A Revised Handbook to the FLORA OF CEYLON

VOLUME III

EBENACEAE
GENTIANACEAE
GESNERIACEAE
LAMIACEAE (LABIATAE)
LECYTHIDACEAE
MARTYNIACEAE
MENYANTHACEAE
MORACEAE
PANDANACEAE
PEDALIACEAE
ROSACEAE
SABIACEAE
SCROPHULARIACEAE
SONNERATIACEAE
SYMPLOCACEAE
UMBELLIFERAE

TT 78–52029/03

A Revised Handbook
to the
FLORA OF CEYLON

VOLUME III

Sponsored jointly by the
University of Peradeniya,
Department of Agriculture, Peradeniya, Sri Lanka,
and the Smithsonian Institution,
Washington, D.C., U.S.A.

General Editor

M.D. DASSANAYAKE

Editorial Board

M.D. DASSANAYAKE and F.R. FOSBERG

Published for the Smithsonian Institution, and the
National Science Foundation, Washington, D.C.,
by Amerind Publishing Co. Pvt. Ltd., New Delhi
1981

September 1981

Published for the Smithsonian Institution, pursuant to an agreement with the National Science Foundation, Washington, D.C., by Amerind Publishing Co. Pvt. Ltd., 66 Janpath, New Delhi 110001

Available from the U.S. Department of Commerce, National Technical Information Service, Springfield, Virginia 22161

Printed at Oxonian Press Pvt. Ltd., Faridabad, India

FOREWORD

The Handbook to the Flora of Ceylon, by Henry Trimen, published in 1893–1900, was in its time, one of the most comprehensive and outstanding floras available for any comparable tropical area. In 1931 A.H.G. Alston added a volume of additions, updating, and corrections to the original five volumes. These six volumes for many years served their purpose very well.

However, the original Handbook was published in a very small edition, and the paper on which it was printed, as was usually the case at the time, was very poor and has deteriorated very badly. Hence the Handbook has for years been absolutely unobtainable, and there are very few copies available even in libraries in Ceylon. Furthermore, botanical science has made substantial progress since the Handbook appeared, and many of Trimen's taxonomic and nomenclatural conclusions are now outdated. Also, with more thorough botanical exploration, new plants have been found to be members of the Ceylon flora. Hence, a new edition of this magnificent work was long overdue.

For quite a number of years Professor B.A. Abeywickrama had in mind a revision of Trimen's Handbook. But heavier and heavier administrative duties consumed his time and there was little opportunity for work on the Ceylon flora, though he did produce, in 1959, an updated Checklist of the Ceylon Flora, which has been most useful to botanists.

Fortuitously, in 1967, the Smithsonian Institution initiated a number of research projects in Ceylon, in cooperation with Ceylonese institutions and scientists. These included an investigation of several problems in plant ecology, with which Prof. Abeywickrama was associated as Co-Principal Investigator. These projects were financed by the Smithsonian using U.S. excess foreign currency under the provisions of Public Law 480.

While we were discussing the ecological investigations, Prof. Abeywickrama wondered if it might not be possible to initiate, using PL-480 support, a project for the revision of Trimen's Handbook to the Flora of Ceylon. I offered to work up a cooperative proposal and submit it to the Smithsonian Special Foreign Currency Program.

This was duly done, approved, and a year's tentative budget authorized. The project was started under the joint auspices of the Smithsonian Institution, the Ceylon Department of Agriculture and the University of Ceylon. I was appointed Principal Investigator. Co-Principal Investigators are Prof. B.A. Abeywickrama, Dr. J.W.L. Peiris, Mr. D.M.A. Jayaweera, Prof. M.D. Dassanayake and Mr. K.L.D. Amaratunga. The plan was to enlist the cooperation of botanists, from wherever available, who were preferably experts in particular families represented in the Ceylon flora. These monographers would be given a period of field work in Ceylon, an opportunity to study the specimens in the Ceylon National Herbariam, in the Royal Botanic Gardens, Peradeniya, with expenses met by the Smithsonian Institution. In return, they would provide updated manuscripts of their families for the revised Handbook, which would then be published by the University of Ceylon, with Smithsonian financing.

This enterprise was initiated in February, 1968, and has been continued without interruption since that date. Quarters for the work, herbarium and library, and other facilities have been furnished by the Division of Systematic Botany of the Department of Agriculture, Peradeniya and by the Botany Department, Faculty of Science, University of Ceylon, Peradeniya. We have enjoyed the cooperation of the U.S. Embassy, Colombo, various Ceylon government departments and agencies, especially the Wildlife Department and the Forest Department, and of many plantations and individuals in all parts of Ceylon, too numerous to enumerate.

Special thanks must be offered to Professor Dieter Mueller-Dombois, of the Botany Department, University of Hawaii, who was for two years Principal Field Investigator for the plant ecology project, and who, on top of his duties in that capacity, supervised the activities of the flora project staff, facilitated the work of the visiting botanists, and acted as finance officer of the project. Without his help, the flora project could not have got started.

Special thanks are also offered to Dr. Marie-Hélène Sachet, Research Botanist, Smithsonian Institution, who, though in no official capacity in the Project, has carried much of the administrative burden, at a sacrifice of her own work. The members of the Flora Project staff at Peradeniya directed by Mr. F.H. Popham, Smithsonian Representative in Ceylon, also deserve great credit for their willing and enthusiastic assistance to the visitors, handling and processing of specimens, typing of labels and manuscripts, and keeping the Project's work going.

The materials on which the flora revisions are based are the visiting botanists' own collections, the herbarium at Peradeniya, personal collections of Mr. K.L.D. Amaratunga and the late Mr. Thomas B. Worthington of Kandy, and materials housed in various foreign herbaria, especially those of Kew, British Museum, and the Indian National Herbarium, Calcutta for the use of which we are grateful to those in charge. A large amount of valuable material was amassed as vouchers for the ecological observations mentioned above and has been utilised by the flora project botanists. Sets of the specimens collected under the auspices of the Smithsonian projects are deposited and permanently available in the U.S. National Herbarium and the Ceylon National Herbarium, Peradeniya. Partial sets are also being deposited in several other Ceylon institutions and in a number of herbaria with tropical interests in other parts of the world.

The resulting revised treatments of the families are to be published, as material accumulates, in volumes of convenient size, without regard to the order of families. Those families, previously published in fascicles 1 and 2, are to be republished, in revised form, in the new format, as manuscripts are received from the authors. An editorial format has been suggested by us, but the content of each revision and the taxonomic conclusions are those of the various authors. An attempt has been made to have the nomenclature in accord with International Code of Botanical Nomenclature, but the application of the Code is, again, the final responsibility of the authors.

A comprehensive index to these volumes will be prepared and published as a separate volume.

It is hoped that, after the Handbook treatments are published, simplified versions can be prepared, suitable for lower school use and for the use of non-botanists.

We also have the hope that the new Handbook will stimulate active interest in the plant resources of Ceylon. Above all, it is hoped that this interest will bring about the establishment of more national parks and nature reserves in all parts of Ceylon, in order that the remarkable Ceylon flora may still have a suitable range of habitats in which to live. By this means, only, the species will be able to survive for the use and pleasure of many future generations of Ceylonese and of visitors from other parts of the world. Without a great increase in such reserves, at the present rate of deforestation and bringing land under agriculture, very many species will surely become extinct in the near future, as some probably already have.

F.R. FOSBERG,
Botanist, Emeritus,
Smithsonian Institution,
Washington, D.C., U.S.A.

CONTENTS

A Revised Handbook
to the
FLORA OF CEYLON

VOLUME III

EBENACEAE

(by A.J.G.H. Kostermans*)

Vent., Tab. Règne vég. t. 443, An VIII. 1799 (excl. sev. gen.); R. Br., Prod. ed. 1. 524. 1810.

Diospyraceae Voigt, Hort. suburb. Calcutt. 343. 1845. Type: *Diospyros* L.

Trees. Bark usually black, rough, brittle and peeling off, rarely (*Diospyros montana*) white and smooth. Branches rarely becoming spines (*D. montana*). Leaves spirally arranged, usually distichous, without stipules, margins entire, nervation pinnate, midrib usually impressed on upper surface. Flowers small, regular, monoecious, dioecious or polygamous-dioecious, the distribution of these different flower kinds rather irregular in the inflorescences. Inflorescences cymose or pseudo-racemiformous, peduncled or sessile, axillary or rami- or cauliflorous or extra-axillary, bracteolate. Flowers solitary (female ones) to numerous. Calyx free, usually enlarged in fruit, consisting of a more or less developed tube and 3–5 (rarely 6) small to medium-sized segments (lobes), only in *D. racemosa* the calyx in bud globose and tearing irregularly; lobes in bud imbricate. Corolla tubular, long or short, constricted at the throat with 3–5 lobes, in bud contort and sinistrorse. Male flower with 4–64 stamens, inserted at the base of the corolla tube or hypogynous, filaments often unequal, sometimes pairwise (or more) connate (filaments forked); anthers narrow, large, acute, dehiscent by lateral slits (in Ceylon species); pollen globose, smooth; ovary rudiment present or reduced to a bunch of hairs or none. Female flower often larger than male one, especially the calyx lobes, which are gradually accrescent, much larger than in the male flower. Staminodes (when present) 4–16, usually less than in the corresponding male flower, but with distinct anthers and filaments; ovary conical to globose, 3–16-celled, superior, with 1–2 ovules in each cell. Fruit indehiscent with thin to thick, fleshy or semiwoody epicarp, usually sessile, apiculate and smooth; dividing septa complete or rudimentary; seeds large, few, surrounded by a juicy usually rather scanty, rather watery, sweet or bitter-sweet pulp. Testa smooth or with distinct pattern; embryo surrounded by copious, dense ruminate or equable endosperm. Seeds always wedge-shaped, in cross section the outer wall concave, the two inner walls formed by two planes, meeting at usually wide

*Krukoff Botanist, Rijksherbarium, Leiden, Netherlands.

angle. Fruiting calyx very characteristic and the best diagnostic; the calyx tube becoming cup-like or not, the lobes either not or hardly enlarged or becoming greatly enlarged and thickened, becoming patent and reflexed to pointing downwards or enrolled, the lateral margins of the lobes also often reflexed, the space in between the lobes sagging and often forming large, marsupial sacs; often this complex pattern of the fruiting calyx becomes flattened out more or less under the ripe fruit.

A family of tropical and subtropical trees, forming only a few genera, represented in all continents. In Ceylon only one genus (*Diospyros*), since *Maba* has been fused with *Diospyros*.

DIOSPYROS

L., Gen. Pl. ed. 6. 550. 1764.

Maba J.R. & G. Forst., Char. Gen. Pl. 121. 1776. Type: *Diospyros kaki* Thunb.

Of the 32 species known (*D. zeylanica* only in sterile condition) the greater part is confined to the lowlands of the wet zone; 6 occur in the dry and intermediate zone. Only one or two go upwards into the lower montane zone. No less than 21 are endemic, and others (like *D. montana*) are represented by forms different from those from outside Ceylon. Many species are rare to extremely rare, one (*D. oppositifolia*) occurring only on Hinidumkande near Hiniduma, others (*D. opaca*) known only from a single specimen, collected over a century ago. Infection of fruit, causing galls (*D. atrata, albiflora*) and the difficulty of extracting the cotyledons from the testa are among the reasons why the species are so rare (cf. Wright in Ann. R. Bot. Gard. Perad. 2. 1904). One species, *D. discolor*, is cultivated occasionally.

Diospyros is the source of ebony wood, of which two kinds are known, the evenly pitch-black one (*D. ebenum*) and the variegated one, called Calamander (Kalumederiya). One of the best suppliers of the latter, *Diospyros quaesita*, is almost extinct.

KEY TO THE SPECIES
(*D. zeylanica* not in the key)

1 Flowers and fruiting calyx trimerous, endosperm equable (not ruminate)
2 Fruiting calyx without cup, consisting only of the calyx lobes
3 Leaves small, less than 12 cm long. Branchlets with adpressed white, silky hairs
4 Leaves thin, long acuminate, reticulate above. Male flowers in sessile clusters. Small trees . . . . **1. D. acuminata**
4 Leaves coriaceous, shortly acuminate, smooth above. Flowers in racemes. Large trees . . . . **2. D. chaetocarpa**

3 Leaves as a rule more than 12 cm long. Branchlets with patent, rusty hairs........ ....3. **D. oblongifolia**

2 Fruiting calyx cup-like with small or obscure teeth (lobes)

5 Leaves attenuate both sides. Rheophyte........................5. **D. rheophytica**

5 Leaves obtuse

6 Lateral nerves 3–4 pairs. Leaves up to 0.5×1 cm............4. **D. nummulariifolia**

6 Lateral nerves more than 4 pairs. Leaves longer

7 Leaves spathulate. Fruiting calyx cup-like..........................6. **D. ferrea**

7 Leaves elliptic. Fruiting calyx flattened..........................7. **D. ebenoides**

1 Flowers and fruiting calyx 4–5-merous. Endosperm equable or ruminate

8 Endosperm equable

9 Calyx of male flowers almost truncate with minute or hardly conspicuous lobes

10 Calyx barrel-shaped, slender. Fruiting calyx disk-like, small. Fruit oblong-ovoid, verruculous..8. **D. oocarpa**

10 Calyx elongate-oblong. Fruiting calyx large, flattish, thick, with strongly reflexed margins and large intersegmental pouches..........................9. **D. quaesita**

9 Calyx of male flowers with distinct lobes

11 Corolla glabrous (or nearly so) outside

12 Clusters of male flowers sessile.................................10. **D. ovalifolia**

12 Clusters of male flowers peduncled

13 Corolla tubular. Stamens very unequal

14 Fruiting calyx lobes broadly spathulate. Fruit globose, 2.5 cm diam......... ....12. **D. koenigii**

14 Fruiting calyx lobes triangular, acute. Fruit depressed globose, 1.5 cm diam.... ....11. **D. ebenum**

13 Corolla short, campanulate or urceolate. Stamens almost equal

15 Leaves thinly chartaceous, 1.5×3–4×10 cm. Trees spiny with smooth, almost white bark...13. **D. montana**

15 Leaves coriaceous, larger. Trees not spiny. Bark rough, black

16 Leaves finely, minutely prominulously areolate-reticulate; lateral nerves erect-patent. Calyx in bud globose, splitting irregularly at anthesis.......... ....14. **D. racemosa**

16 Leaves obscurely or conspicuously more coarsely reticulate. Lateral nerves steep. Calyx in bud with distinct cup and lobes

17 Leaves both sides very conspicuously, rather smoothly, prominently reticulate ....15. **D. albiflora**

17 Leaves with less conspicuous reticulation

18 End bud and inside of calyx black-haired. Fruiting calyx with reflexed, stiff lobes, pointing downward, 4–10 mm long.................16. **D. atrata**

18 End bud and inside of calyx white sericeous. Fruiting calyx with patent, large lobes or these clasping the fruit base, stiff, 10–20 mm long......... ....17. **D. malabarica**

11 Corolla densely pilose outside

19 Lower leaf surface pilose

20 Lower leaf surface with short, rusty hairs. Leaves large, coriaceous........... ....31. **D. opaca**

20 Lower leaf surface with long, yellowish brown, stiff hairs, especially on the veins. Leaves small, thinner.................................18. **D. trichophylla**

19 Leaves glabrous or nearly so

21 Stamens 4 or 5. Fruit pointed

22 Leaves thinly chartaceous, 2×5–4×12.5 cm...............19. **D. attenuata**

22 Leaves coriaceous 2.5 × 12–7.5 × 30 cm..**20. D. acuta**
21 Stamens 12–16. Fruit depressed globose..**21. D. walkeri**
8 Endosperm ruminate
23 Corolla of male flowers urceolate..**22. D. sylvatica**
23 Corolla of male flowers more or less tubular
24 Lower leaf surface pubescent
25 Fruit globose, glabrous or glabrescent..**23. D. melanoxylon**
25 Fruit ellipsoid-ovoid, densely silky rusty hairy..**24. D. hirsuta**
25 Fruit unknown..**31. D. opaca**
24 Lower leaf surface glabrous
26 Male flowers fascicled, sessile; calyx lobes conspicuous
27 Stamens 5, equal..**25. D. moonii**
27 Stamens 8–20, unequal
28 Leaves opposite. Stamens 8..**26. D. oppositifolia**
28 Leaves alternate
29 Fruit glabrous (except apex). Fruiting calyx large, 4-angled, like the cup of an acorn..**27. D. insignis**
29 Fruit with silky, dense, rusty tomentum. Fruiting calyx flattened..**28. D. thwaitesii**
26 Male flowers cymose, pedicellate. Calyx segments small, rather obscure
30 Calyx of female flower with acuminate lobes..**29. D. affinis**
30 Calyx of female flower with rounded, apiculate lobes..**30. D. crumenata**

1. Diospyros acuminata (Thw.) Kosterm., Ceylon J. Sci., Biol. Sci. 12 (2): 92. 1977.

Macreightia acuminata Thw., Enum. Pl. Zeyl. 424. 1864; Beddome, For. Man. in Fl. Sylv. 3: CXLVII. 1869; Clarke in Hook. f., Fl. Br. Ind. 3: 550. 1882; Trimen, Handb. Fl. Ceylon 3: 88. 1895.
Maba acuminata (Thw.) Hiern, Trans. Cambridge Philos. Soc. 12: 112. 1873; Clarke, l.c.; Trimen, l.c.
Ebenus acuminata (Thw.) Kuntze, Rev. Gen. Pl. 2: 408. 1891.
Diospyros imbricata Bakh. Sr., Bull. Jard. Bot. Buitenzorg Ser. 2, 15: 22. 1936. Type: *C.P. 3718* (PDA), Hinidum Pattu.

Tree, 3–5 m high and 5–10 cm diam. Bark smooth or finely fissured, strips 1 cm wide, 0.5–1 mm thick. Live bark 2–4 mm thick, pale reddish to light brown, outside pale reddish-brown. Wood white. Branchlets very slender, apically with a dense tomentum of adpressed, stiff, long silvery silky hairs. Flush initially covered by tiers of imbricate, carinate, acute bracts, the apical ones up to 7 mm long, glabrous. Young leaves with long silky adpressed hairs below, glabrescent. Mature leaves chartaceous to stiffly chartaceous, ovate, lanceolate or ovate-elliptic, 2 × 6–4 × 12 cm (in saplings more lanceolate, with acute base, up to 5 × 19 cm), caudate-acuminate, acumen 2–3 cm long, base rounded or acutish, upper surface prominulously reticulate, midrib slender, flat or slightly impressed, lower surface laxly, prominulously reticulate, midrib slender, prominulous, lateral nerves c. 7 pairs, slender, prominulous, erect-patent, arcuate towards margin, connected by somewhat horizontal, rather widely spaced veins. Petiole slender, 2–5 mm long, pilose. Flowers white with white silky hairs, 1–3 together in the axils of young leaves or fallen leaves, at base surrounded by deciduous, tiered, carinate, acute, at margin densely pilose bracts. Calyx tubular, slender, 4–5 mm long, silky haired, lobes acute, 1–2 mm long. Corolla tube slender, densely silky pilose, up to 1 cm long, lobes glabrous, patent, ovate, fleshy, sometimes reflexed, orifice very small. Stamens 4 or 5. Ovary vestigial, ellipsoid, attenuate, pilose. Female flowers similar to male ones. Fruit globose, 2–2.5 cm diam., sparsely long silky pilose, glabrescent, with remnant of style base; pericarp thin. Seeds brown, smooth, up to 15 mm long, wedge-shaped; endosperm equable. Fruiting calyx not enlarged, no cup; lobes 3, spreading, slender, acute, very hairy.

Distr. Endemic.

Ecol. Wet, evergeen lowland forests in SW Lanka. Locally fairly common.

Note. In one sapling tree the hairs of the flush were much shorter than usual.

Specimens Examined. RATNAPURA DISTRICT: Morapitiya logging area, alt. 100 m, Oct., fl., fr., Kostermans *25695 & 25698* (A, BO, G, K, L, P, PDA, US). GALLE DISTRICT: Kaneliya forest reserve, sapling, *Kostermans 25595A* (G, L); ibid., alt. 100 m, Sep., buds, *Kostermans 25522* (A, BO, G, K, L, P, PDA, US); ibid., May, fl., *Waas 521 A* (L, PDA, US); Hinidumkande (Haycock), alt. 200 m, sapling, *Kostermans 25668 A* (G, L); ibid., Sep., fl., *Kostermans 25559 & 25661* (A, BO, G, K, L, P, PDA, US); ibid., Aug., fl., *Kostermans 25479* (id.); ibid., Aug., fr., *Kostermans 25394 & 25395* (id.); ibid., Sep., young fr., *Kostermans 25662* (id.); Godakande near Hiniduma, alt. 400 m, sapling, *Kostermans 25678 A* (G, L); Hinidum Pattu, buds, *Walker, C.P. 3718*, (BM, K, PDA); Weerakande, alt. 600 m, Sep., ster., *Kostermans 25607* (G, K, L). DISTRICT UNKNOWN: s. loc., fl. *s. coll., C.P. 1353* (K).

2. Diospyros chaetocarpa Kosterm., Ceylon J. Sci., Biol. Sci. 12 (2): 93. 1977.

Macreightia ovalifolia Thw., Enum. Pl. Zeyl. 424. 1864; Beddome, For. Man. in Fl. Sylv. 3: CXLVII. 1869.

Maba ovalifolia (Thw.) Hiern, Trans. Cambridge Philos. Soc. 12: 113. 1873; Clarke in Hook. f., Fl. Br. Ind. 3: 551. 1882; Trimen, Handb. Fl. Ceylon 3: 88. 1895.

Ebenus ovalifolia (Thw.) Kuntze, Rev. Gen. Pl. 2: 408. 1891. Type: *C.P. 3717* (PDA). Hinidum Pattu.

Tree, up to 20 m high and 50 cm diam. Bark black, fissured, strips 5–10 mm wide, 1–2 mm thick, loose. Live bark pale straw to light reddish, finely ridged longitudinally inside, 2–3 mm. Branchlets slender, stiff, apically with sparse, adpressed long, silky hairs. Leaves alternate, stiffly coriaceous, ovate or ovate-elliptic, 1.5 × 3–4 × 10 cm, shortly acuminate; base contracted into the petiole, rounded or acutish; upper surface smooth, dark green (fresh), midrib slender, impressed, lateral nerves filiform, slightly impressed; lower surface pale to yellowish-green (fresh), reticulation lax, prominulous, midrib prominent, sparsely adpressed, long-pilose, lateral nerves 5–6 pairs, prominulous, erect-patent, arcuate. Petiole rather thick, 3–7 mm long. Racemes of male flowers axillary, densely, long, silky, adpressed pilose, up to 2 cm long, few-flowered, initially with many tiered, concave, acute, deciduous bracts at base. Pedicel very short or none. Flowers silvery sericeous, white, c. 0.75 cm long. Calyx broadly tubular, 5–6 mm long, densely long-adpressed pilose; tube very short; lobes 3, acute, slender, almost to the base of the tube. Corolla slightly longer, tubular, bulbous below, lobes ovate-orbicular, obtuse, fleshy, patent, inside glabrous, orifice very small. Stamens 13–16, glabrous, vestigial ovary a tuft of hairs. Female flower not seen. Fruit brown, depressed-globose or globose, up to 1 cm diam., covered with a dense layer of brown hairs, interspaced with long silky hairs, glabrescent. Calyx not enlarged, slender, flat, lobes slender, acute, 5 mm long, covered with long stiff hairs all over. Seeds ellipsoid, smooth, 5 mm long, only one in each fruit, all empty.

Distr. Endemic.

Ecol. Wet, evergreen lowland forests in SW Lanka. On Hinidum and Godakande near Hiniduma, near Eratna and Kuruwita. Everywhere scarce.

Note. The older trees have obscure buttresses. It is not known whether the tree has any appreciable black heartwood. It is like all ebonies called kalumederiya. Only immature fruits are known.

Specimens Examined. GALLE DISTRICT: Hinidumkande (Haycock), Aug., fr., *Kostermans 25483* (A, BO, G, K, L, P, PDA, US); ibid., Sep., ster., *Kostermans 25670* (id.); ibid., Sep., young fr., *Kostermans 25677* (id.); Godakande near Hiniduma, Sep., fl., *Kostermans 25675* (id.); Hinidum Pattu, fl., *s. coll.*, *C.P. 3717* (K, PDA).

3. Diospyros oblongifolia (Thw.) Kosterm., Ceylon J. Sci., Biol. Sci. 12 (2): 103. 1977.

Macreightia oblongifolia Thw., Enum. Pl. Zeyl. 183. 1860 et 423. 1864.

Maba oblongifolia (Thw.) Hiern, Trans. Cambridge Philos. Soc. 12: 112. 1873; Beddome, For. Man. in Fl. Sylv. 3: CXLVII, t. 21, f. 1. 1874; Clarke in Hook. f., Fl. Br. Ind. 3: 550. 1882; Trimen, Handb. Fl. Ceylon 3: 89. 1895.

Ebenus oblongifolia (Thw.) Kuntze, Rev. Gen. Pl. 2: 408. 1891. Type: *C.P. 3396* (PDA).

Tree, up to 8 m high and 20 cm diam. Bark black, brittle, narrowly fissured, 1–2 mm thick, peeling off in small, rectangular pieces. Live bark 3 mm, light brown. Wood white. Branches with copious, spreading, rusty, glossy, 3 mm long hairs. Leaves coriaceous, stiff, oblong-oval to oblong-lanceolate, 3 × 8–7 × 17 cm, shortly acuminate, base rounded. Young leaves pilose on both sides, adult leaves above glabrous, smooth, the slender midrib and laterals impressed, below with prominent, lax venation, the veins hispid, tardily glabrescent, midrib strongly prominent, lateral nerves erect-patent, 8–10 pairs, prominent, towards margin arcuate. Petioles rather thick, 2–5 mm long, densely hispid. Flowers white. Male ones in short, spicate racemes in axils of leaves or fallen leaves, female ones solitary, same position. Calyx pale green, very rusty hairy, male one with short lobes, female ones with lobes almost to the base. Corolla c. 1 cm long; tube cylindrical, slightly bulbous, very hairy with 3 short, rounded, patent lobes. Stamens 12. Pistil in male flower minute, pilose. Ovary in female one pilose, globose, 6-celled; style short, thick, 3-armed. Fruit depressed-globose to globose, smooth, feebly 6-grooved, 6-seeded; septa membranous, transparent, pericarp 1–2 mm thick, brittle, densely rusty fulvous tomentose, the adpressed hairs difficult to rub off. Calyx not enlarged, cup none; lobes slender, acute, 2 mm long with many long bristles along margin. Seed wedge-kidney-shaped, up to 15 mm long and 10 mm diam., black, smooth, round on the back; endosperm equable, testa thin, splitting dorsally.

Distr. Endemic.

Ecol. Wet, evergreen lowland forest of SW Lanka. Locally not uncommon.

Note. A small tree without value as timber tree.

Specimens Examined. GALLE DISTRICT: Kaneliya forest reserve, Aug., fr., *Kostermans 25383 & 25393* (A, BO, G, K, L, P, PDA, US); ibid., May, young fr., *Kostermans 24766 & 24726* (id.); ibid., Sep., buds and fr., *Kostermans 25521 & 25492* (id.); ibid., April, fl., *Waas 500* (K), s. loc., fl., fr., *s. coll., C.P. 3396* (BM, K, PDA).

4. Diospyros nummulariifolia Kosterm., Ceylon J. Sci., Biol. Sci. 12 (2): 102. 1977.

Maba buxifolia var. *microphylla* Thw., Enum. Pl. Zeyl. 183. 1860; Trimen, Handb. Fl. Ceylon 3: 90. 1895. Type: *C.P. 1916* (PDA), *C.P. 477* (BM, p.p.).

Treelet, up to 3 m high and 5–6 cm diam. Bark black, finely fissured, smoothish, strips 2 mm wide, 1 mm thick. Live bark reddish-brown to brown, 3 mm thick. Crown small, very dense. Wood white. Branchlets slender, stiff, apically with a sparse tomentum of short, sericeous, adpressed hairs. Leaves numerous, coriaceous, obovate-cuneate, 4 × 5–5 × 10 mm; apex broad, rounded, base gradually tapered; above smooth, below midrib prominulous, lateral nerves 3–4 pairs, smooth, broad, obscure, prominulous, reticulation lax, obscure, the veinlets broad, smooth. Petiole 1–2 mm long. Fruit brown-red, ellipsoid-ovoid, laterally slightly compressed, 5 × 6–8 mm, apiculate; pericarp thin, brittle. Fruiting calyx cup-shaped, wide, shallow, up to 2 mm deep, the margin entire or slightly wavy, showing the 3 lobes. Seeds usually 2, wedge-kidney-shaped, flattish, 4 mm long, brown, smooth; endosperm equable. Flowers unknown.

Note. Very close to *D. ferrea*, but there is a constant difference: the smaller fruit with very shallow cup and almost entire margin and especially the very small cuneate leaves with few lateral nerves.

It occurs in a limited area, mixed with *D. ferrea*, where in the field it can always be picked out easily.

Specimens Examined. VAVUNIYA DISTRICT: Near Vavuniya, road to Anuradhapura, dry zone, low scrub, July, fr., *Kostermans 25112, 25113, 25231, 25216* (A, BO, G, K, L, P, PDA, US); road Vavuniya to Madhu, July, fr., *Kostermans 25210* (id.); Thunkai near Vavuniya, July, fr., *Kostermans 25264* (id.). BATTICALOA DISTRICT: East coast near Kalmunai, low sea dunes, June, fr., *Kostermans 24349* (id.). COLOMBO DISTRICT: Negombo, Pamunugama, ster., *Simpson 10004* (BM). LOCALITY UNKNOWN: fr., *s. coll. s.n.* (Herb. Willdenow, Berlin); fr., *s. coll. C.P. 1916* (K).

5. Diospyros rheophytica Kosterm., Ceylon J. Sci., Biol. Sci. 12 (2): 104. 1977.

Maba buxifolia var. *angustifolia* Thw., Enum. Pl. Zeyl. 183. 1860 (excl. *Maba angustifolia* Miq.); Clarke in Hook. f., Fl. Br. Ind. 3: 551. 1882; Hiern, Trans. Cambridge Philos. Soc. 12: 117. 1873 (sphalm.: Miq. ex Thw.); Trimen, Handb. Fl. Ceylon 3: 90. 1895; Bakh., Bull. Jard. Bot. Buitenzorg Ser. 2, 15: 54, 57, 62, 197. 1836 (p.p. quoad cit. specim. Ceyl.). Type: *C.P. 1917* (PDA).

Treelet, 1.5–2 m high, diam. 3 cm. Bark smooth, dark brown to black, 1 mm thick, live bark 3 mm, light red. Branchlets slender, horizontal, glabrous, end bud minutely sericeous. Leaves alternate, glabrous, rigidly chartaceous, lanceolate, 1.5 × 4–3 × 10 cm, tapered both ends; upper surface smooth to obscurely laxly reticulate, midrib slender, impressed; lower surface laxly, prominulously reticulate, midrib slender, prominent, young leaves with few, minute adpressed hairs; lateral nerves 6–10 pairs, erect-patent, at some distance from margin arcuately connected. Petiole slender, 3–8 mm long. Female flowers solitary, axillary; calyx campanulate, sericeous except apical part of somewhat reflexed lobes, 3 mm with large, broad triangular teeth, inside of teeth with a densely white pilose band near the margin; corolla tubular, 6–8 mm, densely adpressed-strigose, lobes 2.5 mm, inside glabrous, lanceolate, acute. Ovary densely pilose, ovoid-globose, apiculate, up to 1.5 cm long, smooth; staminodes 6, slender, long. Fruit smooth, yellowish-green, ovoid-globose, up to 1.5 cm long, apiculate; calyx almost disk-shaped, up to 1 cm diam., the triangular, large, woody, thin, acute lobes stretching almost to the centre. Seeds flat-convex or the 2 inner planes making a very wide angle, up to 12 mm long, brownish-black; endosperm equable.

Distr. Endemic.

Ecol. Wet lowland forest of SW Lanka in or on the banks of swiftly running rivulets, liable to inundation.

Note. Close to *D. ebenoides,* differing by the tapered leaves and the tomentum of the inside of the calyx lobes. It is certainly different from *D. ferrea* and from *D. nigrescens,* the latter with copiously strigose branchlets and now named *D. angustifolia* (Miq.) Kosterm.

Specimens Examined. GALLE DISTRICT: road to Hiniduma, near bridge, in rivulet, alt. 50 m, wet, evergreen, fl., *Kostermans 25544* (A, BO, G, K, P, PDA, US); ibid., Aug., buds, *Kostermans 25381* (id.); ibid., June, fl., *Kostermans 25003* (id.); Kaneliya forest reserve, on river bank of swift-running rivulet, Sep., fr., *Kostermans 25602* (id.); Hiniduma, fr., *Walker C.P. 1917* (BM, K, L, PDA).

6. Diospyros ferrea (Willd.) Bakh., Gard. Bull. S. S. 7: 162. 1933; Bakh., Bull. Jard. Bot. Buitenzorg Ser. 2, 15: 50. 1936 (pro minime parte).

Ehretia ferrea Willd., Phytogr. 1: 4, t. 2, fig. 2. 1794; Kosterm., Ceylon J. Sci., Biol. Sci. 12 (2): 98. 1977. Type: *Koenig s.n.* (Herb. Willdenow, Berlin).

Pisonia? buxifolia Rottb., Nye Saml. Kongel. Danske Vidensk. Selsk. Skr. 2: 536, t. 4, fig. 2. 1783.

Ferreola buxifolia (Rottb.) Roxb., Pl. Corom. 1: 35, t. 45. 1795; Roxb., Hort. Beng. 72. 1824 (Ferriola); Willd., Sp. Pl. 4 (2): 798. 1806.

Maba buxifolia (Rottb.) Juss., Ann. Mus. Hist. Nat. Paris 5: 418. 1804; Moon, Cat. 68. 1824; Wight, Ic. Pl. Ind. Or. 3 (1): 4, t. 763. 1843; A. DC. in DC, Prod. 8: 240. 1844 (excl. cit. Burma); Thw., Enum. Pl. Zeyl. 183. 1860 (CP 477; excl. vars.); Hiern, Trans. Cambridge. Philos. Soc. 12: 116. 1873 (p.p., quoad specim. Ceylon.); Beddome, For. Man in Fl. Sylv. 3: CXLVII, t. 19, f. 4. 1869; Clarke in Hook. f., Fl. Br. Ind. 3: 551. 1882 (excl. syn., excl. Kurz); Trimen, Handb. Fl. Ceylon 3: 89. 1895 (excl. vars.); Fischer, Kew Bull. 59. 1932; Anon., Wealth of India 3: 78. 1952.

Ebenus buxifolia (Rottb.) Kuntze, Rev. Gen. Pl. 2: 408. 1891. Type: *Koenig s.n.* (C).

Diospyros ferrea var. *buxifolia* (Rottb.) Bakh., l.c. 57, 59. 1936 & 432. 1941 (p.p., quoad specim. Ceylon & Ind.).

Highulhaenda Hermann., Mus. Zeyl. 21. 1717.

Highulhaenda folio Myrti Burm., Thes. Zeyl. 121. 1737; L., Fl. Zeyl. 202. 1748; Moon, Cat. 68. 1824.

Small tree, rarely up to 7 m high and 15 cm diam., in the drier zones usually not more than 2 m tall. Crown very dense. Bark blackish to dark grey, rather smooth, roughish to the touch, in old trees fissured, rough, black, 0.5–2 mm thick; live bark reddish-brown, pinkish outside, up to 3 mm thick. Branchlets numerous, slender, stiff, apically very sparsely, minutely sericeous. Leaves alternate, stiffly coriaceous, obovate-oblong or spathulate, 6 x 15–10–15 x 30 mm, obtuse to very obscurely, broadly, very shortly obtusely acuminate, often emarginate, base tapered into the 1–2 mm long petiole; above smooth, midrib slender, level with the surface or basal part slightly impressed; below reticulation conspicuous, prominulous, lax, the veins smooth, thick; lateral nerves 5–6 pairs, erect-patent, at some distance from the margin arcuately connected. Flowers 1–3, axillary, white. Male flower: calyx pale yellowish-brown (fresh) cup-shaped or broadly campanulate, 3 mm long, outside densely sericeous; lobes 3, triangular-ovate, acute, 1–1.5 mm long, erect. Corolla tube tubular, slightly swollen about the middle, 5–6 mm long, densely sericeous outside; lobes 3, oblong-ovate, acute, 1–1.5 mm long, patent, glabrous inside. Stamens 6, hypogynous with long, narrow, acute, 2 mm long anthers and 1 mm long, pubescent filaments. Pistil rudiment globose, densely pilose. Female flower as male one; ovary globose, silky-pubescent, style long, densely sericeous, stigmas 3, short. Fruit smooth, glossy, yellow to brown-red (fresh), ovoid to subglobose, 7–9 mm long, initially sericeous around apex, style long persistent. Fruiting calyx cup-shaped, shallow, 2–3 mm deep; lobes 3, depressed-triangular, much broader than high, acute, small. Seeds brown, 3 or less, wedge-shaped, 4 mm long; endosperm equable.

D i s t r. Southern part of peninsular India, ? Tenasserim.

E c o l. Common in the dry and very dry zones, scarcer in the intermediate zone; low altitude.

N o t e. By Bakhuizen lumped with many other species, which will be re-established (cf. Kostermans in Blumea, 1977). Close to *D. nummularifolia* from which it differs by the larger leaves and more numerous lateral nerves. The spathulate leaves distinguish it from the Malesian and Australian *D. littorea.*

S p e c i m e n s E x a m i n e d. JAFFNA DISTRICT: Jaffna, March, Fl., *Simpson 9264* (BM); Point Pedro, March, fr., *Simpson 9274* (BM). MANNAR DISTRICT: Between Pooneryn and Mannar, Oct., fr., *Jayasuriya et al. 10* (L); 4 miles N. of Mankulam, Dec., fr., *Fosberg et al. 53544* (L.); road Madhu-Mannar, May, buds, *Kostermans 24877* (A, BO, G, K, L, P, PDA, US); ibid., May, fl., *Kostermans 24895* (id.); Madhu, May, young fr., *Kostermans 24884* (id.). VAVUNIYA DISTRICT: Vavuniya, June, fr., *Kostermans 25111 & 25112* (id.); ibid., June, young fr., *Kostermans 25126* (id.); ibid., July, fl., *Kostermans 25205* (id.); coastal lagoon Alampil, July, fr., *Meyer 765* (L). PUTTALAM DISTRICT: Puttalam, Palawi junction, Oct., fr., *Kundu et al.*

382 (L). MATALE DISTRICT: Leloya, Oct., fr., *Jayasuriya 327* (L). KANDY DISTRICT: Hunnasgiriya, July, fl., *Kostermans 25279* (A, BO, G, K, L, P, PDA, US); Haragama, alt. 400 m, Aug., fl., *Kostermans 25421* (id.). RATNAPURA DISTRICT: Balangoda, May, fl., *Kostermans 23568* (id.); above Belihuloya, dry scrub, May, buds, *Kostermans 23459* (id.). HAMBANTOTA DISTRICT: Ruhuna Nat. Park, Sep., fr., *Comanor 428* (L); Yala, Aug., fr., *Kostermans 25443* (A, BO, G, K, L, P, PDA, US); Hambantota, Malala Falls, Aug., fr., *Simpson 9952* (BM). TRINCOMALEE DISTRICT: Trincomalee, May, fl., *Simpson 9685* (BM). LOCALITY UNKNOWN: fl., *Walker 263* and *s.n.* (K); fr., *s. coll. C.P. 477* (BM, PDA, p.p.).

7. Diospyros ebenoides Kosterm., Ceylon J. Sci., Biol. Sci. 12 (2): 93. 1977.

Maba buxifolia var. *ebenus* Thw. (non Spr.), Enum. Pl. Zeyl. 183. 1860 (excl. *M. neilgherrensis* Wight); Hiern, Trans. Cambridge Philos. Soc. 12: 117. 1873; Clarke in Hook. f., Fl. Br. Ind. 3: 551. 1882; Trimen, Handb. Fl. Ceylon 3: 90. 1895 (excl. syn.); Bakh., Bull. Jard. Bot. Buitenzorg Ser. 2, 15: 432. 1941 (as var. *ebenus* (Stokes) Thw.).
Diospyros ferrea var. *ebenus* (Thw.) Bakh., l.c. 432 & 438, p.p. (quoad specim. Ceylon., cet. excl.). Type: *C.P. 3395* (PDA).

Tree, up to 4 m tall and 8 cm diam. Bark black, finely fissured, roughish to the touch, 1 mm thick. Live bark 2–3 mm, reddish to light brown. Branchlets slender, glabrous, apically sparsely, minutely adpressed-pilose; end bud sparsely adpressed-pilose. Leaves alternate, glabrous, coriaceous, elliptic or rarely subobovate-elliptic, 1 × 3–2.5 × 6–4 × 9 cm, obtuse, base shortly acute; above obscurely, laxly reticulate; midrib slender, impressed, below laxly prominulously reticulate; lateral nerves 6–10 pairs, erect-patent, at some distance from the margin arcuately connected. Petiole 2–3 mm long, slender. Young leaves pale reddish above, green below (fresh). Female flowers axillary, solitary, sessile on very short inflorescence peduncle; calyx pale green (fresh), campanulate, glabrous, 3 mm long; lobes 3, acute, 1 mm long, erect, inside slightly pilose near the apex; corolla white, tubular, densely white adpressed-pilose, 6–8 mm long; lobes 3, 1.5–2 mm long, oblong, acute. Staminodes 6, consisting of a filament and a long, narrow acute anther. Ovary densely pilose. Fruit globose, 6 mm diam., smooth, yellowish (fresh), rarely apiculate. Fruiting calyx flat or reflexed, 5–7 mm diam.; lobes inconspicuous.

D i s t r. Endemic.

E c o l. Intermediate zone, always near rivulets or where there is water (base of rocks). Is becoming extremely rare, because of extension of agricultural land in the intermediate zone and burning.

N o t e. *Maba buxifolia* var. *ebenus* Thw. is not conspecific with *Maba ebenus* Spr. (which is *Diospyros elliptifolia* (Forst.) Green), nor is it the same as *Diospyros neilgherrensis*. The combination *Maba buxifolia* var. *ebenus* (Stokes) Thw. is an invention of Bakhuizen.

The species differs from *D. ferrea* by the larger, elliptic leaves and the flattened fruiting calyx. It does not occur in the area where *D. ferrea* grows.

The name is inappropriate, as it does not contain ebony and it is not related to *D. ebenum*.

Specimens Examined. BADULLA DISTRICT: Road Bibile to Mahiyangane, Mile 13, along rivulet, June, fl., *Kostermans 25170* (A, BO, G, K, L, P, PDA, US), July, fl., fr., *Kostermans 25304 & 25305* (id.), Aug., fl., fr., *Kostermans 25331, 25350, 25428, 25332 A* (id.). COLOMBO DISTRICT: Mirigama, Kandalama Hill, alt. 100 m, top, base of wet rock, Sep., fl., fr., *Kostermans 25580* (id.); ibid., fl., fr., *s. coll. C.P. 3395* (BM, K, PDA).

8. Diospyros oocarpa Thw., Enum. Pl. Zeyl. 180. 1860 & 423. 1864; Hiern, Trans. Cambridge Philos. Soc. 12: 171. 1873; Clarke in Hook. f., Fl. Br. Ind. 3: 560. 1882; Trimen, Handb. Fl. Ceylon 3: 97. 1895; Alston in Trimen, Handb. Fl. Ceylon 6: 183. 1931; Wright, Ann. R. Bot. Gard. Peradeniya 4: 163, tab. IV (15), VII (1–7). 1904; Worthington, Ceylon Trees 320. 1959. Type: *C.P. 1914* (PDA), Kornegalle (Kurunegala) and Haragam (Haragama) near Kandy.

Diospyros oocarpa Miq. ex Thw., l.c. 423. Type: Pl. *Hohenacher 562*.

Tree, up to 30 m high and 80 cm diam. Base of the straight bole sometimes fluted for 1–2 m. Buttresses small, concave. Bark in medium-sized trees rather smooth, dark brown to grey, peeling off in large irregular pieces leaving scalloped pits; in old trees very rough, blackish, scalloping not so clear. Bark 2–3 mm thick. Live bark yellowish to pale brown, 3–5 mm thick, outside pinkish. Branchlets slender, apically very sparsely, minutely adpressed pilose or glabrous. End bud sericeous. Leaves alternate, thinly coriaceous, glabrous, ovate to ovate-oval, rarely oval, 2×4–5×9 (–6×12) cm, obtusely, shortly, broadly acuminate, base contracted into the petiole, rounded and centrally acutish, very widely, obscurely reticulate in young leaves, old leaves smooth, above the slender midrib impressed, lateral nerves obscure, below paler, midrib prominulous, lateral nerves c. 10 pairs, erect-patent to rather patent, filiform. Petiole slender, 5–9 (–15) mm long, flat, or channelled longitudinally in the centre. Trees dioecious. Male inflorescence a dense sessile or sub-sessile cyme, flowers subtended by a concave round bract. The slender barrel-shaped calyx green, hardly, minutely pilose, 5 mm long. Lobes hardly distinct, up to 0.5 mm long, erect, triangular. Corolla yellow, tubular, tapered in bud, throat narrow, 13 mm long, densely sericeous outside. Lobes 3–4, narrowly oblong, tapered, reflexed. Stamens 9–11–12–14, unequal, epipetalous or hypogynous, unequal, not apiculate, single or in pairs, anthers pilose, 1 mm, filaments glabrous, 0.5–2 mm. Pistil none or a hairy disk. Female inflorescence a sessile cluster of 3 flowers, or flower solitary. Flowers 12 mm long in bud. Calyx brownish black, glabrous or slightly, minutely adpressed-pilose, barrel-shaped, 5–7 mm long. Lobes indistinct as in the male. Corolla greenish-yellow, tapered upwards, 15 mm, orifice narrow. Lobes 3–4, inconspicuous, thick. Staminodes 3, epipetalous, 0.5 mm long, filaments 1/3 anther length. Ovary globose, hairy, 6–8-celled. Stigmas 1–3, short. Fruit oblong-ovoid, up to 2.3 × 3.5 cm, wider at the top (especially when young, cup-shaped); top flat to rounded, apiculate, black, glabrous, papillate when young, adult verruculose and (especially in 1-seeded fruit) strongly longitudinally grooved, 4–6-celled, one seeded. Calyx slightly enlarged, disk-shaped, shallow, 6–11 mm diam., 2 mm deep with thin, entire margin, glabrous, black. Seeds flattened, elliptic with 2 longitudinal grooves, sometimes as broad as long, up to 2.5 cm; testa brown, glossy; endosperm equable. Seedling epigeal, cotyledons detached early. Seedling develops as that of *D. insignis*, epicotyledonary leaf rudiments resemble those of *Garcinia*.

Distr. Also in S. India.

Ecol. Dry and intermediate zone. Rare.

Vern. Kalu kadumberiya (S); Vellai Karunkkali (T).

Note. A species in many characteristics (bark, calyx) different from other Ceylonese ebonies. The timber is variable. Young trees usually with small brown heartwood, sapwood pale pinkish; large trees with much black heartwood, consisting of alternating black and brown layers ("bastard ebony" of commerce). Sapwood of large trees have black lines in the wood. Average weight 45 lb/cu.ft.

Specimens Examined. BADULLA DISTRICT: Mahiyangane, near bridge, Aug., buds, *Kostermans 25342* (A, BO, G, K, L, P, PDA, US). MATALE DISTRICT: Lenadora near Dambulla, alt. 50 m, low hill, dry zone, Aug., buds, *Kostermans 25319* (A, BO, G, K, L, P, PDA, US); N. foot of Matale Hills, Etabendiwena-Dambulla Rd., mile 38/9, June, fl., *Worthington 4799* (K). AMPARAI DISTRICT: Divulane forest, mile 18 Amparai-Kandy Rd., intermediate zone, July, fr., *Kostermans 25311* (A, BO, G, K, L, P, PDA, US). ANURADHAPURA DISTRICT: Mihintale S., alt. 130 m, Mar., fl., *Worthington 4590* (K).

9. Diospyros quaesita Thw., Enum. Pl. Zeyl. 179. 1860; Beddome, For. Man. in Fl. Sylv. 3: CXLV. 1869; Beddome, Ic. 26, t. 128. 1871 (fruit and venation of leaves wrong!); Hiern, Trans. Cambridge Philos. Soc. 12: 174. 1873; Clarke in Hook. f., Fl. Br. Ind. 3: 560. 1882; Trimen, Handb. Fl. Ceylon 3: 97. 1895; Alston in Trimen, Handb. Fl. Ceylon 6: 183. 1931; Wright, Ann. R. Bot. Gard. Peradeniya 2: 166, tab. V (17), VI (5–9). 1904; Worthington, Ceylon Trees 322. 1959. Type: *C.P. 3010* (PDA).

Diospyros hirsuta (non L.) Moon, Cat. 74. 1824.

Tree, dioecious, up to 35 m high and 60 cm diam. with buttresses up to 2 m high and 40 cm out. Bark black, very rough, peeling off in 2–3 mm thick sheets, showing the brown layer underneath. Live bark pale reddish to whitish, 3–5 mm thick. Branchlets rather thick, glabrous. End bud minutely sericeous. Leaves alternate, glabrous, coriaceous to stiffly coriaceous, elliptic-oblong or oblong-lanceolate, 3 × 8–8 × 18 cm, shortly obtusely acuminate, base acute; both surfaces prominulously reticulate or upper surface smooth, midrib slender, deeply channelled, lateral nerves inconspicuous, lower surface with prominent midrib, lateral nerves rather patent, 8–10 pairs with one or two slightly shorter ones in between, at margin arcuate, prominulous, very slender. Petiole rather slender, 8–20 mm long, channelled above. Male inflorescences minutely adpressed-pubescent, raceme-like, dichasial cymes in axils of small caducous bracts or leaves with 3–9 flowers on nodding pedicels on up to 2 cm long peduncles (variable in length). Pedicels slightly pubescent, slender, wider at the top, 6 mm long with two opposite small caducous bracteoles. At base of pedicel a sessile, caducous, 8 mm long, tapered, obtuse bracteole. Calyx green, elongate-oblong, slightly swollen in bud, 9 mm long. Lobes deltoid, pubescent, 1–2 mm long, acute. Corolla yellow, tubular, thick, basal part swollen, 12 mm long, densely sericeous. Lobes 4–5, rotate, broadly ovate, acutish, sericeous outside. Stamens 16 or more on the receptacle, 3 mm long. Anthers slender, almost sessile, attenuate, apex pilose, filaments glabrous. Pistil none or very small. Female flowers solitary, pedunculate (2 cm long) in axil of leaf or caducous bract, in bud 5 mm long, 8 mm diam. Open flower 2 cm long, 3 mm diam., aperture narrow. Calyx brown, adpressed pilose, broadly barrel-shaped. Lobes 5, margins initially slightly recurved, but later very wide intersegmental grooves are formed, these pouches are at first horizontal, but widen gradually and become strongly reflexed, laterally and in between downwards, only the sharp tips remain upright. Corolla yellowish-white to white, densely adpressed pilose, 15 mm long. Lobes 5, obtuse, recurved in old flowers, 5 mm long. Staminodes 5, epipetalous, 2 mm long, or more strongly rudimentary ones; filaments short, glabrous. Ovary globose, pubescent; stigmas 5, short, apex wavy. Fruit subglobose to ovoid-globose, apiculate, 6–7.5 cm long, slightly flattened or impressed, soon glabrous. Calyx at last flattish, 3–4 cm diam., woody, not very thick with strongly reflexed margins and large intersegmental pouches, but all flattened out. Seeds 4–10, elliptic, flattened, up to 3 cm long; testa brown, glossy, striated; endosperm equable. Seedlings epigeal, cotyledons detached early.

Distr. Endemic.

Ecol. Wet, evergreen lowland forests in SW Lanka, up to 500 m alt. Rare.

Vern. Kalumederiya (S).

Note. The timber—calamander—is considered to be one of the best for durability and ornamental purposes. It is black with brown bands. It has become very scarce, not only because of cutting, but also as a result of the very slow growth, the irregular flowering period and the curious mode of development (cf. Wright, l.c.).

Specimens Examined. GALLE DISTRICT: Hinidumkande near Hiniduma, alt. 300 m, Aug., fl., *Kostermans 25373* (A, BO, G, K, L, PDA, US); ibid., 600 m alt., Oct., fr., *Kostermans 25489* (id.). RATNAPURA DISTRICT: cultivated in forestry plot near Rakwana, June, fr., *Kostermans 25009* (id.).

10. Diospyros ovalifolia Wight, Ic. Pl. Ind. Or. 4 (1): (10), t. 1227. 1848; Thw., Enum. Pl. Zeyl. 181. 1860 (*C.P. 1815, 2533*); Beddome, For. Man. in Fl. Sylv. 3: CXLXXX. 1869; Hiern, Trans. Cambridge Philos. Soc. 12: 237. 1873; Clarke in Hook. f., Fl. Br. Ind. 3: 557. 1882; Trimen, Handb. Fl. Ceylon 3: 91. 1895; Wright, Ann. R. Bot. Gard. Peradeniya 2: 134, tab. I (6), XV (1–5). 1904; Worthington, Ceylon Trees 321. 1959. Type: Wight.

Tree, up to 15 m high and 30 cm diam., usually smaller, dioecious, becoming partly deciduous in the very dry northern provinces. Bark with superficial ridges, 1 cm wide, 1–2 mm thick, alternate black (ridges) and brown (fissures) striped longitudinally in old trees. Live bark pale to reddish-brown. Branchlets slender, sparsely, minutely, adpressed-pilose apically. End bud minutely sericeous. Leaves alternate, glabrous, thinly coriaceous, oblanceolate to obovate-oblong to oblong, 1.5 × 4–5 × 13 cm, obtuse or rounded, base gradually acute; both surfaces laxly prominulously reticulate, above very glossy; midrib slender, faintly impressed, laterals prominulous, slender, lower surface with prominent midrib; laterals 6–7 pairs, erect-patent, arcuate, lower pair steeper. Petiole very slender, 3–7 mm long, narrowly channelled above. Male inflorescence a sessile cluster of 3–7 flowers on woody twigs, bracts form a basal involucre. Flowers yellow, 5–10 mm long. Calyx green, sparsely, minutely adpressed-pilose, broadly campanulate. Lobes 4–5, orbicular. Corolla yellow, urceolate, almost glabrous, 5–8 mm long. Lobes 4–5, round, recurved. Stamens 13–20, paired, surrounding a central rudimentary disk-like pilose pistil; anthers glabrous, filaments unequal, much shorter than anthers. Female flowers solitary or in clusters of 3–9 flowers in leaf axils or on bare twigs; buds pilose, surrounded by 2–3 small, deciduous bracts. Calyx broadly campanulate. Lobes 4–5, brown or green, pubescent, 3 mm long, rounded or acutish. Corolla yellow, conical in bud, tubular, lobes 4–5, shorter than tube. Staminodes 1–7, usually 4, sometimes lacking, hypogynous or at base of corolla. Ovary conical, 4-celled, densely pubescent; stigmas 2–4, short, reniform. Fruit subsessile, depressed-globose, 1–2 cm diam., solitary, usually 1-seeded, glabrous, smooth. Calyx very shallow, small, thick. Lobes thick, orbicular, concave, reflexed or patent, up to 5 mm long. Seeds wedge-shaped (or globose in 1-seeded fruit), up to 9 mm long, testa brown, rather smooth; endosperm equable. Seedlings epigeal; cotyledons detached early, testa does not split.

Distr. South India, Ceylon.

Ecol. Common in dry and intermediate zone.

Vern. Kunumella, Habara (S); Vedukunari (T).

Note. Wight's Indian specimens have very large leaves, those of Lanka are always much narrower and smaller. The leaves are distinctly oblanceolate or obovate-oblong, contrarily to most descriptions. Timber brown to dirty white or pinkish, inferior; heartwood small and usually decayed.

Specimens Examined. ANURADHAPURA DISTRICT: Ritigala, W. of N.C.P., Jan., fr., *Worthington 5061* (PDA); Base of Ritigala Kande, Jan., fr., *Waas 422 A* (L, PDA, US); ibid., alt. 400 m, Oct., fr., *Jayasuriya 941* (id.); ibid., alt. 250 m, Jan., fr., *Jayasuriya 1061* (id.); Wilpattu Nat. Park, June, fr., *Wirawan et al. 87* (id.). HAMBANTOTA DISTRICT: Ruhuna Nat. Park near Hambantota, Jan., fr., *Wirawan 813* (L, PDA, US). AMPARAI DISTRICT: Mile 18 Amparai-Kandy Rd., May, ster., *Kostermans 24851* (A, BO, G, K, L, P, PDA, US). MATALE DISTRICT: Nalande, mile 36, alt. 300 m, Nov., fl., *Worthington 6690* (K).

11. Diospyros ebenum Koenig, Physiogr. Salsk. Handl. (Lund) 1: 176. May 1781; L. f., Suppl. Pl. 440. Summer 1781 (p.p.?); Willd., Sp. Pl. 4 (2): 1109. 1805 (p.p., quoad cit. *D. glaberrima* Rottb., caet. exclud.); Moon, Cat. 74. 1824; Thunberg, Fl. Ceil. 6. 1825; ? Roxb., Fl. Ind., ed. Carey 2: 529. 1832; ? Wight, Ic. Pl. Ind. Or. 1: t. 188. 1839; A. DC. in DC., Prod. 8: 234. 1844 (excl. cit. Retz. & Rumph.); Thw., Enum. Pl. Zeyl. 180. 1860 (*CP 1912*, *1913*, *2437*, *2439*); Beddome, Fl. Sylv. 1: t. 65. 1869; Hiern, Trans. Cambridge Philos. Soc. 12: 208. 1873 (p.p., excl. *D. assimilis*); Clarke in Hook. f., Fl. Br. Ind. 3: 558. 1882; Trimen, Cat. 32. 1885; Hort. Zeyl. 49. 1888; Trimen, Handb. Fl. Ceylon 3: 94. 1895 (excl. *D. ebenum* Retz.); Broun, Ind. Forester 25: 275. 1899; Gamble, Man. Ind. Timb., ed. 2: 456. 1902; Gamble, Fl. Pres. Madras 5: 775. 1923; Wright, Ann. R. Bot. Gard. Peradeniya 2: 146, 200, 201, t. IV (16); XIII (1–8); XVIII (8); XIX; XX. 1904; Bailey, Stand. Cycl. Hortic., ed. 3, 2: 106. 1919; Troup, Silvic. Ind. Trees 2: 653. 1921; Pearson & Brown, Ind. Timb. 2: 654. 1931; Bakh., Bull. Jard. Bot. Buitenzorg Ser. 3, 15: 216. 1937 (p.p., excl. Gaertn. f., Sprengel, *D. ebenaster* Retz., *D. reticulata* var. *timoriana* A. DC., *D. nigricans* Dalz., *D. assimilis* Bedd., *D. ebenum* var. *timoriana* (Miq.) Bakh.; quoad specim. Ind. Or. & Ceyl., caeter. exclud.; exclud. many references); Rendle, J. Bot. 71: 179. 1933; Anon., Wealth of India 3: 77. 1952 (p.p., excl. *D. assimilis* Bedd., *D. sapota* Roxb.); Worthington, Ceylon Trees 315. 1959 (excl. cit. Malaya); Howard & Norlindh, J. Arn. Arb. 43: 94–107. 1962 (excl. tab. I); Kosterm., Ceylon J. Sci., Biol. Sci. 12 (2): 23. 1977. Lectotypus propos. (by Howard & Norlindh): *Koenig*, fr. specimen (C). Syn-lectotypes: *Koenig*, fl. male specimen (C).

Diospyros glaberrima Rottb., Acta Haffn. 2: 540, t. 5. 1783. Type: *Koenig*, fr. specim. (C).
Diospyros melanoxylon (non Roxb.) Willd., Sp. Pl. 4: 1109. 1805. Type: *Willd. 19243* (Berlin), fide Hiern.
Diospyros reticulata Wall. (non Willd.) Wall., Cat. 4120 E, fide Hiern.
Diospyros ebenaster (non Retz.) Roxb., Fl. Ind., ed. Carey 2: 529. 1832; (non Retz.) Spach, Hist. Veg. Phan. 9: 407. 1840, t. 135. 1846, fide Hiern.
Diospyros ebenum var. *glaberrima* (L. f.) Bakh., l.c. 219.
Ebenus, Burman, Thes. Zeyl. 91. 1737.
Ebenus, Rumph., Herb. Amb. 3: 4. 1750 (p.p., quoad cit. Trinkanamale).

Tree, up to 30 m high and 90 cm diam., dioecious, polygamous and occasionally monoecious. Buttresses in old trees up to 2 m high, 30 cm out, merging into the fluted base of the straight bole. Bark black to grey-black, rough, peeling off in small rectangular pieces, 3 mm thick, fissured, strips 2–3 cm wide, brittle. Live bark 3–5 mm, light brown, inside yellowish (turning dark quickly after exposure). Branchlets slender, apical part sparsely minutely adpressed pilose or glabrous. End bud densely brownish to whitish sericeous. Leaves alternate, thinly coriaceous, glabrous, oblong-oval to oblong-lanceolate, 2 × 5–6 × 13 cm, with short, broad, obtuse, rather obscure acumen or rounded; base contracted into the 4–8 mm long, glabrous petiole; centre cuneate; venation conspicuous, dense, prominulous and pellucid on both surfaces (beneath less dense); upper surface with slender, impressed midrib (except apical part); laterals faint; lower surface with prominent midrib, lateral nerves 5–8 pairs, slender, prominulous, rather steep, especially the lower pair. Polygamous condition most common (male and hermaphrodite). Male inflorescence an axillary dichasial compressed cyme of 3–15 flowers on 2–5 mm long, pilose slender main peduncle; ramifications very short. Flowers sessile or subsessile; bracts small, acute, pilose, caducous, 2 mm long. Male flower: Buds slender, tubular, tapered. Calyx funnel-shaped, c. 4 mm, lobes 4, green or speckled with white, obtuse; margin ciliate. Corolla glabrous, tubular, yellow to yellowish-white (inside white), throat constricted, c. 10 mm long; lobes 4 (rarely 3), rotate, glabrous, narrow, fleshy, obtuse, 6 mm long. Stamens unequal, 20–30 (usually 16) in 8 epipetalous connate groups of 2's and 3's opposite and alternate with corolla lobes; anthers slender, apiculate; filaments glabrous, except near anther; rudimentary pistil a flat, pilose disk. Female flowers axillary, solitary, subsessile with 2 basal, caducous, small bracts. Calyx cupular, 3 mm; lobes patent or reflexed, 3–4, acute, glabrous, 9 mm long. Corolla 15 mm long with wide mouth; lobes 3–4, rotate, broad, obtuse, reflexed, 9 mm long. Staminodes 8, opposite and alternate with corolla lobes; anthers deltoid, apiculate, pilose at apex; filaments glabrous, 2 mm. Ovary glabrous, 8-celled; style slender, long with 4–5-lobed stigma. Fruit depressed-globose to globose, up to 1.5 cm diam., smooth, glabrous, sometimes apiculate. Fruiting calyx a woody, green cup, inside pilose, the enlarged calyx lobes 8–10 mm long, initially patent, triangular, acute, thick, with a callous rim at their inner base, later completely reflexed, the callous ring forming a thin, upright, 3.4 mm high rim. Seeds 2–8, flattened and wedge-elliptical in shape, slender, up to 15 mm long; testa brown or black, smooth; endosperm equable. Seedling epigeal, cotyledons persistent; epicotyledonary stem does not appear for several weeks or months after full exposure of cotyledons.

Distr. Southern India (not in Malay Peninsula!).

Ecol. Dry and intermediate zone, especially in the east and northeast and south of Jaffna and Anuradhapura Distr. Scattered.

Uses. The ebony is mainly exported to China for making furniture, chopsticks and (formerly) opium pipes, to Europe as a fancy wood. The gummy astringent fruits are a medicine (attenuant, lithontriptic), and eaten in times of famine. It is a fish poison.

Vern. Kaluwara (S); Karunkali (T).

Note. This species supplies the greater part of Ceylon ebony. The ebony in the stem centre may increase from below upwards or may repeatedly increase and decrease upwards. Ebony has been found in saplings and in 2-year old twigs and even in roots. The ebony is jet black or more rarely streaked, very hard and heavy (80–100 lb/cu. ft.). About 1910, 2,635 tons have been sold in Colombo, this is less now. The ebony is obtained by cutting away the sapwood (lying on the ground, termites will do this); attack of borers in ebony has been observed. The proportion of black heartwood is said to increase on bad soils. Growth is slow, a tree of 60 cm diam. is about 200 years old.

The forming of the duramen and the coloured matter has been a puzzle for a long time. It is assumed that they are formed from lignin by an oxidation process into humic acids, comparable to fossilisation of wood (cf. Griffioen in Receuil Trav. bot. néerl. 31: 780, 809. 1934).

Specimens Examined. POLONNARUWA DISTRICT: between Habarana and Polonnaruwa, April, fl., *Jayasuriya et al. 1623* (L, PDA, US); Polonnaruwa, July, fr., *Kostermans 25288* (A, BO, G, K, L, P, PDA, US). BADULLA DISTRICT: Mahiyangane, road along canal to Polonnaruwa, Aug., fr., *Kostermans 25340*, *25341*, *25339*, July, fr., *Kostermans 25288*, *25289* (A, BO, G, K, L, P, PDA, US); road Mahiyangane-Bibile, June, fr., *Kostermans 25168* (id.); Uma Oya, waterfall near bungalow, June, fr., *Kostermans 25184* (id.); Dunhinda waterfall, road Badulla-Bibile, June, fr., *Kostermans 24446* (id.); road to Dunhinda Falls, Oct., fr., *Balakrishnan et al. 859* (L). MATALE DISTRICT: Lenadora near Dambulla, Aug., fl., *Kostermans 25320* (A, BO, G, K, L, P, PDA, US); mile 37/4 Matale road, N of Naula, Aug., fr., *Meyer et al.* 680 (L); Eraula Weva, below Eraula Kanda, E. of Dambulla, July, fr., *Fosberg 55086* (id.). MONERAGALA DISTRICT:

Bibile to Nilage forest, June, ster., *Kostermans 25163 A*, *25167 A* (G, K, L). VAVUNIA DISTRICT: Vavunia, along road, June, fr., *Kostermans 25209* (A, BO, G, K, L, P, PDA, US). KURUNEGALA DISTRICT: Kurunegala hills, July, fl., *Kostermans 25241* (A, BO, G, K, L, P, PDA, US). MANNAR DISTRICT: Madhu, May, after anthesis, *Kostermans 24908* (PDA); ibid., May, fl., *Kostermans 24869* (A, BO, G, K, L, P, PDA, US). TRINCOMALEE DISTRICT: Trincomalee to Kuchchavelli, May, fl., *Kostermans 24804* (PDA); ibid., May, fr., *Kostermans 24801* (A, BO, G, K, L, P, PDA, US); Trincomalee, March, fl., *Worthington 76* (PDA); Trincomalee rd., mile 75/4, April, fl., *Worthington 167* (id.); Trincomalee, below junction Spy Glass rd., March, fl., *Worthington 869* (id.). PUTTALAM DISTRICT: Wilpattu Nat. Park, June, fr., *Cooray et al. 892* (L, PDA, US); Wilpattu, June, fr., *Wirawan et al. 892* (L); 2 miles N. of Puttalam, Sept., fr., *Jayasuriya 6* (L). DISTRICT UNKNOWN: road Kandy-Anuradhapura, July, fr., *Kostermans 25206* (A, BO, G, K, L, P, PDA, US); ibid., June, buds, *Kostermans 25108* (id.).

12. Diospyros koenigii Kosterm., Ceylon J. Sci., Biol. Sci., 12 (2): 100. 1977.

Type: *Kostermans 25355* (L.).

Tree, up to 15 m tall and 50 cm diam. Buttresses high, small, merging into the fluted bole. Bark black, fissured, strips 2 mm wide, wavy, 2 mm thick. Live bark white to light brown, 3 mm thick. Leaves alternate, chartaceous, glabrous, oblong to elliptic, 3 × 7–6 × 13 cm, obtuse or obscurely, broadly, shortly, obtusely acuminate; base contracted into the slender, c. 1 cm long petiole, shortly acutish centrally; upper surface glossy, midrib slender, impressed; lower surface paler, less glossy, the reticulation rather obscure, midrib prominent, lateral nerves 6–8 pairs, very slender, erect-patent, the 2 or 4 lower ones steeper. Fruit axillary or on a very short, slender, pilose, axillary branchlet, sessile, globose, up to 2.5 cm diam. Fruiting calyx green (fresh) consisting of a rahter shallow cup, densely sericeous inside, the 4 lobes completely reflexed, not very woody, broadly spathulate, acute, plicate, with a wide interspace, inside very sparsely, minutely, adpressed pilose, at their inner base a vertical, 2 mm high, thin, wavy rim. Seeds slender, flat-convex, 14 mm long, 4 mm wide, smooth, 2 mm thick; endosperm equable.

Distr. Endemic.

Ecol. Thus far only found in the Gannoruwa jungle behind the Bot. Garden, Peradeniya.

Note. Easily mistaken for *D. ebenum* because of the similar leaves, which are, however, thinner and have a less conspicuous reticulation. The fruits are different and so is the fruiting calyx. It is also near *D. crumenata*, which has a reticulate leaf and different fruit.

Specimens Examined. KANDY DISTRICT: Gannoruwa jungle behind Botanic Garden, Peradeniya, alt. 500 m, rather wet zone, Aug., fr., *Kostermans 25353* (A, BO, G, K, L, P, PDA, US), *25355 & 25434*, (id.), May, young fr., *Kostermans 24035* (id.), cultd. in Bot. Gard. (sub *D. crumenata*), May, young fr., *Kostermans 24858* (id.). LOCALITY UNKNOWN: *s. coll.*, *CP 2437*, *2439* (K).

13. Diospyros montana Roxb., Pl. Corom. 1: 37, tab. 48. 1795 (excl. fruit); Roxb., Fl. Ind., ed. Carey 2: 538. 1832; Poiret in Lam., Enc. 5: 434. 1804; Willd., Sp. Pl. 4 (2): 1110. 1805; A. DC. in DC., Prod. 8: 230. 1844 (*Wallich 4115*); Wight, Ic. Pl. Ind. Or. t. 1225. 1848; Hiern, Trans. Cambridge Philos. Soc. 12: 220 et 221. 1873 (p.p., only var. *montana*); Beddome, For. Man. in Fl. Sylv. 3: CXLIII. 1869; Trimen, Handb. Fl. Ceylon 3: 92. 1895; Gamble, Man. Ind. Timb. 454. 1902, p.p.; Wright, Ann. R. Bot. Gard. Peradeniya 2:

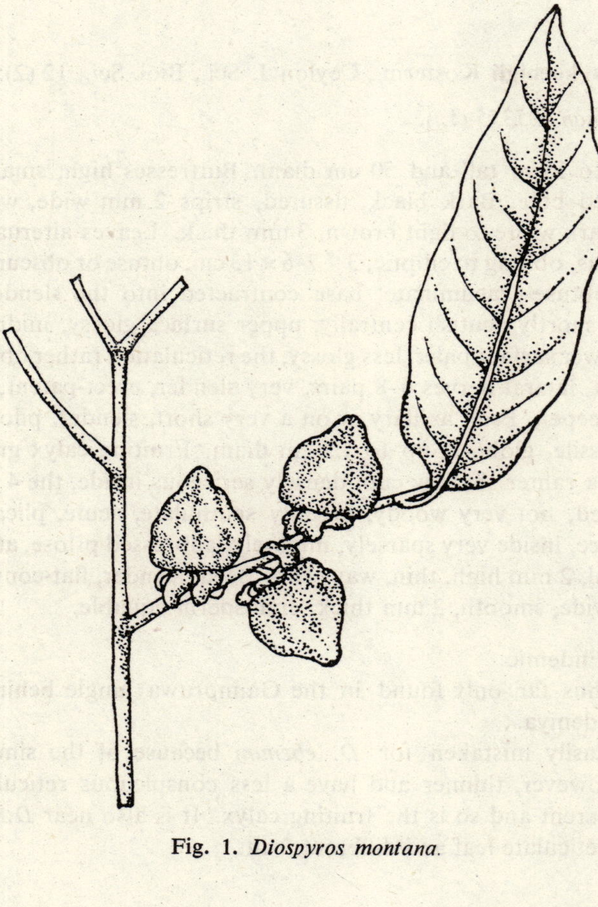

Fig. 1. *Diospyros montana*.

136, 199, 202, tab. V (19); XVI (7–10). 1904; Gamble, Fl. Pres. Madras 776. 1923; Bakh., Bull. Jard. Bot. Buitenzorg Ser. 3, 15: 203. 1937 (p.p., quoad var. *montana* (Roxb.) Hiern); Anon., Wealth of India 3: 84. 1952, p.p.; Worthington, Ceylon Trees 319. 1959. Type: Plate 56 of Roxburgh (Kew), which is tab. 48, Pl. Corom.—**Fig. 1.**

Diospyros montana var. *montana* (Roxb.) Hiern, l.c. 221; Bakh., l.c. 203.
Diospyros cordifolia Auct. (non Roxb.) Thw., Enum. Pl. Zeyl. 178. 1860 and 423. 1864 (Jaffna, *Gardner*.).
Diospyros ebenum Retz., Obs. Bot. 5: 31. 1789. Type: *Koenig s.n.* (C and Lund).

Tree, up to 14 m high and 60 cm diam., usually much smaller. Bole and branches with long spines formed from stumps of branches, hard, stiff, divaricate. Bark smooth, grey to yellowish grey (like guava bark), 1–2 mm thick, in old trees and at base of bole peeling off in stiff irregularly outlined, rather large pieces, 5–10 cm long, 2 mm thick. Live bark 3–5 mm thick, yellow with little greenish-yellow, watery sap. Branchlets slender, glabrous, only those of the flush sparsely minutely pilose. Leaves alternate, chartaceous, oblong to subovate-oblong to subobovate-oblong, 1.5 × 3–4 × 10 cm, very shortly, broadly, obtusely acuminate; base rounded or truncate or subcordate with near the petiole a short cuneate part, both sides smooth; upper surface with midrib level with the surface, laterals very faint; lower surface with slender, prominulous midrib, laterals 6–8 pairs, very slender, rather steep, arcuate. Petiole slender, 5–8 mm long, flattish above. Male inflorescences: Cyme 3-flowered or panicled in axils of small, caducous bracts or leaves. Main peduncle thin, 2–6 mm long. Pedicel thin, 2–5 mm long, basal bract small, narrow, 2 mm long. Pedicels varying in length. Flower bud 4 mm, conical. Mature flower 8 mm long, somewhat bell-shaped. Calyx lobes 4, thin, rounded or deltoid-ovoid, small. Corolla with wide throat, urceolate, white or yellow; lobes 4. Stamens numerous or 16 in equal or unequal pairs, often united to form 8 pairs; anthers slender, apiculate, 2–4 mm long; filaments glabrous, 0–1 mm long. Ovary a small or large disc. Female flower usually solitary on twig or in leaf axil; peduncle 4–12 mm long, glabrous or pubescent with small, caducous bracts. Calyx pilose, lobes 4, ovate-suborbicular, 2–5 mm long, rounded. Corolla white or yellow with wide throat, glabrous, 11 mm long. Lobes 4, rotate. Ovary glabrous, 8-celled; style distinct, bifid at apex. Fruit globose or depressed-globose, or ovoid, usually acute and apiculate, 8–20 mm diam., glabrous, smooth, first yellow, turning to red, ultimately brown; pulp very sweet and bitter. Pericarp thin. Calyx lobes ovate-orbicular, acutish or obtuse, 8 mm long, strongly reflexed, the apex incurved to the pedicel, thin; no tube. Seeds 2–6–8, semilunar wedge-shaped, black, smooth, up to 8 mm long; endosperm equable. Seedling epigeal, cotyledons persistent, ovate, tapered, green, glabrous, 3 × 1.5 cm with flat, 2 mm long petioles. Epicotyledonary system shows delayed development.

D i s t r. South India, Ceylon.

E c o l. Common in the dry zones. Becomes partly deciduous in very dry scrub.

V e r n. Mulkarunkali, Katukanni, Vakkana (T).

N o t e. Closely related to *D. cordifolia* with which it has been combined. Gamble keeps it separate and enumerates the differences. The two species occur in different areas. The Ceylonese one is completely constant in pilosity characters, but the fruits are variable, the most common being globose-ovoid, acute and apiculate. The smallest fruits occur in the NW part of Lanka in a very dry area. These are only 8 mm diam., are compressed, but remain acute. Timber brown or dirty white or faintly red, inferior. Heartwood small and usually decayed. Some specimens have a thinly black-streaked pretty heartwood.

According to Gamble, *D. montana* should have larger fruits than those of *D. cordifolia*, contrarily to the information given in Wealth of India.

Worthington mentioned a dentation in the leaf; this is a pseudodentation, caused by folding.

The plate of Wight is better than that of Roxburgh (the stigmas are not bifid, as they should be). In Roxburgh's plate the calyx under the fruit is adpressed to the fruit, entirely different from the completely reflexed calyx found in our material.

S p e c i m e n s E x a m i n e d. Form with very small fruit (8–12 mm diam., depressed, acute)—VAVUNIYA DISTRICT: Vavuniya, July, fr., *Kostermans 25208* (A, BO, G, K, L, P, PDA, US); road Vavuniya to Madu, July, fr., *Kostermans 25207*, *25223*, *25237* (id.); Parayanalankulam to Vavuniya, very dry scrub, June, fr., *Kostermans 25134*, *25137*, *25139* (id.). HAMBANTOTA DISTRICT: Tissamaharama Resthouse, Feb., fr., *Worthington 5677* (K); Ruhuna National Park, Nov., fr., *Wirawan 709* (L, PDA, US). ANURADHAPURA DISTRICT: Madawachiya, Jan., young fr., *Worthington 4460* (K). Form with larger, globose-ovoid or globose fruit—ANURADHAPURA DISTRICT: rd. side near Anuradhapura, June, fr., *Kostermans 24214* (L, PDA, US). VAVUNIYA DISTRICT: near Vavuniya, July, fr., *Kostermans 25214* (id.). HAMBANTOTA DISTRICT: Ruhuna National Park, Jan., fr., *Wirawan 792* (L, PDA, US). PUTTALAM DISTRICT: Wilpattu National Park, July, fr., *Cooray et al. 1084* (L, P, PDA, US).

14. Diospyros racemosa Roxb., Fl. Ind. 2: 536. 1824; Roxb., Fl. Ind. ed. Carey 2: 536. 1832; Wight, Ic. Pl. Ind. Or. 2: t. 416. 1840 (plate of Roxburgh); A. DC. in DC., Prod. 8: 139. 1844 (leaves sericeous, wrong translation from glossy); Alston in Trimen, Handb. Fl. Ceylon 6: 181. 1931; Bakh., Bull. Jard.

Bot. Buitenzorg Ser. 3, 15: 298. 1937 (omits to cite the first edition of Roxb., Fl. Ind.).

Embryopteris racemosa (Roxb.) G. Don, Gen. Hist. 4: 41. 1838. Type: Roxburgh's plate as reproduced by Wight, Ic. t. 416.

Diospyros toposia Hamilton, Trans. Linn. Soc. London 15: 115. 1827; A. DC., l.c. 237; Thw., Enum. Pl. Zeyl. 179. 1860 (*CP 1911*); Beddome, Ic. 1: 25, tab. 122. 1871 (reticulation badly drawn); Hiern, Trans. Cambridge Philos. Soc. 12: 263. 1873; Clarke in Hook. f., Fl. Br. Ind. 3: 556. 1882; Trimen, Handb. Fl. Ceylon 94. 1895; Wright, Ann. R. Bot. Gard. Peradeniya 2: 142, tab. III (12); X (7–10); XVII. 1904; Brandis, Ind. Trees 432. 1906; Gamble, Fl. Pres. Madras 5: 778. 1923; Bakh., l.c. 298 (p.p., exclud. *D. foveo-reticulata* Merr. & *D. collinsae* Craib); Anon., Wealth of India 3: 87. 1952. Type: Roxburgh's plate (K) (=Wight, Ic. Pl. Ind. Or. t. 416).

Tree, dioecious, up to 20 m high and 40 cm diam. Bark black, or dark greyish brown, rather smooth (except the rough basal part of the bole), 1–2 mm thick, hardly peeling off. Live bark 5 mm, reddish. Branchlets slender, glossy, drying pale brown, glabrous. End bud sparsely, minutely adpressed-pilose. Leaves alternate, thinly coriaceous, glabrous, glossy, elliptic to oblong to lanceolate- or ovate-oblong, 2 × 10–5 × 3 cm, broadly, shortly acuminate; base acutish or rounded, both surfaces very densely, prominulously areolate-reticulate; upper surface with slender, impressed midrib, lateral nerves filiform, faint; lower surface paler, midrib slender, prominent, lateral nerves 8–9 (–17) pairs, erect-patent, arcuate, filiform, prominulous. Petiole slender, glabrous, 5–12 mm long, flat above. Male inflorescence in axil of caducous bract or leaf, in the simplest case a cyme of 3 flowers, in more complex ones the dichotomy is repeated, up to 12 flowers; primary peduncle slender, 8–12 mm, with caducous or persistent light brown hairs, lateral pedicels pilose, up to 5 mm long, slender with a hood-shaped, acute, 2 mm long bract at its base, beneath the calyx on the pedicel a tiny (1.5 mm) pilose bract. Flowers yellow, globose-ovoid in bud, 8 mm long, open flowers 10 mm long, fragrant. Lobes 2–3, small, thick, inside pubescent, may remain closed, though ripe pollen is formed inside. Corolla yellow, urceolate, with a very narrow throat, tapered. Lobes 4, along middle line pilose inside. Stamens numerous, irregular, 18–33, hypogynous or in centre on receptacle, anthers 2 mm, not very apiculate, filaments 0–2 mm, glabrous. Pistil none or a small disk. Female flowers solitary or cymose, on the lower part of young shoots in axil of small bract or leaf. Pedicel 6–10 mm long, slender, with minute, adpressed hairs, slightly enlarged towards flower. Flowers in bud globose-apiculate, pilose, 5 mm, remaining closed for a long time as in the male. Calyx deeply cut into 4 tapered, acute lobes, both sides sericeous. Corolla yellowish white, urceolate to urceolate-tubular. Lobes 4, triangular, acute, short. Staminodes 12–16, in an epipetalous ring, short, anthers broad. Ovary densely pubescent, 4.5–6-celled.

Style 0; stigmas 3–4, short. Fruit ovoid-ellipsoid to subglobose, top flattened, yellowish, smooth, covered with very short brown hairs, glabrescent, 2.5–6 cm high, 2–4 cm diam. Calyx 1.5–2.5 cm diam., flat, rather thin; tube none; lobes triangular-ovate, acute, 5–10 mm long, glabrescent.

Distr. As far as Sylhet in Bangladesh and Khasi Hills in NE India.

Ecol. Moist region only, up to 1000 m alt.

Vern. Kaluwelle (S); Vellai thoverii (T).

Note. A. DC. (l.c.) described the leaves as sericeous, wrongly translating Roxburgh's glossy leaves (he saw no specimen). Bakhuizen kept the name *D. toposia* (although quoting *D. racemosa*); he missed or did not understand that *D. racemosa* had priority, because it was published in the first edition of Roxburgh's Flora Indica, as clearly indicated by Alston.

The fruiting specimen which I collected, had ripe, yellow, sweetish fruits; they are much smaller than those indicated by Wright (only 2.5 cm long, the calyx only 1.5 cm diam.).

Specimens Examined. GALLE DISTRICT: Kaneliya forest near Hiniduma, alt. 200 m, Aug., fr., *Kostermans 25369* (A, BO, G, K, L, P, PDA, US); Hinidumkande (Haycock) near Hiniduma, alt. 200 m, Aug., fl., *Kostermans 25472* (id.). KANDY DISTRICT: Rangala to Corbets Gap, alt. 700 m, May, fl., *Kostermans 23492* (A, BO, G, K, L, P, PDA, US); Galaha Estate, March, fl., *Worthington 3634* (K).

15. Diospyros albiflora Alston in Trimen, Handb. Fl. Ceylon 6: 180. 1931 (excl. cit. Borneo & Java).—**Fig. 2.**

Diospyros embryopteris var. *nervosa* Thw., Enum. Pl. Zeyl. 178. 1860; Trimen, Handb. Fl., Ceylon 3: 93. 1895; Wright, Ann. R. Bot. Gard. Peradeniya 2: 142. 1904; Kosterm., Ceylon J. Sci., Biol. Sci., 12 (2): 92. 1977. Type: *C.P. 1910* (PDA).

Tree, up to 25 m high and 40 cm diam. Buttresses few, up to 1 m high, 50 cm out, thin. Bark rather smooth, dark purplish brown, hard, 0.25–0.5 mm thick. Live bark 5–8 mm thick, brown to pale reddish-brown. Twigs rigid, rather thick, apical part and end bud with adpressed or sub-adpressed black to dark brown hairs. Leaves rigidly coriaceous, glabrous, ovate-lanceolate, rarely ovate, 2.5 × 7–7 × 19 cm, rather shortly acuminate with sharp tip, base rounded and the central part very shortly acutish, both surfaces conspicuously, rather densely, prominulously, smoothly reticulate; on upper surface midrib slender, impressed, on lower surface prominent; lateral nerves 5–7 pairs, arcuately ascendent, prominulous. Petiole rather slender, up to 1.5 cm long, narrowly channelled above. Inflorescence axillary, few-flowered; flowers 3–5 together (male ones) on short pedicels; inflorescence and calyx with a black

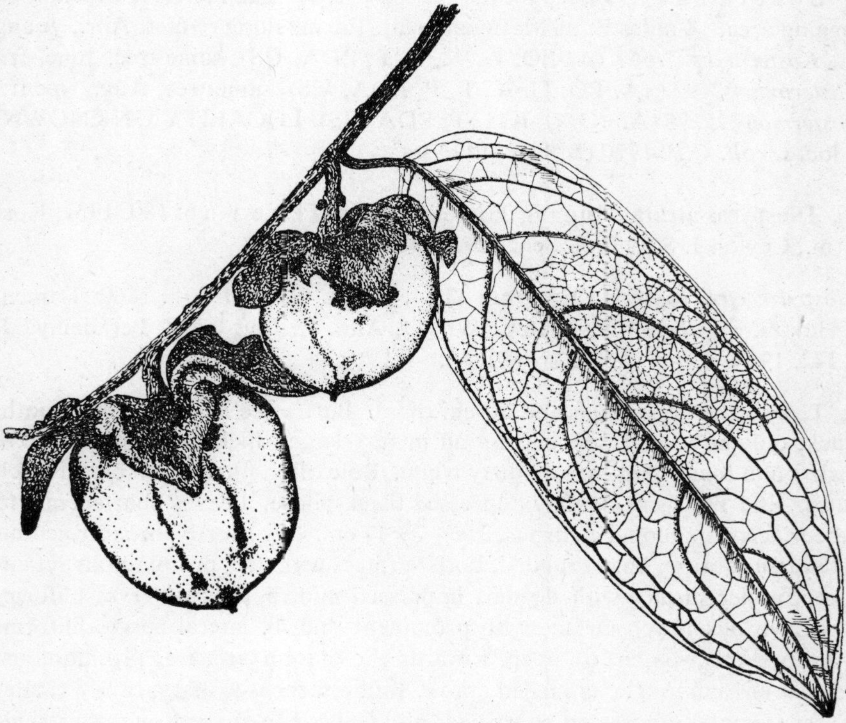

Fig. 2. *Diospyros albiflora* Alston.

tomentum. Flowers white, calyx campanulate, deeply lobed. Female not known. Fruit globose to sub-ellipsoid-globose, 5 cm diam., resembling those of *D. peregrina* with the same, easily-rubbed-off dark rusty scurvy tomentum. Pericarp leathery, 2–3 mm thick. Seeds 4, large, 3 cm long, wedge-shaped; endosperm equable. Base of fruit surrounded by 4 very coriaceous, ovate-triangular, 2.5–3 cm long, acute, at base 2.5 cm wide calyx lobes, where they meet at the base protruding as four sharp ridges. Pedicels rather thick, c. 6 mm long.

Distr. Endemic.

Ecol. Everwet, lowland forest of SW Lanka. Very rare.

Note. After the original collection of Thwaites, only one tree was discovered by me, of which regularly fruits were collected; unluckily the tree was blown over later. The fruits are much bigger than anticipated and resemble strongly those of *D. peregrina*. There are 4 seeds, but I found only one which was perhaps normal; the woody part surrounding the pericarp looks abnormal.

Specimens Examined. KALUTARA DISTRICT: Moropitiya logging area, 14 miles E. of Matugama, alt. 100 m, along rivulet, Apr., young fr., *Kostermans 24661* (A, BO, G, K, L, P, PDA, US), same tree, June, fr., *Kostermans 24977* (A, BO, G, K, L, P, PDA, US), same tree, Aug., ripe fr., *Kostermans 25298* (A, BO, G, K, L, P, PDA, US). LOCALITY UNKNOWN: s. loc. *s. coll. C.P. 1910* (BM, K, PDA).

16. Diospyros atrata Alston in Trimen, Handb. Fl. Ceylon 6: 180. 1931; Kosterm., Ceylon J. Sci., Biol. Sci., 12 (2): 92. 1976.

Diospyros embryopteris var. *atrata* Thw., Enum. Pl. Zeyl. 178. 1860; Trimen, Handb. Fl. Ceylon 3: 93. 1895; Wright, Ann. R. Bot. Gard. Peradeniya 2: 142. 1904. Type: *C.P. 2731* (PDA).

Tree, up to 25 m high and 70 cm diam. Buttresses small. Bark smooth, black, only at base of bole peeling off in very large, thick, black plates. Live bark 7 mm thick, pink, inside dirty white. Bole often fluted with broad thick flutes. End bud very densely adpressed black-pilose. Leaves chartaceous to subcoriaceous, oblong, glabrous, 3×8–7×17 cm, very shortly, broadly acuminate or subobtuse, base rounded, both surfaces with fine, prominulous reticulation; upper surface with slender, impressed midrib, lateral nerves filiform, hardly visible; lower surface with prominent midrib, lateral nerves filiform, 5–9 pairs, erect-patent to steep, towards the margin arcuate, prominulous. Petioles c. 1 cm long, channelled above. Inflorescences axillary, of 3–8 (male) or less (female) flowers on short pedicels. Calyx cup-shaped or rather shallow, thick, 2–4 mm high; lobes fleshy, ovate-triangular, 2–4 mm long, acutish. Corolla pinkish white or white, fleshy, 8–12 mm long, 10–12 mm diam. Corolla tube very short, fleshy; lobes 4, fleshy, ovate, acute, 4 mm long, centre slightly adpressed pilose outside. Stamens numerous, pubescent. Calyx with black hairs on both sides, denser inside. Female flower about same size, ovary globose, styles 4. Fruit globose, up to 7 cm diam., with rusty brown scurvy tomentum, easily rubbed off. Calyx lobes 4–10 mm long, inside black-haired, reflexed, stiffly coriaceous. Seeds 2.5×1.5 cm.

Distr. Endemic.

Ecol. Gannoruwa jungle behind Bot. Gard. Peradeniya, Alutnuwara, Gattekellie, rare. A magnificent tree in the Peradeniya Bot. Garden.

Note. The leaves are less thin than described by Alston; moreover, they are oblong. They resemble strongly those of *D. peregrina*, but are somewhat thinner. The calyx lobes under the fruit are reflexed (in *D. peregrina* clasping the fruit), much thicker. The leaves are glabrous and not black-haired as stated by Trimen.

According to Gamble also occurring in India, which I doubt.

All fruits collected from the tree in the Peradeniya Botanic Garden were seedless, even the mature ones; the septa have become woody and very hard.

Specimens Examined. KANDY DISTRICT: Gannoruwa, jungle behind Bot. Gard. Peradeniya, alt. 500 m, Apr., fl., *Kostermans 23123* (A, BO, G, K, L, P, PDA, US); Bot. Garden, Peradeniya, May, fr., *Kostermans 24917* (id.); ibid., Sep., fr., *Kostermans 24539* (id.); Gattekellie, *de Silva 25* (PDA). LOCALITY UNKNOWN: *s. coll.*, fr., *C.P. 2731* (BM, K, PDA); *Gardner 531* (K); *s. coll., C.P. 1913* (K).

17. Diospyros malabarica (Desr.) Kostel., Allg. med.-pharm. Fl. 3: 109. 1834; Hochreutiner, Bull. Inst. Bot. Buitenzorg 22: 65. 1905 (cum var. *genuina*); Troup, Silvic. Ind. Trees 2: 651, fig. 249. 1921; Alston in Trimen, Handb. Fl. Ceylon 6: 180. 1931; Bakh., Bull. Jard. Bot. Buitenzorg Ser. 3, 15: 325. 1937 (excl. vars. *atrata & nervosa* Thw., forma *atrata & pallida* Bakh., exclus. specim. of Celebes and Java); Worthington, Ceylon Trees 317. 1959; Kosterm., Ceylon J. Sci., Biol. Sci. 12 (2): 103. 1976.

Garcinia malabarica Des. in Lam., Enc. 3: 701. 1719. Type: *Panitsjika maram* Rheede, Hort. Mal. 3: t. 41. 1682.

Diospyros malabarica forma *pallida* Bakh., l.c. 329.

Diospyros peregrina (non Guerke) Anon., Wealth of India 3: 85. fig. 47, col. pl. 1952.

Diospyros embryopteris (non Spr. nec Bojer) Pers., Syn. 2: 624. 1807; Ker-Gawl, Bot. Reg. 6: t. 499. 1820 (excl. *Cavanillea philippensis* Lam.); A. DC. in DC., Prod. 8: 235. 1844; Thw., Enum. Pl. Zeyl. 178. 1860 (excl. vars., *C.P. 1915*); Bedd., Fl. Sylv. 1: t. 69. 1870; Hiern, Trans. Cambridge Philos. Soc. 12: 257. 1873 (excl. vars.); Hiern, J. Bot. 15: 100. 1877; Bentley & Trimen, Med. Pl. t. 168. 1880; Clarke in Hook. f., Fl. Br. Ind. 3: 556. 1884; Wright, Ann. R. Bot. Gard. Peradeniya 2: 138, t. V (20), IX, XVIII (7–8). 1904; Gamble, Man Ind. Timb. ed. 2: 455. 1912.

Embryopteris glutinifera Roxb., Pl. Corom. 49, tab. 70. 1795; Willd., Sp. Pl. 4 (2): 836. 1806; Wight, Ic. Pl. Ind. Or. 3: (4), t. 843, 844. 1848 (*glutinifera* and *glutenefera*). Type: tab. 70. Roxburgh.

Diospyros glutinosa Koenig ex Roxb., Fl. Ind. ed. Carey 2: 533. 1832.

Diospyros globularia (Miq.) Koord. & Val., Bijdr. Booms. Java 1: 43. 1898.

Embryopteris globularia "Roxb." Miquel, Fl. Ind. Bat. 2: 1048. 1856 (sphalm. for *glutinifera*).

Embryopteris glutinifera G. Don, Gen. Syst. 4: 41. 1837 (sphalm.).

Ebenus arbor Ind. etc. Pluk. ex Burm., Thes. Zeyl. 91. 1731 (exclus. Rumph. & Grimm).

Panitsjika maram Rheede, Hort. Mal. 3: 45, t. 41. 1862 (exclus. Piso, Clusius, Bauhin); Hamilton, Trans. Linn. Soc. London 15: 107. 1827.

Dioecious and polygamous trees, up to 14 m high and 70 cm diam. Bole fluted in old trees. Bark black, rough, scaly, strips 5–20 mm wide, 1–3 mm thick, brittle. Live bark 4–10 mm thick, dark beefy red. End bud densely white-silky. Branchlets yellowish, covered with a sparse tomentum of minute, erect hairs (the flush). Leaves alternate, glabrous, thickly to thinly coriaceous, drying yellowish, oblong to lanceolate-oblong, 3 × 9–7 × 20–5 × 22 cm, obtuse or obscurely, broadly, obtusely acuminate; base cantracted into the petiole, the centre acute or cuneate, densely prominulously reticulate both sides; upper surface dark green, glossy, midrib slightly impressed, laterals filiform; below paler, midrib prominent, laterals 8–10 pairs, rather steep, straight, at margin ascendent, very slender, prominulous. Petiole 5–20 mm long, pilose in the flush, soon glabrous, canaliculate above. Male inflorescences in bud globose, apiculate, densely pilose, surrounded by bracts, 3–7-flowered. Mature inflorescence on an up to 2 cm long thickish, densely pilose peduncle; pedicels 4 mm long, subtended by an early deciduous, 8–10 mm long, acuminate, densely sericeous bract; bracteoles on pedicel up to 2 mm long. Flowers 6 mm long, ovoid in bud, mature ones 10–15 mm long. Calyx cup-shaped, silky; lobes 4–5, 3–5 mm long, ovate, acute. Corolla widely tubular, 8–13 mm long with wide throat which is silky on both surfaces; lobes 4–5, yellow, ovate, rounded, 3 mm. Stamens 24, 30, 40, 64 in an epipetalous ring around pistil rudiment or central, unequal in length, usually in pairs, c. 8 mm long, with rectangular attenuate anthers; filaments short, pilose. Rudimentary pistil small or large, a disk or a bunch of hairs. Female inflorescence 1–5-flowered, young same as male one. Peduncle 10 mm long with yellowish hairs which turn rapidly brown-red; flowers in axils of bracts, sometimes only the terminal flower remains. Flowers up to 2 cm long, 3 cm diam. Calyx cup-shaped. Lobes ovate, tapered, 15 mm long and broad (at base), glabrescent. Corolla widely tubular, with wide throat. Lobes up to 18 mm long, tip recurved. Staminodes 4, adnate to tube, short, pilose (sometimes more). Ovary globose, apiculate in bud, yellowish-green, densely pilose. Styles 4, short, pilose, each forked and the tips lobed. Fruit subsessile or shortly pedicelled, globose to depressed-globose; top often depressed, up to 6 cm diam., the multicellular glands dead and changed to a red-orange colour are mealy and easily rubbed off. Pulp transparent, sweet, astringent. Lesions in the thick pericarp often with droplets of hard ruby-red, clear gum. Calyx patent or the lobes somewhat erect, flexible or woody, consisting of a reflexed rim, up to 10 mm high and ovate, acute to hastate lobes, 10–20 mm long, glabrescent inside (the hairs minute, rather short). Seeds 4–8, up to 25 mm long, elliptical-wedge-shaped. Testa brown, smooth, endosperm equable. Seedling epigeal, cotyledons detached early.

Distr. South and Central East India, Siam, Indo-China, Western Malesia (cf. notes).

Ecol. Dry and intermediate zones, but always near rivers and tanks, invariably associated with *Terminalia glabra*.

Vern. Timbiri (Timbeeree) (S); Panichchai (T); Gaub Persimmon (India); Fendu (Hindi).

Note. Timber moderately hard and heavy (50 lb/cu. ft.), used for boat building and construction. Usually the heartwood is rotten; occasionally, when the heartwood is sound, it is variegated black and brown and is as pretty as calamander, although inferior.

Specimens Examined. MATALE DISTRICT: Road Dambulla-Anuradhapura, along river near Dambulla, swampy, low, June, ripe fr., *Kostermans 25104* (A, BO, G, K, L, P, PDA, US), May, young fr., *Kostermans 24501* (id.). POLONNARUWA DISTRICT: Polonnaruwa Ruin area, dry, deciduous forest, low alt., May, young fr., *Kostermans 24293 & 24294* (id.); mile 66, Polonnaruwa-Batticaloa road, June, young fr., *Meyer et al. 124* (L, PDA, US). ANURADHAPURA DISTRICT: Wilpattu Nat. Park, June, fr., *Wirawan et al. 912* (id.). KURUNEGALA DISTRICT: Kurunegala Hills, rocky streamlet, Apr., fr., *Kostermans 24501* (A, BO, G, K, L, P, PDA, US). TRINCOMALEE DISTRICT: Trincomalee-Kuchchavelli road, low, dry scrub, May, after anthesis, *Kostermans 24803* (id.); near Trincomalee, ster., *Koenig s.n.* (L); Kantalai, Apr., fr., *Worthington 2658* (PDA). MONERAGALA DISTRICT: N. of Kataragama, Menik Ganga, Oct., fr., *Wirawan 664* (L, PDA, US). KANDY DISTRICT: Botanic Garden Peradeniya, alt. 400 m, Apr., young fr., *Kostermans 24789* (A, BO, G, K, L, P, PDA, US). BADULLA DISTRICT: Galbokka, between Uraniya-Ekiriyankumbura, Apr., fl., *Jayasuriya 1928* (PDA, US). LOCALITY UNKNOWN: fr., *s. coll. C.P. 2731* (L, P, PDA).

18. Diospyros trichophylla Alston in Trimen, Handb. Fl. Ceylon 6: 181. 1931.

Diospyros pruriens (non Dalz.) Thw., Enum. Pl. Zeyl. 423. 1864; Trimen, Handb. Fl. Ceylon 3: 95. 1895; Wright, Ann. R. Bot. Gard. Peradeniya 2: 151, t. I (1), XIII (9–12). 1904. Type: *C.P. 2836* (PDA).

Dioecious tree up to 8 m high and 10 cm diam., usually much smaller. The end bud and slender branchlets densely covered with long, stiff, 3–5 mm long hairs. Bark black, rather smooth with narrow short longitudinal ridges, 1 mm thick. Live bark reddish brown, 2 mm thick. Young leaves whitish. Leaves alternate, thin, 1.5 × 5–4.5 × 10 cm, obtusely acuminate, base rounded to subcordate, upper surface glabrous, except along midrib, venation lax, faintly pellucid, laterals faint; lower surface with a lax tomentum of long yellowish brown stiff hairs, especially on veins, midrib, lateral nerves and margin, lateral nerves c. 9 pairs, erect-patent, obscure; venation rather lax, obscure. Petiole 2–5 mm, reddish-brown, with long, stiff hairs. Male inflorescence an axillary raceme of 3–4 pedicellate flowers; peduncle reddish-brown with long, stiff hairs, 1 cm long, slender, widening at the flowers. Pedicel short, basal bracts reddish-brown, orbicular, 3 mm long, caducous, very hairy outside. Flowers yellow, 13–25 mm long. Calyx green, cup none. Lobes 4, chartaceous, linear oblong, narrow at base, tapering to acuminate apex, 8 mm long, with yellowish-white, 5 mm long hairs. Corolla yellowish-white, tubular, 13–25 mm long, 2.5 mm diam. at base (base swollen), expanded top 2.5 cm diam.; tube inside hairy. Lobes 4, much longer than tube, rotate, basal parts overlapping, acute, reddish-brown, 13 mm long, 6 mm wide at base. Stamens 12–14, connate into a central column (cf. notes). Pistil 0 or a rudimentary disk. Anthers narrow, acute. Female flower solitary in upper leaf axils, yellow. Peduncle up to 16 mm long, with long stiff hairs, often nutant. Calyx green, accrescent in flower. Lobes 4–5, oblong, acute, 8 mm long, 3 mm wide at base, inside with long white hairs, outside subglabrous. Corolla yellow, tubular, constricted towards middle, 10–28 mm long. Lobes 4–5. Staminodes 5 or more, epipetalous, linear, glabrous; ovary depressed globose, 4-celled, 5 mm diam., densely long white pilose; styles 2, 4 mm long. Stigmas 4, yellow. Fruit solitary, globose, strongly apiculate, 2 cm diam., 2.5 cm high, greenish-yellow with long white hairs. Pericarp thin. No cup. Calyx lobes incurved or patent, thin, 11 mm long, 4 mm wide at base. Seeds 1–4 per fruit, elliptic-wedge-shaped to globose-ovoid, 16 mm long, 10 mm wide and thick. Testa reddish-brown, smooth. Endosperm equable. Seedling epigeal, cotyledons detached early, testa does not split; hypocotyl reddish, densely covered with long, white hairs.

Distr. Endemic.

Ecol. Southern and S. Western Province, wet, evergreen forest, up to 700 m, Gilimale, Sinharaja, Eratna, etc., along rocky streams. Trees may be only a few cm high. Rare.

Note. *D. pruriens* of S. India differs mainly by its sessile flowers. The tree has no black heartwood, only black traces; the timber is white, tinged with yellow, turns red on exposure. It is too small for commercial purposes.

Wright is the only one who saw live flowers. In one case the flower contained an androgynophore with two tiers of stamens, a case unique in *Diospyros*.

Specimens Examined. RATNAPURA DISTRICT: Eratna, Paratala Mukalana, 400 m, very wet, evergreen, Oct., ster., *Kostermans 25691 & 25693* (A, BO, G, K, L, P, PDA, US).

19. Diospyros attenuata Thw., Enum. Pl. Zeyl. 182. 1860; Beddome, For. Man. in Fl. Sylv. 3: CXLVI. 1869; Beddome, Ic. 1: t. 139. 1874; Hiern, Trans. Cambridge Philos. Soc. 12: 182. 1873; Clarke in Hook. f., Fl. Br. Ind. 3: 561. 1882; Trimen, Handb. Fl. Ceylon 3: 96. 1895; Wright, Ann. R. Bot. Gard. Peradeniya 2: 154, t. III (10), XI (10–15). 1904; Alston in Trimen, Handb. Fl. Ceylon 6: 182. 1931; Kosterm., Ceylon J. Sci., Biol. Sci. 12 (2): 92. 1976. Type: *C.P. 3478* (PDA).

Tree, up to 7 m tall and 15 cm diam., usually much smaller, dioecious, bearing flowers when trunk is only 2.5 cm diam., branching low; branches stiff, patent or weeping, slender. Bark black, smooth or finely fissured, 1 mm thick. Live bark 3–4 mm, reddish, inside white. Wood white, smells faintly of passion fruit. Apex of branchlets sparingly puberulous, soon glabrous. Young leaves reddish above, glabrous, except midrib beneath. Mature leaves alternate, glabrous, thinly chartaceous, narrowly ovate to oblong-lanceolate, 2 × 5–4 × 12.5 cm, acuminate to caudate-acuminate with obtuse tip or subacute, base acute or cuneate; both surfaces with conspicuous, dense, prominulous reticulation, feebly pellucid in young leaf; above midrib slender, channelled, lateral nerves hardly differentiated from reticulation; lower surface with prominent, slender midrib; lateral nerves widely spaced, rather patent hardly differentiated from reticulation; accessory lateral nerves in between. Petiole glabrous, slender, 2–4 mm long. Male flowers white, sessile in dense clusters of 1–25 in axils of leaves or fallen leaves. Calyx 2 mm long, 1.5 mm wide, green, pubescent (black hairs); lobes 1–5, deltoid, reflexed, acute, hardly 1 mm long, dark hairy on both surfaces; at base a tiny bract with black hairs on the back. Corolla tubular, dirty white, up to 8 mm long, diam. at base 2 mm, outside densely dark hairy, lobes 4–5, rotate, 2 mm long, subglabrous inside. Stamens epipetalous, 4–6, unequal, glabrous; anthers tapered upwards, sometimes apiculate, 1.6 mm long; filaments white, 0.3–1.2 mm long; pollen smooth. Rudimentary pistil a hairy disk. Female flowers sessile, solitary in leaf axil, in bud 4-angled, dirty yellow, 10 mm long, diam. 5 mm. Calyx 6 mm long, sparsely black-pilose, deeply split into 3–5 deltoid segments with reflexed margin and feeble intersegmental pouch, sparsely pilose. Corolla dirty yellow (red after anthesis), 3–4 mm longer than calyx; lobes 3–5, tapered, acute, rotate with under surface of exposed part with dense pubescence, 4 mm long. Staminodes 4–5, epipetalous, alternate with corolla lobes; anthers glabrous, 1 mm long; filaments yellowish-white, glabrous, 2 mm long. Ovary pear-shaped, 4-celled, covered with stiff, black hairs, 3 mm long. Style short, stigmas 2, black, 1 mm long. Fruit ovoid, up to 2 cm diam. at the base, gradually tapered upwards into a sharp tip, 2–4 cm long, green, with few short hairs. Calyx enlarged to a diameter of up to 1.5 cm; lobes narrow, acute, 8 mm long, thin, green, pilose. Seeds 2–4, elliptical-wedge-shaped or semilunar, up to 2.5 cm long, 1 cm wide and 5 mm thick. Testa brown, smooth; endosperm copious, equable. Seedling epigeal, cotyledons detached early. First epicotyledonary leaves opposite, broadly ovate; apex tapered; base rounded; reticulate, 4–5 cm long, 2.5–3 cm wide.

Distr. Endemic.

Ecol. Wet, evergreen lowland forests in SW Lanka with a very restricted area near Elpitiya and further eastwards. Very rare.

Vern. Kadumberiya.

Note. One of the very rare *Diospyros* species, but locally not uncommon and one with the smallest flowers. Tends to grow in groups. Often many slender boles on one short (20 cm) communal trunk. Fallen fruit shows germinating seeds of which the root pierces the tip of the fruit. The timber is reddish with a little black, decaying heartwood, without commercial value.

Clarke described the leaves as coriaceous, which they are certainly not.

Specimens Examined. GALLE DISTRICT: Beriliya forest near Elpitiya, low alt., Sep., fr., *Kostermans 25639* (A, BO, G, K, L, P, PDA, US); ibid., Sep., fl., *Kostermans 25640* (id.); ibid., Sep., fr., *Kostermans 25641, 25614, 25619* (id.); MATARA DISTRICT: Akuresse, Mar., buds, *Worthington 2529* (PDA); Lower Deniyaya, 150 m alt., ster., *Worthington 2222* (PDA). DISTRICT UNKNOWN: fl., fr. *s. coll. C.P. 3478* (BM, K, PDA).

20. Diospyros acuta Thw., Enum. Pl. Zeyl. 182. 1860; Beddome, For. Man. in Fl. Sylv. 3: CXLI. 1869; Hiern, Trans. Cambridge Philos. Soc. 12: 182. 1873; Clarke in Hook. f., Fl. Br. Ind. 3: 561. 1882; Trimen, Handb. Fl. Ceylon 3: 96. 1895; Wright, Ann. R. Bot. Gard. Peradeniya 2: 157, t. III (11), VII (8–14). 1904; Alston in Trimen, Handb. Fl. Ceylon 6: 182. 1931; Worthington, Ceylon Trees 313. 1959. Type: *C.P. 3476* (PDA).

Monoecious tree, up to 10 m tall and 10 cm diam. usually 4–6 m high and 2–7 cm diam.; flowers when only 3 cm diam. Branchlets slightly pilose, soon glabrous. Bark black, finely fissured, hard, 1 mm thick. Live bark reddish, 3 mm thick, outside glossy dark red with paler stripes. Buds adpressed, rusty pilose. Branchlets stiff, zigzag. Leaves alternate, coriaceous, glabrous, lanceolate-oblong, 2.5 x 12–7.5 x 30 cm, gradually tapered to an acute or obtuse tip, base contracted into the petiole, slightly rounded or acutish; both surfaces smooth, lower one drying a red-brown colour; upper surface with slender, channelled midrib, lateral nerves obscure, reticulation obscure, below midrib strongly prominent, lateral nerves patent to erect-patent, reticulation sometimes subconspicuous. Petiole stout, 1–2.5 cm long, channelled above. Inflorescences adpressed-fulvo-pilose. Flowers white, male ones forming sessile clusters of 1–7 flowers; bracts green, caducous, pubescent, tapered, acute, 7 mm long, 5 mm wide at base. Flowers yellowish, 14 mm long, diam. 3 mm, at lobe height expanded, 10 mm diam. Calyx brown, densely pubescent, 5–7 mm; lobes 4–5, accrescent, lanceolate, acute, 4 mm. Corolla yellowish-brown, pilose, 10–16 mm; lobes 4–5, rotate, very hairy on half the outer surface, acuminate, 4 mm long. Stamens 5, 7, 9, epipetalous, glabrous, free or in pairs, unequal, inner stamens of a pair often very small with little pollen; anther yellowish-brown, apiculate, 3 mm; filament white, 2 mm. Pollen smooth, globose. Pistil 0 or a bunch of hairs. Female inflorescence axillary, sessile, of 1–4 flowers, usually in the upper leaf axils. Bracts larger than in male ones, early caducous, hairy. Flower in bud 15 mm long, 5 mm diam. Calyx 10–12 mm long, brown, both sides hairy. Lobes 5–6, lanceolate, acute, 8 mm long with small, intersegmental groove; margins straight (reflexed according to Hiern). Corolla yellow, 18 mm long, at base 6 mm diam.; open flower 14 mm diam., outside pilose; lobes 5–6, rotate, reflexed, hairy outside, subglabrous inside, apex tapered, 5.5 mm long, base 5.5 mm wide. Staminodes 5, epipetalous, alternate with corolla lobes, anthers yellow, filaments white. Ovary globose, apiculate, densely pilose, 4–6-celled; style 4 mm, densely pilose; stigmas 2–3, reddish-brown, spathulate, glabrous, 1.5 mm long. Fruit ovoid-attenuate, strongly tapered towards stigma, up to 15 × 35 mm. Calyx cup-shaped, 1 cm deep; lobes 10–15 mm long, lanceolate, acute, intersegmental groove and reflexed margins pronounced. Seeds 2–6 per fruit, flattened-elliptic, strongly accuminate at micropyle, up to 7 × 22 mm. Testa brown, smooth; endosperm copious, equable. Seedlings epigeal; cotyledons detached in an early stage.

Distr. Endemic.

Ecol. Very rare, wettest parts of SW Lanka, Vevella, Weerakande, Hewesse, Pasdun Korale, Galle Distr. (Kottawa Arboretum).

Note. Often more than one bole on a communal very low trunk, at base with many small, black, stiff roots. Timber dirty white, turns red on exposure, heavy, dense, no black heartwood. Too small to be of commercial value.

Specimens Examined. GALLE DISTRICT: Kottawa Arboretum near Galle, Aug., fl., fr., *Kostermans 25468*, *25469*, *25470*, *25471* (A, BO, G, K, L, P, PDA, US); Kanneliya forest reserve, Sep., fl., *Jayasuriya 1542* (L, PDA); Buona Vista Hill near Galle, July, fl., *Meijer 260* (K); ibid., Oct., fl., *Balakrishnan 964* (K). LOCALITY UNKNOWN: fl., *Walker C.P. 3476* (BM, K, PDA).

21. Diospyros walkeri (Wight) Guerke in Pflanzenfam. 4 (1): 162. 1897; Alston in Trimen, Handb. Fl. Ceylon 6: 182. 1931; Bakh., Bull. Jard. Bot. Buitenzorg Ser. 3, 15: 279. 1937 (*walkerii*).

Patonia walkerii Wight, Ill. 1: 18. 1831; Bakh., l.c. (sub *D. undulata* Wall. ex Don); Kosterm., Ceylon J. Sci., Biol. Sci. 12 (2): 35. 1976. Type: *Walker s.n.*

Diospyros gardneri Thw., Enum. Pl. Zeyl. 181. 1860; Beddome, For. Man. in Fl. Sylv. 3: CXLVI. 1869; Beddome, Ic. 27, t. 132. 1871 (fruit wrong!); Hiern, Trans. Cambridge Philos. Soc. 12: 29, 34, 53, 75, 214. 1873; Clarke in Hook. f., Fl. Br. Ind. 3: 561. 1882; Trimen, Handb. Fl. Ceylon 3: 96. 1895; Wright, Ann. R. Bot. Gard. Peradeniya 4: 160, tab. V (18), XII. 1904; Lewis, l.c. 10, 24, 88. 1926. Type: *C.P. 1908*, Saffragam and Kornegalle (Kurunegala) Distr., and near Kandy (PDA).

Diospyros undulata Wall. ex Don var. *walkeri* (Wight) Bakh., l.c. 281. 1937. (p.p., excl. cit. specim. Asia meridionalis).

Diospyros ebenaster Moon, Cat. 74. 1824.

Small tree, up to 6 m high and 10 cm diam., rarely up to 15 m high and

20 cm diam., dioecious and polygamous, fruiting already when 3 m high and 3 cm diam. Bark black, finely fissured, strips 2 cm wide, 0.5 mm thick. Live bark straw-yellow, 2 mm, with yellow sap. Wood white. Branchlets rather slender with apically a rather dense tomentum of adpressed hairs. Buds densely, coarsely sericeous. Cut branches turn immediately dark yellow inside. Leaves chartaceous to thinly coriaceous, glabrous, drying brownish-green, oblong, ovate-oblong to lanceolate-oblong, 2.5 × 6–6 × 18 (–9 × 22) cm, distinctly acuminate; acumen often long; base acute, slightly decurrent into petiole; both surfaces very laxly, prominulously, rather obscurely reticulate; above midrib slender, impressed, the surface glossy, slightly wrinkled (dry), lateral nerves very slender, prominulous (often in a groove); below less glossy, midrib slender, prominent, laterals 6–8 pairs, erect-patent, arcuate. Petiole slender, 7–20 mm, soon glabrous, channelled above. Flowers in polygamous trees are: a) with short, accrescent calyx lobes, fertile anthers, abortive pistil, b) with large, deltoid calyx lobes with straight or recurved margins, fertile anthers, abortive pistil, c) with large deltoid calyx segments, with deep intersegmental grooves, fertile anthers, fertile pistil. The enlarged, recurved calyx lobes are an indication of hermaphroditism, but this character is not always present and occurs also in female flowers. Male and hermaphrodite flowers occur separately or in the same inflorescence. Male inflorescence sessile or subsessile pubescent cymes, initially as a red densely pilose bud in a leaf axil with or without peduncle, bearing decussate bracts. The first flower on a peduncle up to 10 mm long, others always sessile. Calyx green, campanulate with neck at base, inside sericeous, 5 mm long, outside very minutely brown pilose. Lobes 4, deltoid, acuminate, margin straight or slightly recurved, 3 mm long. Corolla yellow or white, in buds conical, 10–13 mm long; tube inflated, densely sericeous. Lobes 4, half the tube length, patent, inside glabrous. Stamens 12–16, epipetalous or monadelphous and hypogynous, single or in pairs, inner whorl shorter; anthers slender, apiculate, glabrous or somewhat pilose. Pistil none or a flat disc, showing compartments in cross section. Female flower solitary, axillary, sessile or nearly so. Peduncle 4 mm, with 4 decussate bracts without flowers. Calyx green, widely campanulate, very minutely brown-adpressed-pilose, thick. Lobes 4 (rarely 3), thick, deltoid, 1 cm long, conspicuously, sharply acuminate, margin slightly recurved laterally so as to form a ridge projecting outwards and downwards. Corolla white or yellow. Tube tubular, 15 mm long, densely sericeous. Lobes strongly reflexed or patent, 7–10 mm long. Staminodes 8, epipetalous, 5 mm long, filaments 3 mm, pilose, adnate to corolla tube at base. Ovary globose, densely sericeous, 8-celled; style 5 mm, stigmas 3–4. Fruit yellowish, glabrescent, depressed globose, apiculate, up to 2.5 cm high, 3.5 cm diam., smooth, glossy. Calyx tube hemispherical, 1 cm or more deep, up to 3 cm diam., thick with thin rim, which is either entire or slightly wavy (the lobes) and then these with a very narrow recurved lateral margin, showing the densely

sericeous inner side of the calyx. Seeds 8, thin, up to 18 mm long; testa brown, smooth; endosperm equable. Seedling epigeal; cotyledons persistent; epicotyledonary system late in development.

Distr. Endemic.

Ecol. Wet evergreen lowland forest in SW Lanka.

Vern. Kadum beriya, Kaluwelle (S).

Note. Bakhuizen, l.c., included this in *D. undulata* as a variety. None of his distinctive varietal characters fit. He missed, however, the characters of the entirely different fruiting calyces of the species, that in *D. undulata* being rather shallow with three or four thick, sharp large lobes and a tomentum of rather long, adpressed hairs, that of *D. walkeri* having a very deep calyx with thin rim, with hardly any trace of lobes, the tomentum being extremely minute and moreover the calyx of the fruit has a short, thick neck at the base, not found in *D. undulata*. The leaves of both species are strikingly alike. Also, the shape of the calyx lobes, especially in the female flower, is quite different from that in *D. undulata*; the lobes look similar to the petals found in some Annonaceae (and hence Wight's mistake; he referred it to Annonaceae).

The original orthography *walkerii* of Wight was improved by Guerke to *walkeri*. Bakhuizen quotes this wrongly.

A cut branch turns deep yellow immediately, a character in common with *D. insignis*. The timber is useless because of the small size; it is white, turning yellow with some black streaks.

Specimens Examined. RATNAPURA DISTRICT: road Kalawana to Morapitiya in rivulet, Apr., fr., *Kostermans 24639* (PDA). June, fr., *Kostermans 24947 & 24956* (PDA), Aug., fr., *Kostermans 25360* (PDA), June, ster., *Kostermans 24953* (A, BO, G, K, L, P, PDA, US). GALLE DISTRICT: Kaneliya forest near Hiniduma, treelet 3.5 m, ster., (the hairs of the branchlets are patent!) *Kostermans 25384 & 25661 A* (PDA); ibid., Sep., fr., *Kostermans 25501 & 25524* (A, BO, G, K, L, P, PDA, US); Hiniduma, in rivulet near bridge near Hiniduma, June, fl., *Kostermans 25000* (id.); Kaneliya, Feb., fl., *Jayasuriya 1522* (L, PDA, US). COLOMBO DISTRICT: Labugama watercatchment area, Wagga, Indikada mukalana Reserve, Jan., fr., *Worthington 3471.* (K).

22. Diospyros sylvatica Roxb., Pl. Corom. 37, t. 47. 1895; Roxb., Fl. Ind., ed. Carey 2: 537. 1832; A. DC. in DC., Prod. 8: 231. 1844 (exclus. var. *velutina*); Thw., Enum. Pl. Zeyl. 178. 1860 (*C.P. 2729*); Beddome, For. Man. in Fl. Sylv. 3: CXLIII. 1869; Beddome, Ic. 25, t. 121. 1871; Hiern, Trans. Cambridge Philos. Soc. 12: 29, 33, 41, 65, 67, 161, 287. 1873 (excl. *D. orixensis* Klein ex Willd.); Clarke in Hook. f., Fl. Br. Ind. 3: 559. 1882; Trimen, Handb. Fl. Ceylon 3: 98. 1895; Alston in Trimen, Handb. Fl. Ceylon 6: 183. 1931;

Wright, Ann. R. Bot. Gard. Peradeniya 2: 17, 171, t. II (7); XI (1–9). 1904; Anon., Wealth of India; 3: 86. 1952. Type: tab. 47 of Roxburgh.

Large tree, up to 30 m high and 60 cm diam. (according to Wright up to 60 m and 130 cm diam.), dioecious and polygamous. Buttresses rather small, 50 cm high and out. Bole straight. Bark black to greyish-black, irregularly cracked, strips 2 cm wide, 2 mm thick, rather soft. Live bark 4 mm, yellowish-white to light brown. Wood white; little black heartwood. Branchlets slender, glabrous; only the flush sparsely, minutely, adpressed pilose. End buds minutely sericeous. Leaves alternate, chartaceous, glabrous, oval to narrowly elliptic, 3 × 7–6 × 13 cm, shortly acuminate; base narrowed, acute and slightly decurrent; both surfaces smooth (the lower one with very lax, obscure reticulation); upper surface with slender, impressed midrib, lateral nerves filiform, obscure, prominulous (or so in a groove), below midrib slender, prominent, lateral nerves 7–8 pairs, erect patent, arcuate towards margin. Petiole slender, 5–12 mm long, flat above. In polygamous trees, the flowers male or hermaphrodite, in the latter case with more accessory staminal whorls. In first weeks of flowering no hermaphrodite flowers. Both kinds occur in different clusters or in the same cluster. Flowers nodding, strongly aromatic, visited by bees. Male inflorescence a dichasial cyme of 7–15 flowers in axil of hairy, 6–10 mm long caducous bract or leaf (in the latter case more abundant), up to 2 cm long; main peduncle 7–10 mm long; pedicels of outer flowers up to 2 mm, subsequent ones up to subsessile; bracteoles small, pilose, caducous. Flowers small, bell-shaped. Calyx green, cup-shaped, 3 mm high; lobes 3–4, pilose or ciliate, 1 mm long, apex rounded. Corolla white, urceolate to campanulate, 8 mm long, at apex 8 mm diam.; lobes 3–4, rotate, rounded, glabrous, 3 mm long. Stamens slender, irregular, 13–22, epipetalous, single or in 8–10 fascicles of 2's and 3's; anthers apiculate; filaments white, united or free, glabrous, except near apex. Pistil none or a flat, undulating disc with locular structure. Female flowers solitary or as a cyme of 3 flowers, axillary; main peduncle of varying length, short. Pedicels minute, bracteate. Calyx green, cup-shaped, 12 mm diam.; lobes 3–5, ovate-elliptic, obtuse, 8 mm long. Corolla yellowish-white, 10 mm; lobes 4, rounded, 3 mm long. Staminodes 4, epipetalous, alternating with corolla lobes, glabrous. Ovary globose, tapered, glabrous, 6–8-celled; style with 3 stigmas. Fruit globose, smooth, 15–20 mm diam., often depressed. Calyx tube shallow; lobes suborbicular, spreading, at last reflexed, 6–7 mm long. Pedicel thick, 2 mm long, pilose. Seeds 2–8, ovoid to oblong or globose, up to 14 mm long; testa brown, streaked; endosperm strongly ruminate. Seedling epigeal; cotyledons persistent, the entire or greater part of the endosperm is absorbed before the testa splits. Epicotyledonary stem does not appear until after several weeks or up to 5 months after cotyledons mature.

Distr. South India, Ceylon.

E c o l. Wet and intermediate zone, up to 1300 m alt. Not in the very wet southwest part of Lanka. Rather rare.

V e r n. (doubtful) Sudu kadumberiya (S); Karruppu Thoveria (T).

N o t e. Worthington's plate (Ceylon trees 324. 1959) looks more like *D. ebenum*. The bark is described as conspicuously white, this is only found in *D. montana*. The calyx is said to be 3-pointed. The distribution also does not fit. What is this? The name white-barked ebony is certainly wrong and misleading.

The leaves are smooth and show hardly any reticulation, they are at the most thinly coriaceous.

S p e c i m e n s E x a m i n e d. MONERAGALA DISTRICT: mile 166, Moneragala Rd., June, fl., *Meyer 175* (L, PDA, US). KURUNEGALA DISTRICT: Kurunegala low hill, along dry streamlet, June, ster., *Kostermans 24037 A* (G, K, L), July, fr., *Kostermans 25239* (A, BO, G, K, L, P, PDA, US), Sep., ripe fr., *Kostermans 25540* (id.). KANDY DISTRICT: Botanic Garden Peradeniya, May, fr., *Kostermans 24860*, July, fr., *Kostermans 25259 A* (A, BO, G, K, L, P, PDA, US); Amparawa above Gampola, Aug., fr., *Kostermans 25349* (id.). ANURADHAPURA DISTRICT: Ritigala Reserve, alt. 600 m, May, fr., *Jayasuriya 1727* (L). MATALE DISTRICT: Nalande, mile 32, alt. 300 m, Nov., fr., *Worthington 2412* (K).

23. Diospyros melanoxylon Roxb., Pl. Corom. 1: 36, t. 46. 1795; Roxb., Fl. Ind., ed. Carey 3: 530. 1832; A. DC. in DC., Prod. 8: 224. 1844; Hiern, Trans. Cambridge Philos. Soc. 12: 159. 1873 (excl. *D. roylii* DC.); Clarke in Hook. f., Fl. Br. Ind. 3: 564. 1882; Trimen, Handb. Fl. Ceylon 3: 99. 1895; Alston in Trimen, Handb. Fl. Ceylon 6: 183. 1931; Wright, Ann. R. Bot. Gard. Peradeniya 2: 181, tab. I (5), VII (6–9). 1904; Brandis, Ind. Trees 432. 1906; Gamble, Fl. Pres. Madras 772. 1921; Anon., Wealth of India 3: 31–34, fig. 44–46. 1952; Worthington, Ceylon Trees 318. 1959. Type: Roxburgh, t. 46.

Diospyros wightiana A. DC., l.c. 223; Beddome, Fl. Sylv. 1: t. 67. 1871.
Diospyros dubia Wall. ex A. DC., l.c. 223; Wight, Ic. Pl. Ind. Or. t. 1223. 1848.
Diospyros exsculpta Beddome, Fl. Sylv. 1: t. 66. 1871.
Diospyros melanoxylon var. *beddomei* Clarke, l.c.

Medium to large tree, up to 2 m in diam., dioecious or polygamous. Bark rough, deeply fissured, ridges black, fissures lighter. Live bark 10 mm, brown. Branchlets slender, densely, minutely light rusty woolly, the hairs very slender. Leaves opposite, subopposite or alternate, chartaceous, elliptic to oblong, 1.5 × 4–7 × 14 cm (of rootsuckers up to 10 × 24 cm); obtuse or rounded, base shortly acute, the margins slightly decurrent to the upper side of the petiole. Young leaves densely, minutely light brown woolly on both surfaces, adult ones glabrous or nearly so; both surfaces laxly, prominulously reticulate, above midrib flush with the surface or prominulous, lateral nerves filiform, obscure, lower surface with prominent midrib, lateral nerves 8–9 pairs, erect-patent to rather patent, arcuate, filiform, prominulous. Petioles 6–20 mm, rather slender, apical part flat above. Male inflorescence usually a dichasial cyme of 3–7 flowers in the lower part of a young shoot in a leaf or bract axil, sometimes the lateral flowers are not placed in opposite pairs but at different levels, suggesting a racemose type. Main peduncle slender, 3–1 mm long, densely woolly. Pedicels of first flowers 15 mm long, others short. Bracts and bracteoles long, woolly, gradually decreasing in size upwards. Male flower in bud 5 mm long, pilose, open flower 13 mm. Calyx yellowish-green, campanulate, 5–8 mm, 5 mm diam., sometimes much smaller. Lobes 3–6, acute, 1.5–3 mm long. Corolla yellowish-white, densely sericeous, 12 mm long, tubular; lobes 4, 5 or 6, rotate, rounded or acuminate, 2 mm long. Stamens 8–20 on central disc, single or in groups of 2's and 3's; anthers somewhat apiculate, as long as the glabrous filaments. Pistil none or an apiculate rudiment or a bunch of hairs. Female flowers axillary, solitary; peduncle short, stout, light brown woolly, 1.5 mm long, bracteoles present. Calyx green to brown, deeply lobed, woolly, forming a rather shallow cup. Lobes 4–7, usually 6–7 mm long, 5 mm wide, with tapered apex, broad base, margins laterally reflexed, undulate, at last patent. Corolla yellow, with wide throat, both sides with a dense cover of silky hairs, 9 mm long. Lobes 5, tapered, 3 mm long. Staminodes 8–12, filaments longer than barren anthers. Ovary globose, densely pubescent; stigmas 4, hairy, fleshy. Fruit yellowish, globose, apiculate, glabrescent, up to 4 cm high, 3 cm diam., 2–4-celled, smooth. Calyx 2.5 cm diam., rather flat, pubescent, margin becomes depressed but still shows the acute tip of the lobes, the laterally reflexed margins and the obtriangular intersegmental, pilose part. Seeds 1–4, oval-wedge-shaped, up to 20 mm long; testa red brown; endosperm ruminate. Seedlings probably epigeal with cotyledons detached at an early stage.

Distr. South India, widely distributed.

Ecol. In Leiden in the Persoon Herbarium is a specimen from Ceylon, perhaps dating from Hermann's time. The species was re-discovered in 1888 near Bibile. It was unknown to Thwaites. Flowers were obtained by Wright in 1901. I have found the species near Bibile, where it occurs scattered in fire climax forest and Nilgala, where it is not uncommon, but very scattered. Many trees have been killed by fire.

Vern. Kadumberiya (S).

Note. In Ceylon the tree is best known for its leaves, used as wrappers for cigarettes (beedee leaves). Cigarettes and wrappers are a home industry (road to Nilgala), the leaves of rootsuckers are preferred as they are larger; they are cut in quadrangular pieces and dried. The dried leaves do not tear, nor deteriorate. Timber is one of the oldest known kinds of ebony for export.

Only in old trees enough brown and black variegated wood is found. The trees in Ceylon have very little heartwood. Both sap- and heart-wood are durable. One of the well known trees of the drier areas of South India as far up as Bombay (cf. Wealth of India).

Bakhuizen (l.c. 220) included it erroneously in species with glabrous leaves.

Specimens Examined. MONARAGALA DISTRICT: Bibile, Hill, burnt forest, Aug., ster., *Kostermans 25431* (L); Talawa forest near Bibile, Ekiriyankumbura (Bibile-Batticaloa road) mile 51, alt. 150 m, Aug., young fr., *Worthington 4931* (PDA), July, fl., *Worthington 4055* (K). POLONNARUWA DISTRICT: Mutugala (Polonnaruwa), alt. 50 m, Sep., young fr., *Worthington 6413* (K).

24. Diospyros hirsuta L. f., Suppl. Pl. 440. 1781; A. DC. Prod. 8: 223. 1844 (non Desf.); Thw., Enum. Pl. Zeyl. 181. 1860; Beddome, For. Man. in Fl. Sylv. 3: CXLVI. 1869; Beddome, Ic. 1: 28, t. 137. 1871; Hiern, Trans. Cambridge Philos. Soc. 12. 1873 (excl. synonyms); Clarke in Hook. f., Fl. Br. Ind. 3: 565. 1882; Trimen, Handb. Fl. Ceylon 3: 99. 1895; Wright, Ann. R. Bot. Gard. Peradeniya 2: 179, t. I (2), XV (6–12). 1904; Alston in Trimen, Handb. Fl. Ceylon 6: 184. 1931. Type: *Thunberg s.n.* (lost); lectotypus prop. *C.P. 382* (PDA).

Dioecious, monoecious or polygamous dioecious tree, up to 6 m tall and 10 cm diam., usually smaller. Crown open with long, horizontal branches. Bark black, finely fissured, strips 3 mm wide, 1 mm thick. Live bark reddish, 3 mm thick. Young branchlets subadpressed densely pubescent. Leaves alternate, subcoriaceous, 3.5 × 7–4 × 11–7 × 17 cm (in saplings up to 8 × 20 cm), ovate-lanceolate or elliptic, acuminate to caudate-acuminate, base tapered, shortly acute. Young leaves sparsely adpressed pubescent (especially underneath); mature ones above smooth, glabrous, midrib impressed, lateral nerves sometimes impressed, below (in sicco red brown) sparsely to more densely finely, adpressed pilose, midrib strongly prominent, lateral nerves filiform, rather patent, prominulous, widely spaced, near margin ascendant, in between parallel shorter ones. Petiole 8–11 mm long, densely pubescent. Male inflorescences axillary, dense (up to 20 flowers) sessile cymes, rarely up to 2 cm long. Flowers sessile or subsessile, at base with opposite pair of small, pilose bracteoles. Calyx 2.5 mm, pubescent, lobes 4–5, acute, 1 mm long. Corolla tubular, 5 mm long, yellow (fresh), throat narrow, expanded 5 mm diam.; lobes 4–5, overlapping in bud with exposed half externally densely sericeous, explanate, acute. Stamens 5, epipetalous, alternate with corolla lobes, glabrous, apiculate, 2.5 mm long; filaments short, pollen smooth, circular. Pistil rudimentary. Female inflorescence a sessile cluster (containing bisexual flowers), never solitary, bracteolate. Calyx larger than in male, 7–9 mm long;

lobes silky brown, triangular, 3–4 mm long, reflexed. Corolla yellow (fresh), thick, up to 9 mm long; lobes 5, acute. Staminodes 5 with glabrous anthers, 2 mm long, apiculate, filaments glabrous, 0.3 mm long. Ovary globose, apiculate, 8–10-celled, densely pubescent; styles 2, 2 mm long. Fruit ellipsoid-ovoid to subglobose, up to 3 cm high and 2.5 cm diam., densely, coppery-rusty haired (the hairs coarse, not easily detachable). Fruiting calyx flattish, initially pressed against young fruit, ultimately flat, thick with short, broad, acute lobes, of which the tips are reflexed, 2–3 cm diam., intersegmental pouches rather obscure, densely pilose both sides, glabrescent outside. Pericarp 2 mm. Seeds 1–8, usually 6, elliptic wedge-shaped with beaked micropile, 2 cm long, 1 cm wide; testa brownish black, transversally prominently striate; endosperm horny, ruminate. Seedling epigeal, cotyledons detached early.

Distr. Endemic.

Ecol. Wet lowland forests in SW Lanka, usually along streamlets.

Note. Timber without any black heartwood, dirty white, turning reddish on exposure. Too small for commercial purposes.

Wright mentioned a variety with flat, circular fruit and stout, persistent style from Eratna, Madol and Kadawatte. I have found this near Morapitiya. It is not a variety but a gall-like abnormality, the so-called style is the hardened corolla tube, of which the base is very hairy and inflated and contains rudimentary seeds.

Specimens Examined. KALUTARA DISTRICT: Morapitiya logging area, low, Aug., young fr., *Balasubramaniam 320* (L, PDA); ibid., 14 miles E of Matugama, Apr., fr., *Kostermans 24664* (A, BO, G, K, L, P, PDA, US); ibid., Oct., buds, *Kostermans 25694* (id.), Aug., Galled fr., *Kostermans 25361* (id.). RATNAPURA DISTRICT: Gilimale forest reserve, Oct., fr., *Worthington 3168* (BM, K, PDA). GALLE DISTRICT: Kanneliya forest reserve, alt. 100 m, May, fr., *Kostermans 24671* (A, BO, G, K, L, P, PDA, US); ibid., June, buds, *Kostermans 24989* (id.); ibid., Sep., buds, *Kostermans 25530 & 25332* (id.); ibid., Sep., fr., *Kostermans 25535 & 25536* (id.); ibid., Aug., fr., *Kostermans 25387 & 25390* (id.); ibid., Apr., fl., *Balasubramaniam 263* (L); Kottawa Arboretum, Sep., buds, *Kostermans 25587* (id.), Sep., buds, *Worthington 4107* (PDA); May, fl., *Kostermans 23639* (id.); "Haycock" (Hinidumkande), sapling, *Kostermans 25472* (G, L); Dediyagala forest, alt. 200 m, Mar., fr., *Worthington 2538* (PDA). LOCALITY UNKNOWN: fr., *s. coll., C.P. 382* (BM, K, PDA).

25. Diospyros moonii Thw., Enum. Pl. Zeyl. 182. 1860; Beddome, For. Man. in Fl. Sylv. 3: CXLVI. 1869; Beddome, Ic. 28, t. 138. 1874; Clarke in Hook. f., Fl. Br. Ind. 3: 566. 1882; Trimen, Handb. Fl. Ceylon 3: 101. 1895; Wright, Ann. R. Bot. Gard. Peradeniya 2: 191, t. I (3), XIV (6–10). 1904; Alston in Trimen, Handb. Fl. Ceylon 6: 185. 1931. Type: *C.P. 2833* (PDA).

Diospyros hirsuta forma *moonii* Hiern, l.c. 164 (excl. cit. Helfer 3623).

Tree, 7–20 m tall, and 15–50 cm diam. Bole obscurely fluted and inconspicuously, shortly buttressed. Bark rough, black, deeply fissured, peeling off; strips 3 mm wide, 2 mm thick, red or light red, outside ribbed, inside whitish, turning brown after exposure. Branchlets blackish, smooth, sparsely minutely sericeous towards the apex. End buds sparsely silky. Leaves alternate, coriaceous, young ones sparsely sericeous, especially underneath, mature ones glabrous, ovate-lanceolate to oblong-lanceolate or elliptic, 7 × 16–11 × 28 cm, shortly acuminate, base rounded to acutish; above midrib impressed, lateral nerves filiform, faint, below glabrescent, midrib stout, cylindrical, prominent, lateral nerves up to 9 pairs, widely spaced, rather patent, obscure. Petiole green, stout, 6–15 mm long. Young leaves reddish above. Flowers sessile in axils of leaves or fallen leaves. Male ones in clusters of 4–10, females usually solitary, at base with tapered, acuminate, 3 mm long, deciduous bracts. Male flower with brown, densely silky calyx; lobes 5, broadly triangular, acute, 4 mm long. Corolla cylindrical-urceolate, yellow (fresh), 8 mm long; lobes 4–5, short, triangular. Stamens 5. Pistil rudimentary. Female flowers same size; calyx sericeous both sides; lobes 5, obtusely apiculate, margins recurved, slight intersegmental groove. Corolla yellowish-white (fresh), lobes 5. Staminodes present. Ovary globose, pilose. Fruit globose, or subovoid-globose, sparsely adpressed pilose, glabrescent, solitary or a few together, apex depressed, up to 3–4 cm diam., slightly, smoothly ribbed, 6–10-celled. Fruiting calyx thick, hard, flattened, lobes broadly triangular, acute, densely, minutely sericeous both sides, margins slightly recurved. Seeds elliptic-ovate, flattened, up to 23 mm long; testa brown, transversally striate; endosperm strongly ruminate. Seedling epigeal, detached cotyledons exist a few days as white and shrivelled leaves, by which time the epicotyledonary leaves are green and well developed.

Distr. Endemic.

Ecol. Wet lowland forest of SW Lanka; usually on sandy or rocky soil in rivers or along rivulets, where inundation occurs. Very rare.

Vern. Kadumberiya, Kaluwelle (S).

Note. Wood reddish with occasional black strands; black heartwood lacking. Timber is inferior (for further information cf. Wright).

Specimens Examined. KALUTARA DISTRICT: Kalutara, fl., fr., *s. coll. s.n.* (K). GALLE DISTRICT: Hiniduma, along road near village near bridge in rivulet, June, fl., fr., *Kostermans 25002* (A, BO, G, K, L, P, PDA, US); ibid., Aug., fr., *Kostermans 25381 & 25458* (id.); Kanneliya forest reserve, in sandy rivulet. Aug., ster., & Sep., fr., *Kostermans 25496* (id.); Nagoda, Udugama, Sep., fr., *Worthington 4133* (BM, PDA). LOCALITY UNKNOWN: fr., *s. coll., C.P. 2833* (BM, PDA); young fr., *Walker s.n.* (K).

26. Diospyros oppositifolia Thw., Enum. Pl. Zeyl. 181. 1860; Hiern, Trans. Cambridge Philos. Soc. 12: 157. 1873; Beddome, Ic. 1: 27, t. 131. 1874; Beddome, For. Man. in Fl. Sylv. 3: CXLVI. 1869; Clarke in Hook. f., Fl. Br. Ind. 3: 565. 1882; Trimen, Handb. Fl. Ceylon 3: 100. 1895; Wright, Ann. Bot. Gard. Peradeniya 2: 185, t. 6. 1904; Alston in Trimen, Handb. Fl. Ceylon 6: 184. 1931. Type: *C.P. 3011* (PDA).

Monoecious tree, up to 10 m high and 50 cm diam. Bark very rough, black, peeling off, not deeply fissured; young branchlets fulvous pubescent. Leaves opposite or subopposite, coriaceous, glabrous (young leaves slightly pubescent), ovate or elliptic, 3.5 × 5–7.5 × 15 cm, abruptly very shortly acuminate, base obtuse or rounded or very shortly acutish; both surfaces smooth, lateral nerves very slender, erect-patent, prominulous, widely spaced; in young leaves the numerous connecting, parallel, filiform veins visible. Petiole dark, 2–5 mm long. Male flowers sessile or subsessile in bunches of 1–9 in axils of leaves or fallen leaves, never terminal and placed lower than the female ones, rarely mixed with female flowers. At the base a small caducous bract. Calyx tube narrowly campanulate, 2.5 mm long; lobes 4, narrowly triangular, acuminate, subglabrous, 4 mm long. Corolla yellow, tubular, swollen at base, 15 mm long. Lobes 4, inside red-brown, rotate, acute, 5 mm long, triangular, half of outside pilose. Stamens 4 + 4, usually in pairs, alternating with corolla lobes; anthers yellow, apiculate, hairy at back, 2.5 mm long, filaments white, 1 mm long, pilose, inner ones much shorter or 0. Pistil 0 or minute, pilose. Female flower usually solitary, terminating young shoot, subsessile or on 8–15 mm long finely pubescent peduncle. Calyx green, minutely pilose, 9 mm long, 4 mm diam.; lobes 4 (3) deeply cut, acute, thick, 6 mm long. Corolla yellow, outside minutely white pilose, 16 mm long; lobes 4 (3), rotate, triangular, acute, 10 mm long. Staminodes 4 or 8, filaments 1.5 mm. Ovary almost glabrous, 8-celled. Style thick, 4 thin stigmas. Fruit solitary, terminating shoot, ovoid-attenuate, 3 cm long, 12 mm diam., glabrous, broadly or narrowly based. Fruiting calyx woody, deep, cupular, 18 mm high, 18 mm diam. Lobes triangular-ovate, acute, 6 mm. Seeds 2–5 per fruit, wedge-shaped, outer wall convex, strongly tapered both ends, up to 10 × 28 mm, testa brown, smooth; endosperm horny, equable.

Distr. Endemic.

Ecol. Only known from Hinidumkande (Haycock) near Hiniduma, at 700 m alt. Very rare. It grows solely near the summit of Hinidumkande, where there are less than 50 trees. The prevailing climate is determined by the clouds which hang permanently as a blown-away plume on the top. Hence the name Hinidumkande (hinidum = cloud, kande = hill). The vegetation is xerophytic; there is very little soil.

Note. Timber dirty white, centre of tree usually rotten, no black heartwood.

S p e c i m e n s E x a m i n e d. GALLE DISTRICT: Top of Hinidumkande (Haycock), alt. 700 m, wet, evergreen, Oct., fl., *Kostermans 25717* (A, BO, G, K, L, P, PDA, US); ibid., Sep., fl., *Kostermans 25519* (A, BO, G, K, L, P, PDA, US). LOCALITY UNKNOWN: buds, *s. coll. C.P. 3011* (BM, K, PDA).

27. Diospyros insignis Thw., Enum. Pl. Zeyl. 180. 1860; Beddome, For. Man. in Fl. Sylv. 3: CXLV. 1869; Beddome, Ic. 1: 26, t. 130. 1871; Hiern, Trans. Camb. Philos. Soc. 12: 157. 1873; Clarke in Hook. f., Fl. Br. Ind. 3: 565. 1882; Trimen, Handb. Fl. Ceylon 3: 100. 1895; Wright, Ann. R. Bot. Gard. Peradeniya 2: 181, t. III (3), XIV (1–5). 1904; Worthington, Ceylon Trees 316. 1959. Type: *C.P. 2730* (PDA), syntype: *C.P. 3477* (PDA).—**Fig. 3.**

Tree, up to 6 m high and 5–10 cm diam. (20 cm acc. to Wright), dioecious and polygamous (perhaps monoecious). Bole straight, branched near the top, branches very long, patent. Bark black, finely fissured, 1 mm thick, fissures c. 1 cm apart, running for 10–12 cm, peeling off in large, thin flakes, exposing brown surface. Live bark 3 mm, pale brown, outside pinkish. Young branchlets sericeous, like end bud. Inside of branches turns yellow immediately after cutting. Leaves glabrous, chartaceous to stiffly membraneous-chartaceous, alternate, rarely subopposite, ovate-elliptic to elliptic, 4 × 10–15 × 35 cm, shortly acuminate, base contracted into the petiole and then shortly cuneate. Young leaves pinkish. Both surfaces intricately areolate-reticulate under the lens, upper one with prominulous midrib and laterals (often in a groove), lower one with strongly prominent midrib, lateral nerves prominent, widely spaced, secondary veinlets prominulous, parallel, conspicuous, often horizontal. Petiole stout, glabrous, 1.5–2 cm long. Hermaphrodite and male flowers on same tree, the latter with all gradient stages of pistil development. Hermaphrodite ones in the same cluster or separate. Male clusters axillary, sessile, of 3–20 flowers. Calyx broad, 4 mm long, lobes 4, as long as cup, ovate, acute, almost glabrous outside. Corolla white to yellowish-white, tubular, 1 cm long, slender, throat narrowed; lobes 4, rotate, pilose outside, ovate, acute. Stamens 14–20, epipetalous at base of corolla, single or paired, inner ones shorter, c. 5 mm long, very unequal; anthers apiculate, longer than filaments (3.5–2 mm, inner ones); filaments white, glabrous or pilose, 3–0.5 mm, inner ones long. Pistil an apiculate central disc, 5 mm high. Female flowers in sessile clusters of 1–6. Calyx as in male one, pilose; lobes 4, 5 mm, accrescent in flower with feebly recurved margins and narrow intersegmental pouch. Corolla white to yellowish-white, tubular, 1 cm long, throat narrow; lobes 4, outside silky, tend to become recurved. Staminodes 4–5, epipetalous, sterile anthers apiculate, filaments almost glabrous. Ovary 4-celled, slightly pilose. Stigmas 4, green, after anthesis almost completely enclosed in the accrescent calyx. Fruit green, glabrous (exc. apex), depressed globose, the apex impressed

Fig. 3. *Diospyros insignis* Thw.

and with obscure rounded thick ribs, up to 3 cm long. Calyx considerably thickened and enlarged, like an acorn, the lobes after they become recurved hardly distinguishable, the tips somewhat sharply protruding outward and downward, ultimately the calyx a woody cup of 2 cm deep, the upper part quadrangular, with outward sloping thick, densely rusty pilose walls, 3 cm diam. and 5–10 mm high. Pedicel disc-like, 2–3 mm thick, up to 7 mm diam. Seeds 4–8, ovate-wedge-shaped, 3 cm long, 1–1.5 cm wide, 7 mm thick. Testa deep brown, smooth, obscurely transversally striated; endosperm copious, horny, ruminate. Seedling epigeal, cotyledons detached early, seeds never raised above the ground. First epicotyledonary leaves 1 mm long, topped by two diametrally placed glabrous, thin, bullate, 5 cm long leaves. Wood white, turning dirty after exposure. Heartwood none or little, usually rotten.

Distr. Ceylon and Anaimalai Hills in S. India (?).

Ecol. Wet, evergreen forests in SW Lanka.

Vern. Porawa mara (porawa = axe, mara = to kill) (S); Wal-mederiya (T).

Note. Indicated as a large tree with good timber by Thwaites (repeated by Beddome, etc.). However, as stated by Wright, it is a small, not economic tree with hardly any heartwood.

Reported from the Anaimalai Hills by Beddome, which needs confirmation. The tree does not occur near Kandy, as reported.

The leaves are chartaceous, never coriaceous as stated by Clarke.

Specimens Examined. GALLE DISTRICT: Kanneliya forest near Hiniduma, low alt., wet, evergreen, Sep., fl., *Kostermans 25500* (A, BO, G, K, L, P, PDA, US); ibid., June, fr., *Kostermans 25004 A* (id.); ibid., May, fr., *Kostermans 24767* (id.); ibid., ster., Sep., *Kostermans 25495* (id.); ibid., Aug., fr., *Kostermans 25459* (id.); Kottawa Arbor. near Galle, Sep., fr., *Kostermans 25584, 25585* (id.); Pitigalle, along river ravine, Sep., fl., *Kostermans 25572* (id.); Hinidumkande (Haycock), Aug., fl., *Kostermans 25478* (id.). KALUTARA DISTRICT: Morapitiya road, along rivulet, June, fr., *Kostermans 24950* (id.). KANDY DISTRICT: Kadugannawa, Tampana valley, Mar., fr., *Worthington 5227* (K).

var. **parvifolia** Kosterm., var. nov.—Foliis parvis lanceolatis sensim attenuatis.

Specimens Examined. RATNAPURA DISTRICT: Mareekelle near Ratnapura, low, Oct., buds, *Kostermans 25675* (G, L). GALLE DISTRICT: Hinidumkande, alt. 400 m, Sep., fl., *Kostermans 25666* (A, BO, G, K, L, P, PDA, US), alt. 500 m, Sep.-Oct., buds, fr., *Kostermans 25555, 25707, 25549 A* (id.), alt. 600 m, sapling, *Kostermans 25506* A (G, L); Kanneliya forest reserve, alt. 200 m, Sep., fl., *Kostermans 25595, 25596 A, 25597* (id.); Godakande near Hiniduma, alt. 400 m, sapling *Kostermans 25675 A* (G, L).

28. Diospyros thwaitesii Beddome, Ic. 1: 27, t. 135. 1874; Beddome, For. Man. in Fl. Sylv. 3: CXLIV. 1869; Clarke in Hook. f., Fl. Br. Ind. 3: 566. 1882;

Trimen, Handb. Fl. Ceylon 3: 101. 1895; Wright, Ann. R. Bot. Gard. Peradeniya 2: 188, t. 16 (1–6). 1904; Alston in Trimen, Handb. Fl. Ceylon 6: 185. 1931. Type: *C.P. 3394* (Calcutta).

Diospyros candolleana (non Wight) Thw., Enum. Pl. Zeyl. 181. 1860. Type: *C.P. 3394* (PDA).

Diospyros hirsuta forma *thwaitesii* Hiern, Trans. Cambridge Philos. Soc. 12: 164. 1873.

Monoecious, dioecious or polygamous tree, up to 25 m high and 30 cm diam. with long straight cylindrical bole. Bark black, finely, shallowly fissured, strips 5–30 mm wide, 1–2 mm thick, hardly peeling off. Live bark reddish-brown, paler, 3–4 mm, smelling of passion or grape fruit, outside dark red with wavy lines. End bud and the slender branchlets densely, minutely rusty pubescent. Flush densely dark rusty tomentose. Leaves alternate, chartaceous to subcoriaceous, ovate to lanceolate, 1.5 × 4–5.5 × 14 cm, long-acuminate with sharp tip, base rounded or acutish. Young leaves sparsely adpressed-pilose both sides, more dense on midrib. Adult leaves glabrous, both surfaces smooth, lateral nerves invisible or (in thicker leaves) midrib above prominulous, lateral nerves slightly prominulous with the adjacent tissue, lower surface brown (dried), midrib slender, prominulous, lateral nerves inconspicuous or filiform, widely spaced, at some distance from the margin ascendant, sometimes fairly laxly reticulate. Petioles very slender, pubescent, glabrescent, concave above, 5–12 mm long. In monoecious trees male and female flowers irregularly dispersed on the same or different twigs, polygamous trees with male and hermaphrodite flowers are common. There is a possibility that the trees change their sex in different years. Male flowers 4–15 in sessile axillary clusters, subtended by 2 mm long, acuminate, small bracts, densely rufous pilose outside, glabrous inside. Calyx thick, fleshy, cup-shaped, green, densely ferruginously tomentellous, 4–5 mm long, 3 mm diam. Lobes 5, short, broad, acute, stiff, 2 mm long. Corolla yellowish-white or white, bulbous-tubular, very wide, throat flaring out, 8 mm diam. at the level of the rotate 5 suborbicular-ovate, acute, recurved lobes, 4 mm long, partly covered with a dense layer of rufous hairs, fleshy. Stamens usually 5 epipetalous pairs, alternating with corolla lobes, attached by free or united filaments, or 10–18 in pairs and singles without fixed orientation. Anthers unequal, strongly apiculate, 3 mm long; filaments whitish, glabrous, 1.5 mm long. Pollen smooth. Rudimentary pistil a pubescent apiculate disc. Female flowers axillary, solitary or in a compressed cyme, the solitary ones surrounded by many small, 9 mm long, 5 mm wide bracteoles. Calyx as in male, 10 mm long, lobes as in male, sometimes with an undulating intersegmental groove. Corolla yellowish-white, 6–10 mm, throat very wide. Lobes 5, rotate, half of the inner surface pilose. Staminodes 5, epipetalous, alternate with corolla lobes, filaments attached near top of throat; anthers apiculate with long silky hairs; filaments white, very short. Ovary globose, apiculate, densely pilose, 4-celled. Style 0, stigmas 2. Fruit globose to globose-ovoid, covered with not easy detachable, silky, rusty, dense tomentum, up to 2 × 3 cm. Fruiting calyx not much enlarged, flat, thick, up to 17 mm diam., pouches in between reflexed, segments rather small. Seeds 2 per fruit, oval-elliptic with strongly beaked micropyle, slightly compressed, up to 18 mm long, 10 mm wide and 8 mm thick. Testa brown, transversally striated. Endosperm copious, ruminate. Seedlings epigeal, cotyledons drop after persisting for a few days as a white lamina, meanwhile the epicotyledonary leaves are already deep green.

Distr. Endemic.

Ecol. Wet, evergreen lowland forests of SW Lanka. Rather common, but very scattered.

Note. The anthers do not have the long setae as depicted by Beddome, but the long silky hairs are protruding beyond the tip. The sapwood is dirty white and has the same pleasant passion- or grape-fruit smell as the live bark. With this smell the tree can be easily identified. The heartwood is light red, there are only black traces, no real black heartwood. Because of its small dimensions the timber is not fit for commerce. It is certainly not so rare as indicated by Trimen.

Specimens Examined. GALLE DISTRICT: Hinidumkande (Haycock), alt. 500 m, Sep., fr., *Kostermans 25558* (A, BO, G, K, L, P, PER, US); ibid., ster., *Kostermans 25669 B* (id.); ibid., Aug., buds, *Kostermans 25473* (id.); Kanneliya forest reserve, May, fl., *Kostermans 24750* (id.); ibid., Sep., ster., *Kostermans 25659 A* (id.); ibid., June, fl., *Kostermans 24990* (id.); Beriliya forest near Elpitiya, low, Sep., sapling, *Kostermans 25641 B* (G, K, L); Godakande near Hiniduma, 200 m, Sep., ster., *Kostermans 25570* (id.); Hinidumkande, Aug., buds, *Kostermans 25369* (id.); ibid., Sep., fr., *Kostermans 25505* (id.); Kottawa Arb., sapling, *Kostermans 25620 A* (id.). RATNAPURA DISTRICT: Sripalabaddala, Feb., fr., *Waas 451* (L). s. loc., fl., fr., *s. coll. C.P. 3294* (BM, K, PDA).

29. Diospyros affinis Thw., Enum. Pl. Zeyl. 179. 1860; Beddome, For. Man. in Fl. Sylv. 3: CXLV. 1869; Beddome, Ic. 1: t. 127. 1871; Hiern, Trans. Cambridge Philos. Soc. 12: 169. 1873 (excl. specim. *C.P. 2924* pro parte quoad *D. opaca* Clarke); Clarke in Hook. f., Fl. Br. Ind. 3: 566. 1882; Trimen, Handb. Fl. Ceylon 3: 102. 1895; Alston in Trimen, Handb. Fl. Ceylon 6: 186. 1931; Wright, Ann. R. Bot. Gard. Peradeniya 2: 194. t. I (4), II (7), X (1–6). 1904; Worthington, Ceylon Trees 313. 1959. Type: *C.P. 2924* (PDA), Ooma Oya. —**Fig. 4.5.**

Tree, up to 30 m high and 60 cm diam. Bole straight, fluted, buttresses 1–2 m high, 1 m out, merging into the flutes, dioecious or polygamous. Bark conspicuously fissured and ridged, the ridges wavy, up to 4 cm wide, the

Fig. 4. *Diospyros affinis* Thw.

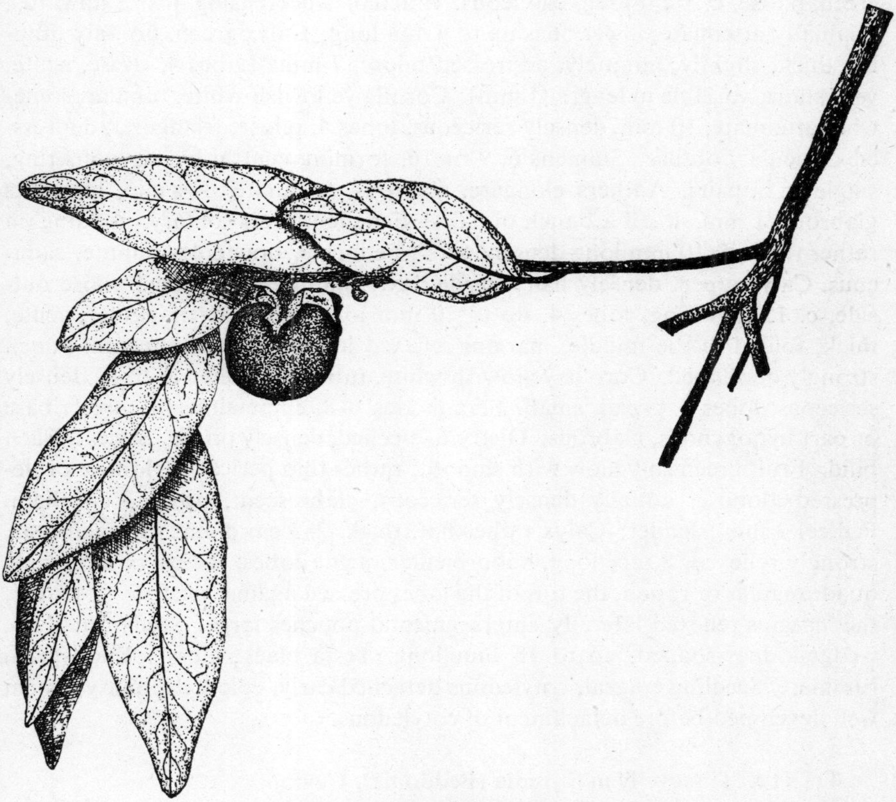

Fig. 5. *Diospyros affinis* Thw.

fissures broad, grey. Dead bark 2–5 mm thick, rough, peeling off in large patches. Live bark 5–6 mm thick, light red. Branchlets slender, apically minutely, sparsely adpressed pilose. End bud densely, minutely sericeous. Leaves alternate, thinly to stiffly coriaceous, young ones below sparsely, minutely adpressed pilose, adult ones glabrous, ovate-oblong to oblong to lanceolate-oblong, 2 × 4–4 × 10 cm, tapered towards the obtuse apex, base acute, both sides densely, prominulously reticulate (above sometimes obscure), above midrib slender, impressed, laterals filiform, sometimes obscure, below midrib slender, prominent, laterals filiform, 8–10 pairs, erect-patent, arcuately ascendant towards margin, often obscure. Petiole slender, 6–10 mm long, flat or slightly concave above. Polygamous (male and hermaphrodite) trees very common, produce fruit slightly smaller than those of female trees. Hermaphrodite flowers same as male ones but wider calyx, number of stamens the same. Male inflorescence cymes of 3–7 flowers in axil of caducous bract or leaf, main peduncle pilose, up to 1 cm long. Pedicels pilose, 2–4 mm. Bracts green, pilose, c. 1 cm long, caducous. Bracteoles decreasing in size upwards, gradually attenuate, lower ones up to 9 mm long. Calyx green, broadly tubular, thick, slightly, minutely, adpressed pilose, 7 mm. Lobes 4, ovate, acute, very short, variable in length (1 mm). Corolla yellowish-white, tubular, somewhat attenuate, 10 mm, densely sericeous, lobes 4, pilose, triangular, depressed orbicular, obtuse. Stamens 6, 9 or 16, forming central hypogynous ring, single or in pairs. Anthers elongate, glabrous, apiculate, 3–4 mm, filaments glabrous, 1 mm. Pistil a bunch of hairs. Female flowers solitary, nodding on rather thick, 5–10 mm long densely pilose peduncle, bracteoles minute, caducous. Calyx green, densely hairy inside, laxly, minutely depressed pilose outside, c. 13 mm long; lobes 4, up to 10 mm long, elongate-triangular, acute, thick, folded in the middle, margins reflexed laterally, intersegmental pouch strongly developed. Corolla yellowish-white, tubular, swollen at base, densely sericeous; lobes 4, broad, small. Staminodes 6–8, epipetalous at corolla base or part hypogynous, glabrous. Ovary 6–8-celled, densely pilose. Styles 2, each bifid. Fruit greenish-yellow with smooth, rather thin pericarp, globose or depressed globose, initially densely sericeous, glabrescent, up to 3 cm diam. Pedicel 3 mm, slender. Calyx rather flat, thick, 2–3 cm diam. with glabrous, strongly reflexed, 8 mm long, suborbicular, acute lobes; the calyx ultimately quadrangular or round, the tips of the lobes pressed against the fruit or patent, the margins reflexed laterally, intersegmental pouches large. Seeds usually 4, wedge-kidney-shaped, up to 18 mm long. Testa black, glossy. Endosperm ruminate. Seedling epigeal, cotyledons detached early, epicotyledonary system well developed before detachment of cotyledons.

Distr. Tinnevelly in S. India (Beddome), Ceylon.

Ecol. Dry and intermediate zone, sometimes fairly common, scattered.

Vern. Kaluwelle (S).

Note. The leaves are never very thick. The pulp is edible and is nicely sweet. The intermittently grey and black longitudinal fissures and ridges of the bark are striking. Bakhuizen (in Bull. Jard. Bot. Buizenz., Sér. 3, 15: 406. 1911) includes in this wrongly *D. opaca* Clarke. Timber: Coloured heartwood little, but of good quality; it is streaked brown. The colouring matter is also abundant in the sapwood.

Specimens Examined. AMPARAI DISTRICT: mile 18 Amparai-Kandy rd., Devulane forest, alt. 50 m, intermediate zone, May, ster., *Kostermans 24842* (A, BO, G, K, L, P, PDA, US), May, fr., *Kostermans 24840* (id.), June, fl., *Kostermans 25174* (id.). BADULLA DISTRICT: Mahiyangane-Bibile rd., dry, low alt., June, fl., *Kostermans 25158*, fr., *25169*, *25171* (A, BO, G, K, L, P, PDA, US), July, fr., *Kostermans 25302* (id.); mile 13 Mahiyangane-Bibile rd., Aug., fr., *Kostermans 25350* (id.); Mahiyangane, near canal to Polonnaruwa, July, fr., *Kostermans 25287* (id.). MATALE DISTRICT: Lenodora near Dambulla, foot of hill, dry zone, Aug., fr., *Kostermans 25316* (A,

BO, G, K, L, P, PDA, US); Elehera rd., mile 3 Naula, dry zone, low atl., Aug., young fr., *Kostermans 25322* (id.). POLONNARUWA DISTRICT: Polonnaruwa, Attanakadawala on canal bank, Oct., young fr., *Jayasuriya 322* (PDA). Also planted along road to Kataragama.

30. Diospyros crumenata Thw., Enum. Pl. Zeyl. 179. 1860; Beddome, For. Man. in Fl. Sylv. 3: CXLV. 1869; Beddome, Ic. 26, tab. 126. 1871; Clarke in Hook. f., Fl. Br. Ind. 3: 567. 1882; Trimen, Handb. Fl. Ceylon 3: 102. 1895; Alston in Trimen, Handb. Fl. Ceylon 6: 186. 1931; Wright, Ann. R. Bot. Gard. Peradeniya 2: 197, t. IV (14), VII (1–5), XVIII (6). 1904; Worthington, Ceylon Trees 314. 1959. Type: *C.P. 2438* (PDA).—**Fig. 6.**

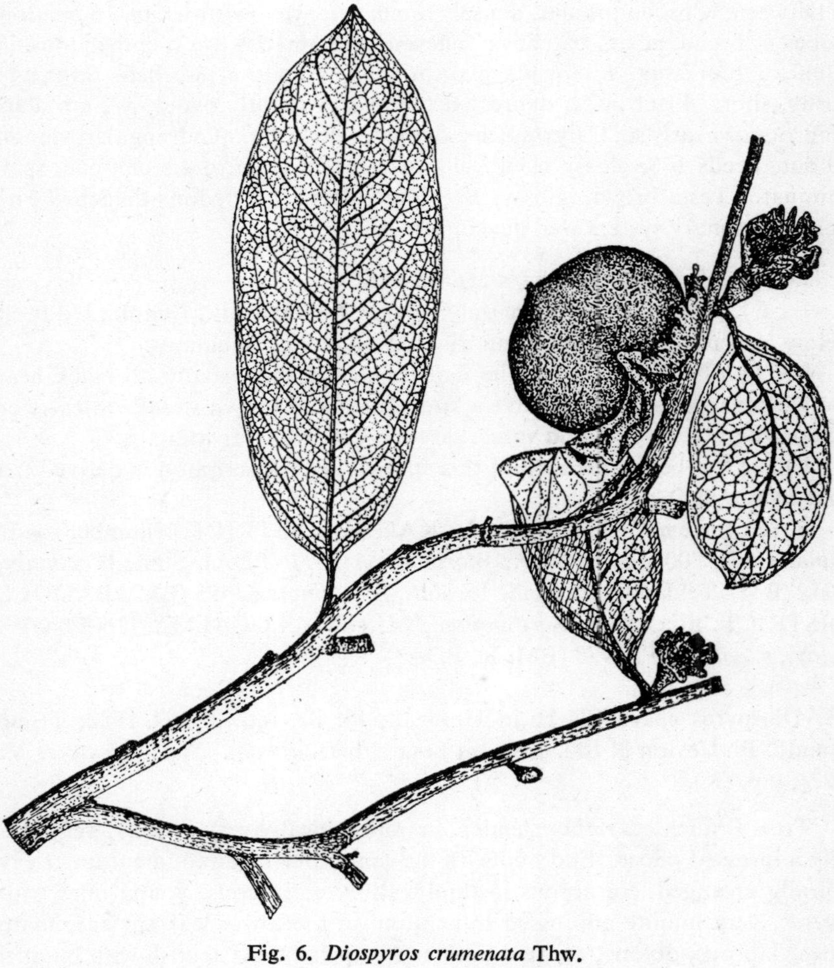

Fig. 6. *Diospyros crumenata* Thw.

Large, dioecious tree. Bark black, rather scaly. Branchlets glabrous. Leaves alternate, subcoriaceous, glabrous, glossy, oblong-ovate, 2.5 × 5–5 × 12 cm, shortly abruptly acuminate, base rounded or acutish, reticulation rather prominent. Petiole 6–10 mm long. Male inflorescences shortly stalked cymes, 3–7-flowered; peduncle 3–10 mm, pilose. Pedicels nodding, short, ultimate ones 1–4 mm. Calyx cup-shaped, densely tomentose, glabrous inside, 5–10 mm long. Lobes 4, obscure or very short, obtuse. Corolla yellow, tubular, 14 mm long. Lobes 4, short, broad, rounded, recurved. Stamens 8–14, hypogynous, 4–6 mm long, anthers 4–5 mm long; filaments glabrous, 0–1 mm long. Pistil 0 or small. Female flower solitary in leaf axil, larger, globose in bud. Peduncle 5–10 mm long; pedicel stout, short. Calyx large, cup-shaped, tomentose both surfaces. Lobes 4, deep, broad, rounded, acute with large deep pouches in between. Corolla tubular, densely tomentose, yellowish-white, 15 mm long. Lobes 4, broad, acute, patent or reflexed. Staminodes 8–12, epipetalous, unequal, 3 mm long. Filaments glabrous. Ovary pilose, 8-celled. Stigmas 4, fleshy, short. Fruit large, depressed globose to slightly ovoid, 3–5 cm diam., glabrous or nearly so. Calyx much enlarged, nearly flat, quadrangular, smoothed out. Seeds 1–8, glossy black, elliptic, flattened, up to 4.5 cm; endosperm ruminate. Testa brown, glossy. Seedling epigeal, cotyledons detached early, epicotyledonary system well developed.

Distr. Endemic? Perhaps also in S India.

Ecol. Rare. Moist region only from 500–1300 m alt., Deltota, Uduwella, below Haputale, Hantane, Gannoruwa, Kalugalle, Ambalawa.

Note. Timber is of medium quality, reddish. Usually no black heartwood, but some specimens have a large black and brown streaked heartwood of considerable beauty and value. Trees are worth cultivating.

I have not been able to find this species. The description is derived from that of Wright.

Specimens Examined. KANDY DISTRICT: Nilambe, road to Galaha, alt. 700 m, Aug., fr., *Worthington 3013* (PDA), Ster., *Worthington 3052* (BM); Teldeniya Rd., mile 15, ster., *Worthington 1628* (BM). BADULLA DISTRICT: Ella, ster., *Worthington 2945* (BM). LOCALITY UNKNOWN: fl., fr., *s. coll., C.P. 2438* (BM, K, PDA).

31. Diospyros opaca Clarke in Hook. f., Fl. Br. Ind. 3: 567. 1882; Trimen, Handb. Fl. Ceylon 3: 102. 1895 (in nota sub *D. affinis*). Type: *Thwaites C.P. 2924*, p.p. (K).

Tree. Branchlets rather slender, smooth, apically very densely, very minutely adpressed pilose. End bud with the same, but pale indumentum. Leaves spirally arranged, coriaceous to thinly, stiffly coriaceous, young ones with a sparse, very minute adpressed tomentum on the lower leaf surface, mature ones glabrous, oblong, tapered both ends, apically less acutish with blunt tip, base subcuneate, slightly decurrent; both surfaces smooth, above dull, midrib not or hardly impressed (more so in younger leaves), lateral nerves filiform, obscure, prominulous, lower surface drying dark red brown, midrib prominulous, lateral nerves c. 5–7 pairs, erect-patent, very faint. Petiole 3–5 mm long, concave above. Male inflorescences axillary, very densely, minutely adpressed grey pilose, consisting of cymes of few flowers, communal peduncle very short or none, the cymes almost sessile. Flowers densely minutely grey adpressed-pilose, sessile. Calyx shortly campanulate, 2.5 mm long, of which the small, 4, triangular teeth are 1 mm long. Corolla 6 mm long, the tube short, the 4 fleshy lobes elongate, acutish, 4 mm long. Stamens about 16, glabrous, anthers lanceolate, acute, c. 1 mm long.

Distr. Endemic.

Ecol. Unknown.

Note. Only known from a single specimen of unknown origin, kept in the Kew Herbarium. The possibility is not excluded, that it does not belong to the Ceylon flora.

Diospyros discolor Willd., Sp. Pl. 4 (2): 1108. 1805; Popenoe, Man. Trop. & Subtr. Fruit 373. 1920; Bakh., Bull. Jard. Bot. Buitenzorg Ser. 2, 15: 146. 1937.

Embryopteris discolor (Willd.) G. Don, Gen. Syst. 4: 41. 1837.
Diospyros philippinensis (Desr.) Guerke in Pflanzenfam. 4 (1): 164. 1897.

Tree, usually not more than 15 m tall and 30 cm diam. Bark blackish or dark brown, rough, peeling off; live bark pinkish. Sapwood grey to pink, heartwood small, black. Branchlets and end bud silvery sericeous. Leaves alternate, rigidly coriaceous, oblong or elliptic-oblong, rarely oblanceolate, 2.5 × 8–12 × 30 cm, acuminate or acute, base obtuse or subcordate, above glabrous, glossy, midrib impressed, beneath initially silver or golden sericeous, glabrescent, glaucous, lateral nerves obscure. Petiole thick, 0.5–1.5 cm long. Inflorescence axillary or pseudo-terminal, cymose or subracemose, often in pseudo-spikes, densely golden or silvery sericeous with large, 7–10 mm long ovate or elliptic, concave, sessile bracts and bracteoles. Male inflorescences shortly pedunculate, 3–7-flowered, female usually 1-flowered. Flowers white, fragrant, 4–5-merous, densely sericeous. Male ones smaller than female ones, shortly pedicelled, calyx fleshy, turbinate-campanulate, deeply 4-lobed, inside glabrous; lobes coriaceous, erect, ovate, obtuse, ciliate, 5–10 mm; corolla twice as long as calyx, up to 1.5 cm, 1.5 cm diam. at apex; tube subcylindrical; lobes pale yellow, fleshy, subequal to the tube, ovate or oblong, revolute to patent up to 1 cm long. Stamens 24–30, hypogynous, pairwise connate, glabrous; filaments up to 2 mm long. Pistil rudiment pilose. Female flowers subsessile; calyx lobes 4, ovate-rotundate or oblong, green, glabrous inside, 1–1.5 cm long; corolla white, trumpet-shaped, 1.5–2 cm long; tube cylindrical, 1 cm long; lobes as long as tube, ovate, obtuse. Staminodes 4–5 or 8–10, glabrous.

Ovary densely sericeous, 8–10-celled; styles 4–5, simple, erect or rather patent. Fruit globose or depressed globose, fleshy, densely rusty or yellowish velutinous, edible, 5–12 cm long, 4–10-seeded; mesocarp fleshy, pinkish, pulp white or cream. Fruiting calyx flattened, adpressed to the fruit base, 2.5–3.5 cm diam.; lobes 1–1.5 cm long, 1–2 cm wide. Seeds ellipsoid, flat-convex, up to 4 cm long; testa rough; endosperm equable.

Distr. Originally from the Philippines, cultivated in all tropical countries.

Vern. Mabola tree, Velvet apple (E).

Note. Many forms have developed in cultivation, also seedless ones. The disagreeable smell of the fruit is located in the skin and disappears when this is removed. The heartwood, if present, is streaked black and brown.

EXCLUDED SPECIES

Diospyros zeylanica Kosterm., Ceylon J. Sci., Biol. Sci. 12 (2): 107. 1977.

This plant represents probably a *Stemonoporus* (Dipterocarpaceae).

GENTIANACEAE

(by L.H. Cramer*)

Juss., Gen. Pl. 141. 1789 ('Gentianae'). Type genus: *Gentiana* L.

Annual or perennial herbs, occasionally twiners, rarely shrubs. Leaves opposite, often sessile, entire, exstipulate. Flowers in 2–3-chotomous, terminal, paniculate or axillary cymes, sometimes in capitate clusters, or occasionally solitary, 4–5-merous. Bracts present, often foliaceous, or absent. Bracteoles rarely present on pedicels. Flowers regular, subregular or irregular, hermaphrodite. Calyx persistent; tube campanulate or very short; lobes imbricate. Corolla gamopetalous, rotate to infundibular; lobes contorted, occasionally with plicate appendages in the sinuses. Stamens as many as and alternate with corolla lobes, inserted on throat or tube of corolla; filaments filiform, sometimes dilated at base or with basal scales; anthers erect, 2-celled, the cells poricidal or longitudinally dehiscent. Disk generally 0, occasionally annular, entire or lobed. Ovary superior, sessile, sometimes stalked, 1-locular, occasionally 2-locular by intrusion of locular wall; placentation parietal or appearing axile; style simple; stigma entire, 2-lobed or 2-lamellate. Fruit a capsule, septicidally 2-valved or indehiscent; seeds angular or globose, rarely compressed, occasionally narrowly winged, with fleshy or oily endosperm.

About 70 genera and 1100 species according to De Wit (Pl. World, 1967); fairly cosmopolitan, more numerously represented in temperate zones and in the montane regions of the tropics. In Ceylon 7 genera are known to occur, represented by 14 species, 3 subspecies (according to me) and 2 varieties; except for 5 endemics among these, the others are all indigenous. No introductions occur. The endemics are chiefly restricted to the highlands where by preference they occur on hilly slopes and usually in wet, black patana. The family is of no economic importance in Ceylon.

The aquatic genus *Nymphoides* Seguier previously placed by Trimen under this family has been transferred to the Menyanthaceae.

Some authorities regard the Gentianaceae as being allied to the Menyanthaceae; others see it as being allied to the Loganiaceae, and perhaps even to the Apocynaceae and the Asclepiadaceae. I follow Cronquist (Evol. & Classification Fl. Pl., 1968) in regarding its affinities as being with the Loganiaceae

*Flora of Ceylon Project. Peradeniya. Present address: University College, Batticaloa.

from which this family is distinguished by its representation by mostly herbs with exstipulate leaves and by a parietal placentation in the ovary.

KEY TO THE GENERA

1 Stem erect, 4-angular to sub-4-angular; leaves usually sessile; ovary without a disk; seeds angular or globose, wingless
2 Corolla regular; all stamens perfect
3 Calyx usually winged dorsally; anther-cells poricidal; ovary 2-locular by intrusion of locular wall, the placentation axile .... **1. Exacum**
3 Calyx smooth; anther-cells longitudinally dehiscent; ovary 1-locular, placentation parietal
4 Cymes capituliform; stigma entire; filaments with a basal scale........ **2. Enicostema**
4 Cymes laxly dichotomous or paniculate; stigma 2-lobed; filaments simple at base
5 Flowers in cymes; corolla lobes without appendages, each with a fringed nectary at base; capsule sessile; erect herbs .... **3. Swertia**
5 Flowers usually solitary; corolla lobes with intermediate plaits, smooth at base; capsule often stalked; usually decumbent herbs .... **4. Gentiana**
2 Corolla subregular to irregular; only 1 stamen perfect
6 Cymes lax; corolla 2-labiate or lobes very unequal, the tube 8–10 mm long, slightly bent outwards at throat; stamens inserted at different levels on corolla tube; capsule oblong .... **5. Canscora**
6 Cymes subfastigiate; corolla subregular, the lobes subequal, the tube 3–4 mm long, straight; stamens inserted at same level on corolla tube; capsule globose...**6. Hoppea**
1 Stem twining, terete; leaves distinctly petiolate; ovary surrounded by disk; seeds trigonous, winged .... **7. Crawfurdia**

1. EXACUM

L., Sp. Pl. 112. 1753; Endl., Gen. Pl. 1: 102. 1838; Griseb., Gen. Sp. Gent. 107. 1839; Benth. & Hook. f., Gen. Pl. 2: 803. 1876; Engl. & Prantl, Pflanzenfam. 4 (2): 63. 1895. Type species: *E. sessile* L. (lectotype).

Annual or perennial, erect, glabrous herbs; stem often branched, more or less 4-angular. Leaves sessile or shortly petiolate, ovate to elliptic or lanceolate, entire, 3–5-nerved at base. Flowers generally in leafy, dichotomous cymes, occasionally solitary by abortion of lateral buds of cyme, 4–5-merous. Bracts foliaceous. Calyx deeply lobed, enlarged and persistent in fruit; lobes dorsally keeled, winged or flat. Corolla rotate, often persistent in fruit; tube short, broadly cylindrical; lobes contorted in bud, ovate to elliptic or oblong, acute or obtuse at apex. Stamens inserted in throat of corolla; filaments very short, dilated at base; anthers attenuate towards apex or uniformly broad, 2-celled, the cells poricidal, rarely laterally dehiscent. Ovary 2-celled by intrusion of the locular wall; placenta large, appearing axile; style long, declinate; stigma capitate or 2-lobed. Capsule globose or ovoid, septicidally 2-valved; seeds numerous, minute, angular.

About 40 species, palaeotropic; 3 species endemic to Ceylon.

KEY TO THE SPECIES

1 Corolla lobes 5, 12–37 mm long; capsule 6–16 mm long; anthers usually gradually attenuate to apex
2 Corolla white, the lobes obliquely erect; flowers usually solitary; leaves broadly ovate to suborbicular, truncate to rounded at base .... **1. E. walkeri**
2 Corolla blue or violet, the lobes spreading; flowers in cymes; leaves elliptic to lanceolate, acute or narrowed at base .... **2. E. trinervium**
1 Corolla lobes 4, 2.5–12 mm long; capsule 3–8 mm long; anthers abruptly narrowed at apex or uniformly broad
3 Calyx lobes dorsally winged; corolla lobes acute, 4.5–12 mm long; anthers 1.5–7 mm long
4 Leaves tapering to base, the petiole obscure; corolla lobes spreading, 11–12 mm long; anther cells poricidal; flowers usually solitary; capsule ovoid, 7–8 mm long .... **3. E. axillare**
4 Leaves cuneate to truncate at base, distinctly petiolate; corolla lobes obliquely erect, 4–8 mm long; anther-cells dehiscing laterally; capsule globose, 4–5 mm long .... **4. E. petiolare**
3 Calyx lobes dorsally flat; corolla lobes obtuse, 2.5–3 mm long; anthers 1–1.5 mm long .... **5. E. sessile**

1. Exacum walkeri Arn., Ann. Sci. Nat. Bot. 2 (11): 176. 1839; Arn., Ann. Nat. Hist. 3: 89. 1839; Griseb., Gen. Sp. Gent. 110. 1839; Griseb. in DC., Prod. 9: 45. 1845; Thw., Enum. Pl. Zeyl. 203. 1860; Clarke, J. Linn. Soc. Bot. 14: 426. 1875; Clarke in Hook. f., Fl. Br. Ind. 4: 96. 1883, excl. syn.; Trimen, Handb. Fl. Ceylon 3: 180. 1895; Abeywick., Ceylon J. Sci., Biol. Sci. 2 (2): 210. 1959. Type: *Walker s.n.*, Ceylon, 1445–1800 m (holotype).

Annual with a short rootstock; stem to 60 cm high, dichotomously branched, subquadrangular at base, 4-angular in upper half. Leaves broadly ovate to suborbicular, rounded or truncate at base, subacute to acute, 3 (–5)-nerved, the nerves prominent beneath, shortly petiolate; petioles 2–4 mm long. Flowers solitary in forks of branches and leaf-axils, very occasionally in terminal, dichasial cymes. Pedicels slightly nodding, 1.5–2.6 cm long. Bracts ovate-lanceolate, 2–11 × 1.25–6 mm, foliaceous above. Calyx lobes 5, lanceolate, to 7 × 1.75 mm, long-acuminate; wings in fruiting calyx 2–3 mm wide, rounded at base. Corolla white; lobes 5, obliquely erect, oblong-elliptic, 12–16 × 6–8 mm, acute. Anthers uniformly broad, 4–5 mm long. Capsule ovoid to elliptic, 8–10 mm long.

Distr. Endemic.

Ecol. In partial shade by moist embankments of country paths, by rocky stream borders and edges of disturbed forests in the uplands and highlands; chiefly in the southeast sector of the Central Province; common. Alt. more or less 900–2500 m. Easily recognized by its white corollae with obliquely erect, acute lobes and a greenish eye. Flowering throughout the year.

Specimens Examined. KANDY DISTRICT: Hantane, *Gardner*

C.P. 38 (PDA); Dolosbage, behind Raxawa Tea Estate, *Theobald & Grupe 2354* (PDA, US); Ambagamuwa, *Gardner C.P. 38* (PDA); Kunadiyaparawita, *Lewis 54* (PDA); Moray Estate, *Kostermans 24247* (L, PDA, US); on way to Adam's Peak, *Fosberg 58044* (PDA, US), along stream banks, *Tirvengadum & Cramer 314* (PDA, US). NUWARA ELIYA DISTRICT: Punduluoya, along stream bank, *Cramer 5148* (K, PDA, US); Pidurutalagala, beside footpath towards top of Peak, *Cramer 3500* (PDA, US); Hakgala, *Bremer 944* (S, PDA, US), summit, *Silva s.n.* (PDA); Horton Plains, along road to Small World's End, *Robyns 6961* (BR, PDA, US), *Fosberg 50063* (PDA, US), *Cramer 5162* (PDA, US). LOCALITY UNKNOWN: *s. coll. s.n.* (PDA); *Thomson 1845* (K); *s. coll. C.P. 1878* (K); *Gardner s.n.* (K).

2. Exacum trinervium (L.) Druce, Bot. Exch. Club Brit. Isles Rep. 3 (5): 418. 1914 ('*trinerve*' sphalm.); Alston in Trimen, Handb. Fl. Ceylon 6: 197. 1931; Alston, Kandy Fl. 55, f. 294. 1938; Abeywick., Ceylon J. Sci., Biol. Sci. 2 (2): 210. 1959.

Chironia trinervia L., Sp. Pl. 189. 1753; Willd., Sp. Pl. ed. 4, 1 (2): 1065. 1798. Type: *Herman s.n.* Ceylon, in Mus. Zeyl. 3: 26 (holotype, BM).

Exacum zeylanicum Roxb., Hort. Beng. 83. 1814; Roxb., Fl. Ind. 1: 414. 1820; Griseb., Gen. Sp. Gent. 110. 1839; Griseb. in DC., Prod. 9: 45. 1845; Arn., Ann. Sci. Nat. Bot. 2 (11): 176. 1839; Arn., Ann. Nat. Hist. 3: 89. 1839 (ut *E. ceylanicum*); Hook., Bot. Mag. 75: t. 4423. 1849; Thw., Enum. Pl. Zeyl. 203. 1860, excl. syn. var. *β*; Hook. f., Fl. Br. Ind. 4: 97. 1883; Trimen, Handb. Fl. Ceylon 3: 181. 1895. Type: same as for *Chironia trinervia* L.

Lisianthus zeylanicus Spreng., Syst. Veg. 1: 586. 1825. Type: same as for *Chironia trinervia* L.

Examination of live and herbarium material reveals that this endemic species is actually a heterogeneous complex of 4 different, infraspecific taxa. These differences are found chiefly in the stem, corolla lobes and anthers. The stem character, angular or cylindrical, is easier to observe in fresh than in pressed material.

KEY TO THE SUBSPECIES

1 Stem 4-angular to sub-4-angular at base; corolla lobes obtuse to subacute, 13–24 mm long, bright to pale blue

2 Anthers 8–13 mm long, more or less straight, attenuate to apex; corolla lobes obtuse; capsule 10–16 mm long; upper half of stem 4-angular

3 Angles of stem wingless; wings of fruiting calyx 2–4 mm wide; leaves at middle of stem narrowly oval-oblong to oval-lanceolate, acute or shortly acuminate, usually 3-nerved at base .... **1.** subsp. **trinervium**

3 Angles of stem narrowly winged; wings of fruiting calyx 4–7 mm wide; leaves at middle of stem broadly elliptic, long acuminate, 5-nerved at base.... **2.** subsp. **ritigalensis**

2 Anthers 4–5 mm long, more or less uniformly broad and abruptly narrowed at apex; corolla lobes subacute; capsule 8–10 mm long; upper half of stem cylindrical to sub-4-angular .... 3. subsp. **pallidum**
1 Stem cylindrical at base; corolla lobes distinctly acute, 23–37 mm long, violet to purplish blue .... 4. subsp. **macranthum**

1. subsp. **trinervium**

Annual with a short rootstock; stem 50–100 cm high, generally slender and branched above, sharply 4-angular throughout, often ribbed at angles. Leaves sessile or subsessile, narrowly oval-oblong to oval-lanceolate, 5.5–9.0 × 1.2–2.0 cm, cuneate to tapering at base, acute to acuminate, faintly undulate at margins, 3-nerved at base, the nerves prominent beneath. Cymes terminal or in the upper leaf-axils, lax. Bracts foliaceous, to 2.6 × 0.7 cm, acuminate. Pedicels 1.0–2.6 cm long. Calyx tube broadly campanulate, 4–4.5 mm long; lobes 5, lanceolate, 7–9 × 2–3 mm, acuminate; wings in fruit 2–4 mm wide, truncate to cordate at base. Corolla tube 4–7 mm long, pale yellowish-green within at throat; lobes 5, obovate, 13–24 × 9–10 mm, obtuse, concave, pale to dark blue. Anthers 8–13 mm long, straight, gradually attenuate to apex. Capsule ovoid, 10–13 × 4–4.5 mm, tipped by persistent style; seeds irregularly rhomboid, 0.5 mm long, angular.

Distr. Endemic. On exposed roadside embankments close to a watercourse or in sloping ground in the wet lowlands; often in clayey-gravelly soil among populations of *Gleichenia linearis* (especially in the Galle district) where it grows in association with *Hedyotis fruticosa*, *Arundina graminifolia* and *Wrightia wallida*; common; rarely in the northeast sector of the Central Province. Alt. sea level—more or less 190 m. The plants are often found as scattered, isolated individuals.

Ecol. Flowering July–April.

Uses. Owing to their attractive colour the flowers are offered in Buddhist temples.

Vern. Binara or Ginihiriya (S).

Note. In general appearance the stem is more slender and the leaves narrower than those of the other varieties.

Specimens Examined. KALUTARA DISTRICT: Agalawatte, in grassy ground, *Cramer 2788* (PDA, US). COLOMBO DISTRICT: without precise locality, *Wight 4359* (PDA). GALLE DISTRICT: Hiniduma, slopes of *Gleichenia* fernland, *Cramer 2691* (PDA, US); Udugama, damp rocky road embankments, *Cramer 5206* (PDA, US). RATNAPURA DISTRICT: Karawita Kande, *s. coll. C.P. 512* (PDA); Delgoda, *Lewis s.n.* (PDA); Gilimale, grassy roadside close to watercourse, *Cramer 5253* (PDA, US); Carney Estate, *Bernardi 15776* (G, US); Eratne, along forest path *Cramer 5014* (K, PDA, US); Sinharaja Forest, along trail, *Meijer 517* (PDA, US); Weddagala, along Dorewella Creek, *Hoogland 11437* (CANB, BR, E, L, PDA), *Waas 1953*

(PDA, US); summit of Bulutota Pass, *Bremer 875* (PDA, S, US). MATALE DISTRICT: Midlands, road embankment on way to Laggala, *Cramer 5139* (K, PDA, US). LOCALITY UNKNOWN: *s. coll. C.P. 512* (G); *s. coll. s.n.* (G); *s. coll. s.n.* (L); *s. coll. s.n.* (K).

2. subsp. **ritigalensis** (Willis) Cramer, stat. nov.

Exacum trinerve var. *ritigalense* Willis, Ann. R. Bot. Gard. (Peradeniya) 3: 282. 1907; Alston in Trimen, Handb. Fl. Ceylon 6: 197. 1931. Type: *Willis* 22 March 1905, Ceylon, Ritigala Summit (holotype, PDA).

Stem to 85 cm high, thickened at base, 4-angular throughout, narrowly winged at angles. Leaves spreading, broadly oval-elliptic, 5.2–16 (–20) × 2.0–5.0 (–6.5) cm, cuneate at base, acuminate to long acuminate, those at middle 5-nerved at base, the nerves prominent beneath. Calyx lobes lanceolate, 11–19 × 4–7 mm, long acuminate; wings in fruit 4–7 mm broad. Corolla lobes broadly obovate, 18–28 × 16–18 mm, obtuse to rounded at apex, pale to dark blue. Anthers 10–13 (–15) mm, straight, gradually attenuate to apex. Capsule ovoid, 11–16 × 6 mm.

Distr. Endemic. In gravelly soil of rocky places and of patanas in the dry intermediate zone towards the central part of the country and in the dry uplands. The type locality is an isolated hill (Ritigala) of the North Central Province; but the plant occurs more commonly in the western sector of the Uva Province and only in isolated pockets in the North Western & Central Provinces. Alt. more or less 400–1200 m.

Ecol. Flowering throughout the year.

Note. The winged character of the stem and the broadly oval-elliptic, 5-nerved leaves at its middle are characteristic. The bluish colour of the corolla varies in depth according to the intensity of sunlight.

Specimens Examined. KURUNEGALA DISTRICT: Ibbagamuwa, summit of Doluwakande, grassy pockets, *Jayasuriya & Balasubramaniam 515* (PDA, US). ANURADHAPURA DISTRICT: Ritigala, along path of great rock slide, *Jayasuriya 1078* (K, PDA, US), in stony banks of gorge on way to summit, *Cramer 5190* (K, PDA, US). KANDY DISTRICT: Alagalla Rock, open grassland, *Kostermans 24041* (L, PDA, US); Nawalapitiya, Nagastenne Group, in patana grassland, *Hoogland 11574* (CANB, BR, PDA, US); Looloowatte, grassland, *Bremer 1043* (PDA, S, US); Hunasgiriya, *de Silva 59* (PDA); between Pussellawa & N'Eliya, in dry patana, *Comanor 330* (PDA, US). MATALE DISTRICT: Wahacotte, *s. coll. s.n.* (G); Laggala, road embankment close to dry patana, *Cramer 5187* (K, PDA, US), Rattota-Illukumbura Rd., *Huber 730* (PDA, US). BADULLA DISTRICT: Diyatalawa, *Mueller-Dombois 17* (PDA, US); Demodera, *de Silva s.n.* (PDA, US); Badulla-Passara Rd., dry rocky steep slope, *Robyns 7320* (BR, PDA, US); Dickwella, on road embankment close to dry patana, *Cramer 5179* (PDA, US); Dickarawa, in dry patana *Cramer 5187* (PDA, US); Yallele Badde, in dry patana, *Fosberg 51827* (PDA, US); Ohiya, steep rocky slope, *Robyns 7314* (BR, PDA, US).

3. subsp. **pallidum** (Trimen) Cramer, stat. nov.

Exacum trinervium var. *pallidum* Trimen, Handb. Fl. Ceylon 3: 181. 1895. Type: *Gardner C.P. 38*, Ceylon, Ramboda (lectotype chosen by me, PDA).
Exacum zeylanicum var. *lewisii* Willis, Ann. R. Bot. Gard. (Peradeniya) 7 (1): 45. 1919. Type: *Lewis 2*, Ceylon, Adam's Peak, Kunadiyaparuwatha (holotype & isotype, PDA).

Stem more or less 40–100 cm high, slender, 4-angular to sub-4-angular at base, much branched and cylindrical to sub-4-angular in the upper half. Leaves spreading, ovate to narrowly ovate-elliptic, 3.0–9.0 × 1.0–2.8 cm, cuneate at base, acuminate, 3 (5–)-nerved at base. Calyx lobes lanceolate, 8–10 cm long, long-acuminate; wings in fruit 3–5 mm wide. Corolla lobes ovate-elliptic to ovate-oblong, 14–16 × 6–8 mm, flat, subacute, pale blue. Anthers 4–5 mm long, uniformly broad and abruptly narrowed at apex. Capsule broadly ovoid, 8–10 × 4–5 mm.

Distr. Endemic. In usually sloping ground, often close to streams, in the wet uplands; common especially in the Knuckles Range, very occasionally in the western sector of the highlands. Alt. more or less 900–1400 m. Individual plants occur in close proximity to each other but never in stands.

Ecol. Flowering February–March, September–December.

Note. The stem branches are more diffuse than in the typical variety and in subsp. *macranthum*. The short, thick anthers with abruptly narrowed tips together with the small, pale blue corollae are distinctive.

Specimens Examined. KANDY DISTRICT: Rangalla, summit of Ridge, *s. coll. s.n.* (PDA), by wayside, *Alston 905* (PDA), along road to Corbet's Gap, *Davidson 8512* (PDA, US); Knuckles Range, *Bremer 1036* (PDA, S, US), Bambarella, in *Gleichenia* fernland, *Cramer 5236* (K, PDA, US); Hantane, *Gardner C.P. 512* (PDA); Moray Estate, in secondary submontane forest, *Fosberg 58062* (PDA, US); slopes of Adam's Peak, *Sohmer 9799* (PDA, US). NUWARA ELIYA DISTRICT: Pidurutalagala, *s. coll. s.n.* (PDA), southern slopes, *Robyns 6946* (BR, PDA, US), in marshy ground, *Cramer 3132* (PDA, US). LOCALITY UNKNOWN: *s. coll. C.P. 38* (G, GOET, K); *Gardner 584* (K); *Mrs. Walker s.n.* (K).

Note. Except for the colour of the flowers (described as white), the type of var. *lewisii* matches the specimens of subsp. *pallidum*. Besides, the type locality of the former includes one where the latter occurs. I consider var. *lewisii*, therefore, as identical with (or at most only a form of) this subspecies.

It is possible that the type was collected from an ageing plant in an exposed situation where the original pale blue colour of the flowers had bleached; the dehisced capsules on the flowering branches seem to indicate this.

4. subsp. **macranthum** (Arn.) Cramer, stat. nov.

Exacum macranthum Arn., Ann. Sci. Nat. Bot. 2 (11): 176. 1839; Arn., Ann. Nat. Hist. 3: 38. 1839; Griseb., Gen. Sp. Gent. 111. 1839; Griseb. in DC., Prod. 9: 46. 1845; Hook., Bot. Mag. 80: t. 4771. 1854; Thw., Enum. Pl. Zeyl. 203. 1860; Trimen, Handb. Fl. Ceylon 3: 181. 1895; Abeywick., Ceylon J. Sci., Biol. Sci. 2 (2): 210. 1959. Type: *Walker s.n.*, Ceylon, 1800 m (holotype).

Exacum ovale Griseb., l.c.; Griseb. in DC., Prod. l.c. Type: *Walker s.n.*, Ceylon, 1800 m (holotype).

Exacum zeylanicum var. *macranthum* Clarke in Hook. f., Fl. Br. Ind. 4: 97. 1883.

Stem more or less 50–100 cm high, cylindrical at base, sub-4-angular in the upper half, often simple. Leaves sessile, occasionally subpetiolate, obliquely erect, ovate-oblong, ovate-lanceolate or suborbicular, 3.6–8 (–9.5) × 1.5–3.2 (–3.6) cm, subcuneate or slightly rounded and 3–5-nerved at base, acute to obtuse, rather thick, the nerves prominent beneath. Calyx lobes linear-lanceolate, to 11 × 4 mm, long acuminate; wings in fruit 2–3 mm wide, rounded at base. Corolla lobes spreading, ovate-lanceolate, 23–37 (–46) mm long, acute, violet to purplish-blue. Anthers 10–13 mm long; capsule ovoid, 9–17 × 5 mm.

Distr. Endemic. In the wet, black patanas of the highlands chiefly in the southeast sector of the Central Province; common, very occasionally in the wet uplands. Alt. more or less (1350) 1700–2350 m.

Ecol. Flowering January–June, September–December.

Note. In exposed situations the leaves are obliquely erect and obtuse to rounded at the apex; in shaded situations (which is rare) they are spread out and tend to be acuminate. The violet or purplish-blue corollae are very striking, and can be spotted from a distance.

Specimens Examined. MATALE DISTRICT: Midlands, road embankment on way to Laggala, *Cramer 5142* (K, PDA, US). KANDY DISTRICT: Corbet's Gap, along road, *Robyns 7111* (PDA, L, US). NUWARA ELIYA DISTRICT: Without precise locality, *Gardner CP 1877* (PDA, GOET), *s. coll. CP 1877* (PDA), *Gardner s. n.* (PDA); Maturata, High Forest, *Willis s.n.* (PDA); North of Pattipola, in patana, *Robyns 6960* (BR, PDA, US) in wet patana, *Robyns 6960* (K, PDA, US); Ambawela Road, in soggy grassland, *Cramer 5166* (K, PDA, US); Horton Plains, trail to Kirigalpotta, in wet patana, *Nowicke & Jayasuriya 232* (K, PDA, US), Reflection Lake, in patana grassland, *Clayton 5484* (K, PDA, US); Road to Ohiya, in grassy area,

Comanor 419 (K, PDA, US), in patana close to Baker's Falls, *Cramer 4450* (PDA, US); near Farr Inn, *Fosberg 53320* (PDA, US); Hakgala, eastern slopes, in rocky crevices, *Robyns 6956* (K, PDA, US), swampy meadow, *Huber 492* (PDA, US); Ambawela, Mipiluwana patana, *Robyns 6959* (PDA, US); Site Eliya, marshy grassland, *Nooteboom 3249* (L, PDA, US). BADULLA DISTRICT: Fort Macdonald, *Silva s.n.* (PDA); Erabedde, along swamp in reforestation, *van Steenis 29527* (L). RATNAPURA DISTRICT: above Belihuloya, *Kostermans 23629* (K). LOCALITY UNKNOWN: *s. coll. C.P. 1877* (Z), *G. Thomas 1845* (K), *Gardner 1847* (K).

Note. *E. ovale* Griseb. Under this name Grisebach distinguished his species from that of *Exacum macranthum* Arn. as having ovate-orbicular, obscurely petiolate or subpetiolate leaves which mark only one point in the variation range of the shape and base of the leaves of this subspecies. Except for such variable characters specimens of the former match those of the latter. I, therefore, regard *E. ovale* as consubspecific with subsp. *macranthum*.

3. **Exacum axillare** Thw., Enum. Pl. Zeyl. 203. 1860; Clarke in Hook. f., Fl. Br. Ind. 4: 96. 1883; Trimen, Handb. Fl. Ceylon 3: 180. 1895; Abeywick., Ceylon J. Sci., Biol. Sci. 2 (2): 210. 1959. Type: *s. coll. C.P. 2998*, Ceylon, Pasdun Korale (holotype, PDA).

Annual, with a shallow rootstock; stem erect, 3.5–45 cm high, 4-angular, narrowly winged at angles, diffusely branched. Leaves ovate-oblong to ovate-elliptic, 1.6–8 × 1–3.5 cm, much tapering to base, acute to acuminate, 3-nerved at base, very thin when dried; petioles obscure, those of lower leaves occasionally to 5 mm long. Flowers usually solitary in forks of branches or in lower leaf-axils, occasionally in axillary or terminal cymes. Bracts lanceolate, to 1.6 × 0.5 cm. Pedicels 6.5–14 mm long, occasionally curved, narrowly winged at angles. Calyx lobes 4, narrowly lanceolate, to 10 × 2 mm, long-acuminate; wings in fruit semicordate-lanceolate, 3.5–4 mm wide, truncate at base, conspicuously nerved. Corolla pale blue, the throat yellow; lobes 4, spreading, elliptic-lanceolate, 11–12 × 4–5 mm, acute to shortly acuminate. Anthers uniformly broad, 4–6 mm long, abruptly narrowed to apex. Capsule narrowly ovoid, to 8 × 4 mm; seeds cubiform to irregularly rhomboid, 0.5 mm long, granulate.

Distr. Endemic. Under shade along stream banks on rocky slopes and on damp road embankments in secondary wet lowland forests chiefly in the western and southern provinces; common. Alt. sea level—more or less 150 m.

Ecol. Often grows in association with *Selaginella* sp. and *Polytrichum* sp. Flowering throughout the year.

Specimens Examined. KALUTARA DISTRICT: Hewesse, *s. coll. C.P. 2998* (PDA); Pelawatte, in damp soil on top of Ranawaka Rock, *Cramer 2803* (PDA, US). GALLE DISTRICT: Riceland, Udugama, *Fer guson 1886*

(PDA); Kanneliya, on road embankment, *Cramer 4541* (PDA, US), *Kundu & Balakrishnan 540* (PDA, US); Hiniduma, Godekande, on rock slopes of Katukitulgala, *Cramer 3346* (PDA, US); Elpitiya, Beraliya Forest, along stream, *Cramer 5215* (K, PDA, US); Mandagala Mukalana, *s. coll. s.n.* (PDA). LOCALITY UNKNOWN: *s. coll. C.P. 2990* (PDA), *s. coll. C.P. 2998* (PDA, GOET), *Moon s.n.* (PDA), *Mrs. Walker 99* (K).

var. **pentamera** Clarke in Hook. f., Fl. Br. Ind. 4: 96. 1883; Trimen, Handb. Fl. Ceylon 3: 180. 1895. Type: *Thwaites C.P. 2998* Ceylon (holotype, PDA).

Stem erect, 28–100 cm high, 4-angular, narrowly winged at angles, diffusely branched. Leaves oval to oval-elliptic, 5.3–11.5 × 2.3–3.3 cm, 3–5-nerved at base. Wings of fruiting calyx 4–5 mm broad, subcordate to rounded at base. Corolla lobes 5, oval-lanceolate, 15–20 × 7–11 mm, subacute to obtuse. Stamens 5, anthers 4 mm long.

Distr. Endemic. Under shade in secondary submontane forests in the Knuckles Range and in secondary wet, evergreen forests of the south central parts of the country; occasional. Alt. more or less 150–300 m.

Ecol. Flowering October–February.

Note. In each individual plant the number of corolla lobes and stamens is constantly 5, and the former are broader than those in the typical variety.

Specimens Examined. MATALE DISTRICT: Corbet's Gap, along road, *Robyns 7111* (BR, K, PDA, US), *Balasubramaniam 2104* (PDA, US), *Alston 905* (BM). RATNAPURA DISTRICT: Karawita Kande, in secondary forest, *Waas 33* (PDA, US), along stream banks, *Cramer 5254* (K, PDA, US).

4. Exacum petiolare Griseb. in DC., Prod. 9: 46. 1845; Wight, Ic. Pl. Ind. Or. 3: t. 1324, f. 2. 1848; Clarke, J. Linn. Soc. Bot. 14: 427. 1875; Clarke in Hook. f., Fl. Br. Ind. 4: 98. 1883, excl. syn.; Trimen, J. Bot. 27: 164. 1889. Type: *Hugel s.n.* ex India orientali (holotype, n.v.)

Exacum pedunculatum var. *petiolare* Trimen, Handb. Fl. Ceylon 3: 182. 1895. Type: same as above.
Exacum pedunculatum auct. non L. 1753, Trimen, Handb. Fl. Ceylon l.c.

Annual with a shallow, fibrous rootstock; stem erect, to 30 cm high, sharply 4-angular, occasionally ribbed at angles, branched in the upper half, the branches usually obliquely erect. Leaves broadly ovate, 1.6–5.8 × 1.2–3.4 cm, cuneate or truncate and 5-nerved at base, subacute to obtuse; petioles 0.3–1.8 cm long. Cymes terminal, occasionally axillary. Pedicels 4–20 mm long, slightly curved. Calyx tube to 3 mm long; lobes 4, lanceolate, 4–5 mm long, very acute; wings in fruit semicordate, 2–3 mm wide, strongly nerved. Corolla pale to bright blue; tube 2.4–3 mm long, inflated at base; lobes 4, obliquely erect, oblong to ovate-oblong, 4.5–6 × 2.5–3 mm, acute to subacute. Stamens alternate with corolla lobes; anthers uniformly broad to apex, 2.75–5 mm long, the lobes dehiscing laterally. Stigma simple. Capsule globose, to 5 mm in diam.

Distr. Under shade on damp, dilapidated walls of ruins and on damp roadside embankments of the northern, north-western and eastern sectors of the dry lowlands; common, especially in the environs of Trincomalee town and Bibile. Alt. sea level—more or less 150 m. Also in Western India and the Deccan.

Ecol. The plant often grows in association with *Canscora decussata*, *Lobelia alsinoides* and *Selaginella* sp. Flowering is seasonal, occurring just after the northeast monsoonal rains; January–March.

Note. My checking of the type of *E. pedunculatum* L. and other representative material in the Herbarium, Kew, with the original description of Linnaeus and that of subsequent authors leaves no doubt as to Trimen's misidentification.

The results of my field work and examination of the more recent material collected for the Flora of Ceylon Project confirm the fact that *E. pedunculatum* does not occur in Ceylon.

Specimens Examined. ANURADHAPURA DISTRICT: Sigiriya, damp grassy places at foot of Rock, *Cramer 3660* (PDA, US), damp walls of Ruins, *Cramer 5127* (K, PDA, US). MATALE DISTRICT: Lenadora, Nalanda, *s. coll. s.n.* (PDA). PUTTALAM DISTRICT: Wilpattu National Park, *Cooray 39* (PDA, US). MONERAGALA DISTRICT: near Bibile, *s. coll. s.n.* (PDA); Bintenne, *Nevill 1877* (PDA). TRINCOMALEE DISTRICT: Mihindupura, Trincomalee-Kanniyai road, *Cramer 5103* (K, PDA, US); Kanniyai, *Ferguson s.n.* (PDA). LOCALITY UNKNOWN: *Moon s.n.* (L); *s. coll. C.P. 1876* (PDA); *Gardner C.P. 1876* (PDA); *s. coll. C.P. 1876* (GOET, L).

5. Exacum sessile L., Sp. Pl. 112. 1753; Vahl, Symb. Pl. 3: 16. 1794; Willd., Sp. Pl. ed. 4, 1 (2): 635. 1798; Griseb., Gen. Sp. Gent. 113. 1839; Griseb. in DC., Prod. 9: 47. 1845; Wight, Ic. Pl. Ind. Or. 3: t. 1324, f. 1. 1848; Clarke in Hook. f., Fl. Br. Ind. 4: 98. 1883; Trimen, Handb. Fl. Ceylon 3: 183. 1895; Abeywick., Ceyl. J. Sci., Biol. Sci. 2 (2): 210. 1959. Type: *Herman*, Ceylon in Mus. Zeyl. 3: specimen on p. 42 (holotype, BM).

Annual with a shallow rootstock; stem erect, to 45 cm high, 4-angular, narrowly winged at angles, dichotomously branched. Leaves sessile, broadly ovate, 3.5–9 × 2.5–7 mm, truncate to rounded at base, obtuse or shortly acuminate, obscurely 3–5-nerved at base. Flowers subsessile, solitary or in cymes at terminal ends of branches. Pedicels 0.75–1.5 mm long, nodding in fruit, slender. Calyx tube 0.5 mm long; lobes 4, ovate, 3–3.5 × 1.5 mm, acute, flat. Corolla tube 1.5 mm long; lobes 4, obovate, 2.5–3 × 2 mm, obtuse, reticulately nerved. Stamens shortly exserted from corolla tube; anthers linear-oblong, 1–1.5 mm long, uniformly broad to apex. Stigma simple. Capsule globose, 2.5–3.5 mm in diam.

D i s t r. In fields and in roadside embankments of the wet lowlands and the wet uplands. Alt. sea level—more or less 1200 m. Formerly rather common, now becoming rare. In spite of persistent efforts I have not been able to locate this plant. Occurs also in the Deccan and in S. India.

E c o l. Flowering September–March, July–August.

S p e c i m e n s E x a m i n e d. KURUNEGALA DISTRICT: Giriulla, *s. coll. C.P. 250* (PDA). KALUTARA DISTRICT: in fields, *Macrae s.n.* (K). RATNAPURA DISTRICT: Kukul Korale, *s. coll. C.P. 250* (PDA); between Delgoda and Karawita, *Lewis & Silva s.n.* (PDA). KANDY DISTRICT: Dolosbage, *s. coll. s.n.* (PDA); Adam's Peak, *s. coll. C.P. 250* (K). NUWARA ELIYA DISTRICT: Pundulu Oya, *Green s.n.* (PDA). LOCALITY UNKNOWN: *s. coll. s.n.* (GOET, G-DC); *s. coll. C.P. 250* (GOET); *Macrae s.n.* (GOET); *Moon s.n.* (K); *Mrs. Walker s.n.* (K).

2. ENICOSTEMA

Blume, Bijdr. 848. 1826; Endl., Gen. Pl. 605. 1838 ('*Henecostema*' sphalm.); Don, Gen. Syst. 4: 200. 1837; Benth. & Hook. f., Gen. Pl. 807. 1876; Gilg in Pflanzenfam. 4 (2): 67. 1895. Type species: *E. littorale* Blume [= *E. axillare* (Lam.) Raynal subsp. *littorale* (Bl.) Raynal.].

Adenema G. Don, Gen. Syst. 4: 201. 1837. Type species: not known.
Slevogtia Reichb., Conspectus 133. 1828. Type species: not known.

Perennial herbs; stem erect, much branched from base, 4-angular. Leaves opposite, sessile, amplexicaul, lanceolate or linear, 3-nerved at base. Flowers subsessile, numerous, in axillary capitate fascicles, the fascicles opposite. Calyx tubular or narrowly campanulate; limb 5-lobed, the lobes unequal. Corolla infundibular; tube short, cylindric, slightly expanded above middle; limb 5-lobed, the lobes contorted, equal. Stamens 5, included in corolla tube below sinuses; filaments filiform, each with a scale at base; anthers erect, oblong, apiculate. Ovary 1-locular; placentation parietal; style short; stigma globose. Capsule oblong, septicidally 2-valved, the valves inflexed on both margins; seeds globose, foveolate.

Three species, from the Antilles through tropical Africa and Madagascar to Asia and Malesia.

Enicostema axillare (Lam.) Raynal subsp. **axillare** Raynal, Adansonia 9: 75. 1969.

Gentiana axillaris Lam., Ill. Gen. 1 (2): 487. 1793, non Rafinesque 1828 nec

Reichb. 1828. Type: *Sonnerat s.n.*, India, Pondicherry (lectotype, P-LA).
Exacum hyssopifolium Willd., Sp. Pl. ed. 4, 1 (2): 640. 1798. Type: *Fabricius s.n.*, specimen 328.30 (lectotype, LINN.).
Cicendia hyssopifolia Wight & Arn., Bot. Mag. 2: 249. 1836.
Adenema hyssopifolia G. Don, Gen. Syst. 4: 201. 1837.
Slevogtia orientalis Griseb. in DC., Prod. 9: 65. 1845; Thw., Enum. Pl. Zeyl. 204. 1860. Types: several specimens (syntypes).
Enicostema verticillare Baill., Hist. Pl. 10: 131. 1891.
Gentiana verticillata L., Syst. Nat. ed. 12: 200. 1767 p.p., non Syst. Nat. ed. 10. 1760.
Enicostema verticillatum auct. non Gilg, 1895. Alston in Trimen, Handb. Fl. Ceylon 6: 198. 1931.
Enicostema littorale auct. non Blume, 1826. Clarke in Hook. f., Fl. Br. Ind. 4: 101. 1883; Trimen, Handb. Fl. Ceylon 3: 185. 1895.

Glabrous perennial; stems many from a long rootstock, to 55 cm high, ribbed at angles, younger parts glaucous green. Leaves oval-oblong to linear-oblong, 3.2–5.0 × 0.7–1.2 cm, tapering to base, subacute, mucronate, the lateral nerves faint, glaucous green. Flowers crowded, erect. Calyx tube 2 mm long; lobes unequal, lanceolate, 1.0–1.5 mm long, acute, with narrowly scarious margins. Corolla white; tube broadly cylindric, 4–5 mm long; lobes oblong-lanceolate, 2.5–3 mm long, curving outwards at apices. Filaments short. Anthers linear, shortly apiculate. Stigma capitate. Capsule ovoid to subglobose, 3–4 × 3 mm, faintly grooved on 2 opposite sides; seeds subglobose to rhomboid, 0.5 mm long, reticulate.

Distr. In open, sandy places among sparse grass close to the beach throughout the dry zone; very common. Alt. sea-level. Also occurs in eastern and southern Africa, and in India.

Ecol. Flowering throughout the year.

Vern. Vellakuru (T).

Note. The plant has a musky odour; this and the lanceolate or oblong leaves with the marginal nerves and the subacute or obtuse apices are distinctive.

Specimens Examined. JAFFNA DISTRICT: Elephant Pass, *Silva s.n.* (PDA); south of Elephant Pass, *Simpson 8037* (PDA), *Gardner C.P. 1871* (PDA). VAVUNIYA DISTRICT: Mullaitivu, flat sandy area near sea, *Townsend 73/94* (K, PDA, US). MANNAR DISTRICT: southwest of Pesalai, *Fosberg 53619* (PDA, US); between Mannar & Thirukatheeswaram, *Kundu & Balakrishnan 614* (PDA, US). PUTTALAM DISTRICT: Lagoon, *Alston 1296* (PDA); Manattivu, along railway track, *Cramer 4652* (PDA, US); Pallugaturai, west of Wilpattu Sanctuary, *Robyns 7334* (BR, PDA, US). HAMBANTOTA DISTRICT: Kirinde, *s. coll. s.n.* (PDA); Ruhuna National Park, Buttawa, *Cooray 15* (PDA, US), dunes and meadows behind beach, *van*

Steenis 19537 (L). BATTICALOA DISTRICT: Keeli-Kudah, seashore, *Waas 2138* (PDA, US). TRINCOMALEE DISTRICT: Foul Point, scrub jungle, *Simpson 9682* (PDA); Kinniyai, in sandy waste ground near coast, *Cramer 4920* (PDA, US); Uppuweli, sandy ground close to beach, *Cramer 4932* (PDA, US). ANURADHAPURA DISTRICT: Hunnuwilagama, Wilpattu, *Sumithraarachchi 8173* (PDA, US). LOCALITY UNKNOWN: *s. coll. s.n.* (L); *s. coll. C.P. 1871* (G); *s. coll. s.n.* (PDA).

Note. Geographically *Enicostema axillare* (Lam.) Raynal is confined to the Old World and is a different plant from *Enicostema verticillatum* (L.) Gilg., a Central American species. Studies on the variations of the former together with those on its distributional ranges led Raynal to distinguish 3 subspecies within this complex, viz. subsp. *littorale* (Malesia), *latiiobium* (East Africa) and *axillare* (Africa and Asia). Our Ceylon species has the distinguishing calyx and leaf characters of the typical subspecies. (Adansonia l.c. p. 74).

3. SWERTIA

L., Sp. Pl. 226. 1753; L., Gen. Pl., ed. 5, 107. 1754; Endl., Gen. Pl. 1: 691. 1839; Benth. & Hook. f., Gen. Pl. 2: 816. 1876; Gilg in Pflanzenfam. 4 (2): 87. 1895 ('*Sweertia*,' sphalm.). Type species: *S. perennis* L. (lectotype).

Ophelia D. Don, Trans. Linn. Soc. London 17: 224. 1837. Type species: not known.

Annual or perennial herbs; stems simple or branched. Leaves opposite, sessile, cauline and basal, the basal ones often petiolate. Cymes dense or lax, in racemose panicles or corymbs; flowers regular, 4–5-merous, blue or yellow. Calyx lobes deeply divided, linear to lanceolate, 1–3-nerved. Corolla rotate; tube very short; lobes contorted, obliquely erect, each at base with 1–2 circular, fringed pits (nectaries). Stamens inserted at base of corolla tube; filaments rather long, filiform or dilated at base; anthers ovate to lanceolate, subsagittate. Ovary 1-locular; parietal placentae 2; style very short; stigma 2-lobed, the lobes very short and recurved. Capsule sessile, valvular; the valves 2; seeds ovate to orbicular, often compressed.

About 100 species, mostly cosmopolitan (fide Willis, Dict. Fl. Pl. ed. 7. 1966), principally mountainous north temperate species; only one species in Ceylon, endemic, restricted to the highlands.

Swertia zeylanica (Griseb.) Walker ex Clarke in Hook. f., Fl. Br. Ind. 4: 127. 1885; Trimen, Handb. Fl. Ceylon 3: 187. 1895; Gilg in Pflanzenfam. 4 (2): 88. 1895; Abeywick., Ceylon J. Sci., Biol. Sci. 2 (2): 211. 1959.

Ophelia zeylanica Griseb., Gen. Sp. Gent. 316. 1839; Griseb. in DC., Prod. 9: 124. 1845; Thw., Enum. Pl. Zeyl. 205. 1860. Type: *Walker s.n.*, Ceylon, 1200–1800 m, in Hb. Hook. (holotype, K).

Annual with a shallow rootstock; stem to 36 cm high, cylindrical to sub-4-angular, grooved on opposite sides, branched above, glabrous. Leaves numerous, thick; lower ones oval-oblong to lanceolate, to 10 × 4 mm, obtuse; upper ones linear-oblong, 0.9–1.4 × 0.2–0.4 cm, subacute, becoming bracteate above. Cymes terminal, trichotomous, flat-topped. Bracts linear-oblong, 3.5 × 1 mm. Pedicels to 8 mm long. Calyx lobes suberect, linear lanceolate to oblong-lanceolate, to 4 × 1 mm, subacute. Corolla ashy-blue or lavender; limb to 8 mm in diam.; lobes oval-oblong, to 10 × 2.5 mm, subacute, reticulately nerved; nectaries orbicular, yellow, pectinately fringed. Capsule erect, linear-oblong to ovate-oblong, 7–8 × 2–3 mm, tipped by persistent style; seeds ovate-rhomboid, 0.75 mm long.

Distr. Endemic. In the wet patanas of the highlands; common. Alt. more or less 1500–2200 m.

Ecol. Flowering seasonal; August–March.

Specimens Examined. NUWARA ELIYA DISTRICT: Hakgala, patana, *Simpson 9039* (K), *s. coll. s.n.* (PDA); Nuwara Eliya, *Gardner C.P. 1875* (PDA); Horton Plains, *s. coll. s.n.* (PDA), along slopes of patana near Farr Inn, *Cramer 3138* (PDA, US), *Fosberg & Sachet 53326* (PDA, US), *Theobald & Krahulik 2735* (PDA, US), *Sohmer et al. 8622* (PDA, US); near World's End, *Fosberg & Jayasuriya 5326* (PDA, US). LOCALITY UNKNOWN: *Col. Walker s.n.* (K); *Maxwell s.n.* (K); *s. coll. C.P. 1875* (PDA); *Champion s.n.* (PDA); *Thompson s.n.* (K); *Gardner s.n.* (K).

4. GENTIANA

[Tournef.] L., Sp. Pl. 227. 1753; L., Gen. Pl. ed. 5, 107. 1754; Endl., Gen. Pl. 1: 100. 1839; Griseb., Gen. Sp. Gent. 210. 1839; Griseb. in DC., Prod. 9: 86. 1839; Benth. & Hook. f., Gen. Pl. 815. 1876. Type species: not known.

Ericoila Borkh., Arch. Bot. 1 (1): 26. 1796. Type species: not known.

Small, perennial or annual herbs; stem usually decumbent, occasionally erect. Leaves sessile or shortly petiolate. Flowers terminal, solitary, sessile or pedicellate, bracteolate or not, 5-merous, rarely 4-merous, regular, blue, violet or purple. Calyx tubular; tube terete, rarely keeled; lobes triangular. Corolla hypocrateriform or infundibular; tube clothed within with minute scales or cilia, or naked; lobes contorted, often with membranous plaits (appendages) in the sinuses. Stamens inserted below middle of corolla tube; filaments linear, equal; anthers ovate or oblong or linear. Ovary 1-locular; parietal placentae 2; style very short or none, the stigma 2-lobed or 2-lamellate, the lobes or lamellae spreading or recurved. Capsule sessile or stipitate, oblong or ellipsoid, 2-valved; seeds indefinite, globose or oblong, often trigonous.

About 500 species, cosmopolitan; only one species in Ceylon represented by a variety.

Gentiana quadrifaria Blume var. **zeylanica** (Griseb.) Kusnez, Trudy Imp. S. Peterburgsk. Bot. Sada. 15: 403. 1904.

Gentiana zeylanica Griseb., Gen. Sp. Gent. 274. 1839; Griseb. in DC., Prod. 9: 108. 1845. Type: *Walker s.n.*, Ceylon, 1800–2400 m, in Hb. Hook. (holotype, K).

Gentiana pedicellata (D. Don) Wall. ex Griseb. var. *orbiculata* Clarke, J. Linn. Soc. Bot. 14: 435. 1875, p.p. quoad descr.

Gentiana pedicellata sensu Thw. non Wall. ex Griseb. 1839.

Gentiana quadrifaria sensu Clarke in Hook. f., Fl. Br. Ind. 4: 111. 1885 non Bl. 1826; Trimen, Handb. Fl. Ceylon 3: 186. 1895 p.p. quoad descr.

Glabrous annual with a shallow rootstock; stem decumbent, to 25 cm long, sub-4-angular, laxly branched from base, the upper branches subfastigiate. Leaves numerous, opposite-decussate, broadly ovate to orbicular, 4–7 × 2.5–5 mm, rounded at base, recurved at apex, shortly aristate, 1-nerved. Flowers terminating branches, giving appearance of a corymb. Bracts 0. Pedicels 1–1.5 mm long. Calyx tube to 5 mm long; lobes lanceolate, 2 mm long, shortly aristate. Corolla indigo-blue; tube to 6 mm long, paler blue; lobes obovate-lanceolate, to 5 mm long, the plaits broadly ovate or broadly 3-angular, 2-fid. Filaments to 3 mm long, flattened, slightly broadened at base. Capsule ovoid, 5–6 mm long, dull yellow with age; stalk to 6 mm long; seeds ovate-oblong, 1 mm long, trigonous, alveolate, pale brown.

Distr. In soggy patana in the highlands; very common. Alt. more or less 1200–2200 m. Also in Bombay and South India.

Ecol. The flowers open fully only in bright sunlight usually between about 9.30 a.m. and 1.00 p.m. The plant often grows in association with *Lycopodium carolinianum*. Flowering throughout the year.

Specimens Examined. NUWARA ELIYA DISTRICT: patana near Nuwara Eliya, *Silva s.n.* (PDA); Pidurutalagala, summit, *Wheeler 12537* (PDA, US), southern slopes, *Robyns 6949* (BR, PDA, US); Lake Gregory, short grassland, *Robyns 6951* (BR, PDA, US), *Cramer 2856* (PDA, US); Mahagastota, in soggy grassland, *Cramer 2877* (PDA, US); Horton Plains, *Willis s.n.* (PDA), patanas, *Simpson s.n.* (K), *van Beusekom 1468* (PDA, US), *Jayasuriya 2399* (PDA, US); in soggy patana behind Farr Inn, *Cramer 4818* (PDA, US), *Gould & Cooray 13861* (PDA, US), beside Ohiya Rd., *Cramer 5159* (K, PDA, US).

Note. Both Clarke and Trimen have provided differing descriptions of Blume's species, the former description referring to the Indo-Ceylon and Java plants, and the latter to the Indo-Ceylon plant only.

5. CANSCORA

Lam., Enc. Méth. 1: 601. 1798; Griseb., Gen. Sp. Gent. 152, 1839; Griseb. in DC., Prod. 9: 64. 1845; Endl., Gen. Pl. 1: 603. 1838; Benth. & Hook. f., Gen. Pl. 811. 1876; Gilg in Pflanzenfam. 4 (2): 76. 1895. Type species: *Canscora perfoliata* Lam.

Pladera Sol. ex Roxb., Fl. Ind. 1: 416. 1820. Type species: *Exacum heteroclitum* L. (lectotype, fide Pfeiffer, Nom. 2: 736. 1864).

Heterocanscora Clarke, J. Linn. Soc. Bot. 14: 431. 1875. Type species: *Canscora schultesii* Wall. ex Griseb.

Slender, annual, glabrous herbs with a short, fibrous rootstock; stem erect, 4-angular, winged at angles. Leaves opposite, sessile or subsessile, 3–5-nerved at base. Cymes terminal, dichotomous, lax; flowers sessile or pedicellate, 4-merous. Calyx tubular; tube striate or winged; lobes 4. Corolla irregular; tube tubular or cylindrical, slightly bent outwards at throat; limb subbilabiate, 3-lobed, the lobes imbricate to subcontorted; upper lip 2-lobed, the lower 1-lobed, the lobe deeply 2-fid. Stamens 4, unequal, the longer one inserted at the union of the lobes of the lower lip, the shorter ones at top of the corolla tube at different levels; anthers erect, oblong to ovate, 2-celled, only 1 anther with the long filament fertile. Ovary 1-locular; style filiform, stigma 2-lobed or 2-lamellate. Capsule oblong, dehiscing septicidally into 2 valves or irregularly; seeds numerous, minute, reticulate to rugose.

About 30 species, palaeotropical (fide Willis, Dict. Fl. Pl. ed. 7, 1966).

KEY TO THE SPECIES

1 Calyx tube striate; lobes long-acuminate
  2 Flowers pedicellate, white or faintly tinged pink; stem branches long, slender or filiform
    3 Upper leaves free from each other, those at lower half of stem broadly ovate; lower lip of corolla 2.5–4.0 mm long; filament of longer stamen uniformly thin to apex.... .... **1. C. diffusa**
    3 Upper leaves connate into perfoliate bracts, those at lower half of stem lanceolate to oblong-lanceolate; lower lip of corolla 5–6 mm long; filament of longer stamen expanded at apex .... **2. C. roxburghii**
  2 Flowers sessile, pink or pinkish-red; stem branches short, broadened towards base.... .... **3. C. heteroclita**
1 Calyx tube winged; lobes shortly acuminate .... **4. C. decussata**

1. Canscora diffusa (Vahl) R. Br., Prod. Fl. Nov. Holl. 451. 1810; Griseb., Gen. Sp. Gent. 152. 1839; Griseb. in DC., Prod. 9: 64. 1845; Thw., Enum. Pl. Zeyl. 204. 1860; Clarke, J. Linn. Soc. Bot. 9: 429. 1875; Clarke in Hook. f., Fl. Br. Ind. 4: 103. 1885; Trimen, Handb. Fl. Ceylon 3: 184. 1895; Gilg in Pflanzenfam. 4 (2): 76. 1895; Abeywick., Ceylon J. Sci., Biol. Sci. 2 (2): 211. 1959.

Gentiana diffusa Vahl, Symb. Bot. 3: 47. 1794. Type: ex India orientali (not extant in Hb. Vahl (C)).

Exacum diffusum Willd., Sp. Pl. ed. 4, 1 (2): 637. 1798. Type: same as for *Gentiana diffusa* Vahl.

Canscora tenella Wight., Ic. Pl. Ind. Or. 4: 7, t. 1327, f. 3. 1850. Type: *Wight* Ic., l.c.

Stem to 30 cm high, diffusely branched from above the base, the branches long, slender or filiform; wings of older parts to 1 mm broad. Leaves few, sessile; lower ones broadly ovate or broadly lanceolate, 1.8–2.4 × 0.8–1.8 cm, truncate to cuneate and 3–5-nerved at base, acute, glabrous. Branches of cyme diffuse. Bracts linear-lanceolate, to 2 × 0.5 mm. Pedicels of lateral flowers slender, 0.75–1.0 mm long, of terminal ones shorter, thicker, all narrowly winged. Calyx tube linear-oblong, to 3.5 mm long, striate; lobes linear-lanceolate, 1.5 mm long, long acuminate. Corolla tube linear, to 8 mm long; limb buff to pale mauve, sometimes pale pink; upper lip to 4 mm across, the lobes oblong; lower lip 2.4–5 mm long. Filament of longer stamen uniformly thin to apex. Style glabrous; stigmatic branches oblong-orbicular, the margins minutely pubescent. Capsule linear-ovoid, 4–6 mm long, the valves recurved.

Distr. In fairly damp places among grass in the eastern part of the dry and intermediate zones and in the dry uplands; rather rare; rarer in the lower uplands. Alt. sea-level—more or less 700 m. Also occurs in tropical Africa, and spread through Southern and South-east Asia to tropical Australia.

Ecol. In the dry zone the plant often grows in association with *Ageratum conyzoides* and *Canscora roxburghii.* Flowering October–February.

Specimens Examined. TRINCOMALEE DISTRICT: Mihindupura, on the road to Kanniyai, *Cramer 5102* (K, PDA, US). MONERAGALA DISTRICT: Bintenne, tank, *s. coll. C.P. 2841* (PDA). KANDY DISTRICT: Hantane, Wariagala, *s. coll. s.n.* (PDA). LOCALITY UNKNOWN: *s. coll. C.P. 2841* (G, GOET).

2. Canscora roxburghii Arn. ex Miq., Anal. Bot. Ind. 3: 10. 1852; Thw., Enum. Pl. Zeyl. 442. 1864; Trimen, Handb. Fl. Ceylon 3: 185. 1895; Abeywick., Ceylon J. Sci., Biol. Sci. 2 (2): 112. 1859. Type: *Roxburgh* Ic. No. 1227 (holotype, K).

Pladera perfoliata sensu Roxb., Fl. Ind. 1: 418. 1820, non quoad basionym.

Canscora perfoliata auctt. non Lam. 1785: Griseb., Gen. Sp. Gent. 155. 1839; Griseb. in DC., Prod. 9: 65. 1845; Thw., Enum. Pl. Zeyl. 2–4. 1860; Wight, Ic. Pl. Ind. Or. 4: 7, t. 1327, f. 2. 1850; Clarke, J. Linn. Soc. Bot. 14: 431. 1875.

Exacum connatum Moon, Cat. 11. 1824 p.p. quoad specimen Jaffnae.

Canscora wallichii Clarke in Hook. f., Fl. Br. Ind. 4: 105. 1885. Type: *Wallich,* Num. List. 4365 (K-WALL.).

Stem to 35 cm high, the branches diffuse; wings 1 mm broad. Lower leaves oval-lanceolate to oblong-lanceolate, 1.2–4 × 0.3–1.2 cm, acute at both ends, occasionally rounded at base, 3-nerved; upper ones connate in pairs into perfoliate bracts, each ovate to suborbicular, to 1.0 × 0.4 cm. Flowers axillary, solitary in the lower axils, mostly in terminal cymes. Pedicels 4–16 mm long, faintly winged. Calyx tube 6–8 mm long, strongly striate; lobes erect, linear-lanceolate, to 3 × 0.75 mm long, acuminate. Corolla tube 9–10 mm long, white; upper lip 10–12 mm across; lobes obovoid, 5–7 mm long; lower lip 5–6 mm long. Anthers faintly pinkish-buff. Style filiform. Capsule linear-cylindrical, to 6 × 2 mm; seeds rhomboid, angular, black.

Distr. Among grass in sandy-gravelly soil on road and quarry embankments in the dry intermediate country and in the northern and eastern parts of the dry zone; common, especially towards the interior around Trincomalee town. Alt. sea-level—more or less 300 m. Also in South India.

Ecol. Flowering January–March.

Specimens Examined. VAVUNIYA DISTRICT: Kuchchaveli, *Gardner & Glennie CP 1874* (PDA). ANURADHAPURA DISTRICT: Ritigala Natural Reserve, along path of great rock slide, *Jayasuriya & Burtt 1137* (PDA, US), in gorge among grass, *Cramer 4913* (PDA, US). TRINCOMALEE DISTRICT: Palaiottu, on grassy railway embankments, *Cramer 5099* (K, PDA, US); Mihindupura, on damp road embankment, *Cramer 5104* (PDA, US). LOCALITY UNKNOWN: *Gardner & Glennie, C.P. 1874* (PDA); *s. coll. C.P. 1874* (K, PDA).

Note. This species has had a confused nomenclatural history owing to its misidentification with *Canscora perfoliata* Lam. by the older authors. Though Roxburgh had described under *Pladera perfoliata* a new species he had misapplied to it a new combination based on an earlier epithet of Lamarck. This is clear from the fact that Lamarck's species based on Rheede's plate (Hort. Mal. 10: t. 52. 1689) is quite a different plant.

Arnott considered his Ceylon specimens of *Canscora roxburghii* to be specifically not different from *C. alata* Wall. ex Griseb. But Miquel clarified the taxonomic status of Arnott's plants considering them conspecific with Roxburgh's species and determining *C. alata* as conspecific with Lamarck's. Since the epithet '*perfoliata*' was already preoccupied for Lamarck's species Arnott rightly adopted that of '*roxburghii*' for Roxburgh's species.

3. **Canscora heteroclita** (L.) Gilg in Pflanzenfam. 4 (2): 76. 1895; Abeywick., Ceylon J. Sci., Biol. Sci. 2 (2): 211. 1959.

Gentiana heteroclita L., Mant. 560. 1767. Type: *Koenig s.n.* (holotype, specimen 328.43 (LINN; isotype, BM).

Exacum heteroclitum Willd., Sp. Pl. ed. 4, 1 (2): 639. 1798. Type: same as above.

Pladera sessiliflora Roxb., Fl. Ind. 1: 417. 1820, nom. illeg.
Canscora sessiliflora Roem. & Schult., Syst. 3, Mant. 230. 1827; Griseb., Gen. Sp. Gent. 150. 1839; Griseb. in DC., Prod., 9: 64. 1845; Thw., Enum. Pl. Zeyl. 204. 1860; Clarke, J. Linn. Soc. Bot. 14: 430. 1875; Clarke in Hook. f., Fl. Br. Ind. 4: 104. 1885; Trimen, Handb. Fl. Ceylon 3: 184. 1895.

Stem erect, to 25 cm high, the branches obliquely erect, short and broadened towards base. Leaves few; lower ones broadly ovate, 6–15 × 5–12 mm, truncate and 3-nerved at base, obtuse, the upper linear-lanceolate, to 3 × 1 mm, 1-nerved. Cymes diffuse, the uppermost ones with only 1 lateral branch. Flowers sessile. Bracts linear-lanceolate, to 1.5 mm long. Calyx tube to 7 mm long, striate, in mature flowers split adaxially to a little above base; lobes linear-lanceolate, to 1.5 mm long, long-acuminate. Corolla pinkish-red; tube to 8 mm long; lobes of upper lip slightly spreading, broadly obovoid, 3 mm long; lobes of lower lip obovate, to 4 mm long. Stigmatic branches linear-oblong. Capsule linear-cylindrical; seeds irregularly rhomboid, 0.25 mm long.

Distr. In exposed, damp, grassy places usually among rocky outcrops in the northern, western and eastern sectors of the dry lowlands, and in the dry intermediate zone; common. Also in South India.

Ecol. In the dry zone the plant usually grows in association with *Dopatrium nudicaule* and *Mitrasacme alsinoides*, occasionally with *Eriocaulon* sp. Flowering October–February.

Note. Citing *Gentiana heteroclita* L. in synonymy, Roxburgh failed to adopt the Linnaean epithet '*heteroclita*'. Hence *Pladera sessiliflora* becomes a nomen superfluum.

Specimens Examined. PUTTALAM DISTRICT: Wilpattu National Park, between office and gate, *Cooray 26 R* (K, PDA, US). KURUNEGALA DISTRICT: Hiripitiya, *F. Lewis s.n.* (PDA). POLONNARUWA DISTRICT: Habarana, beside rocky pool on Dambulla Road, *Cramer 5087* (PDA, US). MATALE DISTRICT: Dambulla, on rocky slope, *Alston 1208* (PDA), *Robyns 7206* (BR, PDA, US), *Cramer 5129* (PDA, US). AMPARAI DISTRICT: on rocky outcrop among short grass, *Cramer 5050* (K, PDA, US); Sashtraveli, on rocky outcrop among sparse grass, *Cramer 5058* (K, PDA, US). BATTICALOA DISTRICT: Muttugoda, in damp, grassy ground beside Pollonnaruwa Road, *Cramer 5075* (PDA, US). LOCALITY UNKNOWN: *s. coll. s.n.* (PDA); *Glenie C.P. 3581* (PDA).

4. Canscora decussata (Roxb.) Roem. & Schult., Mant. 289. 1827; Griseb., Gen. Sp. Gent. 153. 1839; Griseb. in DC., Prod. 9: 64. 1845, excl. syn. *Exacum alatum* Roth.; Thw., Enum. Pl. Zeyl. 204. 1860; Clarke, J. Linn. Soc. Bot. 14: 430. 1875; Clarke in Hook. f., Fl. Br. Ind. 4: 104. 1885; Trimen, Handb. Fl. Ceylon 3: 185. 1895; Abeywick., Ceylon J. Sci., Biol. Sci. 2 (2): 211. 1959.

Pladera decussata Roxb., Fl. Ind. 1: 418. 1820, excl. syn.
Exacum alatum Roth; Spreng., Syst. Veg. 1: 427. 1825; Curtis, Bot. Mag. 58: t. 3066. 1831. Type: Roxburgh Ic. No. 236 (K).

Stem erect, to 45 cm high, branched above. Leaves fairly numerous, narrowly ovate to oblong-lanceolate, 1.2–3.6 × 0.4–1.6 cm, subcordate to rounded and 3-nerved at base, acute, the uppermost ones bracteiform. Pedicels 12–18 mm long, strongly winged. Calyx tube 8–11 mm long; lobes lanceolate, 3–4 × 1 mm, shortly acuminate, dorsally winged. Corolla white; tube 8–10 mm long; lobes of upper lip obovate to orbicular, 5–6 × 4–5 mm, each with 2 green spots towards base; lower lip 3.25–4.5 mm long, the lobes oval-oblong. Stigmatic branches orbicular, hirtellous. Capsule linear-oblong, 6–8 × 2 mm, obtuse at apex, tipped by persistent style.

Distr. In partially shaded or exposed roadside or railway embankments in the northern and southeastern parts of the dry lowlands and in the drier parts of the uplands; common; occasionally in the wet lowlands. Alt. sea-level—more or less 1300 m. Also from tropical Africa through the Mascarene Islands to southern and Southeast Asia.

Ecol. In the dry zone the plant grows often in association with *Ageratum conyzoides* and *Exacum petiolare*; in wetter parts it is occasionally found along with *Lobelia alsinoides*. Flowering February–June.

Specimens Examined. MATALE DISTRICT: Sigiriya, on damp walls of Ruins, *Cramer 5126* (PDA, US). KANDY DISTRICT: Hantane, *s. coll. C.P. 2400* (PDA, US). BADULLA DISTRICT: Ella, paddy fields, *Silva s.n.* (PDA); in dry patana beside Wellawaya Road, *Cramer 4830* (PDA, US); Welimada, roadside embankment *Cramer 5168* (PDA, US). TRINCOMALEE DISTRICT: Palaiottu, on sloping railway embankment among tall grass, *Cramer 5100* (K, PDA, US), *Glenie C.P. 2400* (PDA). COLOMBO DISTRICT: near Gal-modera, *Alston s.n.* (PDA). LOCALITY UNKNOWN: *s. coll. s.n.* (L).

6. HOPPEA

Willd., Ges. Naturf. Freunde Berlin neue Schriften 3: 434. 1801; Benth. & Hook. f., Gen. Pl. 805. 1876. Type species: *H. dichotoma* Willd.

Annual, dwarf herbs; stem often much branched. Leaves small, sessile, ovate or elliptic. Cymes dichotomous, lax or dense; flowers sessile or pedicellate, 4-merous, subregular. Calyx campanulate; tube subventricose or uniformly tubular; lobes short, contorted, equal to unequal. Stamens inserted on corolla tube, 1 perfect and longer than the others; filaments filiform; anthers ovate, erect, the cells parallel. Ovary 1-locular; placentae 2, slightly intruding into loculus; style short; stigma simple, capitate. Capsule subglobose or ellipsoid, 2-valved; seeds minute, reticulate.

Two species in South India, Ceylon and Burma.

Hoppea fastigiata (Griseb.) Clarke in Hook. f., Fl. Br. Ind. 4: 100. 1885; Trimen, Handb. Fl. Ceylon 3: 183. 1895; Abeywick., Ceylon J. Sci., Biol. Sci. 2 (2): 211. 1959.

Cicendia fastigiata Griseb., Gen. Sp. Gent. 158, 1839; Griseb. in DC., Prod. 62. 1845. Type: *Macrae s.n.* Ceylon, Colombo, in fields, in Hb. Benth. (K, n.v.).

Pladera pusilla auct. non Roxb. 1826: Thw., Enum. Pl. Zeyl. 204. 1860.

Pladera fastigiata Clarke, J. Linn. Soc. Bot. 14: 429. 1875. Type: same as for *Cicendia fastigiata.*

Glabrous herb with a shallow rootstock; stem to 12 cm high, 4-angular, narrowly winged at angles. Leaves rotundate to broadly ovate, 5–9 × 5–7 mm, rounded or truncate at base, subacute or obtuse, the lateral nerves invisible. Cymes compact, terminal or axillary, to 3 cm long. Bracts linear-oblong, to 5 × 1 mm, subacute. Calyx tube 1.5–3 mm, 4-angular; lobes erect, oblong-lanceolate, 3 mm long, acuminate, mucronate, the margins thickened. Corolla tube 3 mm long, closely investing ovary; lobes oblong, to 2 × 1.2 mm, mucronulate, concave, white. Filaments very short; anthers subrotundate. Capsule oblong-elliptic, to 3.5 × 2 mm; seeds rhomboid-cuneate to irregularly orbicular, 0.5 mm long, rugulose, jet black.

Distr. Among sparse grass in damp, sandy ground among rocky outcrops; common in the northern, north-central and eastern sectors of the dry lowlands; occasional in the wet low country, the wet intermediate zone and in the lower uplands. Alt. sea level—more or less 125 m. Also in south India and Burma.

Ecol. In the dry lowlands the plant grows in association with other annuals such as *Mitrasacme alsinoides*, *Geniosporum tenuifolium*, *Burmannia pusilla* and *Dopatrium junceum*. Flowering December–March.

Specimens Examined. JAFFNA DISTRICT: Eluthumattuval, *Balasubramaniam 1256* (PDA). MATALE DISTRICT: Digapatana, among sparse grass in soggy ground beside Laggala Rd., *Cramer 4883* (PDA, US). POLONNARUWA DISTRICT: Habarana, in shallow soil of rocky depression, *Cramer 5124* (PDA, US). KANDY DISTRICT: Peradeniya, *s. coll. C.P. 2840* (PDA); Madugoda, among wet rocks, *Simpson 9476* (PDA); COLOMBO DISTRICT: in fields, *Macrae s.n.* (K). KALUTARA DISTRICT: Nagoda, among sparse grass in abandoned airport, *Cramer 2973* (PDA, US). AMPARAI DISTRICT: Lahugala, among sparse grass on rocky outcrop, *Cramer 5047* (PDA, US). BATTICALOA DISTRICT: Sashtraveli, among sparse grass on rocky outcrop, *Cramer 5058* (PDA, US). LOCALITY UN-

KNOWN: *s. coll. s.n.* (K); *s. coll. s.n.* (G, DC); *s. coll. C.P. 2840* (G, GOET).

7. CRAWFURDIA

Wall., Tent. Fl. Nepal. 63, 48, t. 47. 1826; Griseb. in DC., Prod. 9: 120. 1845; Benth. & Hook. f., Gen. Pl. 815. 1876. Type species: not known.

Tripterospermum Blume, Bijdr. 849. 1826. Type species: *T. trinerve* Blume.

Perennial, mountainous herbs; stem dextrorsely twining, terete. Leaves opposite, petiolate, often 3-nerved. Flowers axillary, solitary or fasciculate, fairly large, 5-merous. Calyx tubular; tube 5-winged or smooth; lobes linear, distant. Corolla tubular-campanulate, widened at plicate limb; lobes short or as long as the tube, contorted. Stamens inserted at base of corolla tube; filaments filiform; anthers ovate, sagittate at base. Disk cupular, lobed or entire. Ovary stipitate, 1-locular; style rather long; stigmas 2. Fruit 2-valved or indehiscent, ellipsoid or oblong; seeds trigonous, winged.

About 10 species; southern and eastern Asia.

Both *Crawfurdia* Wall. and *Tripterospermum* Bl. have the same date of publication, and more or less a similar circumscription, though that of *Crawfurdia* is more complete. Bentham and Hooker f. were the first to unite *Tripterospermum* under *Crawfurdia* and hence, according to the Code (Art. 57) their choice must be followed.

Crawfurdia championii (Gardner) Trimen, Handb. Fl. Ceylon 3: 198. 1895.

Tripterospermum championii Gardner, Calcutta J. Nat. Hist. 8: 15. 1847; Abeywick., Ceylon J. Sci., Biol. Sci. 2 (2): 211. 1959. Type: *Gardner C.P. 137*, Ceylon, Horton Plains (PDA, lectotype chosen by me).

Crawfurdia japonica var. *championii* (Gardner) Clarke in Hook. f., Fl. Br. Ind. 4: 108. 1883; Trimen, Handb. Fl. Ceylon 3: 187. 1895.

Crawfurdia fasciculata var. *championii* (Gardner) Thw., Enum. Pl. Zeyl. 204. 1860.

Stem very slender, terete; internodes to 9 cm long. Leaves ovate, 2–3.4 × 1.2–1.8 cm, rounded or truncate and slightly tapering at base, very shortly acuminate, glabrous. Flowers often in opposite pairs; pedicels to 4 mm long. Calyx tube to 7 mm long, with broad sinuses at mouth; lobes linear, to 6 mm long, curving outwards at apex. Corolla tube much exserted, to 2.3 cm long; lobes broadly ovate, 3 mm long, shortly acuminate. Filaments to 1.4 cm long, flattened towards base. Disk entire. Ovary shortly stipitate. Fruit ovoid to obovoid, to 10 mm long.

Distr. Endemic. In the highlands, chiefly at Horton Plains; very rare. I have not encountered this plant at Horton Plains in spite of many trips there;

it is probably extinct owing to clearings. Alt. more or less 2250 m.

Ecol. Flowering February and August.

Note. The description is based on herbarium material.

Specimens Examined. NUWARA ELIYA DISTRICT: Horton Plains, *Gardner C.P. 137* (PDA), turn off to Totapola, *Jowitt s.n.* (PDA), *s. coll. s.n.* (K).

GESNERIACEAE

(by William L. Theobald and Donald A. Grupe*)

Dum., Commentat. 57. 1822.

Acaulescent or caulescent herbs, or slightly woody shrubs, occasionally epiphytic, sometimes 1-foliate and the leaf cotyledonary in origin. Leaves usually opposite, sometimes whorled or alternate, simple, entire or toothed; stipules absent. Inflorescence axillary or terminal, usually a simple or compound dichasium, occasionally solitary, pseudo-racemose, or congested and sub-capitate. Bracts usually small or absent. Flowers nearly always hermaphroditic, often protandrous, usually irregular (regular in *Championia*), often large and showy. Calyx 5-merous, divided to base, or tubular and 5-lobed or parted. Corolla 5-merous, rarely 4 (*Championia*), gamopetalous, usually with distinct tube and 2-lipped; lobes imbricate, adaxial pair usually interior. Stamens usually 2 or 4, inserted on corolla tube; anthers usually connate or connivent in pairs, rarely connivent as 4, or free; staminodia often present. Disk annular, cupular, one-sided, or represented by 1–5 distinct glands. Ovary superior, unilocular with 2 bifid, usually intrusive, parietal placentae, occasionally meeting in center (bilocular); ovules numerous; style simple; stigma simple, or variously lobed. Fruit usually a capsule splitting loculicidally, sometimes septicidal or circumscissile, occasionally a hard or soft, fleshy berry. Seeds numerous, small, sometimes with hair-like appendages at either end.

A large family of about 120 genera and 2000 species, widespread in the tropics and subtropics of the New and Old World. Ten of the 14 taxa represented in Ceylon are apparently endemic, and generally of a rather restricted distribution. Known localities are primarily undisturbed sites in the moist montane and lowland forests of the central and southwestern parts of the island. Most are found in dense shade on either moist rock outcrops or in stream beds. Occasionally they are found exposed in rock crevices or on rock surfaces where there is abundant underground seepage. Nearly all of the taxa are known from only a few collections. They are also apparently unable to cope with the disturbing influences of man on their environment.

The above family description has largely been confined to Old World representatives of the family, the subfamily Cyrtandroideae. Members of this

*Pacific Tropical Botanical Garden, Hawaii.

subfamily have a superior ovary, usually cymose inflorescences, and cotyledons which become unequal after germination. Aside from three species of *Rhynchoglossum* they are all confined to the Old World. The New World representatives, the subfamily Gesnerioideae, have a superior, or more or less inferior ovary, usually racemose inflorescences, and cotyledons which remain equal after germination.

For an excellent series of works on the Gesneriaceae of the Old World see the articles published in the Notes from the Royal Botanic Garden, Edinburgh by B.L. Burtt and colleagues at that institution. These include: Studies in the Gesneriaceae of the Old World, a continuing series begun in volume 21, 1954. Numbers I and II (1954) include a general introduction and discussion of types and lectotypes of certain genera and groups of lower rank, respectively. Number 24 (1963) includes a tentative key to the tribes and genera. At the same institution and in the same journal J.A. Rattner (1963), alone, and in conjunction with H.T. Prentice (1964, 1967), has published a valuable series on chromosome numbers in this family. A translation from Russian of Dr. L.I. Ivanina's studies of the fruit and seeds of representative Gesneriads has also been published in this journal (Application of the carpological method to the taxonomy of Gesneriaceae. Notes R. Bot. Gard. Edinburgh 26: 384–403. 1966).

Four of the five tribes recognized by Burtt are represented in Ceylon; Cyrtandreae (*Rhynchotechum*), Trichosporeae (*Aeschynanthus*), Klugieae (*Rhynchoglossum*, *Epithema*) and Didymocarpeae (*Championia*, *Didymocarpus* and *Chirita*). The last two genera mentioned are the largest in the country with three and five taxa, respectively. The Loxonieae, a small tribe of four genera, is not represented.

KEY TO THE GENERA

1 Inflorescence a simple or compound dichasium, sometimes reduced to 1–2 axillary flowers

2 Fruit dehiscent, oblong to long-linear; pericarp thin, dry; fertile stamens 2 or 4, variously connate or connivent; leaves basal, opposite, or whorled, if alternate, borne along creeping rhizome

3 Fertile stamens 4

4 Corolla 5-merous, red, tubular, limb 2-lipped; stamens didynamous; inflorescence 1–2 flowered in axils of terminal leaves; seeds with hair-like appendage at each end... **1. Aeschynanthus**

4 Corolla 4-merous, white, campanulate-rotate, regular; stamens equal, connivent; inflorescence a compound dichasium, lax, seeds unappendaged...... **4. Championia**

3 Fertile stamens 2

5 Acaulescent herbaceous perennials, or caulescent with a creeping rhizome; corolla tube without yellow stripe and longitudinal ridges within; anthers without tuft of hairs on back.. **2. Didymocarpus**

5 Caulescent herbs or shrubs; corolla tube with a yellow stripe and 2 longitudinal ridges within; anthers with a tuft of hairs on back...................... **3. Chirita**

2 Fruit indehiscent, ovoid; pericarp fleshy; fertile stamens 4, free; leaves alternate, borne at end of erect stem .... 7. **Rhynchotechum**

1 Inflorescence pseudo-racemose, or congested and subcapitiate with a large, solitary, leafy bract

6 Inflorescence pseudo-racemose; bracts inconspicuous, linear; leaves unequal-sided at base; corolla with oblique limb, personate; fertile stamens 4....... 5. **Rhynchoglossum**

6 Inflorescence congested and subcapitate; bract solitary, large, leafy; leaves equal sided at base; corolla obscurely 2-lipped; fertile stamens 2.................... 6. **Epithema**

1. AESCHYNANTHUS

Jack, Trans. Linn. Soc. London 14: 42. 1823, nom. cons.

Trichosporum D. Don, Edinburgh Philos. J. 7: 84. 1822.

Subshrubs, epiphytic or scrambling on rocks, usually with many, slender branches. Leaves opposite or whorled, fleshy or coriaceous, usually entire, veins obscure. Inflorescence terminal or axillary, 1–2 flowered, rarely dichasial or clustered, often appearing umbellate. Calyx tubular or campanulate, 5-lobed, sometimes deeply divided. Corolla showy, tubular-funnelform, curved, 5-lobed; upper lobes erect, lower larger and reflexed. Stamens 4, didynamous, exserted; anthers coherent in pairs by their tips, protandrous. Disk annular. Ovary 1-loculed with 2 deeply intruding, recurved, parietal placentae; style slender, stigma capitate with a horizontal median groove. Fruit a capsule, long-linear, cylindrical, 2-valved, loculicidally dehiscent. Seeds numerous, small, oblong, with a thick appendage at each end, or with a long hair at the apex, and 1 or more hairs at the base.

A genus of about 80 species particularly abundant in the Indomalaysian region and Southern China. Only one representative in Ceylon, *A. ceylanica.* For a discussion of the generic limits of this genus and other members of the tribe Trichosporeae see: B.L. Burtt, Studies in the Gesneriaceae of the Old World XXIX: A reconsideration of generic limits in the tribe Trichosporeae. Notes R. Bot. Gard. Edinburgh 28: 219–225. 1968.

Aeschynanthus ceylanica Gardn., Calcutta J. Nat. Hist. 6: 474. 1846; Trimen. Handb. Fl. Ceylon 3: 272. 1895.—**Fig. 1.**

Stems long, slender, cylindrical, scandent, often rooting at the thickened nodes; bark smooth and shiny, yellowish. Leaves thick and fleshy, scattered, pale green above, whitish-green below, oblanceolate to narrowly elliptical, 4.5–9 cm long, 1.2–2 cm broad, base tapering into petiole, margin entire, apex acute to obtusely accuminate, glabrous; petioles short, thick, 2–6 mm long. Inflorescence pseudoterminal; flowers 1–2 in axils of terminal leaves. Calyx segments divided to near base, linear-lanceolate, apex obtuse, glabrous, c. 6 mm long, 1 mm broad. Corolla tube scarlet, c. 25 mm long, 2.5 mm broad at base, 4–5 mm broad at throat, glandular-pubescent without, yellowish and glandular-pubescent within; throat squarish; lower 3 lobes, c. 4 mm long, 3 mm broad, yellowish-green with a purplish-red blotch near base and scattered purplish-red flecks and stripes in front, green on back, glandular-pubescent on both surfaces, and especially along margins; upper 2 lobes slightly smaller, c. 3 mm long, 3 mm broad, yellow with dull reddish-purple flecks and stripes in front, reddish-green on back, glandular-pubescent on both surfaces and especially along margin. Filaments violet, glandular-pubescent, upper pair c. 22 mm long, lower c. 18 mm long; anthers c. 2.5 mm long, 1.5 mm broad, reddish-purple towards filament, yellowish-green above; sacs parallel, glabrous, staminode glabrous, c. 8 mm long. Ovary glabrous; style c. 18 mm long at maturity; stigma purple, c. 1.5 mm long, 2 mm broad. Capsule straight or curved, c. 15 cm long, 3 mm broad, glabrous, base empty and narrower, gynophore-like. Seeds with one long hair at each end, hairs c. 1 cm long; seeds brown, rugose.

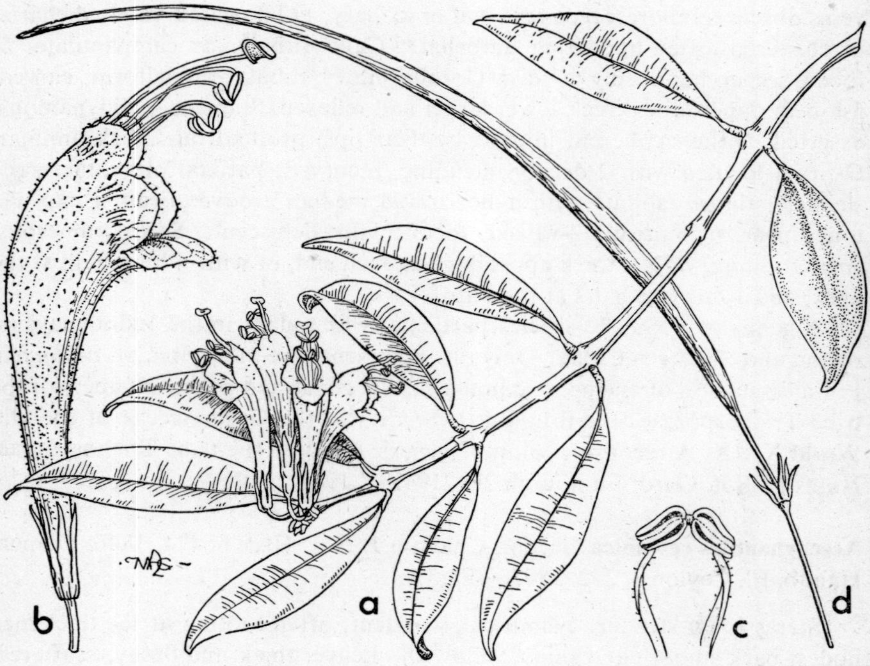

Fig. 1. *Aeschynanthus ceylanica* Gardner. a, flowering branch, $\times \frac{5}{6}$. b, flower, $\times 2\frac{2}{3}$. c, anthers, $\times 3\frac{1}{3}$. d, mature fruit, $\times \frac{9}{10}$.

D i s t r. Endemic. Climbing and rooting over rock surfaces and on trees in the montane forests at elevations above approximately 1300 m. Reported to be common by Trimen.

E c o l. Flowers exhibit extreme protandry with style and stigma deeply included in the tube at anthesis, and only expanding as the stamens begin to wither. Pollinators, unfortunately, were not observed. Flowering March, August–October, and probably throughout the year.

Note. Clarke (DC., Mon. Phan. 5: 26. 1883) has described a variety, var. *pinguis*, from Ceylon and India, but neither Trimen nor we have seen it on the island. The material at Kew from Ceylon (*Mrs. Walker 28, Wight s.n.*), which was cited by Clarke for this variety, is definitely typical *A. ceylanica*. Additional material from India (*Pycarah, Wight s.n.*), which was also referred to this variety by Clarke, is in all probability typical *A. perrottetii*, a species common in southern India. Gamble (Fl. Pres. Madras 2: 985. 1921) notes that var. *pinguis* is a smaller form of *A. perrottetii*. The latter species is apparently very closely allied to *A. ceylanica*, and possibly of identical coloration. It differs somewhat in leaf shape and in having the corolla slightly longer and glabrous (?) without. Cooke (Fl. Pres. Bombay 2: 396. 1958 reprint) thought the two taxa possibly conspecific. *Aeschynanthus ceylanica* may be a smaller flowering subspecies of this taxon, but we hesitate in making the transfer until further evidence is obtained from critical examination of more *A. perrottetii* material.

Specimens Examined. KANDY DISTRICT: Hantane, *Gardner s.n.*, part of *C.P. 1782* (PDA). NUWARA ELIYA DISTRICT: Ramboda, *Gardner 599* (K, PDA); Ramboda, *Thwaites s.n.*, part of *C.P. 1782* (BM, K, PDA) jungle above Hakgala, *de Silva s.n.*, (PDA); Hakgala, *Simpson 9080* (BM); montane forest behind Hakgala Botanic Garden, *Theobald & Grupe 2396* (E, K, PDA, UC, US). LOCALITY UNKNOWN: *Col. Walker s.n.* (K); *Wight s.n.* (K); *Mrs. Walker 28* (K); *Petch s.n.* (PDA).

2. DIDYMOCARPUS

Wall., Edinburgh Philos. J. 1: 378. 1819, nom. cons.

Roettlera Vahl, Enum. 1: 88. 1805, non Willd. 1797.
Henckelia Spreng., Anheit. ed. 2, 2: 402. 1817.
Roettlera subgen. *Didymocarpus* (Wall.) Fritsch in Pflanzenfam. 4, (3B): 148. 1895.

Perennial, acaulescent or caulescent, herbs. Leaves simple, opposite, alternate or basal. Inflorescence a scapose, paniculate, compound dichasium (in ours), axillary. Bracts inconspicuous. Calyx small, 5-lobed or divided. Corolla campanulate; tube curved, oblique at throat; lobes 5, unequal, rounded. Stamens 2, inserted near base of corolla tube; filaments short, curved; anthers connate, applanate, biocular, locules confluent. Staminodia 2, rarely 3, small, swollen at tip. Ovary unilocular with 2 intruding, recurved, parietal placentae; style slender; stigma oblique, subcapitate, capitate, or flattened, expanded and orbicular. Fruit a linear capsule, loculicidally dehiscent. Seeds small, elliptical, smooth or reticulate.

A large and widespread genus of approximately 120 species, extending from tropical Africa and Madagascar east through the Indomalaysian region and Southeast Asia south to Australia. Represented in Ceylon by three species, two of which are endemic, *D. floccosus* and *D. zeylanicus*.

KEY TO THE SPECIES

1 Rhizome erect, short; leaves in basal rosette, usually rugose; petiole winged; corolla tube without prominent dark veins; capsules pubescent
2 Peduncles and calyx villous, corolla tube not distinctly inflated and constricted at throat .... **1. D. humboldtianus**
2 Peduncles and calyx floccose-tomentose, corolla tube distinctly inflated and constricted at throat .... **2. D. floccosus**
1 Rhizome creeping, elongate; leaves scattered along rhizome, not rugose; petiole unwinged; corolla tube with prominent dark veins; capsules glabrous .... **3. D. zeylanicus**

1. Didymocarpus humboldtianus Gardn., Calcutta J. Nat. Hist. 6: 477. 1846; Trimen, Handb. Fl. Ceylon 2: 273. 1895.—**Fig. 2a.**

Didymocarpus primulaefolia Gardn., Calcutta J. Nat. Hist. 6: 478. 1846.
Didymocarpus humboldtianus var. *primulaefolia* (Gardn.) Thw., Enum. Pl. Zeyl. 207. 1860.
Didymocarpus humboldtianus var. *recedens* Clarke, DC., Mon. Phan. 5: 103. 1883.
Roettlera humboldtiana Kuntze, Rev. Gen. Pl. 2: 476. 1891.

Rhizome erect, short, thickened. Leaves basal, numerous, ovate to ovate-oblong, 3.5–23 cm long, 3–11 cm broad, base rounded or obtuse, and then abruptly tapering and decurrent on the petiole, rarely acute or truncated, margin shallowly to coarsely dentate-crenate, apex rounded, surface usually rugose, venation conspicuous, villous-tomentose above and below when young, becoming less so with age above, rarely sparsely pubescent; petiole usually winged to base, c. 1–10 cm long. Inflorescences usually several; peduncles villous, minutely glandular, 5–25 cm long; bracts linear-oblong, villous, 3–8 mm long, c. 1 mm broad; pedicels slender, 3–20 mm long. Calyx very deeply divided, segments linear-oblong, 2–4 mm long, c. 1 mm broad, apex obtuse, villous and glandular. Corolla very variable in size; tube not constricted at throat, 5–7 mm long above, 10–15 mm long below, pure white to pale pink or violet without, pure white, or white with a yellow spot within, glandular-pubescent without, glabrous within; throat 4–7 mm long, 3–5 mm broad; lobes white to pale pink or violet, rounded, pubescent towards back, glabrous in front, lower three lobes 3.5–10 mm long, 5–12 mm broad, upper 2 lobes smaller, 2–6 mm long, 3.5–8 mm broad. Filaments glabrous or sparsely

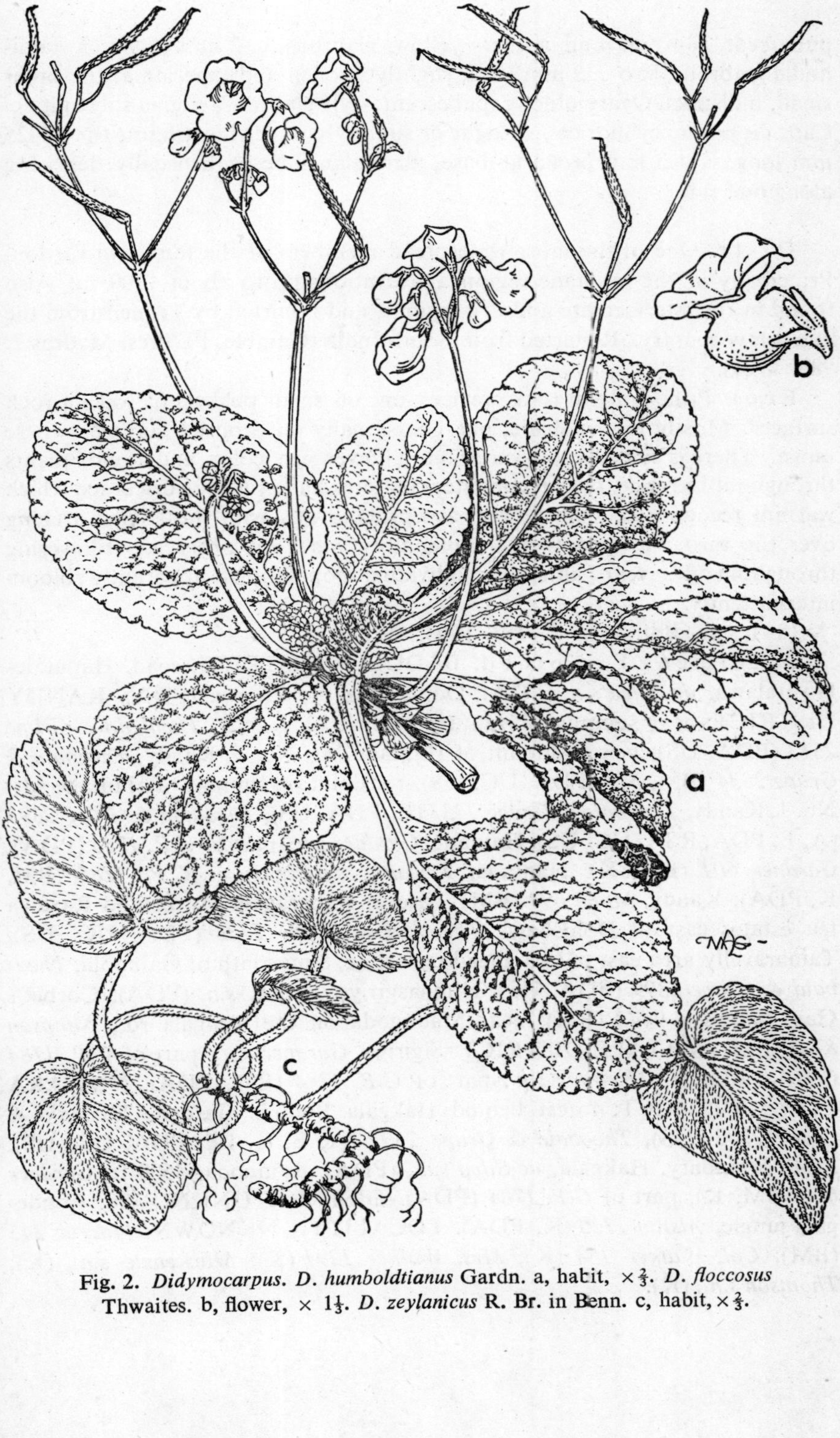

Fig. 2. *Didymocarpus*. *D. humboldtianus* Gardn. a, habit, × ⅔. *D. floccosus* Thwaites. b, flower, × 1⅓. *D. zeylanicus* R. Br. in Benn. c, habit, × ⅔.

pubescent, 3–4 mm long; anthers yellow, glabrous, c. 2 mm broad; 3 staminodia glabrous, two c. 2 mm long, slightly swollen and tapering at tip, other small, indistinct. Ovary oblong, pubescent; style glabrous; stigma subcapitate. Capsule linear, cylindrical, straight or slightly curved, tapering at tip, 15–25 mm long, c. 1.3 mm broad at base, glandular-pubescent, usually dehiscing along one side.

Distr. One of the most widespread members of the family in Ceylon. Principally in the montane region at elevations up to about 1700 m. Also found in the intermediate and dry regions, and reported by Trimen from the moist low country. Reported from South India (Gamble, Fl. Pres. Madras 2: 989. 1921).

Ecol. Principally in rock crevices and on small pockets of soil on rock surfaces. Mostly in the shade, but occasionally on exposed surfaces where moist. There is considerable variation in flower size, color, and size of plants throughout its range, but no consistent patterns have yet been discerned which warrant recognition. The flowers are protandrous with the stigma arching over the throat after anthesis. Pollinators were not observed. Flowering throughout the year, although individual populations apparently bloom intermittently.

Illustr. Bot. Mag. t. 4757. 1853.

Specimens Examined. BADULLA DISTRICT: road, Haputale-Diyatalawa, *de Silva s.n.* (PDA); Ooma Oya, *s. coll. s.n.* (PDA). KANDY DISTRICT: near summit of Hantane Mt. No. 2, Kandy, *Theobald & Grupe 2313* (PDA, US); below summit of Hantane Mt. No. 1, Kandy, *Theobald & Grupe 2334* (BM, NY, PDA, UC, US); rock cliff on summit of Hantane Mt. No. 1, Kandy, *Theobald & Grupe 2337* (E, PDA, US), *Theobald & Grupe 2395* (A, E, PDA, RSA, UC, US); Hantane, *Gardner s.n.*, part of *C.P. 1785* (PDA), *Gardner 601* (BM, K); Hantane, *Thwaites s.n.*, part of *C.P. 1785* (BM, K, PDA); Kandy, *Moon 698* (BM); base of Kabaragala Mt. behind Raxawa tea estates east of Dolosbage, *Theobald & Grupe 2358* (PDA, UC, US); Tamaravelly area east of Craighead tea estate, 8 mi south of Gampola, *Theobald & Grupe 2399* (PDA, US); Hunnasgiriya, *s. coll. s.n.* (PDA); Corbet's Gap, *Simpson 9443* (BM); near Madugoda on the Urugala rd., *Simpson 8786* (BM). MATALE DISTRICT: Sigiriya, *Gardner s.n.*, part of *C.P. 1784* (PDA); Matale, *Thwaites s.n.*, part of *C.P. 1784* (BM, PDA). NUWARA ELIYA DISTRICT: forest behind Hakgala Botanic Garden, *Theobald & Grupe 2302* (US), *Theobald & Grupe 2397* (A, E, K, LE, PDA, UC, US); south boundary, Hakgala, *de Silva s.n.*, (PDA); Ramboda, 4500 ft, *Gardner 600* (BM, K), part of *C.P. 1784* (PDA). DISTRICT UNKNOWN: Condegala jungle, *Alston 1726* (K, PDA). LOCALITY UNKNOWN: *Macrae 245* (BM); *Col. Walker 184* (K); *Mrs. Walker 1291* (K); *Mackenzie s.n.*, (K); *Thomson s.n.*, (K).

2. Didymocarpus floccosus Thw., Enum. Pl. Zeyl. 207. 1860; Trimen, Handb. Fl. Ceylon 3: 274. 1895.—**Fig. 2b.**

Roettlera floccosa Kuntze, Rev. Gen. Pl. 2: 476. 1891.

Rhizome erect, short, stout. Leaves basal, numerous, rhomboid-ovate, 7–15 cm long, 4–9.5 cm broad; base gradually tapering and decurrent on petiole, margin shallowly serrate-crenate, apex obtuse or acute, surface rugose, venation conspicuous, very densely floccose-tomentose above and below, especially when young, becoming less so with age above. Inflorescences usually several, peduncles densely floccose-tomentose, 9–22 cm long; bracts oblong, tomentose, 4–6 mm long, c. 1 mm broad; pedicels slender, tomentose, 5–10 mm long. Calyx very deeply divided, segments oblong, floccose-tomentose, 3–4 mm long, c. 1 mm broad. Corolla tube inflated and then constricted at the throat, c. 6 mm long above, 15 mm long below; white without, white with a yellow spot within, pubescent without, glabrous within; throat c. 3 mm square; lobes pale violet, rounded, pubescent towards back, glabrous in front; lower 3 lobes c. 8 mm long, 8 mm broad, upper 2 lobes c. 6 mm long, 8 mm broad. Filaments glabrous, c. 3.5 mm long; anthers with violets spots, glabrous, c. 2 mm broad; staminodia glabrous, 2 slightly swollen at tip, c. 1.8 mm long; other not swollen, c. 1.2 mm long. Ovary oblong, densely pubescent; style glabrous, stigma capitate. Capsule linear, cylindrical, usually straight, tapering at tip, 12–23 mm long, c. 1.8 mm broad at base, pubescent, dehiscing along one side.

Distr. Endemic. Known only from the vicinity of the type locality at elevations of about 650 m.

Ecol. The only observed population was found growing in rock crevices and on rock faces in a very open area adjacent to a tea plantation. There appeared to be considerable underground seepage. Flowering July–September. No other data available.

Illustr. Bot. Mag. t. 5161. 1860 (*D. primulaefolia*).

Note. As far as it is known this species had only been collected once on the island and at the time of its discovery by Thwaites. We have since rediscovered it, and it is possible that both collections have been made in the same area. Thwaites' type locality was the Raxawa estate, Dolosbage. This estate is presently very small and west of the town. Field work in this area only revealed populations of *D. humboldtianus* at elevations higher than that noted by Thwaites. At lower elevations the area has been completely converted to tea. Our collection, however, was made to the east of Dolosbage but at an elevation equal to that of Thwaites. Within a few miles of this collection of *D. floccosus* one finds typical populations of *D. zeylanicus* and *D. humboldtianus*. The latter is clearly the most closely allied of the two species, but can readily be distinguished by leaf and pubescence characters, flower colour, and the absence of the inflated tube and constricted throat characteristic of *D. floccosus*.

Specimens Examined. KANDY DISTRICT: Raxawa, Dolosbage, *Thwaites s.n., C.P. 3368* (BM, K, PDA). KEGALLE DISTRICT: just east of waterfalls, c. 3 mi east of Dolosbage, *Theobald & Grupe 2403* (A, BO, E, K, LE, NY, PDA, SING, UC, US).

3. Didymocarpus zeylanicus R. Br. in Bennett, Pl. Jav. Rar. 119. 1840; Trimen, Handb. Fl. Ceylon 3: 274. 1875.—**Fig. 2c.**

Didymocarpus longipetiolata Gardn., Calcutta J. Nat. Hist. 6: 475. 1846.
Roettlera zeylanica Kuntze, Rev. Gen. Pl. 2: 477. 1891.

Rhizome prostrate, elongate, branching, not appreciably thickened. Leaves few, scattered along rhizome, usually closely placed, long-petioled, ovate to orbicular, 4–8.5 cm long, 3.5–7.5 cm broad, base cordate to rounded, margin crenate-dentate to serrate, apex acute or obtuse, surface smooth (not rugose), only main vein evident, silky-pubescent above and below; petioles silky-pubescent, 3–20 cm long. Inflorescences usually few, only in axils of last few terminal leaves; peduncles villous to sparsely pubescent, c. 7–15 cm long; bracts lanceolate, pubescent, 2–4 mm long, c. 1 mm broad; pedicels subfiliform, sparsely pubescent, 3–10 mm long. Calyx very deeply divided, segments linear, obtuse, spreading, sparsely pubescent, 3–5 mm long, c. 1 mm broad. Corolla tube slightly gibbous at base, slightly constricted at the throat, c. 11 mm long above, c. 13 mm long below, a diffuse pinkish-violet with prominent darker veins without, colour similar within, very sparsely pubescent without, glabrous within; lobes white to very pale pink, rounded, sparsely pubescent on back, glabrous in front; lower 3 lobes 6–9 mm long, 6–9 mm broad; upper 2 slightly smaller. Filaments glabrous, c. 2.5 mm long; anthers glabrous, c. 2 mm broad; 3 staminodia glabrous, 2 slightly swollen at tip, c. 1.2 mm long, other minute. Ovary oblong, glabrous; style glabrous; stigma oblique, flattened, orbicular, formed from expansion of lower portion, margin entire or fimbriate, c. 2 mm long, 1.5 mm broad. Capsule linear, cylindrical, usually straight or slightly curved, long-tapering at tip, 18–32 mm long, 1–1.5 mm broad at base, glabrous, dehiscing along one side.

Distr. Endemic. Known from the montane region around Adam's Peak north to the Dolosbage area at elevations between 900 and 2000 m.

Ecol. Found in dense shade on rock surfaces near streams, and also in loose soil near rock areas in the forests. Rarely found at base of trees. Populations apparently bloom intermittently throughout the year with all plants within populations blooming at the same time. Flowering December–March (Trimen), and July–August; possibly throughout the year.

Specimens Examined. KANDY DISTRICT: near Carolina Tea

Factory, 1 mi south of turnoff to Watawala on rd. to Hatton, *Theobald & Grupe 2359* (E, PDA, US); near Carolina Tea Factory, *Tunnard* (?) *s.n.* (PDA); Gartmore, Rosamaliya, *de Silva s.n.* (PDA); Malgama, Maskeliya, *Trimen s.n.* (PDA); Adam's Peak, *Gardner s.n.*, part of *C.P. 352* (K, PDA); Ambagamuwa, *Gardner s.n.*, part of *C.P. 352* (PDA). KEGALLE DISTRICT: Windsor Forest east of Dolosbage on rd. to St. Helen's Tea Estate, *Theobald & Grupe 2401* (PDA, US). LOCALITY UNKNOWN: *Thwaites s.n.*, part of *C.P. 352* (K, PDA); *Mrs. Walker s.n.* (K); *Col. Walker s.n.* (K).

3. CHIRITA

Buch.-Ham. ex D. Don, Prod. Fl. Nepal 89. 1825.

Roettlera subgen. *Chirita* (D. Don) Fritsch in Pflanzenfam. 4 (3B): 148. 1895.
Didymocarpus subgen. *Chirita* (D. Don) Chun, Sunyatsenia 6: 290. 1946.

Caulescent herbs or shrubs. Leaves opposite or whorled, petiolate. Inflorescence axillary, 1-flowered, or a simple or compound dichasium. Calyx large, deeply 5-parted or divided. Corolla large, tubular-funnelform, ventricose; lobes 5, rounded, upper 2 slightly connate, 2-lipped. Stamens 2, inserted near base of tube; filaments flattened, geniculate; anthers connate, applanate, with a tuft of hairs on back. Disk annular or lobed, surrounding base of ovary. Ovary unilocular with 2 intruding, recurved, parietal placentae; style slender, stigma oblique, lower lip flattened, thin, enlarged, sometimes bifid, upper reduced or absent. Fruit a capsule, linear, loculicidally 2-valved, seeds numerous.

A large genus of about 80 species known from India and Ceylon east through Southeast Asia and north into Nepal and Southern China. The five taxa found in Ceylon are all endemic and found in the moist lowlands and montane region.

Two of the species, *C. moonii* and *C. walkeri*, are large caulescent shrubs found in rock crevices on steep rock faces. The other two, *C. zeylanica* and *C. angusta*, are caulescent herbs of stream areas in undisturbed forests.

The flowers of all species exhibit a similar floral biology with the anthers united and hinged in such a way that the insect vector is probably dusted with pollen when the geniculate filament arms are disturbed. A broad yellow guide line is also present. The stigma is stabilized between the ridges on the upper part of the tube, and arches over the throat at maturity and after anthesis. The insect vector transmits the pollen upon entering the flower and hitting the stigma. Unfortunately, no pollinators were observed at the time of collection during the rainy season.

Although members of the genus *Chirita* are readily distinguishable from those of *Didymocarpus* on gross morphological grounds, it is difficult to find any one character that is consistent for either group. (See B.L. Burtt, Studies in the Gesneriaceae of the Old World I. General Introduction. Notes R. Bot.

Gard. Edinburgh 21: 185. 1954, and several subsequent articles in this series.)

KEY TO THE SPECIES

1 Caulescent shrubs of rock cliffs; flowers solitary, sometimes 2 in leaf axils, rarely a few-flowered dichasium
  2 Calyx divided to near base; corolla greater than 7 cm long, lobes bluish to mauve-violet .... **1. C. moonii**
  2 Calyx 5-parted, lobes distinct about half way; corolla less than 6 cm long, lobes deep purple.. **2. C. walkeri**
1 Caulescent herbs of stream beds or moist areas; flowers in paniculate compound dichasium
  3 Leaves broadly ovate; corolla reddish-violet to purple................ **3. C. zeylanica**
  3 Leaves ovate-lanceolate or oblong-lanceolate to elliptical; corolla clear blue-violet.... .... **4. C. angusta**

1. Chirita moonii Gardn., Calcutta J. Nat. Hist. 6: 479. 1846; Trimen, Handb. Fl. Ceylon 3: 275. 1895.—**Fig. 3a, b.**

Martynia lanceolata Moon, Cat. 45. 1824, nom. nud.
Roettlera moonii Kuntze, Rev. Gen. Pl. 2: 476. 1891.

Large, caulescent, irregularly branching shrubs, woody at base. Stems 6–12 dm long, slightly thickened at the nodes, silky pubescent when young, becoming glabrate with age below; leaf scars conspicuous. Leaves bright, light green, numerous at ends of branches, usually whorled, rarely opposite, narrowly oblong-lanceolate to elliptical, 7–14.5 cm long, 1.5–4 cm broad, tapering at both ends, margins entire or minutely glandular-denticulate, apex acute or subacuminate, silky-pubescent and shiny above and below, pinnately veined; veins numerous, oblique, conspicuous, especially below; petioles short, silky-pubescent, 0.5–2 cm long. Flowers very large, solitary, or rarely 2 in the upper leaf axils; peduncles silky-pubescent, 2–5 cm long; bracts small, linear, borne on peduncles, 1–2 mm long; c. 0.5 mm broad. Calyx very deeply divided, separate nearly to base, segments oblong-lanceolate, 20–32 mm long, 3–5 mm broad, apex acuminate, keeled, silky-pubesent. Corolla large, 7–9.5 cm long; tube c. 4–6 cm long, c. 2–3 cm broad at throat, whitish towards base with diffuse pale violet without, similar within but with a broad yellow line, 2 longitudinal ridges evident along length of tube both above and below, style within upper ridges, glandular pubescent on both surfaces; throat distinctly oblique; lobes pale bluish-violet to mauve-violet with age; lower 3 lobes spreading, 3 cm long, 3 cm broad; upper 2 lobes slightly smaller. Filaments white, glabrous or sparsely pubescent, swollen at bend, twisted in lower half, lower arm 8–10 mm long, upper arm 5–6 mm long; anthers yellow, c. 3–4 mm across, tuft of hairs on back 2–3 mm long; staminodia 3, 2 hooked, twisted near base, pubescent with a tuft of hairs near swollen apex, c. 3 mm long; 3rd smaller. Ovary glabrous; style glandular-pubescent; stigma 2-lipped, lower lip cuneate, truncated,

c. 5 mm long, 3–4 mm broad, upper very small, deltoid, c. 1.2 mm long, 1.2 mm broad. Capsule linear, glabrous, 7–14 cm long, 3–5 mm broad, persistent on stem.

Fig. 3. *Chirita. C. moonii* Gardn. a, flowering branch, ×⅔. b, calyx, ×⅔. *C. walkeri* Gardn. c, inflorescence, ×⅔. d, calyx, ×⅘.

Distr. Endemic. Known only from a few localities in the vicinity of Kandy at elevations of 500–1500 m.

Ecol. Found in rock crevices on very steep rock cliffs, usually in very open locations. The large shrubs often hang down for a considerable distance. Flowering June–October (Trimen), and possibly at other times.

Illustr. Bot. Mag. t. 4405. 1848.

Note. Although very similar to *C. walkeri* in its habit and vegetative characteristics, the flowers are quite different in size and colour.

Specimens Examined. KANDY DISTRICT: near summit of Hantane Mt. No. 1, Kandy, *Theobald & Grupe 2338* (BM, E, PDA, SING, US); Madugoda-Alutnuvara Rd., *Simpson 9466* (BM); Hantane, *Gardner s.n.*, part of *C.P. 1789* (PDA); *Rangala, Alston 462* (PDA, UC); Doluwa Kande, *s. coll. s.n.* (PDA). KEGALLE DISTRICT: above railroad tracks just west of Kadugannawa, *Theobald & Grupe 2372* (PDA, US); *Theobald & Grupe 2391* (A,

E, K, PDA, UC, US). LOCALITY UNKNOWN: *Mrs. Walker 42* (K); *Macrae 16* (BM).

2. Chirita walkeri Gardn., Calcutta J. Nat. Hist. 6: 480. 1846; Trimen, Handb. Fl. Ceylon 3: 275. 1895.—**Fig. 3c, d.**

Roettlera walkeri Kuntze, Rev. Gen. Pl. 2: 477. 1891.

Large, caulescent, irregularly branching shrubs, woody at base. Stems 6–10 dm long, slightly thickened at nodes, silky-pubescent when young, becoming glabrate with age below; leaf scars conspicuous. Leaves bright, light green, numerous at ends of branches, usually whorled, rarely opposite, narrowly lanceolate to ovate-lanceolate or elliptic, 6–12 cm long, 2–4.5 cm broad, margin entire or minutely glandular-denticulate, apex acute or subacuminate, silky-pubescent and shiny above and below, becoming less so with age above; veins numerous, oblique, conspicuous, especially below; petioles short, silky-pubescent, 0.5–4 cm long. Inflorescence axillary, flowers 1–2 in the leaf axils, or a simple dichasium, rarely compound and few-flowered; peduncles 3–6.5 cm long; pedicles where present 5–9 mm long; bracts small, linear, 2–5 mm long, c. 1 mm broad. Calyx 5-lobed, tube 6–8 mm long, 2–3 mm broad; lobes distinct to about the middle, linear-lanceolate, keeled, apex acuminate, silky-pubescent, 4–14 mm long, 1–2 mm broad. Corolla 4–6 cm long; tube 25–32 mm long, 15–20 mm broad at throat, white with diffuse purple without, similar within but with a broad yellow line below, 2 longitudinal ridges evident along length of tube above and below, style within upper ridges, glandular-pubescent on both surfaces; lobes deep purple, lower 3 lobes 7–10 mm long, 9–14 mm broad; upper 2 lobes slightly smaller; filaments white, glabrous or sparsely pubescent, swollen at bend, twisted in lower half, lower arm c. 5 mm long, upper arm c. 6 mm long; anthers yellow, c. 2 mm broad, tuft of hairs on back c. 1.5 mm long; staminodia 3, 2 hooked, twisted near base, pubescent and with a tuft of hairs near swollen apex, c. 1.5 mm long, 3rd smaller. Ovary glabrous; style glandular-pubescent, stigma 2-lipped, lower lip cuneate, truncated, c. 2 mm long, c. 2 mm broad, upper lip very small, deltoid, 0.5 mm long. Capsule long-linear, glabrous, 4–13 cm long, 1–2 mm broad, persistent on stem.

Distr. Endemic. Subspecies *walkeri* is known only from elevations above 1400 m in the Kandy and Nuwara Eliya districts, while subsp. *parviflora* is only known from elevations below 300 m in the Kalutara District.

Ecol. Occupies habitats identical to those of *C. moonii*. Flowering throughout the year (Trimen).

KEY TO THE SUBSPECIES

1 Inflorescence a simple dichasium or reduced to 1–2 flowers; calyx lobes 10–14 mm long .... **1. subsp. walkeri**

1 Inflorescence usually a few-flowered, compound dichasium, sometimes simple; calyx lobes 4–7 mm long. . . . . 2. subsp. **parviflora**

1. subsp. **walkeri**

Inflorescence axillary, a simple dichasium or reduced to 1–2 flowers; calyx lobes 10–14 mm long; corolla 35–42 mm long.

Distr. This subspecies is known from only a few localities between Kandy and Horton Plains in the Nuwara Eliya District.

Illustr. Bot. Mag. t. 4327. 1847.

Specimens Examined. KANDY DISTRICT: base of Kabaragala Mt. behind Raxawa tea estate east of Dolosbage, *Theobald & Grupe 2357* (E, PDA, UC, US); Raxawa, *Thwaites s.n.*, part of *C.P. 2843* (PDA). NUWARA ELIYA DISTRICT: Pidurutalagala, *Thwaites s.n.*, part of *C.P. 2843* (PDA); Hakgala Bot. Gdn., from a cutting originally brought from Horton Plains in 1903, *s. coll. s.n.* (PDA).

2. subsp. **parviflora** (Clarke) Theobald & Grupe, Ceylon J. Sci., Biol. Sci. 10: 70. 1972.

Chirita walkeri var. *parviflora* Clarke, Mon. Cyrt. 112. 1882; DC., Mon. Phan. 5: 112. 1883.

Inflorescence axillary, usually a few-flowered, compound dichasium, sometimes simple; calyx lobes 4–7 mm long; corolla 32–37 mm long.

Distr. Known from only two localities in the Kalutara District.

Specimens Examined. KALUTARA DISTRICT: Rayigam Korale, *Thwaites s.n.*, part of *C.P. 542* (PDA); Pasdun Korale, *Thwaites s.n.*, part of *C.P. 542* (PDA).

3. Chirita zeylanica Hook., Bot. Mag. t. 4182. 1845; Trimen, Handb. Fl. Ceylon 3: 276. 1895.—**Fig. 4b.**

Chirita communis Gardn., Calcutta J. Nat. Hist. 6: 481. 1846.
Roettlera communis Kuntze, Rev. Gen. Pl. 2: 476. 1891.

Caulescent, often branching, herbs, silky-pubescent, 10–35 cm high. Leaves opposite, bright green, paler beneath, fleshy (drying thin), ovate, 9–15 cm long, 4.5–8.5 cm broad, base rounded or subcordate, margin entire, apex acute, silky-pubescent to strigose above and below, pinnately veined; veins oblique, conspicuous below; petioles pubescent, 2–9 cm long. Inflorescence a paniculate, compound dichasium; peduncles glabrous to sparsely pubescent, 7–13 cm long; bracts ovate, caducous, 2–5 mm long, 2–4 mm broad; pedicels 5–12 mm long. Calyx segments free or weakly coalescent below, broadly ovate-lanceolate to lanceolate, glabrous, 8–17 mm long, 4–6 mm broad. Corolla

2.5–4 cm long; tube 25–30 mm long, 10–15 mm broad at throat, usually pinkish-white to reddish-purple without, white with a broad yellow line within, 2 longitudinal ridges evident along length of tube both above and below, style within upper ridges, glabrous on both surfaces; throat suborbicular in outline; lobes reddish-violet to purple, lower 3 lobes 5–8 mm long, 6–10 mm broad, upper 2 slightly smaller. Filaments white, glabrous to very sparsely pubescent, swollen at bend, twisted in lower half, lower arm 4–5 mm long, upper arm c. 6 mm long; anthers yellow, c. 2 mm broad, tuft of hairs on back c. 1 mm long; staminodia 3, 2 hooked, twisted near base, pubescent and with a tuft of hairs near swollen apex, c. 5 mm long, 3rd smaller. Ovary linear, glabrous; style glandular-pubescent; stigmatic upper lobe not evident, lower lobe greatly expanded laterally into rhomboidal or elliptical flattened plate, 2 mm high, 4 mm broad. Capsule long-linear, glabrous, 5–10 cm long, 1–1.5 mm broad.

Fig. 4. *Chirita. C. angusta* Theobald & Grupe. a, habit, × ⅔. *C. zeylanica* Hook. b, leaf, × ⅔.

D i s t r. Endemic. Rather common in undisturbed montane forests of the moist and intermediate zones at elevations above c. 750.

E c o l. The species appears to be confined to rocky areas in and along stream beds of undisturbed forests; usually in dense shade. It is very closely allied to *C. angusta,* but can readily be distinguished by leaf size and shape, and especially flower colour. Flowering in March and June–August; probably irregularly throughout the year.

Illustr. Bot. Mag. t. 4182. 1845 (colour inaccurate).

Specimens Examined. KANDY DISTRICT: 3 mi from Kotmale on rd. to Nawalapitiya, *Theobald & Grupe 2311* (A, BO, E, K, PDA, SING, UC, US); in forest area near summit of Hantane Mt. No. 2, Kandy, *Theobald & Grupe 2314* (BM, E, LE, PDA, UC, US); Hantane 2300 ft, *Gardner 602* (BM, K); Kandy, *Moon 681* (BM); 28 mi east of Rangala rd. to Looloowatte, *Theobald & Grupe 2316* (E, PDA, UC, US); Sentry Box East Mt. behind Raxawa tea estate, east of Dolosbage, *Theobald & Grupe 2344* (A, E, K, PDA, UC, US). MATALE DISTRICT: Lagalla, *Alston s.n.* (PDA). DISTRICT UNKNOWN: between Weva and Delgoda, *Lewis & de Silva s.n.* (PDA). LOCALITY UNKNOWN: *s. coll. C.P. 1788* (BM, K, PDA); *Macrae 243* (BM); *Mrs. Walker s.n.* (K).

4. Chirita angusta (Clarke) Theobald & Grupe, Ceylon J. Sci., Biol. Sci. 10: 70. 1972.—**Fig. 4a.**

Chirita zeylanica var. *angusta* Clarke, Mon. Cyrt. 112. 1883; DC., Mon. Phan. 5: 112. 1883; Trimen, Handb. Fl. Ceylon 3: 276. 1895.

Caulescent, unbranched or few-branched, herbs, silky-pubescent, 5–25 cm high. Leaves opposite, bright green, paler beneath, fleshy (drying thin), ovate-lanceolate to narrowly oblong-lanceolate or elliptical, 5–14 cm long; 0.7–5.5 cm broad, tapering at base, or rarely acute to obtuse, margin entire, apex acute, silky-pubescent to strigose above and below, pinnately veined; veins oblique, conspicuous below; petioles pubescent, 1–6 cm long. Inflorescence a paniculate, compound dichasium; peduncles glabrous to sparsely pubescent, 6–14 cm long; bracts ovate-lanceolate, cauducous, 3–5 mm long, 2–3 mm broad; pedicels 5–12 mm long. Calyx segments free or only slightly coalescent below, glabrous, narrowly ovate-lanceolate to lanceolate, 7–10 mm long, 2–3 mm broad. Corolla 2.7–4 cm long, tube 18–28 cm long, 8–13 mm broad at throat, white with diffuse clear blue-violet colour towards lobes without, white with a broad yellow line below within, 2 longitudinal ridges evident along length of tube both above and below, style within upper ridges, glabrous on both surfaces; throat angular in outline; lobes clear blue-violet; lower 3 lobes 5–7 mm long, 7–9 mm broad; upper 2 slightly smaller. Filaments white, glabrous to sparsely pubescent, swollen at bend, twisted in lower half; lower arm c. 4 mm long; upper arm c. 6 mm long; anthers yellow, c. 2 mm across, tuft of hairs on back c. 1 mm long; staminodia 3, 2 hooked, twisted near base, pubescent and with a tuft of hairs near swollen apex, c. 4 mm long, 3rd smaller. Ovary glabrous; style glandular-pubescent; stigmatic upper lobe not evident, lower lobe greatly expanded laterally into rhomboidal to elliptical flattened plate, 2 mm high, 4 mm broad. Capsule long-linear, glabrous, 5–7 cm long, c. 1 mm broad.

Distr. Endemic. Forested areas of the low country from the Galle District to the foothills above Ratnapura.

Ecol. Found in habitats similar to those of *C. zeylanica*, but at lower elevations (i.e., on rocks along and in small streams in undisturbed forests, usually dense shade). Flowering in March and August; probably irregularly throughout the year.

Note. This taxon definitely appears distinct enough to warrant specific recognition. Besides the characteristic differences in leaf size and shape there is the distinctive clear blue-violet flower colour not evident in any living populations of *C. zeylanica*, and a more angular corolla throat shape. Both of these characters are not evident in herbarium material.

Specimens Examined. GALLE DISTRICT: 3.1 mi east on Forest Dept. Logging Rd., 3 mi north of Udugama on rd. to Hiniduma, *Theobald & Grupe 2363* (PDA,US). MATARA DISTRICT: Morawaka Korale, *Trimen s.n.* (PDA). RATNAPURA DISTRICT: Sinharaja Forest, Kukul Korale, *Thwaites s.n., C.P. 3437* (BM, K, PDA).

4. CHAMPIONIA

Gardn., Calcutta J. Nat. Hist. 6: 485. 1846.

Low, erect, sometimes branching, caulescent, herbaceous perennials, pubescent. Leaves opposite, petiolate. Inflorescence axillary, a compound dichasium, solitary or rarely in pairs, lax; peduncles slender, shorter than the leaves. Calyx very deeply divided, 5-merous, the segments linear. Corolla 4-merous, regular, rotate, tube extremely short. Stamens 4, equal, inserted on corolla tube; filaments short; anthers connivent, ovate-oblong, 4-locular with 2 locules smaller. Disk absent. Ovary oblong-conical, unilocular with 2 T-shaped parietal placentae; style slender; stigma subcapitate, small. Capsule oblong, pointed, loculicidally 2-valved, each valve splitting into 2 halves. Seeds minute, broadly elliptical, blunt at the ends, reticulate.

A monotypic genus endemic to Ceylon and readily distinguishable by the unusual combination of a 5-merous calyx and a regular, 4-merous corolla.

Championia reticulata Gardn., Calcutta J. Nat. Hist. 6: 485. 1846; Trimen, Handb. Fl. Ceylon 3: 277. 1895.—**Fig. 5.**

Stems unbranched or only sparsely branching from near the base, glabrous to densely pubescent, especially above, becoming glabrate with age below, 6–30 cm high. Leaves slightly fleshy, those of a pair often slightly unequal, oblanceolate-oblong, 5–14 cm long, 1.5–4 cm broad, base cuneate, obtuse, or rounded, margin entire, apex acute or obtuse, glabrous to densely pubescent above, pubescent especially along veins and margin below; petioles glabrous to pubescent, 1–3.5 cm long. Peduncles 4–9 cm long; bracts linear-lanceolate, glabrous to pubescent, 4–12 mm long; pedicels filiform, glabrous or pubescent, 5–20 mm long. Calyx segments linear-lanceolate, 4–5 mm long, c. 1 mm broad, pubescent, persistent. Corolla white, glabrous; tube c. 1 mm long; lobes nearly distinct to base, oblong lanceolate, 6–8 mm long; c. 3 mm broad, apex obtuse. Filaments glabrous, 1–2 mm long; anthers 1 mm long, 1 mm broad, dehiscing on inner face. Style c. 2.5 mm long. Capsule glabrous, 6–8 mm long, 1.5 mm broad.

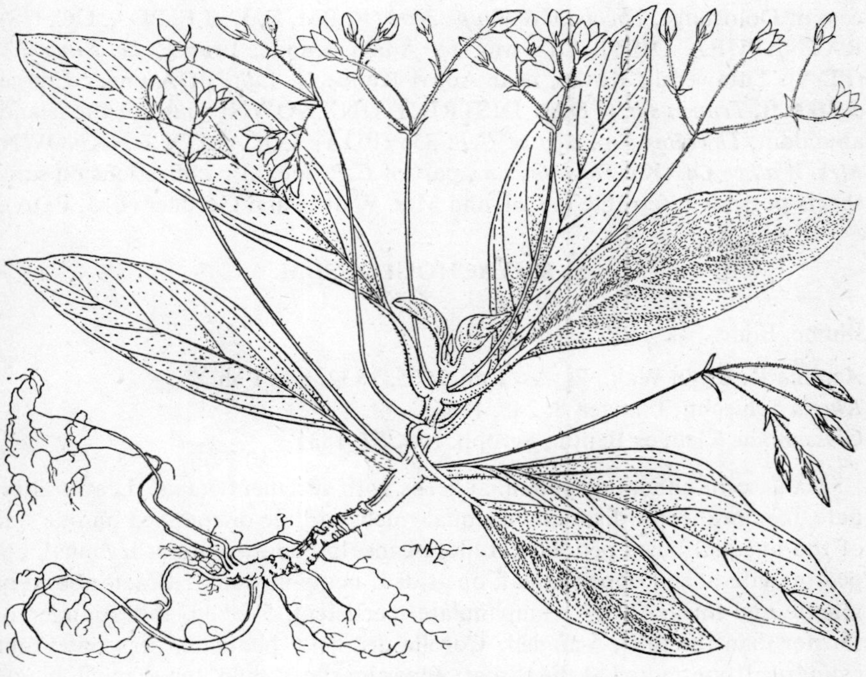

Fig. 5. *Championia reticulata* Gardn. Habit, ×⅔.

Distr. Moist low country from the Galle District north to the foothills of Adam's Peak and the Dolosbage area at elevations up to about 1000 m.

Ecol. In dense shade and loose soil along stream beds in undisturbed forests, also rarely on forest floor near streams. Possibly self-pollinated. Whole populations apparently flower irregularly throughout the year with all individuals blooming at about one time. Flowering February–March and July–August; probably throughout the year at irregular intervals.

Illustr. Clarke, DC., Mon. Phan. 5: t. 15. 1883.

Note. Collections from the Galle District area are generally more pubescent than those near Adam's Peak and the Dolosbage area. There are also possible differences in corolla size and lobe shape, but unfortunately only a few flowers were seen.

Specimens Examined. GALLE DISTRICT: 2.7 mi east along Forest Dept. Logging Rd., 3 mi north of Udugama on rd. to Hiniduma, *Theobald & Grupe 2362* (US); 2.3 mi east along same rd., *Theobald & Grupe 2379* (NY, PDA, UC, US); 5.6 mi east along same rd., *Theobald & Grupe 2383* (E, US, PDA); Hiniduma, *Thwaites s.n.*, part of *C.P. 358* (K). KANDY DISTRICT: Adam's Peak, *Gardner s.n.* (?), part of *C.P. 358, 370* (K, PDA). KEGALLE DISTRICT: on rd. to St. Helen's tea estate, Windsor Forest area east of Dolosbage, *Theobald & Grupe 2402* (A, BM, E, K, LE, PDA, UC, US). RATNAPURA DISTRICT: way to Adavi Kande, Eratne, *de Silva 122* (PDA); Tittaweralu Kotha, near Adavi Kande, *s. coll.* (PDA); near Eratne, c. 1000 ft, *Trimen s.n.* (PDA). DISTRICT UNKNOWN: south of the island, abundant, *Thwaites s.n.*, part of *C.P. 358* (PDA). LOCALITY UNKNOWN: *Mrs. Walker s.n.* (K); *Thwaites s.n.*, part of *C.P. 358*, also collections on same sheets possibly referrable to Col. and Mrs. Walker and Gardner (BM, PDA).

5. RHYNCHOGLOSSUM

Blume, Bijdr. 741. 1826.

Antonia R. Br. in Wall., Pl. As. Rar. 3: 65. 1832, non Pohl.
Klugia Schlecht., Linnaea 8: 248. 1833.
Glossanthus Klein ex Benth., Scroph. Ind. 57. 1835.

Caulescent annual or perennial herbs with succulent stems. Leaves alternate, membraneous, distinctly unequal-sided, cordate or rounded on one side of the unequal base, petiolate. Inflorescence pseudoracemose, terminal, appearing lateral and opposite leaf, one-sided, many-flowered. Bracts absent or minute and linear. Calyx campanulate, persistent, 5-lobed, lobes longer or shorter than the tube, 5-angled. Corolla distinctly bilabiate, personate; tube cylindrical, contracted at the throat; upper lip short, bifid, lower much longer and broader, entire or shallowly 3-lobed. Stamens 2 or 4 (in ours), included, didynamous; anthers coherent in pairs or connate as 4, bilocular, locules confluent. Disk annular or cup-shaped. Ovary unilocular, with 2 parietal placentae; style slender; stigma oblique, sub-capitate, obscurely bilobed. Fruit a capsule, enclosed in persistent calyx, 2-valved, loculicidally dehiscent. Seeds numerous, oblong, reticulate.

A widespread genus of about 12 species extending from India, Ceylon, and China eastwards to the Philippines and New Guinea. Also from Mexico to northern South America. (See B.L. Burtt, Studies on the Gesneriaceae of the

Old World XXIII. *Rhynchoglossum* and *Klugia*. Notes R. Bot. Gard. Edinburgh 24: 167–171. 1962.)

KEY TO THE SPECIES

1 Calyx tube 5-angled, upper angle with rounded, wing-like crest, lobes lanceolate, shorter than the tube; lower lip of corolla usually greater than 16 mm long. . . . . **1. R. notonianum**

1 Calyx tube 5-angled, all angles about equal, narrowly winged, lobes narrowly lanceolate-acuminate, longer than the tube; lower lip of corolla less than 14 mm long. . . . . **2. R. gardneri**

1. Rhynchoglossum notonianum (Wall.) Burtt, Notes R. Bot. Gard. Edinburgh 24: 170. 1962.—**Fig. 6c.**

Wulfenia notoniana Wall., Tent. Fl. Nepal. 46. 1826.
Glossanthus malabaricus Klein ex Benth., Scroph. Ind. 57. 1835.
Glossanthus notonianus R. Br. in Horsfield, Pl. Jav. Rar. 121. 1840.
Klugia notoniana A. DC. in DC., Prod. 9: 276. 1845; Trimen, Handb. Fl. Ceylon 3: 277. 1895.
Klugia glabra Gardn., Calcutta J. Nat. Hist. 6: 489. 1846.
Rhynchoglossum scabrum Dalz. in Hook. J. Bot. Kew Gard. Misc. 2: 140. 1850.
Klugia scabra Dalz. in Dalz. & Gibs., Bombay Fl. 134. 1861.
Klugia notoniana var. *glabra* Clarke, DC., Mon. Phan. 5: 159. 1883.
Klugia notoniana var. *scabra* Clarke, DC., Mon. Phan. 5: 160. 1883.

Erect, branching herbs, sometimes reduced to a single leaf, 10–80 cm high, glabrous, or with a villous line down one side of stem. Leaves bright green above, paler beneath, ovate-oblong in general outline, 6–20 cm long, 3–11 cm broad, base on one side obtuse to acute, other rounded-cordate, margin entire or minutely dentate, somewhat sinuate, apex acuminate, glabrous or scabrous above, glabrous below; petioles glabrous or villous above, 0.5–4 cm long. Peduncles glabrous or villous along one side, 1–4 cm long; bracts linear-filiform, glabrous, 2–5 mm long; pedicels 3–5 mm long. Calyx 8–11 mm long; tube 5-angled, upper angle with rounded wing-like crest on its lower half; crest 3–5 mm high; other angles slightly or equally crested; lobes 5, equal, lanceolate, shorter than the tube, margin ciliated, 2–4 mm long. Corolla tube white, 7–10 mm long, 3–4 mm broad, glabrous without, pubescent within; upper lip white, c. 3 mm high, c. 3 mm broad at base; lower lip deep rich purplish-blue, sometimes slightly paler, with yellow spot at base, glabrous, 16–26 mm long, 15–20 mm broad. Stamens 4; filaments glabrous, one pair c. 6–7 mm long, other c. 1 mm shorter; anthers glabrous, c. 1 mm broad, all 4 weakly connate. Disk cupular. Ovary glabrous; style glabrous; c. 8 mm long. Capsule obovate, membraneous, 8–10 mm long, 5–6 mm broad.

Fig. 6. *Rhynchoglossum*. *R. gardneri* Theobald & Grupe. a, habit, $\times\frac{2}{3}$. b, inflorescence, $\times 1\frac{3}{5}$. *R. notonianum* Burtt. c, inflorescence, $\times 2$.

D i s t r. Known primarily from the montane forests of the moist and intermediate zones, also possibly at lower elevations (var. *glabra*?) of the moist zone. Also common in South India.

E c o l. Usually found in stream beds at edges of forests, or sometimes in open stream areas. No intermediate populations have been found between this species and the closely related, but distinctive, *R. gardneri*.

V e r n. Diya-nilla (S).

S p e c i m e n s E x a m i n e d. BADULLA DISTRICT: Bandarawela, flowers white (?), *Simpson 8654* (BM, PDA); Haputale, *de Silva 9* (PDA, UC). KANDY DISTRICT: at base of Kabaragala Mt. behind Raxawa tea estate east of Dolosbage, *Theobald & Grupe 2356* (E, PDA, UC, US); Raxawa, *Thwaites s.n.*, part of *C.P. 3369* (PDA); Hunnasgiriya, *Thwaites s.n.*, part of *C.P. 1787* (PDA); Dolosbage, *Trimen s.n.* (var. *glabra*) (PDA). NUWARA ELIYA DISTRICT: stream behind Director's home, Hakgala Botanic Garden, *Theobald & Grupe 2303* (A, BM, BO, E, LE, PDA, SING, US); Hakgala Garden, 5000 ft, *Fairchild & Dorsett 326* (UC); jungle behind Hakgala, *de Silva s.n.* (PDA); *de Silva s.n.* (PDA); along rd. from Horton Plains to Agrapatana, *Theobald & Grupe 2309* (E, K, PDA, UC, US); Horton Plains, *Thwaites s.n.*, part of *C.P. 3369* (PDA); Elk Plains, *Gardner s.n.*, part of *C.P. 1787* (PDA); Maturata, *Thwaites s.n.*, part of *C.P. 1787* (PDA); Pundaluoya. *Green s.n.* (var. *glabra*) (PDA). LOCALITY UNKNOWN: *Col. Walker s.n.* (K); *s. coll. C.P. 1787* (BM, K); *s. coll. C.P. 3369* (BM).

2. Rhynchoglossum gardneri Theobald & Grupe, Ceylon J. Sci., Biol. Sci. 10: 70. 1972.—**Fig. 6a, b.**

Klugia zeylanica Gardn., Calcutta J. Nat. Hist. 6: 490. 1846; Trimen, Handb. Fl. Ceylon 3: 278. 1895.

Erect or prostrate herbs, 20–60 cm high, glabrous or with a sparsely villous line down one side of stem. Leaves bright green above, paler beneath, ovate to ovate-oblong in general outline, 4–9.5 cm long, 2.5–6.5 cm broad, base on one side acute to obtuse, other usually distinctly incurved and cordate, margin entire, somewhat sinuate, apex acuminate, glabrous to scabrous above, glabrous below; petioles glabrous or villous above, 0.3–3.5 cm long. Peduncles glabrous to villous along one side, 1–3 cm long; bracts linear-filiform, glabrous, 1–3 mm long; pedicels 2–5 mm long. Calyx 6–9 mm long; tube 5-angled, all angles equal and slightly winged; lobes 5, equal, lanceolate-acuminate, longer than the tube, 4–6 mm long. Corolla tube white, 8–10 mm long, 3–4 mm broad, glabrous without, pubescent within; upper lip white, c. 2–3 mm high, c. 3 mm broad, glabrous; lower lip bright rich blue with a yellow spot at base, 8–14 mm long, 9–14 mm broad, glabrous. Stamens 4; filaments glabrous, one pair c. 5 mm long, other c. 1 mm shorter; anthers glabrous, c. 1 mm broad, all 4 weakly connate. Disk cupular. Ovary glabrous; style glabrous, c. 7 mm long. Capsule ovate, membraneous, 4–5 mm long, 3–4 mm broad.

Distr. Endemic. Known from the moist lowlands of the southern Ratnapura District north into the Kandy District and to elevations of about 1100 m.

Ecol. Found in, or near, stream bed habitats similar to those of *R. notonianum*. Flowering April, July and August; probably throughout the year.

Illustr. Bot. Mag. t. 4620. 1851 (*Klugia notoniana*).

Note. This species has been named in honour of George Gardner, who during the 1840's did much to further our understanding of Ceylon gesneriads, as well as many other Old World taxa.

Specimens Examined. KANDY DISTRICT: forested area near summit of Hantane Mt. No. 2, Kandy, *Theobald & Grupe 2315* (PDA, US); Hantane, *s. coll.*, part of *C.P. 1786* (PDA); Kandy, *Gardner 604* (BM, K); Kandy, *Alston s.n.* (PDA); Raxawa, *s. coll.*, part of *C.P. 1786* (PDA). KEGALLE DISTRICT: on north side of Hwy. A-1 west of Kadugannawa, just below tunnel, *Theobald & Grupe 2370* (A, BO, E, K, LE, PDA, SING, UC, US); 4 mi east of tunnel, *Theobald & Grupe 2371* (BM, NY, PDA, RSA, US); RATNAPURA DISTRICT: on rocks in stream at Hwy marker 65/21 along Hwy. A-17 from Rakwana to Deniyaya, *Theobald & Grupe 2360* (A, E, K, PDA, UC, US); LOCALITY UNKNOWN: *Macrae 242* (BM); *s. coll. 443* (BM, K); *s. coll. C.P. 1786* (BM, K).

6. EPITHEMA

Blume, Bijdr. 737. 1826.

Aikinia R. Br. In Wall., Pl. As. Rar. 3: 65, t. 288. 1832, non Wall. 1832, nec Salisb. ex A. DC. 1830.

Small, succulent herbs, pubescent. Leaves few or solitary, lower alternate and petiolate, upper opposite and usually sessile or short petioled. Inflorescence congested and subcapitate, a scorpioid cyme (monochasium), 1-several in the upper leaf axils, free, or confluent with the petiole, or mid-rib of blade. Bract large, solitary, sessile, 1-sided, concave, subspathaceous. Calyx campanulate, 5-lobed. Corolla small, bilabiate; tube short, terete; lobes 5, unequal. Stamens 2; filaments short, anthers coherent, applanate, short, bilocular. Disk of 1–5 distinct, flattened, yellow glands. Ovary ovoid, unilocular with 2 parietal placentae; placentae stalked, bearing ovules on both sides; style filiform; stigma small, subcapitate. Fruit a capsule, globose, membraneous, dehiscence irregularly circumscissile, enclosed in enlarged calyx. Seeds oblong or elliptical, spirally reticulate.

A genus of about 10 species extending from tropical Africa east through the Indomalaysian area to the Philippines, and Indonesia. Delimitation of species is very difficult in the genus and our sole representative is the most widespread of the group.

Epithema carnosum (G. Don) Benth., Scroph. Ind. 57: 665. 1835.—**Fig. 7.**

Aikinia carnosa G. Don, Gen. Syst. 4: 665. 1838.
Epithema zeylanicum Gardn., Calcutta J. Nat. Hist. 6: 492. 1846.
Epithema carnosum var. *zeylanicum* Clarke, Mon. Cyrt. 178. 1883; DC., Mon., Phan. 5: 178. 1883; Trimen, Handb. Fl. Ceylon 3: 279. 1895.

Stems with a dense cluster of adventitious roots near base, unbranched, or only few-branched from the axil of first leaf, 8–40 cm high, scabrous. Leaves membraneous, light green above, paler beneath, very variable in size, lower leaf often very large, ovate-cordate, 8–20 cm long, 5–15 cm broad, margin entire, serrate or doubly serrate-dentate, apex acute or obtuse, scabrous; upper leaves much smaller, otherwise similar; petioles absent or up to 8 cm long. Peduncles 2–13 cm long; bract broadly ovate to suborbicular, c. 5–10 mm long, 5–10 mm broad, serrate-dentate. Calyx scabrous, lobes lanceolate, c. 3.5 mm long, c. 0.8 mm broad. Corolla tube white without and within, c. 6 mm long, c. 1.5 mm broad, glabrous; lobes blue without and within, glabrous; lower 3 lobes larger, rounded, c. 3 mm long, 3 mm broad; upper lobes slightly fused with distinct H-shaped purplish blotch across lobes, c. 2 mm long, 1 mm broad. Filaments curved near anther, glabrous, 1.5–2 mm long; 2 staminodia curved and swollen at tip, glabrous, c. 1.5 mm long. Capsule small, c. 1.5 mm long, 1.5 mm broad.

Distr. Known from only a few localities in montane forests between Kandy and Dolosbage. Otherwise a very widespread species known from India and Southern China east to the Borneo region.

Ecol. In crevices of damp rocks in the montane forests. The plants are very variable in size and confined to dense shade. All individuals in a population seem to flower at about the same time, and from growth habit they appear to be short-lived. Flowering February and July–September; possibly throughout the year.

Illustr. Wight, Ic. Pl. Ind. Or. 4: t. 1354. 1848 (not representative).

Note. The Ceylon populations may warrant subspecific recognition on the basis of calyx and seed size, but we hesitate making this change until better and more representative collections of *E. carnosum* are available, and chromosome counts are made for populations from both areas.

Specimens Examined. KANDY DISTRICT: Hantane, *s. coll.*, part of *C.P. 2844* (PDA); Raxawa, *Thwaites s.n.*, part of *C.P. 2844* (PDA); Kadugannawa, *Trimen s.n.* (PDA). KEGALLE DISTRICT: north side of

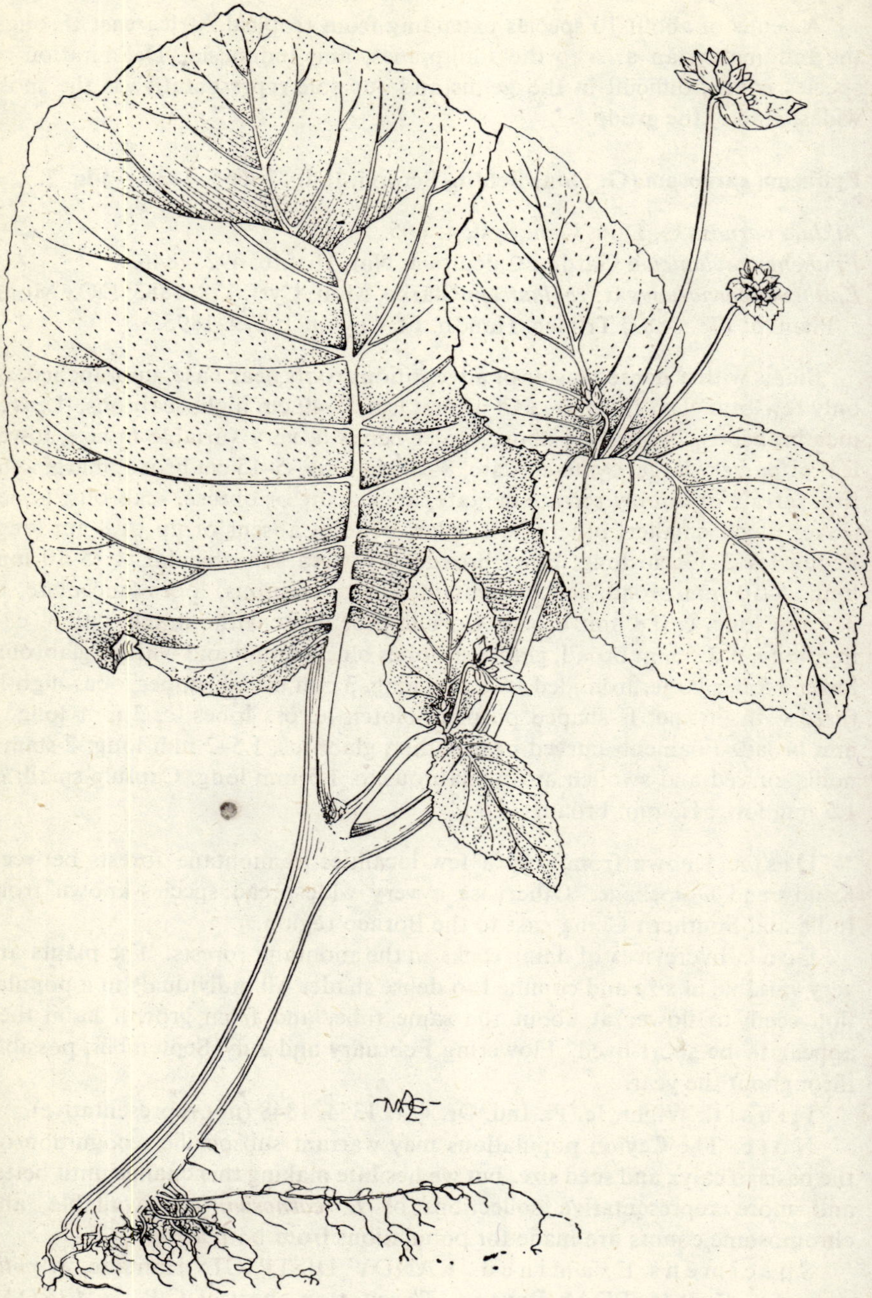

Fig. 7. *Epithema carnosum* Benth. Habit, × ¾.

Hwy. A-1 west of Kadugannawa, just below sharp bend in road on down-grade below tunnel, *Theobald & Grupe 2366* (A, E, K, LE, PDA, UC, US), *Theobald & Grupe 2404* (BM, F, NY, PDA, RSA, US); Jatianpulana, Avisawella, *Simpson 8434* (BM). LOCALITY UNKNOWN: *Mrs. Walker s.n.* (K); *Gardner 606* (K); *Macrae 144, 244* (BM).

7. RHYNCHOTECHUM

Blume, Bijdr. 775. 1826.

Isanthera Nees, Trans. Linn. Soc. London 17: 82. 1834.

Erect, few-branched, caulescent undershrubs, tawny-tomentose. Leaves opposite or alternate. Inflorescence a paniculate, axillary, compound dichasium; bracts narrow; flowers numerous, small. Calyx deeply divided, 5-merous, segments narrow. Corolla campanulate-rotate, 5-lobed, obscurely 2-lipped. Fertile stamens 4, inserted near base of corolla; filaments short, curved; anthers small, 2-celled; cells confluent at apex. Disk very small and annular, or absent. Ovary sessile, ovoid, unilocular with 2 deeply intruding, bifid, recurved, parietal placentae; style slender; stigma small, subcapitate. Fruit a berry, ovoid to globose, wall fleshy, indehiscent. Seeds very small, elliptical, pitted or coarsely-reticulate.

A genus of about 12 species extending from Ceylon and South India to New Guinea and north to the eastern Himalayas and Southern China. (See: B.L. Burtt: Studies on the Gesneriaceae of the Old World XXI. *Rhynchotechum* and *Isanthera*. Notes R. Bot. Gard. Edinburgh 24: 35–39. 1962.)

Rhynchotechum permolle (Nees) Burtt, Notes R. Bot. Gard. Edinburgh 24: 39. 1962.—**Fig. 8.**

Isanthera permollis Nees, Trans. Linn. Soc. London 17: 82. 1834; Trimen, Handb. Fl. Ceylon 3: 280. 1895.
Isanthera floribunda Gardn., Calcutta J. Nat. Hist. 6: 483. 1846.

Stems unbranched or rarely 1–2 branched near base, 20–60 cm high, occasionally taller, tawny, woolly-tomentose above, becoming glabrate with age below; bark whitish, marked with prominent leaf scars. Leaves alternate, closely placed terminally on stem, broadly oblanceolate-oblong, 7–22 cm long, 2.5–9 cm broad, base tapering into petiole, margin serrate, apex acute, dark green above, whitish below, tawny, woolly-tomentose on both surfaces when young, becoming glabrate with age above; petioles 1–3.5 cm long. Inflorescences solitary in the leaf axils, densely woolly-tomentose, peduncles short, 1–15 mm long; bracts linear-lanceolate. Calyx c. 7 mm long, segments lanceolate, c. 5 mm long, woolly-tomentose. Corolla white, glabrous, lower 3 lobes ovate, c. 5 mm long, 4 mm broad, upper 2 only slightly smaller. Filaments

Fig. 8. *Rhynchotechum permolle* Burtt. a, flowering branch, $\times\frac{2}{3}$. b, corolla, $\times 3\frac{1}{3}$. c, flower, $\times 1\frac{3}{5}$. d, anther, $\times 15$. e, pistil, $\times 10$.

glabrous, curved outward and then upward towards staminodium, 1–2 mm long; anthers yellow, glabrous, dehiscing on inner face by a crescent-shaped slit; staminodium very short, swollen at apex. Ovary minutely pubescent, ovoid, c. 2 mm long; tapering into slender style, c. 1.5 mm long; stigma slightly swollen. Fruit ovoid, 4–7 mm long.

Distr. Found in the moist region, both in the central mountains and in the lowlands towards the southern coast at elevations up to about 1300 m. Also found in South India and possibly Burma.

Ecol. Commonly found in dense shade in undisturbed forests especially near small streams. The stems often lie prostrate on the ground for a short distance and root at the nodes. On occasion the plants become very robust with tall stems up to one metre high. The inflorescence and flowers are inconspicuous and no pollinators were observed. Burtt (Notes R. Bot. Gard. Edinburgh 24: 35–39. 1962) reports that *R. discolor* (Maxim) Burtt is readily self-fertile. Flowering June–October and probably throughout the year.

Illustr. Wight, Ic. Pl. Ind. Or. 4: t. 1355. 1848.

Note. Burtt has also very clearly stated the case for the union of *Rhynchotechum* and *Isanthera*, and has also noted the astute nature of Gardner's observations regarding the latter genus (Calcutta J. Nat. Hist. 6: 471–493. 1846). Unfortunately, Burtt had no Ceylon material of *R. permolle* available with anthers in the proper stage for a critical study of dehiscence and determination of the possibility of any real point of difference between the two genera. Critical study of specimens in the wild clearly indicates the same type of valvular dehiscence in *R. permolle* which has been attributed to species of *Rhynchotechum* (sensu stricto).

Specimens Examined. GALLE DISTRICT: 2.7 mi east along Forest Dept. Logging Rd., 3 mi north of Udugama on road to Hiniduma, *Theobald & Grupe 2361* (E, PDA, US). KANDY DISTRICT: approx. 3 mi from Kotmale on rd. to Nawalapitiya, *Theobald & Grupe 2312* (A, BM, E, K, LE, NY, PDA, RSA, UC, US); Kandy, *Moon 234* (BM); Hantane, 2300 ft, *Gardner s.n.* (BM, K); Rangala, *Trimen s.n.* (PDA). NUWARA ELIYA DISTRICT: Maturata, *Thwaites s.n.*, (PDA). DISTRICT UNKNOWN: Ambagamuwa, *Gardner s.n.*, part of *C.P. 1670* (PDA). LOCALITY UNKNOWN: *Thwaites s.n.*, part of *C.P. 1670* (BM, K, PDA); *Col. Walker 43* (K); *Mrs. Walker 1761* (K).

LAMIACEAE (LABIATAE)

(by L.H. Cramer*)

Type genus: *Lamium* L., Sp. Pl. 579. 1753.

Annual or perennial herbs, undershrubs or shrubs, usually with aromatic oil glands or globules; stem and branches usually 4-angular. Leaves opposite, rarely whorled, exstipulate, simple, entire or toothed, rarely lobed. Flowers bisexual, mostly zygomorphic, rarely subregular, in axillary pairs or cymes combined into verticils or terminal spike-like or racemose panicles, or aggregated into heads. Floral leaves often present. Bracts usually reduced, deciduous. Bracteoles small, often absent. Calyx tubular or campanulate, often accrescent, annulate within or not; limb generally unequally 4–5-lobed, or 8–10-toothed, or 2-lipped, the upper lip decurrent on or continuous with the tube. Corolla gamopetalous, sometimes subregular; tube short or long, occasionally annulate within; limb 4–5-lobed, mostly 2-lipped and personate. Stamens 4, (2 in *Salvia*), didynamous, the pairs unequal or subequal in length, the anterior pair the longer, the pairs sometimes connate into a tube; posterior filaments rarely with a hairy crest or projection at base; anthers typically 2-celled, sometimes 1-celled, the cells confluent or divaricate, rarely separated by an elongated connective (e.g. *Salvia*). Disk prominent, ± symmetrical, or with an anterior projection. Ovary superior, 2-carpellary, becoming 4-lobed or 4-partite with partition of each carpel into 2-loculi; style gynobasic, simple, 2-fid above; arms equal or unequal. Fruit of four 1 seeded nutlets enclosed within a persistent calyx; seed small; embryo usually straight, with scanty endosperm.

A cosmopolitan family concentrated chiefly in the Mediterranean region. It comprises over 3000 species in the world distributed among 200 genera. In Ceylon there are 63 species belonging to 22 genera. Of these 54 are indigenous, 12 being endemic, and they occur widely from sea level to c. 2760 m throughout the wet and dry zones; the other species are cultivated. Representatives of various genera, such as *Ocimum*, *Geniosporum* and *Leucas* are commonly prevalent in both the lowlands (wet and dry) and the hill country (uplands and highlands). *Leonotis* is the sole genus with a single species growing only in the dry lowlands, whilst several genera such as *Scutellaria*, *Pogos-*

*Flora of Ceylon Project, Peradeniya. Present address: University College, Batticaloa.

temon, *Plectranthus*, *Coleus*, *Stachys*, *Mentha*, *Prunella* and *Leonurus* are restricted to the uplands above c. 1000 m. Representatives of other genera such as *Ocimum*, *Leucas*, *Anisomeles*, *Hyptis*, *Salvia* and *Stachys* form part of the weed flora occurring in waste places, along roadsides and in cultivated areas. The other genera are not characteristic of a particular zone or type of vegetation. The only exceptions are the genera *Pogostemon*, *Plectranthus* and *Coleus* which form part of the moist undergrowth or field layer of the montane and submontane forests.

The economic importance of the family stems chiefly from the copious presence of volatile oils among many of its genera. Owing to such oils many species in a variety of genera—e.g. *Salvia*—have culinary uses. Species of *Ocimum* are locally used as an insect repellant and of *Thymus* as a culinary herb. *Coleus rotundifolius* is the only plant in the family with edible tubers. Notable, too, is the horticultural value of a few genera such as *Salvia* and *Coleus* whose cultivated species constitute decorative garden plants.

The closest affinities of the family are with the Verbenaceae; but in the latter the ovary is not truly lobed and the style is terminal. Furthermore, almost all members of the Verbenaceae are trees or shrubs and the fruit is usually a drupe or a berry.

According to Briquet's classification (1897, 1917), the family as it occurs in Ceylon is separated into 4 tribes:

1 Style not gynobasic; nutlets with a latero-ventral scar, commonly large, often projecting above half the height of the ovary (*Teucrium*)........................**1. Ajugoideae**
1 Style completely gynobasic; nutlets with a basal scar, often smaller
  2 Seeds more or less transversely placed: embryo with a curved radicle lying on one of the cotyledons (*Scutellaria*)....................................**2. Scutellarioideae**
  2 Seeds erect; embryo with a straight, short (superior) radicle
    3 Stamens erect or spreading and projecting straight forward (*Salvia*, *Clinopodium*, *Pogostemon*, *Stachys*, *Prunella*, *Anisomeles*, *Mentha*, *Thymus*, *Leucas*, *Leonurus*, *Leonotis*, *Dysophylla*)..**3. Stachyoideae**
    3 Stamens declinate (descending), lying upon or enclosed in the lower lip of cordia (*Ocimum*, *Orthosiphon*, *Geniosporum*, *Anisochilus*, *Coleus*, *Plectranthus*, *Hyptis*)..... ....**4. Ocimoideae**

The key which follows deviates from this classification in an attempt to facilitate identification in the field and in the herbarium.

KEY TO THE GENERA

1 Stamens declinate
  2 Verticils 6-more-flowered; upper lip of corolla present; disk generally annular or projected anteriorly; funicular scar of seed basal
    3 Verticils simple
      4 Lower lip of corolla flat
        5 Stamens exserted, anthers 1-celled

6 Verticils 6-flowered; calyx upper lip decurrent on tube, the tube dotted without with sessile glands; filaments appendaged at base with a hairy process or tuft; disk symmetrical or equally lobed; aromatic herbs........................**1. Ocimum**

6 Verticils more than 6 flowered; calyx upper lip continuous with tube, the tube not dotted without with sessile glands; filaments not appendaged; disk projected anteriorly; herbs without a distinctive scent........................**2. Geniosporum**

5 Stamens included; anthers 2-celled

7 Upper lip of calyx continuous with on tube, the tube not dotted without with sessile glands; corolla up to 3 mm long; disk symmetrical.................**3. Basilicum**

7 Upper lip of calyx decurrent on tube; the tube dotted without with sessile glands; corolla much more than 3 mm long; disk projected anteriorly....**4. Orthosiphon**

4 Lower lip of corolla navicular

8 Stamens entirely free; upper lip of calyx usually continuous with tube; members of each pair of young floral leaves laxly imbricate with those of the other..**5. Plectranthus**

8 Stamens fused into a sheath around style; upper lip of calyx decurrent on tube; members of each pair of young floral leaves closely imbricate with those of the other..**6. Coleus**

3 Verticils compound, or flowers in heads

9 Corolla distinctly 2-labiate, the lower lip navicular; stamens included; disk projected anteriorly

10 Calyx accrescent in fruit; tube 10-nerved; verticils distinct, in racemes or panicles

11 Stamens entirely free; upper lip of calyx usually continuous with tube; members of each pair of young floral leaves laxly imbricate with those of the other...**5. Plectranthus**

11 Stamens fused into a sheath around style; upper lip of calyx decurrent on tube; members of each pair of young floral leaves closely imbricate with those of the other..**6. Coleus**

10 Calyx unchanged in fruit, the upper lip deflexed over mouth; tube plain; verticils dense, in tetragonal or cylindrical spikes........................**7. Anisochilus**

9 Corolla subbilabiate, the lower lip saccate; stamens shortly exserted; disk symmetrical...**8. Hyptis**

2 Verticils usually 2-flowered; corolla without upper lip; disk lobed; funicular scar of seed oblique..**21. Teucrium**

1 Stamens erect

12 Verticils symmetrical; upper lip of corolla present; funicular scar of seed basal

13 Calyx tube smooth

14 Leaves opposite; floral leaves absent; calyx 2-lipped, the upper lip before anthesis with a hooded appendage, the lips closed after anthesis; corolla tube longer than calyx, lobes 5; anthers of anterior stamens often 1-celled............**9. Scutellaria**

14 Leaves whorled; floral leaves present, whorled; calyx regular, the lobes spreading after anthesis; corolla tube as long as calyx, lobes 4; anthers of all stamens 2-celled..**10. Eusteralis**

13 Calyx tube nerved

15 Stamens 2; calyx mouth closed after anthesis; anther-cells dimidiate; floral leaves ovate in outline..**11. Salvia**

15 Stamens 4; calyx mouth open after anthesis; anther-cells 1 or 2; floral leaves generally linear or lanceolate in outline

16 Calyx regular or subregular, lobes subequal

17 Corolla limb subbilabiate; upper lip nearly flat, sparsely puberulous without; corolla tube exannulate within; stamens generally exserted

18 Stamens equal; filaments glabrous throughout; floral leaves linear; bracteoles absent; aromatic herbs .... **12. Mentha**
18 Stamens subequal; filaments usually bearded at middle; floral leaves ovate; bracteoles present; herbs without a distinct scent .... **13. Pogostemon**
17 Corolla limb 2-labiate; upper lip hooded, villous-hirtellous without; corolla tube generally annulate within; stamens included
19 Calyx limb 5-lobed or 5-toothed, lobes acute or acicular-acuminate; mouth uniformly level
20 Lobes of calyx equal; one cell of anterior anthers sterile; often aromatic herbs .... **14. Anisomeles**
20 Lobes of calyx subequal or limb subbilabiate; both cells of anterior anthers fertile; herbs without a distinct scent .... **15. Stachys**
19 Calyx limb 8–10 toothed; lobes aristate or spinescent at apices; mouth generally oblique
21 Leaves serrate or crenate; calyx tube 10-nerved, the mouth oblique
22 Floral leaves spinescent at apex; calyx tube dotted with sessile glands without; filaments villous at base .... **16. Leonotis**
22 Floral leaves acuminate at apex; calyx tube not dotted with sessile glands without; filaments glabrous or finely pubescent at base .... **17. Leucas**
21 Leaves deeply incised or lobed; calyx tube 5-nerved, the mouth straight .... **18. Leonurus**
16 Calyx 2-labiate
23 Calyx tube inwardly curved, 13-nerved, pilose-annulate within, not dotted without with sessile glands .... **19. Clinopodium**
23 Calyx tube straight, 5–10 nerved, exannulate within dotted without with sessile glands
24 Leaves ovate in outline; cymes verticillate; bracteoles 0
25 Upper lip of calyx entire; spikes interrupted; filaments entire at apex .... **15. Stachys**
25 Upper lip of calyx 3-dentate; spikes condensed into a head; filaments 2-denticulate at apex .... **20. Prunella**
24 Leaves linear in outline; cymes generally racemose; bracteoles present .... **21. Thymus**
12 Verticils secund; upper lip of corolla absent; funicular scar of seed oblique .... **22. Teucrium**

1. OCIMUM

L., Sp. Pl. 597. 1753; L., Gen. Pl., ed. 5, 651. 1754; Benth. in Wall., Pl. As. Rar. 2: 13. 1831; Benth., Lab. Gen. Sp. 1. 1832 (*Ocymum* sphalm.); Benth. in DC., Prod. 12: 31. 1848; Endl., Gen. Pl. 608. 1838; Benth. & Hook., Gen. Pl. 1171. 1876; Briq. in Pflanzenfam. 4 (3a): 369. 1897. Type species: *O. basilicum* L.

Becium Lindl., Bot. Reg. 28: Misc. 43. 1842. Type species: *B. bicolor* Lindl.

Herbs, undershrubs or shrubs, usually with strongly aromatic oil. Leaves simple, entire to serrate at margins, petiolate. Racemes terminal, simple or branched at base, of cymes combined into 6-flowered, distant verticils. Floral leaves bracteate, caducous. Bracts often petioled, rarely longer than flowers, often deciduous. Calyx ovoid or campanulate, deflexed in fruit, 2-lipped; tube 10-nerved, often glabrous within, the mouth sometimes closed after anthesis by upcurved lower lip; upper lip orbicular or obovate, decurrent at margins on tube, strongly reflexed in fruit; lower lip unequally 4-lobed, anterior ones often aristate. Corolla 2-labiate; tube campanulate, ± equalling calyx, glabrous within; upper lip erect, unequally 4-lobed; lower lip hardly longer, declinate, flattened or concave, entire. Fertile stamens 4, didynamous, exserted; filaments free from each other, anterior ones longest, subglabrous to pubescent, posterior ones at base with a transverse hairy process or hairy tuft; anthers 1-celled, often confluent. Disk tumid, annular or 3–4-lobed. Stylar branches 2, subequal, subulate or flattened. Nutlets obovoid or globose, occasionally finely warty, often mucilaginous when wetted; basal scar small.

About 150 species, pantropic and in warm temperate regions.

KEY TO THE SPECIES

1 Leaves subacuminate to acuminate at apex; mouth of calyx closed after anthesis by upcurved lower lip; median lobes of lower lip of fruiting calyx shorter than upper lip; shrubs or undershrubs .... **1. O. gratissimum**

1 Leaves acute or rounded at apex; mouth of calyx ± open after anthesis; median lobes of lower lip of fruiting calyx as long as or longer than upper lip; herbs occasionally with stems slightly woody at base

2 Verticils moderately close; leaves pubescent on lower surface; calyx tube sparsely or densely pilose without at base; lateral lobes of lower lip of calyx entire; bracts without a receptacle at base

3 Leaves acute at apex; calyx tube hirsute-annulate within; posterior filaments at base with a transverse hairy process; nutlets mucilaginous when wetted

4 Fruiting calyx 4.5–5.5 mm long; corolla 5–7 mm long .... **2. O. canum**

4 Fruiting calyx 6–8 mm long; corolla 7–10 mm long .... **3. O. basilicum**

3 Leaves obtuse at apex; calyx tube completely glabrous within; posterior filaments at base with a hairy tuft; nutlets unchanged when wetted .... **4. O. sanctum**

2 Verticils distant; leaves glabrous or subglabrous on lower surface; calyx tube glabrous without; lateral lobes of lower lip of calyx 2-mucronate or with mucronate serrations; bracts with a concave, glandular receptacle at base .... **5. O. adscendens**

1. Ocimum gratissimum L., Sp. Pl. 1197. 1753 (*Ocymum* sphalm.); Moon, Cat. 44. 1824; Benth., Lab. Gen. Sp. 7, 707. 1832; Benth. in DC., Prod. 12: 34. 1848 (cum syn.); Thw., Enum. Pl. Zeyl. 236. 1860; Hook. f., Fl. Br. Ind. 4: 608. 1885; Trimen, Handb. Fl. Ceylon 3: 367. 1895; Alston, Kandy Fl. 65, f. 348. 1931; Mukerjee, Rec. Bot. Surv. India 14 (1): 20. 1940; Abeywick., Ceylon J. Sci., Biol. Sci. 2 (2): 218. 1959. Type: Hortu Upsalensi, *749.2* (LINN.)

Shrub or undershrub; stem up to 1.90 m high, much branched, obtusely 4-angular, grooved on opposite sides, glabrous and woody towards base, hispidulous above. Leaves ovate to ovate-lanceolate, 4.8–12.8 × 3–9 cm, subcuneate to tapering at base and subdecurrent on petiole, acuminate to subacuminate, glabrescent to sparsely scaberulous above, puberulent on nerves beneath, copiously glandular punctate on both sides, thin; petioles 0.7–6 cm long. Racemes rarely in axils of upper leaves, up to 18.6 cm long; peduncles densely pubescent with recurved hairs. Floral leaves decussate, broadly ovate-lanceolate to oblong-lanceolate, 5–6 × 4 mm, 3-nerved at base, acuminate, ciliate, deciduous. Pedicels 1.5–2.5 mm long, recurved. Calyx campanulate, up to 5 mm long; tube 2–2.5 mm long, sparsely hispidulous or strigose without and dotted with sessile oil globules, glabrous within; upper lip broadly ovate, 2–2.5 × 2–3 mm; median lobes of lower lip shorter than laterals. Corolla 5–7 mm long, greenish white to greenish-yellow; tube infundibular, 2.5–3.5 mm long, broadened above, puberulous without and dotted with sessile oil globules; upper lip 2.5–3.5 mm long; lower one ovate-lanceolate, 2–3 mm long. Transverse process of posterior filaments with gland-tipped hairs. Disk annular. Style glabrous. Nutlets subglobose, 2 mm in diam., faintly foveolate.

Distr. Pantropic; native origin unknown.

Ecol. A common weed usually under shade by roadsides and in waste ground of the lowlands and lower uplands from sea level to c. 530 m, reaching bush-like proportions. This habit, the large leaves with decurrent bases and the conspicuous floral leaves give this plant a distinct facies in the field. The leaves when bruised have a pleasant, apple-like scent. There is no doubt that the plant is well naturalized in Ceylon. Flowering throughout the year.

Vern. Gas-tala (S).

Specimens Examined. POLONNARUWA DISTRICT: Polonnaruwa, Sacred Area, *Ripley 277* (PDA, US), in open, waste ground, *Cramer 3677* (PDA, US). KANDY DISTRICT: Peradeniya, cultivated in Royal Botanic Gardens, *Simpson 9146* (PDA). KURUNEGALA DISTRICT: Gonagama, *s. coll. C.P. 2062* (E, PDA). COLOMBO DISTRICT: Ja-Ela, coconut plantation, *Comanor 1016* (PDA, US). MATARA DISTRICT: Talalla, under shade by roadside, *Cramer 3434* (US). HAMBANTOTA DISTRICT: Yala, Ruhunu National Park, Talgasmankade, under shade in scrub, *Cramer 4123* (E, K, PDA, US), Patanagala, in scrub forest, *Fosberg & Mueller-Dombois 50125* (PDA, US), *Comanor 890* (PDA, US), Block 1, Yala Road, *Cooray 69011101* (PDA, US).

var. **suave** (Willd.) Hook. f., Fl. Br. Ind. 4: 609. 1885; Trimen, Handb. Fl. Ceylon 3: 367. 1895.

Ocimum suave Willd., Enum. Hort. Berol. 629. 1809; Benth., Lab. Gen. Sp. 7. 1832; Benth., in DC., Prod. 12: 35. 1848; Thw., Enum. Pl. Zeyl. 236. 1860. Type from Botanic Gardens, Berlin, cultivated specimen (n.v.).

Shrub; stem up to 1.5 m high, much branched, bluntly 4-angular, pilose-hirsute to hirsute. Leaves up to 12.3 × 4.6 cm, acute at base, densely strigose above, villous-tomentose to hoary-tomentose beneath. Peduncles densely pilose-hirsute. Fruiting calyx densely villous-hirsute without.

Distr. From tropical Africa through Madagascar to India and Ceylon.

Ecol. A common weed in similar habitats as the typical variety but more widespread. Flowering throughout the year.

Note. J.K. Morton (J. Linn. Soc. Bot. 58: 232. 1962) considers *O. gratissimum* and *suave* quite distinct species in West Africa, though he admits "they appear to be connected in Asia by intermediates." Examination of Asian material, however, has shown that there are differences, though only minor, between these two taxa, a greater density of pubescence in the variety being quite obvious. Though in West Africa these taxa are known to have different chromosome numbers (*O. gratissimum* having 2n = 40, and *suave* having 2n = 64), Morton himself admits of chromosome numbers varying from the above from plants not of West Africa. Until a constant difference in numbers between the two taxa can be established I prefer to follow Hooker in reducing *O. suave* to varietal rank under *O. gratissimum*.

Specimens Examined. ANURADHAPURA DISTRICT: Alutoya, *Jayasuriya 677* (PDA, US). MATALE DISTRICT: Matale, by roadside, *Cramer 3025* (PDA, US). KANDY DISTRICT: Kandy, hillside, *Hepper 4491* (K, PDA, US), Ambagamuwa, *s. coll. C.P. 3623* (E, K, PDA); Peradeniya, University Campus, off rail road, *Comanor 333* (K, PDA, US); Kadugannawa, Menikdiwela, in open waste ground, *Cramer 4217* (E, K, NVB, PDA, US); Pussellawa, Sangillipalama, under shade in waste ground, *Cramer 3861* (PDA, US). RATNAPURA DISTRICT: between Balangoda and Belihul Oya, *Comanor 715* (PDA, US). BADULLA DISTRICT: Haputale, in waste ground, *Cramer 3882* (PDA, US).

2. Ocimum canum Sims, Bot. Mag. 51: t. 2452. 1824; Benth., Lab. Gen. Sp. 3, 707. 1832; Benth. in DC., Prod. 12: 32. 1848; Thw., Enum. Pl. Zeyl. 236. 1860; Hook. f., Fl. Br. Ind. 4: 607. 1885; Trimen, Handb. Fl. Ceylon 3: 365. 1895.

Ocimum americanum sensu auctt. Alston in Trimen, Handb. Fl. Ceylon. 6: 234. 1931; Abeywick., Ceylon J. Sci., Biol. Sci. 2 (2): 218. 1959, non L. 1755.

Annual with a shallow rootstock; stem up to 60 cm high, much branched, sub-4-angular, grooved on opposite sides, pubescent, with appressed retrorse hairs. Leaves oval-lanceolate to oval-oblong, 1.7–4.2 × 0.6–2 cm, rounded to acute at base, acute, distinctly serrulate, glabrous to sparsely scaberulous above, sparsely pilose along prominent nerves beneath and copiously dotted with sessile oil glands, thin; petioles 0.5–2.2 cm long, slender, with a few pilose hairs. Racemes to 13.5 cm long; peduncles densely pubescent with recurved hairs. Floral leaves ovate, to 5 × 2 mm, clawed, ciliate, conspicuous at anthesis. Pedicels recurved, to 2 mm long, hispidulous. Calyx campanulate, 4–5 mm long; tube 2 mm long, copiously dotted without with sessile oil globules, hirsute-annulate within above glabrous base; upper lip broadly orbicular, 2 × 2.5–3 mm, apiculate; lower lip 4 mm long, the lateral teeth broadly triangular, shortly aristate, the median ones lanceolate, long-aristate. Corolla 5–6 mm long, white; tube 3 mm long, glabrous on both surfaces; lobes of upper lip ovate-oblong, villous at back; lower lip ovate-oblong, 3 mm long. Filaments glabrous. Style glabrous; branches linear. Fruiting calyx 4.5–5.5 mm long. Nutlets ovoid, 1.5 × 1 mm, subtrigonous, smooth.

Distr. Palaeotropic (cultivated in N. America).

Ecol. A common weed in open, waste, sandy ground along roadsides in the dry lowlands, at c. sea level; occasionally cultivated in the lower uplands. The leaves when bruised have a distinctive, pleasant, camphor-like scent. Flowering throughout the year.

Note. Alston (in Trimen, Handb. Fl. Ceylon 6: 234. 1931) and successive modern authors have misidentified this species as being conspecific with *O. americanum* L. Regarding the identity of *O. americanum* see note below under *O. basilicum* L.

Specimens Examined. JAFFNA DISTRICT: *s. coll. C.P. 3624* (PDA); POLONNARUWA DISTRICT: Polonnaruwa town, under shade along bank of Parakrama Samudra, *Cramer 3675* (PDA, US), Polonnaruwa Government Farm, *Senaratne 3501* (PDA). PUTTALAM DISTRICT: Tabbowa, along bund of tank, *Cramer 4071* (PDA, US). KURUNEGALA DISTRICT: Polgahawela, in waste ground, *Cramer 3275* (PDA, US); Badegama, in waste ground, *Cramer 3591* (PDA, US). KANDY DISTRICT: Kadugannawa, beside foot-path close to habitation, *Cramer 3838* (PDA, US). MATALE DISTRICT: Dambulla, rocky slope, *Cramer 3545* (PDA, US); Sigiriya, at foot of Rock, *Cramer 3662* (PDA, US). MATARA DISTRICT: Polhena, in sandy, waste ground, *Cramer 3443* (PDA, US). HAMBANTOTA DISTRICT: near lagoon *Alston s.n.* (PDA); Iriyagama, *Hendrick s.n.* (PDA); Ruhunu National Park, Patanagala, along edge of rock, *Cramer 4137* (K, PDA, US); east of Veddangewadiya, grass field, *Wirawan 645* (PDA, US). BATTICALOA DISTRICT: Kalkudah, in waste ground, *Cramer 3621* (PDA, US), Batticaloa Town, in waste ground near bridge, *Cramer 3632* (PDA, US).

3. Ocimum basilicum L., Sp. Pl. 597. 1753; Benth., Lab. Gen. Sp. 4. 1832; Benth. in DC., Prod. 12: 32. 1848 (incl. var. *glabratum*); Moon, Cat. 44. 1824; Thw., Enum. Pl. Zeyl. 236. 1860; Hook. f., Fl. Br. Ind. 4: 608. 1885; Trimen, Handb. Fl. Ceylon 3: 366. 1895. Type: West Asia, specimen *749.5* (LINN.)

Ocimum americanum L., Cent. I. Pl. 15. 1755; L., Amoen. Acad. 4: 276. 1759. Type: *Miller*, America (n.v.).

Ocimum album L., Mant. 85. 1767; Moon, Cat. 44. 1824. Type: Hortu Upsalensi, *749.3* (LINN.).

Annual with a short rootstock; stem to 1 m long, much branched, obtusely 4-angular, sparsely or densely villous. Leaves ovate-elliptic, 3.5–6 × 1.7–2.6 cm, cuneate at base, acute, entire to distantly crenulate, puberulous above, pubescent with appressed hairs on nerves beneath and copiously dotted with sessile oil glands; petioles 0.8–1.8 cm long. Racemes to 14 cm long. Floral leaves broadly ovate, to 5.5 × 2.5 mm, caudate-acuminate, long ciliate. Pedicels to 2 mm long. Flowering calyx to 6 mm long; tube 3 mm long, sparsely strigose without, hirsute-annulate within at base; upper lip broadly ovate to orbicular, 4–5 × 3 mm, bluntly acute; lower lip to 5 mm long, the lateral lobes broadly ovate, shortly acuminate, the anterior ones lanceolate, long-acuminate. Fruiting calyx 6–8 mm long. Corolla 7–10 mm long, white, faintly tinged pink; tube glabrous on both surfaces. Filaments glabrous, the posterior ones with a transverse process of tufted hairs. Fruiting calyx 6–8 mm long. Nutlets black, ellipsoid, 2–3.5 × 1.25–1.5 mm, subtrigonous, invested with dense hairs, mucilaginous when wetted.

Distr. Palaeotropic.

Ecol. An uncommon species cultivated in village gardens and especially in Hindu temples, for the fragrant leaves and flowers. The floral leaves and calyces are conspicuously coloured dark purple. This and the fragrant flowers are distinctive. Flowering August–December.

Uses. The dried leaves and flowers are mixed with the holy ash of Hindu temples to give it a fragrant scent.

Note. Various authors have long considered *O. basilicum* and *O. americanum* to be distinct species. Morton, however, (J. Linn. Soc. Bot. 58: 234. 1962) concluded from examination of type material that both plants are conspecific. I have examined the same type material and agree with Morton's position.

Specimens Examined. MANNAR DISTRICT: Mantai, Tirukeeheeswaram Temple, *Balasubramaniam 1266* (PDA). KANDY DISTRICT: Peradeniya, Royal Botanic Gardens, *s. coll. s.n.* (PDA). LOCALITY UNKNOWN: *s. coll. 1861* (K.)

4. Ocimum sanctum L., Mant. 1: 85. 1767; Benth., Lab. Gen. Sp. 11. 1832; Benth. in DC., Prod. 12: 38. 1848; Thw., Enum. Pl. Zeyl. 236. 1860; Hook. f., Fl. Br. Ind. 4: 609. 1885; Trimen, Handb. Fl. Ceylon 3: 366. 1895; Mukerjee, Rec. Bot. Surv. India 14: 19. 1940; Abeywick., Ceylon J. Sci., Biol. Sci. 2 (2): 218. 1959. Type: Hortu Upsalensi, specimen *749.7* (LINN.).

Ocimum inodorum Burm., Fl. Ind. 130. 1768.

Annual or semi-shrubby perennial; stem to 1.05 m high, much branched, obtusely 4-angular, hirsute above, glabrous and often slightly woody below; young branches more densely hirsute with reddish-purple hairs. Leaves oval-oblong, 1.4–3 × 0.9–2.0 cm, obtuse at both ends, undulate, distantly and coarsely serrate, puberulous above, hirtellous beneath, dotted on both surfaces with sessile oil glands; petioles 0.4–1.8 cm long, slender, hirtellous with reddish-purple hairs. Racemes to 12 cm long, generally simple, occasionally branched at base. Floral leaves suborbicular, 2 × 3 mm, cuspidate, ciliate and faintly purplish at margins. Pedicles 2 mm long, villous. Calyx campanulate, 3 mm long; tube 2 mm long, densely dotted with sessile oil globules; upper lip broadly orbicular, 1.5 × 2 mm, apiculate, flat, strongly reflexed in fruit, dull purple; median lobes of lower lip curved upwards. Corolla tubular-campanulate, to 4.5 mm long, reddish-pink; tube 2.5 mm long, dotted with sessile oil glands; lobes of upper lip oblong-orbicular, puberulous at back; lower lip ovate-oblong, 2.5 mm long. Posterior filaments at base with a tuft of simple hairs; anterior ones usually curled upwards at ends. Style glabrous. Nutlets ellipsoid, 1.5 × 1 mm, minutely warty, dark brown, unchanged when wetted.

Distr. Palaeotropic.

Ecol. A common weed along roadsides and in open waste ground, occasionally near dilapidated buildings, in the lowlands and the lower uplands to c. 530 m. Its high erect habit combined with the obtuse ends of the leaves makes it distinctive within the genus. The leaves when bruised have an aromatic scent. Flowering December–August.

Vern. Maduru-tala (S).

Specimens Examined. ANURADHAPURA DISTRICT: Wilpattu National Park, Moderagama Aru, in scrub, *Fosberg 51899* (PDA, US). POLONNARUWA DISTRICT: Polonnaruwa Rest House, grassy bank, *Fosberg 51899* (PDA, US). Sacred Area, sandy soil, *Ripley 350* (PDA, US). KURUNEGALA DISTRICT: Badegama, in open ground, *Cramer 3590* (PDA, US). GALLE DISTRICT: Galle, among beach vegetation close to Park, *Cramer 3428* (PDA, US). HAMBANTOTA DISTRICT: Yala, Ruhunu National Park, Rakinawela, *Cramer 3298* (PDA, US); Patanagala, edges of scrub, *Fosberg 50257* (PDA, US); Koma Wewa, in scrub, *Comanor 410* (PDA, US); between Andunoruwa and Koma Wewa, in deciduous scrub, *Mueller-Dombois 69010714* (PDA, US). RATNAPURA DISTRICT: *Gardner C.P. 249* (CAL, E, K, M, PDA). MATALE DISTRICT: Ritigala, in forest clearing by historic ruins, *Cramer 3526* (PDA, US); Dambulla-Habarana Road, roadside, *Comanor 1027* (PDA, US). BATTICALOA DISTRICT: Kalkudah, in waste ground, *Cramer 3620* (PDA, US); North of Panama, open ground, *Fosberg & Sachet 52976* (PDA, US). TRINCOMALEE DISTRICT: Puliyankulam, *Alston 1330* (PDA).

5. Ocimum adscendens Willd., Sp. Pl. 3: 166. 1800; Benth., Lab. Gen. Sp.9. 1832; Benth. in DC., Prod. 12: 35. 1848; Hook. f., Fl. Br. Ind. 4: 609. 1885;

Trimen, Handb. Fl. Ceylon 3: 366. 1895; Alston in Trimen, Handb. Fl. Ceylon 6: 234. 1931; Abeywick., Ceylon J. Sci., Biol. Sci. 2 (2): 218. 1959. Type: "India Orientalis" (n.v.).

Annual, with a short rootstock; stem erect, 15–30 cm high, obtusely 4-angular, pubescent with retrorse hairs, diffusely branched, with the branches to 18.4 cm long. Leaves oblong to ovate-oblong, 0.7–4.2 × 0.35–1.6 cm, cuneate to tapering to base, obtuse, faintly serrulate, glabrous and copiously dotted with oil globules on both surfaces; petioles 1–2.2 cm long. Racemes 3–5 cm long; verticils distant. Floral leaves linear-oblong, with a concave, glandular receptacle at base. Calyx tube 2.5 mm long, subglabrous to sparsely pubescent without, glabrous within; upper lip broadly ovate-oblong, to 8 mm long in fruit, slightly longer than the lower; lateral lobes of lower lip bimucronate or truncately serrated, median ones spinous-acuminate. Corolla 6–7 mm long, puberulous without, pinkish-white; upper lip 3 mm long, lower equally long. Stamens much exserted, ± twice length of corolla; filaments glabrous. Nutlets globose, 1.5 mm long, smooth, mucilaginous when wetted.

Distr. India and Ceylon.

Ecol. In sandy-gravel soil in open places in the dry lowlands at sea-level. The short, erect stems with diffuse branches and the small, oblong leaves are very distinctive. Flowering August–December.

Specimens Examined. JAFFNA DISTRICT: Kokkawil, in 1897, *s. coll. s.n.* (PDA). VAVUNIYA DISTRICT: Cheddikulam, *Kundu & Balasubramaniam 568* (PDA, US). ANURADHAPURA DISTRICT: Mihintale Hill, *Balasubramaniam & Jayasuriya 1125* (PDA, US). PUTTALAM DISTRICT: Tabbowa, in sandy bund of tank, *Cramer 3232* (PDA, US). HAMBANTOTA DISTRICT: Ruhunu National Park, Patanagala, in scrub, *Cramer 3303* (PDA, US). TRINCOMALEE DISTRICT: Trincomalee, *Glenie C.P. 3999* (E, PDA); Puliyankulam, *Alston 1330* (PDA).

2. GENIOSPORUM

Wall. ex Benth., Bot. Reg. sub t. 1300. 1830; Benth. in Wall., Pl. As. Rar. 2: 18. 1831; Benth., Lab. Gen. Sp. 19. 1832; Benth in DC., Prod. 12: 44. 1848; Benth. & Hook., Gen. Pl. 1172. 1876; Endl., Gen. Pl. 608. 1838; Briq. in Pflanzenfam. 4 (3a): 367. 1897. Type species: *G. colorata* (D. Don) Kuntze.

Perennial herbs; stem erect or procumbent, much branched. Racemes or spikes long, terminal; verticils many-flowered; flowers small. Floral leaves often coloured. Calyx at anthesis campanulate, in fruit tubular, suberect, or declinate; tube glabrous within; limb 5-lobed, the upper lobe broadest, not decurrent at margins on tube, the lateral ones free or united with the upper into an upper lip, the lower ones free or connate into a lower lip. Corolla tube short, campanulate; limb subbilabiate; upper lip broad, shortly 4-lobed or 4-fid; lower one slightly longer, declinate. Stamens 4, declinate; filaments free, toothless. Disk slightly produced anteriorly. Style shortly 2-fid above. Nutlets ellipsoid or oblong, smooth or minutely punctate.

About 25 species, from Africa eastwards through Madagascar to southwest Asia; Indo-China.

KEY TO THE SPECIES AND VARIETIES

1 Floral leaves small, 1.5–4 mm broad, faintly purplish at margins; flowers distinctly pedicellate; calyx tube at least at anthesis with glassy, gland-tipped hairs at base without

2 Leaves oblong-lanceolate or obspathulate, 4–10 mm broad; stem and leaves scabrid or villous-pilose .... **1. G. tenuiflorum**

2 Leaves linear, 1.25–3.5 mm broad; stem and leaves glabrous or sparsely puberulous .... **G. tenuiflorum var. gracile**

1 Floral leaves large, 4–7 mm broad, white at margins; flowers subsessile; calyx tube with simple hairs at base without .... **2. G. elongatum**

1. Geniosporum tenuiflorum (L.) Merr., Philipp. J. Sci. 19 (3): 379. 1921; Alston in Trimen, Handb. Fl. Ceylon 6: 234. 1931; Abeywick., Ceylon J. Sci., Biol. Sci. 2 (2): 218. 1959.

Ocimum tenuiflorum L., Sp. Pl. 597. 1753. Type: Hortu Upsalensi, specimen *749.13* (holotype, LINN.).

Geniosporum prostratum (L.) Benth., Bot. Reg. sub t. 1300. 1830; Benth. in Wall., Pl. As. Rar. 2: 18. 1831; Benth., Lab. Gen. Sp. 21. 1832; Benth. in DC., Prod. 12: 45. 1848; Thw., Enum. Pl. Zeyl. 237. 1860; Hook. f., Fl. Br. Ind. 4: 610. 1885; Trimen, Handb. Fl. Ceylon 3: 368. 1895; Mukerjee, Rec. Bot. Surv. India 14 (1): 33. 1940.

Ocimum prostratum L., Mant. 2: 566. 1771. Type: Hortu Upsalensi, *749.19* (holotype, LINN.).

Lumnitzera prostrata (L.) Spreng., Syst. Veg. 2: 687. 1825.

Herb perennating by a shallow or deep, fibrous rootstock; stem erect, to 35 cm high, diffusely branched from base, or prostrate and ascending, obtusely 4-angular, scabrid or villous-pilose. Leaves often turned upwards, sessile, lanceolate to oblong-lanceolate, 1.8–4.2 × 0.4–1.0 cm, cuneate or tapering to base, acute, distantly serrate from above base, scabrid above, more densely so or pilose beneath, punctate with sessile globules on both surfaces. Racemes to 19.5 cm long; verticils 8–16 flowered. Floral leaves deflexed, broadly ovate, 2.5–4 × 1.5–4.5 mm, cuspidate, faintly purplish at margins, glabrous above, strigose beneath. Pedicels to 2 mm long, slender. Calyx erect, deflexed in fruit; tube cupular-campanulate, 1 mm long, with glassy gland-tipped hairs at base; upper lip broadly ovate, 0.5 × 1 mm, subacuminate; lower lip obovate-oblong,

2 mm long, concave, pubescent without, 2 median lobes of lower lip narrower than the laterals. Corolla pale blue; tube 1.25 mm long; lobes of upper lip ovate-oblong; lower lip 2 mm long, the lobes obovate-oblong. Stamens shortly exserted; anther-cells reniform. Disk annular. Style glabrous. Nutlets ellipsoid, 0.25 mm long, smooth.

Distr. From Central India southwards through the Deccan to Ceylon.

Ecol. In open, sandy ground, chiefly near the shore, in the dry and moist lowlands. The size of this plant is variable, and it should not be confused with the small-leaved plants of the var. *gracile*. Flowering January–July.

Specimens Examined. MANNAR DISTRICT: Madhu road, wayside weed, *Jayasuriya et al. 593* (PDA, US), Ilupaikadavai, in 1890, *s. coll. s.n.* (PDA), ibid., *Alston 1331* (PDA). POLONNARUWA DISTRICT: between Polonnaruwa and Batticaloa, *Townsend 73/258* (K, PDA, US); Gunner's Quoin, open grassy ground, *Cramer 3636* (PDA, US). PUTTALAM DISTRICT: Wilpattu National Park, *Meijer 353* (PDA, US), Kokarewilu, sandy ground, *Cramer 4674* (E, NVB, PDA, US). HAMBANTOTA DISTRICT: Ruhunu National Park, Block 1, Rakina Wewa, near Gonalabbe Lewaya, in tall scrub among grass between clumps, *Fosberg 50279* (PDA, US); near Karugaswela, grassy openings between clumps, *Cooray 68072805* (PDA, US). BADULLA DISTRICT: Bintenne tank, *Gardner C.P. 2079* p.p. (PDA). AMPARAI DISTRICT: Lahugala, in sandy ground, *Cramer 3653* (PDA, US). BATTICALOA DISTRICT: Valaichchenai, sandy ground, *Jayasuriya et al. 593* (PDA, US); Nitabelli, *Ferguson s.n.* (PDA). TRINCOMALEE DISTRICT: Ostenburg, *Ferguson s.n.* (PDA). LOCALITY UNKNOWN: *s. coll. C.P. 2079* p.p. (K, PDA).

var. **gracile** (Benth.) Thw., Enum. Pl. Zeyl. 237. 1860; Hook. f., Fl. Br. Ind. 4: 610. 1885; Trimen, Handb. Fl. Ceylon 3: 368. 1895; Mukerjee, Rec. Bot. Surv. India 14 (1): 34. 1941.

Geniosporum gracile Benth., Lab. Gen. Sp. 21. 1832; Benth. in DC., Prod. 12: 45. 1848. Type: *Macrae s.n.*, Ceylon, Colombo (holotype, K).

Stem more usually prostrate; branches to 36 cm long, subglabrous or sparsely pubescent. Leaves linear, 12–23 × 1.25–3.5 mm, tapering to base. Calyx tube sparsely pubescent; lateral lobes of lower lip ovate, to 0.75 mm broad.

Distr. South India and Ceylon.

Ecol. In similar localities as the typical variety.

Note. The character differences between this and the typical variety are only minor. Besides, the distributional ranges between the two varieties overlap. For these reasons I agree with Thwaites in not considering this taxon worthy of being raised to specific rank.

Specimens Examined. JAFFNA DISTRICT: *Gardner C.P. 2078* p.p. (K, PDA), ibid., 30 miles S.E., *Townsend 73/79* (K, PDA). PUTTALAM DISTRICT: Wilpattu, Periya Nagaswillu, *Mueller-Dombois et al. 69043016* (PDA, US). COLOMBO DISTRICT: Cinnamon Gardens, *Ferguson s.n.* (K, PDA), near sea-shore, *Macrae s.n.* (K). KALUTARA DISTRICT: Nagoda, in sandy ground, *Commanor 2971* (PDA, US). AMPARAI DISTRICT: Pottuvil, near sea-shore, *Ariyaratne s.n.* (PDA). LOCALITY UNKNOWN: *Mrs. Walker s.n.* (K, PDA); *Graham s.n.* (K, PDA); *s. coll. s.n.* (PDA).

2. Geniosporum elongatum Benth., Lab. Gen. Sp. 21. 1832; Benth. in DC., Prod. 12: 45. 1848; Thw., Enum. Pl. Zeyl. 236. 1860; Hook. f., Fl. Br. Ind. 368. 1895 (excl. syn. *Rhinanthus indica* Burm. f.); Alston in Trimen, Handb. Fl. Ceylon 6: 234. 1931; Mukerjee, Rec. Bot. Surv. India 14 (1): 33. 1941; Abeywick., Ceylon J. Sci., Biol. Sci. 2 (2): 218. 1959. Type: *Macrae s.n.*, Ceylon (holotype, n.v.).

Geniosporum indicum sensu Gamble, Fl. Pres. Madr. 718. 1921 [non Briq. in Pflanzenfam. 4 (3a): 367. 1897].

Perennial herb; stem erect, to 67 cm high, much branched at base, unbranched above, slender, acutely 4-angular, pilose-hirsute. Leaves distant, oval-oblong to linear-lanceolate, 1.8–4.6 × 0.5–1.3 cm, narrowed and much tapering to base, acute to sub-acute, distinctly serrate, hirsute and glandular-punctate on both surfaces; petioles obscure, to 5 mm long. Racemes spicate; verticils many-flowered; peduncles villous-hirsute. Flowers subsessile. Upper floral leaves broadly ovate to suborbicular, 3–4 × 5–7 mm, entire, shortly acute, often recurved; lower ones ovate, to 3 × 4 mm, acute; all white, reticulately nerved green. Pedicels obscure. Calyx campanulate, 1.8–2 mm long, hispidulous without; tube 0.75 mm long, 10-nerved; upper lip erect, broadly ovate, slightly longer than the lower, obtuse at apex, ciliolate; lateral lobes of lower lip lanceolate, long-ciliolate, central ones subulate, Corolla shortly tubular, 3.5 mm long; tube 2.5 mm long, slightly bent and broadened above, narrowed below, minutely pubescent without, white; lobes of upper lip shallowly oval-oblong, pinkish-purple; lower lip oval-oblong. 1.5 mm long. Posterior filaments shorter than anterior, hirtellous at base, anterior ones longer, recoiled at apex with age; all flattened, glabrous. Style slightly exserted from corolla tube, shortly 2-fid. Nutlets narrowly oval-oblong, 1 mm long, muricate, dark brown.

Distr. South India and Ceylon.

Ecol. In grassy slopes of the intermediate zone, lower uplands and in the patanas of the uplands, from 150–2000 m; rather common. The floral leaves are very conspicuous and characteristic. In full bloom the spikes afford a pleasant sight. Flowering October–May; July.

Note. *Geniosporum indicum* Briq. was based on *Rhinanthus indica* L., the type of *Razumovia indica* (L.) Alston, a Scrophulariaceous species.

Specimens Examined. MATALE DISTRICT: Sigiriya, *Alston 1725* (PDA), ibid, in grassy slope, *Cramer 3666* (PDA, US). KANDY DISTRICT: Hantane, *s. coll. C.P. 2080* (PDA). NUWARA ELIYA DISTRICT: Padinawala, 63rd Mile Post, Welimada Road, *Balakrishnan 577* (PDA, US), Padinawala, road embankment, among short grass, beside Padinawala-Keppitipola road, *Cramer 4426* (BO, CAL, E, K, NVB, PDA, US). MONERAGALA DISTRICT: Diyaluma, top of Diyaluma Estate, among fire grassland, *Cramer 4490* (CAL, PDA, US). BADULLA DISTRICT: Badulla, *s. coll. s.n.* (PDA); Helahalpe, on patanas, *Cramer 4838* (G, PDA, US).

3. BASILICUM

Moench, Meth. Pl. Suppl. 143. 1802; Kuntze, Rev. Gen. Pl. 2: 512. 1891. Type species: *Ocimum polystachyon* L.

Moschosma Reichb., Consp. 171. 1828, in adnot.; Benth. in Wall., Pl. As. Rar. 2: 13. 1831; Benth., Lab. Gen. Sp. 24. 1832; Benth. in DC., Prod. 12: 48. 1848; Endl., Gen. Pl. 609. 1838; Benth. & Hook., Gen. Pl. 2: 1173. 1876; Briq. in Pflanzenfam. 4 (3a): 368. 1897. Type species: not indicated.

Annual or perennial herbs; stem erect, branched. Leaves petiolate, crenate. Racemes axillary and terminal, spicate; verticils 6–10-flowered, secund, ± laxly arranged. Floral leaves minute. Calyx sublabiate, subdeclinate in fruit; tube ovoid or campanulate, naked at throat; limb 5-toothed, the uppermost tooth often very broad and forming an upper lip continuous with tube, the 2 lateral ones often associated with the 2 anterior ones to form a lower lip. Corolla subbilabiate, tube tubular-campanulate, included in calyx or shortly exserted from it; upper lip 3-lobed, the median lobe notched; lower lip entire, flat or faintly concave. Stamens 4, included, declinate; filaments free, without appendages; anthers 2-celled. Disk symmetrical; style clavate, shortly 2-fid at apex. Nutlets ovoid, compressed, smooth.

Two or three species, palaeotropical.

Basilicum polystachyon (L.) Moench, Meth. Pl. Suppl. 43. 1802 (*polystachion* sphalm); Abeywick., Ceylon J. Sci., Biol. Sci. 2 (2): 218. 1959.

Ocimum polystachyon L., Mant. 2: 567. 1771; Moon, Cat. 44. 1824. Type: "India".

Ocimum tenuiflorum Burm. f., Fl. Ind. 129. 1768 (non L. 1753). Type: Pluk. Ic. in Almagestum botanicum t. 208, f. 4. 1696.

Moschosma polystachyon (L.) Benth. in Wall., Pl. As. Rar. 2: 13. 1831; Benth., Lab. Gen. Sp. 24. 1832; Benth. in DC., Prod. 12: 48. 1848 (*polystachyum* sphalm.); Thw., Enum. Pl. Zeyl. 237. 1860; Hook. f., Fl. Br. Ind. 4: 612. 1885; Trimen, Handb. Fl. Ceylon 3: 369. 1895; Mukerjee, Rec. Bot. Surv. India 14 (1): 35. 1940.

Annual; stem to 1.22 m high, much branched, acutely 4-angular, shallowly furrowed, the angles pustular with tubercles tipped by caducous bristles, the nodes annulate with very small hairs, otherwise glabrous. Leaves broadly ovate to oval-rhomboid, 2.6–7.5 × 2–4.3 cm, truncate to cuneate at base, acuminate, irregularly crenate-serrate, thin, membranous, scaberulous above, glabrous beneath; petioles 1.7–5 cm. Racemes linear, 5–6 cm long; verticils often 6-flowered. Floral leaves obspathulate to oval-oblong, 1 mm long, aristate. Pedicels 1 mm long, hispidulous. Calyx 1.75–2 mm long, longer in fruit; tube subventricose, 1.25 mm long, hispidulous with gland-tipped hairs at base without, somewhat inflated at base in front; upper lip broadly ovate, 0.75 × 1 mm, acuminate; lower lip as long as upper, 4-dentate, the lateral ones triangular-ovate, the anterior ones triangular-lanceolate, acuminate, aristate. Corolla 2.5–3 mm long, mauvish-pink; tube straight, campanulate, 1.75 mm long, glabrous without, thinly pubescent within; upper lip erect, 3-lobed, lobes rounded, central one 2-fid; lower lip slightly longer, oval-oblong, flat, hispidulous ventrally. Stamens didymous, included; anterior filaments longer than posterior, all glabrous; anthers 2-celled, the cells continuous. Style glabrous, shortly 2-fid at apex; branches oblong-oval. Nutlets broadly ellipsoid, 0.8 mm long, compressed, dark brown.

Distr. In tropical Africa, tropical Asia and tropical Australia.

Ecol. A weed growing under shade in moderately damp ground of the dry lowlands, and, occasionally, along the borders of secondary forests, roadside ditches and tanks of the intermediate zone. The branches wither quickly once they are cut, or the plant uprooted. Flowering March–July.

Specimens Examined. ANURADHAPURA DISTRICT: *Alston 329* (PDA), near Medawachchiya, *Simpson 9364* (PDA). PUTTALAM DISTRICT: Tabbowa, under shade beside stream, *Cramer 4074* (E, PDA, US), Pomparipu, under shade in undergrowth of dry forest, *Cramer 4080* (PDA, US). KURUNEGALA DISTRICT: Gonagama, under shade in humus soil overlying depression, *Cramer 3587* (PDA, US), under shade along roadside near culvert, *Cramer 3593* (PDA, US), *Gardner C.P. 2056* (K, PDA). AMPARAI DISTRICT: Lahugala, under shade along borders of secondary forest, *Cramer 3651* (PDA, US); Panama, *Silva s.n.* (PDA); Lahugala Tank, east side 300 yd north of irrigation bungalow, *Mueller-Dombois & Commanor 67072523* (PDA). LOCALITY UNKNOWN: *s. coll. C.P. 2056* (PDA), *s. coll. s.n.* (K).

4. ORTHOSIPHON

Benth., Bot. Reg. sub t. 1300. 1830; Benth. in Wall., Pl. As. Rar. 2: 14. 1831; Benth., Lab. Gen. Sp. 25. 1832; Benth. in DC., Prod. 12: 49. 1848; Endl.,

Gen. Pl. 609. 1838; Benth. & Hook., Gen. Pl. 1174. 1876; Briq. in Pflanzenfam. 4 (3a): 372. 1897; Sleesen, Reinwardtia 5 (1): 37. 1959. Type species: not known.

Clerodendranthus Kudo in Mém. Fac. Agr. Imp. Univ. Taihoku (Bot.) 2 (2): 117. 1929. Type species: *C. stamineus* (Benth.) Kudo.

Perennial herbs or undershrubs with a woody rootstock; stem glabrous or pubescent. Leaves serrate or crenulate, dotted with sessile oil globules beneath. Racemes simple, terminal, elongate; verticils 6-flowered, distant. Calyx tubular-campanulate; deflexed and accrescent in fruit; limb 2-labiate; upper lip broadly ovate or rotundate, the margins decurrent on tube; lower lip 4-lobed, lobes acuminate, aristate. Corolla distinctly exceeding calyx, 2-labiate; tube straight; upper lip 3–4-lobed; lower concave, ± equally long as upper. Stamens didynamous, included in lower lip or much exserted; filaments usually glabrous; anthers 2-celled, the cells reniform. Disk usually enlarged anteriorly. Style at apex minutely capitate or clavate, faintly notched. Nutlets ovoid, smooth or tuberculate; basal scar small.

About 40 species, palaeotropic, mainly in Africa.

KEY TO THE SPECIES

1 Stamens included in lower lip of corolla; leaves crenate-serrate...... **1. O. thymiflorus**
1 Stamens much exserted beyond lower lip of corolla; leaves serrate-dentate........... .... **2. O. aristatus**

1. Orthosiphon thymiflorus (Roth) Sleesen, Reinwardtia 5: 42 & 43. 1959.

Ocimum thymiflorum Roth, Nov. Pl. Sp. 269. 1821. Type: *Heyne s.n.*, India (n.v.).

Orthosiphon tomentosus Benth. in Wall., Pl. As. Rar. 2: 14. 1831; Benth., Lab. Gen. Sp. 27. 1832; Benth. in DC., Prod. 12: 51. 1848 (non De Wildem. 1921); Hook. f., Fl. Br. Ind. 4: 613. 1885. Type: Chingulput, India, *s. coll.*, in Hb. Wt. *2197/58* (holo- & isotype, E).

Orthosiphon glabratus Benth. in Wall., Pl. As. Rar. 2: 14. 1831; Benth., Lab. Gen. Sp. 28. 1832; Benth. in DC., Prod. 12: 50. 1848; Thw., Enum. Pl. Zeyl. 237. 1860; Trimen, Handb. Fl. Ceylon 3: 369. 1895; Abeywick., Ceylon J. Sci., Biol. Sci. 2 (2): 218. 1959. Type: Courtallum, India, Wall. Num. List *27248B* (holotype, K-Wall.).

Perennial herb; stem 32–100 cm high, acute, 4-angular, grooved on opposite sides, slightly woody and glabrous towards base, slightly branched and pubescent above. Leaves ovate to ovate-elliptic, 3.6–7.3 × 2.2–6 cm, truncate to subacute and occasionally unequal-sided at base, acute to subobtuse, coarsely crenate-serrate, scaberulous or puberulous above, more so beneath on nerves and copiously dotted with oil globules; petioles slender, 1–3.5 cm long. Racemes to 15 cm long; verticils lax; peduncles with retrorse hairs occasionally intermixed with erect ones. Floral leaves ovate, to 3 × 2 mm, shortly acute, 3–4-nerved at base, ciliolate, hirtellous without with gland-tipped hairs, reflexed with age, finally deciduous. Calyx tubular-campanulate, oblique-mouthed, membranous; tube 3–4 mm long, pubescent on both surfaces, occasionally hirsute at base, dotted with sessile glands without; upper lip broadly orbicular, 2.5 × 3.5 mm, faintly purplish at margins; lower lip 4 mm long, the lateral lobes shortly ovate, the median ones lanceolate, acuminate-aristate. Corolla to 11 mm long; tube to 7 mm long, pubescent without, whitish; limb pale lilac, the upper lip ovate-oblong, 4 × 2 mm, with the mid-lobe hirtellous without; lower lip broadly orbicular, 4 × 3.5 mm, hirtellous without. Filaments glabrous; anther cells confluent. Style glabrous. Nutlets oblong-orbicular, 1.5 × 1.25 mm, subtrigonous, minutely tuberculate, brown.

Distr. From India to Ceylon, eastwards through Malesia to Indo-China.

Ecol. Under shade in secondary scrub of the dry lowlands and in the drier lower uplands to about 500 m. Flowering January–April.

Note. The six-flowered, lax, verticils might at first sight indicate an *Ocimum* species, but the plant is readily distinguished from an *Ocimum* by the less-branched nature of the stem, the long root stock and the much-exserted corolla. A chemocharacter of the plant is the presence of a lemon-like fragrance in the leaves on bruising.

Specimens Examined. POLONNARUWA DISTRICT: *de Silva s.n.* (PDA). KURUNEGALA DISTRICT: Gonagama, *Gardner C.P. 2059* (PDA). KANDY DISTRICT: Haragama, *s. coll.*, *C.P. 2059* (K, PDA). HAMBANTOTA DISTRICT: Ruhunu National Park, Patanagala, in tall scrub, *Fosberg 50358* (PDA, US), ibid., above mouth of Menik Ganga, *Fosberg 50219* (PDA, US), ibid., Block 2, Kumbukan Oya, *Fosberg et al. 51113* (K, PDA, US), Jamburagala, under shade in secondary scrub, *Cramer 4132* (PDA, US). BADULLA DISTRICT: Bintenne, *Gardner C.P. 2059* (PDA). MONARAGALA DISTRICT: Koslanda, at foot of Diyaluma Falls, *Balasubramaniam s.n.* (PDA). AMPARAI DISTRICT: Nilgala, *s. coll. s.n.* (PDA). LOCALITY UNKNOWN: *Walker s.n.* (PDA), *Mrs. Walker 1846* (K).

2. Orthosiphon aristatus (Bl.) Miq., Fl. Ind. Bat. 2: 943. 1858; Mukerjee, Rec. Bot. Surv. India 14 (1): 26. 1940; Sleesen, Reinwardtia 5: 38. 1959.

Ocimum aristatum Blume, Bijdr. 833. 1825. Type: *Blume s.n.*, Batavia, Java (L).
Orthosiphon stamineus Benth. in Wall., Pl. As. Rar. 2: 15. 1831; Benth., Lab. Gen. Sp. 29. 1832; Benth. in DC., Prod. 12: 52. 1848 (incl. var. *angustifolia* Benth.).

Clerodendranthus stamineus (Benth.) Kudo in Mém. Fac. Sci. Taihoku Imp. Univ. 2 (2): 117. 1929.

Perennial herb; stem diffusely branched, glabrous. Leaves rhomboid to ovate-deltoid, 3.3–7.2 × 1.3–4.2 cm, cuneate at base, acute, serrate from about the middle upwards, glabrous on both sides, glandular-punctate beneath. Racemes 11–21 cm long; peduncles strigulose with purplish hairs; verticils lax. Bracts broadly ovate, 1.25 × 2 mm, ciliolate. Pedicels to 3 mm long. Calyx tube to 3 mm long, 10-nerved, scaberulous on nerves, dotted with oil globules; upper lip broadly ovate, to 3 × 2.5 mm, purplish; lower lip longer, the median lobes acuminate-aristate. Corolla white, strigose-tomentose without; tube 1.3–1.5 cm long, uniformly linear; upper lip orbicular, faintly concave; lower lip as long as upper, navicular. Stamens exserted, filaments to 3.3 cm long, glabrous. Style glabrous, clavate at apex. Nutlets ovoid-oblong, 2 mm long, subtrigonous, glabrous.

Distr. Throughout south-eastern Asia to tropical Australia.

Ecol. Cultivated in gardens as an ornamental on account of its white corollas and extremely elongated stamens.

Specimens Examined. KANDY DISTRICT: Peradeniya, Royal Botanic Gardens, *de Silva 775* (PDA), ibid., *Cramer 4225* (PDA, US). KEGALLE DISTRICT: Kegalle, "*Agricultural Instructor*" *640* (PDA).

5. PLECTRANTHUS

L' Hérit., Stirp. Nov. 84, t. 41. 1788 (nom. cons.); Benth. in Wall., Pl. As. Rar. 2: 16. 1831; Benth., Lab. Gen. Sp. 40. 1832; Benth. in DC., Prod. 12: 55. 1848; Benth. & Hook. f., Gen. Pl. 1175. 1876; Briq. in Pflanzenfam. 4 (3a): 352. 1897. Type species: *P. fructicosus* L' Hérit. (typ. cons.).

Germanea Lam., Enc. 2: 690. 1788. Type species: not known.

Isodon (Schrad. ex Benth.) Kudo in Mém. Fac. Sci. Taihoku Imp. Univ. 2 (2): 118. 1929. Lectotype species: *I. plectranthoides* Kudo nom. illegit. (*Plectranthus rugosus* Wall. ex Benth.).

Annual or perennial herbs or undershrubs; stem sometimes succulent. Leaves opposite, petiolate, serrate or crenate. Panicles terminal, lax, occasionally in axillary, spurious racemes; verticils 6-many-flowered. Floral leaves opposite, laxly imbricate in bud. Bracts small, deciduous. Flowers small, often pedicellate, occasionally subsessile. Calyx campanulate, erect or declinate, 10-nerved, accrescent in fruit; limb often 2-labiate, occasionally ± equally 5-lobed; upper lip equal to or shorter than the lower, entire, or 3-lobed, often deflexed in fruit; lower lip 2–4-lobed. Corolla 2-labiate; tube shortly exserted, straight or decurved at middle, often dorsally subventricose at base; upper lip erect, shortly 4-lobed; lower much longer, entire, naviculate.

Stamens didynamous, inserted at mouth of corolla tube or lower down, exserted from corolla; filaments free; anthers 2-celled. Disk with an anterior tongue-shaped projection. Style shortly 2-fid; branches subequal. Nutlets ovoid or oblong, smooth, granulate or punctulate.

About 250 species, palaeotropical.

KEY TO THE SPECIES

1 Stem ± woody; upper lip of fruiting calyx 3-lobed.
2 Limb of fruiting calyx distinctly 2-labiate; corolla tube obscurely ventricose towards base
3 Leaves along entire length of stem, upper ones petiolate; lateral branches of cymes ± shortened
4 Leaves ovate, entirely crenate; stamens not or hardly exserted beyond lower lip of corolla
5 Stem puberulous or villous above; leaves sparsely or densely scabrid, acute and straight at apex; corolla glabrous within........................**1. P. nigrescens**
5 Stem densely pilose-hirsute above; leaves hirsute above and on nerves beneath, acuminate and curved to a side at apex; corolla sparsely pilose within.......... ....**P. nigrescens** var. **pilosus**
4 Leaves narrowly lanceolate, serrate along upper half of margins; stamens distinctly exserted beyond lower lip of corolla...............................**2. P. walkeri**
3 Leaves chiefly at base of stem, upper ones sessile; lateral branches of cymes diffuse ....**3. P. capillipes**
2 Limb of fruiting calyx subbilabiate; corolla tube distinctly ventricose towards base.. ....**4. P. coetsa** var. **macrae**
1 Stem succulent; upper lip of fruiting calyx entire
6 Leaves sinuate-crenate; lobes of lower lip of calyx aristate............**5. P. subincisus**
6 Leaves crenate; lobes of lower lip of calyx acuminate.
7 Leaves truncate at base; stamens exserted beyond lower lip of corolla; upper lip of fruiting calyx continuous with tube...............................**6. P. gardneri**
7 Leaves subcordate or rounded at base; stamens included in lower lip of corolla; upper lip of fruiting calyx subdecurrent on tube....................**7. P. glabratus**

1. Plectranthus nigrescens Benth., Lab. Gen. Sp. 710. 1835; Benth. in DC., Prod. 12: 58. 1848; Thw., Enum. Pl. Zeyl. 237. 1860 (excl. syn. *P. walkeri* Arn.); Hook. f., Fl. Br. Ind. 4: 617. 1885; Trimen, Handb. Fl. Ceylon 3: 370. 1895; Alston in Trimen, Handb. Fl. Ceylon 6: 235. 1931; Mukerjee, Rec. Bot. Surv. India 14 (1): 41. 1940; Abeywick., Ceyl. J. Sci., Biol. Sri. 2 (2): 219. 1959. Type: *Walker s.n.*, Ceylon (lectotype, K).

Plectranthus hians Benth. in DC., Prod. 12: 57. 1848. Type: *Walker s.n.*, Ceylon, in Hb. Benth. (holotype, K).

Plectranthus scabrellus Benth. in DC., Prod. 12: 58. 1848. Type: *Mackenzie s.n.*, Ceylon, in Hb. Benth. (holotype, K).

var. **nigrescens**

Perennial herb; stem to 98 cm high, slender, acutely 4-angular, glabrescent-puberulous, or pubescent with retrorse hairs, often dull purplish-black. Leaves broadly ovate to ovate-oval or oval-oblong, 1.8–9.5 × 1.2–5 cm, truncate or rounded at base, acute and straight at apex, coarsely crenate-serrate, sparsely or densely scabrid or puberulous on both surfaces, with prominent nerves beneath and dotted with reddish-orange oil globules; petioles 0.8–3 cm long. Racemes terminal, to 21 cm long; cymes lax, 6–10-flowered; peduncles glandular-hispidulous, dotted with reddish-orange oil globules. Floral leaves ovate, 4.6 × 2.5–4 mm, 3-nerved at base, subacute, similarly dotted on both surfaces as leaves, long-ciliate. Pedicels filiform, 3–4 mm long, hirtellous with gland-tipped hairs, dull purple. Calyx 2.25 mm long; tube broadly campanulate, 1.5–2 mm long, 10-nerved, densely hirtellous without with gland-tipped hairs and similarly dotted as leaves; upper lip erect, 3-lobed, flat, becoming broader after anthesis; lower lip 2-lobed, 2 mm long, lobes oval-lanceolate, obtuse. Corolla to 9 mm long, white; tube 5–8.5 mm long, subventricose dorsally, glabrous within; upper lip shortly erect, 4 mm across; lobes oval-oblong, pubescent without, spotted red towards bases; lower lip broadly ovate, naviculate, 4–5 mm long. Stamens occasionally slightly exserted beyond lower lip of corolla; anterior filaments longer than posterior ones, all sparsely villous throughout, pale purple; anther-cells parallel. Style glabrous; branches linear, equal. Nutlets ovoid-oblong, to 1 × 0.5 mm, muricate, brown.

Distr. South India and Ceylon.

Ecol. In shaded roadside ditches in the midlands and in damp undergrowth of submontane and montane forests, often close to streams and water-courses, from about 1000 to 2700 m. This is the commonest labiate species in the Nuwara Eliya district. Dries blackish. Flowering almost throughout the year.

Note. The ovate, scrabrid leaves and the reddish spots on the upper lip of the white corolla are very distinctive. A very variable species as regards the size and shape of its leaves; the pubescence of stem and leaves, except for that of the variety *pilosa*, is also variable.

From examination of type material it is evident that the degree of pubescence of stem and leaves and the dimensions of leaves of *P. hians* and *P. scabrellus* are only variable characters of the same complex species *P. nigrescens.*

Specimens Examined. KANDY DISTRICT: Adam's Peak, under medium shade beside pilgrim path, about midway to summit, *Cramer 3729* (PDA, US); Maskeliya, Rajamally Estate, under shade in secondary forest, *Cramer 4285* (BO, E, K, NVB, PDA, US). NUWARA ELIYA DISTRICT: Ramboda, *Gardner C.P. 425* (PDA), in roadside ditch, under shade, *Cramer 3441* (PDA, US); Pidurutalagala, in open grassy ground among rocks at top of peak, *Cramer 3495* (PDA, US); Punduluoya, under shade of shrubs, along bank of stream, *Cramer 3726* (PDA, US); Hakgala, in undergrowth of secondary forest, close to diverted stream, *Cramer 3466* (PDA, US), in undergrowth of secondary forest midway to peak, *Cramer 3450* (PDA, US); Agrapatna road, *Nowicke & Jayasuriya 254* (PDA, US); Pattipola, in moist embankment under shade, *Cramer 2873* (PDA, US); Horton Plains, Haldumulla road, *Willis s.n.* (PDA, US), under shade on slopes of ravine, *Cramer 3482* (PDA, US), between Little World's End and Great World's End, *Fosberg 50054* (PDA, US); Sita Eliya, *Silva s.n.* (PDA, US); Ambawela, Elk Plains, *s. coll. C.P. 425* (PDA); Pidurutalagala, *Meijer 72* (PDA, US), under shade near beginning of path to hill, *Cramer 3120* (PDA, US). LOCALITY UNKNOWN: *s. coll. C.P. 425* (PDA), *Walker 1834 & 1846* (K), *s. coll. C.P. 2061, 583* (K), *Gardner 651* (K).

var. **pilosus** Cramer, var. nov. Type: *Cramer 4398*, Maskeliya, Ceylon (holotype, PDA, isotype, K).

Foliis basi cuneatis vel truncatis, apice acuminatis atque leviter curvatis, supra atque nervis inferioribus ± pilosis vel hirsutis, petiolis piloso-hirsutis. Nodis pedunculorum inferioribus villoso-caespitosis. Calicis lobis ciliatis.

Perennial herb; stem to 1 m high, obtusely 4-angular, hoary pilose-hirsute above, slightly woody and scabrid towards base. Leaves broadly ovate, 5–11.5 × 2.5–6.3 cm, cuneate or truncate at base, acuminate and slightly curved to a side at apex, ± densely hirsute above and along nerves beneath, dotted in between nerves beneath with sessile oil glands; petioles 1–5.7 cm long, pilose-hirsute. Panicles to 20.5 cm long; lower nodes of peduncles often villously tufted. Lobes of calyx often ciliate. Corolla to 8.5 mm long, white; tube sparsely pilose within. Nutlets ellipsoid, to 1.20 × 1 mm, reticulated, glabrous.

Distr. Endemic.

Ecol. In undergrowth of secondary montane forest from c. 1500–1800 m. Not common. The stem tends to be scandent above if the undergrowth is dense. The plant has a faintly aromatic scent. Flowering October–December; March.

Note. The densely hirsute indumentum of the stem and leaves and the curvature of the leaf apices are quite distinctive. I am satisfied that these characters and the occurrence of the taxon in localities slightly different from that of the typical variety are justifiable reasons for separating it from the typical variety.

Specimens Examined. KANDY DISTRICT: Maskeliya, Rajamally Estate, sloping ground of secondary forest, *Cramer 4398* (BO, E, K, MH, NVB, PDA, US). NUWARA ELIYA DISTRICT: Ramboda, *Gardner C.P. 425* (CAL, E, MH), Sita Eliya, *Silva s.n.* (PDA). BADULLA DISTRICT: Ohiya, in secondary forest along path from Horton Plains, *Cramer 4419* (CAL, MH, PDA, US).

2. Plectranthus walkeri Arn., Nova Acta Phys.-Med. Acad. Caes. Leop. Carol. Nat. Cur. 18: 354. 1836; Benth. in DC., Prod. 12: 58. 1848; Hook. f., Fl. Br. Ind. 4: 617. 1885; Trimen, Handb. Fl. Ceylon 3: 371. 1895; Mukerjee, Rec. Bot. Surv. India 14 (1): 41. 1940; Abeywick., Ceylon J. Sci., Biol. Sci. 2 (2): 219. 1959. Type: *Graham s.n.*, Ceylon (n.v.).

Annual herb; stem erect, to 62 cm high, occasionally decumbent and rooting at lower nodes, slightly branched, grooved on two opposite sides, nodose below, densely strigose and violet-purple above. Leaves narrowly lanceolate, 2.2–5.9 × 0.7–1.5 cm, tapering at base, acute, serrate distally, sparsely scaberulous above, minutely strigose on prominent nerves beneath and dotted with oil globules; petioles 0.4–1 cm long, minutely strigose or villous. Racemes terminal, to 14.4 cm long; cymes lax; peduncles densely strigose. Floral leaves ovate, 5–6 mm long, 3–4-nerved at base, ciliate. Pedicels 2.5–3 mm long, longer in fruit, minutely strigose, violet-red. Calyx to 2.25 mm long, broadly campanulate, slightly declinate in fruit; tube 0.75 mm long, 10-nerved, dorsally incurved above, inflated towards base, minutely strigose without on nerves, glabrous within; upper lip erect, ovate, 1.5 mm long, shortly 3-lobed, with a broader median lobe; lower lip 1.25 mm long, 2-lobed, with oval-oblong lobes, subacute at apex. Corolla 5–7.5 mm long, white; tube narrow, 2–4 mm long, subventricose dorsally above, constricted at base, glabrous on both surfaces; upper lip spreading, 4 mm across, with oval-oblong lobes, the lobes 2 mm long, spotted reddish-violet at base, reflexed with age; lower lip broadly orbicular, 2–2.5 × 2 mm, navicular. Stamens exserted filaments sparsely villous. Disk symmetrical, faintly 4-lobed; stylar branches linear. Nutlets orbicular, 1.25 mm broad, sub-compressed, muricate, dark brown, shining.

Distr. S. India and Ceylon.

Ecol. On semi-exposed or shaded rocky edges of streams in montane forests above 1400 m, usually within the spray of waterfalls or the drip of overhanging rocks. Distinctly seasonal in flowering, and an elegant herb when in full bloom. Rather common. Flowering January–February.

Specimens Examined. KANDY DISTRICT: Adam's Peak, in moderate shade on damp rocks besides stream, *Cramer 3805* (PDA, US), on open rocky ledges within spray of waterfall, *Cramer 4040* (PDA, K, US); Maskeliya, Rajamally, under shade along rocky banks of stream, *Cramer 4047* (PDA, E, NVB, US). NUWARA ELIYA DISTRICT: Ramboda, *Gardner C.P. 90* (PDA). LOCALITY UNKNOWN: *s. coll. C.P. 90* (P), *Walker C.P. 461* (K), *Mackenzie s.n.*, (K), *s. coll. s.n.* (K).

3. Plectranthus capillipes Benth. in DC., Prod. 12: 57. 1848; Thw., Enum. Pl. Zeyl. 237. 1860; Hook. f., Fl. Br. Ind. 4: 619. 1885; Trimen, Handb. Fl. Ceylon 3: 372. 1895; Abeywick., Ceylon J. Sci., Biol. Sci. 2 (2): 219. 1959. Type: *Walker s.n.*, Ceylon, in Hb. Hook. (holotype, K).

Annual; stem with a short rootstock, to 50 cm high, unbranched, bluntly 4-angular, grooved on opposite sides, pilose-hirsute. Leaves few, chiefly at base, broadly-ovate, 2.8–5.6 × 2.4–3.5 cm, rounded or truncate at base, acute, coarsely crenate, pilose-hirsute above, dotted in between reticulations beneath with reddish-orange oil globules; upper ones sessile, becoming bracteate above, lower petiolate; petioles to 1.7 cm long. Panicles terminal, to 37.5 cm long; cymes diffuse; peduncles densely hirtellous with capitate glands. Floral leaves linear to oval-oblong, to 2.5 mm long. Fruiting pedicels to 4.5 mm long, slender, hirtellous with capitate glands. Calyx tubular-campanulate, slightly incurved and declinate in fruit, nerved, dotted with oil globules as in leaves and scabrid without, glabrous within; upper lip erect, the lobes shortly ovate, lower equal, shorter than upper. Corolla to 5 mm long, white, tinged with pink; tube straight, narrow. Stamens exserted. Nutlets ellipsoid, 1–1.5 × 0.5–0.7 mm, subtrigonous, slightly rough, not ridged.

Distr. Endemic.

Ecol. Along roadsides and in open, grassy ground in the lower uplands, 1000–1065 m. Rare. Flowering September–October.

Note. Already in Trimen's time this species was considered rare. The author's attempts to locate this plant have been without avail.

Specimens Examined. NUWARA ELIYA DISTRICT: Ramboda, *Gardner C.P. 2064* (PDA, US). LOCALITY UNKNOWN: *s. coll. C.P. 2064* (PDA), *Walker s.n.* (CAL, PDA), *Gardner C.P. 2064* (K), *Walker 1846* (K), *Walker s.n.* (K).

4. Plectranthus coetsa var. **macraei** (Benth.) Hook. f. ex Gamble, Fl. Pres. Madras 1121. 1924.

Plectranthus macraei Benth., Lab. Gen. Sp. 42. 1832; Benth. in DC., Prod. 12: 59. 1848 (excl. syn. *P. rugosus* var. *tomentosus* Benth.). Type: *Wight s.n.*, Madura hills, India (holotype, K).

Plectranthus coesta sensu Thw., Enum. Pl. Zeyl. 237. 1860. (non Buch.-Ham. ex. D. Don, 1825).

Plectranthus menthoides Benth. in Wall., Pl. As. Rar. 2: 17. 1831; Benth., Lab. Gen. Sp. 42. 1832; Benth. in DC., Prod. 12: 59. 1848; Hook. f., Fl. Br. Ind. 4: 620. 1885 (excl. syn. *P. rugosus* var. *tomentosus* Benth.); Trimen, Handb. Fl. Ceylon 3: 372. 1895; Alston in id. 6: 236. 1931; Abeywick., Ceylon J. Sci., Biol. Sci. 2 (2): 219. 1959. Type: *Wall.* Cat. No. *2744* India (holotype, K-WALL.).

Large perennial herb to 2.6 m high; stem much branched above, somewhat woody and thickened below, to 1.5 cm in diameter towards base, acutely 4-angular, grooved on opposite sides, with a hollowed pith, densely hirsute-tomentose. Leaves ovate to broadly ovate, 3.6–16.2 × 2.9 cm, rounded at base, slightly decurrent on petiole, acuminate, coarsely serrate-crenate, hirtellous and slightly rugose above, densely hoary hirtellous-tomentose beneath and dotted with oil globules in between prominent nerves; petioles 0.3–3 cm long, lanate. Panicles axillary and terminal; verticils dense; peduncles to 24 cm long. Floral leaves ovate, shortly acuminate, persistent, becoming foliaceous below, hoary-tomentose beneath. Pedicels 1.5–3 mm long, slender, hoary-strigose. Calyx deflexed, campanulate; tube 1–1.25 mm long, becoming larger in fruit, 10-nerved, hoary-villous and similarly dotted without as in leaves; limb subbilabiate, lobes ± equal, ovate, 1–1.25 × 0.75–1 mm, ciliolate, purplish at tips. Corolla to 8.25 mm long; tube 3.5–4.5 mm long, slightly ventricose dorsally, minutely strigose on both surfaces, lavender-blue; upper lip erect, broadly obovate, to 4 × 3 mm, the lobes shallowly orbicular; lower lip broadly ovate, 3–4.75 × 3 mm, navicular, minutely strigose without. Stamens included in lower lip; filaments sparsely strigose; anther-cells divergent. Disk annular. Style glabrous; branches 2-fid. Nutlets ellipsoid, 0.75–1 mm long, subtrigonous, sparsely granular, pale brown.

Distr. Deccan peninsula and Ceylon.

Ecol. In open places in the submontane and montane region above c. 1300–2400 m. Fairly common. Flowering November–April.

Note. Panicles scented. The hoary-lanate indumentum of the stem and leaves and the conspicuous panicles of small, blue flowers are very distinctive.

Specimens Examined. NUWARA ELIYA DISTRICT: Horton Plains, along borders of secondary montane forest beside road, *Cramer 3953* (PDA, US), ibid., in undergrowth of forest beside road, *Cramer 2949* (PDA, US); Ambawela, along stream, close to Boragas-Ambawela road, *Trivengadum & Cramer 289* (PDA, US). BADULLA DISTRICT: Hakgala, along borders of Hakgala-Ambawela road, *Cramer 4031* (BO, E, PDA, US), patanas in valley of Fort MacDonald, *Willis s.n.* (PDA, US), by roadside, *Alston 1034* (PDA), *s. coll. s.n.* (MH), Rendapola, along borders of Hakgala road, *Cramer 4032* (PDA, US), in embankments by roadside, *Cramer 2930* (PDA, US), Ramboda, *Gardner C.P. 2063* (PDA, US), Palagalla, *s. coll. C.P. 2063* (PDA, US), Nuwara Eliya, *Macrae s.n.* (K). LOCALITY UNKNOWN: *Col. Walker s.n.* (K), *Gardner 1847* (K), *Mrs. Walker s.n.* (K).

Note. Examination of Ceylon and Indian material of *Plectranthus menthoides* Benth., and of Indian material of *P. coetsa* Buch.-Ham. ex D. Don bears out that the two plants are conspecific except for the minor differences tabulated on page 133.

On the basis of these minor differences *P. menthoides* Benth. is reduced to varietal rank under *P. coetsa* Buch.-Ham. ex D. Don. This position seems to confirm the doubt which already existed in the earlier authors regarding the taxonomic status of *P. menthoides*. Bentham, in De Candolle's Prodromus, l.c., states: "Haec species media est inter *P. coetsam* et *P. macraei* et forte cum

Characters	*P. coetsa*	*P. menthoides*
1. Cymes	lax	dense
2. Indumentum of lower surface of leaf	subglabrous to sparsely scaberulous or sparsely pilose	densely hoary-tomentose
3. Leaf apex	generally acuminate	generally acute
4. Indumentum of stem	subglabrous or finely pubescent with retrorse hairs	densely pubescent with appressed villous hairs

hac pro varietate *P. coetsae* habenda est." Hooker says: "This species is a southern representative of *P. coetsa* and probably only a variety of that plant." Later Trimen himself observed (l.c. 373): "Scarcely distinct from the Himalayan *P. coetsa* Don with which Thwaites unites it."

5. Plectranthus subincisus Benth. in Wall., Pl. As. Rar. 2: 16. 1831; Benth., Lab. Gen. Sp. 36. 1832; Benth. in DC., Prod. 12: 66. 1848; Thw., Enum. Pl. Zeyl. 237. 1860; Hook. f., Fl. Br. Ind. 4: 621. 1885; Trimen, Handb. Fl. Ceylon 3: 371. 1895; Mukerjee, Rec. Bot. Surv. India 14 (1): 49. 1940. Type: *Wall.* Num. List *2737*, Courtallum, India (K-WALL.).

Annual; stem erect, flaccid, moderately thickened at base, slightly tumid at nodes, 4-angular, grooved on two opposite sides, glabrous or sparsely pubescent. Leaves broadly ovate to orbicular-ovate, cordate at base, subacute, sinuate-crenate, sparsely pubescent above, glabrous and dotted with oil globules beneath; petioles to 7.7 cm long. Panicles to 33.5 cm long; cymes racemose; peduncles minutely strigose. Floral leaves to 2.25 mm long, acute. Fruiting pedicels to 6 mm long, slender, minutely strigose. Calyx campanulate, to 6.5 mm long in fruit, declinate; tube sparsely pubescent without and similarly dotted as in leaves; upper lip entire, broadly ovate, subdecurrent on tube; lateral lobes of lower lip oblong-ovate, shortly aristate, median ones ovate-lanceolate, connate in fruit at inner margins above middle, aristate. Nutlets subglobose, 1 mm long, alveolate.

Distr. South India and Ceylon. Exact locality in Ceylon not indicated. I have not encountered this plant. It is probably now extinct.

Specimens Examined. LOCALITY UNKNOWN: *s. coll. s.n.* (PDA).

Note. There is only one unlocalised, poor specimen in the National Herbarium, Peradeniya. No Ceylon specimens are preserved in the Indian herbaria, and none either in the Kew Herbarium or the British Museum. My attempts to locate the plant were without avail. The description is based on type material, and the species is included on the authority of Macrae in De Candolle's Prodromus.

6. Plectranthus gardneri Thw., Enum. Pl. Zeyl. 237. 1860; Trimen, Handb. Fl. Ceylon 3: 371. 1895; Alston in Trimen, Handb. Fl. Ceylon 6: 235. 1931; Abeywick., Ceyl. J. Sci., Biol. Sci. 2 (2): 219. 1959. Type: *Gardner C.P. 16*, Elk Plains, Ceylon (holotype, PDA; isotypes, MH & K).

Plectranthus subincisus Benth. var. *gardneri* (Thw.) Hook. f., Fl. Br. Ind. 4: 622. 1885.

Annual with a shallow rootstock; stem to 60 cm high, much branched from below, obtusely 4-angular, succulent, occasionally tumid at internodes, hirtellous with purplish hairs (fresh). Leaves broadly-ovate to rhomboidovate, 1.5–7.5 × 1.4–6 cm, truncate at base, obtuse, coarsely often doubly crenate, dull green and scabrid with impressed nerves above, hirtellous with purplish hairs along prominent nerves beneath and dotted with orange-red oil globules (fresh); petioles 1.5–2 cm long, hirtellous with purplish hairs. Racemes usually terminal, erect, lax, to 24 cm long; peduncles hirtellous with capitate glands, dull purple. Floral leaves obovate, 2–2.5 × 2–2.5 mm, 3-nerved at base, ciliate with purplish hairs. Pedicels to 3 mm long, hirtellous with capitate glands. Calyx 2.75–3 mm long, broadly campanulate; tube 1–1.25 mm long, to 2.5 mm long in fruit, 10-nerved, scaberulous and similarly dotted as in leaves; upper lip broadly ovate, to 1.25 mm long, obtuse, entire, continuous with tube in fruit; lower one with lanceolate-subulate lobes, the two median ones the longest, all acuminate in fruit. Corolla to 7.75 mm long, pink or sometimes white; tube linear, 3–4 mm long, subventricose dorsally about the middle; upper lip erect, 4–4.5 mm long, concave or arched, 4-lobed, the lobes rounded, 2 upper ones the largest; lower lip ovate-orbicular, to 3.5 × 2 mm, navicular. Stamens much exserted; filaments curled with age, glabrous; anther-cells continuous. Style glabrous; branches linear, equal. Nutlets ellipsoid, to 2 × 1 mm, subtrigonous, shallowly foveolate, dark brown.

Distr. Endemic.

Ecol. Under dense shade in montane forests above c. 1450 m, often close to streams or watercourses. Common and occasionally locally abundant. Flowering almost throughout the year.

Note. The inflorescence in full bloom affords a pretty sight against the forest field. The dull purple under-surface of the leaves and the pinkish corollae with the arched upper lips are distinctive.

This species has been reduced to a variety under *P. subincisus* Benth. by Hooker f. (l.c. 622). It is, however, a valid species distinct from the latter in the crenate or doubly crenate nature of its leaf margins, the acuminate nature of the lobes of the lower lip of the fruiting calyx, and the usually racemose disposition of the verticils.

Specimens Examined. NUWARA ELIYA DISTRICT: Elk Plains, *Gardner C.P. 16* (PDA), Hakgala, on rocks in river below Hakgala, *Silva s.n.* (PDA), under shade on top of damp rock, *Cramer 4185* (K, PDA, US), Pattipola, under shade beside stream, *Cramer 3940* (PDA, US), Horton Plains, at entrance to fog intercept station, *Mueller-Dombois 68051602* (PDA, US), between Little World's End and Greater World's End, in mossy forest, *Fosberg 50054* (PDA, US), *Thomson C.P. 84* (K, PDA), *Harris s.n.* (E), forest back of Farr Inn, *Fosberg & Sachet 53296* (PDA, US), in forest shade, *Balakrishnan 1036* (PDA, US), Ohiya, in damp floor of secondary forest, *Cramer 4415* (CAL, E, K, PDA, US), Pidurutalagala, *Meijer 72* (PDA, US). BADULLA DISTRICT: Haputale, Tangamalai, under shade along stream bank in secondary forest, *Cramer 3819* (PDA, US), under shade of trees along bank of stream in secondary forest, *Cramer 3876* (PDA, US), epiphytic on tree trunk in jungle, *Silva s.n.* (PDA, US), Namunukula, in damp forest floor close to Peak, *Cramer 4566* (PDA, SING, US), *s. coll. s.n.* (PDA), floor of secondary forest, *Cramer 4353* (E, K, NVB, PDA, US). LOCALITY UNKNOWN: *Thomson s.n.* (CAL, K), *Mrs. Walker s.n.* (K), *Gardner 1847* (K), *s. coll. C.P. 16* (BM).

7. **Plectranthus glabratus** (Benth.) Alston in Trimen, Handb. Fl. Ceylon 6: 236. 1931; Abeywick., Ceylon J. Sci., Biol. Sci. 2 (2): 219. 1959.

Coleus glabratus Benth., Lab. Gen. Sp. 58. 1832; Benth. in DC., Prod. 12: 78. 1848. Type: *Wight s.n.*, Madura, India, in Hb. Wt. (holotype, K).
Coleus paniculatus Benth. in Wall., Pl. As. Rar. 2: 16. 1831; Benth., Lab. Gen. Sp. 59. 1832; Benth. in DC., Prod. 12: 79. 1848 (non *P. paniculatus* Jacq., 1800). Type: *Wight s.n.*, India, in Hb. Wt. (holotype, K-WALL., isotype, K).
Plectranthus coleoides Benth. in DC., Prod. 12: 64. 1848; Hook. f., Fl. Br. Ind. 4: 622. 1885; Trimen, Handb. Fl. Ceylon 3: 372. 1895. Type: *Perrotet s.n.*, Nilgiris, India (holotype, G, n.v., isotype, K).

Annual; stem to ± 60 cm high, obtusely 4-angular, glabrous below, puberulous above, succulent. Leaves broadly ovate, 4.5–6.6 × 3–4.2 cm, subcordate or rounded at base, obtuse, irregularly crenate, hispidulous on both surfaces; petioles to 5 cm long. Panicles terminal, to 22 cm long; cymes lax; peduncles hirtellous with gland-tipped hairs. Floral leaves deciduous. Pedicels to 5.5 mm long, hirtellous with capitate glands. Calyx small, much enlarged in fruit, the tube 7.5–8 mm long, striate, subvillous without with gland-tipped hairs, glabrous within; upper lip erect, orbicular, to 3.25 × 3 mm, curved outwards, subdecurrent on tube; lower lip equal to upper in length, the lateral lobes linear-subulate, acute, the median ones acuminate. Corolla puberulous without with gland-tipped hairs; pale violet-purple; lower lip broadly ovate, longer than upper. Stamens included in lower lip of corolla. Nutlets ellipsoid, 1.25–1.75 × 0.75–1.2 mm, subtrigonous, smooth, pale brown.

Distr. South India and Ceylon.

Ecol. Under shade in montane forest, probably close to streams, at c. 1700 m; very rare. My attempts at encountering this plant have failed; Hooker f. does not record this species for Ceylon. Flowering February.

Specimens Examined. NUWARA ELIYA DISTRICT: above Hakgala, *s. coll. C.P. 3942* (K, PDA).

6. COLEUS

Lour., Fl. Cochinch. 2: 272. 1798; Benth. in Wall., Pl. As. Rar. 2: 15. 1831; Benth., Lab. Gen. Sp. 47. 1832; Benth. in DC., Prod. 12: 70. 1848; Benth. & Hook. f., Gen. Pl. 1176. 1876; Endl., Gen. Pl. 689. 1830; Hook. f., Fl. Br. Ind. 4: 625. 1885; Briq. in Pflanzenfam. 4 (3a): 359. 1897. Type species: *C. amboinicus* Lour.

Semisucculent annual or perennial herbs, undershrubs or shrubs, sometimes with underground tubers, often aromatic. Leaves sessile or petiolate, crenate to serrate-crenate, dotted with copious, sessile, globose, pale yellow or amber oil globules. Spikes, racemes or cymose panicles terminal, erect; verticils 6-many-flowered. Pairs of floral leaves imbricate before anthesis in 2 opposite rows, reflexed with age, persistent or deciduous. Calyx ovoid-campanulate, small before anthesis, enlarged and declinate or reflexed in fruit, with the throat often oblique; tube ribbed in fruit; limb subequally 5-lobed, more often 2-labiate; upper lobe or lip broader than the others, often with margins decurrent on tube; the anteriormost lobes of lower lip acute to acuminate, often less than halfway connate along their inner margins. Corolla 2-labiate; tube much exserted beyond the calyx, cymbiform above, sigmoidly or subsigmoidly bent below; upper lip erect, shallowly 3–4-lobed; lower lip elongated, navicular. Stamens 4, declinate, often included in lower lip of corolla; filaments connate into a dorsally split sheath around style with the sheath inserted at mouth of corolla tube; anthers 2-celled, with the cells ± divaricate. Disk produced into an anterior, tongue-shaped projection. Ovary 4-partite to base; style elongated, 2-fid above, with the branches linear, equal in length. Nutlets ovoid or oblong, smooth, granular or 3–5-ribbed above.

About 90 species, palaeotropical.

Hooker f. credits eight species to India and Ceylon with two varieties of *C. malabaricus*. This number is now reduced to seven given that Backer & Bakhuizen (Fl. Java 2: 637. 1965) consider *C. atropurpureus* conspecific with *C. scutellarioides*. Of these seven species, three occur in cultivation. Trimen (1895) records four species for Ceylon, combining *C. malabaricus* var. *mollis* with the typical variety. I have, however, reinstated var. *mollis* to specific rank.

KEY TO THE SPECIES

1 Upper lip of calyx continuous with, or subdecurrent on the tube
2 Verticils dense, pseudo-spicate; stamens much exserted beyond lower lip of corolla; upper lip of calyx continuous with tube .... **1. C. amboinicus**
2 Verticils lax; stamens included in lower lip of corolla; upper lip of calyx subdecurrent on tube
3 Leaves uniformly green; corolla limb mauvish-pink; plants perennating by root tubers .... **2. C. rotundifolius**
3 Leaves variegated; corolla limb blue; plants perennating by a rootstock
4 Leaves more than 4 cm long; upper lobe of calyx longer than the lower median ones .... **3. C. scutellarioides**
4 Leaves not more than 4 cm long; upper lobe of calyx as long as or shorter than the lower median ones .... **C. scutellarioides** var. **crispipilus**
1 Upper lip of calyx distinctly decurrent on the tube
5 Adult stem densely hirsute; young floral leaves silvery-grey; calyx tube hirsutulous-annulate within
6 Decumbent or shortly erect herb; leaves not more than 4.7 cm broad; racemes to 33 cm long; corolla not more than 1.7 cm long; nutlets globose .... **4. C. barbatus**
6 Shrub; leaves more than 4.7 cm broad; racemes to 56 cm long; corolla more than 1.7 cm long; nutlets oblong-orbicular, subtrigonous .... **5. C. grandis**
5 Adult stem glabrous, puberulous or hirsutulous; young floral leaves pale green, or faintly purplish-red; calyx tube exannulate within
7 Stem acutely 4-angular; leaves membranous, acuminate or acute; cymes in branched thyrses; nutlets ellipsoid, subtrigonous, biconvex and orbicular; plants generally not or faintly aromatic
8 Stem tumid above nodes, usually mottled dull purple on sides; verticils annular; nutlets dorsally ribbed; shrubs or undershrubs
9 Angles of stem ± winged; young stem and inflorescence axis glabrous, puberulous, or sparsely hirsutulous; floral leaves deciduous
10 Sigmoid shape of corolla tube weakly defined, the lower lip laterally compressed and keeled; upper lip uniformly lavender .... **6. C. inflatus**
10 Sigmoid shape of corolla tube distinct, the lower lip navicular; upper lip streaked bluish-violet above
11 Inflorescence axis ending in a cyme; inflorescence, including calyx, with eglandular hairs; lateral calyx lobes at anthesis ovate-triangular; nutlets oblong- orbicular, dorsally 3-ribbed
12 Leaves broadly oval-elliptic, usually more than 4.5 cm broad; corolla more than 1.8–2 cm long .... **7. C. malabaricus**
12 Leaves narrowly oval-elliptic, to 4.5 cm broad; corolla not more than 1.8–2 cm long .... **C. malabaricus** var. **leptostachys**
11 Inflorescence axis ending in an aborted bud or flower; inflorescence, including calyx, with gland-tipped hairs; lateral calyx lobes at anthesis truncately semilunar; nutlets narrowly ovoid, dorsally 4-ribbed .... **8. C. kanneliyensis**
9 Angles of stem plain; young stem and peduncles densely hirsutulous; floral leaves ± persistent .... **9. C. mollis**
8 Stem uniformly thickened, uniformly green on sides, verticils secund; nutlets dorsally smooth; herbs .... **10. C. elongatus**

7 Stem obtusely 4-angular; leaves fairly thick, rounded or obtuse; cymes in verticils; nutlets globose; plants strongly aromatic.......................... **11. C. zeylanicus**

1. Coleus amboinicus Lour., Fl. Cochinch. 2: 372. 1790; Alston in Trimen, Handb. Fl. Ceylon 6: 236. 1931; Abeywick., Ceylon J. Sci., Biol. Sci. 2 (2): 219. 1959; Cramer, Kew Bull. 32 (3): 555. 1978. Type: Rumphius Ic. in Herb. Amb. 5: t. 102, f. 2. 1750 (lectotype).

Coleus aromaticus Benth. in Wall., Pl. As. Rar. 2: 15. 1831; Benth. in DC., Prod. 12: 72. 1848; Thw., Enum. Pl. Zeyl. 238. 1860; Hook. f., Fl. Br. Ind. 4: 625. 1885; Trimen, Handb. Fl. Ceylon 3: 374. 1895. Type: *s. lect.*, Patna, India, in Wall. Num. List *2730* (holotype, K-WALL.).
Plectranthus aromaticus (Benth.) Roxb., Fl. Ind. ed. 2, 3: 22. 1832 (non Roxb. 1814).
Plectranthus amboinicus (Lour.) Spreng., Syst. Veg. 2: 690. 1825.

Semi-shrubby aromatic perennial; stem rooting at lower nodes, then ascending or straggling, diffusely branched, the branches to 1.20 m long, rather woody and glabrous towards base, obtusely 4-angular, villous-hirsute in upper parts, dotted with sessile oil glands in between hairs. Leaves broadly ovate, 3.2–10 × 2.6–8.5 cm, rounded to truncate at base, rounded at apex, crenulate, hirsute above and dotted with oil globules as in stem, lanate-tomentose beneath, fleshy; petioles 1.2–3 cm long, villous-hirsute, pinkish-mauve. Pseudo-spikes terminal, to 51 cm long; verticils distant, glomerate, dense, of 12 to more than 20 flowers; peduncles densely villous-hirsute. Floral leaves broadly ovate, to 5 × 3.5 mm, acuminate, concave, hirtellous on both surfaces. Pedicels 1.5–2.5 mm long, hirsute with gland-tipped hairs. Calyx campanulate, 1.5–2 mm long; upper lip erect, ovate-oblong, 2 × 1 mm, ciliate, with margins continuous with tube; lateral lobes of lower lip short, median ones lanceolate, to 1.25 mm long, acuminate. Corolla 10–12 mm long, pink or lilac; tube 3–4 mm long, infundibular above, geniculate below, villous-hirtellous without; upper lip erect, to 4.5 × 3 mm, the lobes suborbicular; lower lip declinate, suborbicular, 5–6 × 4 mm, entire, sparsely hirtellous at back. Filaments exserted beyond lower lip of corolla, glabrous. Style glabrous; branches linear, equal, curled. Nutlets globose.

Distr. Of unknown origin; widely cultivated in the Indo-Malesian region; introduced into Maxico and U.S.A.

Ecol. Introduced into gardens as a medicinal herb, but apparently growing wild on exposed rocks in the lowlands and midlands from c. 100–600 m. Though this is the type species of the genus, the pseudo-spicate nature of its inflorescence is quite original within the genus. The entire plant has a strong aromatic scent much resembling that of turpentine. This and the pseudo-spicate nature of the inflorescence is distinctive in the field. Flowering May–October.

Uses. A decoction of the juice of the plant mixed with that of the onion bulb is given to cattle for dysentery and diarrhoea as well as for loss of appetite.

Vern. Kapparaweliya (S).

Note. The name of this species comes from *Marrubium album amboinicum* Rumph. cited by Loureiro. But since this is pre-Linnaean and a polynomial, Rumphius' authorship cannot be considered for priority.

Specimens Examined. KURUNEGALA DISTRICT: Maraluwawa, under shade in low scrub, close to temple, *Cramer 3850* (PDA, US). KANDY DISTRICT: Kadugannawa, on top of rocks in private garden beside road, *Cramer 4182* (US). LOCALITY UNKNOWN: *Gardner s.n.*, (K), *Walker s.n.* (K).

2. Coleus rotundifolius (Poir.) A. Cheval. & Perr. in A. Cheval., Vég. ut. Afr. Trop. franc. 1: 101, 119. 1905; Cheval., Rev. Int. Bot. Appl. Agric. Trop. 18: 482. 1938; White, Trop. Agric. (Ceylon) 104 (4): 151. 1948.

Germania rotundifolia Poir. in Lam., Enc. Suppl. 2: 763. 1788. Type: *Commerson s.n.*, Mauritio, in Hb. Desfont.

Coleus tuberosus (Blume) Benth., Lab. Gen. Sp. 59. 1832; Alston in Trimen, Handb. Fl. Ceylon 6: 236. 1931.

Plectranthus tuberosus Blume, Bijdr. 838. 1826; Thw., Enum. Pl. Zeyl. 238. 1860. Type: *Blume s.n.*, Java (holotype & isotype. L.).

Coleus parviflorus Benth. in DC., Prod. 12: 72. 1848; Hook. f., Fl. Br. Ind. 4: 625. 1885; Trimen, Handb. Fl. Ceylon 3: 374. 1895.

Herb perennating by means of lateral, tuberous roots; tubers globose-ovoid, to 5 × 3 cm. Young stem erect, to 23 cm high, acutely quadrangular, pubescent with retrorse hairs, the hairs 1-seriate, 3–4-celled; adult stem diffusely branched, rooting at lower nodes. Leaves broadly ovate to ovate-oblong, 3.6–6.6 × 2.6–6 cm, cuneate, rounded or slightly truncate at base and decurrent on petiole, obtuse to rounded at apex, crenate, slightly bullate and scaberulous above, more so on prominent nerves beneath and copiously dotted with amber-coloured oil globules in reticulated excavations between tertiary nerves, uniformly green; petioles 0.8–3 cm long. Racemes simple, to 20 cm long; verticils 6–14-flowered, lax. Floral leaves broadly ovate, 3 × 2.5 mm long, acuminate, dotted beneath with similar globules as on leaves, ciliolate in upper half. Pedicels 1.5–2 mm long, densely scaberulous, pinkish-purple. Calyx broadly campanulate; tube 1 mm long, scaberulous without, sparsely dotted with oil globules between hairs; upper lip erect, broadly ovate, 2.5 × 2 mm, densely scaberulous without, pale purple; lateral lobes of lower lip suborbicular, 0.5 mm long, median ones broadly triangular, 0.8 mm long. Corolla 9–12 mm long; tube sigmoid at base, 4–5 mm long, hirsutulous without and dotted with yellowish oil globules, pale mauvish-pink; upper lip erect, 4–5 mm long, lateral lobes small, central ones larger, white; lower lip ovate-orbicular, 6–7 mm long, navicular, mauve-purple, dotted with oil globules. Filaments glabrous. Style-branches equal, linear. Nutlets not seen.

D i s t r. Native of Tropical Africa, introduced into cultivation in Ceylon, India and Indonesia.

U s e. Cultivated as a food crop in the wet midlands and lowlands. The plant is grown either from tubers of the previous year's crop or from cuttings. The tubers are planted about late July or early August, and the crop is harvested in December–January. The cuttings are made in the ordinary way or from lateral branches which are weighted under a layer of soil. The leaves when bruised have an aromatic odour resembling that of mint. "The tubers have been found to be one of the most paying crops and is also a popular food which can appreciably supplement the diet of the people in Ceylon."—White, l.c. p. 107.

V e r n. Innala (S).

N o t e. This species has had a confused taxonomic history. Apart from the controversy about the generic status of *Coleus*, the identity of this species is in question. Some authors such as Sampson (1936) consider *C. rotundifolius* to be different from *C. parviflorus*. Sampson has reduced *C. rotundifolius* to synonmy under *C. dysentericus* Baker, an African and Malagasy plant of different circumscription not occurring in Asia. From examination of material in the Kew Herbarium I agree with White that *C. rotundifolius* and *C. parviflorus* are identical and different from *C. dysentericus*.

S p e c i m e n s E x a m i n e d. KANDY DISTRICT: Katugastota, Medawela, in private fields by Botota Road, *Cramer 3919* (K, PDA, US); Peradeniya, University Campus, *Cramer 4386* (SING, E, K, NVB, M, PDA, US).

3. Coleus scutellarioides (L.) Benth. in Wall., Pl. As. Rar. 2: 16. 1830; Benth., Lab. Gen. Sp. 53. 1832; Benth. in DC., Prod. 12: 73. 1848.

Ocimum scutellarioides L., Sp. Pl. ed. 2, 834. 1763.
Coleus blumei Benth., Lab. Gen. Sp. 56. 1832; Benth. in DC., Prod. 12: 75. 1848. Type: Java (n.v.).

Perennial herb; stem erect, to 70 cm high, branched, semi-succulent, obtusely 4-angular, finely tomentose. Leaves ovate-deltoid to broadly ovate, 5.6–14.3 × 3.2–9.4 cm, truncate or rounded at base, acute, coarsely crenate, the crenatures faintly 2-fid at apex, scabrid and variously coloured above, hirtellous on prominent nerves beneath, distinctly dotted with oil globules and usually pale green; petioles 2–4 cm long. Racemes terminal, to 26 cm long; verticils lax, 12–16-flowered; peduncles hirtellous. Floral leaves broadly ovate, to 5 × 3.5 mm, long-acuminate, finely tomentose without. Pedicels to 1.5 mm long, slender, sparsely pilose. Calyx obliquely campanulate; tube 1 mm long, hirtellous; upper lip ovate-orbicular, to 2×2 mm, puberulous without and dotted with oil globules as in leaves; lateral lobes of lower lip ovate-rhomboid, 0.5 mm long, median lobes 1.5 mm, acuminate. Corolla to 11 mm long; tube to 4.5 mm long, infundibular and laterally compressed above, narrowed below, hirtellous without and dotted with oil globules, whitish; limb pale to deep blue; upper lip to 1.75 mm long, the median lobes broadly orbicular, lateral ones bluntly ovate; lower lip oval-orbicular, 6×5 mm, sparsely strigose without and dotted with oil globules. Filaments and style glabrous.

Distr. Originally from eastern Asia and Malesia, now pantropical by cultivation.

Ecol. The plant is extensively cultivated in the low country, midlands and uplands, owing to its brightly coloured leaves. A few variations occur in size and colour of the leaves which do not, however, merit separation into distinct varieties. Flowering throughout the year.

Note. This is a very variable species with regard to the shape and colour of the leaves. Until its taxonomy is further elucidated, I prefer to follow Backer & Backhuizen's broad concept of the species (Fl. Java 2: 637. 1965).

Specimens Examined. KANDY DISTRICT: Peradeniya, Botanic Gardens, in flower garden, *Cramer 4228* (PDA, US), in flower bed beside lawn in Botanic Gardens, *Cramer 4221* (PDA, US).

var. **crispipilus** (Merr.) Keng, Gard. Bull. Singapore 24: 56. 1969.

Coleus macranthus var. *crispipilus* Merr., Philipp. J. Sci. 1 (Suppl.) 234. 1906.

Perennial herb; stem erect, to 45 cm high, acutely 4-angular, puberulous with retrorse hairs. Leaves broadly ovate to reniform, 2.6–4×2–3.5 cm, truncate at base, subacute, sparsely pilose and brownish-maroon above with magenta stains along mid-nerve, hirtellous on prominent nerves and pale green beneath with blotches of purple. Racemes or panicles terminal, to 16.5 cm long. Floral leaves ovate-lanceolate, to 2.5×1.5 mm. Corolla pale purplish-blue, to 12 mm long; tube to 5 mm long.

Distr. Cultivated in Ceylon, Malesia and Formosa, etc.

Ecol. Cultivated as an ornamental plant. Flowering throughout the year.

Specimens Examined. KANDY DISTRICT: Peradeniya, Royal Botanic Gardens, border to flower bed, *Cramer 4410* (PDA, US).

4. Coleus barbatus (Andr.) Benth. in Wall., Pl. As. Rar. 2: 15. 1831; Benth., Lab. Gen. Sp. 49, 1711. 1832 (excl. syn. *Germanea forskohlii* Poir. 1821 & *Plectranthus forskohlaei* Willd. 1800); Thw., Enum. Pl. Zeyl. 238. 1860; Hook. f., Fl. Br. Ind. 4: 4: 625. 1885 (excl. eisdem syn.); Trimen, Handb. Fl. Ceylon 3: 373. 1895; Abeywick., Ceylon J. Sci., Biol. Sci. 2 (2): 219. 1959; Cramer in Kew Bull. 32 (3): 555. 1978.

Plectranthus barbatus Andr., Bot. Rep. 9: t. 594. 1809. Type: Andrews Ic. in l.c.

Coleus forskohlii sensu Briq. in Pflanzenfam. 4 (3a): 359. 1897; Alston in Trimen, Handb. Fl. Ceylon 6: 236. 1931; Mukerjee, Rec. Bot. Surv. India 14 (1): 53. 1940 (non Vahl 1790).

Perennial herb; stem generally decumbent, rooting at lower nodes, occasionally erect, obtusely quadrangular, semi-succulent, densely hirsute with uniseriate, multicellular hairs interspersed with gland-tipped ones, dotted with reddish-orange oil globules in between hairs. Leaves oval-oblong, 2.8–8.6 × 1.9–4.7 cm, obliquely cuneate at base, often subdecurrent on petiole, obtuse or rounded at apex, crenate-serrate, pilose on both surfaces, the hairs on the lower surface interspersed with amber-coloured oil globules, thick; petioles 0.6–2.5 cm long. Racemes simple, to 33 cm long; verticils lax, 8–10-flowered; peduncles hispid, the hairs interspersed with oil globules, faintly purplish. Floral leaves much imbricated, ovate-lanceolate, to 1.3 × 1.0 cm, caudate-acuminate, strigose without with silvery hairs, glabrous within, early deciduous. Pedicels 3–4.5 mm long, decurved, hispid. Calyx broadly campanulate; tube 1.5–2.5 mm long, 10-nerved, hispid without with a tuft of longer deflexed hairs towards base, annulate within with hispid hairs; upper lip erect, broadly ovate to ovate-orbicular, 3.5–4 × 2–3 mm, shortly acuminate, subdecurrent on tube, reflexed at apex with age, hirsutulous without, the hairs interspersed with oil globules; lower lip 4.5–5.5 mm long, with ovate-lanceolate lobes, lateral ones 2–3 mm long, median ones 3–4 mm long, all long-acuminate, purplish at tips. Corolla deep purplish-blue, occasionally bluish-white, to 17 mm long; tube 6–8 mm long, expanded above, narrowed and sigmoid below; upper lip shortly ovate, 2.5–3 × 3 mm, dotted with oil globules, lower one stipitate, ovate-orbicular, 8–10 mm long, navicular, sparsely hirsutulous without, the hairs interspersed with oil globules. Stamens unequal; filaments declinate; anther-cells suborbicular. Style glabrous; branches equal, linear. Nutlets globose, to 1.25 mm in diam., granulate, blackish-brown.

Distr. From Tropical East Africa through India to Ceylon.

Ecol. In open, dry, waste places and in shallow soil overlying rocks of the dry lowlands, dry intermediate regions and dry midlands, from sea-level to c. 1500 m. Not cultivated. The erect racemes in full bloom afford an attractive sight. The young imbricate, silver-haired floral leaves are very conspicuous within the genus. The entire inflorescence has a musky scent. Flowering July–December.

Note. *Plectranthus forskohlaei* Willd., the basionym of Briquet's name, was already employed by Vahl (Sym. Bot. 1: 44. 1790) as a superfluous name for *Ocimum hadiense* Forssk., an Arabian plant of a different circumscription correctly placed in *Ocimum*.

Specimens Examined. POLONNARUWA DISTRICT: Gunner's Quoin, in rocky hollows, *Cramer 3639* (PDA, US). MATALE DISTRICT: Dambulla Rock, *Alston 1651* (PDA), *Simpson 9733* (PDA), *Cramer 3282* (PDA, US), Ritigala Summit, *Willis s.n.* (PDA). NUWARA ELIYA DISTRICT: Hakgala, by Hakgala-Palugama Road, *Cramer 4196* (E, PDA, SING, US). BADULLA DISTRICT: Palugama, *s. coll. C.P. 2069* (CAL, PDA). LOCALITY UNKNOWN: *Mrs. Walker s.n.* (K).

5. Coleus grandis Cramer in Kew Bull. 32 (3): 556. 1978. Type: *Cramer 3869* Sita Eliya, Nuwara Eliya District, Ceylon (holotype, PDA, isotype K, US).

Shrub; stem to 1.75 m high, subquadrangular, densely hirsute, the hairs interspersed with reddish-orange oil globules. Leaves ovate-oval to broadly ovate, 6.3–23 × 4.7–18.6 cm, truncate at base, acute, coarsely crenate, hispid on both surfaces; petioles 1–6.8 cm long. Racemes simple, erect, 26–60 cm long; verticils lax, 10-flowered; peduncles stiff, viscous-hispid, dull purple. Floral leaves broadly ovate, to 1.2 × 0.9 cm, acuminate at apex, deciduous. Pedicels 5–7 mm long, stiffly declinate, viscous-hispid, dull purple. Calyx broadly campanulate, dotted without with oil globules as above; tube 3.5–4 mm long, 10-nerved, annulate-hispidulous within, hispidulous without with gland-tipped hairs; upper lip erect, broadly ovate, 4–7 × 4 mm, puberulous without, reflexed above with age; lobes of lower lip ovate-lanceolate, lateral ones 2.5 mm long, median ones 3–3.5 mm long, long acuminate. Corolla 2.2–2.6 cm long, broadened above, laterally compressed, bluish-purple, whitish below; upper lip erect, ovate; lower lip stipitate, oval-oblong, 1.5–2 mm long, navicular. Filaments glabrous. Disk with an anterior prominent tongue-shaped projection. Style glabrous. Nutlets orbicular-oblong, to 2.0 × 1.75 mm, subtrigonous, 3-ribbed above, glabrous, black and shining.

Distr. Tropical East Africa and Ceylon. The country of origin of the plant is not known.

Ecol. Widely grown as a hedge plant in the upper midlands and highlands from c. 1100–2100 m. It thrives well in moist ground. Occasionally it grows as an escape by watercourses close to habitations, but has never been known in the wild. On the other hand nothing is known of the possibility of its having been introduced into the country. The inflorescence has a musky scent; in full bloom the racemes are conspicuous and afford a pretty sight. Cattle are observed not to eat the plant. Flowering September–February.

Note. The identity of this species was arrived at after discussions with Mr. B. Matthews, Kew Herbarium, England, in the course of my work there. According to Mr. Matthews this plant has been observed to be cultivated in Tropical East Africa, too, but has not been described so far. I wish to express my sincere indebtedness to Mr. Matthews for the above discussions.

Specimens Examined. NUWARA ELIYA DISTRICT: Pussellawa, as fence of garden beside Ramboda Road, *Cramer 3869* (PDA, US); at base of ascent to Pidurutalagala, *Cramer 4191* (K, PDA, US). BADULLA DISTRICT: Ambawela, *Comanor 988* (PDA, US), Ohiya, by rail track near Station, *Balakrishnan 482* (PDA, US).

6. Coleus inflatus Benth., Lab. Gen. Sp. 58. 1832; Benth. in DC., Prod. 12: 78. 1849; Thw., Enum. Pl. Zeyl. 238. 1860; Hook. f., Fl. Br. Ind. 4: 626. 1885; Trimen, Handb. Fl. Ceylon 3: 375. 1895; Abeywick., Ceylon J. Sci., Biol. Sci. 2 (2): 219. 1959; Cramer in Kew Bull. 32 (3): 555. 1978. Type: *Macrae s.n.*, Ceylon, in Hb. Benth. (holotype, K).

Coleus benthamianus Arn., Nova Acta Phys. Med. Acad. Caes. Leop. Carol. Nat. Cur. 354. 1836. Type: *Graham s.n.*, Ceylon (holotype).

Perennial herb; stem to 1.85 m high, much branched, acutely quadrangular, tumid above the nodes, winged at angles, warty and to 2 cm thick towards base, glabrous above, transversely lined with very small hairs at nodes, mottled dull purple at sides. Leaves spreading, oval-elliptic, 4–23.5 × 2.4–10 cm, cuneate, narrowed or tapering at base, acuminate, serrate, sparsely scabrid above, pilose on prominent nerves and dotted beneath with reddish-orange oil globules; petioles 0.5–6 cm long, shallowly furrowed and hirsute above, reddish-purple. Panicles terminal, 14–23 cm long; cymes compressed, lax; peduncles subglabrous or hirsutulous, dull maroon. Young floral leaves ovate, to 1.3 × 0.8 cm, fairly conspicuous, soon deciduous. Pedicels 1.0–1.5 mm long, glabrous, dull maroon. Calyx broadly campanulate, to 6 mm long; tube 1.5–1.75 mm long, faintly 10-nerved, glabrous, dotted with oil globules, dull pinkish-purple; upper lip erect, broadly ovate to ovate-lanceolate, 2–4.5 × 2–2.5 mm, with a swollen apical tip; lower lip up to 3.5 mm long, with the lateral lobes ovate-triangular, shortly acuminate, the median ones lanceolate, longa-cuminate. Corolla 2.6–3.1 cm long; tube 2.3–2.7 cm long, laterally subcompressed, expanded above for more than half its length, narrowed and inflexed at base, pale lilac; upper lip erect, broadly oblong, to 6–7 × 3.5–4 mm, concave; lower lip broadly ovate, to 6–7 × 4–5 mm, laterally compressed, keeled, sparsely dotted without with oil globules. Anther-cells divaricate. Disk with a tongue-shaped projection on anterior side. Nutlets ovoid oblong, 3–4 × 1.5–2.25 mm, subtrigonous, dorsally faintly 3-ribbed, dark brown and shining.

Distr. Endemic.

Ecol. A gregarious species in partial shade close to streams in secondary montane forests from c. 1200–2600 m. The panicles are conspicuous and the entire inflorescence is suffused with a pale rose-pink coloration. The lemon-like scent of the inflorescence is distinctive. Flowering almost through the year.

Specimens Examined. KANDY DISTRICT: Rajamally, Moray Estate, under partial shade close to stream in secondary montane forest, *Cramer 4272* (K, PDA, US). NUWARA ELIYA DISTRICT: Ramboda, *Gardner C.P. 2065* (PDA), Nuwara Eliya, *s. coll. C.P. 2065* (K), Park, by road, *Simpson 9120* (PDA); Hakgala, in undergrowth of forest, *Cramer 3914* (PDA, US), jungle, *Silva s.n.* (PDA); Ambewela, under shade close to stream in secondary forest, *Cramer 4248* (BO, PDA, US), in open places in jungle, *Alston 1017* (PDA); Sita Eliya, *de Silva s.n.* (PDA), ibid., *A.W. Silva s.n.* (PDA); Pattipola, under shade in damp sloping ground at edge of jungle, *Cramer 3842* (PDA, US), along road, *van Beusekom 1448* (L, PDA, US); Horton Plains towards Rest House from fork to Diyagama, *Comanor 449* (PDA), *s. coll. C.P. 2065* (PDA), *Simpson 9535* (PDA), Agrapatna Road, roadside thicket, *Nowicke & Jayasuriya 245* (PDA, US), open forest, *Theobald & Krahulik 2764* (PDA); Pidurutalagala, in shallow ravine under shade close to stream, *Cramer 3901* (PDA, US). BADULLA DISTRICT: Ohiya, foot trail to Pattipola, *Balakrishnan 480* (PDA). LOCALITY UNKNOWN: *Wight 745* (M), *Macrae s.n.* (K), *Walker s.n.* (K), *Mackenzie s.n.* (K), *Gardner 649* (K), *Thomson 1845* (K), *Mrs. Walker s.n.* (K), *Simpson s.n.* (BM).

7. **Coleus malabaricus** Benth. in Wall., Pl. As. Rar. 2: 16. 1831; Benth., Lab. Gen. Sp. 47. 1832; Benth. in DC., Prod. 12: 76. 1848; Thw., Enum. Pl. Zeyl. 238. 1860; Hook. f., Fl. Br. Ind. 3: 626. 1885 (excl. syn. *C. ovatus* Benth.); Trimen, Handb. Fl. Ceylon 3: 276. 1895; Abeywick., Ceylon J. Sci., Biol. Sci. 2 (2): 219. 1959; Cramer in Kew Bull. 32 (3): 555. 1978. Type: *s. coll.* in Wall., Num. List *2735A*, 1817. Travancore, India (holotype, K-WALL.).

Coleus macraei Benth., Lab. Gen. Sp. 58. 1832; Benth. in DC., Prod. 12: 77. 1848. Type: *Macrae* Lindl., *s.n.*, Ceylon, in Hb. Benth. (holotype, K).

Coleus walkeri Benth. in DC., Prod. 12: 77. 1848. Type: *Walker s.n.*, Ceylon, in Hb. Benth. (holotype, K).

1. var. **malabaricus** Benth.

Shrubby herb; stem to 2.75 m high, branched, stout, to 3.5 cm in diam. towards base, acutely 4-angular, quadrialate, slightly tumid above nodes, glabrous or hirsutulous, pale green, mottled dull purple. Leaves broadly ovate-elliptic, 5.6–25 × 3.5–13.2 cm, acute or somewhat tapering at base, acuminate, crenate-serrate, scaberulous above, puberulous on nerves beneath and dotted with reddish-orange oil globules; petioles 3.5–6.5 cm long, channelled above, hirsute, pale pinkish-red. Panicles ± pyramidal, to 28 cm long; inflorescence axis ending in a cymelet, with the lateral branches often longer; peduncles glabrous or puberulous with gland-tipped hairs, pinkish-purple. Floral leaves ovate-lanceolate, to 2 × 0.6 cm, lower ones larger; all concave acuminate, ciliolate, glabrous. Pedicels 1.5–2 mm, glabrous or puberulous. Calyx somewhat dilated and deflexed in fruit, dotted with oil globules; tube 1–1.75 mm long, striate, glabrous on both surfaces; upper lip erect, broadly ovate, to 2.5–3 × 2 mm, with a swollen tip; lateral lobes of lower lip ovate-triangular, median ones lanceolate, acuminate. Corolla 1.8–2 mm long, sparsely dotted with oil globules; tube 9–11 mm long, narrowed and sigmoidly bent at base, expanded above, lilac; upper lip erect, broadly spathulate, 5–6.5 mm long, streaked purplish-blue within; lower lip oval-orbicular, 7.5–10 mm long, navicular, glabrous, whitish. Stamens slightly exserted beyond lower lip of corolla; filaments glabrous, white. Disk with an anterior tongue-shaped projection. Style glabrous, branches equal. Nutlets ovoid-oblong, to 1.75 × 1 mm, subtrigonous, 3-ribbed above, glabrous, black and shining.

Distr. South India and Ceylon.

Ecol. Restricted to shaded places near streams in secondary submontane forests, rarely in wet, evergreen forests of the low country, from c. 650–1300 m. A gregarious plant. The indumentum of the petiole, especially of the stem, leaves and peduncles, is variable. The lower surface of the leaves is occasionally a dull purple. The panicles in full bloom are attractive; the flowers are scentless. Flowering almost throughout the year.

Specimens Examined. KANDY DISTRICT: Hantane, *Gardner C.P. 2066* (PDA, US); Ambagamuwa, *s. coll. C.P. 2995* (P), Hunasgiriya, *Silva 66* (PDA, US); Dolosbage, beside stream in secondary wet, intermediate forest, *Cramer 4640* (CAL, E, PDA, US); Maha Oya, north of Corbet's Gap, *Jayasuriya 958* (PDA, US). NUWARA ELIYA DISTRICT: Ramboda, *s. coll. C.P. 17* (PDA, US). RATNAPURA DISTRICT: Balangoda, Maddakande, river bank, *Balakrishnan 301* (PDA, US); Maratenne, along stream, *Jayasuriya & Maxwell 821* (PDA, US); Kuttapitiya Forest, *Waas 1574* (PDA, US). MATARA DISTRICT: Deniyaya, Deniyaya-Bulatala Road, Amingkande, *Sumithraarachchi 327* (PDA, US). BADULLA DISTRICT: Haputale, Tangamalai Reservoir, under shade along stream bank, *Cramer 3820* (PDA, US). LOCALITY UNKNOWN: *s. coll. C.P. 17, 2066* (P), *Mrs. Walker s.n.* (K), *Gardner 1847* (K), *Gardner 656* (K), *Macrae s.n.* (BM), *s. coll. C.P. 17, 3425* (BM), *Macrae 1829* (K), *Walker s.n.* (K).

2. var. **leptostachys** (Benth.) Hook. f., Fl. Br. Ind. 4: 625. 1885.

Coleus leptostachys Benth. in DC., Prod. 12: 77. 1848. Type: *Walker s.n.*, Ceylon, in Hb. Benth. (holotype, K).

Stem puberulous or with depressed, fine hairs. Leaves narrowly oval-elliptic, 6.8–14.5 × 3–4.5 cm, narrowed or tapering to base, acuminate at apex, serrate-crenate, scabrid above, less so or glabrous beneath but with minute, appressed hairs on nerves; petioles to 5.2 cm long. Fruiting panicles to 42 cm long, much branched, the branches to 20.8 cm long, slender. Corolla to 1.2

cm long, glabrous without. Fruiting calyx to 6 mm long; median lobes of lower lip lanceolate, long-acuminate, the lateral ones ovate-lanceolate.

Distr. Endemic.

Note. This variety is known only from the type specimen. I have not encountered this plant.

8. Coleus kanneliyensis Cramer and Balasubramaniam, Kew Bull. 32 (3): 556. 1978. Type: *Cramer 4317*, Kanneliya, Galle District, Ceylon (holotype PDA, isotope E, K, NVB, US).

Perennial herb; stem rooting at lower nodes, ascending to 1.16 m, acutely 4-angular, subquadrialate, pubescent on angles with retrorse hairs, succulent. Leaves ovate to obovate, 3.5–22.5 × 5.4–11.5 cm, cuneate at base and subdecurrent on petiole, acuminate, shallowly crenate, sparsely scaberulous above, equally so on prominent nerves beneath and dotted with amber-coloured oil globules; petioles 1.7–10 cm long, pinkish-red. Panicles terminal, to 10.6 cm long; inflorescence axis ending in an aborted bud or flower, densely hirtellous with gland-tipped hairs; flowers secund, on lateral branches to 9.5 cm long. Floral leaves early deciduous. Calyx declinate, broadly campanulate, to 4 mm long, in fruit 8–10 mm long, dotted with oil globules; tube 1.5–2 mm long, dotted with oil globules; upper lip erect, broadly ovate, to 2 × 2.5 mm, acute with a swollen apex, sparsely hirtellous with gland-tipped hairs; lower lip to 1.5 mm long, longer than the upper one after anthesis, the lateral lobes at anthesis truncately semi-orbicular, the median ones ovate-lanceolate, acute. Corolla to 3.2 cm long; tube to 1.8 cm long, expanded and laterally compressed above, slightly geniculate at base, thinly pubescent, pinkish-white; upper lip spathulate to 12 × 7 mm, longitudinally striped bluish-purple; lower lip broadly ovate, to 12 mm long, sparsely hirtellous. Filaments glabrous. Anterior projection of disk to 1.6 mm long. Style glabrous; branches linear. Nutlets narrowly ellipsoid, to 2.75–3.25 × 1.25–2 mm, subtrigonous, 3-ribbed above, black, shining.

Distr. Endemic.

Ecol. In partial shade by streams in secondary wet-lowland forests to c. 200 m. A gregarious plant. It does not grow to tall proportions; the erect stem easily sets root at the lower nodes. The inflorescence has a faintly musky scent. Flowering June–September. In the forest shade, the leaves of the plant are pale green and are spread out divaricately on the long petioles. Due to the aborted terminal bud, the panicles are arrested in growth unlike in *C. malabaricus* Benth.

Specimens Examined. RATNAPURA DISTRICT: Sinharaja Forest, S.E. of Weddagala, *Meijer 887* (PDA, US); Kelinkande, beside watercourse, in secondary, wet, evergreen forest, *Balasubramaniam 1253* (PDA);

Gongala, beside stream, in secondary, wet, evergreen forest, *Waas 1383* (PDA, US). GALLE DISTRICT: Habarakadde *s. coll.*, *C.P. 171* (PDA); Kanneliya, beside stream, in secondary, wet, evergreen forest, *Cramer 4200* (BO, K, MH, PDA, US), *Balasubramaniam 1044* (PDA).

Note. The locality 'Habarakadde' of *C.P. 17* includes the general area which borders on the Kanneliya Forest. One locality of this area is Hiniduma (including Haycock Hill or 'Hiniduma Kande'), from which the above-documented record probably does not come. I have made a few field trips around Hiniduma itself, including Haycock, and I have failed to encounter any species of *Coleus* there. This record, therefore, probably is representative of an area neighbouring the other areas cited above.

9. Coleus mollis Benth. in DC., Prod. 12: 77. 1848; Cramer in Kew Bull. 32 (3): 560. 1978. Type: *Walker s.n.*, Ceylon, in Hb. Benth. (lectotype, K).

Coleus malabaricus var. *mollis* (Benth.) Hook. f., Fl. Br. Ind. 4: 626. 1885.
Coleus malabaricus sensu Trimen, Handb. Fl. Ceylon 3: 376. 1895, non Benth.

Semi-shrubby perennial; stem to 1.5 m high, branched, tumid above nodes, obtusely 4-angled, glabrous below, densely hirtellous above with gland-tipped hairs, pale green and mottled dull purple on sides. Leaves ovate to oval-oblong, 7–19 × 2.7–8.7 cm, rounded or shortly cuneate at base, caudate-acuminate to long-acuminate, crenate-serrate, scabrid to sparsely pubescent above, equally so on prominent nerves beneath and dotted with reddish oil globules; petioles 1.7–7.6 cm long, upper ones pinkish-red. Panicles terminal, 6–12 cm long; cymes lax; peduncles densely hirtellous with gland-tipped hairs, the hairs multicellular. Floral leaves foliaceous, broadly ovate, 11–15 × 10 mm, acute, hirsutulous, navicular, pinkish-red, ± persistent. Pedicels 2.5–3 mm long; hirtellous, pinkish-red. Calyx deflexed; tube to 2 mm long, in fruit to 6 mm long, hirtellous, dotted with oil globules; upper lip broadly ovate, 2–4 mm long, with a swollen apical tip; lower lip to 3 mm long, lateral lobes shortly ovate-triangular, median ones ovate-lanceolate, acuminate. Corolla 2.5–2.8 cm long, lilac; tube to 1.5 cm long, sigmoid towards base, hirtellous with gland-tipped hairs; upper lip ovate-oblong, to 11 × 6 mm, longitudinally striped bluish-violet. Filaments glabrous. Nutlets oblong-orbicular, 1.75 mm long, subtrigonous, dorsally 3-ribbed.

Distr. Endemic.

Ecol. Under shade along stream banks in secondary montane forests from c. 1400–1500 m. A gregarious plant. The entire inflorescence is suffused pinkish-red and is very distinctive. Flowering March–July.

Specimens Examined. KANDY DISTRICT: Maskeliya, Rajamally, under shade beside stream in forest, *Cramer 4281* (BO, E, K, NBV), close to Rajamally, *Sumitraarachchi 193* (PDA, US); Laxapana, *Simpson 9000* (BM). RATNAPURA DISTRICT: Adam's Peak Sanctuary, Warnagala, under shade beside stream, *Cramer 4694* (K, SING, PDA, US). LOCALITY UNKNOWN: *Walker, s.n.* (K).

Note. *C. mollis* and *C. malabaricus* are two distinct taxa as appear from the characters tabulated below:

Characters	*C. malabaricus*	*C. mollis*
1. Indumentum of stem	puberulous with appressed hairs	densely hirtellous
2. Floral leaves	to 6 mm long, glabrous, early deciduous, slightly concave	11–15 mm long, densely hirsutulous, persistent, navicular
3. Peduncles	to 28 cm long	shorter, to 12 cm long
4. Corolla	to 2 cm long	2.5–2.8 cm long

Both Hooker f. and Trimen had not seen the live plant in which these differences appear more clearly than in dried material.

In addition, there is a difference in the ranges of distribution, *C. malabaricus* occurring in the eastern side of the Kandy district and *C. mollis* on the western side.

10. Coleus elongatus Trimen, J. Bot. 27: 165. 1889; Trimen, Handb. Fl. Ceylon 3: 375. 1895; Willis, Ann. R. Bot. Gard. (Peradeniya) 4: 5. 1907; Cramer, Kew Bull. 30 (1): 35–38. 1975 & 32 (3): 560. 1978. Type: *Trimen*, Ceylon, North Central Province, Ritigala, near summit (holotype, PDA).

Englerastrum elongatum (Trimen) Alston, Kew Bull. 1926: 298. 1926; Alston in Trimen, Handb. Fl. Ceylon 6: 236. 1931; Abeywick., Ceylon J. Sci., Biol. Sci. 2 (2): 219. 1959.

Perennial herb with a short rootstock; stem shortly erect, to ±30 cm high, acutely 4-angular, rather succulent, glabrous and pale brown below, finely tomentose with adpressed hairs and dull purple above, with divaricate branches to 1.75 m long. Leaves ovate to ovate-deltoid, 2.5–4 × 2.1–2.8 cm, truncate at base, subacute to subacuminate, deeply crenate, sparsely scabrid on both surfaces, thin, caducous; petioles 9–18 mm long, slender. Racemes unequal, 2–6 cm long, terminating branches in axils of fallen leaves; peduncles slender, hirtellous; flowers paired, secund. Floral leaves ovate-lanceolate, 1.5–2 × 0.75 mm, acute, hirtellous and dotted without with oil globules. Pedicels to 1 mm long. Calyx subdeclinate, to 3.5 mm long; tube to 1.5 mm long, slightly curved inwards, subgibbous at base anteriorly, hispidulous and copiously dotted with oil globules; upper lip erect, 2 mm long, median lobe ovate-triangular, 1.5 ×

1 mm, acute, lateral ones ovate-deltoid, shorter; lower one 2 mm long, lobes linear-lanceolate, long-acuminate, all sparsely hispidulous without and dotted with oil globules. Corolla 12–13 mm long; tube to 6.5 mm long, sigmoidly bent and whitish at base, expanded and laterally compressed above, finely pubescent without and dotted with oil globules; upper lip obovate-oblong, to 3.5 mm across, bluish-purple, the median lobes orbicular-oblong, large, the lateral ones smaller; lower lip ovate-orbicular, to 7 × 5 mm, finely pubescent and dotted with oil globules without. Filaments glabrous. Style glabrous; branches equal. Nutlets broadly ellipsoid, 0.5 mm long, subtrigonous, shining, dark brown.

Distr. Endemic.

Ecol. In shallow humus soil overlying rocky outcrops in dry forest in the intermediate region, at c. 670 m. Known only from the type locality. Unlike at the time of its discovery, the species now forms a local population. Flowering June–August.

Specimens Examined. ANURADHAPURA DISTRICT: Ritigala, in partial shade in rocky ledge close to peak, *Cramer 4144* (BO, E, K, MH, NBV, PDA), *Jayasuriya 884* (PDA, US).

Note. Alston has apparently overlooked the zygomorphic nature of the calyx in *Englerastrum scandens* (Gurke) Alston to which he related this species. From examination of representative material of both species in the Kew Herbarium there is no doubt that they are congeneric; but according to Briquet's conception of *Englerastrum* (1894) the zygomorphic (bilabiate) nature of the calyx in both rules out their inclusion under this genus. In my opinion they both should be restored to *Coleus*. For a fuller discussion of this subject see Cramer, l.c.

11. Coleus zeylanicus (Benth.) Cramer in Kew Bull. 32 (3): 555. 1978.

Plectranthus zeylanicus Benth., Lab. Gen. Sp. 36. 1832; Benth. in DC., Prod. 12: 66. 1848; Thw., Enum. Pl. Zeyl. 238. 1860; Hook. f., Fl. Br. Ind. 4: 622. 1885; Trimen, Handb. Fl. Ceylon 3: 371. 1885; Abeywick., Ceylon J. Sci., Biol. Sci. 2 (2): 219. 1959.

Annual; stem erect, to 1.20 m high, rooting at lower nodes, branched above, obtusely 4-angular, villous-hirsute, often pale pink in upper parts. Leaves broadly ovate to suborbicular, 2.6–12.5 × 2.2–12.5 cm, rounded to truncate at base, obtuse or rounded, coarsely crenate-serrate, sparsely hirtellous on both surfaces, thick, with prominent nerves beneath forming reticulate depressions dotted with amber-coloured oil globules; petioles to 4 cm long. Panicles terminal, 12–42 cm long; verticils lax; peduncles slender, hirtellous with gland-tipped hairs. Floral leaves sublunar, to 5 × 7 mm, concave, subacute. Pedicels 2–3 mm long, hirtellous. Calyx to 4 mm long, to 1.5 mm in fruit; tube 1.25 mm long, 10-nerved, dotted with oil globules; upper lip broadly ovate, to 2×2.5 mm, abruptly acute, subdecurrent on tube, deflexed over mouth of calyx after anthesis; lower lip 2.25 mm long, lateral lobes triangular-ovate, median ones lanceolate, longer than laterals, acuminate. Corolla 8–10 mm long, pale blue; tube 4–5 mm long, finely pubescent with gland-tipped hairs, whitish; upper lip 4 mm long, shallowly lobed, dotted at back with oil globules; lower lip orbicular, 4–5 mm long, finely pubescent. Style glabrous. Nutlets globose, 0.4 mm broad, smooth.

Distr. Endemic.

Ecol. Cultivated as a medicinal herb in the lowlands and lower uplands. The entire plant has a faintly aromatic scent. Flowering May–October.

Vern. Iriweriya (S).

Uses. The juice of the stem and leaves mixed with honey is drunk as a remedy for diarrhoea.

Specimens Examined. KURUNEGALA DISTRICT: Maraluwawa, in shallow, gravelly soil close to temple, *Cramer 3853* (PDA, US). KANDY DISTRICT: Kadugannawa, cultivated in garden of village houses, *Cramer 3843* (PDA, US). LOCALITY UNKNOWN: *s. coll. C.P. 2083* (K), *Gardner 655* (K).

7. ANISOCHILUS

Wall. in Benth., Bot. Reg. 15, t. 1300. 1830; Benth. in Wall., Pl. As. Rar. 2: 18. 1831 (*Anisochilos* sphalm.); Benth., Lab. Gen. & Sp. 59. 1832; Benth. in DC., Prod. 12: 80. 1848; Endl., Gen. Pl. 610. 1838; Benth. & Hook. f., Gen. Pl. 1177. 1876; Briq. in Pflanzenfam. 4 (3a): 350. 1897. Type species: not known.

Annual or perennial herbs or shrubs. Leaves usually fleshy, with excavations on lower surface in between nerve reticulations. Spikes subtetragonal or cylindrical, usually arranged in panicles; verticils densely imbricate. Floral leaves shorter or longer than flowers, caducous. Flowers small. Calyx suberect, ovoid, inflated above middle, 2-labiate, upper lip (posterior lobe) entire, longer than lower, deflexed over mouth of tube after anthesis; lower lip (anterior lobe) truncate, toothed or 0. Corolla tube slender, exserted beyond calyx, incurved, broadened at throat. Stamens exserted or not, declinate; anther-cells confluent. Disk lobed, the anterior lobe slightly prolonged. Style shortly 2-fid above. Nutlets ovoid, smooth.

About 20 species, in tropical Africa and tropical continental Asia.

KEY TO THE SPECIES

1 Mature stem up to ± 60 cm high; spikes usually more than 1.5 cm long; annual...**1. A. carnosus**

1 Mature stem more than 60 cm high; spikes to 1.5 cm long; perennials
2 Leaves broadly ovate or suborbicular, rounded at base, scaberulous beneath....... ....2. **A. paniculatus**
2 Leaves oval-oblong, acute at base, puberulous or velvety-tomentose beneath....... ....3. **A. velutinus**

1. Anisochilus carnosus (L. f.) Wall. ex Benth. in Wall., Pl. As. Rar. 2: 18. 1831; Benth., Lab. Gen. Sp. 59. 1832; Benth. in DC., Prod. 12: 81. 1848; Thw., Enum. Pl. Zeyl. 238. 1860; Hook. f., Fl. Br. Ind. 4: 627. 1885; Trimen, Handb. Fl. Ceylon 3: 376. 1895; Mukerjee, Rec. Bot. Surv. India 14 (1): 58. 1941; Abeywick., Ceylon J. Sci., Biol. Sci. 2 (2): 219. 1959.

Lavendula carnosa L. f., Suppl. 273. 1781. Type: *Koenig s.n.*, India, Specimen 727/7 (LINN).
Plectranthus dubius Spreng., Syst. Veg. 2: 691. 1825 (nom. illegit.).

Annual; stem erect, to 60 cm high, often slightly branched, obtusely 4-angular, glabrous or with downwardly adpressed, scaberulous hairs, fleshy, dull pinkish-red. Leaves broadly ovate, 1.9–6.6 × 0.8–5.7 cm, truncate, rounded or subcordate at base, subacute or obtuse, crenate-serrate, glabrous or scabrid above, densely so beneath and prominently nerved, dotted with globose, orange-red oil globules on both surfaces, rather fleshy; petioles 0.6–6.6 cm long, puberulous. Panicles diffusely branched; spikes subglobose to oblong-ovoid, 1.5–5.0 cm long, becoming larger and strobilate in fruit. Floral leaves broadly ovate, 4 × 4 mm, truncate at base, similarly dotted as leaves. Calyx tubular; tube to 2 mm long, scaberulous without, glabrous within; upper lip triangular-lanceolate, to 2.5 × 2 mm, acute; lower lip slightly truncate. Corolla pinkish-purple, to 1.4 cm long; tube 4–9 mm long, narrowly tubular, scaberulous and similarly dotted as leaves; upper lip to 4 mm long, lobes shallowly orbicular; lower lip broadly oval, to 3 mm long. Stamens included in lower lip of corolla; filaments glabrous. Style glabrous; branches sub-equal. Nutlets ovoid to suborbicular, 1 mm long, compressed, dark brown and shining.

Distr. Central & South India, Ceylon & Burma.

Ecol. Rooted in niches and hollows of exposed rocks in the dry lowlands and intermediate zone. Fairly common. The pinkish-red coloration of fresh stems is quite distinct. The stems are eaten by goats. Flowering January–March; June.

Note. Sprengel based *P. dubius* on *Lavandula carnosa* L. f. and hence should have employed the Linnaean specific epithet.

Specimens Examined. JAFFNA DISTRICT: Karaveddy, roadside scrub, *Balakrishnan 639* (PDA, US). KURUNEGALA DISTRICT: Tittawela, in fizzures of rocky outcrop near tank, *Cramer 3597* (PDA, US). MATALE DISTRICT: Dambulla, *Gardner C.P. 2055* (K, PDA), *Alston 1332*

(PDA). TRINCOMALEE DISTRICT: Gantalawa, on rocky outcrops, *Jayasuriya 106* (PDA, US); Mutur, Eriyativumalai knoll, on rocky slopes, *Jayasuriya 103* (PDA, US). AMPARAI DISTRICT: Pottuvil, in vicinity of Navataru, *Jayasuriya 2027* (PDA, US); Lahugala, in open rocky crevices close to tank, *Cramer 3652* (PDA, US); Damana, on exposed rocks, *Kostermans 24378* (BO, K, NBV, PDA, US).

2. Anisochilus paniculatus Benth. in DC., Prod. 12: 82. 1848; Thw., Enum. Pl. Zeyl. 238. 1860; Hook. f., Fl. Br. Ind. 4: 62. 1885; Trimen, Handb. Fl. Ceylon 3: 377. 1895; Mukerjee, Rec. Bot. Surv. Ind. 14 (1): 59. 1940; Abeywick., Ceylon J. Sci., Biol. Sci. 2 (2): 219. 1959. Type: *Walker s.n.*, Ceylon, in Hb. Hook. (holotype, K).

Semi-shrubby perennial; stem to 1.72 m high, much branched from base, subquadrangular, marked with long leaf scars and scabrid below, tomentose with deflexed hairs above. Leaves broadly oval to suborbicular, 3.3–10 × 2.6–8.5 cm, rounded at both ends, coarsely crenate, scaberulous on both surfaces, dotted beneath with globose, orange-red oil globules, fleshy; petioles 0.5–2 cm long, fairly thick. Branches of panicles cymose; peduncles 6–14 cm long; spikes ovoid-oblong, 0.6–1.5 cm long. Calyx to 2.75 mm long; tube inflated, 1 mm long; upper lip erect, oblong-ovoid, 1.75 × 1.5 mm, subacute, strongly ciliate, glabrous above, gland-dotted beneath as in leaves. Corolla to 7 mm long, white; tube to 4 mm long, narrowed below, inflated above, glabrous within, hirtellous without and dotted as in calyx; upper lip erect, 3.5 mm long, the lateral lobes very small, the median one larger, suborbicular, finely hirtellous without; lower lip navicular, 3 mm long, slightly deflexed, pubescent without with deflexed hairs and gland-dotted as in calyx. Stamens included in lower lip of corolla; filaments glabrous. Disk projected on anterior side. Style glabrous; branches linear. Nutlets ellipsoid, 0.8 × 0.5 mm, subcompressed, subtrigonous, glabrous, pale brown.

Distr. Central India & Ceylon. On exposed rocks in the midlands to c. 825 m. Up to the present the species is not known from any other area of the country. Rare.

Ecol. As in the others of the genus in general, this species, particularly, has its leaves rolled inwards almost the entire day except the early mornings. Flowering August–September.

Specimens Examined. KANDY DISTRICT: Kadugannawa, *Ferguson s.n.* (PDA, US); Palu Kadugannawa, on bare sloping rock, *Cramer 3839* (PDA, US). LOCALITY UNKNOWN: *Mrs. Walker s.n.* (K).

3. Anisochilus velutinus Trimen, Handb. Fl. Ceylon 3: 377. 1895; Abeywick., Ceylon J. Sci., Biol. Sci. 2 (2): 219. 1959. Type: *s. coll. C.P. 3573*, Gunner's Quoin, Ceylon (holotype, PDA).

Anisochilus suffructicosus sensu Thw., Enum. Pl. Zeyl. 238. 1860 (non Wight, 1850).

Semi-shrubby perennial; stem to 1.75 m high, dichotomously branched, very bluntly 4-angular, sparsely velutinous with retrorse hairs, marked with leaf-scars. Leaves oval to oval-oblong, 4–7 × 1.8–3.2 cm, slightly tapering to base, obtuse, crenate, glabrous or scabrid above, velvety-tomentose beneath and dotted with globose, orange-red oil droplets; petioles 1.8–3 cm long. Panicles terminal, diffusely branched; spikes oblong, 0.7–1.5 cm long. Calyx minutely strigose without; upper lip hooded, ovate-lanceolate, usually obtuse; lower one obsolete. Corolla to 7 mm long, white; tube to 2.5 mm long, villulose without and dotted as in leaves; upper lip short, lobes shallowly ovate; lower one navicular, pale pink. Stamens shortly exserted; filaments glabrous. Nutlets orbicular, 1 × 1 mm, subtrigonous, glabrous, pale yellowish-brown.

Distr. Endemic.

Ecol. On exposed rocks in the dry lowlands; rather rare. Flowering March; July. The flowering season seems to be short, as the spikes are usually encountered in fruit.

Note. Wight's species is distinctly a South Indian plant with a ferruginous-tomentose indumentum and is not found in Ceylon.

Specimens Examined. POLONNARUWA DISTRICT: Mannampitiya, Gunner's Quoin, *s. coll. C.P. 3573* (PDA), open rocky ledges, *Cramer 3640* (PDA, US). ANURADHAPURA DISTRICT: Ritigala, round foot of cairn at summit, *24 Mar. 1905*, *Willis s.n.* (PDA), on rocky ledge, *Cramer 3836* (PDA, US). BADULLA DISTRICT: Maragala, top of rock, *Alston s.n.* (PDA, US). MONERAGALA DISTRICT: Nilgala, *s. coll. s.n.* (PDA). LOCALITY UNKNOWN: *s. coll. s.n.* (PDA).

8. HYPTIS

Jacq., Collect. 1: 101, 103. 1787 (nom. cons.); Benth., Lab. Gen. Sp. 64. 1832; Benth. & Hook. f., Gen. Pl. 1178. 1876; Endl., Gen. Pl. 610. 1838; Briq. in Pflanzenfam. 4 (3a): 333. 1897. Type species: *H. capitata* Jacq. (typ. cons.).

Mesosphaerum R. Br., Hist. Jamaica 257. 1756. (nom. rej.).

Herbs, often aromatic. Leaves opposite, petiolate, unequally serrate. Inflorescence various, often spicate or capitate, or occasionally of few-flowered, cymose racemes. Floral leaves sometimes present. Bracts subulate. Calyx tubular or campanulate, 10-nerved; limb subequally 5-lobed, the lobes subular or toothed. Corolla tube ± length of calyx; limb subbilabiate, spreading, resupinate; upper or posterior lip (now inferior) saccate, 3-lobed, the mid-lobe abruptly deflexed; lower one (now superior) 2-lobed. Stamens 4, declinate, slightly exserted; filaments pubescent; anthers 2-celled, the cells confluent.

Disk symmetrical. Style shortly 2-fid at apex. Nutlets ovoid or oblong, smooth or punctate-rugose.

About 400 species native to South America and the West Indies; of these, about six are adventive to the palaeotropics.

KEY TO THE SPECIES

1 Stem hirsute; base of leaf distinct from petiole; flowers in fascicles or in cymose racemes; calyx tube glabrous within, lobes aristate; corolla blue; aromatic herbs.......... ....**1. H. suaveolens**

1 Stem puberulous; base of leaf decurrent on petiole; flowers in globose heads; calyx tube hirtellous-annulate within; lobes subacute; corolla white; herbs without any noticeable scent..**2. H. capitata**

1. Hyptis suaveolens (L.) Poit., Ann. Mus. Natl. Hist. Nat. 7: 472. t. 29, f. 2. 1806; Benth. in DC., Prod. 12: 126. 1848; Hook. f., Fl. Br. Ind. 4: 630. 1885; Petch, Ann. R. Bot. Gard. (Peradeniya) 5: 338. 1914; Alston in Trimen, Handb. Fl. Ceylon 6: 235. 1930; Mukerjee, Rec. Bot. Surv. Ind. 14 (1): 63. 1940.

Ballota suaveolens L., Syst. Nat. ed. 10. 1100. 1759. Type: *Brown s.n.*, Specimen 737: 6 (LINN).
Bysteropogon graveolens Blume, Bijdr. 824. 1826.

Erect, much-branched undershrub; stem to 2.8 m high, obtusely 4-angular, patently hirsute. Leaves broadly ovate to suborbicular, 3.2–9.5 × 2.6–8.9 cm, lower ones subcordate, upper ones acute at base, all obtusely acute, subrugose and sparsely pilose above, densely so beneath and dotted with oil globules, shallowly crenate-serrulate; petioles distinct, 1.5–4.4 cm long. Flowers axillary, in cymose racemes to 3.6 cm long, or in fascicles. Floral leaves orbicular, to 1.3 × 1.3 cm. Bracts subulate, 1.5 mm long. Calyx deflexed, campanulate; tube 2.5–3 mm long, strongly 10-nerved, hispidulous on nerves, sparsely pilose without and dotted with oil globules at base, glabrous within; mouth oblique, ciliate; lobes erect, subulate-lanceolate, 1.5–2 mm long, aristate. Corolla pale bluish-violet; tube tubular, 4.5–5.5 mm long, subventricose above, puberulous without, glabrous within; lobes of upper lip obliquely oblong, 1.5 mm long, with 2 purplish streaks at base of each; lateral lobes of lower lip obliquely oblong, 1.5 mm long, median one shorter, saccate at base. Stamens inserted towards middle of corolla tube, filaments villous; anther-cells separating with age. Style glabrous; branches unequal, oblong-spathulate. Nutlets oblong, to 3 × 2 mm, compressed, with a central ridge on either face, pubescent with mucilaginous hairs, the areolae white.

Distr. A native of Tropical America, now a pantropic weed by introduction.

Ecol. A fast-spreading weed in open waste places beside roadsides, bunds of tanks and in coconut estates of the dry and wet lowlands and uplands to c. 525 m. It seems to have a preference for dry soils and is becoming obnoxious to coconut plantations particularly. Flowering throughout the year.

Specimens Examined. POLONNARUWA DISTRICT: Giritale, in open waste places close to bank of tank, *Cramer 3684* (PDA, US); Dambulla, on open rocky slope at foot of Rock Temple, *Cramer 3548* (PDA, US). PUTTALAM DISTRICT: Tabbowa, in open grassy bund of tank, *Cramer 4076* (PDA, US). KURUNEGALA DISTRICT: Wenaru Wewa, on bund of tank, *Townsend 73/75* (K, PDA, US); Wariyapola, in coconut plantation, *Cramer 4645* (NVB, PDA, SING, US). KALUTARA DISTRICT: Panadura, waste ground, *Alston s.n.* (PDA). RATNAPURA DISTRICT: Elapata, along roadside between Elapola-Pelawela road, *Cramer 3110* (PDA, US). GALLE DISTRICT: Pitiduwa, Habaraduwa, under shade of palms in coconut grove, *Cramer 3424* (PDA, US). HAMBANTOTA DISTRICT: Kumana, on beach, *Cooray 19* (PDA, US). TRINCOMALEE DISTRICT: Tekiluttu, by roadside, *Cramer 4364* (BO, K, PDA, US). BATTICALOA DISTRICT: Kalkudah, by Valaichchenai Road, *Cramer 4385* (K, PDA, US).

2. Hyptis capitata Jacq., Ic. Pl. Rar. 1: 11, t. 114. 1781–1786; Benth. in DC., Prod. 12: 106. 1848; Mukerjee, Rec. Bot. Surv. 14 (1): 63. 1940; Senaratna, Ceylon J. Sci., Sect. A, 12: 163. 1945. Type: Jacq. Ic., l.c.

Perennial herb or undershrub; stem erect, to 2.80 m high, much branched, robust and somewhat woody below, to 1.2 cm in diam., with a hollow pith, fistular, shallowly furrowed on opposite sides, puberulous. Leaves ovate to oval-elliptic, 4.5–29.5 × 2–9.8 cm, narrowed and decurrent on petiole, acute, serrate-crenate, scabrid to subglabrous on both surfaces, copiously dotted with oil globules beneath; petiole 0.5–4.4 cm long. Heads globose, 1.2–2.5 cm in diam.; peduncles to 4.5 cm long; floral leaves involucrate, oblong to oblong-lanceolate, 6–8 × 2–2.5 mm, scaberulous above, strigose beneath, long-ciliate at basal margins; receptacle covered with long, white hairs. Calyx campanulate; tube 1.5–3 mm long, 10-nerved, hirsute at base; in fruit to 7 mm long, gibbous, reticulated and dotted with oil globules; lobes erect, linear-subulate, 1.5–2 mm long, subacute, scaberulous without. Corolla to 4.5 mm long, white; tube infundibular, 3.75 mm long, sparsely villous without towards mouth; median lobe of upper lip spathulate, 1.5 mm long; lower lip saccate. Stamens shortly exserted; anterior filaments thinly pubescent in lower half, posterior ones villous at base; anther-cells confluent. Style glabrous; branches equal. Nutlets oval-oblong, 1.5 × 1 mm, subtrigonous, blackish-brown, shining; areola basal, white.

Distr. Native of Tropical America, now spread in the Indo-Malesian region.

E c o l. A weed under marginal shade in waste ground close to ditches and watercourses, and along roadsides, in the wet lowlands and lower uplands to c. 600 m. A recent introduction, first collected in 1953. Cattle are known to eat the young plants. Flowering January–September. The dried heads turn black, persist for a long time and make the fruiting branches conspicuous.

U s e s. The dried heads with the peduncles form popular objects of interior decor.

S p e c i m e n s E x a m i n e d. POLONNARUWA DISTRICT: Minneriya, *Comanor 1214* (PDA, US); Polonnaruwa, near Smithsonian camp, *Ripley 402* (PDA, US), *Kundu & Balakrishnan 213* (PDA, US). KURUNEGALA DISTRICT: Wenaruwewa, on bund of tank, *Townsend 73/75* (K, PDA, US). COLOMBO DISTRICT: Veyangoda, *Senaratne s.n.* (PDA). KALUTARA DISTRICT: Panadura, roadside, in coconut plantation, *Comanor 998* (E, PDA); Kalutara, roadside, beside Kalawana-Badureliya road, *Cramer 4288* (K, M, PDA). KANDY DISTRICT: Kadugannawa, Poilkande, under shade close to habitation, *Cramer 4216* (E, K, PDA); Galagedera, *Jayasuriya et al. 1402* (PDA, US); Kandy town, Mahamaya Mawatha, *Jayasuriya & Sumithraarachchi 1403* (PDA, US); Katugastota, Medawela, in marginal shade along Botota road, *Cramer 3920* (PDA, US). MATALE DISTRICT: Dambulla-Trincomalee road, *Comanor 731* (PDA, US). RATNAPURA DISTRICT: between Balangoda and Opanaike, *Comanor 1109* (E, K, PDA); Gilimale, along roadside on Kalawana-Weddagala road, *Cramer 3975* (PDA, US). KEGALLE DISTRICT: Kitulgala, under shade in damp ground close to river, *Cramer 4216* (K, NBV, PDA, US); Colombo-Kandy road, *Balakrishnan 932* (PDA, US). GALLE DISTRICT: Habaraduwa, Hadiwatte, under shade in coconut grove, close to sea, *Cramer 3431* (PDA, US). HAMBANTOTA DISTRICT: Kumana, on beach, *Cooray 19* (PDA, US). AMPARAI DISTRICT: Gal-Oya, dirt road to Siyambalanduwa, *Comanor 585* (PDA, US). Gal-Oya Reservoir, *Comanor s.n.* (K, PDA, US).

9. SCUTELLARIA

L., Sp. Pl. 598. 1753; L., Gen. Pl. ed. 5, 260. 1754; Benth. in Wall., Pl. As. Rar. 1: 66. 1831; Benth. in DC., Prod. 12: 412. 1848; Benth. & Hook. f., Gen. Pl. 1201. 1876; Endl., Gen. Pl. 620. 1838; Briq. in Pflanzenfam. 4 (3a): 225. 1897. Type species: *S. peregrina* L. (lectotype).

Annual or perennial herbs with decumbent, sometimes erect stems. Leaves petiolate, entire or serrate at margins, becoming bracteate above. Spikes or racemes terminal; flowers opposite. Floral leaves 0. Calyx campanulate; limb 2-lipped, closed after anthesis; lips entire, the upper one with a dorsal scale ("scutellum") finally deciduous with the posterior part of limb, the lower one persistent. Corolla tube long, much exserted beyond calyx, recurved at base, expanded above at throat, exannulate within; limb 2-lipped; upper lip erect, hooded, entire or emarginate; lower one spreading, convex or reflexed, 3-lobed, the lateral lobes ± free or united with the median one. Stamens didynamous, included under the corolla upper lip; filaments often pubescent at base; anthers connivent, those of anterior stamens 1-celled. Disk symmetrical. Style very unequally bifid at apex, the posterior branch very short. Nutlets subglobose, areolus basal.

KEY TO THE SPECIES

1 Leaves ovate, truncate or cordate at base; stem and lower surface of leaves pubescent
 2 Bracts shorter than pedicels; flowers lax
  3 Leaves ovate to ovate-lanceolate, finely pubescent or glabrous beneath; racemes 7–29 cm long; corolla tube blue
   4 Leaves acute at apex, the lower surface entirely puberulous or scaberulous to scabrid on nerves; calyx tube glabrate .... **1. S. violacea**
   4 Leaves acuminate at apex, the lower surface entirely glabrous; calyx tube glabrous without .... **S. violacea** var. **glabra**
  3 Leaves broadly ovate to orbicular, hirsute beneath; racemes to 7 cm long; corolla tube whitish .... **S. violacea** var. **hirsuta**
 2 Bracts as long as or longer than pedicels; flowers closely placed .... **2. S. robusta**
1 Leaves linear-oblong, tapering to base; stem and lower surface of leaves glabrous .... **3. S. oblonga**

1. Scutellaria violacea Heyne ex Benth. in Wall., Pl. As. Rar. 1: 66. 1830; Benth., Lab. Gen. Sp. 429. 1833; Benth. in DC., Prod. 12: 418. 1848 (excl. syn. *S. robusta* Benth.); Thw., Enum. Pl. Zeyl. 239. 1860; Hook. f., Fl. Br. Ind. 4: 668. 1885; Trimen, Handb. Fl. Ceylon 3: 382. 1895; Mukerjee, Rec. Bot. Surv. India 14 (1): 145. 1940; Abeywick., Ceyl. J. Sci., Biol. Sci. 2 (2): 219. 1959. Type: *s. coll.*, India, Wall. Num. List 2136/1, in Hb. Heyne, (holotype, K-WALL).

Scutellaria floribunda Benth. in DC., Prod. 12: 418. 1848; Hook. f., Fl. Br. Ind. 4: 668. 1885. Type: *Mrs. Walker s.n.*, Ceylon, in Hb. Hook. (holotype, K).

1. var. **violacea**

Annual; stem usually erect, to 60 cm high, with a rootstock of slender tuberous roots, sometimes decumbent, subquadrangular, pubescent with minutely retrorse hairs. Leaves ovate to oval-lanceolate, 2.0–7.3 × 1.5–4.8 cm, truncate to rounded at base, acute, coarsely crenate-serrate or subdentate, sparsely scaberulous to scabrid above, entirely puberulous beneath or scabrid on nerves; petioles 1.8–5 cm long. Racemes terminal, simple, lax, 7–18.0 cm long; peduncles hirtellous with gland-tipped hairs. Bracts linear-oblong, 1.75–2.5 × 1–2 mm, obtuse, hirtellous without with gland-tipped hairs. Pedicels 3–4 mm long, hirtellous with gland-tipped hairs, dull purple. Calyx tube broadly campanulate, to 2.25 mm long, finely hirtellous without; dorsal scale broadly ovate, 2 mm long, dull purple, finally deciduous; lips truncate. Corolla 1.5–1.7 cm long, bright purplish-blue; tube 10–11 mm long, recurved towards base, gradually widening to throat, villous-annulate within, hirtellous without with gland-tipped hairs; upper lip 5 mm long, the central lobe arched, emarginate; lower lip suborbicular, to 8 mm across, with a white blotch at base. Filaments finely pubescent. Style glabrous; branches linear. Nutlets ovoid-globose, to 1.75 mm long, minutely tuberculate, pale brown.

Distr. South India and Ceylon.

Ecol. Usually under shade in damp vicinity of streams and watercourses of submontane and montane forests of the lower and higher uplands, from c. 600–1400 m. Very common. Variable in the habit of the stem and the size of the leaves which are often purplish beneath. In full bloom this is one of the prettiest of our labiates. Flowering throughout the year.

Note. From examination of the type of *S. floribunda* it is clear that this species, except for the branching of its main peduncle towards the base, is not different from *S. violacea*. The flowers of the former are not exceptionally smaller than those of the latter. These differences being trivial, I concur with Trimen in considering the two taxa conspecific.

Specimens Examined. KANDY DISTRICT: Hunnasgiriya, along bank of forest stream, *Cramer 4504* (PDA, US); Maskeliya, Rajamally Estate, under shade in damp floor of secondary montane forest, *Cramer 4391* (BO, K, MH, NBV, PDA, US); Rajamally, Moray Estate, under partial shade close to stream in secondary forest, *Cramer 4275* (MH, PDA, US); Adam's Peak, *s. coll. C.P. 217* (PDA), in damp undergrowth of secondary forest close to stream, *Cramer 4044* (E, NBV, PDA, US); Girindiella Tea Estate, *Lazarides 7215* (K, PDA, US). NUWARA ELIYA DISTRICT: Labookelle, in sloping ground along bank of stream, by Labookelle-Nuwara Eliya road, *Cramer 3823* (PDA, US); Horton Plains, under shade along border of jungle on way to World's End, *Cramer 4023* (BO, K, PDA, US). RATNAPURA DISTRICT: Dotalugala, under shade among rocks beside stream, *Cramer 4708* (K, PDA, US); Sinharaja Forest, near Heen dola, *Waas 1775* (PDA, US); Adam's Peak Sanctuary, along stream, *Maxwell et al. 920* (PDA, US). BADULLA DISTRICT: Haputale, Tangamalai, under shade, along stream bank in secondary forest, *Cramer 3821* (PDA, US); under shade, close to stream beside road to Adisham, *Cramer 3752* (PDA, US); *Mueller-Dombois 7* (PDA, US); Madawelagama, Pine Nursery, *Balakrishnan & Jayasuriya 814* (PDA, US). LOCALITY UNKNOWN: *Mrs. Walker s.n.* (K), *s. coll. C.P. 2716* (K), *Gardner 1847* (K), *s. coll. C.P. 3370* (K), *s. coll. C.P. 527* (K), *Mrs. Walker s.n.* (K).

2. var. **glabra** Trimen, Cat. 70. 1885; Trimen, Handb. Fl. Ceylon 3: 382. 1895;

Thw., Enum. Pl. Zeyl. 240. 1864 (ut var. β). Type: *s. coll. C.P. 3370*, Ambagamuwa, Ceylon, (holotype, PDA, isotype, K).

Stem completely glabrous. Leaves oval, 3.8–10 × 1.9–5.2 cm, rounded at base, acuminate, coarsely dentate, glabrous to sparsely scaberulous above, completely glabrous beneath; petioles 0.8–3.1 cm long, slender, subglabrous. Racemes simple, 12.5–28.5 cm long; peduncles and pedicels finely puberulous. Calyx tube glabrous or sparsely puberulous without.

D i s t r. Endemic to the lower uplands of Ceylon. Rare.

E c o l. Flowering February.

N o t e. Thwaites' citation is without a legitimate varietal epithet. Hence the need of Trimen's epithet. The description given above is based on type material.

S p e c i m e n s E x a m i n e d. RATNAPURA DISTRICT: Adam's Peak Sanctuary, Warnagala, under shade beside stream, *Cramer 4694* (K, SING, PDA, US). *Maxwell et al. 926* (PDA, US).

3. var. **rotunda** Cramer, nom. nov. Thw., Enum. Pl. Zeyl. 240. 1860 (ut var. γ). Type: *Gardner C.P. 302*, Horton Plains, Nuwara Eliya, Ceylon (holotype, PDA).

Annual; stem erect, to 27 cm high, slender, sometimes rooting at lower nodes, densely hirsute-hispid with spreading hairs. Leaves broadly ovate to rotundate, 1.2–2.6 × 1–2.4 cm, truncate at base, subacute, very hirsute on both surfaces; petioles to 2.4 cm long. Racemes to 7 cm long. Calyx tube hirsute without. Corolla to 8 mm long; tube hirtellous without, white; lower lip to 7 mm across, pale blue. Filaments glabrous.

D i s t r. Endemic to the highlands of Ceylon.

E c o l. On floor of secondary montane forest, often growing among moss, at c. 2400 m; restricted to Horton Plains. Rather rare. Flowering September–October; March–May.

N o t e. Thwaites' variety (in symbol) needs a legitimate name. Agreeing with Thwaites, I am inclined to consider this variety a distinct species. But I am suspending this judgement till I am able to satisfy myself on the constancy of its characters by observing the plant under cultivation.

S p e c i m e n s E x a m i n e d: NUWARA ELIYA DISTRICT: Horton Plains, trail to World's End, *Gould & Cooray 13814* (US, PDA), on floor of dense mossy forest, *Fosberg 50035* (US, PDA), on forest floor close to stream near Little World's End, *Cramer 4600* (CAL, E, K, NBV, PDA, US).

2. Scutellaria robusta Benth. in DC., Prod. 12: 418. 1848; Trimen, Handb. Fl. Ceylon 3: 383. 1895; Abeywick., Ceylon J. Sci., Biol. Sci. 2 (2): 219. 1959. Type: *Walker s.n.*, Ceylon, in Hb. Hook. (holotype, K).

S. violacea var. *hispidior* Benth., Lab. Gen. Sp. 420. 1832; Benth. in DC., Prod. 12: 418. 1848; Hook. f., Fl. Br. Ind. 4: 688. 1885 (excl. syn. *S. wightiana* Benth.).

Perennial herb? Stem to more than 36 cm high, subquadrangular, densely villous-hirsute. Leaves broadly ovate to ovate-oblong, 3.1–7.2 × 1.7–4.5 cm long, subcordate at base, subacuminate, coarsely crenate, glabrous to sparsely scaberulous above, scabrid to hirtellous on nerves beneath; petioles 1.2–4 cm long. Racemes 4–9 cm long; peduncles densely hirsute; flowers closely placed. Bracts broadly ovate, to 5 mm long, hirtellous beneath. Pedicels 3–4 mm long. Calyx tube to 2.5 cm long, hirtellous without. Corolla to 1.5 cm long, dark bright purple; tube widened towards throat, villous-puberulous without.

Distr. Endemic.

Ecol. Close to streams in montane forest at c. 330 m; rare. In spite of several attempts I have not succeeded in locating this species. Flowering October–March.

Note. Hooker f. and Trimen have confused the distribution of this species with that of *S. wightiana* Benth. mistaking the two taxa to be conspecific. Examination of type material of *S. wightiana* showed that it is, generally, a smaller plant and further distinct in the gland-tipped hairs of its peduncle and the white or rose-coloured corollae. *S. wightiana*, a South Indian species, does not appear to occur in Ceylon.

Specimens Examined. KANDY DISTRICT: Adam's Peak, *s. coll. s.n.* (PDA), in jungle, *Alston 970* (PDA). LOCALITY UNKNOWN: *Walker s.n.* (K).

3. Scutellaria oblonga Benth. in Bot. Reg. sub t. 1493. 1832; Benth., Lab. Gen. Sp. 429. 1834; Benth. in DC., Prod. 12: 418. 1848 (cum var.); Thw., Enum. Pl. Zeyl. 240. 1860; Hook. f., Fl. Br. Ind. 4: 668. 1885; Trimen, Handb. Fl. Ceylon 3: 383. 1895; Abeywick., Ceylon J. Sci., Biol. Sci. 2 (2): 219. 1959. Type: *Macrae s.n.*, in Hb. Lindl., Central Province, Ceylon (n.v.).

Perennial? Stem erect, to 42 cm high, occasionally rooting at lower nodes, branched above and dark purple, rather woody at base with leaf scars, acutely 4-angular, glabrous. Leaves linear-oblong to oblong-lanceolate, 2–5.4 × 0.6–1 cm, tapering to base, subacute, distantly crenate-serrate, glabrous, glandular-punctate on both surfaces; petioles 0.4–1 cm long. Racemes 3.4–17 cm long, peduncles glabrous, dull purple. Bracts oblong-lanceolate, 2–4 × 1.5–2 mm, truncate at apex, glabrous on both surfaces, slightly concave. Pedicels 2–3.5 mm long, glabrous. Calyx tube 2 mm long, dotted without with oil globules, pale greenish-purple; lips broadly orbicular; dorsal scale orbicular, 1.5 × 1.5 mm, dull purple. Corolla to 1.4 cm long; tube to 8 mm long, villous without with gland-tipped hairs, whitish-purple; upper lip to 5.5 mm long, the central lobe broadly orbicular, faintly emarginate, purplish-blue, lateral ones ovate, smaller, darker purple; lower lip quadrate-orbicular, 6–8 × 7 mm, faintly 3-lobed, dark purple with a whitish blotch at base. Anterior filaments sparsely villous throughout, posterior ones villous below; anther-cells villulose at margins.

Distr. Endemic.

Ecol. Along streams in montane forests, from c. 1100–2300 m. Common during the time of Trimen, now becoming less common. A pretty plant when in bloom. Flowering throughout the year.

Specimens Examined. KANDY DISTRICT: Hunnasgiriya, *s. coll. C.P. 2057* (MH, PDA); Maskeliya, *May 1891, s. coll. s.n.* (PDA); near Gartmore Estate, *Silva 105* (PDA); Adam's Peak, at foot, *May 1891, s. coll. s.n.* (PDA); Adam's Peak, *s. coll. s.n.* (MH). NUWARA ELIYA DISTRICT: Nuwara Eliya?, *s. coll. s.n.* (PDA); Hakgala, on rocks below river, *30 April 1906, Silva s.n.* (PDA). MATALE DISTRICT: Rattota, beside stream. *Cramer 4871* (PDA, US). LOCALITY UNKNOWN: *Wight 757* (MH).

10. EUSTERALIS

Raf., Fl. Tellur. 2: 95. 1836. Type species: *E. pumila* Raf.

Dysophylla sensu El-Gazzar & Watson ex Airy-Shaw in Taxon 16: 190. 1967 non Bl., nom. illeg.

Annual or perennial herbs; stem erect, aerenchymatous. Leaves verticillate, 3–12 in a whorl, sessile, linear, glabrous or pubescent. Verticils many-flowered, in terminal, dense, narrow spikes. Floral leaves whorled. Flowers very small. Calyx ovoid-campanulate, shortly 4-lobed; throat glabrous within. Corolla tubular, exserted or included in the calyx; limb subequally 4-lobed. Stamens 4, unequal, exserted, straight or declinate; filaments bearded at middle; anthers 2-celled, the cells confluent. Disk unequal. Style 2-fid, glabrous; branches equal. Nutlets ovoid or oblong, smooth or granulate.

21 species, in temperate and tropical regions, chiefly in South-East Asia; locally represented by only 1 species.

El-Gazaar & Watson (1967) confirmed the distinction between the genera *Pogostemon* and *Dysophylla*, and transferred to *Pogostemon* four species of *Dysophylla*, including *D. auricularia* (L.) Blume, the type species. This leaves *Dysophylla* as a monotypic genus, which now needs a new name. In the circumstances, the best procedure would be to conserve *Dysophylla* Benth. with retypification for the sect. *Verticillatae* Benth., against Blume's name. While obviating nomenclatural instability through another generic name, this would obviously create a later homonym. Until such a proposal for conservation is formally approved, I here adopt the next valid synonym, viz., *Eusteralis* Raf., for the unnamed genus.

Eusteralis stellata (Lour.) Majumdar, J. Bombay Natl. Hist. Soc. 74. 385. 1977.

Dysophylla stellata (Lour.) Benth. in Wall., Pl. As. Rar. 1: 30. 1830; Benth., Lab. Gen. Sp. 158. 1833; Benth. in DC., Prod. 12: 158. 1848; Hook. f., Fl. Br. Ind. 4: 640. 1885, quoad nom. in omnibus.

Mentha stellata Lour., Fl. Cochinch. 2: 361. 1790. Type: *Loureiro s.n.*, Cochinchina (holotype, BM).

Dysophylla verticillata (Roxb.) Benth. in Wall., Pl. As. Rar. 1: 30. 1830; Benth., Lab. Gen. Sp. 159. 1833; Benth. in DC., Prod. 12: 157. 1848; Thw., Enum. Pl. Zeyl. 239. 1860 (excl. syn.); Hook. f., Fl. Br. Ind. 4: 639. 1885; Trimen, Handb. Fl. Ceylon 3: 380. 1895; Mukerjee, Rec. Bot. Surv. Ind. 14 (1): 81. 1940; Abeywick., Ceylon J. Sci., Biol. Sci. 2 (2): 219. 1959.

Mentha verticillata Roxb., Hort. Beng. 44. 1814; Roxb., Fl. Ind. ed. 2, 3: 5. 1832 (non. 1759). Type: Roxb. Icon. No. 957, India, based on plant from Calcutta (K, CAL).

Annual with a fibrous rootstock; stem to 85 cm tall, terete, succulent, glabrous, thickened below to 2.3 cm in diam., narrowed and vertically 4-ridged above, much branched; internodes to 9.5 cm long. Leaves verticillate, sessile, spreading, 4–12 in a whorl, becoming reduced in number on branches, linear, 2–16 × 0.2–1.3 cm, narrowed at base, very acute at apex, sharply serrate, glabrous on both sides, glandular-punctate beneath with a prominent mid-nerve. Spikes terminal, linear, 2.0–12 × 0.2–0.8 cm; verticils dense. Floral leaves linear-lanceolate, 1.5–2.5 × 0.5 mm, fairly conspicuous, hirtellous without. Calyx broadly campanulate, 1 mm long; tube slightly expanded above, hispidulous and minutely dotted without with oil globules, glabrous within; lobes subequal, triangular-ovate, 0.5 mm long, acute, long-ciliolate, spreading stellate-wise in fruit. Corolla to 1.75 mm long, pale rose or lilac; tube 1 mm long, glabrous on both sides, whitish; lobes ± equal, ovate, long-ciliolate. Stamens ± equal, exserted; filaments flattened, tufted with villous-moniliform hairs towards middle; anther-cells globose, the cells contiguous. Style flattened; branches capitate. Nutlets globose, to 0.7 mm long, glabrous, dull brown.

D i s t r. From India and Ceylon eastwards to S. China and Japan, and through Malesia southwards to Australia.

E c o l. Along edges of ditches and ponds in the dry lowlands to c. 100 m in altitude. If left undisturbed the plant grows in fairly large populations. In the field its wet habitat and the verticillate phyllotaxy of the leaves might at first glance suggest a species of *Limnophila*; but the verticillate inflorescence and nutlets together with the minute flowers and the filaments bearded at the middle confirm its belonging to *Dysophylla*. Adventitious plants occasionally arise from the base of the stem. Flowering Jaunary–April.

N o t e. This species has had a disturbed nomenclatural history. Placed till recently under *Dysophylla* it was named for a long time *D. verticillata* (Roxb.)

Benth. Later Merrill (Trans. Amer. Philos. Soc. 24 (2): 342. 1935) adopted the name *D. benthamina* Hance. While correctly citing *Mentha stellata* Lour. in synonymy he rejected the valid name *D. stellata* (Lour.) Benth., not recognizing that though Bentham's description was misapplied, his nomenclature was, nevertheless, valid. Bentham's descriptions and citations of *D. stellata*, and those of Hooker f., refer to an entirely different plant of South India (typified from Malabar and Mysore) not occurring in Ceylon, viz. *Dysophylla tomentosa* Dalz. But Bentham's name is perfectly legitimate.

Specimens Examined. POLONNARUWA DISTRICT: Polonnaruwa, in open marshy pond close to Gal Vihara, *Cramer 3682* (PDA, US); Minneriya road, *Townsend 73/223* (E, K, US). MATALE DISTRICT: Sigiriya, in large populations in open marshy ditches close to main drive, *Cramer 3673* (PDA, US). MONERAGALA DISTRICT: Bintenne, causeway, *Gardner C.P. 2082* (CAL, MH, PDA). BATTICALOA DISTRICT: Kolokonaveli, in open shallow pond, *Cramer 3996* (PDA, US); *Simpson 8354* (PDA). LOCALITY UNKNOWN: *s. coll. C.P. 2682* (P).

11. SALVIA

L., Sp. Pl. 23. 1753; L., Gen. Pl. ed. 5, 15. 1754; Benth., Lab. Gen. Sp. 190. 1832; Benth. in DC., Prod. 12: 263. 1848; Benth. & Hook., Gen. Pl. 1194. 1876; Endl., Gen. Pl. 614. 1838. Type species: *S. officinalis* L. (Lectotype).

Annual or perennial herbs. Leaves entire, occasionally lobed. Racemes, spikes or panicles terminal or axillary, verticils 2-many-flowered; flowers variously coloured. Floral leaves usually present, occasionally showy. Bracts present. Calyx tubular-campanulate, pubescent with gland-tipped or simple hairs; limb usually 2-labiate, the upper lip entire or 3-lobed, the lower 2-lobed; mouth closed after anthesis. Corolla usually exserted; tube straight or curved, glabrous or annulate within; limb 2-labiate; upper lip erect, hooded, entire or emarginate; lower spreading, the median lobe larger. Perfect stamens 2, enclosed in upper lip of corolla; filaments short; anther-cells sometimes both fertile, more usually only posterior one fertile, in either case one cell separated from the other by an elongated connective transverse to filament and articulating with it, the connective occasionally with a transverse process; staminodes very small or 0. Disk annular. Style unequally branched at apex. Nutlets ovoid, subtrigonous, smooth.

About 700 species in the temperate and tropical regions, very numerous in tropical America; 3 adventive species in Ceylon.

KEY TO THE SPECIES

1 Stem with a creeping rhizome, tumid above nodes; leaf base decurrent on petiole; calyx tube 10-nerved, with gland-tipped hairs without.......................... **1. S. riparia**

1 Stem erect, plain above nodes; leaf base distinct from petiole; calyx tube more or less than 10-nerved, with simple hairs without
 2 Margins of lower leaves variously lobed; both anther-cells fertile........**2. S. nilotica**
 2 Margins of lower leaves entire; 1 anther-cell fertile
  3 Verticils 8-more-flowered; calyx tube 9-nerved
   4 Leaves ovate; calyx tube strigose without; style glabrous..............**3. S. tiliifolia**
   4 Leaves lanceolate; calyx tube villous or cottony without; style bearded
    5 Leaves narrowly lanceolate; calyx limb 2-labiate; tube densely scabrid within; style bearded along 2 opposite sides..................................**4. S. leucantha**
    5 Leaves broadly lanceolate; calyx limb truncate; tube glabrous within; style bearded along one side..**5. S. farinacea**
  3 Verticils 6-less-flowered; calyx tube 8–13-nerved
   6 Floral leaves and calyx scarlet; calyx tube 8-nerved................**6. S. splendens**
   6 Floral leaves and calyx green; calyx tube 13-nerved.................**7. S. coccinea**

1. Salvia riparia Kunth, Nov. Gen. & Sp. 2: 300. 1818; Epling in Feddes Repert. 10: 16. 1939. Type: *Humboldt & Bonpland s.n.*, Peru (P-n.v.).

Perennial herb with a creeping rhizome; aerial stem rooting at lower nodes, ascending to 1.4 m, diffusely branched, acutely 4-angular, tumid above nodes, pubescent with retrorse hairs. Leaves ovate to subrhomboid, 2–8 × 1.2–3.8 cm, obliquely cuneate at base and decurrent on petiole, acute, bullate and scabrid above, strigose on nerves beneath, finely serrate; petioles 1–2 cm long. Racemes 17–27 cm long; verticils spicate, 1–12-flowered; peduncles viscid-hirtellous and partly puberulous. Floral leaves obliquely ovate, 4–4.5 × 2–2.5 mm, acuminate, hispid with gland-tipped hairs without. Pedicels 2 mm long. Calyx tube 3–4 mm long, 10-nerved, densely hispidulous on nerves with gland-tipped hairs; upper lip broadly ovate, 1.5 mm long, mucronate; lower lip 2-lobed, the lobes ovate, 1.5 mm long, mucronate. Corolla to 9 mm long; tube 3–4 mm long, whitish; limb indigo blue; upper lip 3 mm long, hispidulous without; lower lip 6 × 5 mm, median lobes larger than laterals. Filaments lgabrous; posterior anther-cell fertile; anterior end of connective clavate. Style glabrous; branches plano-concave, one acuminate, the other truncately rotund at apex. Nutlets obovoid, to 2 × 1.5 mm, subtrigonous, glabrous, pale brown; basal scar minute.

D i s t r. From Mexico to Peru; West Indies; Ceylon.

E c o l. An adventive weed along roadsides in partial shade in the lower uplands and dry lowlands from c. 200–300 m. The flowers are very small and easily wither once branches are cut. Flowering March–December.

N o t e. This species has not been encountered before in Ceylon and must, therefore, be credited as a new record for the country. Specimen *Cramer 3599* has been kindly determined by Mr. I.C. Hedge, Royal Botanic Garden, Edinburgh, to whom I tender my sincere thanks.

S p e c i m e n s E x a m i n e d. KURUNEGALA DISTRICT: Galagedera, roadside, under partial shade, *Cramer 3599* (E, K, PDA, US). MATALE DIS-

TRICT: Naula, roadside, *Cramer 3286* (PDA, US). BADULLA DISTRICT: Badulla-Passara Road, roadside, *Balakrishnan & Jayasuriya, 812* (PDA, US); Badulla-Dunhinda Road, *Balasubramaniam 1248* (NVB, PDA).

2. Salvia nilotica Juss. ex Jacq., Hort. Vindob. 3: 48, t. 92. 1776; Alston in Trimen, Handb. Fl. Ceylon 6: 235. 1931. Type: Jacquin, Descr. & Ic.

Salvia abyssinica Jacq., Ic. Pl. Rar. 1: 2, t. 6. 1781 (non L. f. 1781). Type: Jacquin, Descr. & Ic.

Herb with a fibrous root; stem erect, to ±60 cm high, acutely 4-angular, grooved on opposite sides, hirsute. Lower leaves petiolate, runcinate to somewhat pinnately lobed, 7.3–16 × 3.4–8 cm, obtuse at apex; petioles 2.2–4 cm long; upper leaves sessile, ovate, to 5 × 2.5 cm, truncate to subcordate at base, subacute; all crenate-serrate, hirsute on both surfaces, with raised nerves beneath. Racemes terminal and axillary, to 14.5 cm long; verticils 6–8-flowered, dense above; peduncles densely hirsute with gland-tipped hairs. Floral leaves broadly ovate, to 4 × 2.5 mm, aristate. Pedicels 1.5 mm long. Calyx erect, accrescent in fruit; tube 3.5 mm long, 11-nerved, ± hirsute without and dotted with oil globules; upper lip 3-lobed, short; lower one 5 mm long; lobes all lanceolate, spinous-aristate. Corolla to 10 mm long, purplish-blue; tube 5 mm long, glabrous on both surfaces; upper lip 4.5 mm long, emarginate; median lobe of lower lip largest, concave, the lateral ones orbicular. Filaments 1.5 mm long, glabrous; anther-cells both fertile. Style glabrous; branches unequally long and broad. Nutlets ellipsoid, 2 × 1.5 mm, subtrigonous, glabrous.

Distr. Eastern and Southern Africa; adventive in Ceylon.

Ecol. A roadside weed in the uplands to c. 1400 m. The lobed nature of the lower leaves is variable. In spite of a careful search I have not encountered this plant. It is probably extinct in Ceylon owing to clearings. Flowering January–September.

Specimens Examined. NUWARA ELIYA DISTRICT: below Hakgala, Sept., 1890, *s. coll. s.n.* (PDA), Hakgala, cultivated, Aug., 1881, *s. coll. s.n.* (PDA). BADULLA DISTRICT: Palugama, casual weed, Jan., 1888, *s. coll. s.n.* (PDA).

3. Salvia tiliifolia Vahl, Symb. Bot. 3: 7. 1794; Benth. in DC., Prod. 12: 299. 1848; Alston in Trimen, Handb. Fl. Ceylon 6: 235. 1931; Epling in Feddes Repert. 110: 241. 1939. Type: *Bellardi s.n.*, (n.v.).

Salvia fimbriata Kunth, Nov. Gen. & Sp. 2: 199, t. 149. 1818. Type: *Humboldt & Bonpland s.n.*, Venezuela, Caracas (P-n.v.).

Annual with a fibrous rootstock; stem to 1 m high, divaricately branched, acutely 4-angular, shallowly grooved on opposite sides, glabrous below, pubescent above with retrorse hairs. Leaves broadly rhomboid to suborbicular,

2.1–6.5 × 1.8–4 cm, truncate to rounded at base, subacute, crenate-serrate, subbullate and scaberulous above, densely so on nerves beneath; petioles 1.2–3.6 cm long, slender, sulcate and minutely villous. Racemes to 20 cm long; verticils lax, 7–8-flowered; peduncles 2 mm long, slender, pubescent with recurved hairs. Calyx tubular-campanulate; tube to 4 mm long, 5-angular, 9-nerved, strigose on ribs, dotted with oil globules in between ribs, glabrous within; upper lip broadly ovate, 2 mm long, acuminate-apiculate, lower 2 mm long, the 2 lobes acuminate-apiculate; mouth closed after anthesis. Corolla tube linear, to 3.5 mm long, glabrous on both surfaces, whitish-blue; upper lip 2 mm long, minutely villous without, dark blue, lower lip broadly orbicular, 2.5 × 2.5 mm, median lobe large, 2-fid, pale blue, with white streaks towards base, lateral ones smaller. One anther-cell fertile. Style glabrous; branches linear, recurved with age. Nutlets obovoid, 1.5 mm long, subtrigonous, glabrous, pale brown.

Distr. Native of Mexico spreading to North and South America; adventive in Ceylon.

Ecol. An adventive weed along open roadsides and borders of cultivated fields in the wet and dry uplands and the highlands, from c. 1300–2000 m, becoming widespread in the Badulla district. The plant quickly withers on uprooting. Flowering almost throughout the year.

Specimens Examined. NUWARA ELIYA DISTRICT: Hawa Eliya, *Silva s.n.* (PDA), Hakgala, border of cultivated field, *Cramer 3467* (PDA, US). BADULLA DISTRICT: Welimada, waste ground beside Haputale Road, *Cramer 4617* (BO, PDA, US).

4. Salvia leucantha Cav., Ic. & Descr. 1: 16, t. 24. 1791; Benth. in DC., Prod. 12: 321. 1848.

Shrubby herb; stem to ± 1.25 m high, diffusely branched, obtusely 4-angular, young parts tomentose and dull purple. Leaves lanceolate, 4.1–10 × 0.8–2.1 cm, rounded at base, acuminate, scaberulous above, cottony beneath; petioles 0.3–1.7 cm long. Racemes to 24 cm long; verticils 6–8-flowered; peduncles villous-cottony. Floral leaves lanceolate, to 10 × 3 cm, acuminate, villous-shaggy without with lavender hairs. Pedicels 2–3 mm long. Calyx tubular-campanulate, deflexed on pedicel; tube to 6 mm long, 9-nerved, villous-shaggy without, densely scabrid within. Corolla lavender-purple; tube to 12 mm long, slightly bent about the middle; lobes of lower lip incurved. One anther-cell fertile; anterior ends of connective flattened. Style bearded above in 2 opposite rows; branches very unequal, longer one curved outwards. Nutlets not seen.

Distr. Native to Mexico, spread in North America; widely cultivated throughout the world.

Ecol. Cultivated as an ornamental in gardens.

Specimens Examined. KANDY DISTRICT: Peradeniya, Botanic Gardens, *Alston 134* (PDA), *Cramer 4223* (E, PDA, US); Adam's Peak Range, Sitagangula, in garden of dispensary, *Cramer 3808* (PDA, US).

5. Salvia farinacea Benth., Lab. Gen. Sp. 274. 1832; Benth. in DC., Prod. 12: 302. 1848.

Perennial herb; stem to 1 m high, much branched from base. Leaves lanceolate to oblong-lanceolate, to 7.6 × 1.7 cm, acute at base, acute to subacuminate, undulate to distantly serrulate, glabrous above, puberulous along nerves beneath; petioles to 1 cm long. Racemes to 20 cm long; verticils 10–16-flowered, dense; peduncles hoary with appressed hairs. Bracts ovate-oblong, minute, caducous. Pedicels to 3 mm long. Calyx turbinate; tube to 6 mm long, glabrous within, lanate without, 9-nerved; limb truncate. Corolla lavendar-blue; tube 7–8 mm long, ventricose above; upper lip villous without; lower to 1 cm across, lateral lobes broad, each with a white streak towards base, median one smaller. Posterior anther-cell fertile; anterior ends of connectives flattened. Style bearded down one side towards the top; longer branch exserted beyond upper lip of corolla, acuminate, recurved.

Distr. Native to Texas, spread further in North America; cultivated in Ceylon.

Ecol. Cultivated as an ornamental in gardens.

Specimens Examined. KANDY DISTRICT: Kandy, Hutt's Garden, *Macmillan s.n.* (PDA); Peradeniya, Botanic Gardens, *s. coll. s.n.* (PDA), *Cramer 4222* (E, PDA, US).

6. Salvia splendens Sellow ex Roem. & Schult., Mant. 1: 185. 1822; Benth. in DC., Prod. 12: 330. 1848.

Annual; stem to 1 m high, glabrous, nodes purplish with hirsute hairs. Leaves ovate, to 7 × 4.5 cm, rounded at base, acuminate, serrate-crenate, glabrous on both surfaces, glandular-punctate beneath; petioles to 3.2 cm long. Racemes to 19 cm long; verticils 4–6-flowered; peduncles puberulous with scarlet hairs. Floral leaves ovate-lanceolate, to 1.2 × 0.5 cm, acuminate, scarlet, early deciduous. Pedicels to 4 mm long. Calyx broadly campanulate, 1 cm long; tube 8-nerved, hirtellous with scarlet hairs, mouth open after anthesis; lobes ovate-oblong, sharply acute. Corolla scarlet; tube broadly infundibular, to 3.1 cm long, laterally subcompressed, narrowed and whitish towards base; upper lip erect, 1.3 cm long, 2-fid at apex; lower lip shorter, median lobe largest. Stamens exserted beyond upper lip of corolla; filaments 5 mm long, glabrous; anterior end of connectives linear-spathulate. Style slightly curved outwards, glabrous; branches unequal.

D i s t r. Native to Brazil, spread further in South America; cultivated in Ceylon.

E c o l. Cultivated in gardens as an ornamental.

S p e c i m e n s E x a m i n e d. KANDY DISTRICT: Peradeniya, Botanic Gardens, *Cramer 4491* (PDA, US).

7. Salvia coccinea Juss. ex Murr., Commentat. Soc. Regiae Sci. Gott. 1: 86, t. 1. 1778; Epling in Feddes Repert. 110: 133. 1939.

Salvia coccinea var. *pseudococcinea* (Juss. ex Murr.) Gray, Syn. Fl. N. Amer. 21: 368. 1878.

Salvia pseudococcinea Juss. ex Murr. in Commentat. Soc. Regiae Sci. Gott. 1: 86. 1778.

Perennial herb; stem to 1 m high, much branched at base, shaggily hirsute in upper parts. Leaves ovate-deltoid, upper ones to 6 × 4.7 cm, lower smaller; all truncate at base, subacute, shallowly crenate-serrate, scaberulous above, more densely pubescent on nerves beneath; petioles to 6 cm long. Racemes to 10 cm long; verticils 4–6-flowered. Floral leaves ovate-lanceolate, to 1 cm long, green. Calyx deflexed on pedicels, green; tube to 6.5 mm long, 13-nerved, pubescent with retrorse hairs on ribs, mouth closed at anthesis; upper lip broadly ovate, stained brownish-purple in flowers with red or scarlet corolla, entirely green in ones with white corolla. Corolla red, scarlet or white; tube to 1.7 cm long, narrowed at base, gradually broadened above; upper lip entire; median lobe of lower lip large, lateral ones smaller. Stamens inserted on corolla throat, filaments to 4 mm long, glabrous; connectives more than 3 times longer than filaments, glabrous. Style glabrous.

D i s t r. Native to Mexico, widely spread throughout the world.

E c o l. Commonly cultivated in gardens for its conspicuous flowers. The plant has a foetid odour. Flowering throughout the year.

N o t e. This is a variable plant in the indumentum of the leaves and the colour of the flowers. I am not convinced that these differences merit varietal rank and I have, accordingly, merged var. *pseudococcinea* under the typical variety.

S p e c i m e n s E x a m i n e d. KANDY DISTRICT: Royal Botanic Gardens, Peradeniya, *Cramer 4226* (PDA, US). NUWARA ELIYA DISTRICT: precise locality not indicated. *Parsons 1945* (PDA).

12. MENTHA

(Tourn.) L., Sp. Pl. 576. 1753; L., Gen. Pl. ed. 5. 290. 1754; Benth., Lab. Gen. Sp. 168. 1832; Benth. in DC., Prod. 12: 164. 1848; Benth. & Hook. f., Gen. Pl. 613. 1876; Briq. in Pflanzenfam. 4 (3a): 317. 1897. Lectotype species: *M. spicata* L.

Perennial, erect or prostrate, aromatic herbs with creeping rootstocks. Leaves opposite, glandular-punctate. Spikes terminal, dense, cylindrical; verticils many-flowered, axillary. Floral leaves linear. Bracteoles 0. Flowers small. Calyx tubular or campanulate, 10-nerved; tube with a glabrous or villous throat; limb subequally 5-lobed or subbilabiate. Corolla tubular; limb 4-lobed, somewhat irregular, the posterior lobe larger, entire or emarginate. Stamens 4, equal, erect, included or exserted; filaments glabrous; anthers 2-celled, the cells divaricate. Disk symmetrical. Style exserted, 2-fid. Nutlets ovoid, smooth.

About 25 to 30 species, mainly north temperate; one variety native to Ceylon.

Mentha arvensis L. var. **javanica** (Blume) Hook. f., Fl. Br. Ind. 4: 648. 1885.

Mentha javanica Blume, Bijdr. 826. 1826; Benth., Lab. Gen. Sp. 183. 1833; Benth. in DC., Prod. 12: 173. 1848; Trimen, Handb. Fl. Ceylon 3: 381. 1895; Abeywick., Ceylon J. Sci., Biol. Sci. 2 (2): 219. 1959. Type *Blume s.n.*, Buitenzorg, Java (holotype, L).

Mentha arvensis Benth. (non L. 1753) var. *zollingeri* Briq. in Pflanzenfam. 4 (3a): 319. 1897.

Stem erect, to 67 cm high, or prostrate, with spreading branches to 16 cm long, acutely 4-angular, appressed pilose on sides. Leaves lanceolate to oval-lanceolate, $3.5–6.6 \times 0.7–3.3$ cm, acute, sharply serrate, glandular-punctate and pubescent on both surfaces, more pubescent on nerves beneath; petioles 0.6–1 cm long. Verticils dense, in axils of upper leaves. Floral leaves linear-lanceolate, to 4 mm long, acuminate, long-ciliate. Pedicels to 0.5 mm long. Calyx tube campanulate, to 1.25 mm long, striate, dotted without with oil globules and strigose, glabrous within; lobes lanceolate, 1 mm long, acuminate, ciliate. Corolla to 4 mm long, very pale violet; tube 2 mm long, glabrous on both surfaces; upper lip oblong, 1.75 mm long, 2-lobed at apex; lower lip 3-lobed, lobes oblong, to 1.75 mm long, obtuse. Stamens exserted. Style glabrous, much exserted.

Distr. Ceylon and Malesia. The plant was considered very rare by Trimen; it is not possible to locate the plant anymore.

Ecol. Along marshy borders of paddy fields in the lower uplands from c. 700–1000 m. Flowering May–November.

Specimens Examined. NUWARA ELIYA DISTRICT: Pussellawa, *Gardner 663* (PDA), *Gardner CP 2077* (K, MH, PDA). KANDY DISTRICT: Peradeniya, Botanic Gardens, *Alston 1676* (PDA). LOCALITY UNKNOWN: *Walker s.n.* (PDA), *Gardner 1847* (K).

13. POGOSTEMON

Desf., Mém. Mus. Hist. Nat. 2: 154. 1815; Benth., Lab. Gen. Sp. 152. 1833; Benth. in DC., Prod. 12: 151. 1848; Benth. & Hook. f., Gen. Pl. 1179. 1876; Endl., Gen. Pl. 612. 1838; Briq. in Pflanzenfam. 4 (3a): 328. 1897. Type species: *P. plectranthoides* Desf.

Annual or perennial herbs or undershrubs, often strongly scented. Leaves opposite, petioled, entire or serrate. Spikes terminal, occasionally subcapitate, more often paniculate or racemose. Flowers small. Floral leaves ovate. Bracteoles present. Calyx tubular or campanulate; limb subequally 5-lobed, the throat glabrous. Corolla tube included, rarely exserted beyond calyx; limb subbilabiate, spreading, 4-lobed, the 3 posterior lobes united into an upper lip, the anterior one entire, longer than others. Stamens 4, exserted, straight or declinate; filaments usually bearded about middle; anthers 2-celled, the cells confluent. Disk subentire. Style subequally 2-fid at apex; branches often subulate. Nutlets ovoid or oblong, smooth.

About 40 species, from India through Ceylon to South China, Malesia and Japan.

KEY TO THE SPECIES

1 Stem erect or ascending, rather thickened at base
2 Verticils secund, in panicles; stem glabrous or puberulous in upper parts .... **1. P. heyneanus**
2 Verticils annular, in $\pm$ dense spikes; stem pilose-hirsute in upper parts
3 Spikes narrowly cylindrical; flowers sessile; filaments bearded about middle .... **2. P. auricularius**
3 Spikes broadly cylindrical; flowers pedicellate; filaments glabrous throughout .... **3. P. reflexus**
1 Stem prostrate, slender throughout
4 Leaves and calyx glabrous; spikes lax; calyx tube 5-nerved .... **4. P. rupestris**
4 Leaves and calyx scabrid or hirtellous; spikes dense; calyx tube 10-nerved .... **5. P. hirsutus**

1. Pogostemon heyneanus Benth. in Wall., Pl. As. Rar. 1: 31. 1830; Benth., Lab. Gen. Sp. 154. 1833; Benth. in DC., Prod. 12: 153. 1848; Thw., Enum. Pl. Zeyl. 289. 1860; Trimen, Handb. Fl. Ceylon 3: 378. 1895; Mukerjee, Rec. Bot. Surv. India 14 (1): 71. 1940; Abeywick., Ceylon J. Sci., Biol. Sci. 2 (2): 219. 1959. Type: India, *Wall.* Num. List *1532A*, in Hb. Heyne (*Origanum benghalense*), (holotype, K-WALL.).

Pogostemon patchouli Hook. f., Fl. Br. Ind. 4: 633. 1885 [non Pellet 1844, excl. var. *suavis* (Ten.) Hook. f.]. Type: same as that of *P. heyneanus*.

Undershrub; stem ascending, to 1.72 m high, often rather thickened at base, semi-scandent with long horizontal branches, obtusely 4-angular, tumid above nodes, glabrous below, puberulous or strigose in upper parts. Leaves ovate to ovate-rhomboid, 2.8–14.2 × 1.4–8 cm, truncate or cuneate at base, acute to subacuminate, irregularly dentate-crenate, membranous, sparsely scabrid above, more so with long hairs beneath, prominently nerved and occasionally dull purple; petioles 1.2–6 cm long. Panicles terminal and axillary, to 14 cm long; verticils secund, globose, dense; peduncles hirtellous with gland-tipped hairs. Floral leaves ovate-lanceolate, 4–6 × 1–2 mm, acute, hirtellous on lower surface. Bracts smaller. Calyx campanulate; tube 3–4 mm long, obtusely 5-angular, hirtellous without; lobes triangular-ovate, 1–2 mm long, acute, erect. Corolla lilac or white; tube linear, 4–5 mm, glabrous on both surfaces; upper lip 4–5 mm across, the lobes oval-oblong; lower lip to 2 mm long. Filaments copiously bearded about middle, lilac. Disk ± symmetrical. Style glabrous; branches linear-subulate. Nutlets orbicular, 1 mm in diam., compressed, glabrous, pale brown.

Distr. From Western to Southern India, through Ceylon to Malesia. Common.

Ecol. In damp undergrowth of montane forests in the lower and higher uplands from c. 700–2000 m; specially abundant in Hakgala Reserve Forest. Flowering January–October.

Uses. Occasionally cultivated in the lowlands on account of the leaves which are used medicinally against stomach and skin diseases.

Note. The secund, dense spikes are distinctive. Floral leaves, bracteoles and calyces turn scarious on withering.

Specimens Examined. KURUNEGALA DISTRICT: Hiripitiya, cultivated in compound of house near temple, *Cram^r 3583* (PDA, US). KANDY DISTRICT: In jungle, *Alston 20* (PDA), Upper Hantane, under shade along rocky borders of stream, *Cramer 3605* (PDA, US), *s. coll. C.P. 33* (MH, PDA); Nillembe, *s. coll. s.n.* (PDA); Hunnasgiriya, Medamahanuwara, along path in deep shade, *Townsend 73/171* (K, PDA, US). MATALE DISTRICT: Midlands, border of secondary forest, *Cramer 4877* (PDA, US). NUWARA ELIYA DISTRICT: Ginigathena, Adam's Peak Road, roadside shade, *Balakrishnan 1188* (PDA, US); Hakgala, undergrowth of montane forest, *Cramer 4106* (BO, PDA, US), *s. coll. s.n.* (PDA), *Silva s.n.* (PDA). BADULLA DISTRICT: Tangamalai, in undergrowth of secondary forest close to stream, *Cramer 4267* (BO, PDA, US); Namunukula, *Willis s.n.* (PDA), in secondary forest at base of Peak, *Cramer 4740* (PDA, US); Ohiya, in undergrowth of secondary forest, *Cramer 4615* (CAL, SING, PDA, US). LOCALITY UNKNOWN: *Walker s.n.* (PDA), *s. coll. s.n.* (E), *Walker s.n.* (E).

2. Pogostemon auricularius (L.) Hassk., Tijdschr. Natuurl. Gesch. Physiol. 10: 127. 1843; El-Gazzar & Watson, Taxon 16 (3): 188. 1967.

Mentha auricularia L., Mant. 1: 81. 1767. Type: India.

Mentha foetida Burm. f., Fl. Ind. 126. 1768.
Dysophylla auricularia (L.) Blume, Bijdr. 825. 1826; Benth., Lab. Gen. Sp. 158. 1833; Benth. in DC., Prod. 12: 156. 1848; Thw., Enum. Pl. Zeyl. 239. 1860; Hook. f., Fl. Br. Ind. 4: 638. 1885; Trimen, Handb. Fl. Ceylon 3: 380. 1895; Mukerjee, Rec. Bot. Surv. India 14 (1): 79. 1940; Abeywick., Ceylon J. Sci., Biol. Sci. 2 (2): 219. 1959.

Annual; stem rooting at lower nodes, ascending to 65 cm, with straggling branches to 1.30 m long, slightly tumid above nodes, obtusely 4-angular, pilose-hirsute, purplish in upper parts. Leaves oval to oval oblong, 2.4–6 × 1.4–2.2 cm, rounded or cuneate at base, subacute, crenate-serrate, hirsute and glandular-punctate on both surfaces; petioles 5–8 mm long, hirsute. Spikes terminal, linear-cylindrical, to 16.5 cm long; verticils dense; peduncles pilose-hirsute. Floral leaves oval-elliptic, 3–5 × 1.5–2.5 mm, lowermost ones denticulate, the upper ones entire, all densely long-ciliate. Bracteoles oblanceolate or spathulate, 2.5–3 × 0.5–1 mm long, long-ciliate. Flowers sessile. Calyx tube campanulate, 1 mm long, puberulous with gland-tipped hairs; lobes ovate, 0.75 mm long, acute. Corolla to 3.5 mm long, lilac; tube infundibular, 2.5 mm long, slightly exserted from calyx, glabrous; lobes oval-oblong, 1 mm long, acute, sparsely villous at back. Filaments bearded about middle. Style glabrous; branches linear.

Distr. In southern and south-east Asia.

Ecol. Along borders of paddy fields, roadside ditches, and in boggy places among short grass in the wet lowlands and uplands to c. 1000 m. Fairly common. The pilose-hirsute stems and the linear-cylindrical spikes are very distinctive. Flowering October–June.

Note. Formerly belonging to *Dysophylla* Blume, this species is now transferred to *Pogostemon*. For a further discussion of the subject, see note under *Dysophylla*.

Specimens Examined. COLOMBO DISTRICT: Mirigama, *Alston 677* (PDA). KEGALLE DISTRICT: Mawanella, *s. coll. s.n.* (PDA). KANDY DISTRICT: Hantane, *Gardner s.n.* (PDA); Pussellawa, border of paddy field, *Cramer 4056* (PDA, US). KALUTARA DISTRICT: Matugama, in paddy fields, *Kundu & Balakrishnan 457* (PDA, US); between Nagoda and Dodangoda, paddy field banks, *Theobald 2788* (PDA, US).

3. Pogostemon reflexus Benth. in DC., Prod. 12: 155. 1848; Thw., Enum. Pl. Zeyl. 249. 1860; Hook. f., Fl. Br. Ind. 4: 627. 1885; Trimen, Handb. Fl. Ceylon 3: 379. 1895; Abeywick., Ceylon J. Sci., Biol. Sci. 2 (2): 219. 1959. Type: *Walker s.n.*, Ceylon, in Hb. Hort. Soc. Lond. (lectotype, K).

Semi-shrubby herb; stem to 1.90 m high, stout and woody below, much branched, obtusely 4-angular, densely pilose-hirsute, dark purplish in upper parts. Leaves ovate to ovate-rhomboid, 2.2–19 × 1.6–8.8 cm, truncate to rounded at base, subacute, irregularly crenate-serrate, sparsely scabrid or pilose above, densely hirtellous or villous beneath, glandular-punctate on both surfaces; petioles 0.8–5.2 cm long, hirsute. Spikes broadly campanulate, to 24.5 cm long; verticils moderately close to each other, sometimes dense. Floral leaves linear, 2.5–3 mm long. Pedicels 1.5–2 mm long, deflexed. Calyx tubular-campanulate; tube 3.5–4.5 mm, villous-hirsute with purplish hairs; lobes subequal, triangular-lanceolate, 2 mm long, acuminate. Corolla tube 7 mm long, narrowed below, expanded above, villulose without, pale pink; limb to 3 mm across, pinkish-purple, upper lip concave, the lobes ovate-orbicular, hirtellous at back, the lower lip 3 mm long, spreading. Stamens much exserted from corolla; filaments glabrous, occasionally villulose at base. Style glabrous. Nutlets ovoid, 1 mm long, subtrigonous, papillose, dark brown.

Distr. Endemic.

Ecol. In exposed or partly shaded rocky areas of submontane and montane regions from c. 1700-2400 m. The long spicate racemes with the deflexed pinkish-purple flowers are conspicuous and afford a pleasant sight. Rather common. Flowering October–March.

Specimens Examined. KANDY DISTRICT: Hantane, above Wariagala, *s. coll. s.n.* (PDA), Adam's Peak, *Gardner C.P. 154* (MH, PDA), in scrub, *Alston 932* (PDA), *s. coll. s.n.* (MH); Hunnasgiriya, in secondary forest, *Waas 1011* (PDA, US). Adam's Peak, *s. coll. C.P. 154* (K). NUWARA ELIYA DISTRICT: Horton Plains, under shade among rocks along border of forest, *Cramer 3955* (PDA, US). LOCALITY UNKNOWN: *s. coll. C.P. 154* (K).

4. Pogostemon rupestris Benth., Lab. Gen. Sp. 156. 1832 (*P. rupestre* sphalm.); Benth. in DC., Prod. 12: 154. 1848; Thw., Enum. Pl. Zeyl. 239. 1860 (excl. syn.); Hook. f., Fl. Br. Ind. 4: 624. 1885; Trimen, Handb. Fl. Ceylon 3: 379. 1895; Abeywick., Ceylon J. Sci., Biol. Sci. 2 (2): 219. 1959. Type: *Macrae s.n.*, Nuwara Eliya, Ceylon, in Hb. Hort. Soc. Lond. (holotype, K).

Annual; stem rooting at lower nodes, straggling, with branches to 3.20 m long, acutely 4-angular, slightly tumid above nodes, glabrous or puberulous with retrorse hairs, dark purple and shining. Leaves ovate to ovate-lanceolate, 1.7–4.8 × 1.3–3 cm, truncate or rounded at base, acute, crenate-serrate, glabrous on both surfaces, glandular-punctate and purplish beneath; petioles 0.8–2.3 cm long. Spikes to 10 cm long; verticils small, lax; peduncles with slightly retrorse hairs on one side. Floral leaves linear-subulate, 1.5 mm long, glabrous, dark purple. Pedicels 1 mm long. Calyx campanulate-urceolate, 2.5 mm long, glabrous, dotted with oil globules; tube 5-nerved; lobes triangular-ovate, 1 mm long, sparsely ciliate. Corolla pale blue to lilac; tube infundibular, 3 mm long, slightly curved above, glabrous, whitish; limb to 4 mm across, the lobes oval-lanceolate, reflexed with age. Filaments densely bearded at middle. Style glabrous. Nutlets ovoid, 1.25 mm broad, subtrigonous, smooth, black, shining.

Distr. Endemic.

Ecol. Under shade close to streams in the submontane and montane forests from c. 1200–2400 m. The glabrous leaves and the distinctly purplish and shining nature of the stem are distinctive. Common. Flowering February–July.

Specimens Examined. KANDY DISTRICT: Rajamally, Moray Estate, under partial shade close to stream, *Cramer 4277* (BO, K, NBV, PDA, US); Maskeliya, near Gartmore Estate, *Silva s.n.* (PDA), Adam's Peak, *Silva s.n.* (PDA), *Kostermans 24216* (K, NBV, PDA, US), *Cramer 3800* (PDA, US); Ambagamuwa, *Gardner s.n.* (PDA); Dolosbage, behind Raxawa Tea Estate, *Theobald & Grupe 2350* (PDA, US); NUWARA ELIYA DISTRICT: *Gardner C.P. 343* (CAL, MH, PDA); Punduloya, in marginal shade along bank of stream, *Cramer 3725* (PDA, US); Sita Eliya, *Silva s.n.* (PDA); Hakgala, road above Hakgala, *Silva s.n.* (PDA); Pattipola, *s. coll. s.n.* (PDA). LOCALITY UNKNOWN: *s. coll. s.n.*, (MH), *s. coll. s.n.*, (E), *Walker s.n.*, (K), *Gardner C.P. 343* (CAL, MH, PDA).

5. Pogostemon hirsutus Benth., Lab. Gen. Sp. 155. 1832; Benth. in DC., Prod. 12: 154. 1848; Hook. f., Fl. Br. Ind. 4: 635. 1885; Trimen, Handb. Fl. Ceylon 3: 379. 1895; Abeywick., Ceylon J. Sci., Biol. Sci. 2 (2): 219. 1959. Type: *Macrae s.n.*, Ceylon, in Hb. Hort. Soc. Lond. (holotype, K).

Annual; stem prostrate, to 72 cm long, rooting at lower nodes, frequently ascending, obtusely 4-angular, slightly tumid above nodes, glabrous below, hirsute with long, deflexed hairs in upper parts, dull purple. Leaves ovate to ovate-lanceolate, 1–5 × 1–3 cm, truncate to cuneate at base, subacute, serrate, sparsely scabrid above, more densely so on prominent nerves beneath and glandular-punctate; petioles 0.3–1.7 cm long, villous-hispidulous. Spikes oblong-cylindrical, 1.5–6 cm long, dense. Floral leaves 2.5–10 × 0.5–1.5 mm, hirsute beneath. Calyx to 5.5 mm long; tube inflated, 4.5–5 mm long, 10-nerved, hispid with long hairs; lobes ovate to ovate-lanceolate, to 2.5 mm long, acuminate, hispid-ciliate. Corolla bluish-white or lilac; tube linear, to 7 mm long, glabrous; upper lip 4–5 mm across, the mid-lobe smaller than the lateral ones. Filaments bearded about middle with long, pale purple hairs. Style glabrous. Nutlets orbicular, 1.25 mm in diam., sub-trigonous, smooth, black, shining.

Distr. Endemic.

Ecol. Under shade in damp floor of submontane and montane forests, to c. 2400 m, often close to a watercourse. Flowering October–March. Very common.

Note. The hirsute nature of the stem and the dense, hispid spikes are distinctive.

Specimens Examined: KANDY DISTRICT: Corbet's Gap, by stream beside Rangalla-Hunnasgiriya Road, *Cramer 3925* (PDA, US); 'Wallekelly', 5000 ft., *Breckets 321* (CAL). NUWARA ELIYA DISTRICT: Horton Plains, *Gardner C.P. 283* (CAL, MH, PDA), *Willis s.n.* (PDA), in forest shade, *Balakrishnan 1034* (PDA, US), in undergrowth of secondary forest mid-way to Peak, *Cramer 3451* (PDA, US); Elk Plains, in open grassy plain, *Cramer 3509* (PDA, US); Sita Eliya, *Silva s.n.* (PDA); Pattipola, under shade beside stream close to road, *Cramer 3941* (PDA, US); Pidurutalagala, on embankments, under shade of shrubs, *Cramer 3940* (NBV, PDA, US). LOCALITY UNKNOWN: *Gardner s.n.* (PDA), *s. coll. s.n.* (E), *Walker s.n.* (K).

14. ANISOMELES

R. Br., Prod. 503. 1810; Benth. in Wall., Pl. As. Rar. 1: 59. 1831; Benth. in DC., Prod. 12: 455. 1848; Benth. & Hook. f., Gen. Pl. 1207. 1876; Endl., Gen. Pl. 625. 1838; Briq. in Pflanzenfam. 4 (3a): 266. 1896. Type species: not known.

Perennial herbs; stems erect, with a villous or lanate indumentum. Leaves petiolate, serrate-crenate to dentate. Racemes terminal, or verticils axillary, many-flowered; flowers medium-sized, sessile or subsessile. Calyx campanulate, slightly enlarged in fruit; tube 10-nerved; limb equally 5-lobed. Corolla tube subequalling calyx, pilose-annulate within; limb 2-labiate; upper lip erect, subhooded, villous, entire, lower one spreading, 3-lobed, the median lobe larger, emarginate or 2-fid. Stamens 4, didynamous, included under the upper lip of corolla; filaments pubescent at middle; posterior anthers 2-celled, anterior ones dimidiate. Disk equal-sided. Style 2-fid at apex. Nutlets ovoid or oblong, smooth.

About seven species, in tropical and subtropical Asia and tropical Australia; two species in Ceylon.

KEY TO THE SPECIES

1 Stem and leaves villous-pilose; leaves broadly ovate; verticils fairly distant. . **1. A. indica**
1 Stem and leaves lanate; leaves lanceolate or oblong in outline; verticils usually close. . . . . **2. A. malabarica**

1. Anisomeles indica (L.) Kuntze, Rev. Gen. Pl. 512. 1891; Alston in Trimen, Handb. Fl. Ceylon. 6: 236. 1931; Abeywick., Ceylon J. Sci., Biol. Sci. 2 (2): 219. 1959.

Nepeta indica L., Sp. Pl. 571. 1753. Type: "India," Specimen *726.28* (LINN.).

Anisomeles ovata R. Br. in Ait., Hort. Kew. ed. 2, 3: 364. 1811; Benth., Lab. Gen. Sp. 702. 1835; Benth. in DC., Prod. 12: 455. 1848; Thw., Enum. Pl. Zeyl. 240. 1860; Hook. f., in Fl. Br. Ind. 4: 672. 1885; Trimen, Handb. Fl. Ceylon 3: 384. 1895.

Ballota disticha L., Mant. 1: 83. 1767. Type: Surat, India, Specimen *737.7* (LINN.).

Perennial herb; stem erect, to 1.90 m high, divaricately branched, stout, acutely 4-angular, grooved on opposite sides, densely villous-pilose with retrorse hairs. Leaves broadly ovate, 3.2–10 × 3.1–8 cm, cordate-cuneate or truncate at base, acute at apex, coarsely crenate-serrate, villous-pilose on both surfaces; petioles 1–7.5 cm long. Spikes spurious, terminal, 8–38 cm long; verticils distant, dense; peduncles villous-pilose. Floral leaves linear-oblong, or subulate, 7–8 × 1 mm. Bracts subulate, to 6 mm long. Calyx long-campanulate; tube 4–5 mm long, 5-angular, 10-nerved, hirsute; lobes ovate-lanceolate, to 3.5 mm long, acuminate. Corolla to 17 mm long; tube linear, to 7 mm long, broadened above, pectinate-annulate within about the middle, pale greenish-white; upper lip oblong-oval, 6 mm long, pale greenish-white, pilose at back; lower lip longer, the lobes spreading, central one broadly orbicular, deflexed, lateral ones smaller. Filaments bearded in front about middle. Disk more or less 4-lobed. Style glabrous; branches linear, recurved. Nutlets ellipsoid, to 2.5 mm long, subtrigonous, glabrous, shining black; areola basal.

Distr. Common. From India through Ceylon to South China; Malesia to Japan.

Ecol. A weed in waste places and along roadsides in the dry and wet lowlands and the lower uplands, from sea level to c. 700 m. Flowering September–February; July.

Specimens Examined. MANNAR DISTRICT: Madhu Road, under shade close to bank of stream beside road, *Cramer 4090* (PDA, US). POLONNARUWA DISTRICT: Polonnaruwa town, under shade along bank of Parakrama Samudra, *Cramer 3674* (PDA, US). KURUNEGALA DISTRICT: Ganewatte, in Coconut Estate beside road, *Cramer 4097* (E). COLOMBO DISTRICT: Moratuwa, Jubilee Road, *Cooray 69090302* (PDA, US). KANDY DISTRICT: Moramkande, *Alston s.n.* (PDA). GALLE DISTRICT: Point Galle, *Gardner C.P. 2072* (PDA); Pitiduwa, Habaraduwa, under shade in Coconut Estate, *Cramer 3423* (PDA, US). MATARA DISTRICT: Rudawela, in sandy, waste ground close to sea shore, *Cramer 3441* (PDA, US). HAMBANTOTA DISTRICT: Yala, Talgasmankade, under shade in sandy soil close to river, *Cramer 4127* (MH, PDA, US). AMPARAI DISTRICT: Lahugala, in borders of open savannah, *Cramer 3657* (PDA, US). BATTICALOA DISTRICT: Kalkudah, in waste grassy ground beside road, *Cramer 3619* (PDA, US). TRINCOMALEE DISTRICT: Between Mutur and Trincomalee, wayside scrubs, *Jayasuriya et al. 66* (PDA, US),

Trincomalee-Kantalai Road, roadside, *Comanor s.n.* (PDA, US). LOCALITY UNKNOWN: *s. coll. s.n.* (K).

2. Anisomeles malabarica (L.) R. Br. ex Sims, Bot. Mag. 46: t. 2071. 1819 Benth. in Wall., Pl. As. Rar. 1: 59. 1830; Benth., Lab. Gen. Sp. 704. 1835 Benth. in DC., Prod. 12: 456. 1848; Thw., Enum. Pl. Zeyl. 240. 1860; Hook f., Fl. Br. Ind. 4: 673. 1885; Trimen, Handb. Fl. Ceylon 3: 384. 1895; Mukerjee Rec. Bot. Surv. India 14 (1): 153. 1940; Abeywick., Ceylon J. Sci., Biol Sci. 2 (2): 219. 1959.

Nepeta malabarica L., Mant. 566. 1771. Type: Hortu Upsalensi, Specimen *726.26* (LINN.).

Anisomeles intermedia Wight ex Benth., Lab. Gen. Sp. 703. 1835; Benth. in DC., Prod. 12: 456. 1848 (cum syn. et excl. var.). Type: *Wight s.n.*, Madura India (lectotype, MH).

Perennial, semi-shrubby herb; stem to about 2 m high, much branched from base, subquadrangular, thickened below to 1.8 cm in diam., densely lanate. Leaves ovate-lanceolate to oblong-lanceolate, 3.6–16 × 1.3–7 cm narrowed and rounded at base, acute, crenate-serrate, slightly bullate and velvety lanate above, densely so beneath; petioles 0.7–3.5 cm long, stout, lanate. Racemes to 34.5 cm long; verticils close, dense; peduncles densely lanate Floral leaves 8–10 mm long, densely lanate. Bracts linear, to 5 mm long Calyx to 9 mm long; tube to 5 mm long, lanate without, glabrous within lobes lanceolate, to 4 mm long, acuminate. Corolla 1.4–2 cm long; tube to 5 mm long, white, glabrous without; throat pilose towards base of lower lip Upper lip oblong, 4–6 mm long, obtuse at apex, slightly arched, whitish lower lip to 1 cm across, coral pink, with 2 white streaks towards base, pilulose without with gland-tipped hairs, the lateral lobes shallowly rounded the median one larger, broadly orbicular, 2-fid. Style glabrous; branches linear, unequal. Nutlets ovoid, 2.5 × 1.5 mm, trigonous, glabrous, blackish brown and shining.

Distr. Peninsular India, Ceylon, Mauritius and Penang.

Ecol. Along roadsides and rail tracks in the dry lowlands, chiefly close to the coast. Common. Flowering June–March.

Note. Racemes conspicuous, flowers faintly scented.

Specimens Examined. JAFFNA DISTRICT: near Jaffna, *s. coll. s.n.* (PDA). PUTTALAM DISTRICT: Mampuri, waste ground beside Mampuri-Talawila Road, *Cramer 4666* (BO, CAL, SING). BATTICALOA DISTRICT: Batticaloa town, in dense population in sloping damp ground near bridge, *Cramer 3630* (PDA, US); Cheddipalayam, roadside, *Cramer 478* (E, NVB, PDA, US). TRINCOMALEE DISTRICT: Kinniya, in open gravelly ground beside road, *Cramer 4325* (E, K, NVB, PDA, US); Uppuval sandy area with palmyrah palms, *Townsend 73/245* (K, PDA, US); Foul Point sandy coastal area, *Jayasuriya et al. 651* (PDA, US).

15. STACHYS

L., Sp. Pl. 580. 1753; L., Gen. Pl. ed. 5, 253. 1754; Benth., Lab. Gen. Sp. 525. 1834; Benth. & Hook. f., Gen. Pl. 1208. 1876; Endl., Gen. Pl. 625. 1838; Briq. in Pflanzenfam. 4 (3a): 260. 1897. Type species: *S. germanica* L.

Erect perennial herbs or diffuse annuals. Leaves opposite, crenate or serrate. Inflorescence terminal, spicate, or axillary, of 2-many-flowered verticils. Flowers sessile or pedicelled. Calyx tubular-campanulate; tube 5–10-nerved; limb 5-lobed, the lobes subequal or the posterior ones enlarged into ± an upper lip. Corolla tube cylindrical, erect or curved, glabrous or annulate within; limb 2-labiate; posterior lip erect, hooded, entire or slightly emarginate; lower lip spreading, 3-lobed, the median one largest. Stamens didynamous, exserted from corolla tube, included under upper lip of corolla; anthers 2-celled, the cells fertile, parallel or divergent. Disk equal. Style 2-fid at apex. Nutlets ovoid or oblong.

About 200 species, almost world-wide, except Australia and New Zealand.

Stachys arvensis L., Sp. Pl. ed. 2, 814. 1762; Benth., Lab. Gen. Sp. 550. 1834 & 741. 1836; Benth. in DC., Prod. 12: 477. 1848; Alston in Trimen, Handb. Fl. Ceylon 6: 237. 1931; Abeywick., Ceylon J. Sci., Biol. Sci. 2 (2): 220. 1959.

Annual herb with a short rootstock; stem erect, to 17 cm high, acutely 4-angular, sparsely pilose. Leaves opposite, broadly oval-oblong, 1.2–2.6 × 1.2–1.8 cm, truncate to subcordate and 3–5-nerved at base, obtuse, crenate-serrate, ciliate at margins, sparsely pilose and with impressed nerves above, less so beneath; petioles 0 or to 1 cm long. Racemes terminal, to 8 cm long. Floral leaves foliaceous, oblong, to 8 × 5 mm. Flowers subsessile; pedicels 1 mm long. Calyx to 5 mm long; tube to 2.5 mm long, 8-nerved, hispidulous; limb subbilabiate, the upper lip 3-lobed, the lower 2-lobed, all lobes lanceolate, upper ones 2.75 mm long, lower 2.25 mm, acuminate-acicular. Corolla tube to 5.5 mm long, sparsely puberulous; limb pale pinkish-mauve; upper lip erect, 2.5 mm long, hirtellous without; lower one broadly ovate, 3.5 × 3.5 mm, faintly spotted red towards base. Filaments finely pubescent on inner edges. Style branches subequal. Nutlets obovoid, 2 × 1.5 mm, muricate, black, shining.

Distr. Europe and N.W. Asia; naturalised in N. America.

Ecol. A common weed in vegetable fields in the uplands at c. 2000 m. It often grows in association with *Veronica arvensis* L. Flowering February–April.

Note. In external appearance this plant closely resembles *Clinopodium umbrosum* but differs from the latter in its erect stem, racemose verticils and the shorter pubescence of the calyx.

Specimens Examined. NUWARA ELIYA DISTRICT: Nuwara Eliya, *Petch s.n.* (PDA), in carrot field, *Cramer 4439* (BO, CAL, E, K, M, NVB, PDA, US).

16. LEONOTIS

R. Br., Prod. 504. 1810; R. Br. in Ait., Hort. Kew. ed. 2. 409. 1811; Benth., Lab. Gen. Sp. 618. 1834; Benth. in DC., Prod. 12: 534. 1848; Benth. & Hook. f., Gen. Pl. 1214. 1876; Endl., Gen. Pl. 628. 1838; Briq. in Pflanzenfam. 4 (3a): 246. 1897. Type species: *L. leonitis* L.

Shrubs or annual herbs. Leaves dentate, usually petiolate. Verticils dense, many-flowered, the verticils usually few or solitary near apex of stem or branches. Floral leaves similar to foliage leaves but smaller. Calyx tubular, 10-nerved, often incurved towards apex; limb 8–10-lobed, with an oblique mouth, the lobes rigid, often spinescent, ± unequal, the posterior usually large and broader than the others. Corolla tubular; tube usually exserted, cylindric or slightly enlarged, faintly curved, glabrous or imperfectly pilose-annulate within; limb 2-labiate; upper lip erect, elongated, concave, densely villous, the lower spreading, 3-lobed, the median lobe scarcely longer than the lateral ones. Stamens didynamous, ascending under upper lip; anthers 2-celled, the cells divaricate; filaments villous at base. Style subulate at apex, 2-lobed. Nutlets ovoid, triquetrous, obtuse or truncate at apex, glabrous.

About 40 species, all African, except one which is pantropical.

Leonotis nepetifolia (L.) R. Br., Prod. 504. 1810; Benth., Lab. Gen. Sp. 618. 1834; Benth. in DC., Prod. 12: 535. 1848; Thw., Enum. Pl. Zeyl. 241. 1860; Hook. f., Fl. Br. Ind. 4: 691. 1885; Trimen, Handb. Fl. Ceylon 3: 387. 1895; Mukerjee, Rec. Bot. Surv. India 14 (1): 185. 1940.

Phlomis nepetifolia L., Sp. Pl. 586. 1753. Type: Surinam?, *s.n.* (BM).

Large annual; stem erect, to 2.63 m high, somewhat woody at base, obtusely 4-angular with thickened, rounded angles, deeply grooved on opposite sides, finely tomentose, medullated. Upper leaves oval-lanceolate, 5.6–8.5 × 1.5–2.6 cm, acute at base; lower ones ovate to broadly ovate, 4–20 × 2.9–15 cm, broadly rounded or truncate at base; all acute to subacute, coarsely crenate-serrate, glabrous or finely tomentose above, minutely glandular-punctate beneath and puberulous along nerves; petioles 1.5–12 cm long. Verticils 2–8 on a stem, distant, globose, 3–7 cm in diam. Floral leaves linear-lanceolate, 12 × 2 mm, spinously tipped. Calyx 1.6–2.3 cm long; tube 1.0–1.4 cm long, faintly curved during anthesis, more so and accrescent later, finely pubescent below, villous-hirsute above; posterior lobe broadly ovate, 6–8 × 4 mm, spinously acuminate, anterior one 5–6 mm long, 7-toothed, teeth unequal. Corolla 1.9–2.8 cm long, orange-red; tube 1.0–1.4 cm long, glabrous without at base, shaggily villous above, within with 3–4 transverse rows of short hairs at base; upper lip 1.2–1.5 cm long, densely and stiffly pilose at back; lower lip 7–8 mm long, browning with age. Filaments flattened, pilose towards base.

Style branches unequal, linear. Nutlets 3–4 mm long, linear-cuneate, trigonous, black, densely poxed with white glands.

Distr. Tropical Africa and Asia.

Ecol. In waste places, often in hollows in rocks, and by roadsides in the dry and intermediate regions (occasionally in the lower uplands) from sea level to c. 400 m. Common, in places locally abundant. Flowering November–June.

Uses. The dried stems after flowering are popularly used for interior decor.

Note. A very handsome plant when in bloom.

Specimens Examined. VAVUNIYA DISTRICT: Mullaitivu, roadside in jungle, *Simpson 9299* (PDA). PUTTALAM DISTRICT: Wilpattu National Park, Kali Villu, *Mueller-Dombois 34* (PDA, US). MATALE DISTRICT: Dambulla, on rocky slopes at foot of Rock, *Cramer 3556* (PDA, US). KANDY DISTRICT: Kadugannawa, *Marimutthu 8839* (PDA). HAMBANTOTA DISTRICT: Ruhunu National Park, Uraniya, *Comanor 867* (K, PDA); Jamburagala, along borders of gravel road close to Gongala plain, *Cramer 3321* (PDA, US). BADULLA DISTRICT: Bibile, *s. coll. s.n.* (PDA). BATTICALOA DISTRICT: Komari, under shade in thicket by roadside, *Cramer 3648* (PDA, US). LOCALITY UNKNOWN: *Walker s.n.* (K).

17. LEUCAS

R.Br., Prod. 504. 1810; Benth. in Wall., Pl. As. Rar. 1: 60. 1831; Benth., Lab. Gen. Sp. 602. 1834; Benth. in DC., Prod. 12: 523. 1848; Benth. & Hook. f., Gen. Pl. 1213. 1876; Endl., Gen. Pl. 628. 1838; Briq. in Pflanzenfam. 4 (3a): 250. 1896. Type species: *L. flaccida* R. Br.

Perennial, villous or lanate herbs or undershrubs. Leaves entire, more often serrate or dentate. Verticils distant, few to many-flowered. Flowers usually sessile. Calyx tubular or tubular-campanulate, 8–10-nerved, with mouth equal or oblique; limb 8–10-toothed; teeth usually unequal. Corolla usually white; tube straight, slender, annulate or not within; limb 2-lobed; upper lip short, erect, hooded, entire, rarely emarginate at apex, very hirsute without; lower lip longer, spreading, 3-lobed, median one largest. Stamens didynamous, included in upper lip of corolla; filaments glabrous; anthers 2-celled, connivent. Disk symmetrical. Style 2-fid; posterior branch very short. Nutlets ovoid, triquetrous.

About 60 species, pantropic, chiefly in warm parts of Africa and Asia.

KEY TO THE SPECIES

1 Leaves linear or lanceolate, ± sessile; calyx teeth not more than 1 mm long

2 Verticils throughout length of flowering stem, lax; calyx mouth straight, villous-annulate

within; limb 10-toothed .... **1. L. longifolia**

2 Verticils towards ends of flowering stem, dense; calyx mouth oblique, glabrous within; limb generally 7–8-toothed

3 Leaves generally more than 2.7 cm long; calyx teeth mucronate; branches of stem ± stiff

4 Calyx teeth 7–8; mouth glabrous within; leaves generally more than 6 mm broad .... **2. L. zeylanica**

4 Calyx teeth 10; mouth villous-annulate within; leaves to 6 mm broad .... **L. zeylanica** var. **walkeri**

3 Leaves generally not more than 2.7 cm long; calyx teeth moderately long-spinescent, branches of stem wiry .... **L. zeylanica** var. **linearis**

1 Leaves ovate, distinctly petiolate; calyx teeth at least 1.5 mm long

5 Leaves lanate-tomentose beneath; calyx softly villous on nerves

6 Stem ± procumbent, straggling; verticils 6–10-flowered; leaves to 3 cm long; corolla tube exannulate within .... **3. L. mollissima**

6 Stem erect, verticils more than 10-flowered; leaves generally more than 3 cm long; corolla tube villous-annulate within .... **4. L. marrubioides**

5 Leaves scabrid or hirtellous beneath; calyx strigose, hirtellous or hirsute on nerves

7 Verticils lax, generally 2–4-flowered .... **5. L. biflora**

7 Verticils dense, generally more than 4-flowered .... **6. L. angularis**

1. Leucas longifolia Benth., Lab. Gen. Sp. 744. 1835; Benth. in DC., Prod. 12: 527. 1848; Hook. f., Fl. Br. Ind. 4: 664. 1885; Trimen, Handb. Fl. Ceylon 3: 386. 1895; Abeywick., Ceylon J. Sci., Biol. Sci. 2 (2): 220. 1959. Type: *Jacquemont 343*, Poona, India (holotype and isotype, P).

Herb with a perennial rootstock; stem erect, acutely 4-angular, hirsute or retrorsely pubescent below, less hirsute or villous-strigose above, grooved on opposite sides. Leaves subsessile, linear to linear-lanceolate, 2.1–5.8 × 0.2–0.4 cm, tapering to base, subacute, entire, finely pubescent on both surfaces; petioles to 4 mm long. Verticils 4–8-flowered. Floral leaves subulate, to 2 mm long, long-ciliate. Flowers subsessile; pedicels to 1 mm long. Calyx turbinate-campanulate, 5–6 mm long, slightly curved above; mouth straight, villous-annulate within, the hairs exserted, white; teeth minutely triangular. Corolla tube uniform, 6–8 mm long, finely pilose above; upper lip to 4 mm long, lower one 6.5–10 mm long.

Distr. West Indies and Ceylon in the wet lowlands. In spite of many attempts I have not encountered this plant in the field.

Specimens Examined. MATARA DISTRICT: Morawaka Korale, *s. coll. C.P. 2076* (PDA).

Note. From examination of the specimen in the Herbarium, Peradeniya (*C.P. 2076*), Trimen says he is not certain as to the name of this species. The particular specimen has subsequently been determined (author not known) as *L. longifolia*. The characters of this specimen agree with those of the type material in Paris, and of supplementary material seen in Kew Herbarium, duly labelled *L. longifolia*.

2. Leucas zeylanica (L.) R. Br. in Ait., Hort. Kew. ed. 2, 3: 409. 1811; Spreng., Syst. Veg. 2: 742. 1825 (*ceylonica* sphalm.); Benth., Lab. Gen. Sp. 614. 1835; Benth. in DC., Prod. 12: 531. 1848; Hook. f., Fl. Br. Ind. 4: 689. 1885; Trimen, Handb. Fl. Ceylon 3: 387. 1895; Abeywick., Ceylon J. Sci., Biol. Sci. 2 (2): 220. 1959; Mukerjee, Rec. Bot. Surv. India 14 (1): 171. 1940.

Phlomis zeylanica L., Sp. Pl. 586. 1753; R. Br., Prod. 504. 1810. Type: Live plant, Hortu Upsalensi, Specimen *740/14* (LINN.).

Annual; stem to 65 cm high, much branched, branches diffuse, acutely 4-angular, hispidulous. Leaves linear-lanceolate to lanceolate, 3–6 × 0.5–1.3 cm, tapering to base, subacute, entire or shallowly and distantly serrate, scabrid and with impressed nerves above, densely hirtellous and glandular-punctate beneath; petioles 1–4 mm long. Verticils 2–3, usually combined into a terminal head. Floral leaves linear-subulate, 4–7 × 0.5–1 mm, long-ciliate, aristate at tip. Pedicels 0.5 mm long. Calyx tube to 5 mm long, slightly curved, 10–11-nerved, hirtellous, dotted in between hairs with oil globules, mouth oblique, glabrous within; limb 7–8-toothed, 3 posterior ones lanceolate, 1 mm long, anterior ones linear, smaller, all mucronate. Corolla to 17 mm long, white; tube to 8 mm long, glabrous, pectinate-annulate within at middle; upper lip to 4 mm long; lower one broadly spathulate, to 11 × 8 mm, lateral ones small. Style glabrous. Nutlets oblong-cuneate, to 3 × 1.5 mm, truncate at top, trigonous, black, shining, with a white boss at base.

D i s t r. Throughout southern and south-eastern Asia.

E c o l. By roadsides and in waste, sandy places in the dry and wet lowlands and uplands (inclusive of patanas) from sea level to c. 1700 m. A variable plant especially in the size of the leaves. Flowering throughout the year.

V e r n. Geta-tumba (S).

U s e s. The boiled plant is used as the principal ingredient of a decoction taken as a vermifuge. The leaves are also eaten as a vegetable.

S p e c i m e n s E x a m i n e d. POLONNARUWA DISTRICT: Giritale, in open waste places close to bank, *Cramer 3683* (PDA, US); sacred area, in open forest among tall grass, *Fosberg 51882* (PDA, US). PUTTALAM DISTRICT: Wilpattu National Park, Kali Villu, *Mueller-Dombois 67051817* (PDA, US). COLOMBO DISTRICT: Seeduwa, Amandoluwa, in Cinnamon Estate, *Cramer 3411* (PDA, US). KALUTARA DISTRICT: Panadura, in coconut plantation, *Comanor 1000* (PDA, US); east of Horana on highway to Ratnapura, *Theobald and Grupe 2389* (PDA, US). MATALE DISTRICT: Sigiriya, in open, damp grassy places, at foot of rock, *Cramer 3661* (PDA, US); Dambulla, in shallow soil, on open rocky slope at foot of rock, *Cramer 3548* (PDA, US). KANDY DISTRICT: Nawalapitiya-Ginigathena road, *Hepper 4501* (K, PDA, US); NUWARA ELIYA DISTRICT: Rendapola, along roadside, by Fort Macdonald Road, *Cramer 3515* (PDA, US). HAMBANTOTA DISTRICT: Yala, Rakinawela, in damp grassy ground beside pond, *Cramer 3269* (PDA, US); Ruhunu National Park, Block 1, Patanagala, *Cooray 69111802R* (PDA, US); Tissamaharama, in waste places, *Alston 1639* (BM, MH, PDA). BATTICALOA DISTRICT: Kalkudah, around bases of coconut palms, *Cramer 3625* (PDA, US). TRINCOMALEE DISTRICT: Kantalai Tank, *Comanor 757* (K). LOCALITY UNKNOWN: *Mrs. Walker s.n.* (K), *s. coll. C.P. 2076* (B, MH, P, PDA), *Thomson 1845* (K).

1. var. **walkeri** (Benth.) Hook. f., Fl. Br. Ind. 4: 689. 1885; Trimen, Handb. Fl. Ceylon 3: 387. 1895.

Leucas walkeri Benth. in DC., Prod. 12: 531. 1848. Type: *Mrs. Walker s.n.*, Ceylon, in Hb. Hook. (holotype, K).

Annual; stem erect, 12–32 cm high, much branched, slender, acutely 4-angular; internodes to 10.5 cm long, densely villous-hirsute or with appressed, retrorse hairs. Leaves suberect, linear, 1.5–6.6 × 0.2–0.6 cm, rather densely hirsutulous beneath. Verticils rather dense. Calyx tube in fruit 6–6.6 cm long; mouth villous-annulate within, the hairs exserted, white; teeth 10, subulate, mucronate. Nutlets narrowly turbinate, 2–2.5 × 1.5 mm, glabrous.

Distr. Endemic.

Ecol. In dry patanas of the upper lowlands and lower uplands, from c. 700–1500 m. Flowering April–October.

Specimens Examined. NUWARA ELIYA DISTRICT: Hakgala, patana, *Silva s.n.* (PDA). MONERAGALA DISTRICT: Maragala, in patanas, *Alston 1631* (PDA). BADULLA DISTRICT: Boralande, *Mueller-Dombois 12* (PDA, US).

2. var. **linearis** Cramer, var. nov. Type: Gardner *C.P. 2076* (holotype, PDA). Caule erecto, e basi virgato-ramoso, ramis effusis; foliis anguste linearibus, 1.4–2.7 × 0.2–0.4 cm; calicis lobis spinoso-aristatis.

Perennial; stem with a long rootstock to 13.5 cm long, erect, to 15 cm high, virgately branched from base, the branches straggling, wiry, villous. Leaves narrowly linear, 1.4–2.7 × 0.2–0.4 cm, hirsute beneath, especially on nerves. Calyx tube slightly curved, to 5 mm long, teeth 7–8, spinous-aristate. Nutlets ellipsoid, to 2.25 × 1.5 mm, truncate at top, glabrous.

Distr. Endemic.

Ecol. In sandy soil close to the coast in the dry lowlands. Flowering April–September.

Specimens Examined. JAFFNA DISTRICT: *Gardner C.P. 2076* p.p. (PDA). BATTICALOA DISTRICT: Kalkudah, sandy places, near coconut groves, *Simpson 9681* (BM, K), in coconut grove close to coast, *Cramer 4383* (BO, MH, NBV, K, PDA, US); near Vakaneri, *de Alwis s.n.* (PDA). PRECISE LOCALITY UNKNOWN: Jaffna District, *Gardner C.P. 2076* p.p. (PDA).

3. **Leucas mollissima** Wall. ex Benth. in Wall., Pl. As. Rar. 1: 62. 1830; Benth., Lab. Gen. Sp. 607. 1834; Benth. in DC., Prod. 12: 525. 1848; Thw., Enum. Pl. Zeyl. 240. 1860; Hook. f., Fl. Br. Ind. 4: 682. 1885; Trimen, Handb. Fl. Ceylon 3: 385. 1895; Mukerjee, Rec. Bot. Surv. India 14 (1): 183. 1940; Abeywick., Ceylon J. Sci., Biol. Sci. 2 (2): 220. 1959. Type: *s. coll.*, Pundua Hills, December 1824, Wall. Num. List. *2054/2* (holotype, K-WALL.); *s. coll.*, Nag Arguton, Nepal, 1821, Wall. Num. List. *2054/1* (paratype, K-WALL.).

Annual or perennial herb; stem straggling or ascending, much-branched, the branches diffuse, hoary with adpressed hairs; internodes fairly long. Leaves subsessile, ovate to ovate-deltoid, 1.6–5.6 × 0.9–2.6 cm, subacute to truncate at base, acute, serrate, densely appressed-tomentose above, lanate-tomentose beneath. Verticils 6–10-flowered. Floral leaves filiform, to 4 mm long, sericeous. Flowers subsessile. Calyx broadly campanulate; tube 5.5–7 mm long, minutely sericeous within in upper half, densely so without; teeth 10, triangular-subulate, to 1.5 mm long. Corolla to 15 mm long, white; tube linear, 6.5 mm, glabrous without except for the sparsely villous part above, exannulate within; upper lip densely villous at back; lower lip orbicular-spathulate. Nutlets ovoid, 1.5 mm long, smooth.

Distr. India to South China, Malesia. I have not encountered this plant. Trimen says: "Gardner's specimens (*C.P. 2075*) are localised 'Alagoda and Peradeniya,' and may have been cultivated or an introduction."

Specimens Examined. KANDY DISTRICT: Alagoda and Peradeniya, *Gardner C.P. 2075* (PDA).

4. **Leucas marrubioides** Desf., Mém. Mus. Hist. Nat. 11: 6, t. 3, f. 1. 1824; Benth., Lab. Gen. Sp. 611, 744. 1834; Benth. in DC., Prod. 12: 528. 1848; Thw., Enum. Pl. Zeyl. 240. 1860; Hook. f., Fl. Br. Ind. 4: 683. 1885; Trimen, Handb. Fl. Ceylon 3: 385. 1895; Mukerjee, Rec. Bot. Surv. Ind. 14 (1): 181. 1940; Abeywick., Ceylon J. Sci., Biol. Sci. 2 (2): 220. 1959.

Leucas indica Moon, Cat. 44. 1824 (nom. nud.).

Perennial; stem to 1.20 m high, often diffusely branched, obtusely 4-angular, rather woody below, semi-succulent above and densely silky lanate. Leaves broadly ovate, 2.7–10.3 × 1.7–7.4 cm, subtruncate at base, acute, coarsely dentate-serrate, densely pilose, and with faintly impressed nerves above, silky lanate beneath; petioles 0.5–2.3 cm long, silky lanate. Verticils throughout length of flowering branches, dense, more than 10-flowered, rounded.

Floral leaves linear-lanceolate, 4–8 × 0.5–0.75 mm, acuminate, velutinous without. Calyx tubular, 7–10 mm long, 8–10-nerved, densely velutinous without, villous within above; limb 8–10-toothed, the teeth lanceolate-subulate. Corolla white; tube 9–10 mm long, faintly striped, villous without above, pectinate-annulate within; upper lip to 5 mm long; lower lip spathulate, 6–8 × 5 mm. Disk symmetrical. Nutlets ovoid, 1.5 mm long, truncate at top, reticulated, glabrous, brown.

Distr. South India and Ceylon: common, especially in the Uva country.

Ecol. In rocky embankments, often by roads, in the dry lowlands and uplands from c. 100–2000 m. Flowering almost throughout the year.

Specimens Examined. ANURADHAPURA DISTRICT: Minintale, in gravelly slope in compound of sacred enclosure, *Cramer 4149* (BO, E, PDA, US); Minintale hill, shade of scrub, *Balakrishnan & Jayasuriya 119* (PDA, US). PUTTALAM DISTRICT: near Tank, *Gardner C.P. 2071* (E, PDA). KANDY DISTRICT: Hunnasgiriya, along shaded bank of roadside stream, *Cramer 3615* (PDA, US); Pussellawa, Sangilipalama, among rocks close to streamlet, *Cramer 3859* (PDA, US). NUWARA ELIYA DISTRICT: Mulhalkele, gravelly slope in distant jungle, *Cramer 4526* (BO, CAL, PDA, US); Hakgala, along rocky roadside, *Cramer 3253* (PDA, US). BADULLA DISTRICT: patanas, by path to Fort Macdonald, *Silva s.n.* (PDA); hill opposite Hakgala, *Willis s.n.* (PDA); Rendapola, beside rocks on road embankment, *Cramer 3514* (PDA, US); Ambawela, on road embankment among rocks, *Cramer 4117* (PDA, US); Padinawela, along embankment of road, *Cramer 4237* (BO, E, MH, NBV). Palugama, in open road embankment beside road, *Cramer 3753* (PDA, US). LOCALITY UNKNOWN: *s. coll. C.P. 2071* (E, PDA), *Silva s.n.* (PDA).

5. Leucas biflora (Vahl) Benth. in Wall., Pl. As. Rar. 1: 62. 1830; Benth., Lab. Gen. Sp. 610. 1834; Benth. in DC., Prod. 12: 527. 1848; Hook. f., Fl. Br. Ind. 4: 683. 1885 (excl. *L. procumbens* Thw.); Trimen, Handb. Fl. Ceylon 3: 386. 1895; Mukerjee, Rec. Bot. Surv. India 14 (1): 180. 1940; Abeywick., Ceylon J. Sci., Biol. Sci. 2 (2): 220. 1959.

Phlomis biflora Vahl, Symb. Bot. 3: 77. 1794; R. Br., Prod. 504. 1810; Moon, Cat. 44. 1824; Roxb., Fl. Ind. ed. 2, 3: 12. 1832.
Nepeta indica Burm. f., Fl. Ind. 126. 1768 (non L. 1753).

Herb with a perennial rootstock; stem often procumbent, much branched, with the branches to 1.2 m long, acutely 4-angular, finely or densely pubescent with retrorse hairs; internodes to 10 cm long. Leaves broadly ovate, ovate-lanceolate or ovate-deltoid, 1.5–5.2 × 1.1–2.3 cm, truncate or rounded at base, acute or obtuse, coarsely crenate-serrate, finely scabrid or scabrid-hirsute above, densely hirtellous or finely tomentose beneath; petioles 5–6 mm long, slender. Verticils lax, distant, 2–4-flowered (occasionally 6-flowered). Floral leaves subulate, to 2 mm long. Pedicels 2–3 mm long. Calyx tubular-campanulate, 6–8 mm long, 10-nerved, the nerves prominent, finely or densely strigose; teeth 10, narrowly lanceolate, 2–4 mm long, unequal, setaceo-acuminate. Corolla to 16 mm long, white; tube 7–8 mm long, glabrous without, villous-annulate within towards middle; upper lip 6.5–7 mm long; lower one broadly ovate-oblong, 9–12 × 10 mm; palate distinctly bullate. Disk symmetrical. Nutlets ovoid, 2 × 1 mm, smooth, pale brown.

Distr. South India, Ceylon.

Ecol. Widespread on road embankments, waste places, patanas and forests of the lowlands and uplands, from sea level to c. 2700 m. Flowering throughout the year. A very variable plant in regard to the pubescence of stem, leaves and calyx, as well as to the shape and size of the leaves.

Vern. Geta-tumba (S).

Specimens Examined. ANURADHAPURA DISTRICT: Ritigala, in rocky ledge near peak, *Cramer 4145* (BO, PDA, US), *Willis s.n.* (PDA); southern slope, below Weweltenne, *Jayasuriya 1304* (PDA, US), open scrub, *Balakrishnan 630* (PDA, US). KANDY DISTRICT: Kadugannawa, under shade along roadside close to stream, *Cramer 3891* (PDA, US). NUWARA ELIYA DISTRICT: Pidurutalagala, among rocks, *Cramer 3498* (PDA, US); Ambawela, by roadside beside Ambawela Road, *Cramer 4193* (E, PDA, US); Horton Plains, Haldumulla Road, *Willis s.n.* (PDA). GALLE DISTRICT: Pitiduwa, Habaraduwa, under shade of trees by roadside, *Cramer 3427* (PDA, US); *Gardner C.P. 2074* (BM, K, P, PDA); Galle Fort, *Simpson 8613* (PDA). HAMBANTOTA DISTRICT: Yala, Talgasmankade, in sandy soil, close to river, *Cramer 4126* (PDA, US); Ruhunu National Park, Block 1, Padikama, in local forest, *Mueller-Dombois 69010801* (PDA, US). BADULLA DISTRICT: Namunukula, *Willis s.n.* (PDA). MONERAGALA DISTRICT: Bibile, *Silva s.n.* (PDA). AMPARAI DISTRICT: east side of Ulpasse Wewa, *Fosberg & Sachet 52953* (PDA, US). LOCALITY UNKNOWN: *Walker s.n.* (PDA).

6. Leucas angularis Benth. in Wall., Pl. As. Rar. 1: 62. 1830; Benth., Lab. Gen. Sp. 608. 1834; Benth. in DC., Prod. 12: 526. 1848; Hook. f., Fl. Br. Ind. 3: 684. 1885; Trimen, Handb. Fl. Ceylon 3: 385. 1895; Mukerjee, Rec. Bot. Surv. India 14 (1): 181. 1940; Abeywick., Ceylon J. Sci., Biol. Sci. 2 (2): 20. 1959. Type: *s. coll.*, Ceylon, in Hb. Lindl. (holotype).

Leucas pubescens sensu Thw., Enum. Pl. Zeyl. 240. 1860 (non Benth. 1834).

Annual; stem straggling, branches fairly long, acutely 4-angular, densely pubescent or shaggy with retrorse hairs. Leaves ovate-deltoid, 1.8–6.2 × 1.3–3.7 cm, truncate at base, subacute, densely strigose-hirsute with appressed hairs above, tomentose or hirtellous beneath; petioles to 10 cm long. Verticils dense, 4–12-flowered or more. Floral leaves linear-subulate, to 5 mm long. Pedicels 2.5 mm long. Fruiting calyx campanulate, 6.5–8 mm long, 10-nerved, densely strigose on ribs, villous within in upper half; teeth linear-lanceolate, 1.5–4 mm long, long-acuminate, densely ciliate. Corolla tube linear, straight, glabrous.

Distr. South India, Ceylon.

Note. This is an unlocalised plant. There are no Ceylon specimens in the Peradeniya Herbarium, and the only ones seen are those in Kew Herbarium.

Specimens Examined. LOCALITY UNKNOWN: *Walker s.n.* (K).

18. LEONURUS

L., Sp. Pl. 584. 1753; L., Gen. Pl. ed. 5, 254. 1754; Benth. & Hook. f., Gen. Pl. 2: 1210. 1876; Briq. in Pflanzenfam. 4 (3a): 256. 1897. Type species: not known.

Erect, stout herbs. Leaves generally deeply lobed, the lower ones often subpalmatifid. Verticils axillary, dense, 6-many-flowered. Floral leaves narrower than foliage ones, lobed or coarsely dentate. Bracteoles subulate, toothed, teeth ± equal, spinescent. Calyx turbinate; limb 5-toothed, the teeth ± equal, spinescent. Corolla tube expanded at throat, naked or pilose-annulate within; limb 2-lipped, the upper lip erect, entire; lower one 3-lobed, the median lobe obcordate. Stamens didynamous, ascending under upper lip; anthers 2-celled. Disk symmetrical. Style 2-fid at apex. Nutlets triquetrous, truncate at top, smooth.

About 8 species, in temperate Europe and Asia; 1 species widespread in the tropics as a weed, sometimes cultivated.

Leonurus sibiricus L., Sp. Pl. 584. 1753; Benth. in DC., Prod. 12: 501. 1848; Alston in Trimen, Handb. Fl. Ceylon 6: 237. 1931; Mukerjee, Rec. Bot. Surv. India 14 (1): 193. 1940; Abeywick., Ceylon J. Sci., Biol. Sci. 2 (2): 219. 1959. Type specimen: ex Hortu Upsalensi, Specimen *739.7* (LINN.).

Annual; stem erect, to 40 cm high, obtusely 4-angular, grooved on opposite sides, densely pubescent with retrorse hairs. Lower leaves broadly ovate, to 6.8 × 4.2 cm, truncate at base, palmatisect, upper leaves narrower, the lobes less divided, all finely pubescent on both surfaces; petioles to 6.6 cm long, slender. Verticils dense. Floral leaves narrowly ovate, to 2.2 cm long, 3–4-lobed. Flowers sessile. Bracteoles linear, to 7 mm long, spinescent. Calyx turbinate, to 5 mm long, with prominent nerves, puberulous; teeth spinescent. Corolla to 1.4 cm long, puberulous; upper lip densely pubescent on back.

Distr. Native of temperate Asia spread in tropical Asia, Africa and America. Weed in gardens. Rare; I have not encountered this plant.

Ecol. Flowering in April.

Specimens Examined. KANDY DISTRICT: Peradeniya, Royal Botanic Gardens, in 1889, *s. coll. s.n.* (PDA).

19. CLINOPODIUM

L., Sp. Pl. 587. 1753; L., Gen. Pl. ed. 5, 256. 1754; Benth. in DC., Prod. 12: 232. 1848; Benth. & Hook. f., Gen. Pl. 1191. 1876, p.p.; Endl., Gen. Pl. 619. 1838, p.p. Type species: *C. vulgare* L. (Lectotype).

Melissa L. sensu Benth., sect. *Clinopodium* (L.) Benth., Lab. Gen. Sp. 391. 1834.

Satureia L. sensu Briq., sect. *Clinopodium* (L.) Briq. in Pflanzenfam. 4 (3a): 302. 1897.

Perennial herbs, sometimes woody at base. Leaves small, entire, or more often serrate-dentate. Verticils dense, many-flowered, and axillary. Floral leaves linear. Calyx tubular, 13-nerved; tube curved inwards, usually sparsely villous within at throat; limb ± bilabiate; upper lip 3-lobed, lower 2-lobed, longer than upper. Corolla bilabiate; tube exserted beyond calyx, often dilated above; upper lip erect, entire or emarginate, the lower spreading, 3-lobed, the median lobe largest. Stamens 4, included in upper lip of corolla, didynamous; anthers 2-celled, the cells divergent. Disk equal. Style glabrous; branches unequal, the upper subulate, the lower longer and wider. Nutlets ovoid, smooth.

About 10 species, chiefly of north temperate regions; one in Ceylon.

Clinopodium L. (1754) and *Calamintha* Mill. (1754) have always been considered very closely related genera. Briquet (1897) classed them among the 13 genera grouped under his "omnibus" genus *Satureia* L. sensu Briquet. Bentham (1848) separated *Calamintha* as a distinct genus annexing *Clinopodium* to it as a section. *Clinopodium* itself subsequently found generic status as evidenced by Hooker f.'s synonymy under the species *Calamintha umbrosa* (M. Bieb.) Fisch. and Mey. (Fl. Br. Ind. 4: 450. 1885). Brenan, however, (1954) considered *Calamintha* to belong still to *Satureia* L. sensu Briquet. He assessed Bentham's character "not or scarcely bilabiate calyx and subequal teeth" as not being distinctive of *Calamintha* and one that was common to members of the complex *Satureia* L. sensu Briquet. This view is supported by Morton (1965) as regards the South African species.

More recent authors, however, favour keeping *Clinopodium* and *Calamintha* as separate genera. Heywood (Fl. Europea 3: 126. 1972) has keyed out the two genera as follows:

1 Calyx tube straight; inflorescence of pedunculate cymes .... **Calamintha**
1 Calyx tube curved; inflorescence of many-flowered verticillasters .... **Clinopodium**

Taxonomically this position is not definitive yet. More research, therefore, is necessary on the entire *Satureia* complex to confirm the validity of the basis for the separation of *Clinopodium* L. from *Calamintha* Mill. and to elucidate its relationship with the latter and the other members of the complex *Satureia* L. I support Heywood's classification and consider *Clinopodium* to be the representative genus of the original complex *Satureia* in Ceylon.

Clinopodium umbrosum (Bieb.) Koch, Linnaea 21: 673. 1848; Kuntze, Rev. Gen. Pl. 514. 1891; Abeywick., Ceylon J. Sci., Biol. Sci. 2 (2): 219. 1959.

Melissa umbrosa Bieb., Fl. Cauc. 2: 62. 1808; Benth., Lab. Gen. Sp. 392. 1834. Type: *Steven s.n.*, Caucasus, around Anamur & Duschet (LE).

Stem diffusely branched from a perennial rootstock, the branches to 68 cm long, rooting at lower nodes or ascending, slender, acutely 4-angular, densely pubescent on angles with retrorse hairs, purplish in lower parts. Leaves ovate, 1.5–2.8 × 0.9–1.7 cm, rounded at base, acute, coarsely crenate-serrate, sparsely scabrid above, densely so beneath on prominent nerves; petioles 3–6 mm long. Racemes to 15 cm long; verticils globose, 6–15-flowered. Bracts linear-subulate, 3.5–5 mm, long-ciliate. Pedicels 2–4 mm long, hirsute with gland-tipped hairs. Calyx to 6.5 mm long, pilose-hirsute on nerves without, annulate within with retrorse hairs towards top; upper lip shortly triangular-ovate, to 1.5 mm long; lower one 2.5 mm long, the lobes subulate, aristate. Corolla mauvish-pink; tube to 7 mm long, hirtellous without; upper lip orbicular, 2–2.5 × 3 mm; lower one 3–4 × 5–7 mm, lobes orbicular-oblong, the median one 2-fid, with violet-purple blotches and 2 villous ridges of white hairs. Filaments glabrous. Style branches curled inwards with age. Nutlets ovoid-orbicular, 1.25 × 1 mm, glabrous, dark-brown, shining.

Distr. From the Caucasus through Afghanistan to southern and south-eastern Asia.

Ecol. In soggy patanas, along borders of marshy land and slopes of tea estates close to watercourses of the wet uplands, c. 2000–3600 m. Fairly common. The upper leaves are often purplish underneath. The corolla has a scrophulariaceous appearance, but the verticils and the bilabiate calyx distinguish the plant as a member of the Labiatae. Flowering almost throughout the year.

Specimens Examined. NUWARA ELIYA DISTRICT: Nuwara Eliya, Single Tree, among tea bushes on way up to summit, *Cramer 3812* (PDA, US); Sita Eliya, patanas, *Silva s.n.* (PDA); Kandapola, Reserve Forest, moist patanas, *Jayasuriya and Maxwell 816* (PDA, US); Ambawela, in patanas, *Alston 1023* (PDA); Elk plains, between Nuwara Eliya and Kande Ela Reservoir, *Mueller-Dombois 68051810* (PDA, US); Hakgala, under shade in eastern end of Botanic Garden, *Cramer 4189* (PDA, US), patanas, *Wills s.n.* (PDA); Horton Plains, *Willis s.n.* (PDA); half mile below Rest House at Ohiya Road, *Mueller-Dombois 67070914* (PDA, US); on floor of mossy forest, *Fosberg* and *Mueller-Dombois 50032* (PDA, US); beside stream by Ohiya Road, *Cramer 4190* (BO, E, NBV, PDA, US); Moon Plains, in soggy patana beside stream, *Cramer 4462* (E, K, MH, PDA, US). LOCALITY UNKNOWN: *Walker 16* (PDA), *Macrae C.P. 2058* (MH, PDA).

20. PRUNELLA

L., Sp. Pl. 600. 1753; L., Gen. Pl. ed. 5, 261. 1754 (Brunella); Benth., Lab. Gen. Sp. 416. 1834; Benth. in DC., Prod. 12: 409. 1848; Benth. & Hook. f., Gen. Pl. 1203. 1876 (Brunella); Endl., Gen. Pl. 620. 1838; Briq. in Pflanzenfam. 4 (3a): 241. 1897. Type species: *P. vulgaris* L. (lectotype).

Perennial herbs with decumbent or ascending stems. Leaves entire, dentate or pinnatifid. Spikes terminal, dense; verticils 6-flowered or less. Floral leaves imbricate. Calyx tubular-campanulate, dorsally subcompressed, irregularly 10-nerved; limb bilabiate, the posterior lobe broadly tridentate, the anterior lobe incompletely bifid. Corolla tube broad, often exserted, pilose- or squamose-annulate within near base, ventricose above, contracted at throat, limb bilabiate, the posterior lip erect, galeate, entire, the anterior one 3-lobed, the lateral lobes deflexed, the median one concave, crenulate. Stamens didynamous, the anterior ones longer, ascending under the upper lip; filaments, especially the posterior ones, bidenticulate, the lower tooth bearing the anther; anthers connivent in pairs, 2-celled, the cells divaricate. Disk symmetrical. Style glabrous, bifid at apex. Nutlets ovoid or oblong, smooth.

About 5 species, widely spread in temperate regions of both hemispheres, and in tropical uplands.

Prunella vulgaris L., Sp. Pl. 600. 1753; Petch in Ann. R. Bot. Gard. (Peradeniya) 6: 71. 1915; Alston in Trimen, Handb. Fl. Ceylon 6: 236. 1931. Type: Specimen *752.1* (LINN.).

Perennial herb; stem rooting at nodes, ascending, obtusely 4-angular, sparsely pilose-strigose. Leaves ovate-oblong, to 3.2 × 1.3 cm, cuneate at base, obtuse, denticulate, sparsely scabrid on both surfaces; petioles slender, to 1.5 cm long. Spikes to 3.2 cm long. Floral leaves broadly orbicular, to 6 × 8 mm, cuspidate, bristly-ciliate, hirsute without. Pedicels 1 mm long. Calyx tube to 4 mm long, sparsely pilose-hirsute without, glabrous within; upper lip broadly ovate-oblong, to 3 mm long, truncate at top, the median tooth shorter than the lateral ones; lower one 3.5 mm long, the lobes connate at middle along inner margins, lanceolate, aristate. Corolla lilac; tube shortly exserted from calyx, to 10 mm long; upper lip 4 mm long, the median lobe larger than lateral ones. Filaments glabrous. Style shortly exserted from corolla; branches linear, long.

Distr. From Northern and Central America to temperate Europe and Asia; apparently naturalized in the uplands of Ceylon. I have not encountered this plant.

Ecol. By roadsides in the higher uplands at c. 3000 m. Flowering April.

Specimens Examined. PRECISE LOCALITY UNKNOWN: Nuwara Eliya district, *Petch 27* (PDA).

21. THYMUS

L., Sp. Pl. 591. 1753; L., Gen. Pl. ed. 5, 257. 1754; Benth., Lab. Gen. Sp. 340. 1834; Benth. in D.C., Prod. 12: 197. 1848; Benth. & Hook. f., Gen. Pl. 1186. 1876; Briq. in Pflanzenfam. 4 (3a): 311. 1897. Type species: *T. vulgaris* L.

Undershrubs or herbs. Leaves small, linear or elliptic, entire, equal to floral leaves or bracts. Verticils few-flowered, axillary or in terminal spikes. Bracteoles minute. Calyx ovoid, 10–13-nerved, 2-lipped, the posterior lip straight, the anterior one spreading or shortly erect or trifid to the middle, lobes equal or the median one larger. Corolla tube included or exserted, naked within; limb subbilabiate, the posterior lip erect, unevenly emarginate, the anterior one 3-lobed, the lobes equal or the median lobe larger. Stamens 4, exserted or included, straight, subequal or didynamous, the anterior ones larger; anthers 2-celled, the cells parallel or divergent. Disk often symmetrical, entire. Style shortly bifid at apex, the lobes subulate, equal, or the posterior one shorter. Nutlets ovoid or oblong, smooth.

About 50 species, indigenous to the temperate regions of the Old World.

Thymus vulgaris L., Sp. Pl. 591. 1753; Benth. in DC., Prod. 12: 199. 1848. Type: Spain, Specimen *744.6* (LINN.).

Dwarf undershrub; to 40 cm high; stem much branched, obtusely 4-angular, woody and glabrous below, puberulent above with retrorse hairs. Leaves often crowded, sessile, elliptic to linear-spathulate, 15–25 × 1.5–3 mm, tapering to base, acute, glabrous on both surfaces, copiously dotted beneath with yellow oil globules. Verticils in terminal racemes to 7 cm long, or flowers axillary. Floral leaves foliaceous, to 10 mm long. Bracteoles to 5 mm long, acuminate-aristate. Pedicels 1–2 mm long. Calyx tube 3 mm long, strigose without and dotted with globose oil droplets, annulate within with sparsely villous hairs; lobes of lower lip linear, 0.5–3 mm long, acuminate-aristate. Corolla lilac; tube 5–6 mm long, slightly expanded above, puberulent without and sparsely dotted with oil globules, puberulous within with depressed hairs and villous at throat; upper lip 3 mm long, concave; lower lip to 5 × 7 mm, the lobes linear-oblong, the median lobe dotted red at base. Filaments glabrous. Style glabrous.

D i s t r. Native of Europe, cultivated in the uplands of Ceylon.

E c o l. Occasionally cultivated for seasoning food. The plant is intensely fragrant. Flowering March.

S p e c i m e n s E x a m i n e d. NUWARA ELIYA DISTRICT: in compound of house close to western end of Lake Gregory, *Cramer 4444* (BO, CAL, NBV, PDA, US).

22. TEUCRIUM

L., Sp. Pl. 562. 1753; L., Gen. Pl. ed. 5. 247. 1754; Benth. in Wall., Pl. As. Rar. 1: 58. 1830; Benth., Lab. Gen. Sp. 660. 1834; Benth. in DC., Prod. 12: 574. 1848; Benth. & Hook. f., Gen. Pl. 1221. 1876; Endl., Gen. Pl. 631. 1838; Briq. in Pflanzenfam. 4 (3a): 210. 1897. Type species: *T. fructicans* L.

Herbs or shrubs. Leaves entire, serrate or multifid. Floral leaves foliaceous or reduced to bracts. Spikes or racemes terminal or axillary; verticils 2–6-flowered, secund. Calyx tubular or campanulate, 10-nerved; limb 5-lobed, the lobes equal or posterior one larger. Corolla tube included in calyx, rarely exserted, exannulate within; limb 1-lipped; upper lip 0, lower spreading, the upper and lower lateral lobes very short or obsolete, the median one larger or narrower, entire or 2-fid. Stamens didynamous, exserted, erect; anthers 2-celled, with the cells confluent. Disk symmetrical; style unequally 2-fid at apex. Nutlets obovoid, smooth, reticulate or rugose, with the funicular scar oblique or lateral.

About 300 species, cosmopolitan in distribution, but more concentrated in the Mediterranean region; only one species in Ceylon.

Teucrium tomentosum Heyne ex Benth. in Wall., Pl. As. Rar. 1: 58. 1830; Benth., Lab. Gen. Sp. 674. 1834; Benth. in DC., Prod. 12: 582. 1848; Thw., Enum. Pl. Zeyl. 241. 1861; Hook. f., Fl. Br. Ind. 4: 700. 1885; Trimen, Handb. Fl. Ceylon 3: 388. 1895; Mukerjee, Rec. Bot. Surv. India 14 (1): 220. 1941; Abeywick., Ceylon J. Sci., Biol. Sci. 2 (2): 220. 1959. Type: *s. coll.*, India, Wall. Num. List *2025* (holotype, K-WALL.).

Perennial herb; stem to 1.5 mm high, obtusely 4-angular, thickened and glabrous below, much branched and tomentose-puberulous above. Leaves deltoid-ovate to broadly ovate, 2.8–11.5 × 2.2–9 cm, rounded or cuneate at base, acute, usually doubly crenate-serrate, densely scabrid above, densely tomentose beneath with prominent nerves; petals 0.6–3.5 cm long, tomentose. Panicles spreading, to 22 cm long; racemes terminal and axillary. Floral leaves ovate-lanceolate to lanceolate, to 5 × 2.5 mm, acuminate, 3-nerved, ciliolate.

Calyx declinate, to 7 mm long; tube campanulate, to 5.5 mm long, faintly gibbous dorsally at base, densely hirtellous without with gland-tipped hairs, villous-annulate within, the hairs filling the tube after anthesis; upper lip erect, broadly ovate, to 2 × 2.5 mm, mucronate; lateral lobes of lower lip triangular-ovate, 2 mm long, median ones lanceolate-subulate, to 3 mm long, acute. Corolla white, 1.4–1.5 cm long; tube narrow, to 6 mm long, slightly incurved, hirtellous without, scabrid within at mouth with whitish hairs along plane of lower lip; upper lateral lobes of lower lip rounded, lower ones ovate-lanceolate, the median one longer than the lateral lobes, broadly ovate, to 5 mm long, navicular. Filaments much exserted beyond lower lip of corolla, glabrous, thinly scabrid at base. Disk 4-lobed; style glabrous; branches linear. Nutlets globose, to 1.5 × 1 mm, trigonous, shallowly reticulate.

Distr. South India; Ceylon.

Ecol. In clayey-loamy soil of grassy hillsides in the uplands and highlands. Altitude c. 730–1400 m. The plant tends to be gregarious. Flowering February–June; September–October.

Specimens Examined. KANDY DISTRICT: Galagama, *Walker C.P. 212* p.p. (K), Hantane, *Gardner C.P. 212* p.p. (PDA), Hunnasgiriya, grassy hillside, *Cramer 3614* (PDA, US). NUWARA ELIYA DISTRICT: Hakgala, Fort Macdonald, patana, *s. coll. s.n.* BADULLA DISTRICT: Erabedde, on sloping hill beside Boralande Road, *Cramer 3739* (PDA, US); Bintenne, *Silva s.n.* LOCALITY UNKNOWN: *Walker s.n.* (PDA), *Gardner s.n.* (K).

LECYTHIDACEAE

(by William Macnae and F. Raymond Fosberg*)

Trees or shrubs, easily recognizable in the field because of the leaves tufted at the end of bare branchlets; leaves large; stipules present; flowers in racemes or spikes, with highly characteristic arrangement of stamens in which filaments are fused into a short tube and with a tendency towards gigantism in the case of species in which one or more whorls of stamens become staminodal; ovary with several loculi and several ovules in each loculus; fruits large and often diagnostic.

Trees of the genus *Barringtonia* are or have been in the past common, even abundant in Ceylon. Now well-grown trees are not common. Schimper in his famous "Indo-Malayische Strandflora" recognized that a "Barringtonia formation" occurred behind but associated with mangrove forests. The species concerned is *B. racemosa*, which in Southwest Ceylon is conspicuous in such areas but in Central Ceylon this species also occurs along the margins of streams and the banks of paddy fields. In all situations it is heavily pollarded. Another species, *B. asiatica*, is also associated with maritime forests. It occurs in the fringe of trees which lie behind a sandy shore. In Ceylon coconut palms occupy this habitat and presumably the trees of *B. asiatica* have been eliminated except in the gardens of a few homes or alongside certain streets and roadways. These trees, some of which are old, are presumably cared for on account of their showy flowers. The third species, *B. acutangula*, is associated with tanks and pans and is less liable to mutilation.

Careya arborea is a tree characteristic of savanna country where fire is a controlling feature. The cannon ball tree, *Couroupita*, is commonly planted around Colombo and Kandy.

*The manuscript for the Lecythidaceae was submitted in draft form, very incomplete and in a somewhat different format than is usual in botanical literature. Professor Macnae died before the treatment could be brought into final form. Fosberg largely rewrote it, inserting much additional material, and brought into final form for use in the Flora of Ceylon. The descriptive material and taxonomic concepts are those originally included by Macnae. The literature citations, nomenclature, and keys have been somewhat rearranged and edited.

KEY TO THE GENERA

1 Flowers in thick, stiff terminal spikes, sessile .... **2. Careya**
1 Flowers in racemes
 2 Racemes branched, borne on trunk and larger branches, stamens fused into a curved fleshy androphore, fruit globose, cannon-ball-like .... **Couroupita**
 2 Racemes simple, pendant, terminal or axillary on small branches, stamens connate at base into a short thick tube, fruit small, less than 10 cm in greatest dimension, or if larger, square and tapering apically .... **1. Barringtonia**

1. BARRINGTONIA

J.R. & G. Forst., Char. Gen. 75. 1776, nom. cons.; Payens, Blumea 15: 157–263. 1967 (revision of genus).

Agasta Miers, Trans. Linn. Soc. London II Bot. 1 : 59, t. 11, 1875.
Butonica Lam., Enc. Meth. Bot. 1 (2) : 521. 1785.
Stravadium Juss., Gen. Pl. 326. 1789.

Shrubs or trees with fissured bark; leaves spirally arranged, usually large (reduced leaves or cataphylls may be present on the flush growth at the end of a shoot, and always occur on seedlings); leaf scars distinct on twigs; stipules present, minute and soon falling; flowers in racemes, terminal or lateral on the twigs, sometimes very long; calyx of 2 or 4 segments adnate to ovary; petals 4 or 5, adnate to the staminal tube; stamens very numerous, connate at the base into a thick tube, filaments very long, inner whorls reduced to shorter bristle-like staminodes, anthers basifixed, 2-celled, dehiscing laterally, often before the bud opens; ovary 2–4-celled with many pendulous ovules in each cell, style long, simple; fruit fibrous or leathery, indehiscent, 1-celled; seed 1, large, embryo without cotyledons but with a spiral of minute scales towards the apex.

KEY TO THE SPECIES

1 Leaves stiff, subcoriaceous, glossy, glabrous, with 6–10 pairs of nerves; flowers with stamens 8–10 cm long, fruit square, 8–10 cm across, with fibrous or corky pericarp .... **1. B. asiatica**
1 Leaves chartaceous, flowers in long pendent racemes, stamens less than 4 cm long, fruit fleshy, not more than 3–4 cm wide
 2 Sepals united into a cap which ruptures irregularly, stamens in 5–6 whorls, fruit ovoid, 5–8 cm long .... **2. B. racemosa**
 2 Sepals free at least above, rounded to ovate, stamens in 3 whorls, fruit somewhat quadrangular, 2–6 cm long .... **3. B. acutangula**

1. Barringtonia asiatica (L.) Kurz, For. Rep. Pegu App. A, IXV, B. 52 in clavi. 1877; Alston in Trimen, Handb. Fl. Ceylon 6: 189. 1931; Payens, Blumea 15: 184. 1967 (full bibliography).

Mammea asiatica L., Sp. Pl. 512. 1753.
Barringtonia speciosa J.R. & G. Forst., Char. Gen. 76, pl. 38. 1776; Thw., Enum. Pl. Zeyl. 119. 1859; Trimen, Handb. Fl. Ceylon 2: 189. 1894.
Agasta indica Miers, Trans. Linn. Soc. London Bot. 1 : 63. 1875.

Tree up to 30 m tall; twigs with large leaf scars; leaves large, 15–40 × 10–18 cm, oblong-obovate, entire, crowded, tapering to base with almost no petiole, shining on both surfaces, nerves 6–10 pairs, marginal vein distinct; raceme terminal or occasionally lateral, bracts sessile, oval, pedicels around 5 cm, opening buds around 4 cm; calyx tube opening into two unequal segments; petals 4, elliptic, convex; staminal tube with 6 whorls, inner one staminodal, filaments up to 10 cm; ovary 4-celled, tetragonous; fruit large, 10 cm, quadrangular, bluntly pointed, ovoid, pericarp very thick, fibrous and spongy, seed over 5 cm.

Distr. From the Comoro Is. and Madagascar eastwards to Tahiti and from Vietnam to North Queensland. It has been introduced widely to tropical islands, such as Hawaii, the West Indies and St. Helena.

Ecol. Seashores, common as a street tree and in shoreward gardens in Colombo and south-western coastal districts; some of these trees are probably growing truly wild, and were not planted.

Vern. Diya-midella (S).

Specimens Examined. COLOMBO DISTRICT: Mount Lavinia, *Robyns 7258* (PDA, US). GALLE DISTRICT: Galle, *Worthington 2523* (K), *Macnae s.n.* in 1968 (US). MATARA DISTRICT: Dondra Head, beach area. *Sohmer et al.* 8866 (US). KANDY DISTRICT: Peradeniya, Royal Botanic Garden, *s. coll. s.n.* (PDA). TRINCOMALEE DISTRICT: Trincomalee, *Worthington 1655* (K). All of these specimens were from probably planted trees.

2. Barringtonia racemosa (L.) Spreng., Syst. Veg. 3: 127. 1826; Thw., Enum. Pl. Zeyl. 119. 1859, incl. vars.; Kurz, J. Asiat. Soc. Bengal 46 (2): 70. 1877; Clarke in Hook. f., Fl. Br. Ind. 2: 507. 1879; Knuth in Pflanzenr. 105, IV, 219: 17. 1939; Payens, Blumea 15: 192. 1967 (full bibliography).

Eugenia racemosa L., Sp. Pl. 471. 1753.
Butonica racemosa (L.) Juss. ex Miers, Trans. Linn. Soc. London Bot. 1: 66, pl. 13. 1875.
Butonica ceylanica Miers, Trans. Linn. Soc. London Bot. 1: 77. 1875.
Barringtonia ceylanica (Miers) Gardner ex Clarke in Hook. f., Fl. Br. Ind. 2: 508. 1879; Niedz. in Pflanzenfam. 3: 7. 1897; Knuth in Pflanzenr. 105, IV, 219: 15. 1939; Trimen, Handb. Fl. Ceylon 2: 190. 1894 (*zeylanica*).
Barringtonia zeylanica "Bl." Alston in Trimen, Handb. Fl. Ceylon 6: 120. 1931 (sphalm.).

A small tree but capable of reaching 20 m or more; leaves tufted at the ends of stout twigs, subsessile, 15 × 4 cm, obovate, oval, tapering to base, glabrous and shiny; racemes long, pendulous, mostly terminal, 25–30 cm, pedicels 10–15 mm, bracts triangular; calyx tube rupturing into 2–5 unequal lobes; petals 4, oblong or oblong-oval, spreading, pink to red in colour; stamens in 5 or 6 whorls of which the innermost one is staminodal; ovary 2–4-celled; fruit ovoid, 5–8 × 2–4 cm.

Distr. From Natal in Southern Africa north to Kenya, thence eastwards to Fiji, Samoa and Caroline Islands, and from Ryukyu Islands and Guam to Queensland.

Ecol. Moist low country, on the shores of backwaters, estuaries, lakes and rivers, but also inland in similar situations. Flowers all year, coastal trees usually with pinkish filaments, inland trees with crimson filaments. The leaves of inland trees are usually smaller than those of coastal trees.

Vern. Goda-midella (S).

Note. The mode of occurrence of *Barringtonia racemosa* in Ceylon has given rise to some confusion. Thwaites recognized two varieties, one (α) being found in estuarine localities and the other (β) being found along inland rivers. Miers (1875) recognized the latter as *B. ceylanica* citing minor differences which could be accounted for by the richer soils of estuaries. Kurz (1877) reduced it to *B. racemosa* again. Clarke in Hooker's Flora of British India lists both with comment that Kurz had united *B. ceylanica* with *B. racemosa*. Trimen lists *B. zeylanica* but comments that he never found it in the living state. Alston (1931) gives a few recognition marks, but these equally apply to *B. racemosa*. Payens (1967) finally unites the two species. We agree. From our experience in Ceylon it is clear that this species is very variable, but the differences may well be due to differences in the growth potential between inland and estuarine environments. The trees have been much modified by pollarding. We never saw a fully grown unpollarded tree.

Specimens Examined. COLOMBO DISTRICT: Gampaha, *de Silva s.n.* (PDA). KALUTARA DISTRICT: Alutgama, 36 mi south of Colombo, *Worthington 2515* (K); Thiriwanaketiya, *Alston 1246* (PDA); Hadigalla, *Waas 911* (PDA). KURUNEGALA DISTRICT: Potuhera, 400 ft, *Worthington 152* (K). KEGALLE DISTRICT: Kitulgala, 200 ft, *Worthington 401* (K); Rambukkana, *E.L.J. s.n.* (PDA). RATNAPURA DISTRICT: Near Pallebedda town, stream along Pelmadulla-Hambantota Road, near mile marker 20/4, *Sohmer et al.* 8818 (US); Avissawella-Dehiowita Road 2 mile marker, roadside, edge of paddy field, *Waas* 1 (US); Ratnapura, *s. coll. C.P. 2682* (PDA); Carney, Adam's Peak sanctuary, c. 400 m, *Hepper et al. 4527* (US). GALLE DISTRICT: Hiniduma, along sandy banks of Gin Ganga, *Sohmer & Waas 10341* (US); Bank of Gin Ganga at Nelluwa, 30 m, *B. & K. Bremer 825* (US); Galle, *Gardner 305 C.P. 2682* (PDA); West of Gin Ganga, Hiniduma, 100 ft, *Worthington 2274* (K); Dedduwa Lake, Bentota, *Worthington 2475* (K); Banks of Bentota Ganga near Alutgama, *Macnae s.n.* (US, PDA); Neluwa, *Balakrishnan 239* (US); Dediyagala, 600 ft, *Worthington 2561* (K). TRINCOMALEE DISTRICT: Trincomalee, *s. coll. C.P. 2682* (PDA). DISTRICT UNKNOWN: *s. coll. C.P. 3610* (PDA); *C.P. 2682* (PDA); Dambemula Tenna Mukalana, Kahigammana, *de Silva s.n.* in 1927 (PDA).

3. **Barringtonia acutangula** (L.) Gaertn., Fruct. 2: 97, Pl. 101. 1791; Trimen, Handb. Fl. Ceylon 2: 191. 1894; Payens, Blumea 15: 226. 1967 (full bibliography).

Eugenia acutangula L., Fl. Zeyl. 85. 1748; L., Sp. Pl. 471. 1753.
Stavadium acutangulum (L.) Miers, Trans. Linn. Soc. London II, Bot. 1 : 80. 1875.

A small tree up to 20 m; leaves 6–15 cm × 2–6 cm, elliptic, glabrous, or slightly hairy beneath, subsessile, tapering to base, acute or acuminate at the tip, rather thin; racemes terminal, pendulous, up to 50 cm long, densely flowered; flowers about 2–3 cm in diameter; sepals 4, free, semicircular, petals 4 or 5; stamens in 3 whorls, inner one staminodal; ovary usually 2-celled; fruit ovoid, 2–6 cm × 1–2.5 cm, bluntly quadrangular.

Distr. Widespread from Afghanistan to North Australia. Payens recognizes two varieties of this species; all Ceylon specimens belong to the typical variety.

Ecol. Around tanks and irrigation channels in the dry region, on soil which may occasionally be flooded, common.

Vern. Era-midella (S); Adampu (T).

Specimens Examined. JAFFNA DISTRICT: Jaffna, *s. coll. C.P. 1592* (PDA). PUTTALAM DISTRICT: Puttalam, *Worthington 234* (K); Chilaw-Puttalam Road, 4 miles from Chilaw, *Simpson 8203* (PDA). TRINCOMALEE DISTRICT: Kantalai Tank, *s. coll. s.n.* (PDA); Near Trincomalee, to NW, *Bernardi 15291* (PDA), Nilaveli, 15 ft, *Worthington 6358* (K). COLOMBO DISTRICT: s. loc, 400 ft, *Worthington 2047* (K); Kalatuwawa Catchment, 360 ft, *Worthington 3534* (K). KEGALLE DISTRICT: Atale, Peradeniya Estate, 8 mi south of Kegalle, road to Ratnapura, *Meijer 391* (PDA, US); Colombo-Kandy 43/2 400 ft, *Worthington 4526* (K). MONERAGALA DISTRICT (?): s. loc, 130 ft. *Worthington 2047* (K). MATALE DISTRICT: Nalanda, 900 ft, *Worthington 6358* (K). BATTICALOA DISTRICT: Batticaloa, *Macnae s.n.* (US, PDA), *Gardner C.P. 1592* (PDA); Verragal Ferry, close to Trikonamadu, *Balakrishnan 368* (PDA, US), *369* (PDA, US); Ruhuna National Park, 1 mile upstream from Kumana, *Cooray 68060602R* (PDA, US). AMPARAI DISTRICT: East of Maha Oya, culvert 27/6, *Maxwell & Jayasuriya 740* (PDA, US); 3 miles NW of Maha Oya, 50 m, *Davidse & Sumith-*

raarachchi 9003 (US); Amparai Road, along Eggalare Oya, *Kostermans 24362* (US). POLONNARUWA DISTRICT: Thamankaduwa, *Waas* 607 (US); Polonnaruwa-Batticaloa Road, 1 mi from Mahaweli Ganga, *Meijer & Balakrishnan* 118 (US, PDA). ANURADHAPURA DISTRICT: Maradankadawela, Jaffna Road, *Waas 236* (US). KURUNEGALA DISTRICT: Narammala, *Waas 295* (US).

2. CAREYA

Roxb., Hort. Beng. 54. 1814; Knuth in Pflanzenfam. 105, IV. 219a: 50. 1939.

Trees up to 20 m; leaves spirally arranged, crowded at the tips of the twigs; flowers in crowded erect spikes; calyx tube adnate to ovary, and not produced beyond it, segments 4, imbricate; petals 4, imbricate; stamens in many whorls, very numerous, connate at the base; ovary inferior, 4-celled, with numerous ovules in 2 rows, style long, simple; fruit large, fleshy, indehiscent, with numerous seeds immersed in the flesh; embryo with cotyledons.

Careya arborea Roxb., Hort. Beng. 54. 1814; Roxb., Pl. Corom. 3: 14, pl. 218. 1819; Roxb., Fl. Ind. 2: 638. 1832; Buch.-Ham., Trans. Linn. Soc. London 15: 97. 1827; DC., Prod. 3: 295. 1834; Wight & Arn., Prod. 1: 334. 1834; Wight, Ill. 20, pl. 99–100. 1841; Thw., Enum. Pl. Zeyl. 119. 1849; Miq., Fl. Ind. Bat. 1 (1): 494. 1855; Miers, Trans. Linn. Soc. Lond. II, 1: 97. 1875; Kurz, J. Asiat. Soc. Bengal 46 (2): 71. 1877; Kurz, For. Fl. Burma 1: 499. 1877; Clarke in Hook. f., Fl. Br. Ind. 2: 511. 1879; Ridley, Fl. Mal. Pen. 760. 1922; Knuth in Pflanzenr. 105, IV, 219: 51, f. 10. 1939.

Careya coccinea sensu Alston in Trimen, Handb. Fl. Ceylon. 6 : 120. 1931, non A. Chev., Cat. Saig. 64. 1919.

A small or middle-sized tree, up to 20 m with very thick rough dark grey bark and a small rounded crown; young branches with prominent leaf scars, growing parts quite glabrous; leaves large, 15–30 × 5–15 cm, usually sessile, broadly ovate, tapering to base, obtuse or rounded at apex, denticulate or crenate, thick, smooth and shining on both sides, pale green; flowers large, 10 cm in diameter, sessile, crowded on very thick, swollen terminal spikes, each with a central oval bract and two lateral linear ones; calyx tube about 2–3 cm, campanulate, glabrous, segments rounded, stiff, erect; petals 5–7 cm, ovate, obtuse, or acute, margin often revolute; filaments about as long as petals, spreading; fruit about 8 cm, green, glabrous, crowned with persistent calyx segments and style, solid with several seeds immersed in flesh.

Distr. This species occurs from Afghanistan through India to Ceylon and east to Thailand.

Ecol. On exposed patana land, common; trees are resistant to fire. This is the Patana "Oak".

Vern. Kahata (S); Kachaddai (T).

Note. Rheede tot Drakenstein published an early treatment and plate of this in his remarkable Hortus Malabaricus 3: 35, pl. 36. 1682.

Specimens Examined. COLOMBO DISTRICT: Colombo Zoo, 40 ft, *Worthington 6683* (K). GALLE DISTRICT: Hiyare, *Silva 223* (PDA). KANDY DISTRICT: Peradeniya, *Cooray 67111608R* (PDA, US), *Mueller-Dombois & Cooray 67111325* (PDA, US); Nilambe, *Worthington 2784* (K). MATALE DISTRICT: Dambulla, 7 mi SE of Kandalama Nikawatuna, *Sumithraarachchi 459* (PDA, US). BADULLA DISTRICT: road from Ohiya to Haldumulla, 5000 ft, *Balakrishnan 531* (PDA, US). AMPARAI DISTRICT: Bibile, culvert 42/6, *Worthington 4950* (K); Bibile-Batticaloa, culvert 50/7, *Worthington 6309* (K); "43½ m Bibile", 590 ft, *Worthington 5113* (K). ANURADHAPURA DISTRICT: 6.5 mi NE of Anuradhapura, *Fosberg & Balakrishnan 53458* (US, PDA, K). DISTRICT UNKNOWN: s. loc. (*Worthington*) *471* (K)?; s. loc. *s. coll. CP 3169* (PDA). No flowering specimens seen.

COUROUPITA

Couroupita surinamensis Mart. ex Berg, Linnaea 27: 462. 1856; Knuth in Pflanzenr. 105, IV, 219a: 47. 1939.

A tall tree with relatively smooth bark; leaves crowded at the apices of the twigs and branches, 10–15 × 6–10 cm, oblong-ovate, apex blunt, almost acute, basally with a short petiole. Flowers in racemes on trunk and lower branches, with short sepals, whitish pink petals each about 5 × 2.5 cm; the androphore large, spathulate, whitish; fruit larger than a cricket ball, very conspicuous—hence cannon ball tree.

Distr. A native of South America, introduced and cultivated everywhere.

Vern. Sal (S).

Specimens Examined. COLOMBO DISTRICT: Colombo, 20 ft, *Worthington 1811* (K). KEGALLE DISTRICT: Mawanella, *Worthington 2479* (K). KANDY DISTRICT: Peradeniya, Royal Botanic Garden, *Worthington 555* (K). All planted. DISTRICT UNKNOWN: *s. coll. s.n.* (PDA); *s. coll. s.n.* (PDA).

MARTYNIACEAE

(by William L. Theobald and Donald A. Grupe*)

Stapf in Pflanzenfam. 4, (3B): 265. 1895.

Annual or perennial herbs, viscid-pubescent. Leaves alternate or opposite, simple; stipules absent. Inflorescence terminal, racemose; bract solitary, deciduous, borne on the pedicel; bractlets at base of calyx 1–2. Flowers hermaphroditic, irregular. Calyx 5-merous, free or partly united, sometimes spathaceous. Corolla 5-merous, gamopetalous; tube cylindrical at base, campanulate or funnelform above, often ventricose and oblique; limb 5-lobed, generally somewhat 2-lipped; lobes imbricate, adaxial pair exterior. Stamens 2 (in ours) or 4, inserted on corolla tube near base of expanded portion; anthers connate or connivent in pairs, locules divaricate. Disk present, annular, glandular. Ovary superior, unilocular with 2 intrusive, parietal placentae, often cohering to form false septa; ovules few to many; style slender; stigma 2-lipped, lips flattened. Fruit beaked, a capsule with a fleshy, deciduous exocarp and a woody persistent endocarp, crested along median line above and occasionally below; style persistent and splitting at maturity into 2 horn-like processes. Seeds black, sculptured, endosperm thin or none, embryo straight, cotyledons large, fleshy.

A small family native to the New World and consisting of about 5 genera and approximately 16 species. Several members of the family, including the sole representative in Ceylon, *Martynia annua*, have become naturalized in some parts of the Old World. The family has sometimes been included in the Pedaliaceae (e.g., Trimen; Hooker, Flora of British India), but can be separated from the latter by the characteristic fruit, the terminal, racemose inflorescence, and the parietal placentation. The latest revision of the family is that of Van Eseltine—A preliminary study of the Unicorn plants (Martyniaceae). New York Agric. Exp. Sta. Tech. Bull. No. 149. 41 pp. 1929. For a discussion of the nomenclatural history and typification of the family, see Hevly—Nomenclatural History and Typification of *Martynia* and *Proboscidea* (Martyniaceae). Taxon 18: 527–534. 1969.

*Pacific Tropical Botanical Garden, Hawaii.

MARTYNIA

(Houstoun in Martyn) L., Sp. Pl. 618. 1753, emend. Stapf in Pflanzenfam. 4 (3B): 265–269. 1895.

Carpoceras A. Rich. in Desf. & Mirb., Bull. Sci. Nat. Geol. 21: 98. 1830.
Disteria Raf., Fl. Tellur. 4: 68. 1838.
Vatkea O. Hoffm., Verh. Bot. Vereins Prov. Brandenburg 73: 45. 1880.

Erect, often branching, annual or perennial herbs, densely viscid-pubescent. Leaves opposite, broad, palmately veined, long petioled. Inflorescence a short, terminal raceme; flowers in the axil of a large, deciduous bract, with 2 membraneous bractlets at the base of the calyx. Calyx of 5 distinct sepals, upper 3 segments narrower than the lower 2. Corolla tube broadened almost from base, obliquely campanulate, ventricose; lobes unequal, lower lobe much the larger. Stamens 2, staminoidia 2–3, 3rd very rudimentary when present; anther locules widely divergent. Ovary unilocular, appearing 4-locular due to intruding T-shaped placentae meeting at the middle. Fruit exocarp thin, deciduous, falling off in 2 pieces; endocarp woody, remaining attached to pedicel, with 8 longitudinal ridges (4 on each side), crested along the dorsal central suture, crowned by 2 claw-like, very sharp, short hooks. Seeds oblong.

A monotypic genus of the New World, now common in waste areas and along roadsides in many parts of the Old World. For a listing of relevant literature see the discussion following the above family description.

Martynia annua (Houstoun in Martyn) L., Sp. Pl. 618. 1753, non L., Syst. Nat. 113. 1759.—**Fig. 1.**

Martynia diandra Gloxin, Obs. 14, t. 1. 1785; Trimen, Handb. Fl. Ceylon 3: 285. 1895.
Martynia angulosa Lam., Enc. 2: 112. 1786.
Carpoceras angulata A. Rich. in Desf. & Mirb., Bull. Sci. Nat. Geol. 21: 98. 1830.
Disteria angulosa Raf., Fl. Tellur. 4: 68. 1838.
Vatkea diandra O. Hoffm., Verh. Bot. Vereins Prov. Brandenburg 73: 45. 1880.

Plants 0.3–2 m high, somewhat woody toward base; stems sub-terete in transection and up to 1.5 cm in diameter. Leaves opposite; broadly ovate to deltoid in outline, 7–23 cm long, 7–20 cm broad, base cordate, margin repand-dentate, apex acute; petiole fistulose, 7–18 cm long. Racemes 10–20 flowered, 4–10 cm long; pedicels 5–25 mm long; bracts and bractlets usually pale pink, membranous, distinctly veined; bracts broadly ovate, 1.2–2.5 cm long, 0.7–1.3 cm broad; bractlets 2, ovate-oblong, 0.6–1.5 cm long, 0.4–1 cm broad. Calyx yellowish to greenish-white, upper 3 sepals narrowly ovate-oblong,

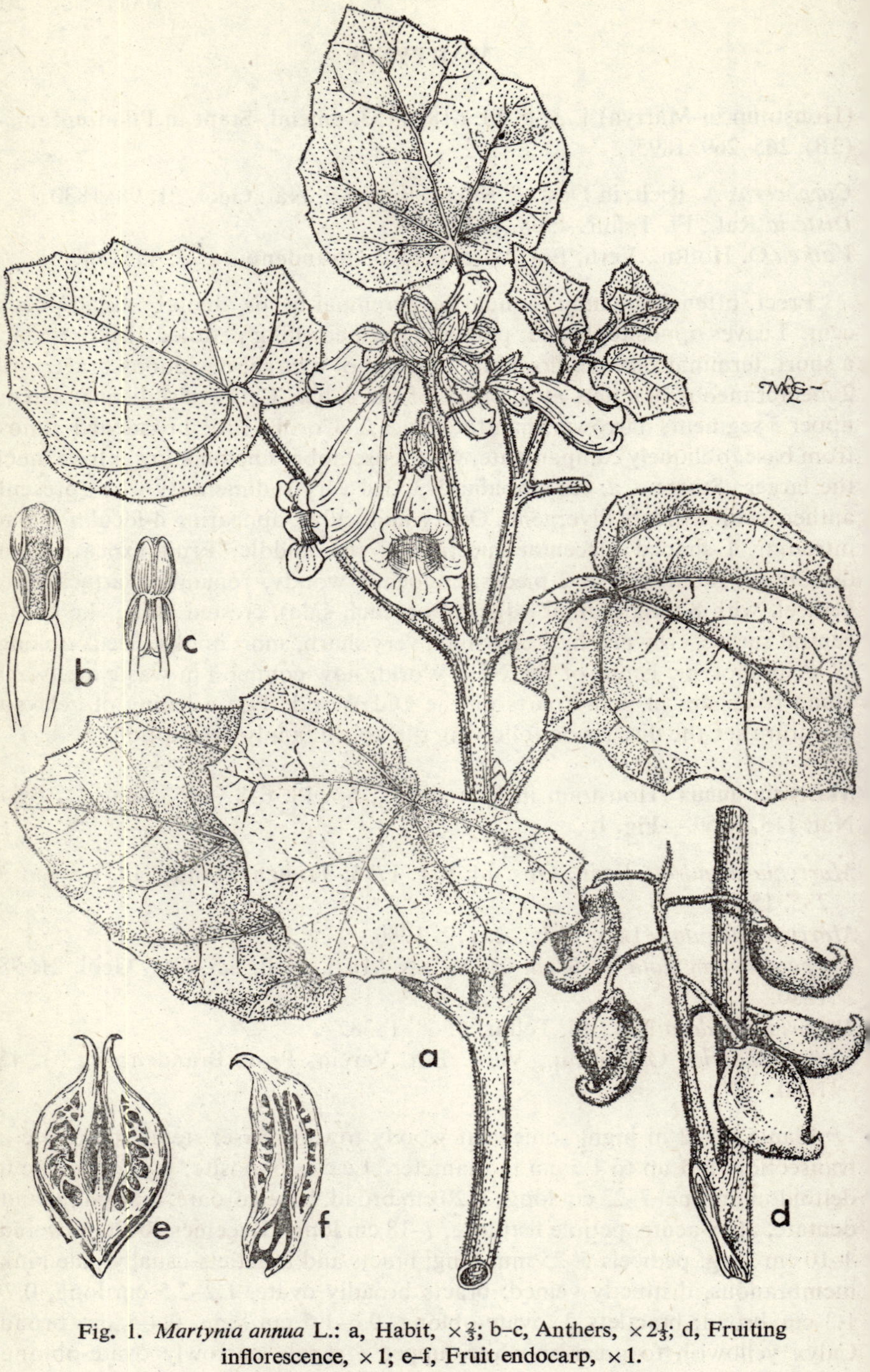

Fig. 1. *Martynia annua* L.: a, Habit, $\times \frac{2}{3}$; b–c, Anthers, $\times 2\frac{1}{3}$; d, Fruiting inflorescence, $\times 1$; e–f, Fruit endocarp, $\times 1$.

9–13 mm long, 3–4 mm broad, lower 2 broadly ovate, 10–13 mm long, 4–7 mm broad. Corolla glandular; tube 3–4 cm long, diffuse pale to reddish-pink without, white to pale pink within, flecked reddish-purple; lower lobe c. 1 cm long, marked by a larger purple blotch with a yellow blotch above; other 4 lobes smaller, broadly rounded, c. equal in size, each marked by a dark reddish-purple blotch. Filaments white, glabrous, 10–15 mm long. Fruit ovoid, flattened somewhat dorso-ventrally; deciduous exocarp green, densely viscid-pubescent; endocarp black, very hard, 2–3 cm long, 1.5–2 cm broad, the recurved claws c. 5 mm long.

Distr. Widely scattered throughout the dry regions of Ceylon. Native of Mexico and Central America now naturalized in many parts of the Old World.

Ecol. Locally common in waste areas, refuse dumps, and along roadsides. Flowering throughout the year.

Vern. Naga—darani (S), Naga—tali (T), Tigers' Claws (E).

Illustr. Bot. Reg. 23: t. 2001. 1837 (*M. diandra*).

Specimens Examined. BADULLA DISTRICT: Ekiriyankumbura, Uva, *Trimen s.n.* (PDA). JAFFNA DISTRICT: along Hwy. 9 near hwy. marker 171, just north of Elephant Pass, *Theobald & Grupe 2332* (US). KANDY DISTRICT: Royal Botanic Gardens, Peradeniya, Herbarium Grounds, *Trimen s.n.* (PDA). MATALE DISTRICT: Dambulla, *Simpson 8083* (BM). TRINCOMALEE DISTRICT: waste area behind sandy beach, approx. 5 mi north of Trincomalee on rd. to Nilaveli, *Theobald & Grupe 2328* (PDA, UC, US). VAVUNIYA DISTRICT: roadside ditch, Mankulam, .1 mi north of inter-section of Hwy. 9 and rd. to Odduchuddan, *Theobald & Grupe 2329* (PDA, US).

MENYANTHACEAE

(by L.H. Cramer*)

Dumort., Anal. Fam. Pl. 20. 1829 ('Menyanthideae'). Type genus: *Menyanthes* L.

Annual or perennial aquatic or marsh herbs without internal phloem; stem rhizomatic or with a tufted rootstock. Leaves radical, submerged and/or floating, the floating leaves single to many, alternate, entire or 3-foliolate, occasionally peltate; petiole sheathing at base. Flowers regular, hermaphrodite, often heterostylous, 5–6-merous, solitary, racemose, fasciculate or in many-flowered cymes. Calyx shallowly or deeply lobed. Corolla lobed, the lobes valvate or induplicate-valvate, the margins entire or the inner surface fimbriate or crested. Stamens isomerous, inserted at base of corolla tube or alternate with the corolla lobes; anthers 2-celled. Hypogynous nectaries usually present. Ovary superior, 1-celled; parietal placentae 2; style short or long; stigmatic lobes 2. Fruit capsular, with irregular dehiscence, rarely fleshy, indehiscent; seeds numerous, winged or not, with copious, cellular endosperm.

Five genera, mostly aquatic, with about 33 representative species in the temperate regions and the tropics of both worlds. The only representative genus in Ceylon is *Nymphoides* Sequier. This genus is difficult of study in the herbarium as dried material is usually devoid of the lower vegetative parts. Further, the corollae are caducous, those still in dried material having shrivelled into fragments. Plants of the genus, therefore, need to be studied from fresh material.

There is a difference of opinion on the systematic position of *Nymphoides*. Some modern authors following the older authorities still retain the genus in the Gentianaceae. I consider it to belong more to the Menyanthaceae as it bears the following characters of the family: the aquatic or marsh habit, lack of internal phloem, alternate leaves, valvate aestivation of the corolla, fimbriate or crested corolla lobes, hypogynous nectiferous glands, evalvular capsule with irregular dehiscence and presence of cellular endosperm in the seeds.

In maintaining this position I have followed the opinions of Dumortier (1829), Hutchinson (1959), Engler (1964), Airy-Shaw (1966), Cronquist (1968) and monographers such as Raynal (1969). All these authors have raised to

*Flora of Ceylon Project, Peradeniya.

family rank either the original subfamily Menyanthoideae (Gilg, 1895) or the tribe Menyantheae (Bentham & Hooker f., 1876).

NYMPHOIDES

Seguier, Pl. Veron. 3: 121. 1754; Hill, Br. Herbal 77. 1756–1757; Kuntze, Rev. Gen. Pl. 2: 429. 1891 ('*Nymphodes*' sphalm.). Type species: *Menyanthes trifoliata* L.

Limnanthemum Gmel. Novi Comment. Acad. Sci. Imp. Petrop. 14 (1): 527. 1770; Endl., Gen. Pl. 606. 1838; Griseb., Gen. Sp. Gent. 341. 1839; Griseb. in DC., Prod. 9: 138. 1845; Benth. & Hook. f., Gen. Pl. 819. 1876; Gilg in Pflanzenfam. 4 (2): 107. 1895. Type species: *L. peltatum* Gmel.

Aquatic or marsh herbs with a rhizomatic stem. Floating leaves single or several, entire, ovate or orbicular, with a deep sinus at base, entire, or obscurely sinuate, occasionally dotted beneath with glands (hydathodes). Flowers fasciculate, nodal or lateral on petioloid stem, occasionally homostylous, usually 5–6 – (– 8) merous. Calyx deeply lobed, persistent. Corolla hairy in the throat; lobes valvate. Hypogynous nectaries 4–5. Fruit ellipsoid, ripening under water; seeds few to many.

About 20 species, in the tropics and temperate regions of the world.

KEY TO THE SPECIES

1 Petioloid stem with a single node and single floating leaf, the leaf glandless beneath; corolla white; seeds glabrous

2 Only floating leaves present, the leaves 3.5–24 cm broad; fascicles of more or less 15–24 flowers; corolla lobes 5–7, the limb 1.4–4.2 cm in diam.; seeds smooth or tuberculate

3 Corolla lobes densely fimbriate on inner surface; petioloid stems thick; style heterostylous; seeds 25-more in a capsule, smooth .... **1. N. indica**

3 Corolla lobes with a median longitudinal crest on inner surface; petioloid stems slender; style homostylous; seeds 5–10 in a capsule, tuberculate .... **2. N. hydrophylla**

2 Floating and radical leaves present, the former 1–3 cm broad; fascicles of 4–8 flowers; corolla lobes 4, the limb to 9 mm in diam.; seeds muriculate .... **3. N. parviflora**

1 Petioloid stems with several nodes and alternate leaves; leaves dotted with black glands beneath; corolla yellow; seeds densely puberulous .... **4. N. aurantiaca**

1. Nymphoides indica (L.) Kuntze, Rev. Gen. Pl. 2: 429. 1891; Abeywick., Ceylon J. Sci., Biol. Sci. 2 (2): 211. 1959.

Menyanthes indica L., Sp. Pl. 145. 1753; Roxb., Fl. Ind. 2: 31. 1824. Type: *Herman s.n.* Ceylon, in Mus. Zeyl. (holotype, BM).

Limnanthemum indicum Griseb., Gen. Sp. Gent. 343. 1839; Griseb. in DC., Prod. 9: 139. 1845; Thw., Enum. Pl. Zeyl. 205. 1860, incl. var.; Clarke in Fl. Br. Ind. 4: 131. 1883; Gilg in Pflanzenfam. 4 (2): 108. 1895; Trimen, Handb. Fl. Ceylon 3: 188. 1895.

Villarsia indica Vent., Choix. Pl. 9. 1803.
Limnanthemum wightianum Griseb., Gen. Sp. Gent. 344. 1839; Griseb. in DC., Prod. 9: 139. 1845. Type: *Wight s.n.*, ex India, Madras, in Hb. Hook. (lectotype, K).

Perennial herb; rhizome oblique, thick; scale-like leaves numerous, ovate; fibrous roots to 45 cm long. Petioloid branches to 1.36 m long, simple, flexible. Floating leaves oval-orbicular to orbicular, 6.5–27 × 3.5–24 cm, deeply cordate at base, coarsely crenate, with the nerves obscure above, prominent on glandless lower surface, rather thick; petioles to 16 cm long, stout. Umbels of more than 20 flowers. Bracts sheathing, oblong-lanceolate, to 7 × 3 mm, fairly thick. Pedicels erect, 5.5–12 cm long. Calyx lobes linear-oblong, to 7 × 2 mm, obtuse, the margins hyaline, reflexed with age. Corolla white; tube to 2 mm long, the throat yellow; limb 2.4–4.2 cm in diam.; lobes 5–7, linear-lanceolate to oblong-lanceolate, 12–16 mm long, subacute at apex, densely fimbriate, on inner surface with multicellular, cottony outgrowths. Filaments broadened at base. Style to 5 mm long. Disk glands 7–8, each suborbicular, ciliolate at apex with papillose hairs. Capsule obovoid, 6–8 × 4.5–5 mm; seeds globose, 1.25 mm in diam., occasionally lenticular, smooth, pale yellowish-white, shining.

Distr. In streams and tanks of the north-central, north-western and southern sectors of the dry lowlands; common. Alt. sea level to more or less 150 m. Palaeotropic.

Ecol. Flowering throughout the year.

Vern. Olu, Maha-ambata (S).

Note. The thick leaves and fimbriate corolla lobes are distinctive.

Specimens Examined. ANURADHAPURA DISTRICT: Siyambalewa, margin of tank, *Hepper & Jayasuriya 4666* (K, PDA, US). PUTTALAM DISTRICT: Wilpattu, Kali Villu, *Cooray 16R* (K, PDA, US); Kottukachchiya, border of tank, *Cramer 2655* (PDA, US). MATALE DISTRICT: Innamaduwa, border of tank, *Cramer 4896* (PDA, US). KALUTARA DISTRICT: Koboduwa, in paddy fields, *Cramer 3220* (PDA, US). MATARA DISTRICT: Borale, Weligama, *Alston 1298* (PDA). KURUNEGALA DISTRICT: Kurunegala tank, *Alston 1298* (PDA); Nikaweratiya tank, *Senaratna 10069* (PDA). HAMBANTOTA DISTRICT: Nabadagas Wewa, Yala, *Fosberg et al. 51079* (K, PDA, US). LOCALITY UNKNOWN: *s. coll. C.P. 3152* (PDA); *Macrae s.n.* (K), *s. coll. s.n.* (G, DC); *Macrae s.n.* (L); *Walker s.n.* (K).

2. Nymphoides hydrophylla (Lour.) Kuntze, Rev. Gen. Pl. 2: 429. 1891; Abeywick., Ceylon J. Sci., Biol. Sci. 2 (2): 211. 1959.

Menyanthes hydrophylla Lour., Fl. Cochinch. 1: 129. 1790. Type: Cochin China.

Menyanthes cristata Roxb., Pl. Corom. 2: 3, t. 105. 1798. Type: Roxburgh Ic. in Pl. Cor., l.c.

Villarsia cristata Spreng., Syst. Veg. 1: 582. 1824.

Limnanthemum cristatum Griseb., Gen. Sp. Gent. 342. 1839; Griseb. in DC., Prod. 9: 139. 1845, excl. syn. *Nedel-Ambel* Rheede; Thw., Eum. Pl. Zeyl. 2–5. 1860; Clarke in J. Linn. Soc. Bot. 14: 150. 1875, excl. syn. *L. aurantiacum* Dalz.; Clarke in Hook. f., Fl. Br. Ind. 4: 131. 1883; Gilg in Pflanzenfam. 4 (2): 108. 1895.

Perennial herb; rhizome short, erect; petioloid branches to 85 cm long, slender. Floating leaves oval-orbicular, 4.0–13.0 × 3.5–10.0 cm, deeply cordate-hastate at base, the lobes rounded, faintly crenate, pale green above, purplish, glandless and with prominent nerves beneath; petioles to 2.5 cm long. Umbels of more than 20 flowers. Bracts ovate, 2.5 mm long. Pedicels 2.8–6.0 cm long, slender. Calyx lobes linear-oblong, 4–5 × 1.5 mm, subacute. Corolla tube 2 mm long, yellow within; limb 1.4–2.0 cm in diam.; lobes 5–6, oval-oblong, 7.0–10.0 × 4.0–5.0 mm, obtuse, with a longitudinal, undulate crest along centre on upper surface, the margins undulate. Filaments short. Style very short; stigmatic branches 2-fid. Capsule broadly ovoid, 3.0–4.0 × 3.5–4.0 mm; testa bulging out over seeds; seeds 5–10, globose, 2 mm in diam., occasionally lenticular, tuberculate, yellowish-white.

Distr. In ponds and stagnant pools in the dry lowlands from the northern to the southern province; very common, especially in the Trincomalee District. Alt. sea level—more or less 150 m. Also in India, Malaysia and South China.

Vern. Heen-olu (S).

Note. In general this is a smaller plant than that of *Nymphoides indica*. The centrally flanged or crested corolla lobes are distinctive.

Specimens Examined. JAFFNA DISTRICT: Kaddaikadu, *Bernardi 14263* (PDA, US). VAVUNIYA DISTRICT: between Mannar and Medawachchiya, *Robyns 7342* (BR, PDA, US); south of Mullaitivu, *Meijer 744* (PDA, US). ANURADHAPURA DISTRICT: Wilpattu National Park, *van Beusekom 1607* (PDA, US); Dambulla-Madatugama rd., *Hepper & Jayasuriya 4606* (K, PDA, US); Kitulhitiyawa, in villu, *Robyns 7265* (BR, PDA, US). POLONNARUWA DISTRICT: Hunukiriyawa, in small tank, *Cramer 5159* (K, PDA, US). TRINCOMALEE DISTRICT: Kannamkulam, *Cramer 3002* (PDA, US); Muthur, in stagnant water, *Robyns 7269* (BR, PDA, US); along jeep track to Kantalai, *Robyns 7278* (BR, PDA, US); Kuchchaveli, tank, *Kostermans 24824* (L, PDA, US). BATTICALOA DISTRICT: Kalkudah, in pond of Kennedy Estate, *Cramer 2734* (PDA, US). HAMBANTOTA DISTRICT: Nabadagas Wewa, in abandoned reservoir, *Fosberg et al. 51079* (PDA, US). MATALE DISTRICT: Sigiriya, in tank, *Nowicke et al. 360* (PDA, US). LOCALITY UNKNOWN: *Gardner C.P. 1870* (PDA).

3. Nymphoides parviflora (Griseb.) Kuntze, Rev. Gen. Pl. 2: 429. 1891; Abeywick., Ceylon J. Sci., Biol. Sci. 2 (2): 211. 1959.

Limnanthemum parvifolium Griseb. in DC., Prod. 9: 141. 1845; Clarke, J. Linn. Soc. Bot. 14: 450. 1875; Clarke in Hook. f., Fl. Br. Ind. 4: 132. 1883; Trimen, Handb. Fl. Ceylon 3: 189. 1895. Type: *Wallich*, Burma, Tavoy, Num. List 4351 (holotype, K-Wall.).

Menyanthes campestris Moon, Cat. 13. 1824, nom. nud.

Limnanthemum moonii Thw., Enum. Pl. Zeyl. 205. 1860. Type: *Moon C.P. 2842*, Ceylon, Kalutara (lectotype, PDA).

Limnanthemum parvifolium var. *moonii* Clarke in Hook. f., Fl. Br. Ind. 4: 132. 1883.

Aquatic or terrestrial herb; rhizome small, erect, with slender, fibrous roots; petioloid branches 7.0–36.0 cm long, very slender. Radical leaves broadly spathulate or deltoid, 7–11 × 6–11 mm, cuneate at base, obtuse; petioles 2.8–3.5 cm long, broad; floating leaves broadly oval to cordate-orbicular or obovate, 1.0–4.2 × 1.0–3 cm, with a deep, triangular sinus; petioles to 2–5 mm long. Fascicles of 4–8 flowers. Pedicels 4–11 mm long. Calyx lobes oblong-lanceolate, 1.75–3.0 mm long, faintly 3-nerved. Corolla white; limb to 8–11 mm in diam.; lobes 4, oblong-lanceolate, 3.5 mm long, subacute, densely fimbriate on the upper surface. Disk of 4–5 minute glands, each minutely penicillate at top. Style very short. Capsule obovoid to broadly turbinate, 2–3.5 mm long, apiculate, longer than calyx; seeds 15 or more, lenticulate, to 1 mm in diam., muriculate, white.

Distr. In shallow ponds, paddy fields or wet places close to the coast in the western and southern sectors of the lowlands; rather rare. Alt. sea level. Also spread from South India to Malaysia and tropical Australia.

Ecol. Flowering May–December.

Vern. Bin-olu (S).

Note. The seeds are seen to bulge out against the semi-transparent wall of the capsule.

Specimens Examined. KALUTARA DISTRICT: Hewesse, *s. coll. s.n.* (PDA), *Moon C.P. 2842* (PDA); Koboduwa, in flooded paddy fields, *Cramer 3218* (PDA, US). GALLE DISTRICT: northwest of Hiniduma, *Alston 2345* (PDA). MATARA DISTRICT: Weligama, *s. coll. C.P. 2842* (PDA). LOCALITY UNKNOWN: *Col. Walker s.n.* (K); *s. coll. s.n.* (K); *Kelaart s.n.* (K); *s. coll. C.P. 2842* (PDA).

Note. This species is variable in the size of its parts, especially in that of the leaves. Two specimens of *C.P. 2842* (one of Moon) in the Herbarium, Peradeniya, appear dwarfed and show in their parts little quantitative differences from that of Wallich's type material. Till further material is available I prefer, for the present, to regard the two former plants as conspecific with, and as a form of, the latter.

The dwarfed forms tend to occur in a terrestrial habitat. My study of herbarium material excludes all doubt that no distinct variety could be recognised in this species.

4. Nymphoides aurantiaca (Dalz.) Kuntze, Rev. Gen. Fl. 1; 429. 1891; Abeywick., Ceylon J. Sci., Biol. Sci. 2 (2): 211. 1959.

Limnanthemum aurantiacum Dalz. in Hooker's J. Bot. Kew Gard. Misc. 2: 136. 1850; Clarke in Hook. f., Fl. Br. Ind. 4: 132. 1883; Trimen, Handb. Fl. Ceylon 3: 190. 1895. Type: *Dalzell s.n.* India, Bombay (lectotype, K).
Menyanthes biflora Moon, Cat. 13. 1824, nom. nud.
Limnanthemum biflorum Thw., Enum. Pl. Zeyl. 205. 1860. Type: *Moon, C.P. 1869* Ceylon, Kalutara (lectotype, PDA).
Limnanthemum forbesianum sensu Griseb., Gen. Sp. Gent. 345. 1839; Griseb. in DC., Prod. 9: 139. 1845; Clarke, J. Linn. Soc. Bot. 14: 450. 1875; Clarke in Hook. f., Fl. Br. Ind. 4: 132. 1883, quoad specimen Macrae.

Perennial herb; rhizome long. Petioloid branches few, to 43 cm long. Leaves on flowering stems alternate, oval-orbicular, 2.2–5.2 × 2–4.7 cm, deeply cordate-hastate at base with a narrow sinus, rounded at apex, pale green above, dull purplish-brown beneath and dotted with blackish glands; petioles 1.5–6.5 cm long. Fascicles lax, of 2–4 flowers. Bracts oblong-lanceolate, to 6 × 2 mm, acute, dotted with reddish glands. Corolla tube 5–6 mm long; limb to 1.5 cm in diam., orange; lobes 5, ovate-oblong, 4–5 mm long, fimbriate, each at base with a small tuft of multicellular, plumose hairs. Filaments 1 mm long; anther-cells linear. Disk glands minute. Capsule depressed-globose, 4–6 × 3.5–4.5 mm, shorter than calyx lobes, apiculate; seeds globose, 2 mm in diam., densely pubescent with soft hairs.

Distr. In shallow ponds close to the coast in the Western Province (wet lowlands); rare. Alt. sea level. Also in Bombay and in South India.

Ecol. The plant comes into flower at the onset of the southwest monsoon, and continues to be in flower and fruit from May–December.

Note. The purplish-red colour of the fresh stem, petioles, pedicels and calyx is distinctive.

Specimens Examined. KALUTARA DISTRICT: Pasdun Korale & Kalutara, *Gardner C.P. 1869* (PDA), *Moon C.P. 1869* (PDA); Nagoda, in shallow pond beside Convent Road, *Cramer 2787* (PDA, US); in fields, *Macrae s.n.* (K). COLOMBO DISTRICT: Western Colombo, *Ferguson C.P. 1869* (K). LOCALITY UNKNOWN: *s. coll. s.n.* (PDA); *Col. Walker s.n.* (K).

Note. Of the syntypes cited by Grisebach only the Macrae specimen (Ceylon) is conspecific with our plant. Other than Macrae's observational error regarding the colour of the corollae (described as white), the following characters in Grisebach's protologue do not refer to the Ceylon plant; the limited length of the leaves (to 1/2 inch.), the stalked gland at the base of each corolla lobe and the punctate vestiture of the seeds. These refer to Forbes' plant (Africa). Clarke, however, cites the Macrae specimen only under Grisebach's name.

MORACEAE

(by E.J.H. Corner*)

Lindl., Veg. Kingdom 266. 1847; Metcalfe and Chalk, Anatomy of the Dicotyledons 2: 1259. 1950; Corner, Gard. Bull. Singapore 19: 187. 1962; Hutchins., Genera of Flowering Plants 2: 151. 1967; Berg, Meded. Bot. Mus. Herb. Rijks. Univ. Utrecht n. 386. 1974.

Trees, shrubs, climbers, or herbaceous (*Dorstenia*, *Fatoua*). Latex present, generally milky. Leaves pinnate, palmate, incised, or simple, stipulate. Inflorescences axillary, typically paired, unisexual or bisexual, monoecious or dioecious, paniculate, racemose, spicate, capitate, or urceolate. Flowers small, unisexual, apetalous; tepals 8–0, generally 4, free or gamophyllous, imbricate or valvate, persistent; stamens isomerous and antitepalous, or reduced to 3, 2, or 1; filaments straight, free or connate, or inflexed; anthers large, mucronate, to small, bilobate and non-mucronate, varying crescentic to turbinate (with transverse, equatorial dehiscence); pistillode present in male flower or not; ovary 1-, rarely 2-, locular, superior to inferior, or immersed in sockets in the inflorescence-axis; styles 2 or 1 with two stigmatic arms, or one stigmatic arm; stigma not capitate or peltate; ovule 1, anatropous or campylotropous, apical or subapical. Fruits drupaceous, discrete or more or less connate in large fleshy syncarps. Seeds large or small, invested in endocarp; testa membranous or disintegrated. Embryos various, curved or straight; radicle long or short, incumbent or accumbent; cotyledons plicate, conduplicate, or plane, foliaceous or thickened, equal to very unequal. Lamina with or without cystoliths.

Genera 53 (75, Hutchinson); species c. 1400 (c. 1850, Hutchinson). Ceylon, 8 genera, 33 species; one genus and several species introduced.

This large tropical family is represented in Ceylon by fragments of widespread genera. Two, *Dorstenia indica* and *Ficus racemosa*, have African affinity and this may be so with *Ficus exasperata*, but the rest appear to have been derived from Indo-Malaysia. There are five endemics, namely *Artocarpus nobilis*, *Broussonetia zeylanica*, *Ficus costata*, *F. fergusoni*, and *F. diversiformis*, none of which seems originally to have been rare or local. Seven species are shared with India, particularly peninsular India, namely *Ficus amplissima*, *F. arnottiana*, *F. benghalensis*, *F. mollis*, *F. trimeni*, *F. tsjahela* and *Plecosper-*

*Botany School, Cambridge.

mum spinosum, to which must be added *F. exasperata* and *D. indica.* Of the remainder, most are species which trend round the periphery of the lowland forest of western Malaysia with the result that the Moraceous element in Ceylon is Indo-Chinese rather than Malayan; the point is emphasized by the recent discovery of *F. pubilimba* in Ceylon. There are merely three species in common between Ceylon and Singapore, namely *Antiaris toxicaria, Ficus laevis* and *F. microcarpa.* The result is a varied, if rather poor, element in the Ceylon flora.

All the conspicuous trees have been illustrated by Worthington (1959). I have limited illustration to botanical details because it is through these that the intricate classification of the family has been derived. It has, indeed, been overdone. If the immense diversity in *Ficus* is maintained, then it follows that many of the smaller genera in other tribes must be reduced to subgeneric rank (Corner, 1962); thus there arise the varying generic quota of the family.

I have continued to use the word 'seed' for the structure usually so-called. Actually it is the true seed enclosed in the thin, often lignified, endocarp.

KEY TO THE TRIBES

1 Flowers inside urceolate receptacles (syconia) closed by bracts; styles not extruded; blastophagous with sterile insect-inhabited flowers. Stamens (when 2 or more) with introrse anthers; filaments straight in bud .... **3. Ficeae**
1 Flowers not enclosed, not blastophagous. Stamens mostly extrorse
 2 Inflorescences bisexual, discoid. Male flowers superficial round the immersed female flowers. Herbaceous .... **2. Dorstenieae**
 2 Inflorescences unisexual. Trees, shrubs or climbers
  3 Female inflorescence racemose, spicate with slender axis, or 1-flowered; ovaries free, superior. Fruits not forming syncarps. Male inflorescence racemose, spicate or capitate, often with a sterile side; stamens with the filaments inflexed in bud; pistillode generally present .... **4. Moreae**
  3 Female inflorescence capitate, thickly spicate or discoid, generally forming a syncarp; ovaries often immersed. Male flowers mostly without pistillode
   4 Male inflorescence discoid with an involucre of bracts; female similar or 1-flowered. Filaments straight in bud .... **5. Olmedieae**
   4 Inflorescences different; female thickly spicate to capitate-globose, not 1-flowered .... **1. Artocarpeae**

KEY TO THE GENERA

1 Herb. Inflorescences discoid with marginal tentacles .... **5. Dorstenia**
1 Woody plants
 2 Flowers in syconia .... **6. Ficus**
 2 Flowers exposed
  3 Leaves rather small with few or no transverse intercostal veins
   4 Shrubs or small trees, often spinous .... **7. Streblus**
   4 Climbers, usually spinous. Flowers in heads. Leaves entire.
    5 Female tepals conspicuous, decussate; styles –2.5 mm long. Male peduncle–10

mm long; filaments straight in bud, anthers not exsert. Syncarp red; seeds numerous, 4–5 mm long, with ligneous coat .... **3. Maclura**

5 Female tepals minute; styles -5 mm long. Male peduncle 12–20 mm long; filaments inflexed in bud, anthers exsert. Syncarp with 1–2 larger seeds, not ligneous .... **4. Plecospermum**

3 Leaves with many transverse intercostals

6 Leaves spirally arranged; stipule-scars annular. Fruit massive, spinous-warted .... **1. Artocarpus**

6 Leaves distichous

7 Base of lamina not or scarcely cordate, edge entire. Inflorescences and syncarp subglobose .... **Artocarpus gomezianus**

7 Base of lamina more or less cordate, edge denticulate or entire

8 Lamina 4–10 cm long, denticulate. Male inflorescence spicate; female subglobose, many-flowered .... **2. Broussonetia**

8 Lamina larger. Male inflorescences discoid, often recurved

9 Lamina 20–50 cm long. Syncarps many-seeded. Cultivated .... **9. Castilla**

9 Lamina smaller. Fruit 1-seeded. Wild .... **8. Antiaris**

1. Tribe ARTOCARPEAE

1. ARTOCARPUS

J.R. et G. Forst., Char. Gen. Pl. 101, t. 51. 1776; Jarrett, J. Arnold Arbor. 40:113–368. 1959; id. 41: 73–140. 1960; Corner, Gard. Bull. Singapore 19: 233. 1962; Sharma, Phytomorphology 15: 185. 1965.

Trees, monoecious, with thick, sticky white latex. Leaves pinnate, pinnatifid or simple. Inflorescences unisexual, pedunculate, thickly spicate to clavate or capitate, covered with the minute sessile flowers and bracteoles. Male flowers with one stamen; perianth tubular, 2–4 partite or merely perforate. Female flowers variously connate; perianth tubular, thin-walled round the ovary, often thick-walled round the style. Syncarp more or less massive, fleshy, pulpy, or with pulpy pericarps, usually many-seeded. Seeds medium-sized to large; embryo straight or slightly curved; cotyledons thick, fleshy, equal or unequal, not folded.

Species c. 50, Ceylon, India and South China to the Solomon Islands, absent from Australia and New Caledonia.

Subgenus ARTOCARPUS

Leaves spirally arranged; stipules amplexicaul, leaving annular scars. Female flowers with the perianths connate in the middle portion, usually with projecting thickened apex round the style. Syncarp covered with conical processes forming a tough rind, the individual fruits enclosed in the persistent, more or less pulpy, saccate parts of the perianth.—*A. heterophyllus*, *A. incisus*, *A. nobilis*.

Subgenus PSEUDOJACA

Trec., Ann. Sci. Nat. Bot. ser. 3, 8: 117. 1847; Jarrett, l.c.

Leaves alternate; stipules small, not amplexicaul, leaving paired scars on the twig. Female flowers with saccate perianth, the apical part usually small or minute. Syncarps often smooth, becoming entirely fleshy or pulpy.—*A. gomezianus.*

N o t e. Only one species of each subgenus is native in Ceylon; they mark a limit of this large genus.

KEY TO THE SPECIES

1 Leaves distichous with paired stipule-scars. Syncarps velvety, globose, 5–8 cm wide. Male heads 1 cm wide, globose .... **1. A. gomezianus**
1 Leaves spirally arranged with annular stipule-scars. Syncarps warted or somewhat spinous, larger. Male heads cylindric, larger
 2 Cauliflorous with very massive syncarps. Leaf 4–10 cm wide, entire .... **2. A. heterophyllus**
 2 Not cauliflorous. Leaf larger
  3 Leaf pinnatifid, very large. Tree of cultivation .... **3. A. incisus**
  3 Leaf entire (lobed in sapling). Wild tree .... **4. A. nobilis**

1. Artocarpus gomezianus Wall. ex Trec., Ann. Sci. Nat. Bot. ser. 3, 8: 118. 1847; Jarrett, J. Arnold Arbor. 41: 88. 1960.

Subsp. **zeylanicus** Jarrett, l.c. 90; Trimen, Handb. Fl. Ceylon 4: 99. 1898 (as *A. lakoocha* Roxb. var. *gomeziana* Wall.); Worthington, Ceylon Trees Pl. 420. 1959 (as *A. gomezianus*).—**Fig. 1.**

Tree –20 m high, evergreen (?); bark grey, somewhat scaly. Twigs 2–4 mm thick, the leaves distichous, pubescent when young. Stipules small, bearing a short scar on either side of the petiole. Lamina 11–25 × 7–16 cm, elliptic, varying ovate to oblong, apex acuminate, base widely cuneate to subcordate, shiny, the sides upcurved, smooth; lateral veins 10–13 pairs, whitish beneath; petiole 1–3 cm long. Inflorescences axillary, mostly solitary; bracteoles peltate. Male heads 8–11 mm wide, globose; peduncles 5–18 mm long; perianth with 3–4 nearly free tepals; stamen 0.5 mm long. Female head stouter; stigmata bifid, exserted to 0.3 mm. Syncarp 2–3 cm wide, subglobose, green then yellow to orange-buff, pulpy; peduncle 13–20 mm long.

D i s t r. Peninsular India, Ceylon.

E c o l. In lowland and submontane forest to 1100 m alt.; flowering March/April and September, fruiting June and February.

U s e s. Fruit edible; timber.

V e r n. Monkey Jak (E); Kana-gona (S).

N o t e. This is a westerly ally of true *A. gomezianus* which occurs in

Indo-Malaya; it differs in the smaller syncarps and smaller male inflorescences with relatively long peduncles.

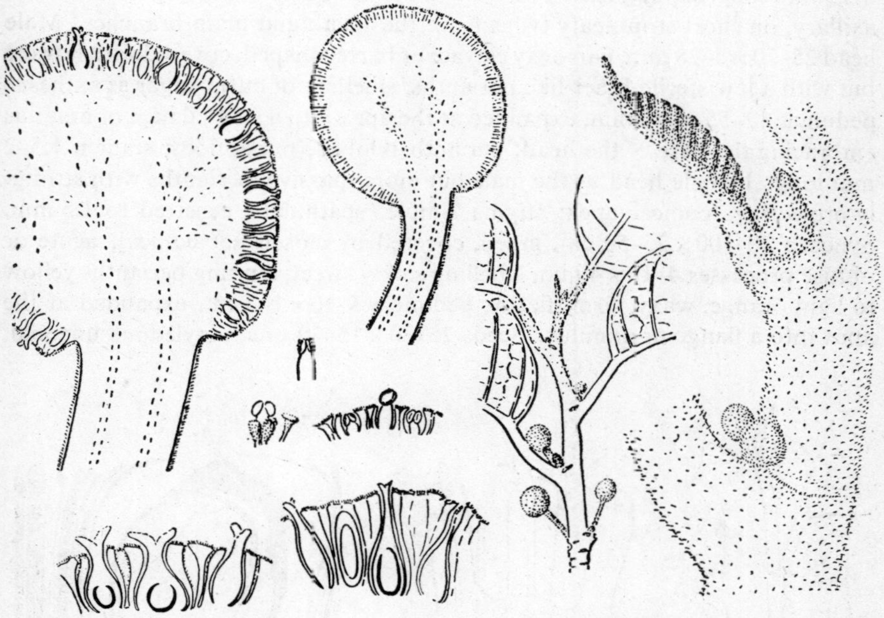

Fig. 1. *Artocarpus gomezianus* ssp. *zeylanicus.* Twig with male inflorescences, × 1; bud with stipule scars, × 5; female and male heads in section, × 5; flowers × 12; young syncarp with very young fruits, × 5.

Specimens Examined. GALLE DISTRICT: Galle, *Champion s.n.* (CGE), *Worthington 2462* (K); Hiniduma, *Worthington 2312* (K). NUWARA ELIYA DISTRICT: Nuwara Eliya, *Champion s.n.* (CGE). KANDY DISTRICT: Kadugannawa, Kolugala, *Hancock 1379* (BM); Kadugannawa, *Worthington 1426* (K); Peradeniya, *Worthington 6744* (K). RATNAPURA DISTRICT: Ratnapura, *s. coll. C.P. 2831* (PDA). LOCALITY UNKNOWN: *s. coll. C.P. 2463* (K); *Macrae 692* (BM, CGE); *s. coll. C.P. 2232* (PDA).

2. Artocarpus heterophyllus Lam., Enc. 3: 210. 1789; Corner, Gard. Bull. Singapore 10: 56. 1939; Jarrett, J. Arnold Arbor. 40: 334. 1959; Worthington, Ceylon Trees pl. 422. 1959 (as *A. integra*).—**Fig. 2.**

Artocarpus integrifolia sensu Trimen, Handb. Fl. Ceylon 4: 99. 1898.

Evergreen tree of cultivation; cauliflorous with massive syncarps; bark often reddish-brown; leaves withering burnt-orange, throughout the year. Twigs 2–6 mm thick, glabrous. Lamina 5–25 × 3.5–12 cm, elliptic to obovate, obtuse to subacuminate, base rather suddenly cuneate and subdecurrent, entire, glabrous; lateral veins 6–10 pairs, with parallel intercostals; petiole 8–25 mm long; sapling leaves with 1–3 pairs of lobes. Inflorescences solitary, axillary, on short stout leafy twigs from the trunk and main branches. Male head 25–70 × 8–28 mm, narrowly clavate or barrel-shaped, covered with flowers but with a few sterile bract-like perianths, smelling of burnt sugar at anthesis; peduncle 12–55 × 2–3 mm, expanded at the apex into a small flange or annulus embracing the base of the head; perianth bilobed, puberulous; stamen 1.5–2 mm long. Female head as the male but more massive; perianths with strongly projecting, conical apex; stigma simple, spathulate, exserted to 1.5 mm. Syncarp 30–100 × 25–50 cm, green, covered by closely set conical, acute or obtuse processes 4–10 × 4 mm, smelling sickly sweet; fruiting perianths yellow to light orange, waxy, firm, fleshy; peduncle 5–10 × 1–3 cm, expanded at the apex into a flange or annulus. Seeds 25–30 × 15–20 mm; cotyledons unequal.

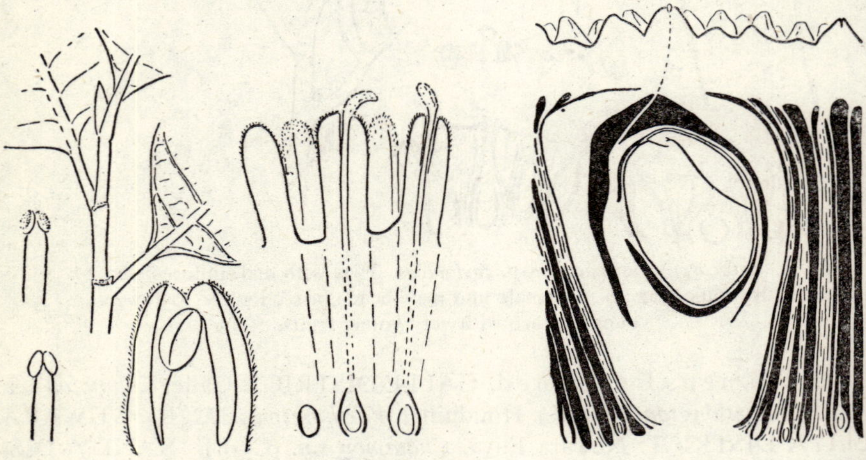

Fig. 2. *Artocarpus heterophyllus*. Twig with annular stipule-scars, × 1; stamens, × 5; male flower-bud in section, × 25; female flowers soon after pollination, × 5; ripe fruit with seed in the syncarp-wall, × 1.

Distr. Widely cultivated throughout the tropics, especially in Southeast Asia, common in Ceylon; origin unknown, possibly South India.

Ecol. Fruit-crop March–April.

Uses. Fruit; timber.

Vern. Jak (E); Kos (S); Pla (T).

Note. Though one of the best known tropical trees, the Jak has been confused botanically with its Malaysian ally the Chempedak to which the name *A. integer* belongs. Both are distinguished by the massive, cauliflorous syncarps, but that of the Jak is one of the most massive fruiting structures in the plant kingdom. There are said to be two forms in Ceylon; 'waraka' has firm pulp in the ripe fruit whereas 'wela' has soft pulp (as in *A. integer*).

Collections. None noted for this very common and well-known species.

3. Artocarpus incisus L. f., Suppl. 411. 1781; Corner, Gard. Bull. Singapore 10: 280. 1939.

Rademachia incisa Thunb., Kongl. Vetensk. Akad. Handl. 37: 253. 1776.
Artocarpus communis J. R. & G. Forst., Char. Gen. 101. 1776; Jarrett, J. Arnold Arbor. 40: 387. 1959.
Artocarpus altilis (Parkinson) Fosberg, J. Wash. Acad. Sci. 31: 95. 1941; Fosberg, Brittonia 12: 101. 1960.
Sitodium altile Parkinson, Journ. Voy. Endeavour 45. 1773.

Evergreen tree with stout twigs and large pinnatifid leaves, cultivated. Twigs 10–20 mm thick; stipules 10–25 cm long. Lamina up to 60 × 20 cm with 1–5 or more pairs of pinnate lobes; petiole 30–60 mm long, stout. Male head 12–30 × 1.5–3 cm, cylindric to clavate, pendent, becoming yellow; peduncle 3–6 cm long; perianth bilobed; stamen 1.5–2.5 mm long. Female head stiffly upright; stigma simple or bifid, exserted to 2 mm. Syncarps 15–30 cm wide, oblong or rounded, green to yellow, set with conical processes –15 × 5 mm, varying simply areolate in seedless varieties; fruiting perianth not fleshy; peduncle 4–13 cm long. Seeds 25 × 20 × 15 mm; cotyledons somewhat unequal.

Distr. Widely cultivated in the tropics, probably originating from New Guinea.

Ecol. Planted in parks and gardens.

Vern. Breadfruit tree (E); Rata-del (S); Era-pla (T).

Note. This well-known tree occurs in two forms, that with seeds and conical processes on the syncarp and the seedless without such processes. According to Worthington the seedless form is more general in Ceylon; in place of the seeds in this form the fruiting perianths develop into starchy edible tissue.

The breadfruit tree has three possible botanical names. I use that by which it has been known longest. The other names, *A. communis* and *A. altilis*, are considered correct by some and incorrect by others.

Collections. None noted.

4. Artocarpus nobilis Thw., Enum. Pl. Zeyl. 262. 1861; King, Ann. R. Bot. Gard. Calcutta 2: 12, t. 10. 1889; Trimen, Handb. Fl. Ceylon 4: 98. 1898; Jarrett, J. Arnold Arbor. 40: 360. 1959; Worthington, Ceylon Trees pl. 423. 1959.

Evergreen tree –25 m high with heavy crown of large, stiff leaves; bark grey with orange scale-scars, fissured by lenticels. Twigs 10–12 mm thick, hispid-scabrid. Stipules 3–12 cm long, hairy. Lamina 14–32 × 8–23 cm, usually obovate, shortly acuminate, base cuneate, margin strongly sinuate and inrolled, smooth; lateral veins 10–13 pairs; petiole 2–3 cm long; sapling leaves pinnately lobed. Inflorescences axillary, solitary or paired. Male head 7–13 × 1.5 cm, cylindric, covered by flowers and bracts; peduncle 3–7 cm long; perianth bilobed; stamen 1.8 mm long. Female head stouter, covered by peltate bracts; stigma simple, exserted to 1 mm. Syncarp up to 20 × 10 cm, cylindric, covered by short, close, rigid, subcylindric processes 1 × 1–1.5 mm, with persistent peltate bracts; peduncle 10–15 cm long; fruiting perianths not fleshy (?). Seeds 8 × 7 mm.

Distr. Ceylon.

Ecol. Frequent in the wet zone forest up to 700 m alt., and in villages; flowering in June.

Uses. Timber, edible seeds.

Vern. Del, Bedi-del (S); Arsini-pla (T).

Note. This fine tree holds an anomalous position in subgen. *Artocarpus*. At the southwest extremity of the subgeneric range it has many points in common with the New Guinea endemic *A. sepicanus* at the other end. Between lies the section *Duricarpus* of Indo-Malaya with which both seem related. It is also noteworthy that this common Ceylon tree has not been found in south India.

Specimens Examined. KANDY DISTRICT: Peradeniya, *s. coll. C.P. 2818* (PDA); Kandy, *Worthington 6749* (K). LOCALITY UNKNOWN: *Koenig s.n.* (BM); *Oltmans 109* (L); *Rouin s.n.* (L).

2. BROUSSONETIA

L'Herit. ex Vent., Tabl. Regn. Veg. 3: 547. 1799; Corner, Gard. Bull. Singapore 19: 233. 1962, with synonyms.

Dioecious unarmed trees, shrubs, or climbers, often (? always) deciduous. Leaves spirally arranged, opposite, or distichous, palmately lobed to undivided, generally cordate to rounded or widely cuneate at the base, dentate to entire, membranous; intercostals numerous; stipules membranous. Bracts and perianth without immersed yellow glands. Male inflorescence racemose to spicate, with a sterile groove; bracts ovate to clavate, truncate, or umbonate, the interfloral bracts generally stalked, those bordering the sterile groove sessile; flowers shortly pedicellate to sessile; tepals 4, joined in the lower half, valvate; filaments 4, inflexed in bud; pistillode spicate, minute. Female inflorescence capitate, globose, without a sterile groove; bracts as in the male, becoming stipitate in fruit; perianth utricular with 2–4 small lobes or teeth; ovary sessile or becoming stipitate in fruit (sect. *Broussonetia*); style 1 (–2), long. Syncarp globose, 10–25 mm wide, the flowers not connate, ripening orange-red, more or less pulpy. Drupes thinly pulpy, invested by the thinly pulpy, yellow to red, perianth. Seeds 1.7–2.5 mm long, ovoid-compressed, faintly keeled; endocarp crustaceous to ligneous, smooth or asperate. Embryo curved: cotyledons equal, thinly fleshy, conduplicate to flat; radicle long, incumbent. Cystoliths hypergenous, hypogenous, or none, often as cystolith-hairs; microscopic gland-hairs 1–3-celled, rather large, capitate; hairs smooth or papillate, often hooked (especially microscopic hairs); sphaerocrystal-cells in the lower epidermis along the veins, or none.

Species about 7, Ceylon to Japan and New Guinea.

Section ALLAEANTHUS

(Thw.) Corner, Gard. Bull. Singapore 19: 234. 1962.

Allaeanthus Thw., Hook. J. Bot. Kew Gard. Misc. 6: 302. 1854.

Drupe sessile. Seed compressed, smooth, ligneous, with simple keel. Cotyledons conduplicate or flat. Cystoliths on the lower side of the leaf. Leaves distichous.

Species 3, Ceylon, Assam to the Philippines and Celebes (? Borneo).

Note. The differences between *Allaeanthus* and *Broussonetia* are so slight that I have reduced them to one genus. It is related with the climber *Malaisia*, both genera having the male inflorescence as in *Streblus* and the female as in a small *Artocarpus*.

Broussonetia zeylanica (Thw.) Corner, Gard. Bull. Singapore 19: 235. 1962. —**Fig. 3.**

Allaeanthus zeylanicus Thw., Hook. J. Bot. Kew Gard. Misc. 6: 302, f. 9b. 1854; Bur. in DC., Prod. 17: 223. 1873; Trimen, Handb. Fl. Ceylon 4: 103. 1898; Alston, Kandy Fl. 35, f. 185. 1938; Worthington, Ceylon Trees pl. 425. 1959.

Deciduous tree to 18 m high; bark smooth, grey, fibrous; latex white; leaves distichous, withering amber-yellow. Twigs, young leaves and inflorescences pubescent with soft white hairs –0.5 mm long. Twigs 1–2 mm thick, fuscous brown. Stipules 6–7 mm long, ovate-lanceolate, denticulate, membranous, subpersistent on young shoots. Lamina 4–10.5 × 1.5–4 cm, ovate to somewhat elliptic or lanceolate, apex acuminate, –15 mm long, base cordate to rounded, acutely serrate, sparsely puberulous above, hirsute beneath along the main veins, then glabrous, smooth, lateral veins 7–15 pairs, oblique; intercostals –12, transverse, regular; basal veins usually short; petiole 8–10 mm long. Male spikes 25–60 mm long with a sterile strip on one side, 1–2 axillary on short shoots from the long twigs, yellow, before or with the new foliage;

peduncle 4–5 mm long; bracts 1–1.3 mm long, ovate, acute, hairy. Female heads 5–8 mm wide, with short peduncle 2–5 mm long; bracts 1–1.3 mm long, 2–3 mm in fruit, subpeltate-clavate, with minutely hooked hairs; perianth saccate with 4 short lobes; style 1, rarely 2, 6–8 mm long, puberulous. Syncarp 20 mm wide. Seeds 2.2–2.6 mm long, smooth. Embryo with corrugate-foliaceous cotyledons (Bureau).

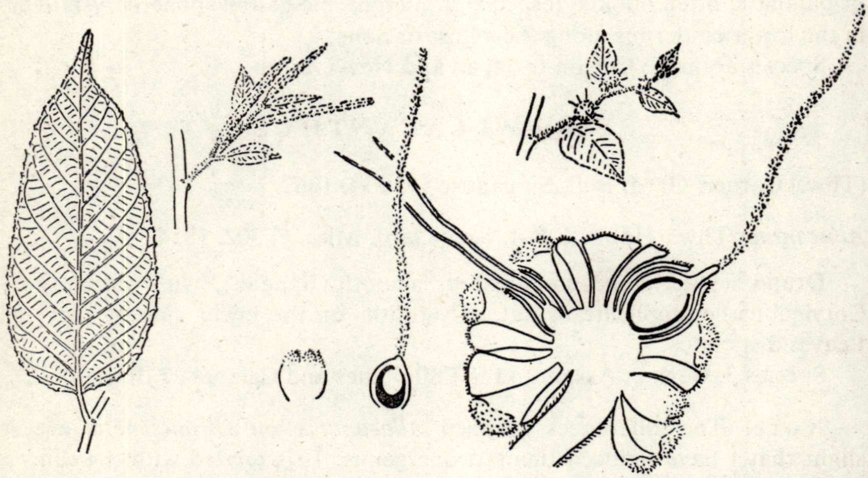

Fig. 3. *Broussonetia zeylanica.* Leaf and twigs with male and female inflorescences, × ½; female perianth, ovary and syncarp, × 6.

D i s t r. Ceylon.

E c o l. In moist regions, rather rare, mostly by rivers; flowering February and August.

V e r n. Alandu (S).

N o t e. This graceful tree, the very tough bark-fibres of which were used for string, seems to have become rare as its lowland forest-habitat has disappeared. No recent material has been available to dissect the flowers and no collection in which the seeds contain embryos. Bureau's description of the cotyledons needs verification. They are flat in the allied Philippine tree *B. luzonica*, which differs in the entire leaves and stipules as well as in the longer male spikes. However, the cotyledons are corrugate-conduplicate in the climbing *B. kurzii* which connects the two other species of this section from Assam to Indochina.

S p e c i m e n s E x a m i n e d. MATALE DISTRICT: Nalanda, *Worthington 1091*, *6686* (K). MONERAGALA DISTRICT: Bibile, *Worthington 2983*, *5432*, *5438* (K). KANDY DISTRICT: Kadugannawa, *Worthington 1441* (K). AMPARAI DISTRICT: Mahaoya, *Worthington 5312* (K). KURUNEGALA DISTRICT: Omeragala, *Worthington 864* (K). LOCALITY UNKNOWN: *Macrae 358*, *359* (CGE).

3. MACLURA

Nutt., Gen. Amer. Pl. 2: 233. 1818, gen. conserv.; Corner, Gard. Bull. Singapore 19: 235. 1962, with generic synonymy.

Dioecious trees, shrubs, climbers, generally with axillary spines. Leaves spirally arranged to distichous, simple, without glands. Male inflorescences paniculate, spicate with a sterile groove, or capitate. Female inflorescences capitate or shortly spicate. Bracts and/or tepals with 2–7 immersed yellow glands. Bracts spathulate to clavate or none. Male flowers 4-merous; tepals free; anthers small, extrorse to laterally dehiscent; filaments inflexed or straight in bud; pistillode subulate or as a slender sterile ovary or none. Female flowers sessile, free or connate proximally; tepals 4, decussate, obtuse, free or gamophyllous to the middle, fleshy in fruit; ovary free or sunk in sockets in the receptacle; stigmatic arms 1–2, short, or long. Drupes very thinly pulpy, enclosed in the fleshy perianth, generally not projecting, often concrescent below into a fleshy syncarp. Seed compressed, with more or less ligneous endocarp 2.5–10 mm long. Embryo curved; radicle long, asymmetrically incumbent to accumbent; cotyledons equal, folded or not, rather thin. Cystoliths amphigenous or none; microscopic gland-hairs 4-celled, capitate; generally with sphaerocrystal cells in the lower epidermis, at least along the veins; hypodermis none; hairs smooth.

Species 12, temperate and tropical, America, Africa, Asia.

Section CUDRANIA

(Trec.) Corner, Gard. Bull. Singapore 19: 237. 1962.

Vanieria Lour., Fl. Cochinch. 564. 1790.
Cudrania Trec., Ann. Sci. Nat. Bot. ser. 3, 8: 122. 1847, gen. conserv.
Cudranus Rumph. ex Miq., Fl. Ind. Bat. 1, 2: 290. 1859.
Maclura subgen. *Leptosura* Blume, Ann. Mus. Bot. Lugduno Batavum 2: 83. 1856.

Trees, shrubs, climbers with leafless axillary thorns in place of short shoots. Male and female inflorescences capitate, bracteate; bracts with 2 immersed yellow glands; tepals free or gamophyllous in the lower half, with 2–7 immersed yellow glands. Male flowers with the filaments straight in bud; pistillode subulate. Female flowers with the ovaries free or in sockets of the receptacle; style short, simple or with two short arms. Syncarp fleshy, –40 mm wide, the drupes generally concealed. Seeds 3–9 mm long. Cotyledons broad with infolded margins and obliquely incumbent radicle, or small and flat with accumbent radicle. Cystoliths none.

Species 5, Asia, Australasia.

Ser. Connatae

Corner, Gard. Bull. Singapore 19: 238. 1962.

Female flowers connate in their lower parts, the ovaries sunk in receptacular sockets, tepals joined in the lower part; style simple, rarely with a minute tooth-like branch. Anthers scarcely exsert, the filaments shorter than the anthers or scarcely exceeding them. Seeds 3–5 mm long. Cotyledons with infolded margins and obliquely incumbent radicle.

Species 3.

Notes. The syncarp of *Maclura* distinguishes it as one of the main genera of Moraceae. The free tepals become fleshy, as in *Morus*, and the endocarps, forming the seed-coats, are compressed and ligneous. Because of the straight filaments of the stamens in bud, *Cudrania* is usually separated and classified artificially with *Artocarpus*; in the Indian *M.* (*Cudrania*) *fruticosa* Roxb. the filaments have inflexed tips and prove that this criterion is fallible.

Maclura cochinchinensis (Lour.) Corner, Gard. Bull. Singapore 19: 239. 1962. —**Fig. 4.**

Vanieria cochinchinensis Lour., Fl. Cochinch. 564. 1790.
Cudrania javanensis Trec., Ann. Sci. Nat. Bot. ser. 3, 8: 123. 1847; Trimen, Handb. Fl. Ceylon 4: 98. 1898.
Plecospermum cuneifolium Thw., Hook. J. Bot. Kew Gard. Misc. 6: 303. 1854.
Maclura javanica Blume, Ann. Mus. Bot. Lugduno Batavum 2: 84. 1856.
Cudrania cochinchinensis (Lour.) Backer, Blumea 6: 302. 1947–8.

Much branched thorny shrub developing long scrambling sprays, or becoming a woody liane with stems –3 cm thick; latex white (? or yellowish); inner root-bark orange. Thorns –5 cm long, unbranched, straight or slightly decurved, developing at every node on long shoots. Twigs, stipules, petioles, and underside of the main veins thinly fulvous or whitish hairy (hairs –0.5 mm long) or appressedly hairy, glabrescent or glabrous from the first. Twigs 1–2.5 mm thick, pale ochraceous to brownish, developing a short shoot from the base of the spine. Stipules –3 mm long, caducous. Lamina 4–11 × 2.5–5 cm, elliptic, acute or subacuminate, often mucronate, base cuneate, entire, thinly to distinctly coriaceous, smooth, drying greyish green; on upper shoots, more or less obovate or lanceolate-obovate, obtuse, and diminishing to 0.5 × 0.3 cm; lateral veins 6–9 (–14) pairs, 4–7 in the upper obovate leaves; intercostals 0–1, ill-developed; basal veins short; glands none; petiole 3–13 (–20) mm long, becoming raised on a slight cauline knob. Inflorescences axillary, paired or solitary. Male heads 7–10 mm wide, cream-yellow; peduncles 3–10 mm long; tepals free, closely to thinly hairy; anthers not exsert. Female

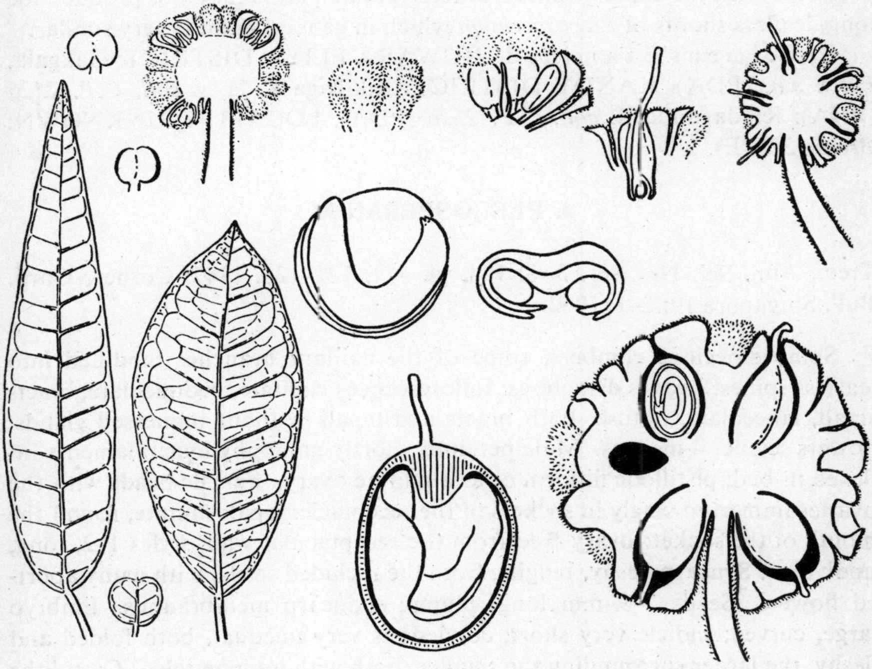

Fig. 4. *Maclura cochinchinensis*. Sapling leaf (left), adult leaf (right) and small leaf of a new shoot, × ½; male and female inflorescences and syncarp in l.s., × 4; male and female flowers (in section), stamens, fruit (in section) and embryos, × 8.

heads 6–8 mm wide at anthesis; tepals joined in the lower part and fused with adjacent flowers; style 1.5–2.5 mm long. Syncarp 15–20 mm wide, ripening yellow to orange and red, succulent; peduncle 3–12 mm long. Seeds 4–5 × 3.5–4.5 × 2–2.5 mm, brown, the hard wall c. 0.15 mm thick.

D i s t r. Ceylon to Japan, New Britain, Australia and New Caledonia.

E c o l. Lowland forest and to 1800 m alt., not common, flowering in April.

U s e s. Fruits edible; wood for dyeing silk, yellow or green.

N o t e. This widespread species is variable. In Malaysia there are all degrees of hairiness from the glabrous to the closely hairy var. *pubescens* (Trecul) Corner, which may not occur in Ceylon. The variety in leaf-shape must be studied in living plants. It seems that seedlings grow into bushes to 1 m high in open ground and have very small, more or less orbicular and obtuse leaves 1–1.5 cm long. Then, on developing long scrambling sprays, the spines appear and the leaves become oblong lanceolate, to 12 × 2 cm with up to 18 pairs of lateral veins. Finally the shorter elliptic or obovate leaves are produced with the axillary inflorescences. Adult plants may not produce the long, leafless shoots of *Plecospermum* which in general habit is very similar.

Specimens Examined. NUWARA ELIYA DISTRICT: Hakgala, *Petch s.n.* (PDA). KANDY DISTRICT: Rikillagaskada, *s. coll. C.P. 2526* (PDA); Kindagodile, *s. coll. C.P. 2526* (PDA). LOCALITY UNKNOWN: *Brodie 225* (E).

4. PLECOSPERMUM

Trec., Ann. Sci. Nat. ser. 3, 8: 124, pl. 4, f. 121–126. 1847; Corner, Gard. Bull. Singapore 19: 241. 1962.

Spiny dioecious climbers, some of the axillary branches modified into leafless spines. Leaves distichous. Inflorescences capitate, pedunculate; bracts small, lanceolate, obtuse, both bracts and tepals without immersed glands; flowers sessile, 4-merous. Male perianth shortly gamophyllous; filaments inflexed in bud; pistillode filiform or as a minute ovary. Female heads with the ovaries immersed singly in sockets of the receptacle; tepals minute, round the mouth of the socket; ovary free from the receptacular wall; styles 1–2, long, undivided. Syncarp fleshy, bulging from the included seeds, with many aborted flowers. Seeds 7–9 mm long, plump; endocarp membranous. Embryo large, curved; radicle very short; cotyledons very unequal, both folded and fleshy, the larger surrounding the smaller, both with long petioles. Cystoliths none.

Species 2 (?3), Ceylon, India, Burma, Indo-China, Thailand, Andaman Isl.

This problematic relic seems related with *Malaisia* but is more advanced with capitate male inflorescence as well as female. The distribution is peripheral to the rain-forest of the Malayan region.

Plecospermum spinosum Trec., l.c.; Hook. f., Fl. Br. Ind. 5: 491. 1888; Trimen, Handb. Fl. Ceylon 4: 103. 1898; Alston, Kandy Fl. 34, f. 183. 1938. **—Fig. 5.**

Trophis spinosa Roxb. ex Willd., Sp. Pl. 4: 734. 1806.
Trophis aculeata Roth, Nov. Sp. Pl. 368. 1821.
Batis spinosa Roxb., Fl. Ind. 3: 762. 1832.

Massive woody climber, at first shrub-like; trunk –15 cm thick; bark fuscous brown, fissured, rugged, transversely cracked; latex white; new shoots reaching 2–3 m long, with long internodes, scale-leaves and more or less reflexed axillary spines; foliage leaves on the short laterals, laxly spiral, glabrous or puberulous. Stipules minute. Lamina 5–9 × 2.5–3.5 cm, elliptic to obovate, acute or subobtuse, base cuneate, entire, subcoriaceous, smooth;

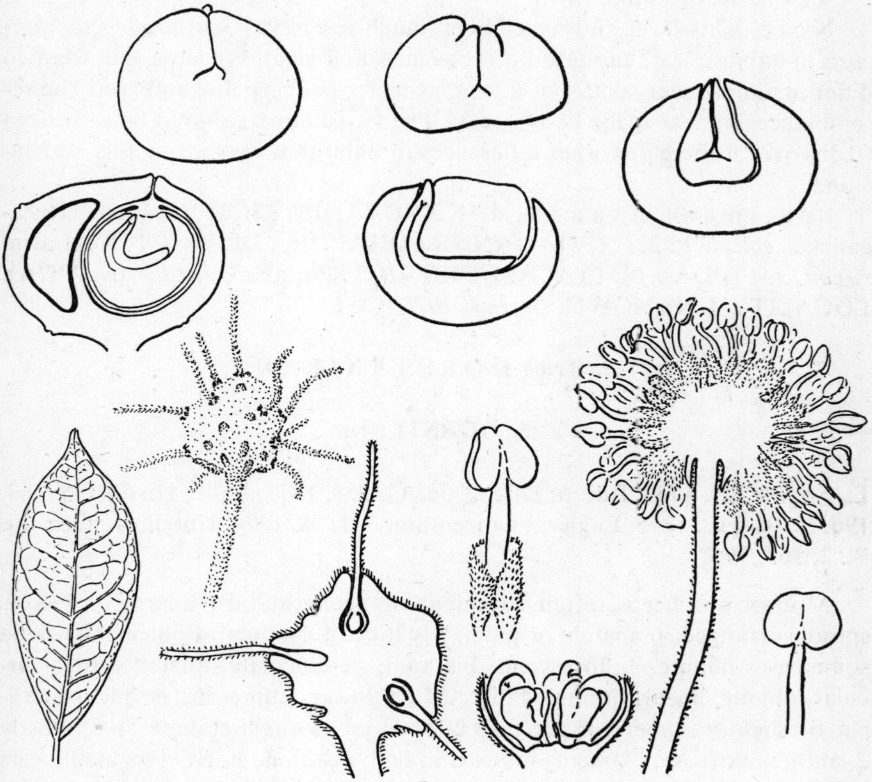

Fig. 5. *Plecospermum spinosum.* Leaf, × ½; female head (at anthesis), × 4; and in l.s., × 8; male head (in section), × 4; male flower-bud (in section) and stamen, × 8; syncarp (in section, upper left), × 2; embryos, × 4.

lateral veins 4–7 pairs; petiole 4–9 mm long. Male inflorescence 10 mm wide (overall) at anthesis, –7 mm wide in bud; peduncle 12–20 × 1 mm, puberulous or glabrous; bracts at the base of the head 1 mm long, several, lanceolate or subulate, puberulous; bracteoles scattered among the flowers, similar or smaller; flowers sessile; perianth green, puberulous, the lobes ovate-acute; anthers with lateral dehiscence. Female inflorescence as the male; style 1, 3–5 mm long, hairy, projecting. Syncarp 10–15 mm wide, bluish-green, subglobose, faintly knobbly, closely brownish velutinate, 1–2 seeded, the wall 1–1.5 mm thick without woody tissue; pericarp 0.2 mm thick, with thin fibrous endocarp.

Distr. Ceylon, Peninsular India.

Ecol. Frequent in low country—1000 m alt., flowering February–April.

V e r n. Katu-timbol (S).

N o t e. This large thorny climber much resembles *Maclura cochinchinensis* in habitat, leaf, capitate inflorescences and small syncarps, but the two differ in almost every detail of flower, syncarp, pericarp and embryo. The resemblances appear to be convergent. The two climbers should be compared to discover if there are other differences in habit that may prove this conclusion.

S p e c i m e n s E x a m i n e d. KANDY DISTRICT: Hantane, Haragama, *s. coll. C.P. 2212* (PDA). ANURADHAPURA DISTRICT: Kekirawa, *s. coll. s.n.* (PDA). PUTTALAM DISTRICT: Puttalam, *s. coll. s.n.* (PDA). LOCALITY UNKNOWN: *Pearson 1078* (CGE).

2. Tribe DORSTENIEAE

5. DORSTENIA

L., Sp. Pl. 121. 1753; Bur. in DC., Prod. 17: 258. 1873; Baill., Hist. Pl. 6: 144, 196, f. 108–113. 1875; Engler, Pflanzenfam. 3, 1: 78. 1894; Hutchins., Gen. Fl. Pl. 2: 163. 1967.

Monoecious herbs, often succulent or pachycaulous, unarmed. Leaves spirally arranged, pinnately or palmately lobed to simple; stipules in pairs, in some cases minute. Inflorescence bisexual, pedunculate, dilated into a circular, oblong, linear, or angled disc. Male flowers numerous, sessile, superficial or slightly immersed; perianth 2-lobed or toothed; stamens 1–3, mostly 2, anthers extrorse, filaments inflexed in bud; pistillode none. Female flowers numerous, generally surrounded by 3–4 males; perianth slight or none; ovary immersed in a hollow in the receptacle; style short, with two short stigmatic arms. Syncarp fleshy, extruding the small crustaceous seeds. Embryo transverse in the lignified endocarp; radicle incumbent; cotyledons thin, equal, flat or slightly recurved.

Chiefly in lowland tropics, 150–200 species, mostly America and Africa, Arabia (1), India (1).

Dorstenia indica Wight, lc. Pl. Ind. Or. 6: 1964. 1853; Bur. in DC., Prod. 17: 272. 1873; Hook. f., Fl. Br. Ind. 5: 494. 1890; Trimen, Handb. Fl. Ceylon 4: 102. 1898; Gamble, Fl. Pres. Madras 3: 1370. 1928; Fyson, Fl. South Ind. Hill Stations 1: 542, f. 474. 1932.—**Fig. 6.**

Herb, tufted, stems fleshy, tapering, curved-ascending, –45 cm high, unbranched but proliferating from the base, rooting from the underside; young parts shortly white hairy; leaves spirally arranged; stipules apparently none; latex white. Lamina 2.5–12.5 × 1.3–4.5 cm, elliptic to sublanceolate, apex subacute, base subrotund or cuneate, margin dentate to sinuate, membranous;

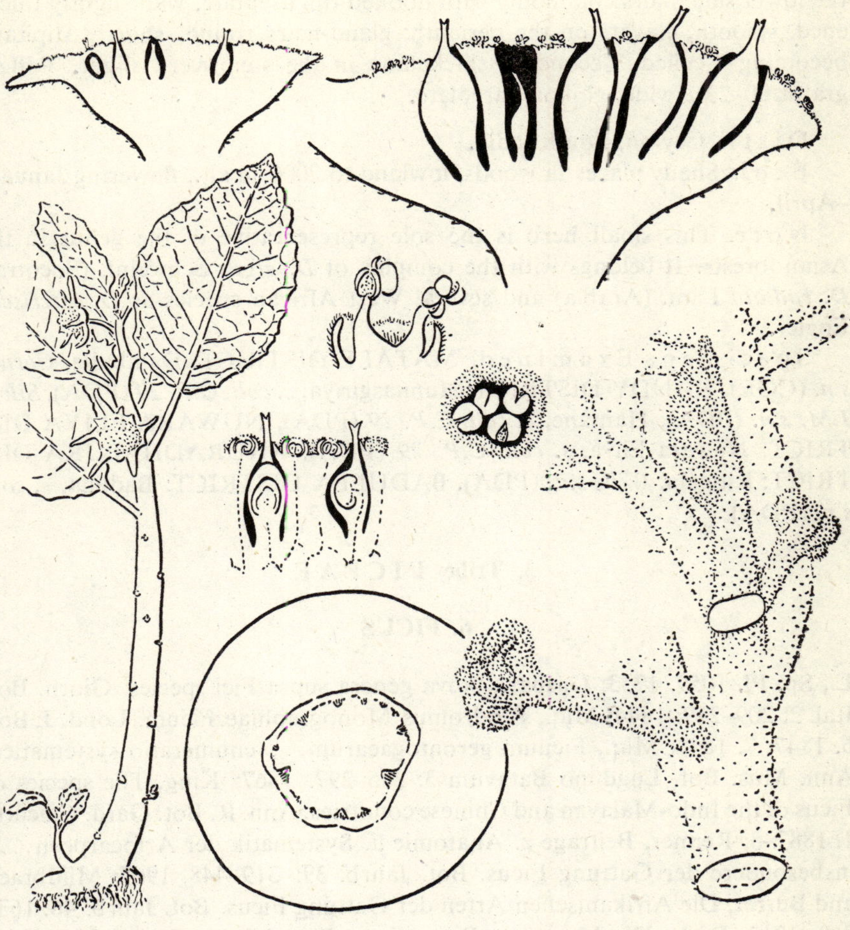

Fig. 6. *Dorstenia indica*. Small plant, × ½; stem in t.s., × 2½; apex of stem with young inflorescences, × 2½; capitulum (in section) at female anthesis (left) and male (right), with fruits forming, × 2½; female flowers (in section) with male flowers in bud, × 10; male flower in section and viewed from above, × 25.

lateral veins 5–9 pairs, reticulation lax; petiole 6–20 mm long. Inflorescences solitary, axillary, often somewhat decurved, green, set with 5–12 linear marginal lobes –7 mm long; peduncle 2–3.5 cm long, dilating into the disc 8–16 mm wide. Male flowers numerous, very small, 0.5 mm high; perianth with 2 slight lobes, puberulous; stamens 1–2 (–3). Female flowers immersed in the disc, opening before the male; perianth vaguely 2-lobed; ovary stalked, stigma bifid. Seed 2 mm, crustaceous, minutely papillate. Leaf without cystoliths; epidermal cells undulate (to subrectangular above); stomata only on the lower side; hairs commonly with hooked tip, aseptate, wall slightly thickened, smooth, clavate on the perianth; gland-hairs round, shortly stipitate, becoming 4-celled. Secondary thickening in the stem very slight. Pollen-grains 20–22 μ wide, globose, areolate.

Distr. Ceylon, South India.

Ecol. Shady places in woods, lowland to 2000 m alt., flowering January –April.

Note. This small herb is the sole representative of the genus in the Asian forests. It belongs with the complex of *D. gigas* Schweinf. (Socotra), *D. radiata* Lam. (Arabia) and several West African species as *D. asteriscus* Engler.

Specimens Examined. MATALE DISTRICT: Dambulla, *Corner s.n.* (CGE). KANDY DISTRICT: Hunnasgiriya, *s. coll. C.P. 29* (PDA), *Silva, J.M. s.n.* (PDA); Hantane, *s. coll. C.P. 29* (PDA). NUWARA ELIYA DISTRICT: Nuwara Eliya, *s. coll. C.P. 29* (PDA). ANURADHAPURA DISTRICT: Ritigala, *Wilis s.n.* (PDA). BADULLA DISTRICT: Badulla, *s. coll. s.n.* (PDA).

3. Tribe FICEAE

6. FICUS

L., Sp. Pl. 1059. 1753; Gaspar., Nova genera supra Fici species. Giorn. Bot. Ital. 2: 209–219. 1844; Miq., Prodromus Monographiae Ficum. Lond. J. Bot. 6. 1847, 7. 1848; Miq., Ficuum gerontogaearum....enumeratio systematica. Ann. Mus. Bot. Lugduno Batavum 3: 285–297. 1867; King, The species of Ficus of the Indo-Malayan and Chinese countries. Ann. R. Bot. Gard. Calcutta 1. 1887–8; Renner, Beitrage z. Anatomie u. Systematik der Artocarpeen.... insbesondere der Gattung Ficus. Bot. Jahrb. 39: 319–448. 1907; Mildbraed and Burret, Die Afrikanischen Arten der Gattung Ficus. Bot. Jahrb. 46: 163–269. 1912; Diels, Die Moraceen Papuasiens. Bot. Jahrb. 67: 144–235. 1935; Elmer, A fascicle of Sorsogon Figs. Leafl. Philipp. Bot. 9: 3427–3431. 1937; Sata, A Monographic Study of the Genus Ficus. Contr. Inst. Hort. Ec. Bot. Fac. Agric. Taihoku Imp. Univ. n. 32. 1944; Vreede, Topography of the laticiferous system in the genus Ficus. Ann. Bot. Gard. Buitenzorg 51: 125–149. 1949; Corner, An introduction to the distribution of Ficus. Reinwardtia 4: 15–45. 1958; Corner, Taxonomic notes on Ficus, Asia and Australasia. Gard. Bull. Singapore 17: 368–485. 1960, 18: 1–69. 1960, 18: 83–97. 1961, 19: 385–401. 1962; Corner, Check-list of Ficus in Asia and Australasia with keys to identification. Gard. Bull. Singapore 21: 1–186. 1965; Corner, Ficus in the Solomon Islands and its bearing on the post-Jurassic history of Melanesia. Philos. Trans. R. Soc. B. 253: 23–159. 1967; Wiebes, Provisional host catalogue of fig-wasps (Hymenoptera, Chalcidoidea). Zool. Verh. Leiden 83: 1–44.

1966; Hill, Figs of Hong Kong. Hong Kong University Press. 1–130. 1967; Condit, Ficus, the exotic species. University of California, Division of Agricultural Sciences 1–363. 1969; Galil and Eisikowitsch, Further studies of the pollination ecology of *Ficus sycomorus*. Tijdschr. Ent. 112: 1–13. 1969.

Woody plants with latex. Stipules free and paired or connate. Leaves simple or palmately lobed, varying pinnately and palmately lobed in saplings, symmetric or asymmetric, dentate or entire, generally with glands on the underside in the axils of the lateral or basal nerves or abaxial at the apex of the petiole, in some cases on the twig at the base of the petiole. Inflorescences paired, varying solitary or clustered, axillary to cauliflorous, urceolate and containing the flowers, the orifice closed by small bracts, with few to many external bracts; bracteoles between the flowers present or not; monoecious or dioecious, strongly protogynous. Flowers unisexual, small to minute, of three kinds (male, gall- and female), in some cases also with neuter flowers (sterile male). Perianth with 2–8, free or joined, red to white, tepals, varying wholly gamophyllous and saccate or cupular to none. Stamens 1–7, free or, if 2, often with the filaments shortly joined; anther generally exsert, often mucronate when large, 4-celled with longitudinal introrse dehiscence or 2-celled with apical crescentic or equatorial transverse dehiscence: pollen-grains small, smooth. Ovary unilocular: style single, subapical to gynobasic; stigma bifid to simple, often shortly infundibuliform in gall-flowers; ovule 1, anatropous, integuments 2. Male flowers ostiolar in one or several rows or disperse within the fig, in some cases with a functional gall-ovary, with a vestigial pistillode or without. Gall-flowers disperse, as modified female flowers, each containing one developing insect, sterile. Fruit (from each female flower) as a drupelet, thinly pulpy, the woody endocarp forming a pyrene 0.5–5 mm long. Seed (true seed) with membranous integuments, albuminous or practically exalbuminous; embryo curved with conduplicate cotyledons in larger seeds to straight with flat cotyledons in small seeds; epigeal. Inflorescence fruit (fig) more or less fleshy, rarely dehiscent, often with sclerotic cells.

Tropics and subtropics, c. 1000 spp.; c. 450 spp. in Asia and Australasia; 23 spp. in Ceylon with several introduced.

The fig-trees of tropical towns and villages are a part of Asian life. In Ceylon, this is manifest by the abundance of the Sacred Bo (*F. religiosa*), of *F. benghalensis*, *F. tsjahela*, and *F. amplissima*, while *F. mollis* frequents the ancient ruins. Symbiotic with the genus and part of its character are the small fig-wasps (Hymenoptera-Chalcidoidea) on which the propagation of the wild figs depends. Introduced species of *Ficus*, such as *F. elastica* and *F. benjamina* in Ceylon, are seedless unless the specific fig-wasp has also been introduced; and this seems to have been the case long ago with *F. religiosa*. The insects escape when the figs begin to ripen and the abundance of these insects in April is said to be a feature of Kandy. How the females deliberately carry pollen, stuffed into special thoracic pockets, has been discovered only very recently (Galil, 1969).

It is not difficult to recognise the relatively few species of *Ficus* in Ceylon from their general habit and from the venation of the leaves. This point is emphasised in the illustrations; the developmental significance of the different kinds of venation is explained elsewhere (Corner, 1967). A special form of leaf is the cordate, derived from the elliptic by the curtailing of apical growth of the lamina and expansion of the base with additional basal veins. The Ceylon flora presents three parallel examples of this, namely *F. religiosa* and *F. tsjahela* (or *F. virens*), *F. arnottiana* and *F. costata*, and *F. benghalensis* and *F. fergusonii*. The cordate-leafed ally of *F. amplissima* is *F. rumphii* of Indomalaysia. *F. benjamina* and *F. trimenii* show the transition from transverse veining, as in *F. drupacea*, to the parallel side-veins of *F. elastica*.

The botanical classification of *Ficus* is intricate and depends on details of the microscopic flowers. It is explained in the following conspectus of the Ceylon species. The key on p. 235 is artificial. The species are arranged in the order of my check-list (Corner, 1965). Wherever it is known, the main pollinating insect is mentioned on the authority of Wiebes (1966).

Subgeneric Classification of Ficus

1. Subgen. **Urostigma** (Gasp.) Miq., Ann. Mus. Bot. Lugduno Batavum 3: 285. 1867; Corner, Gard. Bull. Singapore 17: 370. 1960; id. 21: 6, 99, 100. 1965.

Monoecious with male, female and gall-flowers in the same syconium; usually with interfloral bracts; without external lateral bracts. Stamen 1. Style simple. Seeds smooth. Sclerotic cells in an outer and an inner layer in the fig-wall. Usually epiphytic stranglers with entire coriaceous leaves; leaf-gland at the back of the petiole-apex.

sect. **Urostigma** Corner, l.c. 371. 1960.

Male flowers ostiolar. Ovaries entirely red-brown or in the upper half. Cystoliths only on the underside of the lamina. Petiole usually articulate to the lamina, separating in fallen leaves.

ser. **Religiosae** Miq., l.c. 287; Corner, l.c. 371. 1960.

Fig without internal bristles.—*F. religiosa*, *F. tsjahela*.

ser. **Caulobotryeae** (Miq.) Corner, l.c. 371. 1960.

Figs with abundant chaffy-vesicular internal bristles.—*F. virens*, *F. caulocarpa*.

sect. **Leucogyne** Corner, l.c. 371. 1960.

Male flowers disperse. Ovaries white. Cystoliths only on the upper side of the leaf.—*F. amplissima*.

sect. **Conosycea** (Miq.) Corner, l.c. 371. 1960.

Male flowers disperse. Ovaries with a red mark at the base or along the stylar side. Cystoliths on both sides of the lamina or only on the upper side. Petiole not articulate. Lamina with distinct intercostal venation or with the secondary lateral veins almost as prominent as the primary.

subsect. **Conosycea** Corner, l.c. 372. 1960.

Venation with transverse intercostals.

ser. **Validae** Miq., l.c. 285; Corner, l.c. 372. 1960.

Figs pedunculate.—*F. arnottiana, F. costata.*

ser. **Drupaceae** Corner, l.c. 372. 1960.

Figs sessile.—*F. drupacea, F. mollis, F. benghalensis, F. fergusoni, F. pubilimba.*

subsect. **Dictyoneuron** Corner, l.c. 373. 1960.

Leaf without transverse intercostals, the secondary lateral veins much less conspicuous than the primary.—*F. talboti.*

subsect. **Benjamina** (Miq.) Corner, l.c. 374. 1960.

Leaf without transverse intercostals, the secondary lateral veins almost as prominent as the primary.

ser. **Benjamineae** Miq., l.c. 287; Corner, l.c. 374. 1960.

Leaves thinly coriaceous, drying grey-green to light brownish with the veins distinctly raised on both sides.—*F. benjamina.*

ser. **Callophylleae** Corner, l.c. 374. 1960.

Leaves rather thickly coriaceous, drying dark brown with the petiole blackish; basal veins often elongate.—*F. microcarpa, F. trimenii.*

sect. **Stilpnophyllum** Endl., Gen. Pl. 4, 2: 35. 1847; Corner, l.c. 24. 1962.

Leaf with the secondary lateral veins as prominent as the primary. Fig thickly and shortly pedunculate, the basal bracts very early caducous. Cystoliths on the upper side of the lamina.—*F. elastica.*

2. Subgen. **Pharmacosycea** Miq., Ann. Mus. Bot. Lugduno Batavum 3: 299. 1867; Corner, Gard. Bull. Singapore 17: 405. 1960; id. 21: 28, 100, 111. 1965.

As subgen. *Urostigma* but the stamens 1–3, the stigma bifid, the sclerotic cells diffuse throughout the fig-wall or none. Usually tall buttressed independent trees; leaf-glands in the axils of the basal nerves on the underside of the lamina.

ser. **Vasculosae** Corner, l.c. 406. 1960.

Leaves drying pale green to brownish, the reticulations often raised on one or both surfaces.—*F. callosa.*

ser. **Nervosae** Corner, l.c. 407. 1960.

Leaves drying brown.—*F. nervosa.*

3. Subgen. **Sycomorus** (Gasp.) Miq., Ann. Mus. Bot. Lugduno Batavum 3: 295. 1867.—*Sycomorus* Gasp., Giorn. Bot. Ital. 2: 219. 1844.

As subgen. *Urostigma* but without interfloral bracts; male flowers ostiolar, sessile, compressed, the two stamens with more or less connate filaments; gall- and female flowers with laciniate-denticulate tepals and simple stigma. Leaves often dentate, more or less plicate in bud; leaf-glands in the axils of the basal veins. Buttressed trees, usually cauliflorous.—*F. racemosa.*

4. Subgen. **Ficus** Corner, Gard. Bull. Singapore 17: 4–6. 1960; id. 21: 35, 99, 116. 1965.

Dioecious; gall-plants with male and gall-flowers in the figs, seed-plants with neuter and female flowers. Interfloral bracts absent. Lateral bracts often on the outside of the fig. Sclerotic cells, when present in the fig-wall, usually only as an inner layer near the flowers. Leaves often dentate; leaf-glands various. Seeds often with distinctive form or marking. Habit various.

sect. **Ficus** Corner, l.c. 417. 1960.

Male flowers pedicellate with 2 stamens. Stigma often bifid. Trees or shrubs.—*F. carica* (? occasionally grown in Ceylon).

sect. **Rhizocladus** Endl., Gen. Pl. 4, 2: 34. 1847; Corner, l.c. 18: 3. 1960; 21: 50, 116, 123. 1965.

Root-climbers. Male flowers with 2 stamens. Female flowers usually with bifid stigma. Seeds oblong, generally compressed and keeled.

ser. **Plagiostigmaticae** Corner, l.c. 18: 3. 1960.

Male flowers ostiolar, anthers mucronate.— *F. pumila*, *F. laevis*.

ser. **Distichae** Corner, l.c. 18: 5. 1960.

Male flowers disperse, anthers not mucronate.—*F. diversiformis*.

sect. **Sycidium** Miq., Hook. Lond. J. Bot. 7: 228. 1848; Corner, Gard. Bull. Singapore 17: 443. 1960; id. 21: 62, 117, 130. 1965.

Male flowers ostiolar, mostly with 1 stamen. Seed lenticular to shortly oblong. Leaves often dentate. Trees, shrubs, or epiphytic stranglers.

subsect. **Varinga** (Miq.) Corner, l.c. 17: 446. 1960.

Male flowers without a normal gall-ovary. Seeds shortly oblong.

ser. **Heterophylleae** Corner, l.c. 446. 1960.

Figs pedunculate without lateral bracts. Seeds smooth.—*F. heterophylla*.

ser. **Exasperatae** Corner, l.c. 17: 446. 1960.

Figs pedicellate with small scattered lateral bracts. Seeds reticulately pitted.—*F. exasperata*.

subsect. **Palaeomorphe** (King) Corner, l.c. 17: 446. 1960.

Male flowers with a gall-ovary. Seed shortly oblong, usually with a short apical keel. Figs pedicellate. Epiphytic stranglers.—*F. tinctoria*.

sect. **Sycocarpus** Miq., Ann. Sci. Nat. Bot. ser. 3, 1: 33. 1844; Miq., Fl. Ind. Bat. 1, 2: 322. 1859; Corner, Gard. Bull. Singapore 18: 37. 1960, 21: 84, 116, 147. 1965.

Male flowers with 1–2 stamens. Perianth saccate or cupular, not divided into tepals. Seed often rough with prominent hilum. Leaves usually dentate. Twigs often with a gland at the node. Trees or shrubs.—*F. hispida*.

KEY TO THE SPECIES

1 Climbers or creepers (not strangling figs). Figs stalked
2 Cultivated. Leaves strongly veined beneath. Figs pear-shaped, 3–6 cm long.......... ....**21. F. pumila**
2 Wild. Figs round, smaller
3 Leaf ovate, 5–20 cm wide. Fig 2–3 cm wide..........................**22. F. laevis**
3 Leaf and fig much smaller..**23. F. diversiformis**
1 Trees, shrubs, or strangling figs
4 Small strangling shrub. Leaves scabrid, often toothed or lobed. Figs solitary, stalked, axillary..**24. F. heterophylla**
4 Trees or stranglers
5 Cauliflorous
6 Leaves opposite, scabrid, hairy, denticulate. Figs in long sprays, ripening yellow. Small tree or bush.. ....**27. F. hispida**
6 Leaves spirally arranged, smooth, entire. Figs red. Large buttressed tree......... ....**20. F. racemosa**
5 Not cauliflorous
7 Leaves very scabrid, denticulate, basal veins elongate. Fig stalked, yellow, with a few small lateral bracts..**25. F. exasperata**
7 Not so. Leaves entire (at least in the adult state)
8 Leaves with numerous, very close side-veins; basal veins not elongate. Strangling figs, often cultivated
9 Aerial roots round the trunk copious. Leaf rather large; stipules red. Figs yellow, with short thick stalk...**17. F. elastica**
9 Aerial roots often few or none round the basketing trunk. Figs pink, red or purple
10 Leaf thin, acuminate, small; petiole slender. Fig sessile or stalked-pyriform.... ....**14. F. benjamina**
10 Leaf coriaceous, subacute; petiole rather thick, drying black. Wild, rare...... ....**16. F. trimenii**
8 Leaf with the side-veins distinctly spaced; basal veins elongate or not
11 Fig stalked
12 Fig with small bracts at the base of the stalk. Leaves rather small and narrow
13 Strangler. Leaf asymmetric. Fig scabrid hairy.................**26. F. tinctoria**
13 Large buttressed tree. Leaf symmetric. Fig smooth...........**19. F. nervosa**
12 Fig with small bracts at the top of the stalk
14 Leaf ovate-cordate. Small tree (? strangler)................**6. F. arnottiana**
14 Leaf elliptic, the base cuneate to subcordate
15 Large buttressed tree. Leaves large, shiny, stiff, scabrid beneath. Figs greenish yellow..**18. F. callosa**
15 Stranglers
16 Petiole not articulate to the lamina. Fig orange to red, 12–15 mm wide. Rare..**7. F. costata**
16 Petiole articulate, separating from the lamina in fallen leaves. Figs pink to purple-black
17 Lateral veins 9–16 pairs. Figs 5–8 mm wide..............**4. F. caulocarpa**
17 Lateral veins 5–11 pairs. Figs 8–12 mm wide................**3. F. virens**
11 Figs sessile. Stranglers, but some cultivated as trees
18 Leaf ovate-cordate with long tip; petiole articulate..............**1. F. religiosa**

18 Not so
19 Leaf densely white tomentose beneath, often obtuse. Figs tomentose, small..9. **F. mollis**
19 Not so
20 Figs 12–30 mm wide, ripening red
21 Leaf rather obtuse, often ovate to cordate and hairy beneath; basal veins elongate
22 Developing pillar-roots from the branches. Figs 20–30 mm wide, basal bracts conspicuous. Cultivated..10. **F. benghalensis**
22 Without pillar-roots. Figs smaller. Wild..........................12. **F. pubilimba**
21 Leaf acute, not ovate; basal veins short
23 Buds and young shoots brown hairy. Figs rather oblong..........8. **F. drupacea**
23 Young parts whitish puberulous, soon glabrous. Figs subglobose...11. **F. fergusoni**
20 Figs smaller
24 Copious aerial roots from the branches, developing pillar-roots. Leaf rather thick, acute to obtuse, with prominent basal veins; cystoliths on both sides of the lamina. Figs pink to purple-black...15. **F. microcarpa**
24 Without such habit
25 Leaves thinly coriaceous; cystoliths only on the underside; petiole articulate to the lamina. Figs pink to purple
26 Figs 8–12 mm wide, with glistening internal bristles...............3. **F. virens**
26 Figs smaller, without internal bristles.................................2. **F. tsjahela**
25 Petiole not articulate. Cystoliths only on the upper side of the lamina
27 Leaf rather thickly coriaceous, inclined to lanceolate-elliptic, acute or sub-acuminate. Figs 10–12 mm wide, pink to purple...............5. **F. amplissima**
27 Leaf thinly coriaceous, elliptic, acuminate. Figs smaller (? red). Rare...13. **F. talboti**

1. Subgenus UROSTIGMA

1. **Ficus religiosa** L., Sp. Pl. 1059. 1753; King, Ann. R. Bot. Gard. Calcutta 1: 55, pl. 67A. 1887; Trimen, Handb. Fl. Ceylon 4: 90. 1898; Alston, Kandy Fl. 34, f. 179. 1938; Corner, Wayside Trees of Malaya 683, pl. 204, 206. 1940; Corner, Gard. Bull. Singapore 21: 6. 1965; Worthington, Ceylon Trees f. 414. 1959; Gupta, Botanica 14: 84. 1963; Wiebes, Zool. Verh. Leiden 83: 28. 1966; Galil and Eisikowitch, Phytomorphology 18: 356. 1968; Condit, Ficus Exot. Species 99. 1969.

An evergreen or deciduous banyan or tree, without aerial roots from the branches. Leaves spirally arranged; new leaves pink. Glabrous except for the sericeous or puberulous stipules and basal bracts. Twigs 3–4 mm thick. Stipules –15 mm, but larger on opening shoots. Lamina 7–26 × 4–16 cm, ovate-cordate or ovate with truncate base, caudate-acuminate with the tip 25–90 mm long, thinly coriaceous, smooth, edge often uneven or sinuous; lateral veins 6–9 pairs at a wide angle, usually with several zigzag and obscure inter-costals; basal veins 2 (–3) pairs; petiole 35–130 mm long, mostly longer than the lamina, articulate but, in some cases, rather obscurely. Figs axillary, paired, sessile, ripening pink, purple, and black; basal bracts 3, 3–5 mm long, ovate-obtuse, silky-puberulous, glabrescent; body 8–11 mm wide (10–15 mm, living), subglobose, the orifice closed by 3 apical bracts in a disc 2–3 mm wide; internal bristles none. Male flowers ostiolar, sessile in one ring; tepals 2, reddish, free, ovate-lanceolate. Gall- and female flowers sessile; tepals 3–4, red, free, short; ovary yellowish to ochraceous, red-brown at the apex or in the upper half, sessile or with a stalk –1.5 mm long, somewhat indurate and subangular. Lamina with cystoliths only on the lower side; stomata superficial.

Distr. Himalayan forest from Rawalpindi eastwards, Yunnan, Cochin China, North Thailand.

Ecol. Submontane forest; cultivated widely.

Uses. Bark for dyeing, tanning, and fibre; leaves for silkworm.

Vern. Sacred Bo (E); Bo (S); Arachu (T); Peepul (Hindi).

Insects. *Blastophaga quadraticeps* Mayr., and other genera (Wiebes).

Note. In the wild state *F. religiosa* is now rare but it has prospered in its holiness and been carried into most tropical lands. A sprig of the tree under which the Buddha attained enlightenment was brought to Ceylon in the year 288 B.C. It has survived in the temple of Anuradhapura by means of its ability to supplant the aging trunk with new shoots and this tree is acclaimed as the oldest in historical record. Its insects appear to have accompanied it. In the drier regions *F. religiosa* is deciduous for short periods from December to March. Botanically its nearest ally is the Indo-Malaysian *F. saxophila* Blume, but *F. arnottiana* may be mistaken for it.

Collections. None noted for this common species.

2. Ficus tsjahela Burm. f., Fl. Ind. 227. 1768 (*F. tjakela* auct.); King, Ann. R. Bot. Gard. Calcutta 1: 57, pl. 70. 1887; Trimen, Handb. Fl. Ceylon 4: 91. 1898; Alston, Kandy Fl. 34, f. 180. 1938; Worthington, Ceylon Trees pl. 417. 1959; Corner, Gard. Bull. Singapore 21: 7. 1965.

A deciduous banyan, without aerial roots from the branches. Leaves spirally arranged. Glabrous. Twigs 2–4 mm thick, dark brown. Stipules small over the resting buds, but leafy, –11 cm long on opening shoots. Lamina 9–21 × 3.5–11.5 cm, oblong-elliptic, varying ovate-elliptic, shortly acuminate, the tip –15 mm long, base rounded to widely cuneate, nitid, smooth, thinly coriaceous, drying greenish-grey to olivaceous; lateral veins (7–) 8–14 pairs, –3 vague intercostals, the veins and reticulations slightly, but distinctly, raised on both sides of the lamina; basal veins 1 (–2) pairs, short; petiole –70 × –2 mm, articulate to the lamina. Figs sessile, axillary, paired, and in clusters of 3–6 on the twigs behind the leaves, ripening greyish-purple; basal bracts 3, 2–2.5 × 3–3.5 mm, ovate-rounded, obtuse, often split, body 5–7 mm wide, depressed-globose, the small orifice plane; internal bristles none; fig-wall thin. Male flowers ostiolar in one ring, sessile; tepals 2–3, red, free or shortly joined.

Gall-flowers sessile or with white pedicels; tepals 3–4 (5), red, lanceolate, free or shortly joined; ovary dark red, more or less stalked. Female flowers as the gall-, but sessile; tepals 3, lanceolate. Lamina with cystoliths only on the lower side; stomata superficial.

Distr. Peninsular India; Ceylon.

Ecol. Moist regions below 700 m, also cultivated.

Vern. Kiri-pela (S).

Insects. Not known.

Note. This common tree in Ceylon is distinguished by its very small, sessile figs. It is often mistaken for *F. caulocarpa* with stalked figs and *F. virens* with larger figs, both of which species have copious glistening internal bristles. It must not be confused with *F. tsiela*, the correct name of which is *F. amplissima* (lanceolate, dull leaves with cystoliths on the upper side). Closely allied to *F. tsjahela* is *F. densifolia* Miq. of Réunion.

Specimens Examined. KANDY DISTRICT: Hakkinda, *Alston s.n.* (PDA); Hantane, *s. coll. C.P. 3083* (PDA); Kolugala, *Worthington 1497* (K); Peradeniya, *Worthington 2373* (K); Kadugannawa, *Worthington 6277, 6271* (K). COLOMBO DISTRICT: Labugama, *Worthington 3494* (K). BADULLA DISTRICT: Namunukula, *Worthington 5089* (K). GALLE DISTRICT: Deniyaya, *Worthington 2218* (K). LOCALITY UNKNOWN: *Vahl s.n.* (C); *Koenig s.n.* (BM); *van Royen 131* (L).

3. Ficus virens Ait., Hort. Kew 3: 451. 1789; Corner, Gard. Bull. Singapore 17: 376. 1960; id. 21: 9. 1965; Wiebes, Zool. Verh. Leiden 83: 34. 1966; Condit, Ficus Exot. Sp. 105. 1969.—**Fig. 7.**

Ficus infectoria Willd. sensu Roxb., Fl. Ind. 3: 551. 1882; King, Ann. R. Bot. Gard. Calcutta 1: 59. 1887; Trimen, Handb. Fl. Ceylon 4: 92. 1898.
Ficus lucescens Blume sensu Alston, Kandy Fl. 34, f. 181. 1938.
Ficus tsiela sensu Worthington, Ceylon Trees pl. 416. 1959.

Large deciduous banyan, with few or numerous tufts of aerial roots from the branches. Leaves spirally arranged, glabrous or the stipules finely puberulous. Twigs 3–5 mm thick, brown. Stipules –15 mm long, –8 cm long and leafy on opening shoots. Lamina 8–19 × 3.5–9 cm, ovate-elliptic to elliptic, shortly and often obtusely acuminate with the tip –12 mm long, base subcordate, rounded, subtruncate to cuneate, edge often sinuous, smooth, thinly coriaceous, drying dull brown; lateral veins 7–12 pairs, 1–4 mostly zigzag intercostals, slightly raised below as the reticulations; basal veins 1–2 pairs, not elongate; petiole 25–65 × 2 mm, articulate. Figs axillary, paired or on the twigs just below the leaves, (? also on short burrs on the older twigs), ripening whitish then pink to purple and black; peduncle 1–6 mm long, occasionally hairy; basal bracts 3, 1.5–3 × 2–4 mm, ovate, subacute or obtuse,

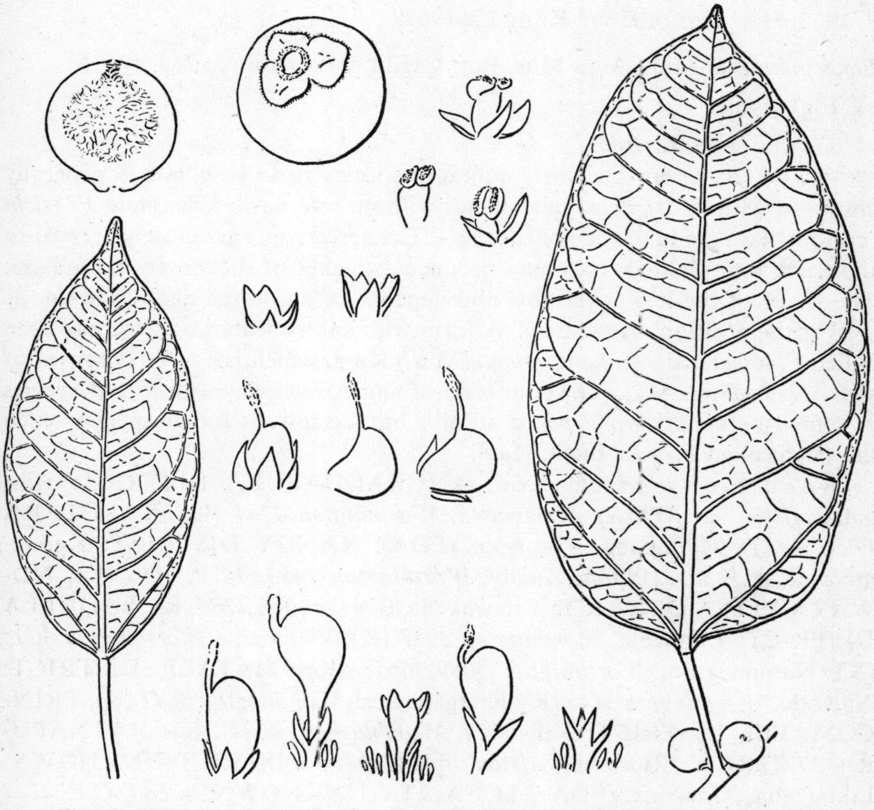

Fig. 7. *Ficus virens*. Leaves, × ½; figs, × 3; flowers (gall-flowers below), × 10.

glabrous or puberulous, persistent; body 7–12 mm wide, (12–20 mm, living), globose, the small orifice closed by 3 flat apical bracts in a disc 1 mm wide; internal bristles abundant, white, chaffy-vesicular. Male flowers ostiolar, sessile, in 2–3 rings; tepals 2–3, ovate, acute, free or shortly gamophyllous. Gall-flowers sessile or shortly pedicellate: tepals 3–4, reddish, spathulate to linear-lanceolate, free (rarely 2 joined); ovary sessile or stalked, red-brown. Female flowers sessile; tepals 3–4, free; ovary red-brown, sessile or stalked. Lamina with cystoliths only on the underside; stomata superficial.

var. **virens**

Figs pedunculate.

Distr. Ceylon, India, to the Solomon Islands.

Ecol. Coastal and in monsoon forest, often in parks and villages.

Insects. *Blastophaga coronata* Grandi and other genera (Wiebes).

var. **sublanceolata** (Miq.) Corner, Gard. Bull. Singapore 17: 377. 1960; id. 21: 9. 1965; Hill, Figs of Hong Kong 22. 1967.

Ficus infectoria Miq., Ann. Mus. Bot. Lugduno Batavum 3: 264. 1867.

Figs sessile.

Distr. as var. *virens.*

Note. *F. virens* is a fairly common species in Ceylon but is generally confused with *F. tsjahela* (smaller figs without internal bristles) and *F. tsiela* (cystoliths on the upperside of the leaf). The aerial roots are closely pressed to the trunk and planted trees may produce a thicket of suckers from the base. The shape of the leaf is variable and depends often on the age of the tree as well as the position on the twig. A form with subtruncate leaf-base has been called *F. infectoria* var. *lambertiana* (Miq.) King, which is a later synonym of var. *aegeirophylla* Miq. The confusion of names over this widespread species is immense. It is known as *pakor* in India but it is not the Indian species which has been called *F. lacor* Buch. Ham.

Specimens Examined. ANURADHAPURA DISTRICT: Ritigala, *Alston s.n.* (PDA); Kalawawa, *Worthington 4254* (K). KALUTARA DISTRICT: Matugama, *Alston s.n.* (PDA). KANDY DISTRICT: Kadugannawa, *W.F. s.n.* (PDA); Kandy, *Worthington 1453, 1479, 5991* (K). NUWARA ELIYA DISTRICT: Talawakelle, *Worthington 1781* (K). BADULLA DISTRICT: Lunugala, *Worthington 2967* (K); Welimada, *Worthington 3003* (K); Namunukula, *Worthington 5089, 6655* (K). MATALE DISTRICT: Nalanda, *Worthington 3448* (K); Kongahawela, *Worthington 6841* (K). TRINCOMALEE DISTRICT: Kiliveddy, *Worthington 4887* (K). RATNAPURA DISTRICT: Rakvana, *Alston 1695* (PDA). DISTRICT UNKNOWN: Lihinigalla, *Alston s.n.* (PDA). LOCALITY UNKNOWN: *s. coll. C.P. 2223* (PDA).

4. Ficus caulocarpa Miq., Ann. Mus. Bot. Lugduno Batavum 3: 268. 1867; Corner, Gard. Bull. S.S. 10: 283. 1939; Corner, Gard. Bull. Singapore 21: 10. 1965; Corner, Wayside Trees Malaya 675, pl. 201. 1940; Corner, Philos. Trans. R. Soc. B. 253: 63, f. 6. 1967.—**Fig. 8.**

Ficus infectoria var. *caulocarpa* (Miq.) King, Ann. R. Bot. Gard. Calcutta 1: 63, pl. 79. 1887.

A very large, deciduous banyan, developing a reticulate root-trunk, often with bunches of aerial roots from the trunk and bases of the main branches; bark pale pinkish-brown. Leaves spirally arranged. Glabrous except the shortly silky stipules, or the twigs and peduncles puberulous. Twigs 3–8 mm thick, dark brown. Stipules –15 mm long, –10 × 1.2 cm and leafy on opening shoots. Lamina 6–22 × 3–8.5 cm, elliptic to oblong, varying slightly obovate, more or less acuminate with the tip –15 mm, base rounded to narrowly cuneate,

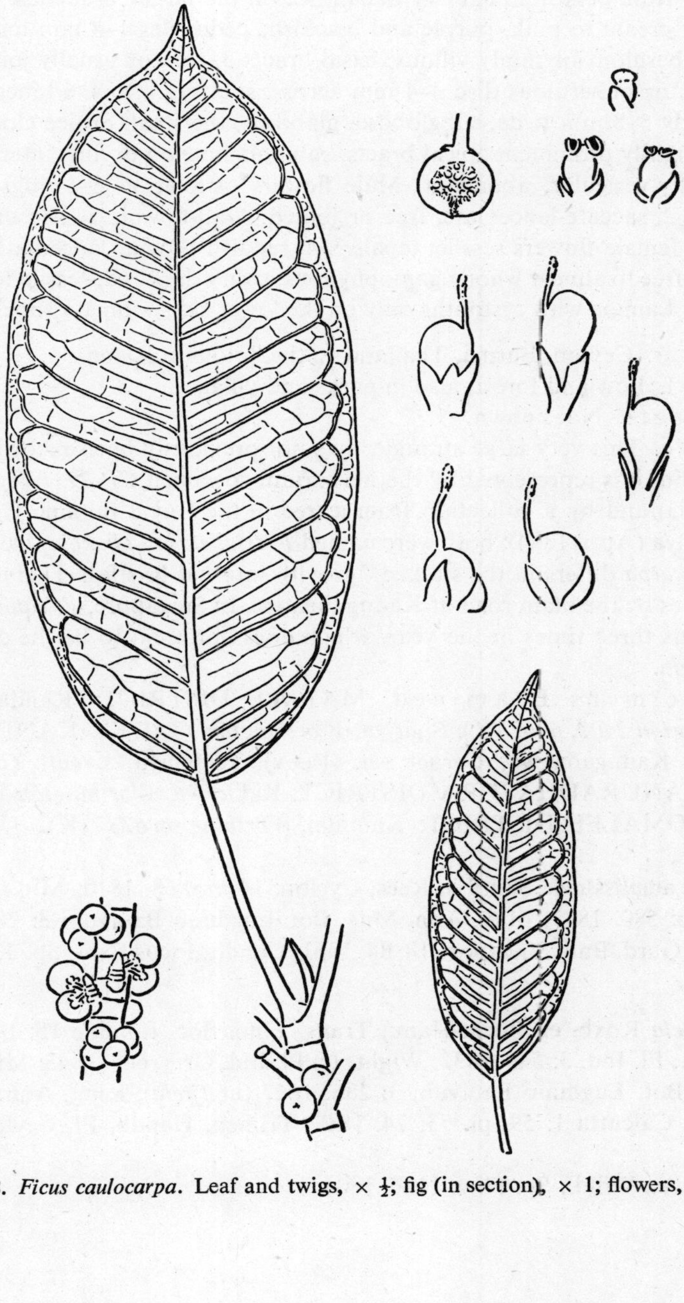

Fig. 8. *Ficus caulocarpa*. Leaf and twigs, × ½; fig (in section), × 1; flowers, × 10.

thinly but firmly coriaceous, drying dull dark brown; lateral veins 9–16 pairs, slightly raised below, 0–3 intercostals, the reticulations barely visible; basal veins 1–2 pairs, short; petiole 15–80 × 1–2.5 mm, articulate. Figs in groups of 2–8 in the axils of old leaves and on the twigs just behind (developing in crops from persistent axillary buds), not on the larger branches, ripening white or cream to pink, purple and blackish; peduncles 1–4 mm long, glabrous, puberulous or thinly villous; basal bracts 3, free or usually joined in a glabrous or puberulous disc 3–4 mm across, splitting into 4–6 lobes, persistent; body 5–8 mm wide, subglobose, glabrous, the small orifice closed by 3 flat or slightly prominent apical bracts, subumbonate on drying; internal bristles chaffy-vesicular, abundant. Male flowers ostiolar in 2–3 rings, sessile; tepals 1–2, saccate-lanceolate, free or joined. Gall-flowers sessile or shortly stalked, female flowers sessile; tepals 3–4, reddish-brown, lanceolate or spathulate, free to almost wholly gamophyllous; ovary dark red, sessile to shortly stalked. Lamina with cystoliths only on the lower side; stomata superficial.

Distr. Ceylon, Burma, Thailand to the Solomon Islands.

Ecol. Lowland forest, also in parks and villages.

Insects. Not known.

Note. This very large strangler has not previously been recorded from Ceylon but it is represented by the herbarium specimen *CP 2931* (Teldeniya, July 1853) and by a collection from a tree in the Royal Botanic Garden at Peradeniya (April 1884); both were named *F. tsjahela* (*F. tsjakela*) from which *F. caulocarpa* differs in the stalked figs with internal bristles. There are two large trees by the main road at Kadugannawa. In Singapore, *F. caulocarpa* is deciduous three times in the year, which does not seem to be the case with *F. tsjahela*.

Specimens Examined. MATALE DISTRICT: Kandanuwara, *Worthington 2402*, *6367* (K); Sigiriya, *Worthington 4350* (K). KANDY DISTRICT: Kadugannawa, *Corner s.n.* (PDA); Teldeniya, *s. coll. C.P. 2931* (PDA). ANURADHAPURA DISTRICT: Kekirawa, *Worthington 5667* (K). TRINCOMALEE DISTRICT: Kantalai, *Worthington 6737* (K).

5. Ficus amplissima. Smith in Rees, Cyclop. 14: no. 68. 1810; Miq., London J. Bot. 6: 580. 1847; Miq., Ann. Mus. Bot. Lugduno Batavum 3: 287. 1867; Corner, Gard. Bull. Singapore 18: 84. 1961; Condit, Ficus Exot. Sp. 135. 1969. **—Fig. 9.**

Ficus tsiela Roxb. ex Buch.-Ham., Trans. Linn. Soc. London 15: 149. 1826; Roxb., Fl. Ind. 3: 549. 1832; Wight, Ic. Pl. Ind. Or. t. 668. 1843; Miq., Ann. Mus. Bot. Lugduno Batavum 3: 286. 1867 (ut *tjiela*); King, Ann. R. Bot. Gard. Calcutta 1: 59, pl. 73, 74. 1887; Trimen, Handb. Fl. Ceylon 4: 92. 1898.

Ficus indica Willd., Sp. Pl. 4: 1146. 1806.

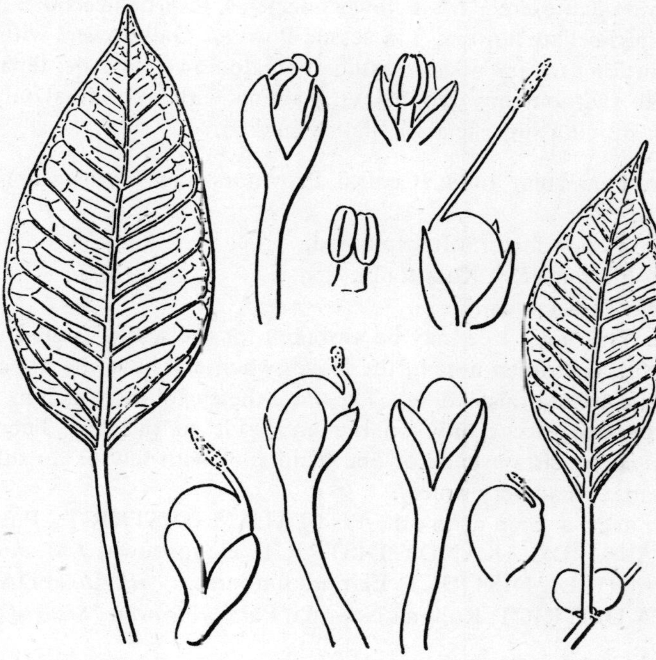

Fig. 9. *Ficus amplissima*. Leaves, × ½; flowers, × 10.

Urostigma pseudobenjamineum Miq., *U. pseudotsiela* Miq., London J. Bot. 6: 566. 1847; Miq., Ann. Mus. Bot. Lugduno Batavum 3: 286. 1867 (ut *tjiela*).
Ficus pseudobenjamina Miq., Ann. Mus. Bot. Lugduno Batavum 3: 286. 1867; Rheede, Hort. Mal. III, 85, t. 63. 1680 (*tsiela*); Burm. f., Fl. Ind. 226. 1768 (ut *F. indica* Burm.).

A large deciduous banyan, rarely developing aerial roots from the branches: bark greenish grey: latex white, copious. Leaves spirally arranged. Glabrous or the young twigs, stipules, and figs puberulous to villous with white hairs –0.5 mm long. Twigs 2.5–4 mm thick, pale. Stipules –25 mm long. Lamina 5–12 × 2–6 cm, ovate-elliptic to lanceolate-elliptic, acute or shortly and obtusely cuspidate, base rounded to cuneate, coriaceous, smooth, drying grey-green to brownish; lateral veins 6–10 pairs, oblique, venation faint, mostly without intercostals; basal veins 1 pair, short or, in ovate leaves, –1/3 lamina; petiole 25–70 mm long, not disarticulating from the lamina. Figs axillary and on the twigs below the leaves, sessile, ripening pink to purple; basal bracts 3, 2–2.5 × 3–3.5 mm, ovate, glabrous or puberulous, broader than long, obtuse, more or less concealed; body 10–12 mm wide, subglobose or obovoid, the plane orifice closed by 3 flat apical scales; internal bristles none. Tepals (2–) 3, reddish, ovate-acute, free. Male flowers disperse, few; pedicels 0–2 mm long; tepals 2 in pedicellate flowers, 3 in sessile flowers. Gall-flowers with pedicels 0.5–2.5 mm long; ovary white, sessile. Female flowers sessile; tepals 3 (–4); ovary white (without any red mark). Lamina with cystoliths only on the upper side; stomata superficial or slightly sunken.

Distr. Peninsular India (Central Provinces and southwards), Ceylon, Maldive Isl.

Ecol. In open forest; often planted.

Vern. Ela-nuga (S); Kalatti (T).

Insect. Not known.

Note. This large tree may be mistaken for the sessile-fruited variety of *F. virens* but, as Trimen noted, 'the leaves when dry have the upper surface covered with minute raised dots'. These are the cystoliths which in *F. virens*, as in *F. tsjahela*, occur only on the underside of the leaf. The trunk of *F. amplissima* consists usually of one main root with few or no subsidiaries and the limbs spread very widely.

Specimens Examined. PUTTALAM DISTRICT: Puttalam, *s. coll. C.P. 2218* (PDA). KANDY DISTRICT: Gannoruwa, *J.M. de Silva s.n.* (PDA). BADULLA DISTRICT: Ekiriankumbura, *s. coll. s.n.* (PDA). HAMBANTOTA DISTRICT: Ruhuna National Park, *Wirawan 788, 814* (PDA).

6. Ficus arnottiana (Miq.) Miq., Ann. Mus. Bot. Lugduno Batavum 3: 287. 1867; King, Ann. R. Bot. Gard. Calcutta 1: 56, pl. 68. 1887; Trimen, Handb. Fl. Ceylon 4: 90. 1898; Worthington, Ceylon Trees f. 400. 1959; Corner, Gard. Bull. Singapore 21: 11. 1965; Wiebes, Zool. Verh. Leiden 83: 9. 1966.—**Fig. 10.**

Urostigma arnottianum. Miq., London J. Bot. 6: 564. 1847.

Urostigma courtallense Miq., id. 564.

Urostigma caulobotryum Miq., id. 568 pr. p. receptaculorum (alt. p.—*F. tsjahela*).

Pogonotrophe emodi Miq., London J. Bot. 7: 73. 1848.

Ficus emodi (Miq.) Miq., Ann. Mus. Bot. Lugduno Batavum 3: 278, 293. 1867; King, Sp. Fic. 181. 1888.

Ficus arnottiana (Miq.) Miq. var. *courtallensis* (Miq.) Miq., Ann. Mus. Bot. Lugduno Batavum 3: 287. 1967; King, Ann. R. Bot. Gard. Calcutta 1: 56, pl. 68B. 1887.

A small deciduous banyan or independent tree to 10 m high, without aerial roots from the branches; bark grey-brown. Leaves spirally arranged; veins and petioles bright pink; new leaves brownish-pink. Glabrous. Twigs 3–5 (–6) mm wide. Stipule 12–40 mm long, glabrous, caducous, –6 cm and leafy on young shoots. Lamina 6–14 × 5–9 cm (–25 × 14.5 cm in saplings), ovate-cor-

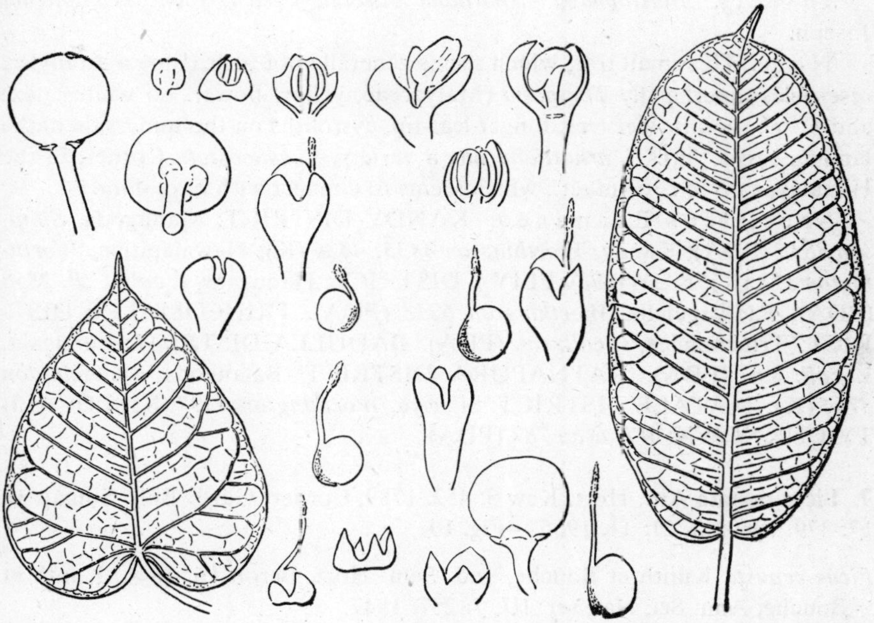

Fig. 10. *Ficus arnottiana* (left) and *F. costata* (right); leaves, × ½; figs of *F. arnottiana* × 3, of *F. costata* × 1; flowers, × 10.

date, rarely short-ellipsoid, acuminate with a tip –22 mm long (–40 mm in sapling), base deeply cordate, entire, thinly to rather thickly coriaceous, undulate, drying dull greyish-brownish, smooth; lateral veins 5–7 pairs (–10 in saplings), scarcely raised below, venation as in *F. religiosa*; basal veins 2–3 pairs, the larger pair 1/3–1/2 lamina; petiole 30–70 mm (–170 mm in sapling), not articulate. Figs in axillary pairs and on the twigs below the leaves, glabrous, ripening yellow-brown, red (?) or purple with green spots (King); peduncle 2–5 × 1 mm: basal bracts 3, 1 × 2–2.5 mm, blunt, thin, eventually caducous; body 5–8 mm wide, globose, not pedicellate, orifice small, plane; internal bristles none. Tepals red, glabrous, rather fleshy, more or less gamophyllous, the perianth 3–4 lobed. Male flowers sessile, round the orifice and sparsely scattered in the interior of fig; stamen 1. Gall- and female flowers sessile or shortly stalked; ovary sessile, straw-white with a slight red patch on stylar side; style slender. Lamina with crstoliths on both sides; stomata superficial; cuticle slightly striate round stomata.

Distr. India (South, Central), Ceylon.

Ecol. Open forest in the dry zone below 1000 m alt., on poor grassy hillsides and rocky places.

Vern. Patana Bo (E); Kaputu-Bo and Kaudu-Bo (S).

Insects. *Blastophaga arnottiana* Joseph, *Terastiozoon incompletum* Joseph.

Note. This small tree, which seems generally not to become a strangler, resembles superficially *Thespesia* (Malvaceae, yellow flower, no white latex) and *F. religiosa* (larger tree, longer leaf-tip, cystoliths on the underside of the lamina, sessile fig). *F. arnottiana* has a variety, v. *subcostata* Corner, in the Himalayas up to 2300 m alt., which seems to connect with *F. costata.*

Specimens Examined. KANDY DISTRICT: Madugoda, *Simpson 8815* (PDA); Kandy, *Worthington 4435, 4436* (K); Nawalapitiya, *Worthington 378* (K). NUWARA ELIYA DISTRICT: Hanugam, *s. coll. C.P. 2856* (PDA); Padiyapelella, *Worthington 6213* (PDA). TRINCOMALEE DISTRICT: Trincomalee, *s. coll. s.n.* (PDA). BADULLA DISTRICT: Lunugala, *s. coll. s.n.* (PDA). RATNAPURA DISTRICT: Balangoda, *Worthington 3754* (K). MATALE DISTRICT: Sigiriya, *Worthington 4337* (K). LOCALITY UNKNOWN: *Wirawan 787* (PDA).

7. **Ficus costata** Ait., Hort. Kew 3: 452. 1789; Corner, Gard. Bull. Singapore 17: 379. 1960; id. 21: 11. 1965.—**Fig. 10.**

Ficus venusta Kunth et Bouche, Ind. Sem. Hort. Berol. 16. 1846; Kunth et Bouche, Ann. Sci. Nat. ser. III, 7: 237. 1847.

Ficus caudiculata Trimen, J. Bot. 23: 242. 1885; Trimen, Handb. Fl. Ceylon 4: 88. 1898; King, Ann. R. Bot. Gard. Calcutta 1: 48, pl. 58A. 1887; Worthington, Ceylon Trees pl. 405. 1887.

Ficus mooniana King, Ann. R. Bot. Gard. Calcutta 1: 57, pl. 69. 1887; Trimen, Handb. Fl. Ceylon 4: 91. 1898; Alston, Kandy Fl. 34, f. 177. 1931.

Urostigma wightianum Miq. var. *majus* Thw., Enum. Pl. Zeyl. 265. 1864.

A large tree (? an epiphytic strangler). Leaves spirally arranged. Glabrous or the stipules puberulous. Twigs 5–7 mm thick, stout, pale brown-ochraceous. Stipules small, leafy on opening shoots. Lamina 10–18 × 4–8.5 cm (–25 × 15 cm in saplings), oblong-elliptic, shortly acuminate, the tip –13 mm long, base rounded-subcordate to rounded-cuneate, smooth, coriaceous, rather stiff, drying dark brown; lateral veins 10–16 pairs, 3–5 zigzag intercostals, reticulations finely raised on both sides of the lamina, minutely brown-areolate beneath; basal veins 2 pairs, short; petiole 14–70 × 2–3.5 mm (–90 mm in saplings), not articulate. Figs axillary, paired, mostly on the twigs just below the leaves, ripening red (Trimen), orange-red (Worthington), at first minutely brownish-scurfy; peduncle 5–12 × 1.5 mm, not thickened into an annulus at the apex; basal bracts 2–3, 1.5–2.5 × 2–4.5 mm, ovate, obtuse, thin, persistent (? caducous eventually), sometimes shortly connate; body 12–15 mm wide (? mature), subglobose, smooth, but drying much wrinkled, the orifice slightly depressed; no internal bristles. Tepals 3, red, fleshy, more or less gamophyllous, the perianth cupular and 3-lobed in female flowers. Male flowers disperse, sessile or pedicellate; anther submucronate. Gall-flowers with red pedicels; ovary sessile, reddish on the stylar side. Female flowers sessile; ovary reddish on the stylar side; stigma simple. Lamina with cystoliths on both sides or none below; cuticle of lower epidermis gyroso-plicate below as in *F. annulata*; stomata superficial.

Distr. Ceylon, Annam (pr. Tourane).

Ecol. Lowland and submontane forest in the wet zone, rare.

Insecs. Not known.

Note. This is close to *F. arnottiana* and differs in leaf-characters, though it appears that saplings may have almost identical, long-petiolate, ovate-cordate leaves. The figs are larger and differently coloured at maturity but the difference in size is bridged by the specimens of *F. arnottiana* from North India. The specimen from Annam (environs de Tourane, 2. vi. 09, leg. Ch. d'Alleizette, det. *F. infectoria*) is fertile and agrees exactly with the Ceylon specimens; it may have been planted.

Specimens Examined. MATALE DISTRICT: Matale, *s. coll. s.n.* (PDA). KANDY DISTRICT: Deltota, *s. coll. s.n.* (PDA); Nawalapitiya, *Worthington 3154* (K). RATNAPURA DISTRICT: Ratnapura, *s. coll. s.n.* (PDA). GALLE DISTRICT: Deniyaya, *Worthington 2196* (K). DISTRICT UNKNOWN: Delgoda, *F. Lewis* and *J.M.S. s.n.* (PDA); Paregodde, *U.J. s.n.* (PDA).

8. Ficus drupacea Thunb., Diss. Ficus 6, 11. 1786; Corner, Gard. Bull. Singapore 17: 380. 1960; id. 21: 13. 1965; Corner, Philos. Trans. R. Soc. B. 253: 64, f. 7. 1967; Wiebes, Zool. Verh. Leiden 83: 15. 1966; Condit, Ficus Exot. Sp. 184. 1969.

Ficus pilosa Reinw. ex Blume, Bijdr. 446. 1825; King, Ann. R. Bot. Gard. Calcutta 1: 21, pl. 16. 1887.

var. **pubescens** (Roth) Corner, Gard. Bull. Singapore 17: 381. 1960.—**Fig. 11.**

Ficus mysorensis Heyne ex Roth and v. *pubescens* Roth in Roehm. et Schult., Syst. Veg. 1: 508. 1817; King, Ann. R. Bot. Gard. Calcutta 1: 19, pl. 14. 1887; Trimen, Handb. Fl. Ceylon 4: 86. 1898; Worthington, Ceylon Trees pl. 411. 1959; Condit, Ficus Exot. Sp. 86. 1969.
Ficus cotoniaefolia sensu Alston in Trimen, Handb. Fl. Ceylon 6: 268. 1931; Alston, Kandy Fl. 34, f. 178. 1938.

A large banyan, abundant descending aerial roots round the trunk of the host-tree, without tufts of aerial roots from the branches; bark grey or brownish, smooth. Leaves spirally arranged to subdistichous. All parts (except the fig-body) at first densely pilose or strigose with stiff ferruginous spreading hairs 2–4 mm long, glabrescent, especially the upperside of the lamina, often

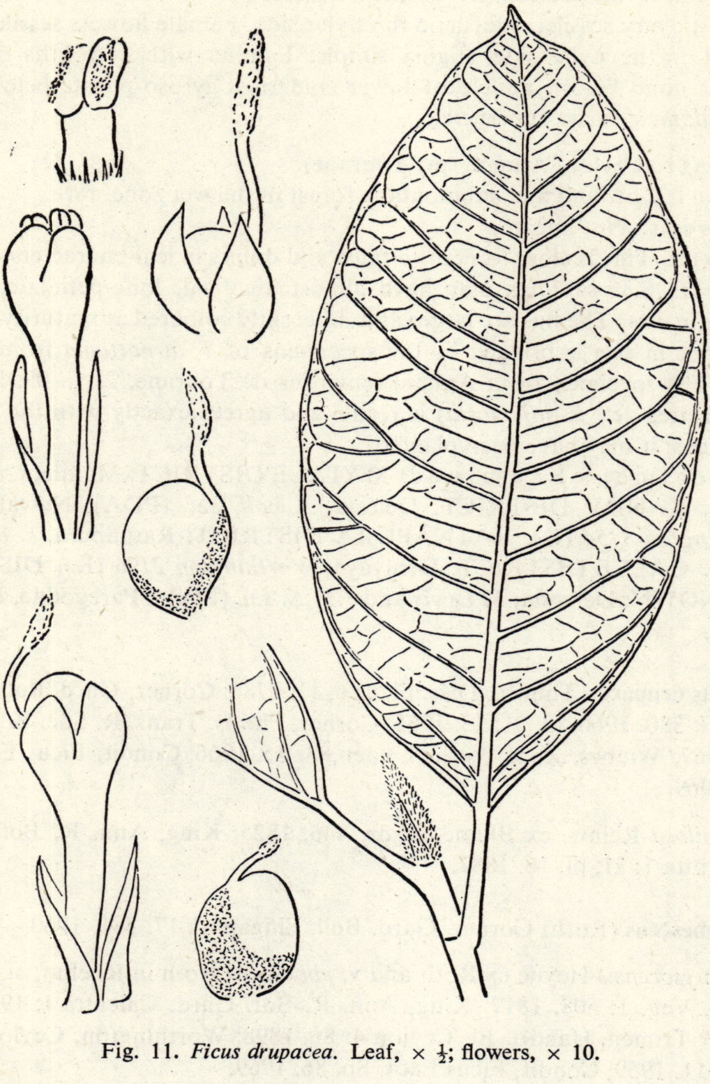

Fig. 11. *Ficus drupacea*. Leaf, × ½; flowers, × 10.

becoming wholly glabrous except the distal node; often with a thin coat of pale undulate hairs on the underside of the lamina. Twigs 3–6 mm thick, dark brown. Stipules 12–20 mm long. Lamina 8–26 × 3–13 cm (–35 × 16 cm in saplings), elliptic, oblong-elliptic, or obovate, shortly acuminate, base narrowly to widely cuneate, rounded or subcordate (especially in saplings), thinly coriaceous, drying pale fuscous to light greyish-brown; lateral veins 9–15 pairs, 3–7 more or less regular intercostals, distinctly raised above and below;

basal veins 3–5 pairs or 1–2 in small leaves, not elongate; petiole 10–45 × 2–5 mm, rather short. Figs sessile, axillary, paired, at first covered with a stout conical tomentose stipule, ripening yellow-ochre or orange to dull red; basal bracts 2–6 × 4–8 mm long and wide, generally very narrow and small, crescentic, hidden by the fig-base, densely brown pilose, one bract often minute or absent; body 15–35 × 12–20 mm, oblong, the orifice closed by 3–4 small umbonate apical bracts, not forming a flat disc; internal bristles none or few, short; wall thick. Tepals free. Male flowers few, pedicels –4 mm long; tepals 2–3; filament bristly at the base; anther wide. Gall-flowers with pedicels 0.2–3.5 mm; tepals (2–) 3. Female flowers sessile; tepals 3–4. Lamina with cystoliths on both sides, often sparse below; stomata shallowly sunken or almost superficial.

Distr. Ceylon, India, Bangladesh, Laos.

Ecol. Low country, especially in the dry zone.

Vern. Bu-nuga (S).

Insects. *Blastophaga belgaumensis* Joseph and other genera (Wiebes).

Note. *F. drupacea* is one of the very widely distributed strangling figs (Ceylon to the Solomon Islands). Its western form, often called *F. mysorensis*, has somewhat larger leaves with subcordate base and larger basal bracts to the fig, which may be more hairy. However some collections from Ceylon are exactly intermediate. As with all such stranglers, the leaves are larger on the sapling and lower branches of mature trees, and the figs associated with them are larger. The species is readily known from the brown woolly young shoots; when the buds are dormant, a brown hairy young leaf can generally be found inside.

Specimens Examined. KURUNEGALA DISTRICT: Kurunegala, *Alston 693* (PDA); Ibbagamuwa, *Worthington 4033* (K). KEGALLE DISTRICT: Kegalle, *Alston 876* (PDA). KANDY DISTRICT: Teldeniya, *Alston 1083* (PDA); Kandy, *Worthington 6793, 6795* (K); Kadugannawa, *Worthington 6278* (K). ANURADHAPURA DISTRICT: Anuradhapura, *s. coll. C.P. 2222* (PDA). NUWARA ELIYA DISTRICT: Attabagie, Pusselawa, *Worthington 2843* (K). MATALE DISTRICT: Sigiriya, *Worthington 4342* (K). BADULLA DISTRICT: Namunukula, *Worthington 5409* (K). LOCALITY UNKNOWN: *Macrae 419* (CGE); *Walker 231* (BM).

9. Ficus mollis Vahl, Symb. Bot. 1: 82. 1790; Vahl, Enum. 2: 192. 1806; Worthington, Ceylon Trees f. 410. 1959; Corner, Gard. Bull. Singapore 21: 14. 1965.—**Fig. 12.**

Ficus tomentosa Roxb. ex Willd., Sp. Pl. 4: 1136. 1806; Roxb., Fl. Ind. 3: 550. 1832; Wight, Ic. Pl. Ind. Or. t. 647. 1843; Miq., Ann. Mus. Bot. Lugduno Batavum 3: 285. 1867 (incl. var. *brevifolia* Miq. et var. *connivens* Miq.);

King, Ann. R. Bot. Gard. Calcutta 1: 22, pl. 18. 1887; Hook f., Fl. Br. Ind. 5: 501. 1888; Trimen, Handb. Fl. Ceylon 4: 87. 1898.

A deciduous banyan or independent tree with few or no aerial roots from the branches. Leaves spirally arranged. Twigs, petioles, underside of leaves, basal bracts, and figs densely tomentose with whitish to pale ferruginous undulate hairs, 1–1.5 mm long on the twigs and petioles, –3 mm long on the stipules; upper side of lamina soon glabrous. Twigs 3–5 mm thick. Stipules short, densely appressedly brown-strigose, caducous. Lamina 6–14 × 4–8.5 cm, reducing –3 × 2 cm on upper twigs, –21 × 14 cm on saplings, broadly elliptic to obovate-elliptic, subrotund in small leaves, obtuse, subacute or subacuminate, base rather deeply cordate to rounded, entire, smooth above, subcoriaceous, drying light brown; lateral veins 5–6 pairs, strongly raised

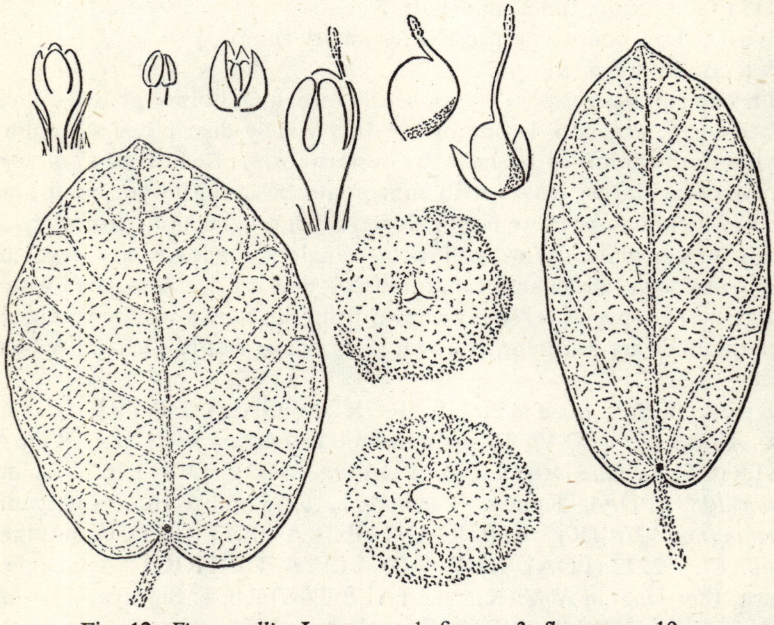

Fig. 12. *Ficus mollis*. Leaves, × ½; figs, × 3; flowers, × 10.

below, –10 intercostals; basal veins 3–4 pairs, 1/3–1/2 the lamina; petiole 10–50 mm long, often with a bare subglandular patch on the back at the junction with the midrib. Figs sessile, axillary, paired; basal bracts 3, 4–5 mm long, half-covering the fig-body, ovate, acute, often split, villous; body 8–9 mm wide, depressed-globose, the orifice closed by 3 glabrous apical bracts in a disc 2 mm wide: internal bristles sparse to abundant. Tepals 3–4, reddish, free. Male flowers disperse; pedicel 0–0.7 mm long. Gall-flowers sessile or pedicellate –0.7 mm. Female flowers sessile. Lamina with cystoliths only on the upperside; stomata superficial or slightly sunken; hairs simple, aseptate, smooth, undulate.

Distr. India (Central, South and Northwest), Ceylon.

Ecol. Lowland forest and open rocky places, in dry regions.

Vern. Wal-aralu (S).

Insects. Not known.

Note. This fairly common species, readily known from the white and softly downy undersides of the leaves, occupies a rather isolated position in classification but seems to be related with *F. benghalensis*.

Specimens Examined. MATALE DISTRICT: Dambulla, *Alston 1006* (PDA), *Worthington 1647, 2801* (K); Sigiriya, *Worthington 4342, 6343, 6709* (K); Nalanda, *Worthington 6397* (K). HAMBANTOTA DISTRICT: Yala, *Corner s.n.* (PDA). LOCALITY UNKNOWN: *s. coll. C.P. 2226* (PDA); *Wirawan 786* (PDA).

10. Ficus benghalensis L., Sp. Pl. 1059. 1753; King, Ann. R. Bot. Gard. Calcutta 1: 18, pl. 13. 1887; Trimen, Handb. Fl. Ceylon 4: 86. 1898; Worthington, Ceylon Trees pl. 402. 1959; Corner, Gard. Bull. Singapore 17: 381. 1960; Hill, Figs of Hong Kong 28. 1967; Condit, Ficus Exot. Sp. 176. 1969.—**Fig. 13.**

Ficus indica L., Sp. Pl. 1060. 1753, non sensu King.
Ficus cotonaeifolia Vahl., Enum. Pl. 2: 189. 1806.

A large banyan, rooting from the spreading branches, deciduous to evergreen. Leaves spirally arranged. Twigs, stipules, petioles, underside of lamina; and figs white puberulous, slowly glabrescent. Twigs 3–7 mm thick, light brownish-ochre. Stipules stout. Lamina 10–30 × 7–20 cm, ovate, obtuse, base cordate, rarely cuneate, firmly coriaceous; lateral veins 5–6 (–7) pairs, –10 intercostals, reticulations regular and raised on both sides; basal veins 2–4 pairs, 1/3–1/2 lamina; petiole 15–70 × 3–5 mm, stout. Figs sessile, axillary, paired, ripening orange to red; basal bracts 3–7 × 10–14 mm, umbonate at the base, obtuse; body 14–17 mm wide (15–20 × 21–25 mm, living), depressed-globose, the plane orifice closed by 3 flat or subumbonate apical bracts in a disc 3–4 mm wide; internal bristles none. Tepals 2–3 in male flowers, 3–4 in gall- and female flowers, free. Male flowers with pedicels 0.5–2.8 mm long, anther shortly mucronate. Gall-flowers with pedicels 1.5–2.3 mm long. Female flowers sessile. Lamina with cystoliths abundant on the upper side, few or none below; stomata deeply sunken.

Distr. Pakistan, India, now widely cultivated.

Ecol. Lowland forest, extending into regions of drought and slight frost and there deciduous; common in towns and villages.

Vern. Banyan (E); Bor (H); Maha-nuga (S); Al or Arla (T).

Insects. *Blastophaga masoni* Saunders and other genera (Wiebes).

N o t e. This most massive strangler is the holy banyan of India and the 'Ficus indica' of the classics, which name has unfortunately been so much misused that botanically it needs to be abandoned. It is recognised from the massive pillar-roots, the blunt, somewhat cordate and softly hairy leaves, and the sessile crimson figs. Instead of suckering from the base as *F. religiosa*, it expands by dropping new pillar-roots from the branches and there seems no

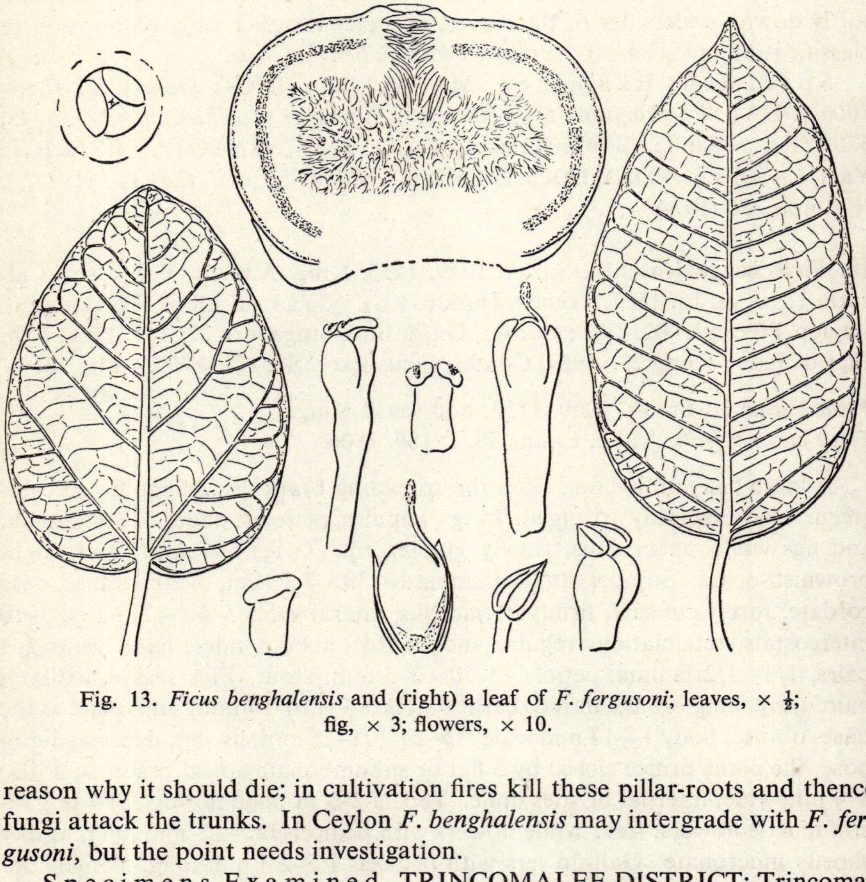

Fig. 13. *Ficus benghalensis* and (right) a leaf of *F. fergusoni*; leaves, × ½; fig, × 3; flowers, × 10.

reason why it should die; in cultivation fires kill these pillar-roots and thence fungi attack the trunks. In Ceylon *F. benghalensis* may intergrade with *F. fergusoni*, but the point needs investigation.

S p e c i m e n s E x a m i n e d. TRINCOMALEE DISTRICT: Trincomalee, *Alston 540* (PDA); *Worthington 6629* (K). COLOMBO DISTRICT: Colombo, *Coert 1463* (PDA). JAFFNA DISTRICT: Jaffna, *s. coll. C.P. 2225* (PDA). MANNAR DISTRICT: Murungan, *Worthington 4467* (K). BADULLA DISTRICT: Diyaluma Falls, *Worthington 5675* (K). BATTICALOA DISTRICT: Erauur, *Worthington 6303* (K). KALUTARA DISTRICT: Kalutara, *Worthington 6592* (K). KURUNEGALA DISTRICT: Kurunegala tank, *Worthington 6797* (K).

var. **krishnae** C. DC., Bull. Herb. Boiss. 760. 1902 (ut *F. krishnae* C. DC.); C. DC., Bot. Mag. t. 8092. 1906; Biswas, Current Sci. 3: 424–5, f. 1–3. 1935; Puri, J.R. Asiat. Soc. Bengal 12: 7. 1946; Puri. Wealth of India 4: 26, f. 10. 1956.

As var. *benghalensis* but with most or all of the leaves having the side of the lamina in its lower half reflexed and connate to form a cup.

D i s t r. Cultivated, original somewhere in India.

V e r n. Krishna Bor, Krishna's Cup (E).

N o t e. This is now regarded as a bud-sport, and most plants have been distributed from the Calcutta Botanical Garden. The plants may gradually change their leaves back to the normal form of var. *benghalensis*. Ten per cent of the seeds are said to breed true, the remainder being var. *benghalensis* (Biswas, 1935).

11. Ficus fergusoni (King) Worthington, Ceylon Trees f. 407. 1959; Corner, Gard. Bull. Singapore 21: 14. 1968.—**Fig. 13.**

Ficus altissima Blume var. *fergusoni* King, Ann. R. Bot. Gard. Calcutta 1: 31, pl. 31. 1887; Trimen, Handb. Fl. Ceylon 4: 87. 1898.

A banyan, developing aerial roots on or near the trunk. Leaves spirally arranged. Young parts whitish puberulous, soon glabrous. Twigs 5–7 mm thick, brownish ochraceous. Stipules thinly appressedly hairy, caducous. Lamina 10–22 × 5–12 cm, elliptic or ovate-elliptic, subacute or shortly acuminate, base subcordate, rounded or cuneate, rather thickly coriaceous, drying brown; lateral veins 8–12 pairs, venation as in *F. benghalensis*; basal veins 2–3 pairs, –1/3 lamina, mostlys hort; petiole 20–50 × 2.5–3.5 mm. Figs sessile, axillary, paired, ripening orange-red; basal bracts 3, 2–3, 3–4, or 5–7 mm long, ovate, obtuse, broader than long, free; body 13–20 mm wide, subglobose or obovoid, with 3 flat apical bracts in a small disc 2–3 mm wide. Tepals 3. Flowers as in *F. benghalensis*. Lamina with cystoliths chiefly on the upper side.

D i s t r. Ceylon.

E c o l. Moist low country up to 1000 m alt., frequent in tea-estates.

V e r n. Nuga, Kos-gona (S); Al or Arla (T).

I n s e c t s. Not known.

N o t e. The short fig with distinct basal bracts distinguishes this species from *F. altissima*, which does not occur in Ceylon, and allies it with *F. benghalensis*. From both it differs in the more elliptic leaf with cuneate base and short basal veins. The leaf is always acute, unlike that of *F. benghalensis*, and in habit *F. fergusoni* is a taller tree without spreading pillar-roots. The fig, however, ripens orange red as in *F. altissima*. It is conceivably the Ceylon relic of the ancestor of this alliance (subgen. *Urostigma* ser. *Drupaceae* subser. *Indicae*).

Specimens Examined. NUWARA ELIYA DISTRICT: Talawakele, *Worthington 1783* (K). KEGALLE DISTRICT: Dolosbage, *Worthington 1907* (K). KANDY DISTRICT: Dunally, Galaha, *Worthington 2651* (K); Deltota, *Worthington 6587*, *6588* (K); Kandy, *Worthington 6743* (K). MATALE DISTRICT: Matale, *s. coll. CP 2221* (PDA); Kandanuwara, *Worthington 6368* (K). DISTRICT UNKNOWN: Mottana Pattana, *Worthington 1478* (K). LOCALITY UNKNOWN: *Weerawan 758*, *770* (PDA).

12. Ficus pubilimba Merr., J. Arnold Arbor. 23: 159. 1942; Corner, Wayside Trees of Malaya 680. 1940 ("Province Wellesley Fig.").

A large banyan, without aerial pillar-roots from the branches; bark pinkish-grey. Leaves laxly spirally arranged. Twigs, petioles, underside of main veins, stipules and young figs white villous-puberulous with soft, erect hairs – 0.7 mm long, often brownish on the twigs, sometimes the leaves glabrous, rarely wholly glabrous. Twigs 2–5 mm thick, brown. Stipules – 13 mm long, – 40 mm on opening shoots. Lamina 6–23 × 3.5–10 cm, elliptic or ovate-elliptic, shortly and rather obtusely subacuminate, base rounded to cuneate, thinly coriaceous, drying pale brown; lateral nerves 4–7 pairs, oblique, raised below, 2–5 intercostals, the regular reticulations raised on both sides; basal nerves 1 (– 2) pairs, 1/4 – 1/3 lamina, rarely more; petiole 15–36 × 2–3 mm. Figs sessile, axillary, paired, ripening red to dark purple; basal bracts 3, 1–4 × 1.5–7 mm, ovate to crescentic, obtuse, pubescent, concealed; body 9–13 mm wide (20–25 mm, living), subglobose, glabrescent, the orifice closed by 3 flat apical bracts in a slightly depressed disc 1.5–2 mm wide; internal bristles none. Tepals (3 –) 4, free. Male flowers abundant, pedicellate. Gall- and female flowers similar, sessile or pedicellate. Lamina without cystoliths or a few above; stomata sunken.

Distr. Ceylon, Burma, Indo-China, Hainan, Thailand, North Malaya.

Ecol. Wild in lowland forest at Yala.

Note. This species is very close to *F. benghalensis* but has smaller figs with small, concealed basal bracts and seems to lack the pillar-roots. The Ceylon collection has leaves very much the same, but in Thailand, Indo-China and north Malaya the leaf-base is generally more or less cuneate.

Specimens Examined. HAMBANTOTA DISTRICT: Katagamua tank, Yala, *Weerawan 783* (PDA).

13. Ficus talboti King, Ann. R. Bot. Gard. Calcutta 1: 51, pl. 63. 1887; Talbot, For. Fl. ombay 2: 512. 1911; Corner, Gard. Bull. Singapore 21: 19. 1965.—**Fig. 14.**

Ficus pierrei Gagnep., Lec. Notul. Syst. 4: 92. 1927; Gagnep., Fl. Gen. I.-C. 5: 763. 1928.

Large banyan, without copious aerial roots. Leaves laxly spiral. Glabrous or the stipules hairy with white to brownish hairs –0.5 mm long. Twigs 1.5–3 mm thick, brown. Stipules –12 mm long. Lamina 6–12 × 2.5–5.5 cm, –17 × 7 cm (? in saplings or on lower branches), elliptic or ovate-elliptic, acuminate with the tip –12 (–20) mm long, base widely to narrowly cuneate or rounded, thinly coriaceous, drying brown: lateral veins 4–8 (–10) pairs, at a wide angle, slightly raised below, no intercostals; basal veins 1 pair, –1/4 (–1/3) the lamina; petiole 10–35 × 1.5 mm. Figs axillary, paired, sessile or with a short, glabrous or puberulous peduncle –1.5 × 1 mm; basal bracts 3, 1–2 mm long, not concealed, obtuse, glabrous or appressedly puberulous; body 6–8 mm long and wide, somewhat obconic and subtruncate, glabrous or

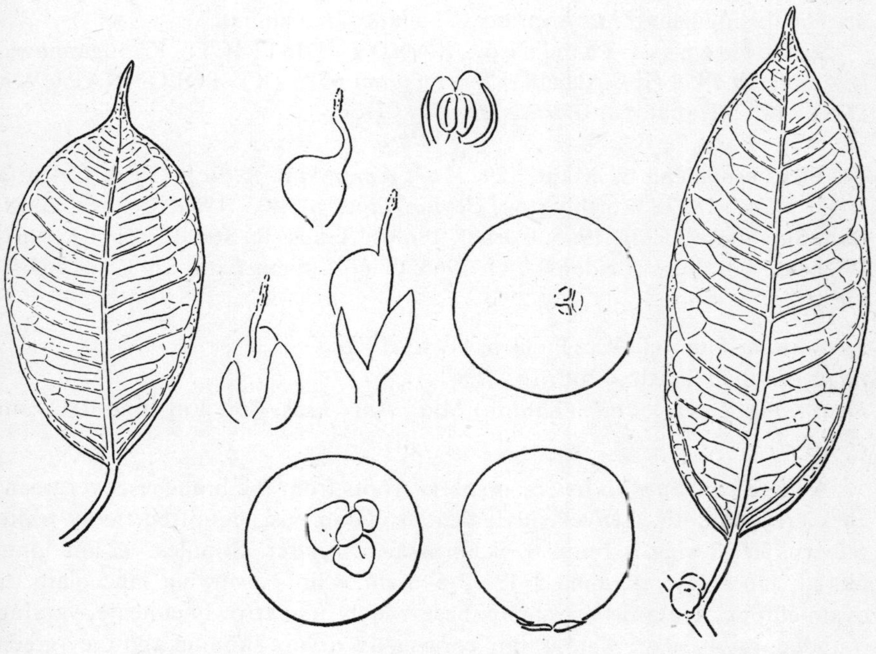

Fig. 14. *Ficus talboti*. Leaves, × ½; figs, × 3; flowers, × 10.

puberulous, the slightly sunken orifice closed by 3 flat apical bracts in a disc 1 mm wide; internal bristles few, minute, brownish. Tepals 3–4, red, with white edges, free, spathulate in male flowers to ovate-subacute in gall- and female. Male flowers disperse, few, shorter than the sessile or shortly pedicellate gall-flowers. Female flowers similar: gall- and fertile ovaries yellowish-white, without a red mark, rather crustaceous; stigma simple. Lamina with cystoliths only on the upper side, or very sparse below; stomata slightly to deeply sunken.

Distr. Ceylon, South India, Burma, Laos, Cambodia, North Thailand.

Ecol. Lowland and submontane forest (Dekanda, Alagalla, Habarane, Kadugannawa).

Note. This indistinctive species in the herbarium (*Worthington 1458, 5917, 6576*) has a peculiar distribution with a wide gap between Southern India and Burma: yet I cannot find any botanical difference between the collections. It resembles *F. microcarpa* but seems to lack the copious aerial roots and has much thinner, acuminate leaves, with spreading lateral nerves and the cystoliths very sparse or absent from the lower side. This last feature, together with the absence of red colour on the ovary, leads me to think that the species may be allied with *F. amplissima* and *F. rumphii*, in which case its geographical distribution is not inept. The form with shortly pedunculate fig may be distinguishable as a variety (Thailand, Indochina).

Specimens Examined. KANDY DISTRICT: Kadugannawa, *Worthington 1458* (K); Alagalla, *Worthington 6576* (K). POLONNARUWA DISTRICT: Habarana, *Worthington 5917* (K).

14. Ficus benjamina L., Mant. 129. 1767; King, Ann. R. Bot. Gard. Calcutta 1: 43, pl. 52. 1887; Worthington, Ceylon Trees pl. 403. 1959; Corner, Gard. Bull. Singapore 21: 21. 1965; Corner, Philos. Trans. R. Soc. B. 253: 67. 1967; Wiebes, Zool. Verh. Leiden 83: 11. 1966; Condit, Ficus Exot. Sp. 128. 1969.— **Fig. 15.**

Ficus nitida Thunb., Diss. Ficus 5, 11, 15. 1786.
Ficus pyrifolia Salisb., Prod. 16. 1796.
Ficus retusa L. var. *nitida* (Thunb.) Miq., Ann. Mus. Bot. Lugduno Batavum 3: 267. 1867.

A large banyan with few or no aerial roots from the branches, evergreen; bark grey, smooth. Leaves subdistichous. Glabrous, less often thinly white puberulous. Twigs 1–2 mm thick, brownish, slender. Stipules –12 mm long, paired, lanceolate. Lamina 3–12 × 1.5–6 cm, elliptic, varying lanceolate or ovate-elliptic, acute or acuminate, base widely to narrowly cuneate, varying rounded, rarely subcordate, thinly coriaceous, drying subnitid and grey-green to olivaceous brownish; lateral veins 6–11 (–14) pairs, the secondary nerves almost as prominent, raised on both sides of the lamina, as the reticulations; basal veins 1 (–2) pairs, short; petiole 4–16 × 1–1.5 mm. Figs sessile, axillary, paired, ripening yellow to orange and dark red, rarely pink to purple; basal bracts 2–3, 0.5–1.5 mm long, crescentic, minute, concealed beneath the fig-body, often unequal; body 8–12 × 7–10 mm, ellipsoid, ovoid, obovoid, or subglobose, sometimes with a slight thick pedicel, orifice closed by 3 small flat apical bracts in a disc 1.5–2 mm wide, often with a slight rim; internal bristles none. Tepals 3 (male) or 3–4 (gall- and female), free. Male flowers abundant, pedicel 0.8–2 mm: anther small. Gall-flowers with pedicels 0.2–2.8 mm long.

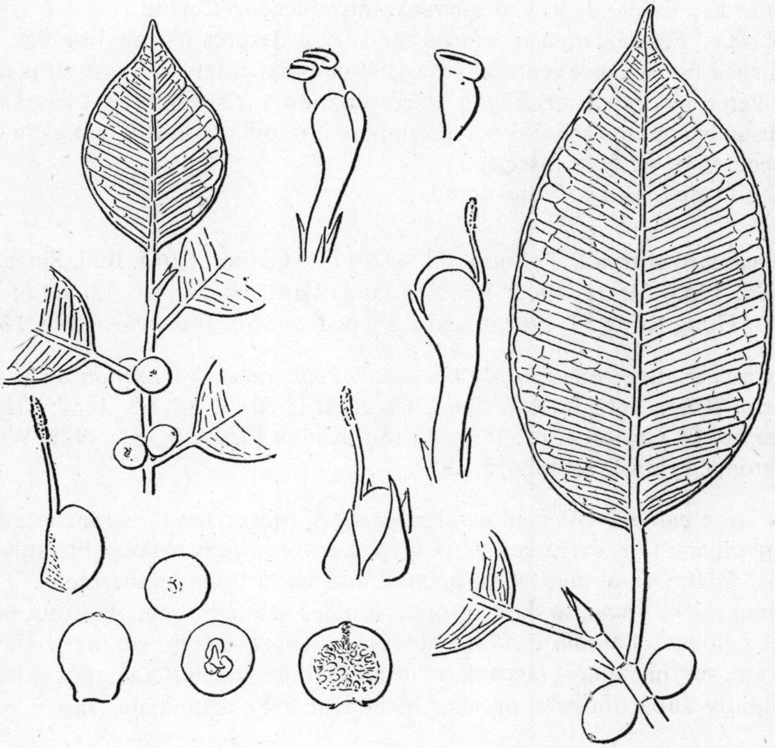

Fig. 15. *Ficus benjamina*. Twigs, × ½; figs, × 1; flowers, × 10.

Female flowers sessile. Lamina with cystoliths on both sides; stomata superficial.

D i s t r. India and South China to the Solomon Islands; introduced to Ceylon, planted for shade.

V e r n. Walu-nuga (S).

I n s e c t s. *Blastophaga adempta* Wiebes and other genera (Wiebes).

var. **nuda** (Miq.) Barrett, Amer. Midl. Naturalist 45: 127. 1951.

Urostigma nudum Miq., London J. Bot. 6: 584. 1847.
Urostigma benjamineum var. *nudum* Miq., Pl. Jungh. 50. 1851.
Ficus comosa Roxb., Pl. Corom. 2: 14, Pl. 125. 1798.
Ficus benjamina L. var. *comosa* (Roxb.) Kurz, For. Fl. Burma 2: 446. 1877; King, Ann. R. Bot. Gard. Calcutta 1: 44, pl. 52B. 1887.

Fig 15–20 × 12–16 mm (–25 mm wide, living), subglobose and sessile to pyriform with a short thick pedicel; basal bracts minute, concealed, or one enlarged 2–2.5 × 6–8 mm.

D i s t r. Evidently as v. *benjamina*; introduced to Ceylon.

N o t e. *F. benjamina* is one of the very widespread strangling figs, distinguished by its small acute leaves with fine parallel lateral veins. It is often mistaken for *F. microcarpa* with laxer venation, often prominent basal veins and usually copious aerial roots developing into pillar-roots. The figs in Ceylon seem to be without insects.

C o l l e c t i o n s. None noted.

15. Ficus microcarpa L. f., Suppl. Pl. 442. 1781; Corner, Gard. Bull. Singapore 17: 397. 1960, 21: 22. 1965; Wiebes, Zool. Verh. Leiden 83: 22. 1966; Hill, Figs of Hong Kong 32. 1967; Condit, Ficus Exot. Sp. 146. 1969.—**Fig. 16.**

Ficus retusa auct., *Ficus nitida* auct. and *Ficus retusa* var. *nitida* auct. (non vera): King, Ann. R. Bot. Gard. Calcutta 1: 50, pl. 61, 62. 1887; Trimen, Handb. Fl. Ceylon 4: 89. 1898; Alston, Kandy Fl. 34, f. 182. 1938; Worthington, Ceylon Trees pl. 415. 1959.

A large banyan with copious slender aerial roots from the branches, developing pillar-roots, evergreen; bark grey. Leaves spirally arranged to subdistichous. Glabrous or the twigs, petioles and basal bracts puberulous. Twigs 2–3 mm thick, brown to dark brown. Stipules –15 mm long, glabrous, or the edges ciliolate. Lamina 3–11 × 1.5–6 cm, (–18 × 8 cm in saplings), elliptic-obovate, varying almost lanceolate, often slightly inequilateral, apex subacute to bluntly subacuminate or shortly and acutely acuminate, rarely retuse

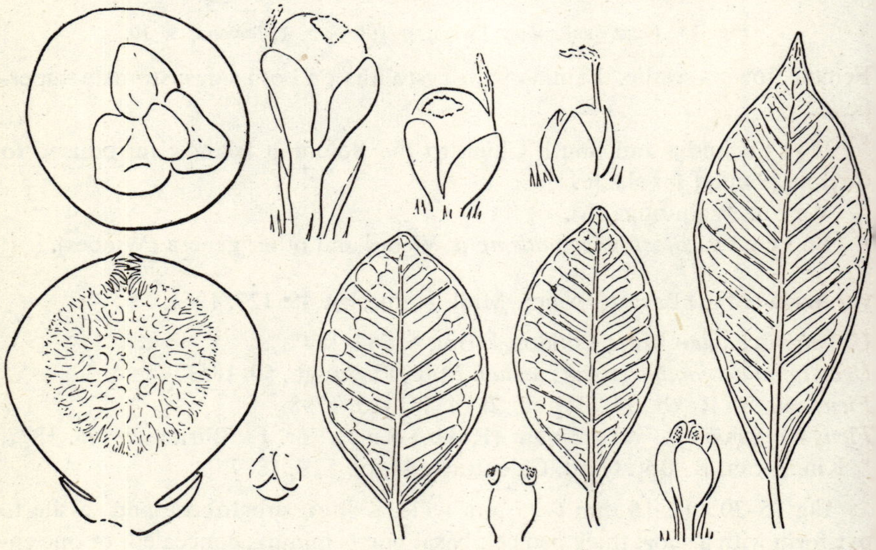

Fig. 16. *Ficus microcarpa*. Leaves, × ½; figs, × 3; flowers, × 10.

(? apex damaged), base widely to narrowly cuneate, thinly coriaceous, smooth, the midrib generally bifurcate just before the tip: lateral veins 5–9 pairs, at an acute angle, slightly raised below, not or rarely prominent; basal veins 1 (–2) pairs, not elongate or reaching 1/3 (–1/2) lamina; petiole 3–20 × 1.5–2 mm. Figs paired, axillary, sessile, ripening pink to mauve, purple and black (not yellow or red); basal bracts 1.5–3.5 × 1.5–2.5 mm, ovate-lanceolate, obtuse to subacute, often split, not keeled but often gibbous at the base; body 5–9 × 6–11 mm, pyriform to subglobose, or 9–11 mm long and wide, the plane orifice closed by 2–3 flat apical bracts in a disc 2–2.5 mm wide, finally slightly gaping; internal bristles short, 0.2–0.5 mm long, abundant to sparse. Tepals 3 (–4), spathulate-obovate, free. Male flowers few, sessile or substipitate, anther barely exsert. Gall-flowers with pedicels –1.2 mm long. Female flowers sessile. Lamina with cystoliths on both sides; stomata superficial to shallowly sunken, the outer pore often constricted.

Distr. Ceylon, India, South China, Ryukyu Isl. to New Britain.

Ecol. Coastal, estuarine, and on tidal river-banks, rarely in lowland inland forest, but not infrequently in rocky places; commonly planted and extending to 2,000 m alt. in montane forest.

Insects. *Blastophaga saundersii* Grandi and other genera (Wiebes).

Note. There has been much error in the name of this species. *F. retusa* L., under which it has usually passed, is a west Malaysian species of different alliance, and *F. nitida* or *F. retusa* v. *nitida* is a synonym of *F. benjamina*. *F. microcarpa* is one of the few banyans that develop pillar-roots in the manner of *F. benghalensis*. It appears to enjoy a much greater ecological range in Ceylon than in most countries where it is wild, because it reaches to the cloud-forest at Hakgala as a form with small, obovate, obtuse and rather thick, congested leaves, as noted by Trimen. A detail, distinctive of *F. microcarpa*, is the presence of short internal bristles between the flowers.

Specimens Examined. KANDY DISTRICT: Alagalla, *Worthington 312, 807* (K); Kandy, *Worthington 6571* (K); Loolokondera, *Worthington 2782* (K). NUWARA ELIYA DISTRICT: Hakgala, *A.W.S. s.n.* (PDA), *Worthington 5378* (K). BADULLA DISTRICT: Haldummulla, *Worthington 4857* (K). COLOMBO DISTRICT: Labugama, *W.F. s.n.* (PDA). TRINCOMALEE DISTRICT: Muttur, *Worthington 1395, 1661* (K); Trincomalee, *Worthington 6375* (K). MATALE DISTRICT: Matale, *Worthington 4300* (K). BATTICALOA DISTRICT: Batticaloa, *Worthington 5304* (K). DISTRICT UNKNOWN: Mulliyavalai, *Alston 1447* (PDA); Kalobologa, *Worthington 1971* (K); Pitiyakanda, *Worthington 2392* (K). LOCALITY UNKNOWN: *Gardner 824* (BM); *s. coll. C.P. 14, 2537* (PDA).

16. Ficus trimenii King, J. Bot. 23: 242. 1885; King, Ann. R. Bot. Gard. Calcutta 1: 46, pl. 55. 1887; Trimen, Handb. Fl. Ceylon 4: 88. 1898; Corner, Gard. Bull. Singapore 21: 24. 1965.—**Fig. 17.**

A large banyan with few aerial roots. Leaves laxly spiral to subdistichous. Glabrous, or the twigs at first puberulous. Twigs 2–3 mm thick. Stipules 9–25 mm long, puberulous. Lamina 7–12 × 4–6 cm, elliptic, subacute, base cuneate, smooth, coriaceous, drying brown; lateral veins 9–13 pairs, crowded, at a wide angle, the secondary veins not quite as strong as the primaries, no intercostals; basal veins 1 (–2) pairs, short; petiole 10–18 mm long, drying black. Figs axillary, paired, sessile, puberulous then glabrous; basal bracts 3, 2–3 mm long, ovate, subacute, often split; body 10–12 mm wide, subglobose, the

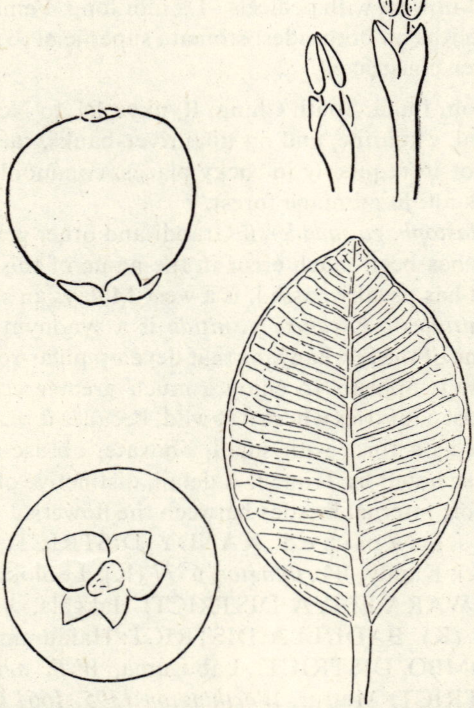

Fig. 17. *Ficus trimenii.* Leaf, × ½; fig, × 2; male flowers, × 10.

orifice closed by 3 flat apical bracts in a disc 2 mm wide, separating at maturity; internal bristles none. Tepals 3–4, free. Male flowers disperse, abundant; pedicels 0.5–1.5 mm; stamen 1, anther exsert. Gall-flowers pedicellate. Female flowers sessile. Lamina with cystoliths on both sides, stomata sunken.

D i s t r. Peninsular India (Canara), Ceylon.

E c o l. Lowland-forest, rare.

I n s e c t s. Not known.

N o t e. Superficially this resembles *F. microcarpa* L. f., but the venation is that of *F. benjamina* L., and its alliance is with *F. callophylla* Bl. of west Malaysia. The tree at the Royal Botanic Gardens of Peradeniya, mentioned by Trimen, still flourishes but there have been no recent collections of wild plants. The early collections were from Kadugannawa, where it seems no longer to exist, and from Ekiriankumbura, Uva.

S p e c i m e n s E x a m i n e d. KANDY DISTRICT: Kadugannawa, *W.F. s.n.* (PDA). LOCALITY UNKNOWN: *s. coll. C.P. 2220* (PDA).

17. Ficus elastica Roxb. ex Hornem., Hort. Bot. Hafn. Suppl. 7. 1819; King, Ann. R. Bot. Gard. Calcutta 1: 45, pl. 54. 1887; Corner, Wayside Trees of Malaya 677, pl. 202, 203. 1940; Corner, Gard. Bull. Singapore 21: 24. 1965; Worthington, Ceylon Trees pl. 406. 1959; Wiebes, Zool. Verh. Leiden 83: 16. 1966; Hill, Figs of Hong Kong 36. 1967; Condit, Ficus Exot. Sp. 122. 1969.— **Fig. 18.**

A large evergreen banyan, developing copious aerial roots from the parts of the branches near the trunk, not developing pillar-roots; midrib, petioles, stipules commonly red on young shoots. Leaves spirally arranged. Glabrous or the twigs, stipules, and the fig-peduncles finely puberulous. Twigs 3–5 mm thick, –8 mm on lower branches. Stipules very large on lower branches and opening shoots, shorter on the upper twigs, often pink. Lamina 10–15 × 4–7.5 cm on upper branches (–40 × 22 cm on lower branches), elliptic varying subobovate, rarely almost lanceolate, acuminate with a short tip –10 mm (–25 mm on lower branches), base cuneate (broadly to almost rounded on lower branches), coriaceous; lateral veins 15–20 (–26) pairs, scarcely raised beneath, without intercostals, the secondaries almost as well-developed as the primaries; basal veins 1 (–2) pairs, short; petiole 25–50 × 2–3 mm, (–80 mm on lower branches), drying blackish. Figs axillary and on the twigs below the leaves, paired, ripening yellow-ochre (? mottled red); peduncle (1–) 3–5 × 4–6 mm, short, thick, obconic, with a slightly tumid entire upper margin; basal bracts 3, 3 mm long, very early caducous (within the stipular bud-sheath); body 9–12 × 8–9 mm, shortly ellipsoid, the plane or slightly prominent orifice closed by 3–4 apical bracts in a disc 2–2.5 mm wide, internal bracts imbricating: internal bristles none. Male flower disperse, shortly pedicellate; tepals (3–) 4, free, eventually spreading; anther 4-celled, dehiscing longitudinally. Gall-flowers sessile or shortly pedicellate, female flowers sessile; tepals 4, free; ovaries not immersed; stigma simple. Seed subrugose. Lamina with cystoliths abundant on the upperside, sparse below; stomata deeply sunken.

D i s t r. Assam, Burma, North Malaya, Java; widely cultivated, introduced to Ceylon.

V e r n. India Rubber Tree (E).

I n s e c t s. *Blastophaga clavigera* Mayr.

N o t e. This tree has been occasionally planted in Ceylon. It is a strange relic without close allies in Southeast Asia but related apparently, as seems clear from the venation and microscopic structure of the leaf, with subgen. *Malvanthera* of Melanesia and Australia. The figs appear to have no basal

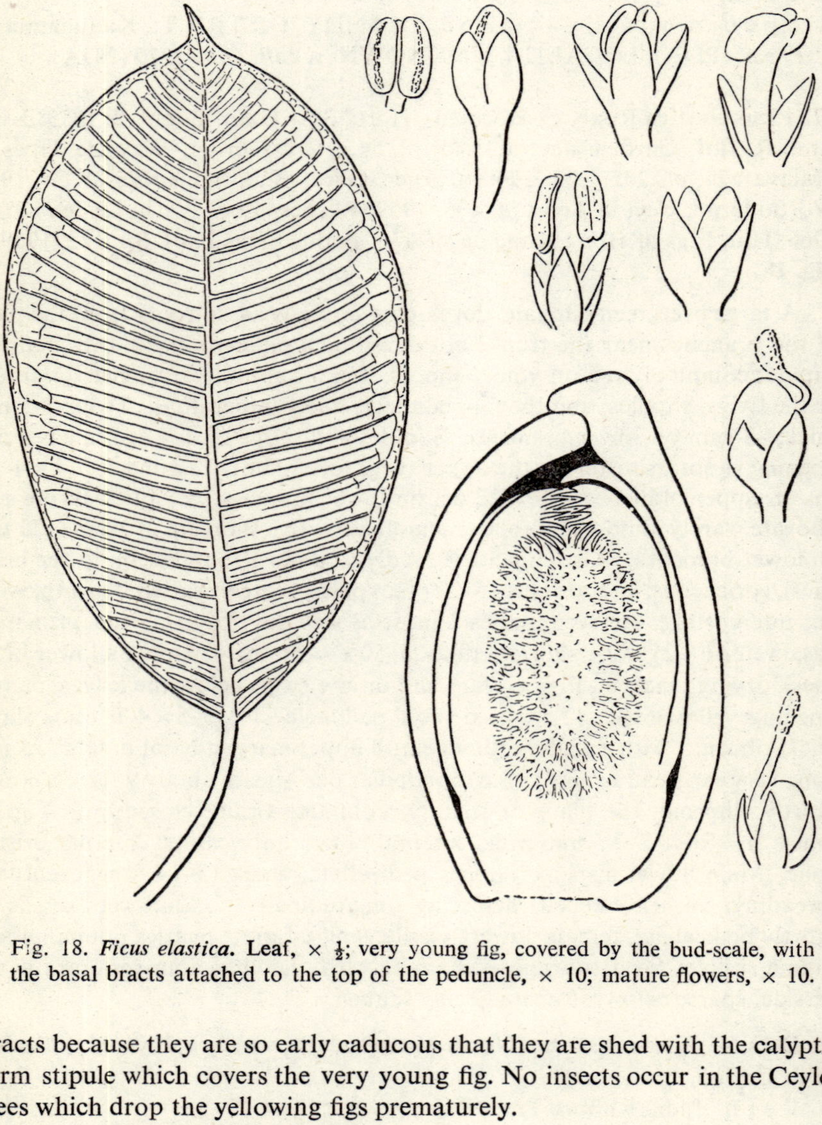

Fig. 18. *Ficus elastica.* Leaf, × ½; very young fig, covered by the bud-scale, with the basal bracts attached to the top of the peduncle, × 10; mature flowers, × 10.

bracts because they are so early caducous that they are shed with the calyptriform stipule which covers the very young fig. No insects occur in the Ceylon trees which drop the yellowing figs prematurely.

C o l l e c t i o n s. None noted.

2. Subgenus PHARMACOSYCEA

18. Ficus callosa Willd., Acta Acad. Berol. 102. 1798; King, Ann. R. Bot. Gard. Calcutta 1: 64, pl. 85. 1887; Trimen, Handb. Fl. Ceylon 4: 93. 1898; Worthington, Ceylon Trees pl. 404. 1959; Corner, Gard. Bull. Singapore 21: 29. 1965; Condit, Ficus Exot. Sp. 205. 1969.—**Fig. 19.**

Ficus cinerascens Thw., Enum. pl. Zeyl. 266. 1864.

Tree –45 m high, becoming buttressed; leaves spirally arranged; bark grey, smooth; latex serous. Young parts finely puberulous, soon glabrous. Twigs 3–8 mm thick, pale ochraceous to brown, stout. Stipules 12–30 mm long, –55 mm on opening shoots. Lamina 10–30 × 5–15 cm, elliptic, varying ovate to

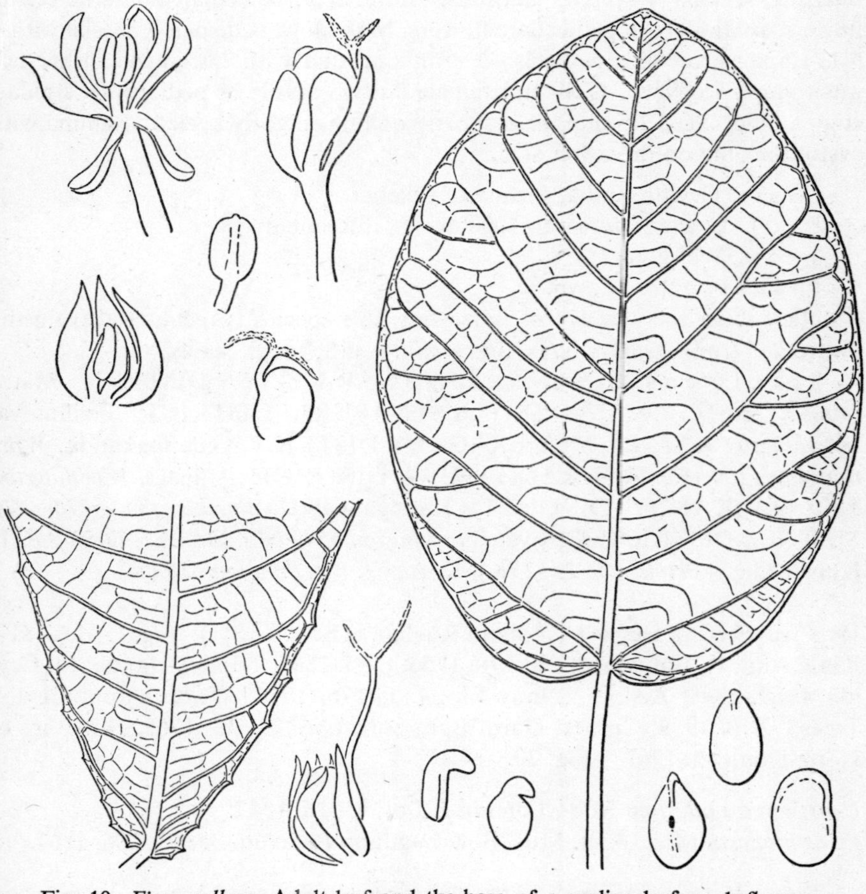

Fig. 19. *Ficus callosa.* Adult leaf and the base of a sapling leaf, × ½; flowers, seeds and embryos, × 10.

obovate, obtuse, subacute, or obtusely subacuminate, base cordate to rounded and widely cuneate, entire, thinly coriaceous, often nitid, finely scabrid beneath, drying grey-green to pale brownish; lateral veins 7–16 pairs, –7 fairly distinct intercostals, slightly raised on both sides; basal veins 2–3 pairs, short, no glands; petiole 20–70 × 2–4 mm; sapling leaves –50 × 23 cm, elliptic or oblong-elliptic or obovate, often with 1–3 short obtuse lobes on each side at or above the middle or merely sinuate, base with 2–8 minute teeth on each side, 20–30 lateral nerves, petiole –10 cm long. Figs axillary, solitary, rarely paired, puberulous or finely villous, glabrescent, ripening yellow; peduncle 2–19 mm; basal bracts 3, 1.5–4 mm long, persistent; pedicel 1.5–5.5 mm; body 15–25 mm wide, subglobose, the orifice closed with 3 apical bracts separating to show the internal bracts; internal bristles none; fig-wall thick, densely sclerotic. Tepals 4–6, red, glabrous, entire, free, lanceolate-acute in sessile flowers, spathulate in pedicellate flowers. Male flowers disperse, sessile with 2 (–1) stamens, or with pedicels –2 mm long and with 1 stamen; anther not mucronate; no ovary. Gall- and female flowers sessile or pedicellate, similar, stigma bifid. Seed 1.5 mm long, shortly oblong, slightly keeled. Lamina with cystoliths only on the lower side.

Distr. Ceylon, India, Burma to Celebes.

Ecol. Lowland forest to 1000 m alt., infrequent.

Vern. Wal-gona (S).

Insects. Not known.

Note. In the absence of fallen figs, this species is recognised from the buttressed trunk, watery latex, and shining, stiff, rough leaves.

Specimens Examined. NUWARA ELIYA DISTRICT: Maturata, *s. coll. C.P. 2562* (PDA). RATNAPURA DISTRICT: Belihuloya, *Worthington 4851* (K). MONERAGALA DISTRICT: Galapitakanda, *Worthington 5100* (K). ANURADHAPURA DISTRICT: Ritigala, *Worthington 5160* (K). KANDY DISTRICT: Kadugannawa, *Worthington 6276* (K). MATALE DISTRICT: Sigiriya, *Worthington 6349* (K). GALLE DISTRICT: Kuneli Ela, *Worthington 2312* (K); Kottawa, *Worthington 618* (K).

19. Ficus nervosa Heyne ex Roth in Roehm. et Schult., Syst. Veg. 1: 513. 1817; King, Ann. R. Bot. Gard. Calcutta 1: 53, t. 65. 1887; Trimen, Handb. Fl. Ceylon 4: 89. 1898; Alston, Kandy Fl. 34, f. 176. 1938; Worthington, Ceylon Trees f. 412. 1959; Corner, Gard. Bull. Singapore 21: 31. 1965; Hill, Figs of Hong Kong 42. 1967.—**Fig. 20.**

Urostigma modestum Miq., London J. Bot. 6: 586. 1847.
Ficus modesta Miq., Ann. Mus. Bot. Lugduno Batavum 3: 263, 286. 1867.

Tree –35 m, buttressed –6 m, evergreen, with dense crown: latex scant, white, yellowish on exposure. Leaves spirally arranged, pinkish brown when young. Glabrous, except the stipules, or thinly pubescent at first with hairs –0.5 mm long. Twigs 2–4 mm, fuscous: pith orange-brown. Stipules –13 mm, finely appressedly silky, caducous. Lamina 6–17 × 2.2–7 cm, elliptic, often narrowly subacute to acuminate, base cuneate to somewhat round, entire, coriaceous thinly, often bullate, drying brown, smooth; lateral veins 7–11 pairs, prominent below, 0–3 intercostals, often vague; basal veins 1 pair, not prominent; petiole 7–25 × 1.5–2.5 mm. Figs axillary and on the twigs behind the leaves, paired or solitary, glabrous or thinly appressedly hairy at first, ripening red; peduncle 0–2.5 mm; basal bracts 3, c. 1 mm long; pedicel 3–12 mm; body 8–15 mm wide, subpyriform, hard, the orifice closed by 3 small flat scales; bracteoles subtending the flowers; internal bristles 0; sclerotic cells abundant in the rather thick wall. Male flowers disperse, mostly round the

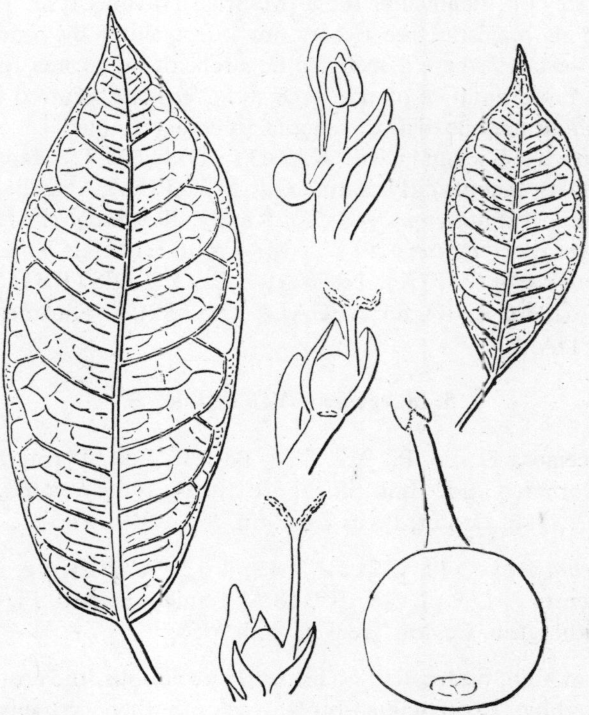

Fig. 20. *Ficus nervosa.* Leaf of v. *nervosa* (left) and v. *minor* (right), × ½; fig, × 3; flowers, × 10.

orifice, sessile to pedicellate; tepals 2–3, red, free; stamen 1 (–2), anther mucronate. Gall- and female flowers sessile to pedicellate; tepals 3–5, red, free or shortly gamophyllous, ovate-lanceolate, glabrous; ovary sessile, yellowish-white, no red; stigma bifid. Lamina with cystoliths only on the lower side.

Distr. Ceylon, India, Burma, South China, Formosa, Tonkin, Annam.
Ecol. Forest tree of the wet zone to 1600 m alt.; frequent.
Vern. Kala-maduwa (S).
Insects. Not known.

var. **minor** King, Ann. R. Bot. Gard. Calcutta 1: 53. 1887.

Urostigma modestum Miq. f. *brevifolia* Miq., Fl. Ind. Bat. I. 2: 353. 1859; Thw., Enum. Pl. Zeyl. 266. 1864.

Rather persistently puberulous on the twigs, petioles, and underside of the midrib. Lamina –10 × 4.5 cm, small, acuminate: lateral nerves 6–7 pairs: petiole 10–22 × 1–1.5 mm, slender.

Distr. Ceylon, Peninsular India (Madras, Travancore).
Note. This majestic tree occurs not infrequently by roads and in the mountain tea-estates where it seems to be a relic of the former forest. It is the westernmost representative of the large ser. *Nervosae*, centred in Melanesia. How far v. *minor* may be distinct remains to be determined.
Specimens Examined. KANDY DISTRICT: Hantane, *s. coll. C.P. 2219* (PDA); Hunnasgiriya *s. coll. C.P. 2219* (PDA); Alagalla, *Worthington 15* (K); Galaha, *Worthington 2488* (K); Kandy, *Worthington 5497*, *6397* (K); Kadugannawa, *Worthington 6297* (K). MATALE DISTRICT: Kandanuwara, Matale, *Worthington 2382* (K). NUWARA ELIYA DISTRICT: Rangbodde, *Worthington 6138* (K). KURUNEGALA DISTRICT: Kurunegala, in 1886, *s. coll. s.n.* (PDA).

3. Subgenus SYCOMORUS

20. Ficus racemosa L., Sp. Pl. 922. 1753; Barrett, Bull. Torrey Bot. Club 73: 312. 1946; Corner, Gard. Bull. Singapore 21: 34. 1965; Wiebes, Zool. Verh. Leiden 86: 27. 1966; Condit, Ficus Exot. Sp. 196. 1969.–**Fig. 21.**

Ficus glomerata Roxb., Pl. Corom. 2: 13, f. 123. 1798; King, Ann. R. Bot. Gard. Calcutta 2: 173. pl. 218, 219. 1888; Trimen, Handb. Fl. Ceylon 4: 96. 1898; Worthington, Ceylon Trees pl. 408. 1959.

Tree –30 m high, buttressed, occasionally deciduous, the crown often irregular and shabby; bark pinkish-brown, smooth then coarsely flaky; latex copious, cream buff. Young shoots and figs finely white hairy, soon glabrous; young leaves pale green. Twigs 1.5–3 mm thick. Stipules –12 mm long, –30 mm long on opening shoots, often persistent on young twigs. Lamina 6–19 × 3.5–8.5 cm, elliptic varying subovate, shortly oblong or somewhat lanceolate, tapered to a subacute or subacuminate apex, base broadly to narrowly cuneate, rarely subcordate, subcoriaceous, smooth, entire (dentate in saplings); lateral veins 4–8 pairs, –7 intercostals slightly raised below; basal veins 1 pair, short, often with a slight axillary gland; petiole 15–70 mm long, becoming brown scurfy. Figs in large clusters on the main branches and trunk, on branching leafless twigs –25 cm long, 2.5 cm thick at the base, ripening rose-red; peduncle 3–12 mm long, stout; basal bracts 1–2 mm long, ovate-triangular, obtuse; pedicel 0–5 mm long; body 20–30 mm wide (35–50 mm, living), subglobose to pyriform, often lenticellate-verrucose, orifice plane or slightly sunken, closed by 5–6 apical bracts; internal bristles none; sclerotic cells none in the thick, soft wall. Perianth with 3 (–4) dentate-lacerate lobes joined below, red, glabrous. Male flowers ositolar in 2–3 rings, sessile, much compressed; stamens (1–) 2, rarely

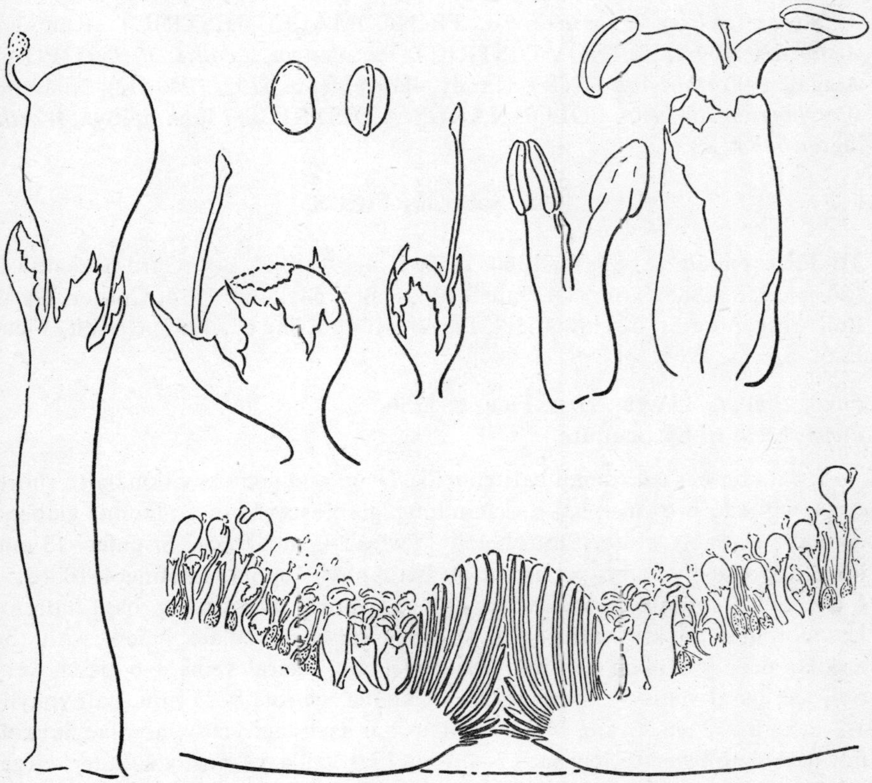

Fig. 21. *Ficus racemosa.* Section through the orifice of a mature fig with the male flower in three circles, the sessile female flowers stippled, × 3; flowers, × 10.

3, often with an abortive ovary. Gall-flowers long-stalked; ovary dark red; style short. Female flowers sessile or very shortly stalked among the gall-flowers; ovary sessile or substipitate, red-spotted; style 2–3 mm long, glabrous, simple. Seed 1 mm long, lenticular, smooth, not or scarcely keeled. Lamina with cystoliths only on the lower side.

D i s t r. Ceylon, Pakistan, India, South China to New Guinea and Queensland.

E c o l. Riverside tree of the lowlands; common.

V e r n. Attikka (S); Atti (T).

I n s e c t s. *Ceratosolen fusciceps* Mayr and other genera (Wiebes).

N o t e. This is the one species of the African subgen. *Sycomorus* which penetrates into Asia. It is recognised from the clusters of large, red figs on the trunk and branches. In West Pakistan and India it is deciduous and this may be the case in the dry zone of Ceylon. The figs are scarcely edible, being tenanted usually by many insects.

S p e c i m e n s E x a m i n e d. TRINCOMALEE DISTRICT: Kantalai, *Alston 519* (PDA). KANDY DISTRICT: Peradeniya, *s. coll. C.P. 3305* (PDA); Alagalla, *Worthington 51* (K); Kandy, *Worthington 5152, 7040* (K); Nilambe, *Worthington 3047* (K). POLONNARUWA DISTRICT: Manampitiya, *Worthington 6288* (K).

4. Subgenus FICUS

21. Ficus pumila L., Sp. Pl. 1060. 1753; King, Ann. R. Bot. Gard. Calcutta 2: 124. pl. 158. 1888; Trulzoch, Jahrb. Wiss. Bot. 54: 1–69. 1914; Corner, Gard. Bull. Singapore 21: 52. 1965; Hill, Figs of Hong Kong 67. 1967; Condit, Ficus Exot. Sp. 75. 1969.

Ficus stipulata Thunb., Diss. Ficus 8. 1786.
Ficus repens of horticulture.

Root-climber with small bathyphylls. Twigs and petioles villous with short, erect, white to brownish hairs –1 mm long, glabrescent slowly; lamina glabrescent above, finely white-villous below. Twigs 2–3 mm thick. Stipules –13 mm long, appressedly hairy, persistent on bathyphyll shoots. Lamina 4–10 × 2.5–6 cm, ovate-cordate to somewhat oblong, obtuse to subacute, base more or less cordate, thickly coriaceous, strongly raised-reticulate below with the areolae deeply sunken and minutely foveolate; lateral veins 3–6 pairs, very oblique; basal veins 3–5 pairs, 1/2–2/3 lamina; petiole 8–25 mm; bathyphylls 1.2–3.5 × 0.7–2 cm, ovate, blunt, cordate but asymmetrically, areolae flat and not foveolate beneath, petioles 2–4 mm. Figs axillary, mostly solitary, large, thinly villous, glabrescent, ripening purple to black; peduncle (0–) 4–20 × 3–5 mm; basal bracts 5–7 mm long, ovate-lanceolate, appressedly hairy; body 3–6 × 2.5–3.5 cm, subglobose (gall-figs) to pyriform, obovoid or turbinate (seed-figs), subtruncate and umbonate at the apex, narrowed to the base; internal bristles abundant, short; wall thick, without sclerotic cells. Male flowers in many rows, filling the distal half of the fig, 10–17 mm long; stamens 2–3. Gall-flowers sessile or with white pedicels –7 mm long; tepals 4–6; ovary sessile, dark red-brown. Female flowers sessile or with white pedicels –12 mm long;

ovary sessile, white; style long, simple, strongly lateral. Seed 2.5–1.2–1.5 mm, subterete, not or scarcely keeled, slightly thickened at each end, often curved. Lamina with cystoliths only on the lower side.

D i s t r. Japan, Ryukyu Isl., Formosa, China, Indo-China (North); introduced to Ceylon.

E c o l. Often cultivated as a creeper on walls.

U s e s. Ornamental; seed-figs edible (? unknown in Ceylon).

I n s e c t s. Not known.

N o t e. This is a frequent garden climber, easily known from the large glaucous figs which are sterile in Ceylon, as in other countries where it has been introduced.

S p e c i m e n s E x a m i n e d. LOCALITY UNKNOWN: *Worthington 6479, 6579* (K).

22. Ficus laevis Blume, Bijdr. 437. 1825; King, Ann. R. Bot. Gard. Calcutta 2: 128, pl. 161. 1888; Corner, Gard. Bull. Singapore 21: 53. 1965; Wiebes, Zool. Verh. Leiden 83: 20. 1966.—**Fig. 22.**

Large root-climber, internodes elongate. Leaves spirally arranged. Glabrous, or with sericeous stipules and thinly appressedly hairy petioles and underside of midrib, or with rather sparse, spreading, white to brownish hairs –1.5 mm long on the twigs, petioles, and underside of the midrib. Twigs 2–4 mm thick. Stipules –15 mm. Lamina 8–25 × 5–20 cm, ovate to elliptic, suddenly acuminate, the tip –20 mm long, base shallowly cordate to rounded-subtruncate and cuneate, entire, membranous, smooth; lateral veins 4–6 pairs, 3–8 intercostals rather lax and slightly raised below, areolae flat; basal veins 2 (–3) pairs, 1/2–2/3 lamina, with 2 slight glands; petiole 35–100 mm long, often with a slight gland at the node; bathyphylls ovate-elliptic, cordate, often with overlapping basal lobes, rather long petiolate. Figs axillary, solitary, occasionally paired; peduncle 10–32 mm; basal bracts 1.5–2 mm long, ovate, acute; pedicel 0–10 mm long; body 20–25 mm wide, globose or pyriform, the plane or slightly depressed orifice closed by many small appressed apical bracts in a rosette 3–4 mm wide; internal bristles abundant; no sclerotic cells in the fig-wall. Tepals 4, red, free, glabrous. Male flowers in one ring, sessile; stamens 2 (–3), the filaments free, the anthers shortly mucronate. Gall- and female flowers sessile or with white pedicels –2.7 mm long; gall-ovary dull red-brown or red-spotted, female ovary white or red-spotted, sessile; gall-style short, female style long and shortly bifid. Seed 1.7–2 mm long, rather plump, slightly keeled at one or both ends; embryo straight. Leaves with cystoliths only on the lower side; no hypodermis; vascular bundles not trabeculate, except in the main veins.

D i s t r. Ceylon, India, Burma, South China to Sumatra, Java and Borneo.

E c o l. Lowland and mountain forest to 1500 m alt., uncommon, possibly rare in Ceylon.

I n s e c t s. *Blastophaga gomberti* Grandi (in the Indian *F. laevis* var. *macrocarpa*).

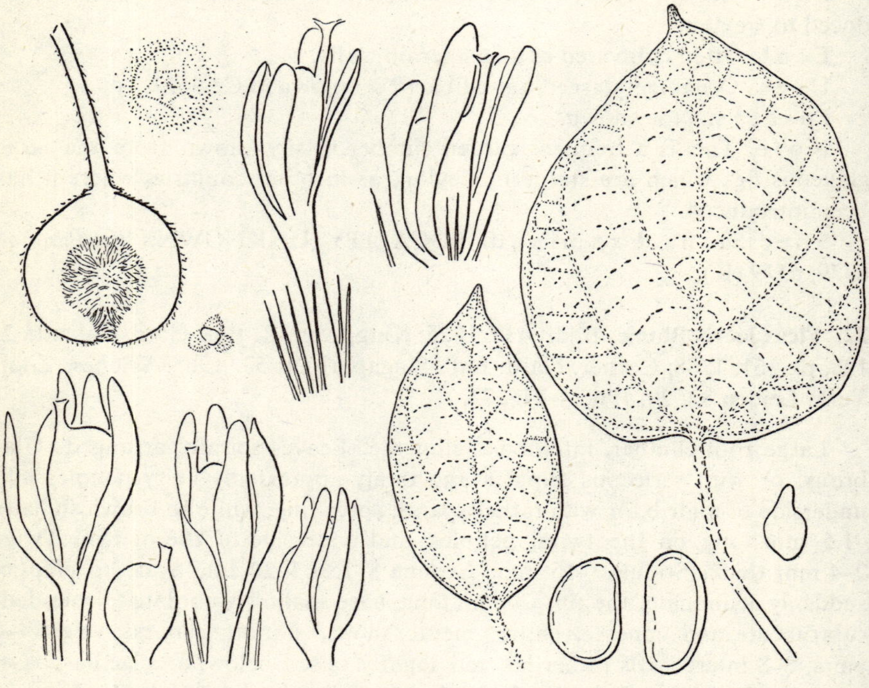

Fig. 22. *Ficus laevis*. Leaves, × ½; fig and basal bracts, × 1; orifice of fig, × 3; flowers, seeds and embryo, × 10.

var. **tomentosa** King, Ann. R. Bot. Gard. Calcutta 2: 128. 1888.

Pogonotrophe dasyphylla Miq., *P. ceylanica* Miq., London J. Bot. 7: 74, 75. 1848.

Ficus ceylanica Miq., Ann. Mus. Bot. Lugduno Batavum 3: 293. 1867; Trimen, Cat. 84. 1885.

Ficus laevis Blume var. *dasyphylla* (Miq.) King, l.c.; Trimen, Handb. Fl. Ceylon 4: 95. 1898.

Twigs, petioles, underside of the main veins, and figs densely villous with stiff spreading brown hairs to 1.5 mm long, in some cases the hairs appressed or hooked on the underside of the veins, the areolae smooth and flat; stipules appressedly silky.

D i s t r. Ceylon, Malaya, Sumatra.

E c o l. Wet zone forest, to 1300 m alt., uncommon; fruiting March–June (Trimen).

I n s e c t s. Unknown.

N o t e. This large, lax climber seems to occur mainly as var. *tomentosa* in Ceylon, but the old collection *C.P. 379* (Adam's Peak) seems to be var. *laevis*. In South India there occurs a variety, var. *macrocarpa* (Miq.) Corner, with large cauliflorous figs 3–6 cm wide when living, but this has not been found in Ceylon.

S p e c i m e n s E x a m i n e d. RATNAPURA DISTRICT: Karawita Kande, *s. coll. s.n.* (PDA). KANDY DISTRICT: Knuckles, *W.F. s.n.* (PDA); Adam's Peak, *s. coll. C.P. 233* (PDA).

23. Ficus diversiformis Miq., London J. Bot. 7: 441. 1848; Miq., Ann. Mus. Bot. Lugduno Batavum 3: 281, 294. 1867.—**Fig. 23.**

Ficus stipulata Moon, Cat. 74. 1824, non Thunb.
Ficus thwaitesii Miq., Ann. Mus. Bot. Lugduno Batavum 3: 229, 294. 1867; King, Ann. R. Bot. Gard. Calcutta 2: 125. pl. 159 B. 1888; Trimen, Handb Fl. Ceylon 4: 95. 1898.

Creeping shrub, rooting at the nodes. Leaves laxly spiral. Glabrous or the twigs, petioles, underside of the veins, and figs thinly puberulous with

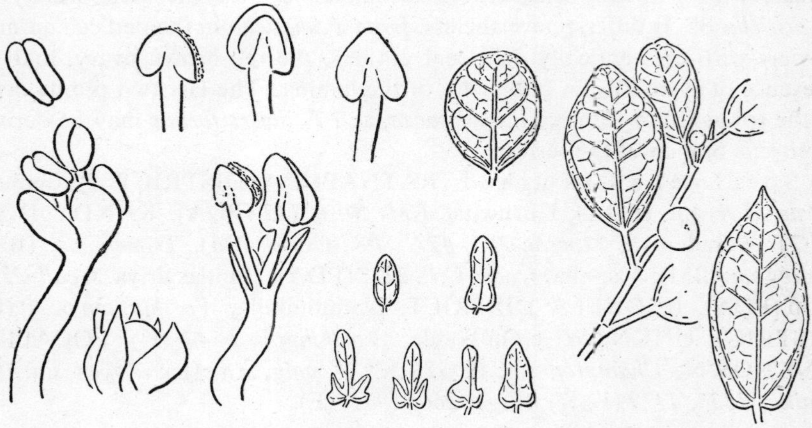

Fig. 23. *Ficus diversiformis*. Adult leaves and (lower centre) bathyphylls, × ½; male and gall-flowers, × 10.

erect white hairs –0.5 mm long, to thinly villous with brownish hairs. Twigs 1–1.5 mm thick, fuscous brown. Stipules short, caducous. Lamina 1.5–5.5 (–10) × 0.9–3 (–5) cm, obovate, blunt, varying elliptic, subacute, to ovate and suborbicular, base cuneate to subcordate, entire, coriaceous, drying fuscous brown and areolate beneath, smooth; lateral veins 2–3 (–4) pairs, rather strongly raised below, no intercostals, areolae brown; basal veins 1 (–2) pairs, 1/2–2/3 the lamina; petiole 3–10 mm long, slender; bathyphylls 6–15 mm long, elliptic, obtuse to hastate with 1–2 pairs of basal lobes, petiole 1–2 mm. Fig axillary, solitary, (? colour); peduncle 3–12 mm; basal bracts 3, in a collar, 0.5–1 mm long, ovate, acute, persistent; pedicel 0–6 mm long; body 10–13 mm wide, subglobose, subscabrid, the orifice plane and small; internal bristles none; fig-wall without sclerotic cells, or very few and scattered in the inner layer of gall-figs. Tepals 3–4, dark red, free, glabrous, fleshy. Male flowers disperse, with glabrous red pedicels 1–2 mm long; tepals blunt; stamens 2 with red filaments, the anthers with crescentic dehiscence over the apex. Gall-flowers sessile; tepals lanceolate; ovary more or less sessile, dark red-brown, not angular. Female flowers as the gall; style glabrous. Seed 2 mm long, subcompressed, slightly and bluntly keeled, smooth. Lamina with cystoliths on both sides, sparse above; hairs aseptate.

Distr. Ceylon.

Ecol. Creeping over the ground, rocks and trunks, chiefly in the mountains, 500–1600 m alt.

Insects. Not known.

Note. Because of the disperse arrangement of the male flowers and the non-mucronate anthers, this little climber fits in ser. *Distichae* of sect. *Rhizocladus*, where it has considerable resemblance with the west Malaysian *F. disticha* Bl. It differs, nevertheless, from *F. disticha* in the pedicellate male flowers with crescentically dehiscent anthers, the red-brown ovary, and the presence of cystoliths on both sides of the lamina. The last two points occur in the robust Indo-Chinese *F. hederacea*, and *F. diversiformis* may be derived from this branch of the series.

Specimens Examined. RATNAPURA DISTRICT: Ratnapura, *A. de Alwis s.n.* (PDA); Kuruwita, *J.M. Silva 112* (PDA). KANDY DISTRICT: Peradeniya, *Macrae 283*, *427*, *808* (CGE, BM), *Trimen s.n.* (BM), *Ridley s.n.* (BM); Alagalla, *s. coll. C.P. 2217* (PDA); Hunnasgiriya, *s. coll. 2224*, *3116* (PDA). BADULLA DISTRICT: Namunukula, *Worthington 6540* (K). DISTRICT UNKNOWN: Galbodde, *Worthington 6540* (K). LOCALITY UNKNOWN: *Champion s.n.* (CGE, K); *Koenig*, *s.n.* (L); *Royen s.n.* (C); *Walker 1338*, *1179* (E, K); *Wight 2665* (*941*) (E).

24. Ficus heterophylla L. f., Suppl. Pl. 442. 1781; King, Ann. R. Bot. Gard. Calcutta 2: 75, pl. 94. 1888; Trimen, Handb. Fl. Ceylon 4: 93. 1898; Corner, Gard. Bull. Singapore 21: 73. 1965; Wiebes, Zool. Verh. Leiden 83: 19. 1966. **—Fig. 24.**

Shrub –3 m, often prostrate. Leaves distichous. Twigs 1–2 mm thick, becoming reddish-brown, thinly to closely pubescent with short white spreading hairs –0.5 mm long, often subscabrid, varying almost glabrous. Stipules short, single, caducous, intrapetiolar. Lamina 5–14 × 1.5–5 cm, oblong or narrowly elliptic to subovate, varying lanceolate, acute to shortly acuminate, base narrowly to widely cuneate or rounded, membranous to subcoriaceous, dentate to subentire, thinly to harshly scabrid-hairy on both sides; lateral veins 4–7 pairs (–12 on lanceolate leaves), 0–2 intercostals; basal veins 1 pair, 1/4–1/2 the lamina, with 1–2 small basal glands; petiole 5–55 mm long, slender, puberulous; juvenile leaves varying pinnately lobed with 1–4 lobes on each side, –9 cm wide, or palmately 3-lobed. Figs solitary, axillary, rarely paired, ripening orange to dark red; peduncle 4–12 mm; basal bracts 3, in a collar,

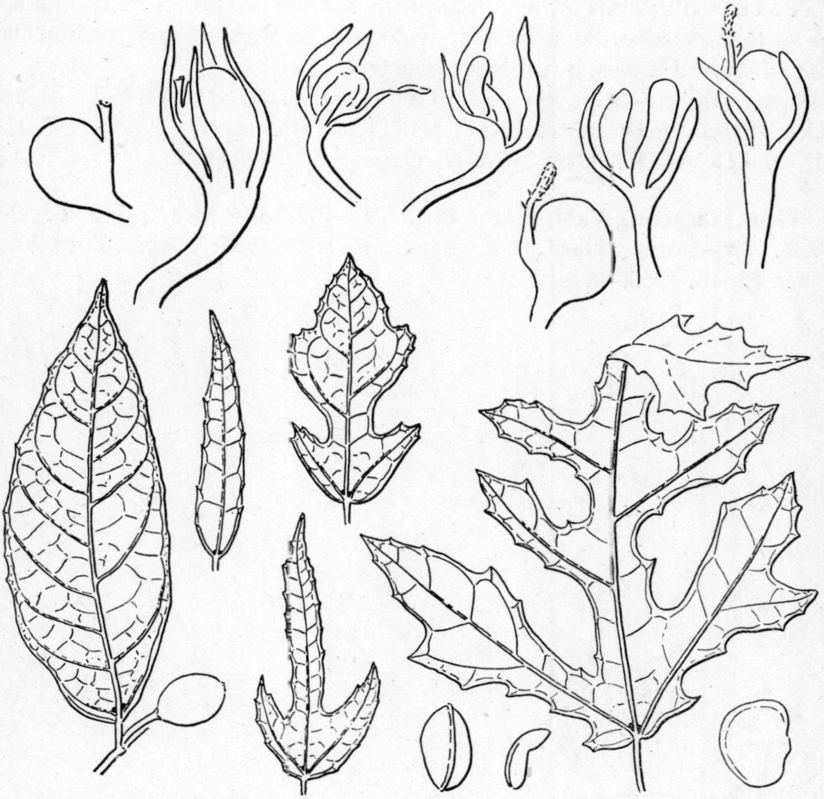

Fig. 24. *Ficus heterophylla.* Leaves, × ½; gall-flowers (left), seed-flowers (right), male flowers, seeds and embryo, × 10.

ovate-subacute, small; pedicel 0–10 mm; body 12–22 × 8–17 mm, ellipsoid, scabridulous then smooth, drying umbonate, no lateral bracts, the orifice closed by 4–6 small umbonate apical bracts; internal bristles none; sclerotic cells none. Tepals white, free, entire, glabrous or slightly puberulous. Male flowers in 2–3 rings, pedicellate –1.5 mm; tepals 3 (–4), linear-oblong; no gall-ovary. Gall-flowers pedicellate –2.5 mm; tepals 4 (–5), lanceolate; ovary sessile to shortly stalked. Female flowers as the gall, tepals oblong and blunt, style glabrous or puberulous upwards. Seed 1–1.3 × 0.6–0.8 mm, subcompressed, shortly oblong, slightly keeled, smooth. Leaf with cystoliths on both sides, abundant, the larger papillate.

Distr. Ceylon, India, Burma, South China to Java and Borneo.

Ecol. In open places, particularly the flood-margins of rivers.

Vern. Wal-ehetu (S).

Insects. Not known.

Note. Collections of this species from Ceylon are few. It may be a species of the dry zone. As a small sprawling shrub with solitary, pedunculate figs and scabrid leaves, it is easily recognized.

Specimens Examined. TRINCOMALEE DISTRICT: Trincomalee, *Alston 556* (PDA). KANDY DISTRICT: Peradeniya, *s. coll. C.P. 2228* (PDA). LOCALITY UNKNOWN: *Gardner 826, 829* (BM).

25. Ficus exasperata Vahl, Enum. Pl. 2: 197, 402. 1806; Miq., London J. Bot. 7: 231. 1848; Corner, Gard. Bull. Singapore 21: 74. 1965; Wiebes, Zool. Verh. Leiden 83: 16. 1966.—**Fig. 25.**

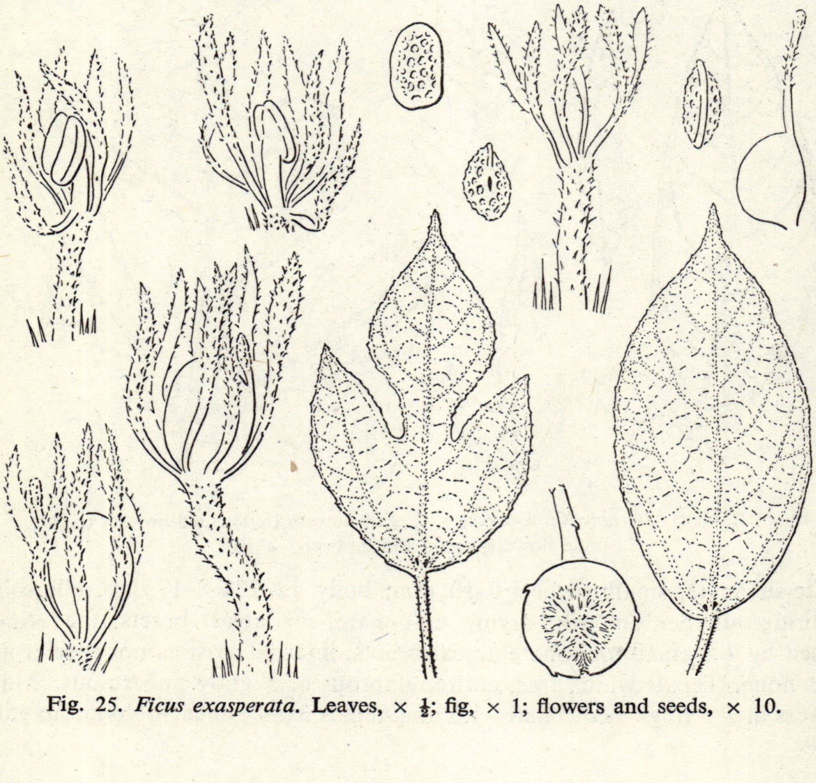

Fig. 25. *Ficus exasperata*. Leaves, × ½; fig, × 1; flowers and seeds, × 10.

Ficus serrata Forsk., Fl. Aeg. Arab. 179. 1775, non L.
Ficus asperrima Roxb., Fl. Ind. 3: 554. 1832; King, Ann. R. Bot. Gard. Calcutta 2: 80, pl. 100. 1888; Trimen, Handb. Fl. Ceylon 4: 94. 1898; Alston, Kandy Fl. 34, f. 174. 1938; Worthington, Ceylon Trees pl. 401. 1959.

Small to medium-sized, deciduous tree, –25 m high. Leaves laxly spiral to opposite or subdistichous. All parts coarsely and harshly scabrid with stout white hairs –0.5 mm long, with pustulate base (no long hairs). Twigs 1.5–3 mm thick, light to dark brown. Stipules short, paired, caducous. Lamina 6–19 × 3–9 cm, narrowly to widely elliptic, varying ovate to obovate or oblong-lanceolate, acute to acuminate, the tip –15 mm long, base cuneate, more or less symmetric, denticulate or sinuate-crenate to serrate, scabrid on both sides, becoming stiffly coriaceous, hard, drying grey-green, fuscous or brownish; lateral veins 3–5 pairs, oblique, –6 intercostals, raised below, faint above; basal veins 1–2 pairs, 1/2–2/3 the lamina, 2 basal glands; petiole 4–65 mm long; sapling-leaves palmately 3-lobed. Figs axillary, mostly solitary, harshly scabrid, ripening yellow (yellow to purple, King); pedicel 4–8 mm long, with 2–3 small scattered lateral bracts, sometimes more or less aggregated into a collar; body 10–15 mm wide (15–20 mm, King), subglobose with scattered small lateral bracts, apical bracts projecting 1–2 mm; internal bristles copious, white; fig-wall without sclerotic cells. Tepals white, free, rarely 1–2 joined, puberulous or hispidulous upwards, exceeding the ovary. Male flowers ostiolar, sessile, in 1–2 rings; tepals 3–4 (–6) oblong-spathulate; stamen 1, no ovary. Gall-flowers sessile to pedicellate; tepals 4–6, lanceolate; ovary white, sessile; styles pink. Female flowers as the gall; tepals 4–7, linear-spathulate; style puberulous. Seed 1.2–1.4 × 0.8–0.9 mm, shortly oblong, slightly keeled, distinctly reticulate. Lamina with cystoliths on both sides, those on the upper side mostly as microscopic spicate hairs, sometimes also on the lower side.

Distr. East Africa, Arabia, Central and South India, Ceylon.

Ecol. Lowland regions up to 1000 m, common.

Uses. Leaves formerly used for sand-paper.

Vern. Bu-thediya, Sewan-mediya (S).

Insects. *Blastophaga gestroi* Grandi and other genera (Wiebes).

Note. This common tree is one of the few species of *Ficus* which occur both in Africa and Asia. The reticulately-pitted seed and the hispidulous tepals at once distinguish it in the alliance of subsect. *Varinga* of Sect. *Sycidium*. Its origin is problematic.

Specimens Examined. KANDY DISTRICT: Peradeniya, *Alston 1364* (PDA); Madugoda, *Simpson 8820* (BM); Kandy, *Worthington 6959* (K); Gelioya-Galaha, *Wirawan 755* (PDA). DISTRICT UNKNOWN: Nambidde, *s. coll. C.P. 2229* (PDA). LOCALITY UNKNOWN: *Macrae 357* (CGE).

26. **Ficus tinctoria** Forst. f., Prod. Fl. Austral. 76. 1786; Corner, Gard. Bull.

Singapore 17: 475. 1960; id. 21: 74. 1965; Corner, Philos. Trans. R. Soc. B. 253: 111. 1967; Wiebes, Zool. Verh. Leiden 83: 32. 1966; Hill, Figs of Hong Kong 74. 1967; Condit, Ficus Exot. Sp. 92. 1969.

ssp. **parasitica** (Willd.) Corner, Gard. Bull. Singapore 17: 476. 1960; id. 21: 75. 1965.—**Fig. 26.**

Ficus parasitica Willd., Mem. Acad. Roy. Sci. Hist. (Berlin) 2: 102. 1798; Trimen, Handb. Fl. Ceylon 4: 85. 1898; Alston, Kandy Fl. 34, f. 175. 1938; Worthington, Ceylon Trees pl. 413. 1959.

Ficus gibbosa Bl. var. *parasitica* (Willd.) King, Ann. R. Bot. Gard. Calcutta 1: 6, pl. 2 a–b. 1887.

Large strangling fig with copious interlacing and anastomosing aerial roots basketing the host; bark grey; latex white. Glabrous to somewhat hairy. Leaves distichous. Twigs 1.5–3 mm thick, pale. Stipules –18 mm long, straight, conical. Lamina 6–15 × 3.5–7 cm, elliptic-lanceolate to ovate-elliptic or subrhombic, acute to strongly acuminate, base cuneate and generally unequal, entire to vaguely lobed, sinuate, or toothed, thinly coriaceous, scabrid or smooth, often hairy beneath; lateral veins 4–6 pairs; basal veins 1 pair, not or shortly elongate, with a gland on the broader side of the lamina: petiole 5–20 mm long. Figs axillary, ripening golden-yellow (gall) or red (seed), scabridulous; peduncle 0; pedicel 1.5–10 mm, with 3 small basal bracts 1–2 mm long; body 8–12 mm wide, pyriform-subglobose, no lateral bracts, orifice

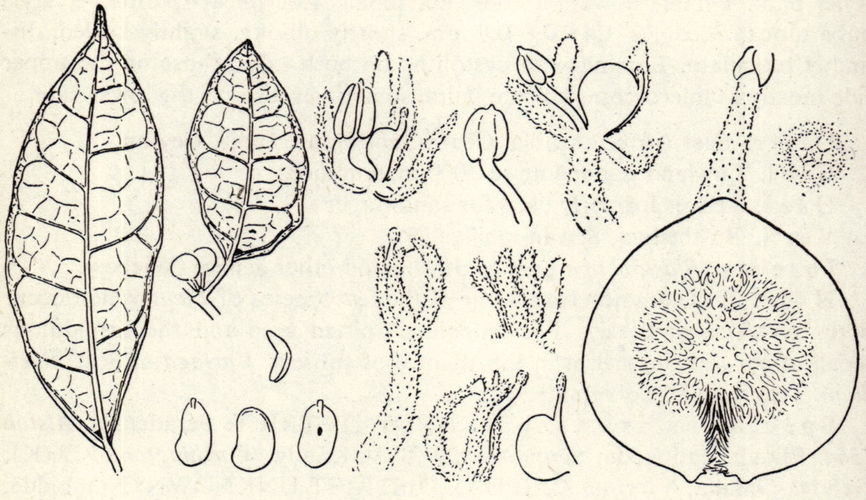

Fig. 26. *Ficus tinctoria* ssp. *parasitica*. Leaves, × ½; fig (in section) with the orifice (surface-view), × 3; male flowers and (below) seed-flowers, seeds and embryo, × 10.

plane, (surrounded by a slight rim, when living), closed by several small apical scales; internal bristles minute, abundant; sclerotic cells abundant in the inner wall of the gall-fig, few or none in the seed-fig. Tepals 3–4 (–5), white, puberulous, oblong or lanceolate, free; flower-pedicels puberulous. Male flowers ostiolar in 1–2 (rarely more) rows, sessile or pedicellate; stamen 1, mucronate, with a gall-ovary. Gall-flowers sessile or pedicellate (–2.5 mm); ovary white, sessile. Female flowers as the gall-flowers. Seed 1–1.3 mm long, oblong, compressed, not or slightly keeled. Leaf with cystoliths on both sides; stomata superficial.

Distr. Ceylon, India, Burma, South China, Indo-China, Andaman and Nicobar Isl.

Ecol. Low country to 1000 m alt., common.

Vern. Wal-ehetu, Gas-netul, Gas-anguna (S).

Insects. Not known.

Note. The continental ssp. *parasitica* is the western part of a great complex that stretches through the Malay Archipelago, as ssp. *gibbosa*, to Polynesia where it occurs as ssp. *tinctoria*. The distinctions are keyed out in my check-list of *Ficus* (Corner, 1965). The banyan-habit is facultative in ssp. *tinctoria* and becomes more pronounced westwards; ssp. *gibbosa* has a root-trunk of many descending roots as in *F. elastica*, with little or no anastomosis, but in ssp. *parasitica* there are also anastomosing and basketing roots. These details of habit require to be studied more carefully. As this epiphytic strangling habit is perfected, there comes a remarkable difference in the stomata; they are deeply sunken in ssp. *tinctoria* and ssp. *gibbosa*, but they are wide and superficial in ssp. *parasitica*; where these last two overlap in Indo-China, intermediates occur which are almost impossible to classify. Thus, while Ceylon has an entity, *F. parasitica*, well distinct from *F. tinctoria*, a complete transition can be found through the intermediate range of ssp. *gibbosa*. *F. tinctoria* and its eastern ally *F. virgata* are the only species of *Ficus* outside subgen. *Urostigma* which are stranglers.

Specimens Examined. KANDY DISTRICT: Peradeniya, *s. coll. C.P. 2227* (PDA); Kadugannawa, *Worthington 1163*, *6844* (K). MATALE DISTRICT: Sigiriya, *Worthington 6345* (K). LOCALITY UNKNOWN: *Champion s.n.* (CGE, K); *Koenig s.n.* (L); *Macrae 429* (BM); *Walker 1172* (PDA); *Gardner 825* (PDA).

27. Ficus hispida L. f., Suppl. Pl. 442. 1781; King, Ann. R. Bot. Gard. Calcutta 2: 116, pl. 154, 155. 1888; Trimen, Handb. Fl. Ceylon 4: 94. 1898; Alston, Kandy Fl. 34, f. 173. 1938; Worthington, Ceylon Trees pl. 409. 1959; Corner, Gard. Bull. Singapore 21: 89. 1965; Wiebes, Zool. Verh. Leiden 83: 19. 1966; Hill, Figs of Hong Kong 82. 1967; Condit, Ficus Exot. Sp. 81. 1969. — **Fig. 27.**

Ficus oppositifolia Roxb., Pl. Corom. 2: 14, pl. 124. 1798.
Ficus daemonum Koen. ap. Vahl, Enum. Pl. 2: 198. 1806.

Tree –15 m high, often shrub-like, laxly branched; bark grey, smooth; latex white. Leaves decussate, varying spirally arranged. All parts hispid with short white to pale brown hairs, 0.5–1.5 mm long, varying appressedly hairy to thinly hairy. Twigs 3–8 mm thick, hollow. Stipules 11–25 mm long, caducous. Lamina 7–35 × 2.6–16 cm (–40 × 18 cm, sapling), oblong-elliptic to ovate-elliptic, acuminate with the tip 10–25 mm long, base cordate to cuneate, symmetrical, crenate-serrate to subentire, membranous, scabrid on both sides, varying almost smooth, drying grey-green; lateral veins 4–9 pairs, curved ascending, 6–12 intercostals raised below; basal veins 1–3 pairs, short or elon-

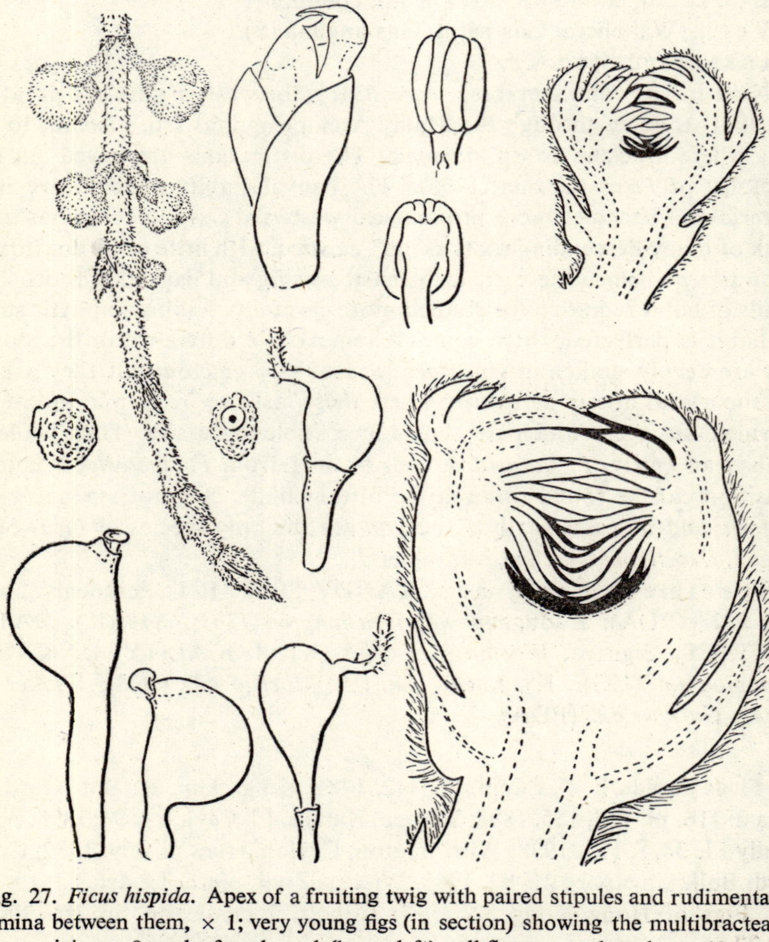

Fig. 27. *Ficus hispida.* Apex of a fruiting twig with paired stipules and rudimentary lamina between them, × 1; very young figs (in section) showing the multibracteate origin, × 8; male, female and (lower left) gall-flowers, and seeds, × 10.

gate 1/3–1/2 lamina; petiole 1–10 (–14) cm × 2–4 mm, with a subnodal gland. Figs axillary to cauliflorous on leafless hanging twigs arising in clusters from the trunk and main branches, up to 1.5 m long, sometimes trailing on the ground or becoming geocarpic, thinly hispid, ripening pale yellow; peduncle 5–15 mm long; basal bracts 3, 1–1.5 mm long, subtriangular; pedicel 0–6 mm; body 15–25 mm wide (–25 × 35 mm, living), depressed globose, with a few appressed lateral bracts 2–4 mm wide, sometimes none, often faintly ribbed towards the plane or slightly depressed orifice closed by 5–6 umbonate peripheral bracts and numerous small inner bracts; internal bristles 0; sclerotic cells none, or in the gall-figs as a thin layer. Male flowers in 1–2 rows; stamen 1. Gall-flowers sessile or stalked, the perianth covering the ovary. Female flowers sessile or pedicellate; perianth as a collar or short tube round the stalk of the red-brown ovary; style hispid. Seed c. 1 mm long, slightly keeled, tuberculate, with prominent hilum. Leaf with cystoliths only on the lower side.

Distr. Ceylon, India, South China to New Guinea and Queensland.

Ecol. In secondary forest, open ground and river banks, up to 1200 m alt., common.

Uses. Latex acrid; fruit, seed and bark emetic.

Vern. Kota-dimbula (S).

Insects. *Ceratosolen solmsi* Mayr and other genera (Wiebes).

Note. This common small tree, easily recognised from the opposite hispid leaves and strings or sprays of figs from the trunk and branches, is the sole representative of the large, predominantly cauliflorous sect. *Sycocarpus* (subgen. *Ficus*) of Malaysia and Melanesia. It is not, however, a species of high dipterocarp forest and is absent from the central part of Malaysia. It varies much in hairiness and in the way in which the figs are borne; sometimes the fruiting sprays become truly geocarpic, but this form seems not to occur in Ceylon.

Specimens Examined. KANDY DISTRICT: Alagalla, *Worthington 38* (K); Kandy, *Worthington 6330* (K); between Galaha and Deltota, *Wirawan 760* (PDA). ANURADHAPURA DISTRICT: Ritigala, *Worthington 5376* (K). LOCALITY UNKNOWN: *Brodie 174* (Herb. Lugd. Bat), *68 & 81* (L); *Royen s.n.* (L); *s. coll. C.P. 2230* (PDA).

4. Tribe MOREAE

7. STREBLUS

Lour., Fl. Cochinch. 615. 1790; Corner, Gard. Bull. Singapore 19: 215. 1962 (with generic synonymy); Corner, Phytomorphology 28: 1. 1975.

Trees, unarmed or spinous, monoecious or dioecious. Leaves distichous, mostly oblong-elliptic to lanceolate, denticulate or entire, mostly short-petiolate, without glands. Inflorescences axillary, bisexual or unisexual, pedunculate, varying cymose, racemose, spicate, and capitate, or the female flower solitary. Male flower with a pistillode, 3–5 merous, mostly 4-merous; tepals more or less valvate, free or shortly joined; filaments inflexed in bud; anthers small, reniform, extrorse. Female flowers 4-merous; tepals decussate, imbricate, free or more or less joined; ovary superior. Drupe rather large, 5–14 mm wide, thinly fleshy, often with a thickened fleshy base, dehiscent or not, the thin tepals investing the drupe or not. Seed large, round, invested by the thin membranous endocarp; embryo various. Cystoliths amphigenous, hypergenous, or none, not papillate except sect. *Streblus*; microscopic gland-hairs various; hairs simple, aseptate, smooth except sect. *Streblus*; stomata superficial.

Species 22, Asia, Australasia; 3 in Ceylon.

Section PHYLLOCHLAMYS

(Bur.) Corner, Gard. Bull. Singapore 19: 217. 1962.

Phyllochlamys Bur. in DC., Prod. 17: 217. 1873.

Male inflorescence capitate or very shortly racemose, flowers pedicellate. Female flower solitary. Drupe with thick fleshy base, ? dehiscent, much exceeded by the enlarged foliaceous tepals. Dioecious, spinous. Species 1; *S. taxoides*.

Section STREBLUS

Male inflorescence capitate, flowers very shortly pedicellate. Female flower solitary. Drupe fleshy, indehiscent, the tepals just covering the drupe, then reflexed. Dioecious (except *S. tonkinensis*), unarmed. Species 2; *S. asper*.

Section TAXOTROPHIS

(Blume) Corner, Gard. Bull. Singapore 19: 218. 1962.

Taxotrophis Blume, Ann. Mus. Bot. Lugduno Batavum 2: 77. 1856.

Male and female inflorescences racemose with pedicellate flowers. Drupe apparently indehiscent and without (or with slight) fleshy base, the tepals not enlarged or just covering the drupe. Dioecious, spinous. Species 3; *S. zeylanicus*.

Note. *Streblus* is a difficult genus of advanced species separated by intricate details. It has been customary to distinguish eight genera but I have reduced them to sections of the one because the differences are no greater, often less, than those which distinguish the subgenera and sections of *Ficus*. In Ceylon, *Streblus* reaches a western limit and only the widespread *S. asper* is common.

KEY TO THE SPECIES

1 Leaves rough. Male flowers in small heads. Female flowers solitary, giving a small orange drupe. Not spiny...... ....**1. S. asper**

1 Leaves smooth. Generally spiny

2 Latex white. Male inflorescences capitate or very shortly racemose. Female flower solitary. Fruiting tepals foliaceous, longer than the fruit...............**2. S. taxoides**

2 Latex orange. Male racemes 1–2 cm long. Female inflorescence with 2–6 flowers. Fruiting tepals not enlarged or just covering the ovary...............**3. S. zeylanicus**

1. Streblus asper Lour., Fl. Cochinch. 615. 1790; Hook. f., Fl. Br. Ind. 5: 489. 1888; Trimen, Handb. Fl. Ceylon 4: 101. 1898; Alston, Kandy Fl. 35, f. 184. 1938; Corner, Wayside Trees of Malaya 692, f. 255. 1952; Worthington, Ceylon Trees pl. 424. 1959.—**Fig. 28.**

Trophis aspera Retz., Obs. 5: 30. 1789; Roxb., Fl. Ind. 3: 761. 1832; Wight, J. Bot. 1: 62, t. 121. 1834.

Trophis cochinchinensis Poir., Enc. 8: 123. 1808.

Trophis aculeata Roth, Nov. Sp. Pl. 868. 1821.

Epicarpurus orientalis Blume, Bijdr. 488. 1825; Wight, Ic. Pl. Ind. Or. t. 1961. 1853.

Calius lactescens Blanco, Fl. Filip. 698. 1837; Merr., Sp. Blanc. 123. 1918.

Streblus lactescens Blume, Ann. Mus. Bot. Lugduno Batavum 2: 80. 1856.

Achymus pallens Soland. ex Blume, Ann. Mus. Bot. Lugduno Batavum 2: 79. 1856.

Cudrania crenata C.H. Wright, J. Linn. Soc. 26: 469. 1899.

Vanieria crenata (C.H. Wright) Chun, J. Arnold Arbor. 8: 21. 1927.

Diplothorax tonkinensis Gagnep., Bull. Soc. Bot. France 75: 98. 1928.

Unarmed, evergreen bush or small tree –14 m high; bark grey, inner bark whitish, thick; latex white, copious; branches often drooping, straggling, becoming densely twiggy, the lower branches even prostrate; young plants often (? always) beginning by creeping on the ground with prostrate twigs, like a creeping *Ficus*. Twigs 1–1.5 mm thick, brown, rough with short, white to brownish, spreading hairs 0.3 mm long. Stipules 2–5 mm long, lanceolate, appressedly hairy, caducous, often subpersistent on young twigs. Lamina 1.2–13 × 0.6–6.5 cm, elliptic to obovate, varying lanceolate, acute or subacute and minutely mucronate, attenuate to the narrowly subcordate or subcuneate, often slightly asymmetric, base, serrate, dentate, or crenulate (at least towards the apex), slightly hairy at first, soon glabrous, harshly scabrid on both sides, varying almost smooth above, rigid to coriaceous and hard, drying brittle and brown; lateral veins 4–7 (–8) pairs, raised below, 0–2 intercostals, the veins often slightly impressed above; basal veins 1 (–2) pairs, short; glands none; petiole 1–3 (–5) mm long, hairy as the twigs. Male capitula 6–10 mm wide, 5–15 flowered; peduncle 3–15 × 0.5 mm, sparsely puberulous to glabrous, with 1–2 small bracts at the base, occasionally one on the stalk, and a few small

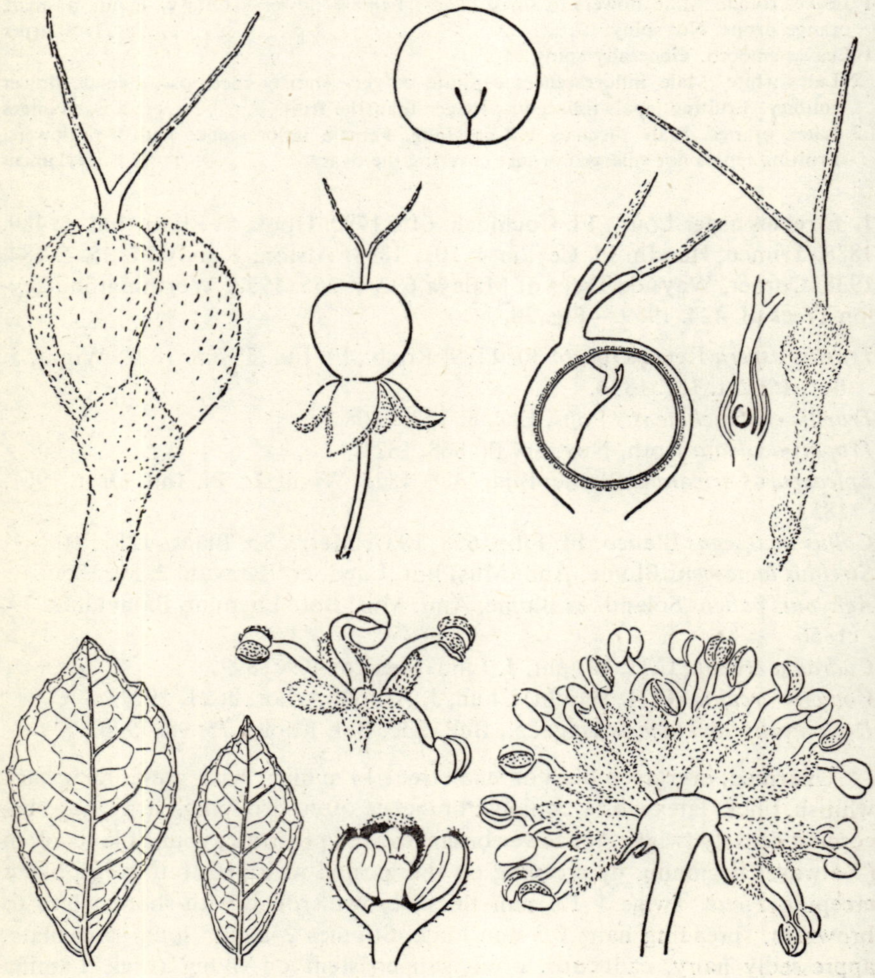

Fig. 28. *Streblus asper*. Leaves, × ½; ripe fruit with reflexed tepals (upper centre), × 2; immature fruit (upper left) and female flower (upper right), × 5; fruit, flower in section, and embryo, × 5; male inflorescence (in section) and male flower, × 5; male flower bud (section), × 8.

bracts at the apex. Male flowers very shortly stalked, fragrant; perianth 1 mm long, puberulous; stamens white. Female flowers green, solitary, at the top of the thinly puberulous peduncle 1–4 × 0.5 mm; bracteoles 2, 1–2 mm long, ovate-acute, appressed to the perianth, puberulous; tepals 2 mm long, ovate-acute or the inner pair obtuse, puberulous; style 1 mm long, then lengthening to 2–3 mm, the stigmatic arms 3–4 mm long at anthesis, lengthening to 6–12 mm long, glabrous. Drupe 6–8 mm long, globose or obscurely 2-humped at the apex, yellow to orange, at first invested by the enlarged tepals 5–8 mm long, then exposed at maturity and the tepals reflexed; peduncle 7–27 mm long. Seed 4–5 mm wide, globose, greyish white. Embryo curved; cotyledons thick, equal or very unequal; radicle short, incumbent. Cystoliths on both sides of the lamina.

Distr. Ceylon, India, Burma, South China, Hainan, Indo-China, Thailand, Andaman Isl., Malaya, Sumatra, Java, Bali, Lombok, Sumbawa, Celebes, Philippines.

Ecol. In strongly seasonal monsoon country, chiefly in secondary forest, by rice-fields, and in open places, lowland.

Uses. Drupe edible, sweet; leaves eaten by cattle and goats.

Vern. Crooked Rough-Bush (E); Géta-netul (S); Patpiray, Pirasu (T).

Note. This common small tree with rough leaves marks the more seasonal climates and is absent from the dense rain-forest. Ecologically it needs investigation because it seems that seedlings develop into prostrate plants before becoming erect; this creeping stage is often seen by rice-fields. Though generally dioecious, there are records of monoecious trees and there is var. *monoicus* (Gagnep.) Corner from Laos which is intermediate to the monoecious *S. tonkinensis* (Dub. et Eberh.) Corner.

Cardenolide glycosides, different from those in *Antiaris*, have recently been investigated from the root-bark of *S. asper*; see Khare, Bhatnagar, Schindler, and Reichstein, Helvet. Chim. Acta 45: 1515–1546. 1962.

Specimens Examined. MONERAGALA DISTRICT: Bibile, *J.M. Silva s.n.* (PDA). BADULLA DISTRICT: Hora Bora Wewa, *Worthington 1540* (PDA); Ella, *Worthington 2940* (PDA). KURUNEGALA DISTRICT: Kurunegala, *Worthington 860* (PDA). KANDY DISTRICT: Haragama, *s. coll. C.P. 2214* (PDA). POLONNARUWA DISTRICT: Minneriya, *Worthington 539* (K); Manampitiya and Topa Wewa, *Worthington 6287* (K). ANURADHAPURA DISTRICT: Yanoya, *Worthington 734* (K); Habarana, *Worthington 1384* (K); Anuradhapura, *Worthington 4252* (K). MATALE DISTRICT: Sigiriya, *Worthington 2057* (PDA). JAFFNA DISTRICT: Tampalai, *Worthington 5307* (K).

2. Streblus taxoides (Heyne) Kurz, For. Fl. Burma 2: 465. 1877; Corner, Gard. Bull Singapore 19: 225. 1962.— **Fig. 29.**

Trophis taxoides Heyne ap. Roth, Nov. Pl. Ind. Or. 368. 1821.
Trophis taxiformis Spreng., Syst. Veg. 3: 902. 1826.
Trophis spinosa Roxb., Fl. Ind. 3: 762. 1832.
Epicarpurus timorensis Decne, Nouv. Ann. Mus. Hist. Nat. 3: 499, t. 21. 1834; Span., Linnaea 15: 344. 1841.
Epicarpurus spinosus (Roxb.) Wight, Ic. Pl. Ind. Or. 6: 7, t. 1962. 1835. (partim).
Taxotrophis roxburghii Blume, Ann. Mus. Bot. Lugduno Batavum 2: 78. 1856.
Phyllochlamys spinosa (Roxb.) Bur. in DC., Prod. 17: 218. 1873; Hook. f., Fl. Br. Ind. 5: 488. 1888; Trimen, Handb. Fl. Ceylon 4: 101. 1898.
Streblus microphyllus Kurz, Prel. Rep. For. Pegu App. A, cxviii. 1875, App. B, 84; Kurz, For. Fl. Burma 2: 464. 1877.
Streblus taxoides Kurz var. *microphylla* Kurz, For. Fl. Burma 2: 465. 1877.
Phyllochlamys wallichii King ap. Hook. f., Fl. Br. Ind. 5: 489. 1888; Corner, Wayside Trees of Malaya 690. 1952.

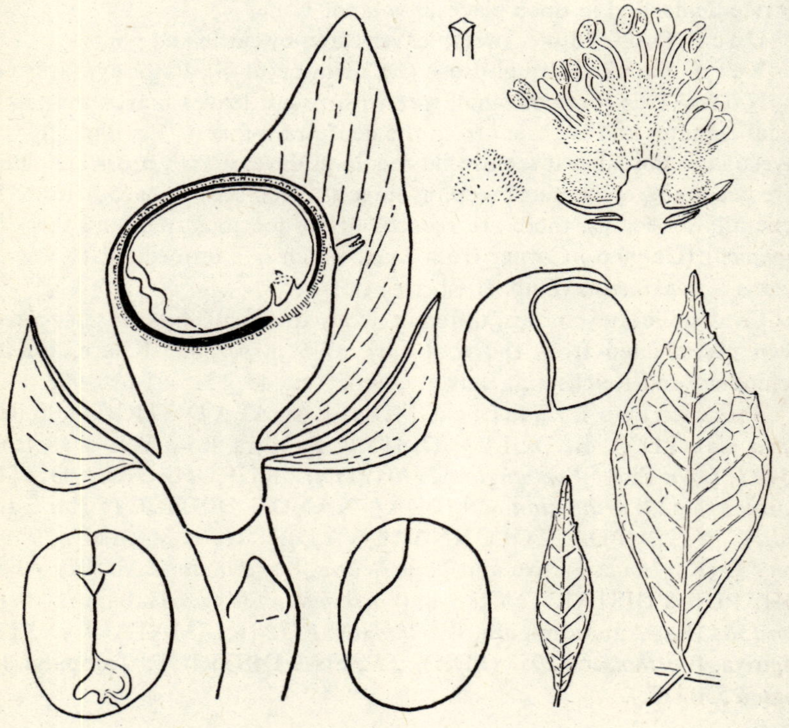

Fig. 29. *Streblus taxoides*. Leaves, × ½; male inflorescences (section) with bract and pistillode, × 5; fruit (in section), × 4; embryos, × 4.

Phyllochlamys taxoides (Heyne) Koorders, Exk. Fl. Java 2: 89. 1912.
Phyllochlamys taxoides var. *parvifolia* Merr., Philipp. J. Sci. 17: 247. 1920.
Phyllochlamys tridentata Gagnep., Fl. Gen. I.-C. 5: 714. 1928.
Taxotrophis poilanei Gagnep., id. 701.

Small, evergreen, very twiggy tree; latex white; twigs with woody spines –1 cm long; trunk dark grey, often with short spiny twigs; glabrous or rarely the petioles and underside of the lamina hairy. Stipules 2–5 mm long, triangular, acute or acuminate, often sharply pointed, caducous. Lamina (2–) 5–10 (–18) × 0.7–6.5 cm, elliptic, varying ovate to obovate, often asymmetric, apex acute to long acuminate and often tridentate, base cuneate or slightly rounded, dentate, especially in the distal half, smooth, coriaceous; lateral veins 8–12 pairs, inarching; basal veins very short; glands none; petiole 2–5 mm long. Male inflorescences 2–5 mm wide, 1–4 axillary, capitate with 8–14 flowers sessile except the central ones pedicellate, or shortly racemose and the flowers with pedicels –1.5 mm long; peduncle very short. Female flowers solitary, peduncles 3–6 mm long; tepals 2–5 mm long; style 1 mm long, the stigmatic arms 0.5–1.5 mm long. Drupe 5–10 mm wide, ellipsoid, asymmetric, very fleshy at the base, covered and exceeded by the foliaceous, narrow or broad, much enlarged tepals –20 mm long. Embryo with short radicle and thick fleshy subequal cotyledons. Cystoliths on the upperside of the leaf, sometimes also on the lower side along the veins.

Distr. Ceylon, India, Burma, Indo-China, Hainan, Thailand, Malaya, Java, Timor, Celebes (southeast), Mindoro, Palawan.

Ecol. In lowland forest in rocky or dry places, frequent.

Vern. Fig-lime (E); Gongotu (S).

Note. This small spiny tree is very variable in leaf-size and toothing, but it does not seem possible to distinguish varieties.

Specimens Examined. MONERAGALA DISTRICT: Bibile, *J.M. Silva s.n.* (PDA). KANDY DISTRICT: Haragama, *s. coll. C.P. 1074* (PDA); Urugala, *Alston 480* (PDA). POLONNARUWA DISTRICT: Mineriya, *s. coll. C.P. 1074* (PDA). ANURADHAPURA DISTRICT: Ritigala, *Willis s.n.* (PDA), *Worthington 291* (K). KURUNEGALA DISTRICT: Kurunegala, *Worthington 4015* (K).

3. Streblus zeylanicus (Thw.) Kurz, For. Fl. Burma 2: 464. 1877; Corner, Gard. Bull. Singapore 19: 229. 1962.—**Fig. 30.**

Epicarpurus zeylanicus Thw., Hook. J. Bot. Kew Gard. Misc. 4: 1. 1852, id. 3: t. 11. 1851.

Taxotrophis zeylanica Thw., Enum. Pl. Zeyl. 264. 1864; Hook. f., Fl. Br. Ind. 5: 488. 1888; Trimen, Handb. Fl. Ceylon 4: 100. 1898; Hutchins., Kew Bull. 149. 1918.

Diplocos zeylanica (Thw.) Bur. in DC., Prod. 17: 215. 1873.
Taxotrophis caudata Hutchins., Kew Bull. 149. 1918.

Small spiny tree, gnarled, bushy, glabrous or the young parts puberulous; spines 0.5–2 cm long, axillary; latex orange. Twigs 3–5 mm thick. Stipules 3 mm long, connate, intra-petiolar, ovate obtuse, early caducous. Lamina 6–12 (–14) × 2–4 (–5) cm, elliptic, varying somewhat ovate or obovate, apex acuminate, base widely cuneate to rounded, margin dentate, rarely entire, smooth; lateral veins (6–) 7–9 pairs; basal veins short; petiole 2–5 mm long. Inflorescences axillary, solitary, occasionally paired, unbranched, peduncles rather short. Male inflorescence 1–2 cm long, greenish, with a sterile band along one

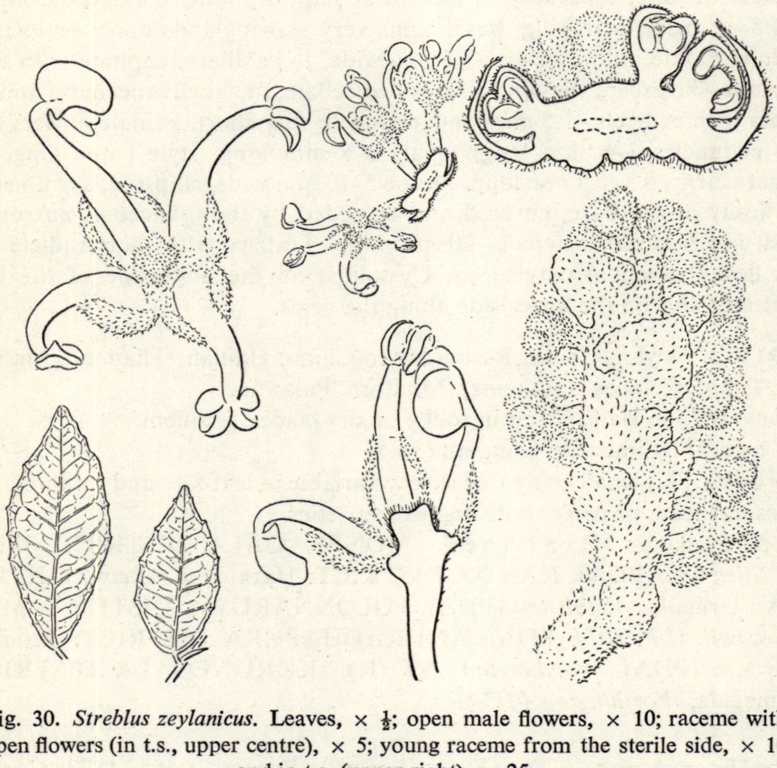

Fig. 30. *Streblus zeylanicus*. Leaves, × ½; open male flowers, × 10; raceme with open flowers (in t.s., upper centre), × 5; young raceme from the sterile side, × 10 and in t.s. (upper right), × 25.

side, the rachis flattened; bracts along the margins of the sterile band, 1–1.5 mm wide, convex-subreniform, puberulous. Male flowers greenish, with pedicels 1–2.5 mm long; perianth with 4 segments, puberulous; stamens 4; pistillode quadrate-columnar, hairy. Female flowers 2–6 per inflorescence, with pedicels 5–10 mm long, elongating 10–20 mm in fruit; perianth as the male; ovary with short style, 1 mm long, and 2 ligulate stigmata 2 mm long. Fruit

8–10 × 6–7 mm, asymmetric, slightly fleshy at the base, perianth not enlarging or enlarging to cover the fruit. Seed 5 mm wide; embryo with relatively long curved radicle and subequal, much folded cotyledons. Cystoliths on the upper side of the leaf.

Distr. Ceylon, India, Burma, South China, Indo-China; uncommon in dry forest.

Note. Sterile trees resemble *S. taxoides* but have usually a more widely cuneate or rounded base to the lamina and, as it seems, orange latex. *S. spinosus* (Bl.) Corner, from Sumatra and Java, is very close.

Specimens Examined. KURUNEGALA DISTRICT: Kurunegala, *s. coll. C.P. 2213* (PDA). MATALE DISTRICT: Nalanda, *Worthington 994* (K); Galewela, *Worthington 957* (K). DISTRICT UNKNOWN: Hayama, *s. coll. C.P. 2213* (PDA).

5. Tribe OLMEDIEAE

8. ANTIARIS

Lesch., Ann. Mus. Hist. Nat. Paris 16: 478, t. 22. 1810 (gen. conserv.); Trec., Ann. Sci. Nat. ser. 3, 8: 142, t. 6, f. 158–168. 1847; Bur. in DC., Prod. 17: 28. 1873; Corner, Gard. Bull. Singapore 19: 244. 1962.

Ipo Pers., Syn. 2: 566. 1807.
Antscha Horsfield, Verh. Bat. Gen. 7: 8. 1814.
Toxicaria Aepnel ex Steud., Nomencl. 434. 1821.
Lepurandra Nimmo, Grah. Cat. Pl. Bombay 193. 1839.

Monoecious deciduous trees; bark smooth, grey; leaves distichous. Inflorescences axillary, pedunculate, in groups of 2–4, the male below the female on the same twig. Male head many-flowered, discoid, involucrate with the small bracts in a single series; interfloral bracts none; flowers sessile; tepals 4–5, free, imbricate; stamens 3–8, mostly 4; anthers introrse; filaments very short or practically none; pistillode none. Female inflorescence 1-flowered; bracts few, small, scattered; tepals 4–5, minute; ovary inferior, concrescent with the receptacle; style very short, with 2 long stigmatic arms; ovule immersed, apical, transversely anatropous. Drupe globose to ellipsoid, fleshy; endocarp thinly crustaceous. Seed rather large. Embryo straight; radicle very short; cotyledons thick, equal, not folded. Cystoliths none; appressed hairs papillate, erect hairs smooth; microscopic gland-hairs 2–4 celled, capitate.

Species 4, lowland tropical forest; 4 species in Africa (2 endemic, Madagascar), 1 species in Asia and Australasia.

The affinity of *Antiaris* seems to be with the American *Olmedia* and *Pseudolmedia*, and this is supported by the recent intensive work on the glycosides of the family. The geographical evidence in *Antiaris* itself also supports an

Australasian derivation across the Pacific. Though there are three other species in the African region, these are clearly specialities, and the evidence of *A. toxicaria* is that it has suffered diminution in fruit-size from the east to the west of its range (Corner, 1962). *Antiaropsis*, in spite of its name and location in New Guinea, is very far removed from *Antiaris*, as relic of some much earlier stage in the tribe.

Antiaris toxicaria Lesch., l.c.; Hook., Comp. Bot. Mag. 1: 11, t. 17. 1835; Trimen, Handb. Fl. Ceylon 4: 97. 1898; Renner, Bot. Jahrb. 39: 374. 1908; Corner, Wayside Trees of Malaya 647. 1940; Worthington, Ceylon Trees pl. 419. 1959; Bisset, Planta Medica 10: 143. 1962; Wehrli, Schindler, and Reichstein, Helvet. Chim. Acta 45: 1083. 1962; Wehrli, Helvet. Chim. Acta 45: 1206. 1962.—**Fig. 31.**

Ipo toxicaria Pers., l.c.
Toxicaria macassariensis Aepnel, l.c.
Lepurandra saccidora Nimmo, l.c.
Antiaris innoxia Blume, Rumphia 1: 172, t. 54. 1835.
Antiaris dubia Span. ex Hook., Comp. Bot. Mag. 1: 310. 1835.

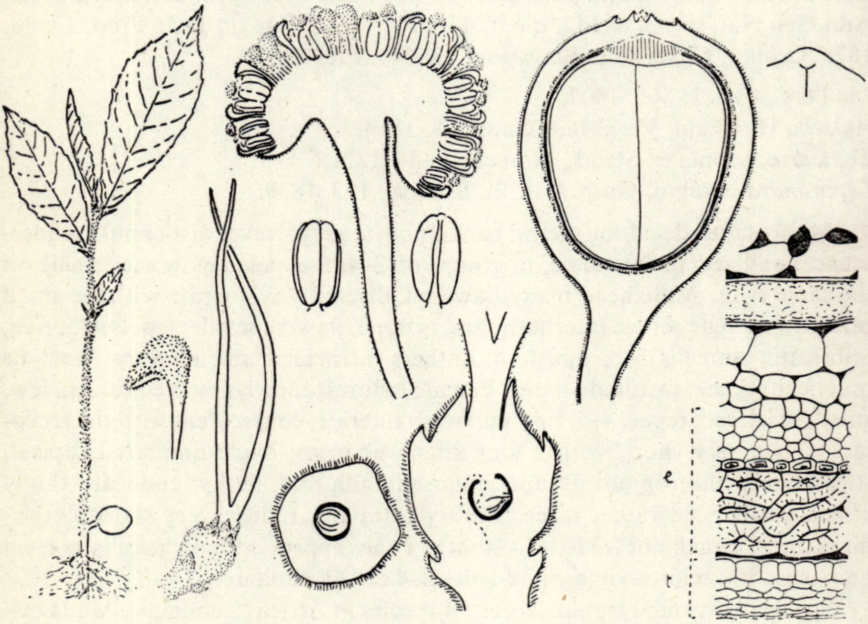

Fig. 31. *Antiaris toxicaria*. Seedling, with two scale-leaves on the epicotyl, × ½; male head (l.s.), × 2; stamens and male tepal, × 10; female flower, × 2, and in section, × 10; fruit (section) with the radicle-end of the embryo, × 2; testa (t) and endocarp (e) in section, × 250.

Antiaris saccidora (Nimmo) Dalz. in Hook. J. Bot. Kew Gard. Misc. 232. 1851.
Antiaris rufa Miq., Fl. Ind. Bat. Suppl. 423. 1861.
Antiaris zeylanica Seem., Bonplandia 10: 4. 1862; Seem., Fl. Vit. 253. 1868.

Tree –50 m high, often (? always) deciduous, monopodial with rather short, but widespread branches, the conical crown at length spreading and irregular; trunk –1 m thick, buttressed 1–3 m high; bark greyish white, smooth, then slightly fissured and lenticellate-pustulate; inner bark whitish, turning red-brown on exposure, fibrous; wood pallid white; latex creamy-white, turning dirty brownish and granular on exposure. Leaves distichous on the branches. Twigs, petioles, and underside of the veins hispid with brown hairs 1–2 mm long, at least in saplings, becoming closely and minutely appressedly hairy (hairs –0.5 mm long) to practically glabrous in adult trees. Twigs 2–5 mm thick, brownish, the terminal bud often oblique. Stipules –11 mm long, densely appressedly brown hairy, paired, somewhat intrapetiolar, caducous. Lamina 5–20 × 2.5–10 cm (–25 cm long in saplings), elliptic, subovate with cordate base, or elliptic obovate with cuneate base, very variable, often somewhat asymmetric, acute or subacuminate, denticulate (sapling) to entire, membranous to thinly coriaceous, often subnitid, the upperside often thinly appressedly hairy and subscabrid, or setose-scabrid (sapling); lateral veins 7–12 pairs, curved ascending; intercostals –8, distinct, scarcely prominent beneath; basal veins 1–3 pairs, short; no glands; petiole 3–11 mm long, always short. Inflorescences borne on the new twigs with the young leaves, minutely appressedly hairy, green. Male head discoid to reflexed pulvinate, 1–2 cm wide; peduncle –15 × 1 mm; involucral bracts linear, short; flowers 2 mm high; tepals spathulate-cucullate; anthers yellow. Female inflorescence 1-flowered, subpedunculate, with several small scattered lanceolate bracts; tepals minute, lanceolate, free; stigmatic arms 12–15 mm long. Drupe 12–25 × 11–20 mm, ellipsoid to pyriform, densely brown velutinate, ripening red to crimson, purple, and finally black, on the twigs below the leaves; peduncle 2–10 mm long; stigmatic arms caducous; pulpy wall 1.5 mm thick; endocarp crustaceous, thin. Seed 10–16 mm long. Seedlings with hypogeal cotyledons; epicotyl with 3–4 minute scale-leaves before the spirally arranged foliage leaves; roots yellowish.

Distr. Africa, Asia, Australasia.

var. **toxicaria**

Lamina mostly elliptic-obovate, acuminate, base cordate to cuneate, often somewhat asymmetric, thinly to rather strongly coriaceous, subscabrid to smooth. Male head –20 mm wide. Drupe (excl. peduncle) 12–25 × 11–20 mm.

D i s t r. Ceylon, India, Burma, South China, Hainan, Indo-China, Thailand, Andaman Isl., Malaya, Sumatra, Java, Borneo, Celebes.

E c o l. Lowland forest -1500 m alt., occasional in villages, flowering August and September.

U s e s. Fruit edible; bark for dyeing and cloth-making.

V e r n. Riti (S); Netavil (T).

N o t e. This is the most widespread species of Moraceae and may be said to characterize the lowland tropical rain-forest of the Old World. It shows a cline in the size of the fruit and, to some extent, of the leaves from Fiji to Africa, the Melanesian trees (or shrubs) having the largest leaves and fruits (20–26 mm wide, when dried). In Ceylon, as in Southeast Asia, the dried fruits are 11–20 mm wide, and in Africa they are 7–12 mm. In western and central Malaysia many trees have in their latex the notorious arrow poison, due to the presence of various cardiac glucosides, especially antiarin. Ceylon trees, according to Worthington, are harmless. Like the related *Castilla*, the trees shed their lateral branches with clean scars.

S p e c i m e n s E x a m i n e d. KURUNEGALA DISTRICT: Kurunegala, *s. coll. C.P. 2231* (PDA). KANDY DISTRICT: Kadugannawa, *Worthington 820* (K). MATALE DISTRICT: Nalanda, *Worthington 980* (K); Matale, *Worthington 1083* (K). MONERAGALA DISTRICT: Bibile, *Worthington 2981* (K).

9. CASTILLA

Cerv., Gaz. Lit. Mexico Suppl. 7. 1794; Trec., Ann. Sci. Nat. ser. 3, 8: 137, t. 5, f. 139–148. 1847; Baill., Hist. Pl. 6: 154, 204, t. 122. 1875; Engler, Pflanzenfam. 3, 1: 84, f. 63. 1894; Pittier, Contr. U.S. Natl. Herb. 13: 247. 1910; Hutchins., Gen. Fl. Pl. 2: 170. 1967; Berg, Flora Neotropica Monograph n. 7, Olmedieae, Brosimeae (Moraceae). 1972.

Castilloa auct.

Monoecious or dioecious trees, unarmed; leaves distichous, short-petiolate, without glands; stipules more or less amplexicaul, leaving annular scars. Inflorescences unisexual, axillary, involucrate, discoid with the distal face fertile, or the male more or less urceolate-compressed to flabellate or conduplicate, many-flowered. Male flowers naked, dispersed among the bracteoles; filaments straight in bud; pistillode none. Female flowers without bracts, many not setting fruit; perianth tubular, 4-dentate, projecting, surrounding the style and the upper part of the semi-inferior ovary; style short, with 2 rather short stigmatic arms; ovule subapical. Syncarp of conic projecting perianths dry or fleshy; endocarps thin, sclerotic, more or less embedded in the receptacle, thickened and woody round the hilum. Seeds large; embryo straight; radicle short, superior; cotyledons very fleshy, plano-convex, equal or subequal. Lamina without cystoliths.

About 5 (–10) species, Tropical America from South Mexico to Bolivia and Peru, following the lowlands of the main mountain ranges; 1 species introduced to Malaysia.

The dioecious state is peculiar in this genus. There are male and bisexual plants, but it is not yet certain whether the male are merely young bisexual trees. In these the axillary female inflorescence is accompanied by two smaller male inflorescences, one on each side of it. In the male trees, the male inflorescences are larger and unaccompanied by female inflorescences, but these male trees are always small or young trees; furthermore, even monoecious trees begin with a male phase. This genus also shows the regular abscission of secondary branches from the main branches which arise from the trunk and acquire, thereby, a lanky appearance; see Went, Ann. Jard. Bot. Buitenzorg 14: 1. 1897.

Castilla elastica Cerv., l.c.; Hook f., Trans. Linn. Soc. London 2: 206. 1886; Worthington, Ceylon Trees pl. 418. 1959; Woodson, Ann. Missouri Bot. Gard. 47: 140, f. 50. 1960; Berg, l.c. 94. 1972.—**Fig. 32.**

Tree –30 m high, buttressed; bark light grey, lenticellate in shallow longitudinal fissures; latex copious, white; vegetative branches wide-spreading with large leaves; flowering branches short, caducous, reaching 1–3 m long before abscission, small-leafed; wood very hard. Twigs, petioles, stipules, and underside of the lamina villous with spreading, less often appressed, stiff, golden hairs 1–2 mm long, the upperside of the lamina thinly appressedly hairy. Twigs 4–7 mm thick, axillary buds rather large (? compound). Stipules 3–6 cm long. Lamina 20–57 × 1–25 cm on the long branches, oblong or oblong-obovate, acuminate, base cordate and often somewhat unequal, stiffly membranous to coriaceous, finely and closely ciliate-denticulate to subentire; lateral veins 16–22 pairs, –12 intercostals close and slightly raised below; basal veins 2–3 pairs, short; petiole 7–25 mm long, short and stout. Male heads usually bigeminate in groups of 4, turnip-shaped, compressed, conduplicate-flabellate, 20–37 mm wide; peduncle 2–15 mm long; complemental male heads attending the female ? clavate. Female heads smaller at anthesis than the males, subsessile. Syncarps 5–9 cm wide, 15–25 mm thick, orange-red, pulpy. Seeds 8–10 × 6–7 mm; pale straw colour.

D i s t r. South Mexico to Colombia; introduced to Ceylon.

U s e s. Planted formerly for rubber; used as a shade tree for cocoa and coffee; scattered in towns and villages; a fine avenue on the main road north of Koholanwela; difficult to axe.

V e r n. Panama Rubber.

N o t e. According to the chemical researches of Brauchli, Schindler

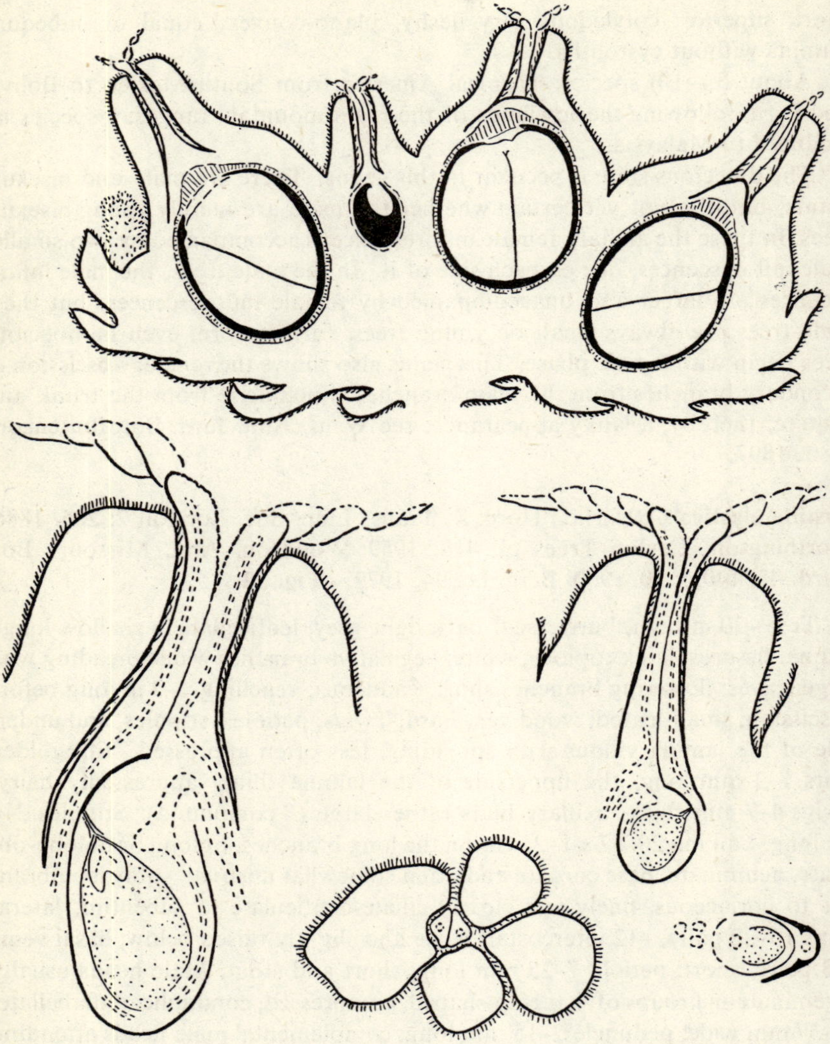

Fig. 32. *Castilla elastica*. Syncarp (in section), × 3; female flowers shortly after pollination and with young embryo, perianth (t.s.) and ovary (t.s.), × 10.

and Reichstein (cited under *Antiaris toxicaria*), the glycosides of *Castilla elastica* differ very considerably from those of *Antiaris* and *Streblus asper*, which are the other Moraceae examined in comparable detail. The seeds of *C. elastica* contain principally cymarin, periplocymarin, and k-strophanthin-β, as in *Strophanthus kombe*. The latex is much poorer in glycosides than the seeds; the characteristic glycosides of *Antiaris* are lacking.

Collections. None noted.

PANDANACEAE

(by Benjamin C. Stone*)

Strictly dioecious woody-stemmed trees, shrubs, or climbers; leaves trifariously or quadrifariously (*Sararanga*) spiralled. Stems erect or ascending, usually branching, sympodial in most cases, annulate by the leaf scars, epidermis at first green. The stems later developing a thin grayish or brownish cortex overlying green inner layer; adventitious roots emitted, especially at base of trunk or on older branches, but sometimes on young aerial branches, descending (and sometimes branching), functioning as prop-roots; or if early aborted, sometimes remaining as blunt spines or protuberances; in a few epiphytic species of *Pandanus*, remaining within the persistent leaf base and often flattened; in the climbers (*Freycinetia*) functioning as clasping roots. Prop-roots often set with longitudinal rows of small spines. Rootcaps (of aerial roots) with persistent tissue layers. Leaves linear ensiform, rarely narrowly elliptic or ovate, bifacial, entire with serratures (prickles), the apex sometimes splitting in older leaves; in cross-section often M-shaped; the single midrib usually obvious, thickly sheathed by sclerenchyma, evident from just above the leaf base attachment to the apex, usually set with prickles along its length or part of it, the prickles decreasing in size nearing the leaf apex, always antrorse but those near leaf base sometimes retrorse, or absent. Margins usually prickly-serrate, occasionally entire or prickly near apex only or near base and apex only, usually larger at leafbase, smaller and more crowded at apex, antrorse, or, very rarely, retrorse but then only at the base, or often divaricate. Twin lateral pleats often well distinguished, smooth or in some species prickly-serrate, near the apex only.

Inflorescences strictly unisexual (staminate or pistillate, rarely the former with pistillodia or the latter with staminodia), terminal on ordinary or on specialized branches, the stem growth sympodial, or branching with site of old inflorescence overtopped or remaining in the "notch" formed by the new branches; peduncular scars thus often seemingly lateral. Vegetative axis converted to inflorescence peduncle; basal foliar appendages leaflike, later (upper ones) intermediate to bracts, apical spathaceous, coloured or white, reduced, often somewhat different in shape, navicular at base, soft to coriaceous,

*Kuala Lumpur, Malaysia.

margins and midrib commonly serrulate at least distally; spadix solitary or spadices several, spicate, axillary, up to 13 or 15 (always an odd number). Pistillate inflorescence erect, later nodding and finally pendulous as fruit ripens. Staminate inflorescence usually nodding to pendulous from the start. Pistillate inflorescence in *Sararanga* branched paniculate. Pistillate spadix formed of 1-celled carpels ripening as drupes, or of 2-many-celled phalanges (syncarps) of carpels, ripening as polydrupes, or of 1-several-celled carpels ripening as berries (*Freycinetia*). Each carpel tipped by a stigma, this of various forms, but always grooved on the edge, the surface papillose by glandular hairs, which also line the interior of the ovarian chamber; one or a few vascular bundles connect the stigmatic groove to the placenta. Style absent or not very clearly distinct, sometimes spiniform. Stigma round, ovate, reniform, deltoid linear, or variously shaped. Ovule single, anatropous (*Pandanus*, *Sararanga*) or several to many in each carpel (*Freycinetia*). Pericarp of carpel thin, fleshy to pulpy, or at apex indurated; upper mesocarp often medullose, sometimes cavernose, lower mesocarp fleshy-fibrous. Endocarp osseous, well-developed (*Pandanus*), this but firm (*Sararanga*), or membranous and evanescent (*Freycinetia*). Seeds albuminous; seedcoat membranous (*Pandanus*) or somewhat thickened (*Freycinetia*), sometimes with a strophiole, usually with a raphe, often larger (up to 5 cm long) in *Pandanus*, small and semilunate (in *Freycinetia* and *Sararanga*).

Staminate inflorescence usually multi-spicate, or in a few cases with a single terminal spike, in *Sararanga* richly paniculate. Stamens singly attached to spike axis or adnate in various types of phalanges, racemose, umbellate, etc., borne on a main column (stemonophore), sometimes pendent from edge or underside of peltate disc (*Pandanus*), or subsessile and subtended by a cupular organ (? bracteal), very rarely surrounding a pistillode (esp. in *Freycinetia*). Filament simple or sometimes bulbous at base, abruptly narrowed at apex. Anthers elongate, 4-celled, commonly white or cream, often yellow, occasionally pink or reddish, blunt or more commonly tipped by a short or long apiculation, ± hastulate at base. Pollen subspherical to ellipsoid, the exine usually ± minutely spinulose, or smooth; pore apical; grains dry or loosely adherent with a light oil, white, yellow, pink, or orange. Calyx and corolla totally lacking.

Genera 3: *Pandanus* (600–700 species), *Freycinetia* (c. 200 species) and *Sararanga* (2 species only). The first two present in Ceylon. Paleotropical, from W. Africa to Polynesia, Australia, New Zealand, N. to Ryukyu Isl. and Japan (southern tip).

A clearly defined family whose relationships are not very clear, probably fairly ancient. Perhaps related nearly to Palms, but not very close to *Typha* (which is often given as a relative). Perhaps related to Agavaceae. The genera are very distinct. *Pandanus* occupies the largest range, from W. Africa to Australia and Polynesia; *Freycinetia* is enclosed within this area, except for the extension into New Zealand. *Sararanga* is limited to the Philippines, New Guinea, the Admiralty Group (Manus) and the Solomon Islands.

KEY TO THE GENERA

1 Woody climbers, with clasping adventitious roots. Carpels many-ovulate. Staminal filament microscopically papillate by protruding epidermal cells. Anthers blunt. Leaf midrib basally unarmed; leaf base margins membranous, forming auricles. Fruits small, fleshy.. **1. Freycinetia**

1 Trees or shrubs, never climbing; adventitious roots not clasping. Carpels 1-ovulate. Staminal filaments smooth; anthers usually apiculate. Leaf midrib usually prickly at base (dorsally); auricles lacking. Fruits usually rather large, drupaceous... **2. Pandanus**

1. FREYCINETIA Gaud.

Woody climbers, usually ascending tree-trunks, sometimes scrambling over rocks; adventitious roots clasping, often flattened, wiry and tough. Leaves linear-ensiform to elliptic or ovate, rarely over 2 m long and usually much smaller, midrib distinct but twin lateral pleats obscure or absent, margins prickly or entire, midrib prickly at apex, almost never so at base; sheath margins clearly differentiated as membranous auricles, thin to translucent, green or purplish, with a very few widely spaced nerves; auricles more or less deciduous; entire or minutely prickly, rounded to lobelike at apex. Inflorescences terminal on normal shoots or on specialized efoliate lateral shoots, pseudo-umbellate or rarely racemose or sometimes solitary. Bracts leaflike below, above shorter, broader, thicker, softer, fleshy, coloured or white; peduncles glabrous (or scabrid in pistillate plants only). Pistillate spikes composed of carpels or syncarps, stigmas minute, apical, flat, crowded or coalescent; ovules several to many per carpel. Staminodia occasionally surrounding each carpel or syncarp. Staminate spike of singly attached stamens, filament papillose by the protuberant epidermal cells, anther oblong to linear, blunt. Pollen grains usually smooth, in at least one species minutely spinulose. Type species: *F. arborea* Gaud. (Hawaii).

Freycinetia is a genus of between 180–200 species; from Ceylon, its westernmost outpost, it occurs E. to Cambodia and Viet Nam, Taiwan, Malaysia, Indonesia, Philippines, Micronesia, Melanesia, N.E. Australia, and Polynesia (incl. New Zealand). It is clearly a derived genus, with its climbing habit, reduced inflorescences, small and numerous seeds, probable bat pollination syndrome (perhaps also rats), sometimes very small ovate leaves, and presence of staminodia and pistillodia as rather frequent. The fruits are probably bird-dispersed, although mammals (bats, monkeys) may also assist. The pollen grains are mostly smooth, in contrast to those of *Pandanus*.

Chromosome number 2n = 56 approx. or n = 25 approx., counts in only 3 species [Ref. C.H. Cheah and B.C. Stone, Bot. Jahrb. Syst. 93 (4): 498–529. 1973].

KEY TO THE SPECIES

1 Leaves clearly lanceolate to short-ensiform, usually 15–30 (–35) cm long. Longitudinal nerves (adult leaves) 15–30. Auricles rounded, subentire. Bracts creamy or yellowish. Pistillate heads 1–3, ellipsoid, up to 3–4 cm long; peduncles smooth, glabrous; berries mostly 3–4-celled (a few up to 6-celled), apex bluntly truncate, red when ripe. Staminate spadix 0.9–1.8 cm long, anthers 0.6–0.7 mm long. . . . . **1. F. pycnophylla**

1 Leaves linear-ensiform, mostly 50–70 cm long. Longitudinal nerves (adult leaves) 40–42. Auricles (on vegetative branches) toothed and lobelike at apex. Bracts red. Pistillate heads usually 2 or 3, oblong-cylindric, to 6–7 cm long. Pistillate peduncles minutely appressed-hispidulous. Berries mostly 2–3-celled, apex narrow tapered but minutely truncate, red when ripe. Staminate spadix 3 cm long, anthers 1.25–1.5 mm long. . . . . **2. F. walkeri**

Section SARAWAKENSES B.C. Stone

1. Freycinetia pycnophylla Solms, Linnaea 42: 91. 1878; Stone, Webbia 23: 591–596. 1969. Syntypes: *Walker*, *Thwaites* (*C.P. 366*). Lectotype: the pistillate plant of the 2 syntypes cited.—**Fig. 1.**

Freycinetia angustifolia sensu Thw., Enum. Pl. Zeyl. 327. 1861, non Blume.
Pandanus scandens Moon, Cat. 67. 1824.

A slender woody-stemmed climber, the main stems clasping trees by the adventitious roots; branches ± descending; leafy stems to 7 cm diam., leaf scars 7–10 mm apart, or smaller and with more crowded leaf scars. Leaves trifariously spiralled, linear-lanceolate, attenuate-acute, mostly 15–30 (–35) cm long (but much smaller in seedlings and juveniles), base sheathing, with membranous auricles 15–28 mm long, 5–8 mm wide, entire, rounded at apex, finally caducous; margins along their length minutely denticulate with prickles up to 0.5 mm long, 1–5 mm apart, often reduced to mere nubs or obsolete especially along the middle of the blade; midrib similarly prickly along distal half or 2/3 of leaf; all prickles antrorse. Longitudinal nerves 15–30 per leaf. Upper surface darker. Staminate inflorescence terminal, bracteate, the bracts deltoid-ovate to ovate-lanceolate, to 4 × 1.8 cm, cream or yellow; spadices usually 3 (rarely up to 6), pseudoumbellate; peduncle to 15 × 1.5 mm; spadix oblong, 9–18 × 4.5–5 mm, pale yellow, of numerous stamens; filaments very slender, 2 mm long, anthers orbicular, blunt, 0.6–0.7 mm long. Pistillate inflorescence terminal, bracteate, bracts as in the male, spadices 3 (rarely up to 6), peduncles glabrous, about 2 cm long, 1–3 mm thick; heads oblong-ellipsoid, in flowering state 13–15 × 5 mm, in fruit up to 4 × 2 cm; berries numerous, scarlet red when ripe, oblong, apex truncate, rimmed, stigmas 3 or 4 (rarely to 6 or as few as 2), black, round, 0.3 mm wide; pileus low-domed—

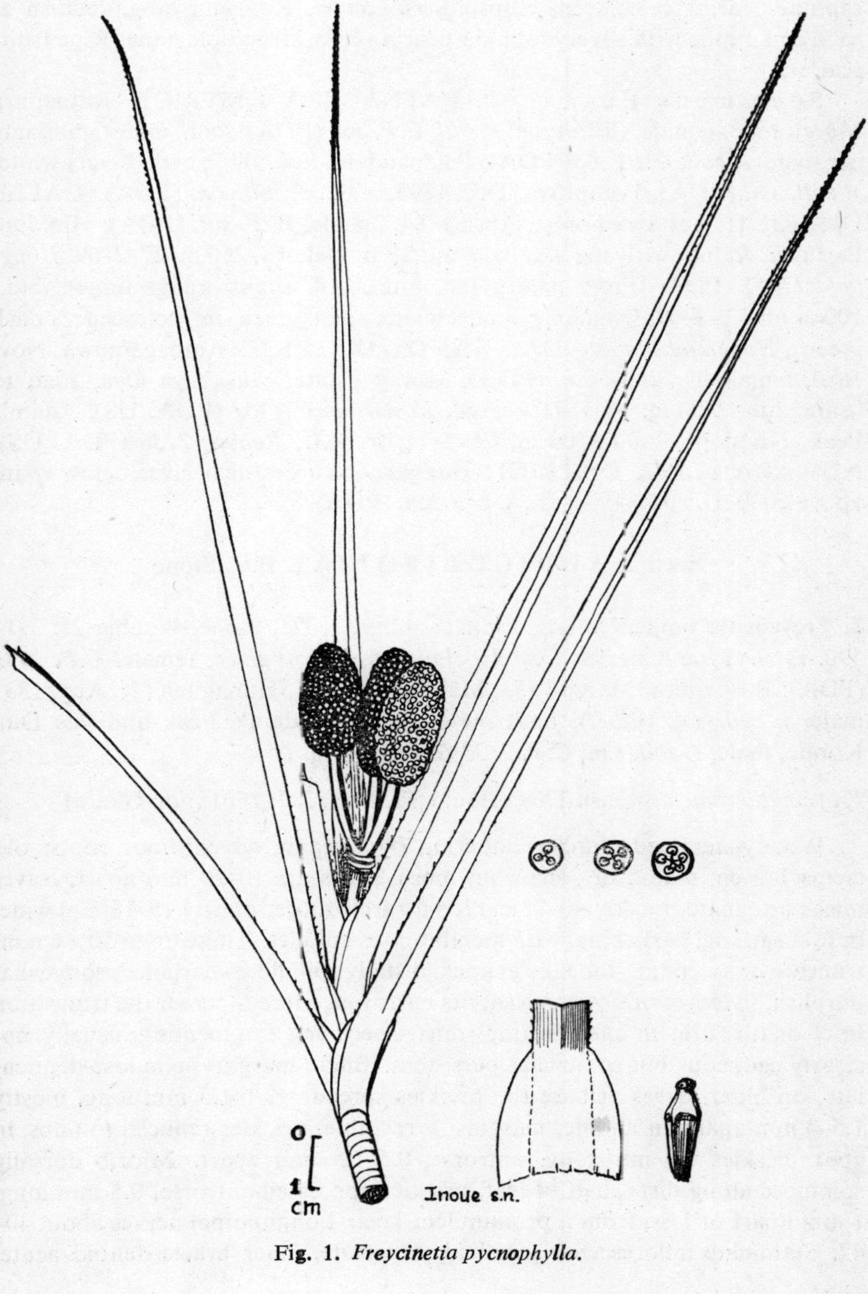

Fig. 1. *Freycinetia pycnophylla.*

pyramidal, somewhat angulate; berry (in fruit) 6–7 mm long, 2–2.5 mm wide, apex coriaceous, lower part fleshy, pericarp membranous, dotted with silvery raphide-bearing cells; seeds ellipsoid-semilunate, 1.25 mm long, reddish at each end, raphe with silvery raphide-bearing cells, strophiole none. Ripe fruits scarlet.

Specimens Examined. RATNAPURA DISTRICT: Ratnapura March 1853, female, (lectotype) *s. coll. C.P. 366* (PDA); ibid., male (staminate paratype) *s. coll. C.P. 366* (PDA); Palabaddale, Feb. 1892, male bracts white, *s. coll. s.n.* (PDA); Potupitiya, Dec. 1893, male, *s. coll. s.n.* (PDA). GALLE DISTRICT: Kottawa Forest, April 1884, female, *W.F. s.n.* (PDA); Hinidum Pattuwa, Kalubowitiyana Kanda, 4 mi. SE of Neluwa, 250 m, 27–2–69, *Hoogland 11455* (PDA, leaves pale green, imm. infl. slight orange tinge); ibid., 100 m alt., 1–3–69, female fl. (bracts white, spad. green, infructescences dark green), *Hoogland 11468* (PDA). KANDY DISTRICT: Ambagamuwa, Nov. 1880, female fls., *W.F. s.n.* (PDA); Moray Estate, Maskeliya Oya, road to fishing hut, 2600 m, 16–5–71, fr. red. *Kostermans 24161* (PDA, US); Adam's Peak, NE slope, 1000–1500 m, 26–2–71, fr. *A.G. Robyns 7236* (PDA, US). NUWARA ELIYA DISTRICT: Hakgala, on trees near river below road, above 5000 ft., 30–4–1906, fr., *s. coll. s.n.* (PDA).

Section AURICULIFOLIAE B.C. Stone

2. Freycinetia walkeri Solms, Linnaea 42: 92. 1878; Stone, Webbia 25: 591–596. 1969. Type material: Ceylon, Hantane, *Col. Walker*, female, *C.P. 2333* (PDA); Ratnapura, March 1853, Male, *s. coll. s.n.*, Hunaguna (?), Aug. 1837 male, *s. coll. s.n.* (PDA), both on one sheet; Adam's Peak and Pas Dun Korale, male, *s. coll. s.n.*; *C.P. 2333* (PDA).—**Fig. 2.**

Freycinetia radicans sensu Thw., Enum. Pl. Zeyl. 327. 1861 (non Gaud.)

Woody-stemmed climber adhering by clasping adventitious roots; old stems 1–2 cm diam., or somewhat more, leaf-scars 10–20 mm apart; leaves linear attenuate, mostly 40–75 cm long, rarely more, mostly 15–18 mm wide; base sheathing, darkening, with membranous auricles. These up to 50 × 6 mm, truncate or ascending-lobelike at apex, distally spinulose-margined, somewhat purplish, in leaves of flowering shoots narrower, more tapered, the truncation brief or obscure; in age splitting transversely and fragmenting, usually not clearly caducous but somewhat persistent. Blade margins spinulose-denticulate, on older leaves at base the prickles spreading, 1–1.3 mm long, mostly 1.5–4 mm apart; in middle, margins ± revolute, prickles reduced to nubs; in apex prickles 0.5 mm long, antrorse, 0.5–1.5 mm apart. Midrib dorsally spinulose along distal half of leaf, prickles appressed-antrorse, 0.5 mm long, 1 mm apart or less, from a prominulent keel. Longitudinal nerves about 40–42. Staminate inflorescence terminal, bracteate, inner bracts deltoid acute,

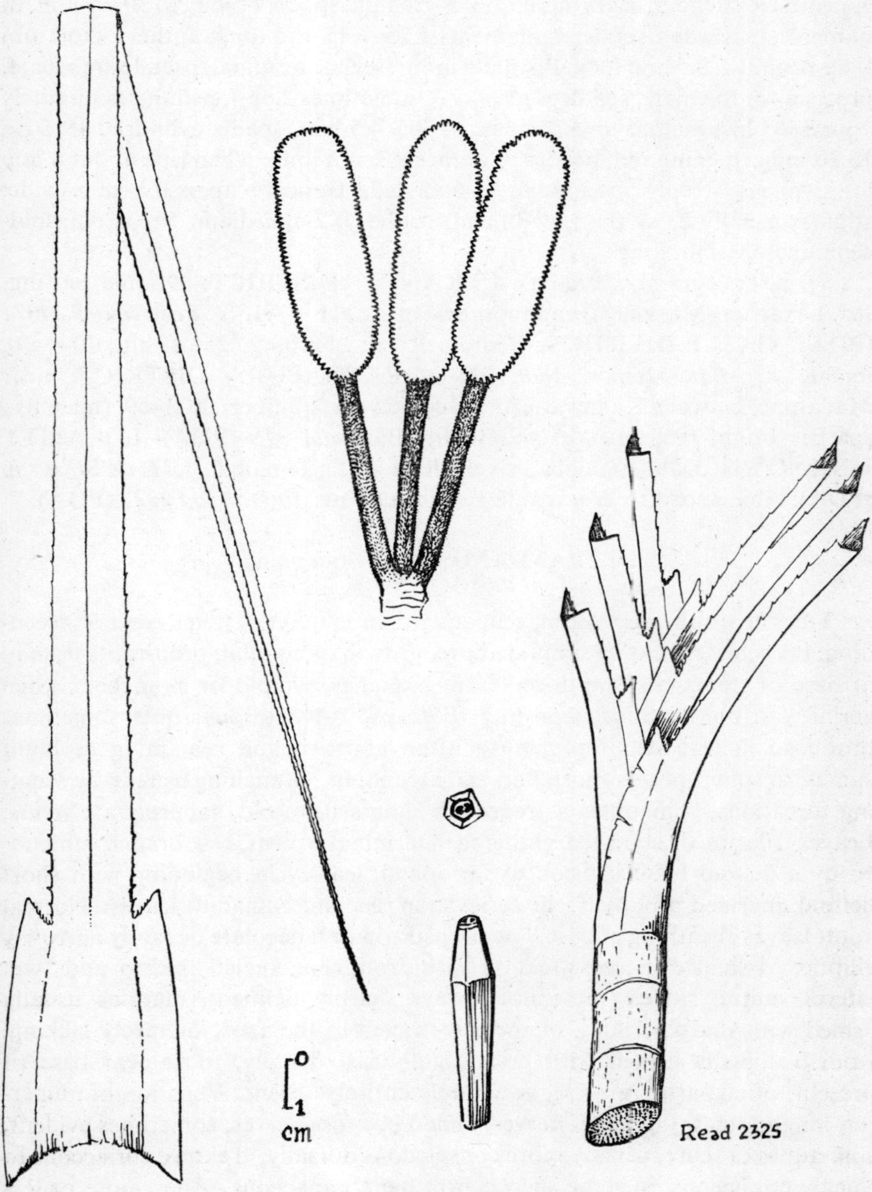

Fig. 2. *Freycinetia walkeri.*

bright red, to 10 × 4 cm, somewhat fleshy; spadices pseudoumbellate, usually 3, peduncles slender, glabrous c. 45 × 2–2.5 mm; spadix oblong to 30 × 5 mm, of numerous crowded stamens; filaments 1.25–1.45 mm long; anthers short oblong-rounded, 0.5 mm long. Pistillate inflorescence terminal, pseudoumbellate, bracts as in the male, spadices usually 3, sometimes 2 or 4, peduncles minutely appressed-hispidulous toward apex, to 50 × 4.5 mm, spadix cylindric, 45–60 × 16.20 mm, ripening red; berries at anthesis 2 mm long, when ripe about 8 mm long, pileus steeply pyramidal, 5–6-angled, truncate apex 0.5 mm wide; stigmas usually 2, rarely 3, confluent, round, 0.2 mm diam. Seeds ellipsoid-semilunate, 1 mm long.

Specimens Examined: KANDY DISTRICT: Double cutting, Rd. Laxapana-Maskeliya, common, 900 m alt., 11–5–71, fr. *Kostermans 24071* (PDA). GALLE DISTRICT: Bentota River, climbing, 250 ft alt., 10–4–70, female fl., *Balakrishnan 280* (PDA). RATNAPURA DISTRICT: near Madampe, between Kahawatte and Gadakewela, climber, 20–1–69, (male fls., spathes bright red, spadices yellowish), *Wirawan 818* (PDA). LOCALITY UNKNOWN: Kalugammana, riverside, 1–2–27, female fl. *J.M. de Silva s.n.* (PDA); Along road to Weragama Estate, 29 Nov. 1969, *Read 2325* (PDA).

2. PANDANUS L. ex Stickman

Trees or shrubs, terrestrial, semiaquatic or epiphytic; stems erect or ascending, branched (or rarely simple) above, growth sympodial, prop-roots usually at base of stems, also produced from branches, simple or branched, when aerial with conspicuous persisting rootcaps. Adventitious roots sometimes produced in leaf axils (epiphytes), often abortive and remaining as blunt spines; in some species short, flat, and ascending. Branching usually by seeming furcations, sometimes as regrowth shoots from old, subprostrate stems. Leaves trifariously spiralled, gathered in terminal crowns; new branch subtended by a deltoid bifacial-biconcave prophyll, leaf cycle beginning with short deltoid unarmed prophylls, the series soon reaching full adult leaves. Normal adult leaves sheathing at base, linear-ensiform or lanceolate or rarely narrowly elliptic, M-shaped in cross-section with prominent keeled midrib and twin lateral ventral ridges (these not always sharply defined), margins usually armed with sharp prickles, or these restricted to the apex, or rarely lacking. Midrib dorsally armed with prickles, at least distally, those near base (if present) often partly retrorse, very rarely entirely absent. Venation of numerous longitudinal ± parallel nerves, joined by cross-nerves, sometimes evident, sometimes obscure, usually more conspicuous dorsally. Texture coriaceous to thickly coriaceous, in some species firm membranaceous. Apex more or less prolonged into triquetrous acumen or flagellum. Epidermis cuticle often waxy, or glaucous, especially dorsally. Inflorescence terminal on normal or, rarely, on specialized branches, strictly unisexual. Staminate inflorescences ephemeral, bracteate, racemoso-spicate, in some species a single terminal spike; spikes axillary; bracts spathaceous, white or coloured. Spike composed of single stamens or of staminate phalanges, these various, racemose, subumbellate, branched, etc. Filaments smooth, slender to bulbous; anthers oblong to linear, 4-celled, at base often ± hastulate, apex blunt or more commonly apiculate by extension of the connective. Pollen grains usually with minutely spinulose exine. Pistillate inflorescence at first erect, later nodding, in fruit often ± pendulous, cephalia solitary or spicately disposed, bracteate as in the male; cephalia in flowering state more or less hidden by bracts; composed of free or adnate carpels ripening as drupes or as syncarps; carpels one-ovulate, ovule within a finally hard-osseous endocarp; mesocarp fibrous and medullose, often cavernose and containing a viscid fluid or gel when submature; lower mesocarp fibrous and fleshy; pericarp more or less indurated above, fleshy below or fleshy throughout, often oily or starchy, sometimes edible, usually brightly coloured; style obscure or partly developed; stigma round, reniform, deltoid, linear, or various; sometimes spiniform, sometimes forked or dentate, consisting of glandular hairs which harden and sclerify later, grooved at base; placenta parietal; ovule anatropous; seed more or less fusiform or teardrop-shaped, albumen white, embryo minute, sublateral. Germination often epigeal, cotyledon remaining internal, hypocotyl also, epicotyl developing into shoot leader and emitting roots.

Chromosome number $2n=60$, in many species; a single tetraploid known so far (*P. spiralis*); cultivars sometimes with slight aneuploid anomalies.

A large complex genus of 600–700 species, exclusively Old World, though pollen of a fossil form genus *Pandaniidites* has been reported from the Miocene of Texas and North-Central North America. Living members occur in West Africa, Madagascar (over 75 species), India, Burma, Thailand, S. China, Malaya, throughout Indonesia, Philippines, Taiwan, Okinawa and Southernmost Japan (Amami O-Shima), throughout Micronesia, Melanesia, and Polynesia, but not in New Zealand, richly developed in New Guinea and N. Australia, also in New Caledonia, reaching North to Hawaii.

From the monographer's point of view *Pandanus* is subdivided into about eight subgenera; these in turn are subdivided into sections; the details of this infrageneric classification are described elsewhere (Stone).

Only four species seem to be indigenous in Ceylon; they belong to three subgenera. Several other species are in cultivation, some of them very abundant.

The production of minor economic products—plaited or woven mats, hats, handbags, etc., flavourings (from leaves of "rampeh" *P. amaryllifolius*), perfumed oils (Keora oil, from male flowers of *P. odoratissimus* and perhaps also from *P. kaida*), make these plants useful as well as ornamental, and most villages and towns have planted specimens. Several species are used as "living hedges" or marker rows in paddy fields, etc. Some species (not in Ceylon) produce edible fruits and/or seeds; one of these, from the Andaman Islands (*P. leram*) is found cultivated in Peradeniya.

Because of the strict distinction of the sexes, and also because some species are rarely or never found in flower or fruit, four keys are provided; one for sterile collections, one for staminate collections, one for pistillate collections, and one for wild-growing species only.

KEY TO THE SPECIES

I. Artificial key based mainly on leaves

1 Leaf apex with twin lateral ventral pleats prickly
2 Wild plants; leaf apex gradually ± narrowly attenuate. Stigmas spiniform on the free drupes. Staminate and pistillate infl. spicate. Stamens singly attached to spike axis, anthers much elongated. Wild .... **5. P. thwaitesii**
2 Cultivated, rare; leaf apex rounded-acuminate. Stigmas round to ovate. Stamens in small fascicles, anthers oblong, short, blunt. Cultivated .... **9. P. polycephalus**
1 Leaf apex with unarmed lateral ventral pleats. Stigmas various. Stamens in phalanges
3 Leaves musky scented, armed only at extreme apex prickly, green. Cultivated shrubs .... **6. P. amaryllifolius**
3 Leaves unscented, prickly at least at extreme apex, or unarmed
4 Leaves entirely unarmed; cultivated shrubs .... **11. P. tectorius var. laevis**
4 Leaves prickly along margins as well as apex, also usually at base
5 Leaves red-marginate, underside gray glaucous .... **P. utilis**
5 Not so
6 Leaves striped green and yellow or green and white, rarely almost completely albino; cultivated shrubs .... **11. P. tectorius** var. **sanderi** etc.
6 Leaves not striped or variegated; mostly wild plants
7 Leaf prickles commonly 1–3 mm long, green or greenish, minutely dark-tipped, all or nearly all antrorse; dorsal surface somewhat greenish glaucous or semi-glossy green, paler than ventral surface. Cultivated shrubs
8 Leaf apex attenuate flagelliform. Common, cultivated .... **11. P. tectorius**
8 Leaf apex narrowly acute, not drawn out. Rare, cultivated .... **7. P. cephalotus**
7 Prickles usually 3–10 mm long (or more)
9 Prickles pale to white; leafbases never reddish-purple. Bracts white to cream
10 Prickles white or whitish, markedly paler than the leaf-surfaces; leaf attitude arching and also usually somewhat twisting, apex flagelliform, curving. Underside, esp. at base, with 2 clearly demarcated very pale-glaucous stripes one on either side of the midrib. Maritime species of ocean beaches, but also cultivated, pistillate phalanges 5–15-celled, staminate phalanges racemose; pistillate heads solitary .... **4. P. odoratissimus**
10 Prickles pale greenish but not white; leaf attitude more stiffly erect, or abruptly bent or decurved toward apex; flagelliform apex straighter. Underside not so strongly glaucous. Cultivated sp. of rice fields, etc. Pistillate phalanges 1–4-celled, staminate phalanges subumbellate. Pistillate heads solitary or sometimes spicate .... **3. P. kaida**
9 Prickles greenish-white, dark-tipped; leafbases usually reddish-purple or bronzy-brownish
11 Pistillate head solitary, of simple drupes with forking rostriform styles. Staminate phalanges subumbellate. Bracts pale dull yellow. Wild along stream-banks or beds, also cultivated .... **2. P. ceylanicus**

11 Pistillate head solitary, oblong, or several together; drupes with simple or forked styles. Staminate phalanges umbellate
12 Pistillate head oblong, usually solitary. Trunk tall, un- or sparsely branched. Bracts straw-coloured, stiff. Wild and cultivated.......................... **1. P. furcatus**
12 Pistillate heads 5–9-together, ellipsoid; trunk branched, crown candelabriform. Bracts cream-white, thin fleshy. Cultivated only.......................... **P. nepalensis**

II. Key based on staminate flowers
(Species not known to flower are omitted)

1 Spikes racemose-spicate, composed of singly attached stamens; filaments very much shorter than the narrow elongated apiculate anthers. Wild and cultivated..**5. P. thwaitesii**
1 Not as above; the stamens joined in phalanges with a column (stemonophore)
2 Anthers blunt, nonapiculate. Stamens few per phalange, crowded subumbellate. Cultivated...**9. P. polycephalus**
2 Anthers mucronate-apiculate by extension of the connective
3 Stamens umbellate at apex of column
4 Bracts and spikes yellow or buff
5 Stamens 7–9 per phalange; spikes about 7 × 2 cm or less; bracts of infl. small, up to c. 18 × 3 cm...**2. P. ceylanicus**
5 Stamens 8–14 per phalange; spikes to 15 × 3 cm or larger; bracts larger, stiff...... ....**1. P. furcatus**
[see also: **P. nepalensis**]
4 Bracts and spikes white or cream coloured
6 Filaments forked, or some forked and some simple, slender. Leaves red-marginate ....**P. utilis**
6 Filaments simple. Leaves not red-marginate
7 Filaments 3–5 mm long; stamens c. 20–25 per phalange. Massive trees with long stout prop-roots. Cultivated.......................................**10. P. leram**
7 Filaments c. 1 mm long. Stamens fewer, commonly 5–15 (–23) per phalange. Small trees
8 Anther with obvious mucro 0.5 mm long; column relatively stout....**3. P. kaida**
8 Anther nearly obtuse, mucro minute or obsolete; column slender............. ....**7. P. cephalotus**
3 Stamens racemose along the column (at least its upper portion)
9 Leaves with large white prickles; wild maritime sp...............**4. P. odoratissimus**
9 Leaves with small greenish prickles; cultivated sp.........**11. P. tectorius** and vars.

III. Key based on pistillate collections

1 Carpels simple, ripening as 1-seeded drupes, (rarely 2–3-seeded)
2 Stigma ± linear on dorsal side of spiniform style. Heads racemose-spicate. Placentation dorsal...**5. P. thwaitesii**
2 Stigma on ventral side of style, or flat or oblique on carpel apex. Placentation ventral
3 Some styles, at least of drupes near the peduncle, forked; styles ± rostriform; stigma ± deltoid, ventral
4 Pistillate heads solitary
5 Pistillate heads small subglobose, up to 15 × 15 cm; styles simple or if forked then scarcely so (barely notched).................................**2. P. ceylanicus**
5 Pistillate heads large oblong, to 30–45 cm long; styles fairly often forked, if so with slender sharp horns...**1. P. furcatus**
4 Pistillate heads racemose-spicate, ellipsoid...........................**P. nepalensis**
3 Styles ± round-reniform, flat or nearly so, apical or subapical

6 Fruit-heads small, red-orange fleshy; racemose-spicate; leaf apex rounded-acuminate; lateral pleats prickly.......................................**9. P. polycephalus**
6 Fruit heads greenish; leaf apex acute, lateral pleats smooth..........**7. P. cephalotus**
1 Carpels adnate in 2-several-celled phalanges
7 Phalanges mostly 2-celled (1–4-celled); cultivated........................**3. P. kaida**
7 Phalanges mostly 5–15-celled
8 Carpels ± concentric
9 Phalanges with abrupt expanded 'shoulders' laterally (but these evanescent)........ ....**4. P. odoratissimus**
9 Phalanges gently expanded or with tapered sides..................**11. P. tectorius**
8 Carpels ± linear-seriate
10 Small trees; leaves red-marginate; phalanges small, 4–5 cm long..........**P. utilis**
10 Massive trees; leaves not red-marginate; phalanges to 15 cm long......**10. P. leram**

IV. Key to wild-growing species

1 Leaf apex with twin lateral ventral pleats prickly. Inland sp............**5. P. thwaitesii**
1 Leaf apex with lateral pleats smooth, unarmed
2 Leaf-marginal and midrib teeth large, white, undersurface markedly glaucous; leaf apex long extended flagelliform. Fruit head composed of polycarpellate phalanges. Maritime sp..**4. P. odoratissimus**
2 Leaf-marginal and midrib teeth green or greenish or brownish, never white. Inland spp.
3 Rheophyte or subaquatic shrub along streams, riverbanks, sometimes in paddy fields; leaves rarely more than 120 cm long and 3.5 cm wide. Fruit head up to 15 cm diam. ....**2. P. ceylanicus**
3 Terrestrial massive shrub or tree; leaves commonly 2 m long or more, 5–10 cm wide or more..**1. P. furcatus**

Section RYKIA (DeVriese) Kurz

1. Pandanus furcatus Roxb., Hort. Beng. 71. 1814; Roxb., Fl. Ind. 3: 744. 1832.—**Fig. 3.**

Rykia furcata (Roxb.) DeVriese, Kl. Akad. Vet. Verh. 203. 1854.

Arborescent, up to 4 m tall or more; not or scarcely branched; leaves comparatively large, linear, apex gradually attenuate; lamina coriaceous, 2 m long or more to 8.5 cm wide; margins and midrib (beneath) armed with sharp prickles, ventral pleats smooth unarmed. Margins, towards the base (but leaf-base not seen) with small antrorse prickles, about 1.5–1.7 mm long, 5–8 mm apart; towards the apex the prickles more crowded (1–3 mm apart) but not much smaller except at extreme apex where they are 0.5 mm long; midrib prickles similar to those of adjacent margin, but much more remote, up to 30–35 mm apart, except along the apex, where their spacing is alike also. Reticulations not or but little evident. Longitudinal nerves conspicuous, c. 128/1f, prominulent on both surfaces but especially so abaxially.

Specimens Examined: KURUNEGALA DISTRICT: Kurunegala, foot of hill, rare, clumps to 4 m tall in a rather, moist place, June 1973

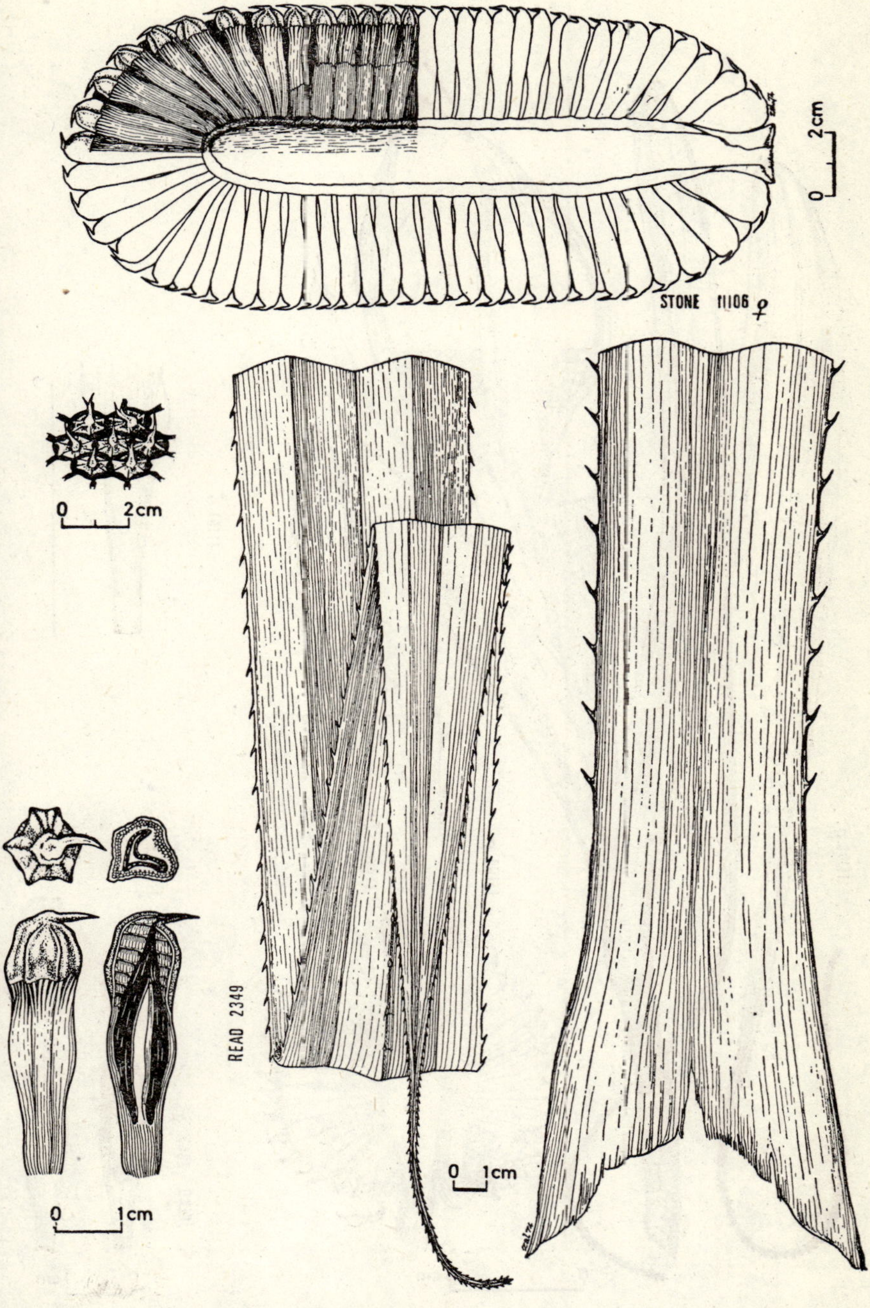

Fig. 3. *Pandanus furcatus*.

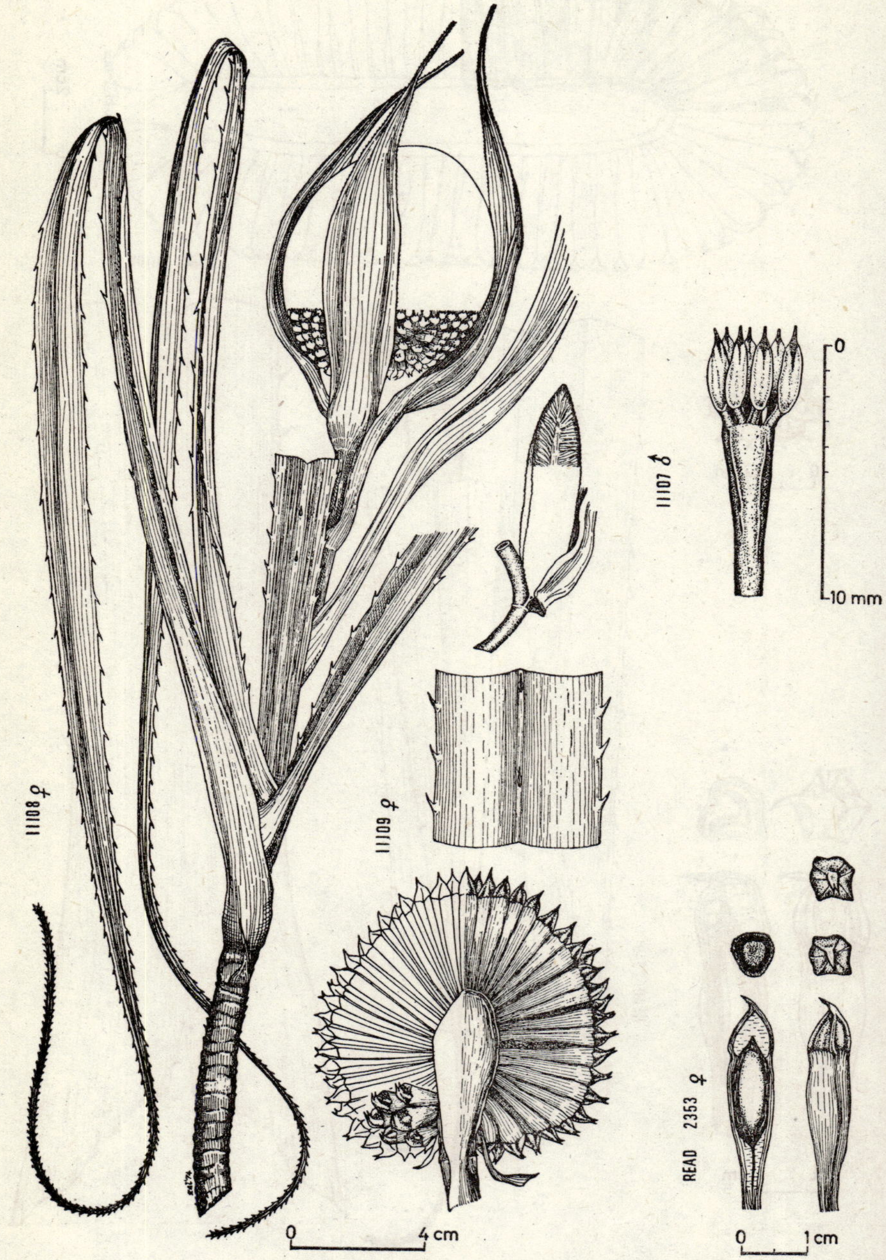

Fig. 4. *Pandanus ceylanicus.*

Kostermans s.n. (KLU, sterile). KANDY DISTRICT: Uduwatta Kelle along road, 12 Dec. 1969, *Read 2349* (PDA); Peradeniya, Botanic Gardens, 500 m., April 1973, female, *Stone 11, 106*; *11313* (KLU).

Note. In the absence of fruits the Kostermans collection cannot be identified with absolute certainty. It is however apparently indigenous, according to the collector, and it is obviously not one of the hitherto-reported Ceylon species. It is at least a new record. The fruit illustration is based on presumably cultivated plants.

2. Pandanus ceylanicus Solms, Linnaea 42: 16. 1878. Type materials: Ceylon, Central Province, Nov. 1861, female, *Thwaites. C.P. 2734* (PDA) (lectotype); Ratnapura, Ekneligoda, March 1853, female, *Thwaites C.P. 2734* (PDA); without locality, male infl., *s. coll. C.P. 2734* (PDA).—**Fig. 4.**

Pandanus furcatus sensu Thw., Enum. Pl. Zeyl. 327. 1864, non Roxb.
Pandanus zeylanicus (sphalm. err.) Hook. f., Fl. Br. Ind. 6: 484. 1898; Trimen, Handb. Fl. Ceylon 4: 339. 1898.
Pandanus porrittianus Martelli, Soc. Bot. Ital. Proc. 301. 1904; Martelli, Webbia 4 (1): 28. 1913; 4 (2): Tav. 28, fig. 38–40. 1914. Syn. nov.

Erect or ascending or partly prostrate shrubs, rather sparsely branched; stems emitting slender prop-roots up to 40 × 1.5 cm, smooth or with a few whorls of blunt protuberances, occasionally branched, mostly from the lower part of the trunk. Stems pale greyish-tan, annulate by leafscars about 1–1.5 cm apart, inner cortex green, main trunk reaching 2–3 m in length, up to about 6 cm in width, comparatively smooth or the older parts set with scattered blunt aborted adventitious rootlets. Leafy branches 1.5–4 cm diam., leafscars 1.5–4 mm apart. Leaves linear-ensiform, coriaceous, straight or gently curved, slightly twisted, mostly 60–120 cm long, 2–3.5 cm wide, apex gradually attenuate, sheathing base slightly expanded, narrowly marginate, in older leaves with a slight to dark coppery or purplish-black tint, otherwise the blades green, darker above, paler and duller beneath, also striate by the 44–74 longitudinal darker nerves; margins sharply prickly along their whole length, at the base the prickles 3–5 mm long, antrorse and slightly arcuate, pale whitish green, or the tips brownish; (7–) 10–25 (–40) mm apart; in midregion, the prickles more appressed, slightly shorter (1.5–3 mm long), slightly closer together; near apex, the prickles 0.5–1 mm long, 1–3 mm apart. Midrib prominent, rounded at base, glossier than leafblade, acute distally, in oldest leaves unarmed near the base, but with prickles along the distal half; in younger leaves, and in juvenile plants, armed near the base with sharp retrorsely hooked pale whitish-green prickles similar in size to those on the adjacent margin, but more widely spaced, (15–) 20–50 (–70) mm apart; beyond the middle, the prickles antrorse, similar to those of the adjacent margin but more widely spaced; at apex, similar to those of adjacent margin in size, orientation, and spacing. Staminate inflorescence terminal, racemose-spicate, at first suberect, later drooping and pendulous, the main rachis firm, fleshy, white, 24–28 cm long, 9 mm wide, trigonous, bearing about 9 to 11 spathaceous bracts, the outer ones with green leaflike apices, the inner ones pale yellow, narrowly ovate, navicular, to c. 18 × 3 cm, the final one or two almost linear, all thinly membranous and slightly coriaceous, unarmed except along the apex. Staminate spikes 5 to 7, sessile or pedicellate (on pedicel 18 × 3 mm), about 7 × 2 cm or less, ± trigonous, composed of numerous staminate phalanges, each of these with a narrowly obconic column c. 2.5–4 (–5) mm long, 1.5–2 mm wide at apex, narrowed to base, pale yellow, bearing usually 7 or 9 anthers subumbellately disposed, filaments to 1 mm long, abruptly narrowed just under the anther; anthers almost white, 2.5 mm long, narrowly oblong, rounded-acute at apex and apiculate, apicules spiniform, straight or curved, 0.4–0.5 mm long. Pistillate inflorescence terminal, bracteate, the peduncle bearing a solitary ovoid-ellipsoid head, at anthesis 3–4 cm long, 3 cm diam., surrounded by pale yellow spathes; carpels pale green, stigmas white; carpels at anthesis 6–9 mm long; spathes similar to those of staminate infl., but comparatively longer or narrower. Peduncle 20–30 cm long, 10–15 mm thick, trigonous, at first cream, later green. Fruiting heads subglobose, up to 15 cm diam., dark green, composed of approximately 500–1000 drupes, nodding; peduncles persistent for a time on the stems after fruits have fallen. Drupes 1-celled, oblanceoloid, 23–30 mm long, up to 8–9 mm wide, 5–6-angled, free apex the upper sixth, c. 5 mm long, pileus pyramidal, produced into the acute, ± antrorsely deflected rostriform-spiniform style, this 2–3 mm long; stigma brown, narrowly deltoid, grooved at base, simple or, on a few drupes near the peduncle, briefly forked. Endocarp osseous brown, slightly above the middle of the drupe, about 10–12 mm long, acuminoid at apex, subtruncate at base, the walls c. 1 mm thick, seed one. Apical mesocarp pithy-fibrous, lower mesocarp fibrous, about 7–10 mm long. Seed, albumen white, ellipsoid, c. 7 × 3 mm.

Specimens Examined. RATNAPURA DISTRICT: bed of the Billahool (= Belihul) Oya, 26 Feb. female infl. *s. coll. 1882* (PDA); N. of Eheliyagoda, road to Talduwa, border of paddy field, 29 Nov. 1969, male, *Read 2305* (PDA). KANDY DISTRICT: Kandy, rocky banks of Mahaweli Ganga, 1500 ft. alt., 7 April 1973, male, *B.C. Stone 11107*, female, *11108*, frts., *11109* (PDA, US, KLU); Kandy to Hunnasgiriya, 4 mi. from Kandy, as paddy divider, 1969, *Read 2353* (PDA). KALUTARA DISTRICT: Kunendrana, Bulathsinhala, Agalawatte, 1969, *Jayasinghe* (+*Read 2344*) (PDA).

Section KAIDA St. John

3. Pandanus kaida Kurz, J.R. Asiat. Soc. Bengal 38: 148. 1869; Stone, Malayan Agriculturist 9: 34–44. 1970. Holotype: Siam, coll. *Teysmann* (BO).—**Fig. 5.**

Pandanus candelabrum sensu Kurz in J. Bot. 5: 127. 1867, non Beauvois.
Pandanus kaida, in Trimen, Handb. Fl. Ceylon 4: 341. 1898.
Pandanus siamensis Williams, Bull. Herb. Boiss. Ser. II, 4: 220. 1904.

A large shrub, in cultivation usually kept short, but when freely growing becoming treelike with an erect trunk 2–4 m tall, branched above; prop-roots basal, rather few and short. Mature leaves to 3–4 m long, rarely longer, and up to 9 cm wide, commonly 2–3 m × 6–8 cm, linear-ensiform, gradually narrowed to the long flagelliform apex, coriaceous, green above and below, not or scarcely glaucous, erect-ascending, rather stiff, the tips of older leaves abruptly downturned. Margins and midrib (beneath) stoutly armed. Margins near base with forward-curved sharp teeth to 5–6 mm long and 3–12 mm apart; near the middle of the blade, with similar but more appressed, more crowded, slightly shorter teeth; near apex, the teeth much smaller, 0.5 mm long, but wide-spaced, mostly 5–10 mm apart. Midrib teeth nearest the leaf-base blunt, short, erect or slightly reflexed, 10–20 mm apart, these followed by a series of longer, sharper, reflexed teeth to 6 mm long; but from the middle of the leaf and beyond, all midrib teeth forwardly curved, shorter (c. 4 mm) and mostly 10–15 mm apart; near leaf apex the midrib teeth similar in size and shape to those of adjacent margin, but somewhat more distant, longitudinal veins fairly conspicuous, especially on the undersurface, in larger mature leaves totalling about 60–65 on either side of the midrib, which is slightly salient. Cross veins nearly invisible. Inflorescences terminal and bracteate. Female infl. pedunculate, the peduncle up to 40–50 cm long, at the apex nearly 2 cm thick, trigonal, bearing several (3–5) foliaceous bracts. Cephalia solitary or sometimes 2, 3 (or rarely up to 7!) borne together spicately, each subtended by several (when solitary or terminal) or one, (when lateral) green to whitish spathes with leaflike apices (lower spathes) or acute, white apices (upper spathes) similar to those of the male inflorescence. Staminate inflorescence a large raceme or spike or spikes, each lateral spike 8–11 cm long, 2.5–3 cm thick, subsessile, subtended by a white spathe. Spikes composed of very numerous rather densely crowded stemonophores, these subterete, about 7–8 mm long, 1.5–2 mm thick, nude to the apex and there bearing up to 23 erect, crowded, white stamens, each on a short filament 0.5–2 mm long, the anther linear or slightly curved, 3.6–4 mm long, hastate at base, acute at apex with an apiculum 0.5–0.7 mm long; all filaments closely crowded at apex of stemonophore. Pollen grains minutely spinulose. Cephalia (terminal or solitary) oblong-ellipsoid, rounded-trigonal, about one and a half times longer than wide, up to 23 × 15 cm, but sometimes smaller and more globular (when lateral on spicate inflorescences), finally pendulous, composed of numerous mostly 1–4-celled phalanges. Phalanges cuneate, up to 45–50 mm long, 10–20 (–24) mm thick, usually 5–7 angled, apex low-pyramidal, centrally produced into the 1–4 short, stocky styles, these erect or oblique, commonly 1–2 mm high

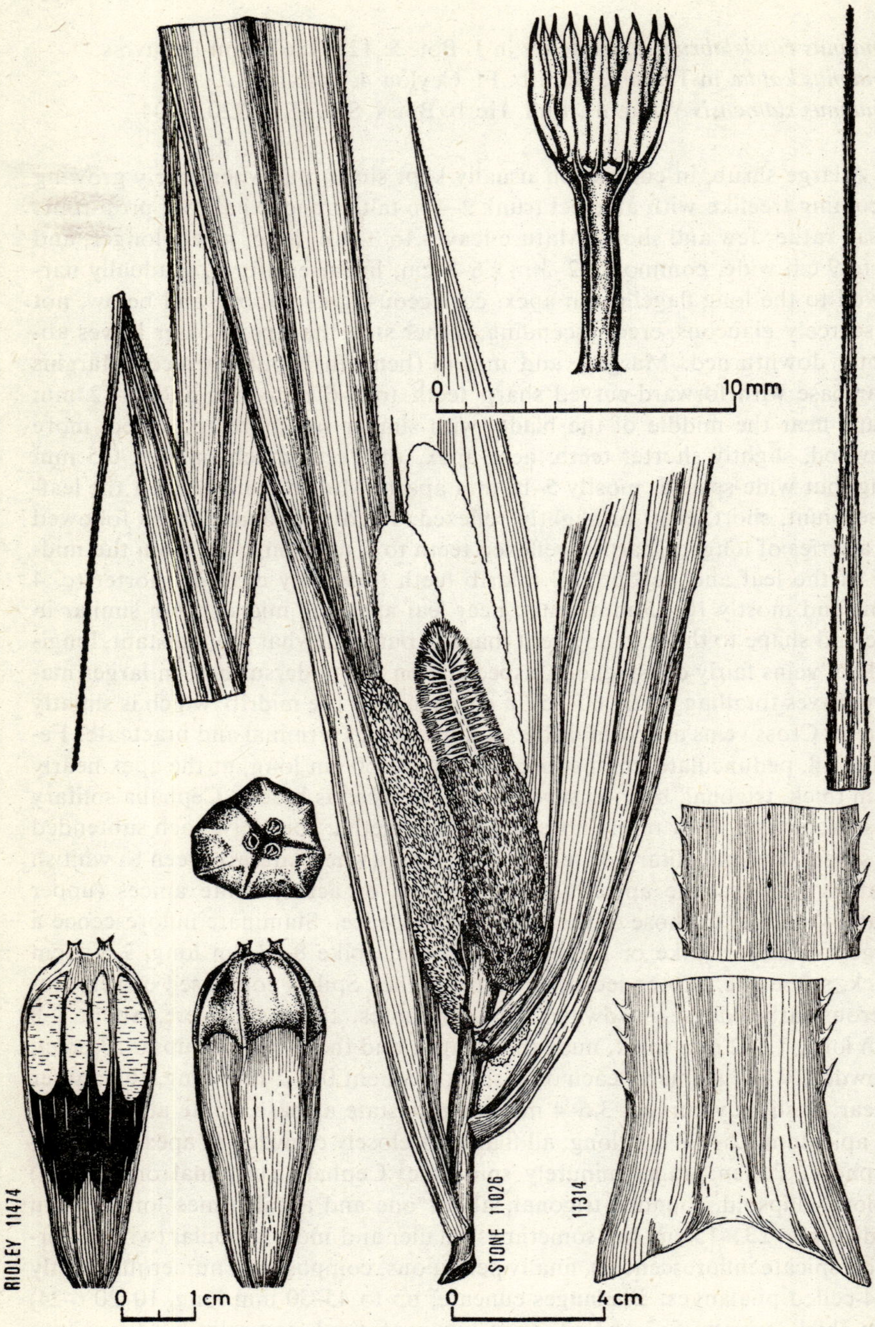

Fig. 5. *Pandanus kaida.*

and 1–3 mm broad, the stigmatic surface brown, centrally grooved, facing each other (i.e. centripetal) when 2 or more together, often wider than long (usu. 2 mm wide), rather close-set, rarely over 5 mm apart; individual carpel apices barely distinguishable by the shallow apical sutures (less than 1 mm deep) separating them. Endocarp subcentral, reddish, osseous, 1–4-celled (usually 2-celled), 12–15 mm long, the outer walls 1–2 mm thick, the seed chambers about 8–10 mm long. Upper mesocarp cavernose (as many chambers as carpels), medullose pithy and, peripherally, slightly fibrous, within, about 13–15 mm long and to 7 mm wide; lower mesocarp fibrous-fleshy. Pericarp at apex firm, up to 2 mm thick; below, very thin.

Vern. Watta-keyiya or watta-keyiya-arulu (S).

Specimens Examined. COLOMBO DISTRICT: Kanuwana near Colombo, Oct. 1884, female, *s. coll. s.n.* (PDA). KANDY DISTRICT: Kadugannawa, 1200 ft., 21 April 1973, female, *Stone 11314* (KLU, PDA, US). LOCALITY UNKNOWN: *s. coll. H82* male infl. (PDA).

Note. I have seen no specimens of *P. kaida* from anywhere, in Ceylon or elsewhere, which could be construed unequivocally as wild. Nonetheless its occurrence as a wild species in Ceylon is not impossible; but its usefulness, whether as a paddy-field divider, or as a source of textile leaf materials, or for other purposes, has made it a widely planted species. Its original home may never be known in detail, but it does not extend beyond the area bounded by India on the west and the eastern perimeter of the continent of Asia, from Malaya to S. China, in the east. I describe it fully in this treatment not because I am certain it is an indigenous species but because it is so similar when not flowering or fruiting to *P. ceylanicus* that it can easily be confused with it, and the two species are often found growing together.

It will be noted that I include *P. kaida* in Sect. *Kaida*. This is a new disposition, as it was formerly attributed to Sect. *Pandanus* or, earlier, to Sect. *Hombronia* (by Martelli). In fact it constitutes the type of a new section but it is certainly closely related to Sect. *Rykia*, and is definitely not close to Sect. *Pandanus*.

Section PANDANUS

4. Pandanus odoratissimus L. f., Suppl. Pl. 64. 1781; St. John, Taxon 12 (5): 201–204. 1963. Type: Ceylon, 1777–8, *O. Thunberg* (U) (not seen).—**Figs. 6, 7.**

Pandanus verus Rumph., Herb. Amb. 4: 139. t. 74. 1743; Kurz, J. Bot. Br. & For. 5: 125. 1867, in part.
Pandanus fascicularis Lam., Enc. 1: 372. 1783.
Pandanus leucanthus Hassk., Fl. Bot. Ztg. 25 (2), Beibl. 14. 1842.
Hasskarlia leucacantha Walp., Ann. 1: 753. 1849.

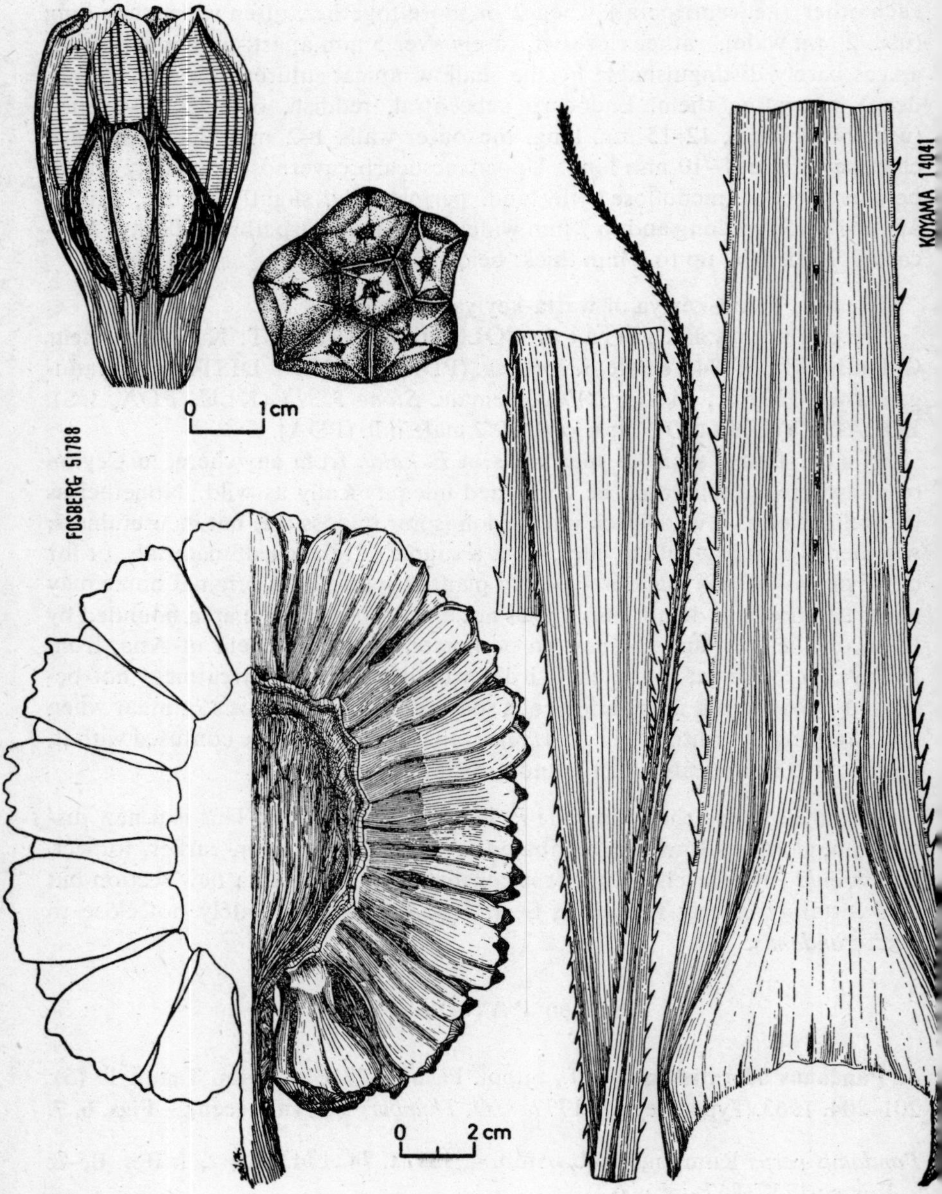

Fig. 6. *Pandanus odoratissimus.*

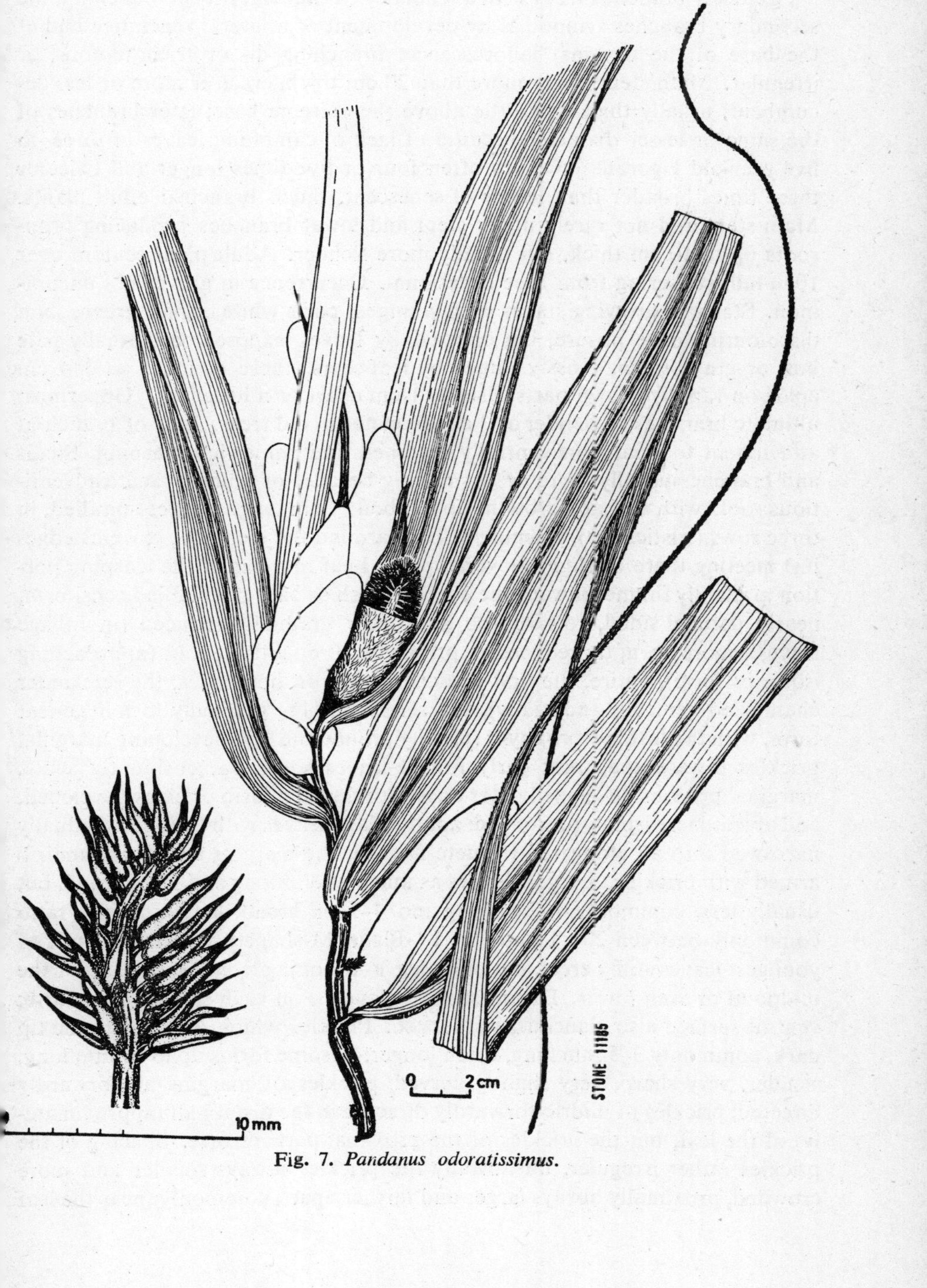

Fig. 7. *Pandanus odoratissimus.*

Coarsely branched trees with a tendency to candelabriform branching, the secondary branches sympodial by development of a lateral vegetative bud at the base of the terminal inflorescence; branching di- or trichotomous, or irregular. Main stem rarely more than 20 cm thick, erect or more or less decumbent, usually thickest a little above the extreme base; later branches of the same or lesser diameter because of lack of cambium; leaves of three- to five-year-old vigorous juveniles often four or five times longer and twice or three times broader than leaves of senescent, much branched adult plants. Main stem and not rarely upper stem and lower branches producing prop-roots up to 10 cm thick, but usually more slender. Adult plant seldom over 10 m tall; suckering from base occasional. Occurrence in groves not uncommon. Stems in growing stages with youngest parts white or pale green, soon discolouring on exposure, but covered by leaves; exposed part usually pale buff or gray-brown, grossly ringed by leaf-scars, these as much as 3–4 cm apart on rapidly-grown parts, usually 1 cm apart or a little less. Uppermost ultimate branches as slender as 3 cm diameter in old trees. Base of branch at attachment to stem dorsiventrally flattened and somewhat clasping. Stems and branches usually more or less prickly because of blunt or acute adventitious roots with arrested growth, up to about 1 cm long. Leaves spiralled, in three rows (tristichous), clasping at base around the stem, the extreme edges just meeting there or slightly overlapping. Leaf margins of the clasping portion gradually thinner toward the edge, whitish or pale green. Blade ensiform, nearly parallel-sided, toward the apex very gradually reduced (in foliage leaves), more abruptly reduced in prophylls. Prophylls deltoid (approaching isosceles form), entire, the first one channelled on both sides, the remainder channelled ventrally (adaxially), the series merging gradually to foliage-leaf form, with the last few prophylls (usually around the 9th) developing marginal prickles toward the apex; early foliage-leaves short, rather bluntly acute, marginal prickles developed; later ones with dorsal midrib prickles developed, and apices increasingly prolonged; adult foliage-leaves with the apex gradually narrowed into an elongated triquete flagella, the margins and dorsal midrib armed with prickles. Largest leaves as much as 3 m long and 9 cm broad, but usually less, commonly 1–2 m long and 4–7 cm broad; length : width ratio commonly between 20 : 1 and 30 : 1. Blade M-shaped in section. Apex of youngest leaves stiffly erect, but older leaves drooping from near the tip or the midpoint or even lower. Dorsal surface glaucous on each side or the midrib; ventral surface also glaucous, but less so. Prickles white, or the extreme tip dark, commonly 3–5 mm long, often longer; in some forms up to 10 mm long, slender, very sharp, very slightly curved; prickles of margins all forwardly directed; prickles of midrib forwardly directed in the distal half (approximately) of the leaf, but the prickles of the proximal part retrorse. Spacing of the prickles rather irregular, but distally the prickles always smaller and more crowded, proximally always larger and farther apart; commonly (near the leaf base) the prickles 1–3 cm apart. (Some forms probably for genetic reasons unarmed.) Parallel veins mostly 40–160, depending on age of leaf, age of tree, etc., visible but not prominent along most of the blade but less or not so toward the base. Transverse lines not at all evident, unless under microscope. Coloured pigmentation absent, i.e. leaves never reddish, purplish, etc.; but white- or yellow-striped variegated forms may perhaps exist. Male inflorescence only on male trees, pedunculate, bracteate, the bracts similar in colour and form to those of the female, but the infloresence a raceme of spikes (an odd number, usually 5, 7, 9, or 11), each in the axil of a bract, each consisting of a fleshy central axis from which emerge numerous lateral phalanges of stamens about 13 mm long, bearing about 19–23 stamens on short filaments 0.5–2 mm long, with anthers 2–3.5 mm long, narrowly sagittate, with an apiculus 0.5 mm long; the basal part of the phalange is about 3 mm long up to the first filament; phalange, receptacle and major axis clearly fasciculate. Overall length of inflorescence 30 to 60 cm; bracts subtending spikes mostly less than 30 cm long, last ones only about 6 cm long. Inflorescence ephemeral, decaying very rapidly after maturing. Female inflorescence borne only on female trees, terminal, pedunculate, the peduncle 10–30 (rarely more) cm long, trigonal, bracteate, the lower bracts leaflike, upper bracts whitish-yellow, successively shorter, more or less navicular, the midrib and margins with reduced prickles, texture subfleshy, the last bracts much reduced to linear-lanceolate shape and much shorter than the syncarp; these bracts all soon withering and more or less persistent but at last fragmenting and falling. Inflorescence a cephalium, globose to ellipsoid, up to about 30 × 20 cm, slightly rounded-trigonal, mostly 15–20 cm × 12–18 cm; composed of a variable number (26–143, usually about 50–70) of carpellate phalanges, these in turn composed of mostly 4–10 concentrically arranged fused carpels (phalanges of only 1–3 carpels or of 11–14 carpels rarely also occur, the former only near the cephalium apex, the latter only near the cephalium base). Carpels each terminating in a single U- or V-shaped stigma; carpel apex truncate, low-rounded, or low-pyramidal, sometimes somewhat acute. Phalanges laterally mostly smooth, rarely with 1 or 2 intercarpellary grooves; apically with slight grooves between the carpel tips. At maturity the phalanges red-orange to vermilion, sometimes the carpel tips still greenish; median part of phalange rather abruptly expanded into fleshy shoulders, these not evident in dried or insect-damaged or animal-gnawed specimens, but always obvious in fresh fully ripe fruit. Phalanges mostly 3–8 cm long, 2.5–4.5 cm wide, pentagonal or hexagonal but sometimes slightly exceeding these measurements; phalanges from cephalium base (peduncular end) always curved, shorter, and with more carpels, than those from lateral or apical part of cephalium; phalanges from apex of cephalium usually longest and with fewest carpels, the carpel tips more expanded and sometimes more acute. Young phalanges greenish-white, later yellow, finally orange then reddish to vermilion. Phalanges free of each other but tightly crowded; eventually falling from the cephalium receptacle, apical ones first, leaving a pentagonal or hexagonal scar. Cephalium when mature more or less drooping; when young usually suberect. Phalange externally covered by a pericarp; within this, upper mesocarp lacunose, each carpel apex with a chamber traversed around the periphery by longitudinal fibres and crossed by thin medullar plates of tissue (aerenchyma); lower mesocarp fibrous, the fibres traversing a pulpy tissue; endocarp central, osseous, reddish-brown, unitary, each carpel represented by one ovular chamber, this fusiform, a few fibres traversing the endocarp, some central in the mesocarp and leading to the stigma, these entering the ovular chamber along one side; upper part of ovular chamber filled with soft white endosperm; embryo basal; germinating downward. Outermost seeds often not developing.

Vern. Watta-keyiya (S).

Specimens Examined. JAFFNA DISTRICT: North shore of Jaffna Lagoon, 10 mi. SE of Navatkuli, sandy island, 5 m tall, semi dry sand at margin of Palmyra palm grove, 18–2–71, female, *T. Koyama et al. 14041* (PDA). BATTICALOA DISTRICT: Kalkuddah Bay, 16 mi, N. of Batticaloa, near Resthouse at beach in front of Mt. Somanader's sandy beach, first row of woody veg. facing the sea, more or less sea level, 6 m tall, spreading over 100 m, gregarious, densely busy, 5 frts. in var. stages of ripeness, 20–4–68, *D. Mueller-Dombois 68/4/20/13* (PDA); ibid., 16 April 1973, male fls., *B.C. Stone 11, 185*, Young pistillate heads, *11, 186* (PDA). GALLE DISTRICT: Hikkaduwa, along strand, August 1969, male, *R.W. Read 2390*, female, *2391* (PDA). COLOMBO DISTRICT: Colombo, Dec. 1968, *M. Samaraweera s.n.* (KLU). KALUTARA DISTRICT: W. Coast, 30 mi. S. of Colombo, Maggona, 20 Dec. 1969, *Fosberg 51787, 51788, 51789* (KLU, PDA, US). LOCALITY UNKNOWN: (2 sheets) annotated "Mudu-Keyeya" *s. coll. C.P. 3739* (PDA).

Section ACROSTIGMA Kurz

5. Pandanus thwaitesii Martelli, Webbia 1: 369. 1904.—**Fig. 8.**

Pandanus humilis Moon, non Rumph., Cat. 67. 1824; Thw., Enum. Pl. Zeyl. 327. 1864; non Lour.

Pandanus foetidus var. *racemosus* Trimen, Handb. Fl. Ceylon 4: 340. 1898.

Shrubs, the stems erect or decumbent-ascending, dichotomous but sparsely branched, to 2–4 m long, to 7–8 cm diam., stems pale grey to whitish, rather smooth, leaf scars distant; prop-roots few, basal, rather short. Leaves well-spaced, the insertion often partly visible, linear, gradually attenuate, up to 220–280 × 3.5–4.2 cm; apical twin pleats denticulate along distal one-fifth of blade, the prickles 0.8 mm long, erect-antrorse, mostly 1–3 mm apart. Margins prickly along their entire length, the prickles all antrorse, sharp, straw to buff-coloured, very sharp; at base the prickles 4 mm long, straight or slightly bent,

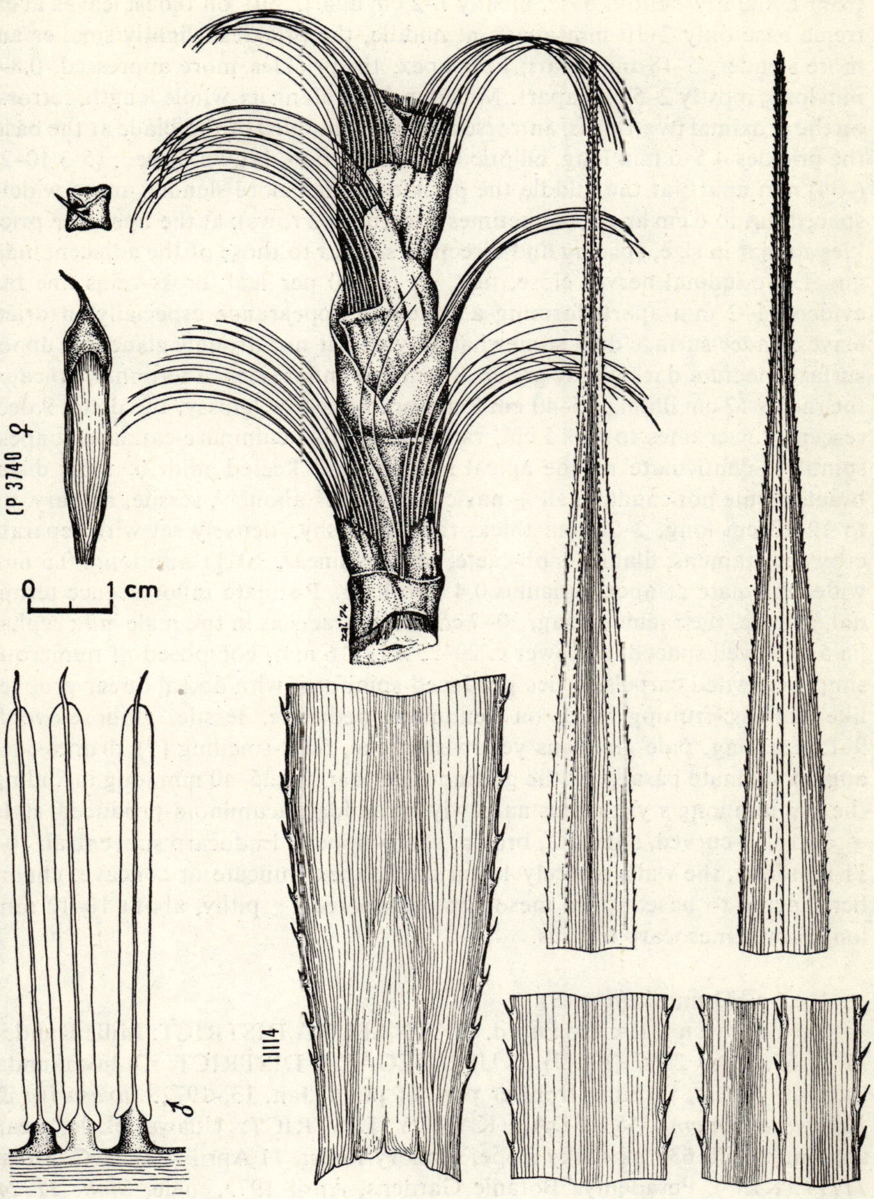

Fig. 8. *Pandanus thwaitesii*.

from a slightly deltoid base, mostly 1–2 cm apart, but on oldest leaves at extreme base only 2–10 mm apart; at middle, the prickles slightly smaller and more slender, 5–18 mm apart; near apex, the prickles more appressed, 0.8–1 mm long, mostly 2–5 mm apart. Midrib prickly along its whole length, retrorse on the proximal two-thirds, antrorse on the distal third of the blade at the base, the prickles 4.5–6 mm long, elliptic in cross-section, brown tipped, (5–) 10–25 (–34) mm apart; at the middle the prickles rather more slender, more widely spaced, up to 6 cm apart, (sometimes present in 2 rows); at the apex, the prickles similar in size, spacing and direction similar to those of the adjacent margin. Longitudinal nerves close, fine, up to 100 per leaf; cross-veins fine but evident, 1–2 mm apart, forming a tessellate appearance especially in dried leaves; under-surface dull, somewhat greyish but not strongly glaucous; upper surface medium dark glossy green. Staminate inflorescence terminal, spicate, the rachis 12 cm diam., 35–40 cm long; bracts yellow, glossy, the distal 9 decrescent, lower ones to 40 × 5 cm, rather abruptly acuminate-caudate at apex, spinulose-denticulate on the apical margins and keeled midrib; most distal bracts, acute not caudate, all ± navicular; spikes about 9, sessile, axillary, up to 12–13 cm long, 2–2.5 cm thick, rachilla fleshy, densely set with separate crowded stamens; filaments obsolete; anthers linear, 10–11 mm long, 0.5 mm wide, apiculate at apex, apiculus 0.4 mm long. Pistillate inflorescence terminal, spicate, the rachis zigzag, 20–? cm long, bracts as in the male infl.; cephalia 5 to 7, well spaced, in flower c. 20–25 × 12–16 mm, composed of numerous simple crowded carpels, styles produced spiniform with dorsal linear groove-like stigmas; fruiting heads on pendulous peduncle, sessile, globose-ovoid, 9–12 cm long, pale glaucous yellowish green, fetid-smelling (?); drupes 5–6-angled, connate basally by the glutinous pericarpia; 35–40 mm long including the 5–7 mm long style; pileus narrowly pyramidal, acuminoid-produced; style ± antrorse-curved, subulate, brown, sharp-tipped. Endocarp subcentral, 10–11 mm long, the walls scarcely 1 mm thick, apex truncate or concave, chamber tapered to base; apical mesocarp chambered, ± pithy, about 10–12 mm long; lower mesocarp fibrous.

Vern. Dumu-keyiya (S).

Specimens Examined. RATNAPURA DISTRICT: Mille Kanda, 11/12/69, *Read 2360* (KLU). KURUNEGALA DISTRICT: Doluwakanda summit, 580 m, forming frontier pop. of pond, Jan. 13, 1972, *Jayasuriya & Balasubramaniam 558* (PDA). KANDY DISTRICT: Udawattekele, near summit of hill, 630 m, rocky slope, semidry forest, 11 April 1973, *B.C. Stone 11114* (KLU); Peradeniya Botanic Gardens, April 1973, male, *Stone 11324* (KLU), *Stone 11315* (KLU). COLOMBO DISTRICT: Bandiyamulle near Gampaha, 21 April 1973, *Stone 11318* (KLU). On same sheet: Heneratgoda, March 1882, female infl. & lf.; ibid., male spikes and bracts; Mirigama, May 1896, (drupes) *Wright* (PDA). LOCALITY UNKNOWN: 3 sheets: Central

Prov., Dec. 1861, one male, one of leaves only, one of pistillate fl. heads, a leaf, and a packet with some matured drupes, this last the lectotype, *s. coll. C.P. 3740* (PDA).

Species in Cultivation only

6. Pandanus amaryllifolius Roxb., Hort. Beng. 71. 1814; Roxb., Fl. Ind. 3: 743. 1832.

A shrub cultivated for the musky-scented leaves commonly used in cooking; undoubtedly introduced from Indonesia centuries ago. Juvenile leaves may be used as well as the adult leaves. The latter are about 115 cm long, 7–8 cm wide, apex rather abruptly rounded-acute, not at all caudate, the lateral pleats obsolete; midrib at base rarely with a few short distant retrorse prickles, but usually unarmed; margins unarmed except at extreme apex where, as on the ultimate portion of the midrib, there may be a few minute prickles; sheathing base green.

Vern. Pandan rampeh (An Indonesian name long used in Ceylon).

Specimens Examined. KANDY DISTRICT: Walegalle, near Kandy, cult. by a house, 19 April 1973, *Stone 11305* (KLU).

Note. This species does not flower; consequently it cannot be assigned to a section of the genus. Roxburgh indicates the source as Amboina.

7. Pandanus cephalotus B.C. Stone, Bull. Maurit. Inst. 7: 13–16. 1972.

Cultivated in the Peradeniya Botanic Gardens (S-810, S-810b).

Specimens Examined. KANDY DISTRICT: Peradeniya Botanic Gardens, 5 April 1973, *Stone 11103* and *11104* (KLU, PDA).

8. Pandanus cfr. **krauelianus** Schum.

In the Peradeniya Botanic Gardens, 2 trees.

9. Pandanus polycephalus Lam., Enc. 1: 372. 1783.

In the Peradeniya Gardens; at least two trees seen flourishing in 1973.

Specimens Examined. KANDY DISTRICT: Peradeniya Bot. Gardens, 5 Sept. 1922, *s. coll. 355* (PDA).

10. Pandanus leram Jones ex Fontana, Asiat. Res. 3: 163. 1792.

Cultivated at Peradeniya and at Heneratgoda and nearby.

Specimens Examined. KANDY DISTRICT: Peradeniya Botanic Gardens, 5 April 1973, *Stone 11101 & 11102* (KLU, PDA). COLOMBO DISTRICT: Gampaha, near Heneratgoda, 21 April 1973, *Stone 11319* (KLU).

11. Pandanus tectorius Parkinson ex "Z" (ed.) in Der Naturforscher, 4: 220–258. 1774; St. John, Biol. J. Linn. Soc. 4 (4): 305–310. 1972.

var. **laevis** Warb. in Pflanzenr. 3 (IV. 9): 48. 1900.

Pandanus laevis Kunth, Enum. Pl. 3: 100. 1841.
Pandanus moschatus Miq., Fl. Ind. Bat. 3: 165. 1855.
Pandanus spurius Miq., CV 'Putat' in St. John, Pac. Sci. 19 (2): 232, fig. 223. 1965, with further synonyms.

Description. see St. John 1965, cited above.

This cultivar, of which only males have been collected in Ceylon, is to be found occasionally in cultivation in Ceylon.

Specimens Examined. COLOMBO DISTRICT: Queen's House, Colombo, Feb. 1883, male infl., *s. coll. s.n.* (PDA). KANDY DISTRICT: Peradeniya Botanic Gardens, male infl., lvs., *s. coll. s.n.* (PDA).

var. **sanderi** (Hort.)

Pandanus sanderi Hort. in Gard. Chron. 3 Ser. 23, 243 fig. 93. 1898; Warb., in Pflanzenr. 3 (IV. 9): 87. 1900.

A cultivar planted for its variegated striped leaves, occasional as in house-yards and around the lake in Kandy.

This is probably not distinct from *P. variegatus* Miq. (Anal. Bot. 28. 1851), from *P. baptistii* Hort. (Hort. Belg. 19: 166 fig. 35. 1893), and from *P. veitchii* Hort. in Dallière, Pl. Orn. in Gard. Chron. 2: 349. 1868.

Of course the wild Ceylon species are in cultivation in various places, including the Peradeniya Gardens, as well. The above listing may not be exhaustive but includes all Pandans seen by the writer in two trips to Ceylon in 1968 and 1973.

PEDALIACEAE

(by William L. Theobald and Donald A. Grupe*)

R. Br., Prod. 519. 1810.

Annual or perennial herbs, rarely shrubs, covered with mucilage-glands (at least on young parts). Leaves opposite or the upper alternate, usually simple, sometimes palmately compound; stipules absent. Inflorescence axillary, usually a single flower, sometimes a simple dichasium; conspicuous and characteristic nectarial glands (metamorphosed flowers) at the base of the pedicels. Flowers hermaphroditic, usually irregular. Calyx 5-merous, deeply divided or lobed. Corolla 5-merous, gamopetalous; tube campanulate, funnelform, or tubular, often ventricose and oblique; limb 5-lobed, 2-lipped to nearly regular; lobes imbricate. Stamens 4, very rarely 2, didynamous; anthers free or connate in pairs, locules parallel or divergent; connective usually gland tipped; staminodium small or absent. Disk present, annular or asymmetrical. Ovary superior, bilocular with axile placentation, often appearing 4-locular due to incomplete false septa, sometimes the loculi again divided by spurious septa; ovules 1-many; style slender; stigma usually 2-lipped. Fruit variable, an indehiscent nut or a dehiscent capsule, often with spines or hooks. Seeds 1-many per locule.

An Old World family of about 12 genera and 50 species, especially abundant in the tropics and subtropics of Africa, including South Africa and Madagascar. It also extends eastwards into India, Southeast Asia and Indonesia. Some have become naturalized in the New World.

Most are herbaceous shore or desert plants. Of the four species found in Ceylon only two are apparently native, *Pedalium murex* and *Sesamum prostratum*. Both of these are shore plants with *Pedalium* also being found far inland along the edge of tanks (reservoirs). *Sesamum indicum* has probably been introduced into the Dry Zone where it is now widely cultivated and has become naturalized. Trimen also believed it not to be indigenous to the island. It is often confused with *S. radiatum*, another introduced species known from more moist areas around Colombo.

The family is probably most closely allied to the New World family Martyniaceae of which there is only one introduced species in Ceylon, *Martynia*

*Pacific Tropical Botanical Garden, Hawaii.

annua. Members of the Pedaliaceae can be distinguished from this family by their axillary flowers, axile placentation, nectarial glands, and characteristic fruit shapes.

KEY TO THE GENERA

1 Fruit indehiscent, subpyramidal with 4 radiating spines; corolla subequally 5-lobed; stems round, subsucculent .... **1. Pedalium**
1 Fruit a dehiscent capsule, ovate-oblong to oblong, unarmed; corolla distinctly 2-lipped; stems obtusely-quadrangular, not succulent .... **2. Sesamum**

1. PEDALIUM

L., Syst. Nat. ed. 10, 1123. 1759.

Prostrate, usually branching, annual herbs; stems subsucculent. Leaves opposite or alternate, simple, toothed or lobed, petiolate. Flowers axillary, solitary, with nectarial glands at the base of pedicels. Calyx 5-merous, deeply divided. Corolla tubular-funnelform, gradually tapering from the base, slender; limb spreading, subequally 5-lobed; lobes rounded. Stamens 4, didynamous; included; anthers free; locules divergent. Disk small, annular. Ovary bilocular; locules undivided; ovules 2 per locule; style slender; stigma 2-lipped. Fruit indehiscent, hard, subpyramidal, 4-angled, rounded to acute at apex with a spreading spine at the base of each of the 4 angles, abruptly contracted below spines. Seeds 1–2 in each locule, pendulous, oblong, black.

A monotypic, widespread genus known from tropical East and West Africa, Socotra, and Madagascar east through India and Ceylon to Indonesia.

Pedalium murex L., Syst. Nat. ed. 10, 1123. 1759; Trimen, Handb. Fl. Ceylon 3: 285. 1895.—**Fig. 1E.**

Stems thick, subsucculent, branching throughout, 15–60 cm long, slightly rough with scaly glands or hairs. Leaves somewhat fleshy, pale green, glaucous, broadly obovate to elliptical in outline, 1.5–4.5 cm long, 1.2–3.5 cm broad, base acute, margin irregularly toothed or lobed, apex broadly rounded or truncated, glabrous above, with minute scaly-glands below; petioles 1–3 cm long. Pedicels slender, curved. Calyx deeply divided, segments lanceolate-oblong, acute, c. 2 mm long, c. 0.5 mm broad. Corolla pale yellow, opening in morning, glabrous, 2–2.5 cm long; tube c. 2 cm long, glandular-pubescent within; lobes rounded, c. 5 mm long, 5 mm broad. Filaments glandular hairy at base. Fruit glabrous or sparsely glandular, rugose or tuberculate on the face; 10–15 mm long; body 5–8 mm broad, excluding the spines; spines c. 3 mm long.

Fig. 1. *Sesamum indicum* L. A, Habit, $\times\frac{2}{3}$. B, Lower leaf, $\times\frac{2}{3}$. *Sesamum radiatum* Schumach. & Thonn. C, mature fruit, $\times\frac{2}{3}$. *Sesamum prostratum* Retz. D. habit, $\times 5/6$. *Pedalium murex* L. E, mature fruit, $\times 1$.

Distr. Along sandy beaches of the dry zones on both the east and west coasts of the island. It has been collected inland at the edge of a tank (reservoir) at Anuradhapura and along roadside near Jaffna. Also known from Africa east through India to Indonesia.

Ecol. The prostrate plants spread out and cover an area of up to 1 metre in diameter. The flowers were observed to open in the morning and close late in the afternoon. No pollinators were observed. They are often found at the edge of beaches or in open grassy areas a short distance from the shore. They are reported to be saline soil indicators in sand or on limestone. Flowering February–August, and probably throughout the year.

Vern. Et-nerenchi (S); Peiru-ar anai-nerinchi (T).

Illustr. Pedaliaceae. Fl. Trop. East Africa p. 5. 1953.

Specimens Examined. ANURADHAPURA DISTRICT: Nuwera Wewa tank, Anuradhapura, Simpson *8054* (BM). BATTICALOA DISTRICT: Batticaloa, *s. coll. C.P. 1779* p.p. (PDA). JAFFNA DISTRICT: along roadside, Hwy. A-9 just north of Elephant Pass, near Hwy. marker 171, *Theobald & Grupe 2330* (PDA, US); Jaffna, *Trimen s.n.* (PDA). PUTTALAM DISTRICT: Chilaw, *Trimen s.n.* (PDA). TRINCOMALEE DISTRICT: along edge of sandy beach below fort at Trincomalee, *Theobald & Grupe 2323* (A, K, L, PDA, UC, US); Trincomalee, *Simpson 8482* (BM); along sandy beach approx. 5 mi. north of Trincomalee on rd. to Nilaveli, *Theobald & Grupe 2327* (A, E, NY, PDA, UC, US).

2. SESAMUM

L., Sp. Pl. 634. 1753; L., Gen. Pl. ed. 5. 282. 1754.

Erect or prostrate herbs with very short, white mucilagenous hairs, and often longer, articulate hairs. Leaves opposite below; alternate above, simple, entire, toothed or lobed, sometimes divided, or compound, often varying on the same plant; usually petiolate. Flowers axillary, solitary, or few and fascicled; shortly pedicellate. Calyx 5-merous, small, persistent, deeply divided or parted. Corolla 5-merous, tube campanulate-ventricose, oblique, distinctly 2-lipped, lower lobe the longest. Stamens 4, didynamous, included; filaments often with a ring of hairs at point of insertion; anthers free, locules parallel; staminoidium minute or absent. Disk annular. Ovary bilocular, appearing 4-locular by presence of false septum; ovules numerous; style slender; stigma 2-lipped. Fruit a capsule, ovate-oblong to oblong, beaked or obtuse-rounded, 4-angled, longitudinally grooved, unarmed, loculicidally 2-valved, appearing 4-loculed. Seeds numerous, obovate, compressed, smooth or rugose.

A genus of about 30 species, particularly abundant in tropical and South Africa, and eastward to Indonesia.

KEY TO THE SPECIES

1 Erect branching herbs; leaves variable, large and long petioled below, smaller and short petioled above; capsule longer than 15 mm
2 Capsule with a conspicuous, subulate beak; mature seed finely reticulate or smooth; lower leaves often deeply divided or compound; corolla usually white, rarely pinkish-violet..1. **S. indicum**
2 Capsule rounded or obtuse, not beaked; seeds with transverse ridges radiating from center; lower leaves never deeply divided or compound; corolla pinkish-violet...2. **S. radiatum**
1 Prostrate branching herbs; leaves usually similar throughout; capsule less than 12 mm long..3. **S. prostratum**

1. Sesamum indicum L., Sp. Pl. 634. 1753; Trimen, Handb. Fl. Ceylon 3: 285. 1895.—**Fig. 1 A, B.**

Sesamum orientale L., Sp. Pl. 634. 1753.

Strong smelling, usually branching, erect annual herbs up to 1 m high; stems and branches obtusely quadrangular in upper part, furrowed. Leaves very variable; lower simple or palmately compound, 6–15 cm long, 3–10 cm broad, in less robust specimens ovate, base rounded, obtuse, or acute; margin sub-entire to irregularly dentate-serrate, apex acute or obtuse; in larger specimens 3-lobed, parted, or palmately compound; long petioled; upper leaves shorter petioled, ovate-oblong to oblong-lanceolate or linear, 3–6 cm long, 0.5–3 cm broad, base rounded to acute, margin usually entire or only shallowly toothed, apex acute; petioles of lower leaves 3–8 cm long, on upper leaves reduced to only a few mm. Pedicels 1–3 mm long. Calyx deeply divided, segments lanceolate-oblong, 3–5 mm long, c. 1 mm broad. Corolla usually white, sometimes pale pinkish-violet, often spotted yellow within and with a yellow blotch on lower lip, 2.5 to c. 3.5 cm long. Filaments glabrous, ring of hairs evident on tube at point of insertion, one pair c. 8 mm long, other c. 6 mm long; anthers glabrous or sparsely pubescent along margins, c. 2 mm long. Ovary densely pubescent, style glabrous; lips of stigma lanceolate. Capsule erect, oblong, 15–28 mm long (beak included), 5–7 mm broad, densely villous-pubescent, apex rounded and abruptly contracted into a conspicuous subulate beak. Seeds c. 2 mm long, pale yellow, brown, or black, finely reticulate or almost smooth when ripe.

Distr. Widely cultivated and naturalized in the dry zone, especially along roadsides and in abandoned fields and waste areas. It is a plant of ancient cultivation and its area of origin is not known with certainty, although probably Africa or possibly India.

Ecol. A plant confined apparently to the dry zone with no known populations from the areas around Colombo where *S. radiatum* is known. Flowering May–August and possibly throughout the year.

Vern. Tel-tala (S); Ella (T); Gingelly (E).

Illustr. Bot. Mag. t. 1688. 1814.

Note. It has often been confused with *S. radiatum* but can readily be distinguished on the basis of its beaked capsule, finely reticulate or smooth seeds, and usually deeply divided or compound lower leaves.

Specimens Examined. JAFFNA DISTRICT: along roadside near Hwy. marker 171, just north of Elephant Pass on Hwy. A-9, *Theobald & Grupe* 2331 (A, E, K, L, LE, NY, PDA, UC, US); Jaffna, *Gardner s.n. C.P. 2023* p.p. (PDA). MATALE DISTRICT: Pelwehera, Dambulla, *Alston 2402* (K, PDA).

2. Sesamum radiatum Schumacher in Schumacher & Thonn., Beskr. Guin. Pl. 282. 1827; Alston in Trimen, Handb. Fl. Ceylon 6: 221. 1931.—**Fig. 1C.**

Sesamum occidentale Regel & Heer, Ind. Sem. Hort. Turic. 1842, ex DC., Prod. 9: 250. 1845; Trimen, Handb. Fl. Ceylon 3: 286. 1895.

Strong smelling, usually branching, erect annual herbs, 20–120 cm high; stems and branches obtusely quadrangular in upper part, furrowed. Leaves variable, the lower ovate-elliptical, never deeply divided or compound, long petioled, the upper ovate to narrowly oblong or elliptical, 3–13 cm long, 1–8 cm broad, base rounded to cuneate, margin entire to coarsely dentate, apex acute or obtuse; petioles of lower leaves 2–6 cm long, on upper leaves usually reduced to only a few mm. Pedicels 4–7 mm long. Calyx very deeply divided, segments lanceolate, 4–7 mm long, c. 2 mm broad. Corolla pinkish-violet to whitish-violet without, similar and spotted purple within, densely pubescent without, glabrous or sparsely pubescent within, 3–4 cm long. Ovary densely tomentose; style glabrous; lips of stigma lanceolate. Capsule erect, oblong, 22–30 cm long, 7–9 mm broad, densely pubescent, apex rounded or very obtuse, not conspicuously beaked. Seeds c. 2 mm long, brown or black, with transverse ridges emanating from a central field on the broad surfaces, especially the flatter one.

Distr. Known from the vicinity of Colombo north to Negombo. Trimen reported it to be "a common weed in waste places about Colombo and elsewhere," and noted that it "no doubt spread from the Botanic Gardens." It is native to tropical West Africa and has become widespread in the Old World tropics, including India, Southeast Asia and Indonesia.

Ecol. It is primarily a weedy species of the more moist areas around Colombo. It has not been reported from the dry zone where *S. indicum* is widely cultivated. Flowering May–August, and probably throughout the year.

Note. It has often been confused with *S. indicum*. However, it can readily be separated on the basis of unbeaked capsule, ridged seed surface, and absence of deeply parted or compound lower leaves.

Specimens Examined. COLOMBO DISTRICT: along railroad tracks 0.3 miles north of Seeduwa Rd. crossing below Negombo Airport, *Theobald & Grupe 2375* (A, E, K, L, LE, NY, PDA, UC, US); Katunayaka, near Negombo, *Alston 2387* (PDA); Colombo, *s. coll. C.P. 3852* p.p. (PDA). KANDY DISTRICT: Royal Botanic Gardens, Herb. Ground, *Trimen s.n.* (PDA); Herb. Ground, *s. coll. s.n.* (PDA); near head office, Peradeniya Gardens, *T. Appuhany s.n.* (PDA).

3. Sesamum prostratum Retz., Obs. 4: 28. 1779; Alston in Trimen, Handb. Fl. Ceylon 6: 221. 1931.—**Fig. 1D.**

Strong smelling, abundantly branched, prostrate herbs; stems and branches obtusely quadrangular in upper part, furrowed, villous-tomentose. Leaves small, broadly ovate to obovate, rarely shallowly 3-lobed, 6–17 mm long, 6–16 mm broad; base rounded, margin crenate-dentate, apex rounded, sparsely pubescent on upper surface, densely covered with villous hairs on veins below, numerous white mucilagenous hairs between veins below; petioles 1–2 mm long. Pedicels 1–2 mm long. Calyx lobes nearly free to base, lanceolate, c. 5 mm long, c. 1 mm broad. Corolla reddish to pinkish-violet without, similar with purple flecks and spots within, densely pubescent without, 20–30 mm long. Capsule ovate-oblong, 6–12 mm long, 4–9 mm broad, densely villous and with white mucilagenous hairs, apex rounded and truncated, beaked. Seeds c. 2 mm long, black, reticulate-pitted.

Distr. Known only from the south-east coast at Panama, on the sea shore. Also found along coasts of South India.

Ecol. Flowering May–August, and possibly throughout the year.

Illustr. Wight, Ic. Pl. Ind. Or. 4; t. 1346. 1848.

Specimens Examined. AMPARAI DISTRICT: Panama, on the sea shore, common, *F. Lewis s.n.* (PDA).

ROSACEAE*

(by Deva Duttun Tirvengadum†)

Type genus: *Rosa* L.

Trees, shrubs or herbs, sometimes straggling or climbing. Leaves alternate, simple or compound, usually stipulate. Inflorescence various, from single flowers to corymbose, racemose or paniculate. Flowers mostly bisexual (rarely polygamous), regular. Calyx tube free, or adnate to ovary; calyx lobes 4 to 6 (– to 10), mostly 5, sometimes with an epicalyx of bracteoles. Disk lining the calyx tube. Petals free, borne on the calyx tube, rarely absent. Stamens 5 (sometimes 4), or numerous; anthers didymous, 2-locular, dehiscing lengthwise. Carpels 1 to many, free or connate and adnate to the calyx tube (ovary superior, semi-inferior or inferior), borne on a convex or concave receptacle; ovules often 2 (rarely 1 or several) in each carpel or locule, anatropous; styles free or connate, terminal, lateral or basal. Fruit various, superior to inferior, drupaceous (Prunoideae), pomaceous (Maloideae), follicular (Spiraeoideae), berry-like (*Photinia*) or of an indefinite number of achenes or drupelets (Rosoideae). Seeds erect or pendulous, mostly without endosperm.

A family of 124 genera and about 3375 species. Widely distributed in all parts of the world, mostly in the North temperate regions. In Ceylon, it is represented by 12 genera with 30 species. Four species are endemic and two species doubtfully endemic. Most Ceylon Rosaceae are montane plants, with the exception of *Prunus walkerii* which extends to the lowland rain forests (Gilimale) and *Rubus indicus* which occurs at an elevation of 300 m and occasionally at lower altitudes in the Southern Province. They are found in varied habitats: heavy jungles, primary or secondary montane forests, open patanas, marshy places, in the shade or in open thickets, often close to streams or rivers. Herbaceous species, such as *Potentilla polyphylla* and *Potentilla sundaica* are restricted to certain localities, which are fast disappearing due to agricultural activities and colonisation by man. *Sanguisorba indicum*, a very rare species collected only twice in Ceylon, seems to have disappeared from the flora. Many genera are of economic importance (*Eriobotrya*, *Pyrus*, *Malus*, *Rubus*, *Fragaria*, *Prunus*). In Ceylon, a few genera produce edible fruits of

*This revision was carried out at the Laboratoire de Phanérogamie, Paris.

†Present address: Mauritius Institute, Port-Louis, Mauritius.

a rather poor quality: pears, peaches, apples, strawberries and raspberries. *Prunus domestica* (common plum), *Prunus armeniaca* (common apricot) and *Prunus amygdalus* (almond) have not been met with, since recorded by Moon (Cat. 37. 1824). A few ornamental species recorded in the Peradeniya Herbarium by some well preserved specimens have not been seen nor collected recently: *Raphiolepis indica*, *Prinsepia utilis*, *Quillaja saponaria*, *Cotoneaster frigida* and *Chrysobalanus icaco*. *Rubus multiflorus* Thunberg (Fl. Ceil. 7. 1825), mentioned by Alston as unknown to him, has not been recorded anywhere in Ceylon, as far as I know. According to Trimen, there is no wild species of *Rosa* (rose) in Ceylon. A few cultivated species introduced from China, namely: *R. indica* (= *R. odorata* (Andr.) Sweet), *R. semperflorens* (= *R. chinensis* Jacq.) and *R. banksiae* Ait. were, however, reported by Moon (l.c.).

Rosaceae when first established by A.L. de Jussieu (1789) was a small family with about 40 genera. Focke (1888—in Pflanzenfam. 3 (3) : 1) recognised 89 genera which he divided into 6 subfamilies: Spiraeoideae, Pomoideae, Rosoideae, Neuradeoideae, Prunoideae and Chrysobalanoideae and 13 tribes. These subfamilies have been generally maintained (Rehder, Man. Trees & Shrubs, ed. 2: 322. 1940; Schulze-Menz, Syll. Pflanzenfam. (2): 209. 1964). Hutchinson, (Fam. Fl. Pl. 1: 149. 1959; Gen. Fl. Pl. 1: 174. 1964), however, established another classification, thereby distinguishing 20 tribes, without grouping them into subfamilies.

Here the classification of Focke is followed. The genera represented in the Ceylon flora may be arranged as follows:

A. KEY TO THE SUBFAMILIES

1 Fruit dehiscent (1 to 5 follicles) .... 1. SPIRAEOIDEAE
1 Fruit indehiscent, dry or fleshy
 2 Ovary inferior or semi-inferior. The 2 to 5 carpels more or less connate and adnate to the cup-shaped calyx tube. Fruit a pome or berry-like .... 2. MALOIDEAE
 2 Ovary superior. Carpels 1 to many, usually free. Fruit a drupe or aggregate of drupelets or achenes
  3 Carpels usually many. Calyx persistent. Leaves often compound. Ovules 1, rarely 2 .... 3. ROSOIDEAE
  3 Carpels usually 1. Calyx usually deciduous. Leaves always simple. Ovules 2 .... 4. PRUNOIDEAE

B. KEY TO THE GENERA ACCORDING TO SUBFAMILIES
(See also keys C & D)

1 SPIRAEOIDEAE
 Stipules absent. Carpels 5 .... **1. Spiraea**
2 MALOIDEAE
 1 Flowers in compound corymbs or panicles
  2 Flowers in corymbs. Seeds rather small .... **2. Photinia**

2 Flowers in panicles. Seeds large .... **3. Eriobotrya**
1 Flowers in umbel-like racemes
3 Styles free (cultivated and naturalized) .... **4. Pyrus**
3 Styles connate at base (cultivated) .... **Malus** Mill.

3 ROSOIDEAE

1 Carpels borne on a convex receptacle
2 Carpels 2-ovulate becoming drupelets. Style terminal. Epicalyx absent .... **5. Rubus**
2 Carpels 1-ovulate becoming achenes. Style lateral or terminal. Epicalyx present
3 Fruit receptacle fleshy, very juicy .... **6. Fragaria**
3 Fruit receptacle spongy and slightly juicy or dry
4 Receptacle spongy, slightly juicy, deciduous. Epicalyx larger than calyx .... **7. Duchesnea**
4 Receptacle dry, persistent. Epicalyx usually smaller than calyx .... **8. Potentilla**
1 Carpels enclosed in concave receptacle (calyx tube)
5 Fruit of one or 2 achenes. Calyx tube hardened in fruit. Style terminal
6 Sepals 5. Calyx tube topped with a ring of hooked spines. Petals present .... **9. Agrimonia**
6 Sepals 4. Calyx tube not spiny, but with 4 narrow wings. Petals absent .... **10. Sanguisorba**
5 Fruit of 1 achene. Calyx tube membranous, not hardened in fruit. Style basal .... **11. Alchemilla**

4 PRUNOIDEAE

Leaves glandular at base or on petiole .... **12. Prunus**

C. KEY TO THE GENERA
(for plants with flowers)

1 Herbs
2 Calyx subtended by an epicalyx
3 Sepals with a 4 (–5) bracteate epicalyx (seemingly 8 (–10). Apetalous. Leaves simple, palmatilobed .... **11. Alchemilla**
3 Sepals with a 5-bracteate epicalyx. Petals present. Leaves pinnately or palmately compound
4 Receptacle after anthesis not enlarged, persistent. Flowers yellow, in terminal, lax, leafy corymbs or panicles. Leaves imparipinnate or palmately compound .... **8. Potentilla**
4 Receptacle after anthesis slightly or strongly enlarged, deciduous. Flowers white (cymes) or yellow (in latter case solitary). Leaflets palmately compound
5 Inflorescence few-flowered. Epicalyx smaller than the calyx. Petals white. Receptacle strongly enlarged after anthesis. Leaflets 3 .... **6. Fragaria**
5 Flowers solitary. Epicalyx larger than the calyx. Petals yellow. Receptacle slightly enlarged after anthesis. Leaflets 3 (–5) .... **7. Duchesnea**
2 Calyx without an epicalyx
6 Flowers in spike-like racemes, bisexual. Calyx tube spiny .... **9. Agrimonia**
6 Flowers capitate, bisexual or polygamous. Calyx tube not spiny (winged in fruit) .... **10. Sanguisorba**
1 Shrubs or trees
7 Shrubs or undershrubs, mostly sarmentose. Branches covered with prickles. Carpels many (more than 5), inserted on a convex receptacle .... **5. Rubus**
7 Erect shrubs or trees. Prickles usually absent. Carpels few (1 to 5) rarely more, included in the calyx tube

8 Stipules absent. Carpels 5. Ovules several .... **1. Spiraea**
8 Stipules present. Carpels 1 to 5, rarely more. Ovules 2
9 Glands present at base of leaf or on petiole. Carpels usually 1, rarely 2 or more.. .... **12. Prunus**
9 Glands absent. Carpels 2 (–5)
10 Flowers in umbel-like clusters. Ovary inferior
11 Styles free .... **4. Pyrus**
11 Styles connate at base .... **Malus** Mill.
10 Flowers in panicles or corymbs. Ovary inferior or half-inferior
12 Flowers in panicles. Ovary inferior .... **3. Eriobotrya**
12 Flowers in corymbs. Ovary half-inferior .... **2. Photinia**

D. KEY TO THE GENERA
(for plants with fruits)

1 Herbaceous plants
2 Fruit consisting of numerous achenes. Epicalyx present
3 Receptacle fleshy, or spongy, deciduous. Leaves 3 (–5) foliolate
4 Receptacle fleshy. Epicalyx smaller than calyx .... **6. Fragaria**
4 Receptacle spongy. Epicalyx larger than calyx .... **7. Duchesnea**
3 Receptacle dry, persistent. Leaves 3 to 9 foliolate
5 Calyx tube not winged in fruit. Leaves palmate, with 3 (–5) leaflets, or imparipinnate .... **8. Potentilla**
5 Calyx tube winged in fruit. Leaves pinnate, up to 9 foliolate....... **10. Sanguisorba**
2 Fruit consisting of 1 or 2 achenes. Epicalyx present or absent
6 Achenes solitary. Epicalyx present. Calyx tube not hardened and spiny in fruit.... .... **11. Alchemilla**
6 Achenes 1 or 2. Epicalyx absent. Calyx tube hardened and spiny in fruit.......... .... **9. Agrimonia**
1 Ligneous or subligneous plants
7 Fruit dehiscent when ripe, consisting of 5 follicles. Stipules absent .... **1. Spiraea**
7 Fruit indehiscent, simple or aggregate. Stipules present
8 Fruit drupe-like, superior
9 Drupelets many, borne on a convex receptacle. Straggling shrubs with prickles.... .... **5. Rubus**
9 A true drupe, enclosed in the calyx tube. Mostly trees, without prickles. . **12. Prunus**
8 Fruit a pome or berry-like, inferior or half-inferior
10 Fruits in panicles or corymbs
11 Fruits in panicles, inferior. Seeds large (sometimes as large as the fruit)......... .... **3. Eriobotrya**
11 Fruits in corymbs, half-inferior. Seeds much smaller than the fruit... **2. Photinia**
10 Fruits in umbel-like racemes
12 Styles free. Stone cells present .... **4. Pyrus**
12 Styles connate at base. Stone cells absent .... **Malus** Mill.

1. SPIRAEA

L., Sp. Pl. 489. 1753; L., Gen. Pl. ed. 5. 216. 1754. Type species: *Spiraea salicifolia* L.

Evergreen or more or less deciduous shrubs. Leaves simple, serrate, exstipulate. Flowers in umbel-like racemes or corymbs, mostly bisexual, white or pinkish. Calyx tube campanulate-turbinate; calyx lobes 5, persistent. Petals 5 or numerous. Stamens 15 to 60. Carpels 5, free, enclosed in the calyx tube; ovary superior, 1-locular; ovules several; styles (sub) terminal. Fruits follicles, dehiscing along the inner suture. Seeds several; endosperm thin or absent.

A genus of about 120 species, mostly of the Northern Hemisphere, from East Europe to North Asia. One species is cultivated and almost naturalized in Ceylon at high elevation.

Most of the species, including numerous hybrids, are popular garden shrubs and much cultivated on account of their decorative flowers.

Spiraea cantoniensis Lour., Fl. Cochinch. 1: 322. 1790; Merr., Trans. Amer. Philos. Soc. n.s. 24 (2): 177. 1935; Rehder, Man. Trees & Shrubs, ed. 2. 333. 1940; Bailey, Man. Cult. Pl., Rev. ed. 497–498. 1949. Type: China, Canton (P).

Spiraea corymbosa Roxb., Fl. Ind. ed. 2, 2: 512. 1832, non Rafinesque 1814. See Rehder for complete synonymy.

Shrub, about 2 m tall. Young branches glabrous. Leaves oblong-lanceolate to rhombic-lanceolate, 3 to 5 cm long, up to 1.6 cm broad, cuneate at base, acute or short-acute at tip, irregularly incised-serrate, glabrous, dark green above, pale bluish-green beneath, persistent or more or less deciduous. Petiole 5 to 8 mm long. Inflorescence in few or dense umbellate corymbs, glabrous. Pedicel about 1 cm long. Calyx lobes triangular, pubescent at apex inside. Petals 5 or numerous, suborbicular to oval, longer than stamens, glabrous, white. Stamens 20 to 24 in normal flowers. Ovary glabrous. Follicles 5, with terminal spreading styles.

D i s t r. In China and Japan. Cultivated in N.W. India, Thailand, in temperate regions of America and Europe and in the Mascarene Islands. In Ceylon, two forms of *Spiraea cantoniensis* are cultivated as ornamental shrubs at high elevation, the simple and the double-flowered states. The simple-flowered form is often found growing spontaneously among hedges, along borders of vegetable gardens in Nuwara Eliya.

f. **cantoniensis**

Leaves rhombic-lanceolate, persistent. Flowers numerous. Petals 5.

I l l u s t r. Lindley, Bot. Reg. n.s. 7 (30): Pl. 10. 1841 (sub. *S. Reevesiana*).
S p e c i m e n s E x a m i n e d. NUWARA ELIYA DISTRICT: Nuwar

Eliya, growing as hedge in compound of Grosvenor Hotel, *Tirvengadum & Cramer 61* (G, K, L, MAU, P, PDA, US).

f. **lanceata** (Zab.) Zabel in Beissner et al., Handb. Laubh.—Ben. 152. 1903.
Spiraea cantoniensis var. *lanceata* Zabel, Strauch. Spiraēn 41. 1893.

Leaves oblong-lanceolate, more or less deciduous. Flowers in groups of 4 to 10. Petals numerous.

Illustr. Verdier in Hortic. France 121, t. 11. 1855 (sub. *S. reevesiana*, fl. pleno).

Specimens Examined. NUWARA ELIYA DISTRICT: Hakgala, cultivated in the Botanical Garden, *Tirvengadum & Jayasuriya 130* (G, K, L, MAU, P, PDA, US).

Note. I have not collected the following specimens recorded in the Peradeniya Herbarium: *Spiraea salicifolia* L. and the hybrid *Spiraea vanhouttei* Zab. They are probably cultivated as ornamental plants. A third specimen bearing the nomenclature *Neillia opulifolia* and recorded as *Spiraea opulifolia* L. in Moon (1824—Cat. 40: 321) is probably *Physocarpus opulifolia* (L.) Max.

2. PHOTINIA

Lindl., Trans. Linn. Soc. London 13: 103, t. 10. 1821. Type species: *Photinia serrulata* Lindl.

Evergreen or deciduous shrubs or trees. Leaves simple, entire or serrate, stipulate. Flowers in terminal corymbs, bisexual, often white. Calyx tube turbinate, adnate to ovary; calyx lobes 5, persistent. Petals 5. Stamens about 20. Carpels 2 to 3, connate; ovary half-inferior, 2 to 3-locular, with 2 ovules in each locule; styles 2 to 3, more or less connate at base. Fruit a berry-like pome, topped with the persistent, connivent calyx lobes; pericarp more or less fleshy; endocarp leathery. Seeds much smaller than the fruit, without endosperm.

A genus of about 60 species in Asia, North and Central America. About 50 species in Asia, Central and Eastern Himalaya: from Nepal to Sikkim to Bhutan, Assam (Khasi Hills), Shillong, South India (Nilgiris), South China (Yunnan), North Vietnam and North Laos. Also in the Malay Archipelago (Sumatra, Java, Timor), Martaban and Lesser Sunda Islands.

Photinia integrifolia Lindl., Trans. Linn. Soc. Lond. 13: 103. 1821; Hook. f., Fl. Br. Ind. 2: 381. 1878; Vidal, Adansonia, n.s. 5 (2): 226. 1965. Type: Nepal, *Wallich 669* (Holotype, K; Isotype, P).
Pyrus integerrima Wall. ex D. Don, Prod. Fl. Nepal: 237. 1825.

Eriobotrya integrifolia (Lindl.) Kurz, J. Asiat. Soc. Bengal 45: 304. 1877.
See Vidal for complete synonymy.

Evergreen shrub or tree. Young branches quite glabrous. Leaves coriaceous, oblanceolate or oval oblong, 8 to 12 (–to 15) cm long, 3 to 4 (–to 6) cm broad, cuneate or rounded at base, subacuminate at apex, entire, glabrous, shining above, paler beneath. Petiole 1 to 1.5 (–to 3) cm long. Stipules triangular or subulate, 2 to 4 mm long, deciduous. Inflorescence in terminal corymbose panicles, spreading, glabrous or more or less pubescent. Pedicel 1 to 3 mm long. Calyx lobes rounded or triangular, glabrous on both surfaces. Petals ovate, slightly villous at the base inside, white. Stamens 15 to 20. Ovary densely pilose at apex. Fruit globose, up to 5 mm in diameter, with calyx lobes inflexed at the top and crowned with persistent bifid style.

Distr. In Central and Eastern Himalaya (Nepal, Sikkim, Bhutan, Assam), South China (Yunnan) and North Vietnam.

var. **integrifolia**

Petiole 1 to 1.5 cm long. Leaves 8 to 12 cm long, 3 to 4 cm broad.

Distr. That of the species.

var. **Sublanceolata** Miq., Fl. Ind. Bat. 1 (1): 387. 1857. Type: Java, *Horsfield 432* (Holotype, K; Isotype, BM).

Photinia notoniana Wight & Arn., Prod. 1: 302. 1834; Thw., Enum. Pl. Zeyl. 100. 1859; Trimen, Handb. Fl. Ceylon 2: 142. 1894. Type: India, Nilgiris, *Wight 1014* (Holotype, K; Isotype, P).
Photinia notoniana var. *ceylanica* Hook. f., Fl. Br. Ind. 2: 381. 1878. Type: Ceylon, *Walker s.n.* (K).
Photinia notoniana var. *macrophylla* Hook. f., Fl. Br. Ind. 2: 381. 1878. Type: India, Khasia, *Hooker f. & Thomson s.n.* (K).
See Vidal for complete synonymy.

Petiole 2 to 3 cm long. Leaves 10 to 15 cm long, 3 to 6 cm broad.

Distr. In S. India (Pulney Hills & Nilgiris), N.E. India (Khasi Hills, Assam). In Ceylon, it is widely distributed in the upper montane zone to the highest elevation, up to 2500 m; it is occasionally found at somewhat lower elevation.

Ecol. Found in jungles, along mountain paths, summit of hills, and in secondary montane forests, often near streams. The new flush is coral red, old leaves burnt orange colour. The flowers are pinkish white and fragrant. Flowering from March to April. The little berry-like fruit has a harsh bitter flavour.

V e r n. Lunu-warala (S); Anreepawlaycody maram (T).

I l l u s t r. Wight, Ill. Pl. 86. 1831; Wight, Ic. Pl. Ind. Or. 3: Pl. 991. 1845; Select. Neilgherry Pl. 1: Pl. 64. 1846; Worthington, Ceylon Trees pl. 225. 1959.

N o t e. The only species of *Photinia* found in Ceylon has been invariably described as *P. notoniana* W. & A. and *P. notoniana* W. & A. var. *ceylanica* Miquel, by Trimen and Hooker f. respectively. The latter, while describing *P. integrifolia* Lindl. had great difficulty in distinguishing it from *P. notoniana* and expressed doubt as to their specific differences, the best character being the very short petiole and uniformly oblanceolate smaller leaf of the former species. Vidal (l.c.), however, pointed out the lack of distinct characters between the various species of *Photinia* with entire leaves, and subsequently put *P. notoniana* var. *ceylanica* under *P. integrifolia*.

The conspecificity of *P. integrifolia* and *P. notoniana* has also been expressed by Kalkman (Blumea 21(2): 421. 1973). In agreement with both Vidal and Kalkman I accept the reduction of the Ceylon species to varietal status and maintain the epithet chosen by the latter, namely, var. *sublanceolata* Miq.

S p e c i m e n s E x a m i n e d. KANDY DISTRICT: Hunnasgiriya, *Tirvengadum, Cramer & Balasubramaniam 73* (G, K, L, MAU, P, PDA, US); Corbet's Gap, 1300 m alt., *Tirvengadum, Cramer & Balasubramaniam 70* (G, K, L, MAU, P, PDA, US); 4000 ft alt., beyond Rangalla, *Worthington 5475* (BM, PDA); Madulkele, Knuckles jungle, *Worthington 1969* (BM, PDA); Norton Road, 1100 m alt., *Worthington 6557* (PDA); NUWARA ELIYA DISTRICT: Nuwara Eliya, 7000 ft alt., *Gardner 274* (BM, K); Sita Eliya, 6000 ft alt., *Worthington 6762* (PDA); Rambodda?, *s. coll. C.P. 136* (MH, P, PDA); Pidurutalagala, *Walker Arnott s.n.* (PDA); Hakgala 6000 ft alt., *Worthington 5734* (PDA); *5794* (BM, PDA); *Silva s.n.* (PDA); Ambawela, *s. coll. C.P. 136* (BM, K, MH, P, PDA); Totapolankande, 2600 m alt., *Tirvengadum & Cramer 110, 284* (G, K, L, MAU, P, PDA, US); Horton Plains, 7000 ft alt., *Worthington 1689* (BM, PDA); *s. coll. C.P. 136* or *135*? (K); *Vladik 710, 893* (PDA, US); *Van Beusekom 1503* (PDA, US); *Meijer, Dassanayake & Balasubramaniam 618* (K, PDA, US); *J.C.W. s.n.* (PDA). BADULLA DISTRICT: Top of Namunukula, *Walker Arnott s.n.* (PDA). LOCALITY UNKNOWN: Regio temp., 6000 ft alt., *Thomson s.n.* (P); *Thomson* in 1845 s.n. (K); *Wight s.n.* (K); *Mrs. G. Walker s.n.* (K); *Walker s.n.* (K).

3. ERIOBOTRYA

Lindl., Trans. Linn. Soc. London 13: 102. 1821.

Photinia Benth. & Hook. f., Gen. Pl. 1: 627. 1865, pro parte. Type species: *Eriobotrya japonica* (Thunb.) Lindley.

Evergreen trees or shrubs. Leaves simple, dentate, with strong, straight nerves ending in the teeth, stipulate. Flowers in broad, terminal panicles, bisexual, usually woolly pubescent, white or whitish. Calyx tube obconical, more or less adnate to ovary; calyx lobes 5, persistent. Petals 5. Stamens about 20. Carpels 2 to 5, connate; ovary inferior, 2 to 5-locular; ovules 2 in each locule; styles 2 to 5, connate at base. Fruit pome-like, with persistent incurved calyx teeth; pericarp more or less fleshy; endocarp thin. Seeds large, without endosperm.

A genus of about 30 species in Asia, in the Himalayan regions, North and Northeast India, Bangladesh, South and Central China, South Japan, Southeast Asia and the Malay Peninsula. One species cultivated and naturalized in Ceylon.

For a revision of this genus, see Vidal (1965—Adansonia n.s. 5: 537–580).

Eriobotrya japonica (Thunb.) Lindl., Trans. Linn. Soc. 13: 102. 1821; Hook. f., Fl. Br. Ind. 2: 372. 1878; Nakai, J. Arn. Arb. 5: 69. 1924; Vidal, Adansonia n.s. 5 (4): 550. 1965. Type: Japan, *Thunberg s.n.* (UPS).

Mespilus japonica Thunb., Fl. Japan 206. 1784; Moon, Cat. 40. 1824, nomen nudum.

Cratageus bibas Lour., Fl. Cochinch. 319. 1790; ed. Willd. 391. 1793.

Tree up to 6 m tall. Young branches rusty tomentose. Leaves elliptic oblong to broadly oblong-lanceolate, 15 to 25 cm long, 4 to 7 cm broad, cuneate at base, acute at tip, remotely dentate, lustrous and dark green above, finely rusty tomentose beneath. Petiole 0.5 to 1 cm long. Stipules subulate, about 1 cm long, persistent. Inflorescence of densely flowered, terminal panicles, with stout branches, rusty tomentose. Flowers more or less sessile. Calyx lobes ovate, subacute, glabrous inside. Petals ovate, tomentose at the base inside, white. Stamens 20. Ovary densely pilose. Fruit globose or ovoid, 3 to 4 cm in diameter, with persistent calyx lobes at apex, yellow when ripe. Seeds sometimes nearly as large as the fruit.

Distr. Mostly in Japan and China. Cultivated in South India, Southeast Asia, South America, South Europe, North Africa, Madagascar and in the Mascarene islands.

Vern. Nocotahmarum (T); Loquat, sometimes also called "Japan plum" (E).

Illustr. Wight, Ic. Pl. Ind. Or. 1: Pl. 226/931. 1839.

Note. The flower is fragrant and the fruit is edible and has an agreeable acid taste.

Specimens Examined. KANDY DISTRICT: Madulkele, Hatale Bungalow, altitude 2500 ft, *Worthington 2004* (PDA); Peradeniya Royal Botanic Garden, May 1888, *s. coll.* (PDA); Nawalapitiya, garden at Emboolpitiya, *Worthington 381* (PDA). NUWARA ELIYA DISTRICT: Hakgala Botanic Garden, behind compound of office, *Tirvengadum & Jayasuriya 132* (G, K, L, MAU, P, PDA, US).

4. PYRUS

L., Sp. Pl. 479. 1753; L., Gen. Pl. ed. 5. 214. 1754, pro parte. Type species: *Pyrus communis* L.

Deciduous or semi-evergreen trees or shrubs (sometimes spinescent). Leaves simple, serrulate or entire, stipulate. Flowers umbelliform or in clusters, normally bisexual, white, rarely pink, appearing before or with the leaves. Calyx tube urn-shaped, adnate to ovary; calyx lobes 5, persistent or deciduous. Petals 5. Stamens 20 to 30. Carpels 2 to 5, connate; ovary inferior, 2 to 5-locular; ovules 2 in each locule; styles 2 to 5, free but closely constricted at base by the disk. Fruit an often pyriform pome; pericarp fleshy with copious stone or grit-cells; endocarp cartilaginous. Seeds without endosperm.

A genus of about 20 species from Europe to East Asia, South to North Africa, Persia and the Himalayas. In Ceylon, the only introduced species *Pyrus communis* is now naturalized at high elevation, mostly in the montane zone.

Pyrus communis L., Sp. Pl. 479. 1753; Moon, Cat. 40. 1824; Hook. f., Fl. Br. Ind. 2: 374. 1878; Rehder, Man. Trees & Shrubs, ed. 2. 403. 1940; Bailey, Man. Cult. Pl., Rev. ed. 514. 1949. Type: *Linné s.n.* (LINN). Microfiche IDC card 334: 647.

Small tree 5 to 7 m tall. Young branches glabrous or slightly pubescent. Leaves broadly ovate or ovate-orbicular, 4 to 8 cm long, up to 5 cm broad, rounded at base, acute or shortly acuminate at tip, finely serrulate, glabrous or pubescent when young, drying black, paler beneath. Petiole 1.5 to 5 cm long. Inflorescence in corymbose or umbellate racemes or few-flowered clusters, almost glabrous. Pedicel 1.5 to 5 cm long. Calyx lobes broadly ovate, acuminate, slightly pubescent outside, villous at base inside. Petals suborbicular, spreading, glabrous, white or pink. Stamens 20, in 2 rows, outer ones longer. Ovary slightly pubescent at apex. Fruit pyriform, with persistent calyx lobes, very warty.

Distr. Cultivated abundantly in the district of Nuwara Eliya. It is often found as an escape, close to orchards and cultivations.

Vern. Common pear (E).

Note. The fruit is of a poor quality and does not look attractive, being very warty. The flesh is very gritty and hard.

Specimens Examined. NUWARA ELIYA DISTRICT: Cultivated in orchard, besides Municipal Tourist Rest House, *Tirvengadum & Cramer 80* (MAU, PDA, US); Hakgala, Botanic Garden, *Tirvengadum & Jayasuriya 141* (MAU, PDA, US).

MALUS Mill.

The genus *Malus* which is sometimes held as a subgenus of *Pyrus* is distinguished from the latter by its more or less pubescent leaf-surfaces, acute serratures, connate styles (at base) and the absence of grit-cells in the fruit.

The species of *Malus* (apple tree) that is cultivated for economic purposes in Ceylon probably belongs to *Malus sylvestris* Mill. (*Pyrus Malus* L.; *Malus communis* DC.).

5. RUBUS

L., Sp. Pl. 492. 1753; L., Gen. Pl. ed. 5. 217. 1754. Type species: *Rubus fructicosus* L.

Shrubs or low undershrubs, often straggling or scrambling. Branches mostly sarmentose, mostly armed with straight or hooked prickles, rarely glandular. Leaves simple or compound, stipulate. Flowers in axillary and terminal panicles or corymbose cymes or in axillary glomerules or clusters, rarely solitary, bisexual, white or pink. Calyx expanded, with a short tube, often shallowly campanulate; calyx lobes 5, imbricate, persistent. Petals 5, mostly unguiculate, white or pink. Stamens indefinite, mostly bi-seriate. Carpels numerous, inserted on a more or less prominent convex receptacle; ovules 2 in each ovary; styles terminal. Fruit an aggregate of few or numerous drupelets, arranged on a rather dry or spongy receptacle. Stones cartilaginous, rugose. Seeds without endosperm.

A very large and variable genus with more than 400 species, including many natural hybrids, nearly worldwide, particularly abundant in the Northern Hemisphere. It is represented by 14 species in Ceylon, one of which is probably endemic.

Of all the Ceylon species, *R. indicus* is the most widely distributed, common and abundant at the same time. Besides, it is the only species which occurs at high elevation and also at lower altitude in the southeast regions of the island. The other species are distributed mostly in the Central Province at altitudes ranging from 800 m to the highest elevated portions of the island. Of these, the commonest is *R. ellipticus* often found growing in varied habitats, sometimes constituting a real pest due to its gregariousness for shade and space to neighbouring plants. *R. fairholmianus* and *R. gardnerianus* are fairly common, the former at altitudes ranging from 1600 to 2800 m. *R. rugosus* is rather rare and seems to prefer cool and shady places at high altitudes in montane forests. *R. niveus* occurs mostly in the lower montane zone. *R. leucocarpus* var. *tomentosa* is fairly widespread whereas *R. leucocarpus* var. *leucocarpus* is localised at high altitude in jungles and forests around Nuwara Eliya. *R. glomeratus* is extremely rare. *R. micropetalus* recorded at various localities by Trimen is now rare. *R. rosifolius* is naturalized in Kandy District, whereas *R. coronarius*, found growing as an escape in Sevastine and spontaneously in the Peradeniya University campus is often cultivated at lower altitude, as an ornamental plant. *R. lineatus* and *R. sorbifolius* have not been seen by recent collectors. They are probably cultivated as garden varieties. Specimens of *R. moluccanus*, (probably a variety of this very variable taxon), have been collected in the montane zone, at about 1800 m alt. and recorded for the first time in Ceylon since Alston.

Found in open sunny places, clearings or open areas in forests and jungles, along roadsides or embankments, often along edges of forests, ravines or riverbanks, climbing or straggling over dense vegetation; sometimes in shady areas along forest or jungle paths or borders of tea estates. All species are found in the regions with a year-round rainfall, except *R. indicus*, which is also found in the monsoon area in the southeastern part of the island.

Some species of *Rubus* are exceedingly polymorphous and hybridize readily. This makes it difficult to define well marked species. Trimen, for example, recorded only 4 species and 3 varieties of *Rubus* in Ceylon, whereas Alston recognised 8 species and 1 variety. Here the arrangement of Focke (1911—Sp. Rub., Biblioth. Bot. 17 (72): 9) and the revision of Alston (1931—in Trimen, Handb. Fl. Ceylon 6, Suppl. 101) are closely followed.

KEY TO THE SPECIES
(systematic)

1 Stipules joined to the petiole; persistent, linear or lanceolate. Leaves compound, pinnate (subg. *Idaeobatus*)
2 Receptacle very prominent. Flowers solitary or arranged in few-flowered cymes
3 Branches provided with stipitate glands. Carpels very numerous, about 400 .... **1. R. sorbifolius**
3 Branches not provided with stipitate glands. Carpels about 200
4 Stipules linear. Flowers in 2 or 3-flowered cymes, often reduced to 1 flower, up to 2.5 cm in diameter. Petals 5 (–10), not numerous .... **2. R. rosifolius**
5 Flowers pentamerous .... var. **rosifolius**
5 Flowers with extra petals, usually 10 .... var. **commersonii**
4 Stipules ovate lanceolate. Flowers always solitary, up to 4 cm in diameter. Petals numerous .... **3. R. coronarius**
2 Receptacle less prominent, rather low. Flowers mostly arranged in many-flowered panicles (a 3-flowered corymbose cyme in *R. leucocarpus*)
6 Branches not provided with dense, long bristles. Leaves imparipinnate, 5 to 9-foliolate. Leaflets acute or acuminate, tomentose or glabrous
7 Leaflets 5 to 9, ovate oblong or elliptic, acute with greyish white tomentum underneath. Calyx not armed with prickles .... **4. R. niveus**
7 Leaflets 5 to 7, ovate lanceolate, acuminate, glabrous or with silvery white tomentum underneath. Calyx armed with prickles .... **5. R. leucocarpus**
8 Leaflets mostly 7, membranous, glabrous .... var. **leucocarpus**
8 Leaflets 5 to 7, coriaceous, with white tomentum underneath (turning buff-coloured with age) .... var. **tomentosa**

6 Branches provided with dense, long bristles. Leaves 3-foliolate. Leaflets obtuse, with grey tomentum underneath..6. **R. ellipticus**

1 Stipules joined to the stem, deciduous or semipersistent, mostly incised. Leaves mostly simple, sometimes compound (subg. *Malachobatus*)

9 Leaves simple

10 Bracts boat-shaped, sub-entire or dentate at apex. Calyx with a distinct hyaline margin. Leaves subacutely 5-lobed, lobulate..........................**7. R. indicus**

10 Bracts laciniate or pinnatifid or palmatifid. Margin of calyx not hyaline. Leaves 5 to 7-lobed, rarely lobulate

11 Bracts pinnatifid or pectinately laciniate with linear segments, persistent

12 Leaves membranous. Tomentum thin and greenish...........**8. R. moluccanus**

12 Leaves more or less coriaceous. Tomentum thick fulvous or cinnamomeous

13 Tomentum cinnamomeous. Petals very small, less than half the length of the calyx lobes..**9. R. micropetalus**

13 Tomentum fulvous. Petals nearly as long as the calyx lobes............. ....**10. R. glomeratus**

11 Bracts palmatifid or palmately laciniate, rarely entire, rather deciduous, occasionally persistent

14 Leaves 5 to 7-lobed, lobulate. Tomentum thick and whitish. Inflorescence condensed and many-flowered..........................**11. R. fairholmianus**

14 Leaves 5 (–3) lobed, not lobulate or scarcely lobulate. Tomentum thin or thick, hispid brown or tawny. Inflorescence few-flowered

15 Leaves 5-lobed, sometimes obscurely 3-lobed, not lobulate, shallowly cordate, membranous. Tomentum tawny. Inflorescence in lax racemes............. ....**12. R. gardnerianus**

15 Leaves 5-lobed, scarcely lobulate, deeply cordate, subcoriaceous to coriaceous. Tomentum hispid brown. Inflorescence in compact racemes or panicles ....**13. R. rugosus**

9 Leaves compound palmate...**14. R. lineatus**

16 Leaflets with silvery white tomentum underneath....................var. **lineatus**

16 Leaflets glabrous underneath...var. **glabrior**

KEY TO THE SPECIES
(artificial)

1 Leaves compound

2 Leaves compound palmate...**14. R. lineatus**

3 Leaflets silvery tomentose underneath..................................var. **lineatus**

3 Leaflets glabrous, except on nerves beneath..........................var. **glabrior**

2 Leaves compound pinnate

4 Branches covered with dense, long reddish bristles or glands

5 Branches provided with stipitate glands. Inflorescence in corymbose cymes. Carpels very numerous (about 400), inserted on a prominent receptacle. Leaflets mostly 7, acuminate..**1. R. sorbifolius**

5 Branches not provided with stipitate glands, but with bristles. Inflorescence in many-flowered panicles. Carpels less than 100, inserted on a low receptacle. Leaflets 3, obtuse..**6. R. ellipticus**

4 Branches not covered with long, reddish bristles or glands

6 Flowers white, 2.5 to 4 cm in diameter. Stem not glaucous

7 Inflorescence in 2 or 3-flowered cymes. Flowers pentamerous, or with extra petals (about 10). Leaflets pilose on both surfaces........................**2. R. rosifolius**

8 Petals 5 .... var. **rosifolius**
8 Petals more than 5, usually 10 .... var. **commersonii**
7 Flowers solitary, with numerous petals (more than 10). Leaflets slightly pubescent above, glabrous beneath .... **3. R. coronarius**
6 Flowers pink, less than 2 cm in diameter. Stem glaucous
9 Inflorescence in cymose panicles, sometimes reduced to 1 or 2 flowers. Leaflets ovate oblong or elliptic, acute, covered with greyish white tomentum beneath. Petiole slender, not perfectly furrowed above .... **4. R. niveus**
9 Inflorescence in 3-flowered corymbose cymes, rarely reduced to 1 or 2 flowers. Leaflets ovate acuminate, glabrous or with white tomentum beneath. Petiole stout, perfectly furrowed above .... **5. R. leucocarpus**
10 Leaflets glabrous underneath .... var. **leucocarpus**
10 Leaflets with whitish tomentum underneath .... var. **tomentosa**
1 Leaves simple
11 Leaves hastate or oval oblong, usually longer than broad
12 Old stem ferruginous hirsute. Leaves 5-lobed .... **7. R. indicus**
12 Old stem covered with whitish tomentum or glabrous. Leaves 3 to 7-lobed
13 Leaves 5 to 7-lobed, with whitish tomentum underneath .... **11. R. fairholmianus**
13 Leaves 3 to 5-lobed, with greenish or cinnamomeous tomentum underneath
14 Leaves distinctly 5-lobed. Tomentum thin, greenish .... **8. R. moluccanus**
14 Leaves 3 (–5) lobed, the 2 basal lobes when present almost incipient. Tomentum thick, cinnamomeous .... **9. R. micropetalus**
11 Leaves broadly ovate or orbicular, usually as long as broad
15 Leaves with fulvous tomentum underneath. Fruit small .... **10. R. glomeratus**
15 Leaves with tawny, pale greyish or dark brown tomentum underneath. Fruit large
16 Leaves often shallowly cordate to truncate at base, distantly rugose with large areoles above. Tomentum tawny .... **12. R. gardnerianus**
16 Leaves deeply cordate at base, minutely rugose with very small areoles above. Tomentum greyish or dark brown .... **13. R. rugosus**

1. **Rubus sorbifolius** Maxim., Bull. Acad. Imp. Sci. St. Pétersb. 17: 158. 1871; Mél. Biol. Bull. Acad. Imp. Sci. St. Pétersb. 8: 390. 1871; Thuan, Fl. Thailand 2 (1): 49. 1970. Syntypes: Japan, Kyushu Archipelago, Higo Province, Mt. Higo-San, 1863, *Maximowicz s.n.*; India, Khasi Hills, *Hooker f. & Thomson s.n.* Lectotype: *Maximowicz s.n.* (P).

Rubus asper Wall. ex Don, Prod. Fl. Nepal 234. 1825 (non Presl, 1822). Type: Nepal, *Wallich 741* (Holotype, K; Isotype, BM).
Rubus rosaefolius auct. non Smith: Hook. f., Fl. Br. Ind. 2: 341. 1878, pro parte.

Shrub or liana. Stem stout, terete, glabrescent. Young branches almost villous, covered with numerous crimson coloured stipitate glands. Prickles few, scattered, hooked. Leaves compound, imparipinnate. Leaflets 7, sometimes 1 to 3 in upper part, membranous; lateral leaflets opposite, ovate lanceolate, 2 to 7 cm long, 1 to 2 cm broad, rounded to obtuse at base, acuminate at tip, doubly incise-serrate, both surfaces with a few scattered hairs; petiolule short; terminal leaflet larger and long-acuminate. Petiole + rachis up to 10 cm long, glandular-villous and sparingly prickly. Stipules lanceolate, long-acuminate, 2 to 7 cm long, joined to the petiole, persistent. Inflorescence in 3-flowered, sometimes 1 to 2-flowered corymbose cymes, glandular-pubescent and prickly. Flowers up to 1 cm across. Pedicel 2 to 3 cm long. Calyx lobes ovate-lanceolate, caudate, densely white pubescent inside and on the margins, reflexed in fruit. Petals obovate or oblong, glabrous, white. Carpels about 400, small, inserted on a very prominent conical, villous receptacle. Fruit oblong, up to 1 cm in diameter, vertically placed on the receptacle, reddish; drupelets numerous. Stones deeply pitted.

Distr. In India, Vietnam, Laos, Thailand, China, Formosa (Taiwan), Korea, Japan and Indonesia. In Ceylon, recorded in Nuwara Eliya.

Illustr. Focke, Sp. Rub. Biblioth. Bot. 17 (72): 158. f. 67. 1911 (Sub. *R. asper*).

Note. This plant is closely related to *R. rosifolius*. It can easily be identified by the presence of stipitate glands and the inflorescence which is usually a 3-flowered cyme (rarely reduced to 1 or 2 flowers).

The fruit is edible.

Specimens Examined. NUWARA ELIYA DISTRICT: Nuwara Eliya, cultivated, Coll. *J. Petch 840* (BM, K); 27.2.1927, *s. coll. s.n.* (PDA).

2. Rubus rosifolius Smith, Pl. Ic. Hact. Inéd. 3: t. 60. 1791, as "rosaefolius"; Moon, Cat. 40. 1824; Hook. f., Fl. Br. Ind. 2: 341. 1878; Focke, Sp. Rub., Biblioth. Bot. 17 (72): 153. f. 65. 1911; Bot. Jahrb. 54: 71. 1916; Van Royen, Gen. Rubus, Phanerog. Monogr. 2: 34, f. 6. 1969. Type: From Mauritius, *Commerson s.n.* Holotype (LINN); Isotypes (MPU, P).

Rubus indotibetanus Koidzumi, Fl. Symb. Or. As. 65. 1930.

See Vidal (1968—Fl. Camb. Laos. & Vietn. 7: 22) for complete synonymy.

Shrub rooting at nodes and branching profusely in the apical part. Stem weak, erect, or subscandent, slightly pilose. Young branches densely pilose. Prickles few, distant, weak. Leaves compound, imparipinnate. Leaflets 5 to 7, in the inflorescence reduced to 3, membranous; lateral leaflets opposite, ovate-lanceolate, 2 to 4 cm long, 1 to 2 cm broad, rounded or obtuse to truncate, sometimes asymmetric at base, acute or acuminate at tip, regularly doubly and deeply incise-serrate, both surfaces green and pilose; petiolule very short or sessile; terminal leaflet larger, usually with 1 lobe on one side at base, tapering towards the acute tip, sometimes the upper part of leaflets merging with the terminal leaflet. Petiole + rachis up to 10 cm long, in the inflorescence as short as 1.5 cm, channelled, densely pilose and sparingly prickly. Stipules linear, up to 1 cm long, borne a little way off the base of the petiole, persistent. Inflorescence in 2 to 3-flowered axillary and terminal cymes, often reduced to a single flower, pilose. Flowers up to 2.5 cm across. Pedicel up to 3 cm long. Calyx lobes lanceolate-caudate, subulate at tip, softly tomentose inside, almost glabrous at the basal part, pilose outside. Petals 5 to 10 or more, but not numerous, obovate to orbicular, glabrous inside, pilose outside on basal part, white. Carpels about 200, small, inserted on a very prominent, conical, more or less glabrous receptacle. Fruit globose; mature fruit subglobular, up to 2 cm in diameter, bright red; drupelets numerous. Stones deeply pitted.

var. **rosifolius**

Petals 5 (flower single).

Distr. A very widespread species, almost cosmopolitan. Naturalized and cultivated in the tropics and warm temperate regions. In India, Tibet (Himalaya), Burma, South China, Vietnam, Cambodia, North Vietnam, Indonesia, South Japan, South Africa, Cameroon, Australia, Java, Philippines (Manila), New Guinea, Madagascar, Mascarene Islands, Mauritius (native), Rodrigues and Reunion, Seychelles and the West Indies. In Ceylon, it is naturalized and localised in Kandy District, at an altitude of about 800 m.

Ecol. Usually found growing in large groups, in shady moist areas, along streams or borders of tea estates, in open places or cleared slopes, along roadsides among bushes on patna margins. The leaves have a dewy appearance due to the abundance of glandular structures present on both surfaces of the leaflets. Flower white. Flowering from January to March. Van Royen reports that it flowers the whole year round in New Guinea.

Vern. "Rada-Kaetambila" (S), according to Moon l.c. Rose-leaved bramble (Curtis) or wild raspberry (E).

Illustr. Curtis, Bot. Mag. 113: Pl. 6970. 1887.

Note. The whole plant is easily recognisable by its shapely and evergreen leaves and its attractive, bright red fruits. The flower is fragrant and ornamental. The fruit, though insipid to the taste, is copiously produced. During the season, it is reported that tons of this wild berry are collected by the tea planters for making jam.

There had been a bit of doubt as to the origin of this plant, but authentic specimens of Commerson seen at the Linnean Society, in Montpellier and in Paris, confirm Mauritius as the place of origin.

For a good discussion on the origin of this plant, see Van Royen l.c.

Specimens Examined. KANDY DISTRICT: Madulkele, Hatale 2300 ft alt., *Worthington 2024* (PDA); between Panwila and Madulkele, *Tirvengadum, Cramer & Balasubramaniam 345* (MAU, PDA, US), *350* (G, K, L, MAU, P, PDA, US); Peradeniya, Plant Collector's Garden, March 22nd 1920, *T.P. s.n.* (PDA). LOCALITY UNKNOWN: *s. coll. s.n.* (M.H.).

var. **commersonii** (Poiret) Tirvengadum, comb. nov. Type: Based on *Rubus commersonii* Poir.

Rubus commersonii Poir. in Lam., Enc. 6: 240. 1804. Type: Java, *Commerson s.n.* (P).

Rubus rosaefolius Smith var. *coronarius* auct. non Sims: Focke, Sp. Rub. Biblioth. Bot. 17 (72): 155. 1911.

Flowers with extra petals (usually twice that of var. *rosifolius*).

Distr. It is rather rare and has been recorded in Java (origin), Mauritius, Brazil and China. I have not seen any specimen in Ceylon, but it is probably cultivated as an ornamental plant.

Note. This species collected by Commerson in Java bears great affinity with *R. rosifolius*, from which it differs by its extra petals. In the herbarium specimens examined, the flowers never reach the size and shape of *R. rosifolius* var. *coronarius* (= *R. coronarius* (Sims) Sweet) with which it has often been confused.

See further discussion which follows under *R. coronarius*.

3. Rubus coronarius (Sims) Sweet, Hort. Brit.: 144. 1827; Bailey, Man. Cult. Pl. Rev. ed.: 522. 1949. Type: Based on *Rubus rosaefolius* J.E. Sm. var. *coronarius* Sims.

Rubus rosaefolius Smith var. *coronarius* Sims., Bot. Mag. 43: Pl. 1783. 1816; Focke, Sp. Rub., Biblioth. Bot. 17 (72): 155. 1911. Type: Sims, Pl. 1783.

Undershrub up to 1 m tall, with a long rootstock, often stoloniferous. Stem woody, erect or inclined, puberulent. Young branches arising in axils of foliar stipules, hirtellous. Prickles numerous, strong, with broad base. Leaves compound, imparipinnate. Leaflets 5 to 7, the lower pair more distant, always 3 in the upper leaf next the flower, subcoriaceous; lateral leaflets opposite, elliptic, up to 3.5 cm long, 1.5 cm broad, more or less rounded or obtuse, often asymmetric at base. acute at tip; serrate, sometimes incised, slightly hirtellous above, glabrous beneath, petiolulate, rarely sessile; terminal leaflet elliptic to ovate-lanceolate, up to 6 cm long, 3 cm broad, petiolule up to 9 mm long. Petiole + rachis up to 5 cm long. Stipules ovate-lanceolate, up to 1 cm long, lower half adnate to base of petiole, persistent. Flowers terminal, solitary, probably sterile, always opposite alternate leaf, 3 to 4 cm across. Pedicel 2 to 5 cm long, the apical part fitting in a sort of spherical depression at base of receptacle. Calyx lobes 5, sometimes 10, outer ones ovate, caudate, slightly hirtellous outside, tomentose inside and at the margins; inner ones finely tomentose on both surfaces. Petals numerous, orbicular, puberulent on both surfaces, white. Carpels about 200, small, inserted on a very prominent, villous receptacle. Fruit not seen.

D i s t r. In Malaya (origin?), Java, Philippines, South America (Brazil), San Domingo, Cuba, Puerto Rico. Cultivated in India, China, Vietnam, Thailand, Japan, Madeira, Madagascar. In Ceylon, naturalized in the montane zone at an alt. of 1500 m. Cultivated at lower altitudes.

E c o l. Found growing in any soil and propagates itself rapidly by suckers which it puts forth abundantly. The habit of the plant varies from a low shrub about 15 cm tall to an undershrub which can reach 1 m. In its low form it frequently puts forth suckers annually which give rise to a close cluster of stems giving the appearance of a single stem much branched at the base. Flower white. The flowering season seems to have a very long range. It was seen in flower between October and February. I have not seen any fruit either in the cultivated or in the naturalized ones in Ceylon. It is frequently cultivated for ornament, on account of its elegant, white flower which almost rivals a rose.

V e r n. Double flowered China (Sweet), Brier-rose or Salem-rose according to Bailey l.c.

I l l u s t r. Sims, Bot. Mag. 43: Pl. 1783 (Sub. *R. rosae-folius* var. *coronarius*).

N o t e. The origin of this plant is rather doubtful. Sims (l.c.) stated it was introduced from the Prince of Wales Islands (Penang) in the East Indies. Later it had also been called *Rubus chinensis*, but again its origin from China is exceedingly doubtful.

It has often been considered as a variety of *R. rosifolius*. With the abundance of material now available, a number of specific characters have been listed which clearly distinguish it from the latter and more precisely from *R. rosifolius* var. *commersonii*, which differs from *R. rosifolius* only in its extra petals. The most striking character is the flower, which is full double, like a small rose.

Further evidence is the fact that no fructification has been recorded so far, since Sims (l.c.) stated "the parts of fructification seem to remain so perfect that fruit may very well be expected, though we do not find that any has yet been perfected." On the other hand, *R. rosifolius* var. *commersonii* (= *R. commersonii* Poir.) has a reddish fruit the size of a raspberry. The Paris herbarium material of *R. coronarius* from India, China and Japan, besides seems to confirm this. All the specimens seen from those regions bear flowers and not fruits; whereas specimens of *R. rosifolius* var. *commersonii* from Java and South America show both flowers and fruits.

S p e c i m e n s E x a m i n e d. KANDY DISTRICT: Peradeniya University Campus, *Tirvengadum & Balasubramaniam 211* (G, K, L, MAU, P, PDA, US); Maskeliya, Moray Estate, *Tirvengadum, Cramer & Balasubramaniam 311* (G, K, L, MAU, P, PDA, US). MATALE DISTRICT: Sevastine, 1,300 m alt., *Tirvengadum, Dassanayake & Jayasuriya 15* (K, L, MAU, P, PDA, US). NUWARA ELIYA DISTRICT: Horton Plains, near Farr Inn, *Tirvenga-*

dum & Cramer 104 (G, K, L, MAU, P, PDA, US). RATNAPURA DISTRICT: Kurupanawala, *Tirvengadum, Cramer & Balasubramaniam 182* (MAU, P, PDA, US).

4. Rubus niveus Thunb., Dissert. Rub. 9, f. 3. 1813; Focke, Sp. Rub., Biblioth. Bot. 17 (72); 182. f. 75. 1911; Alston in Trimen, Handb. Fl. Ceylon 6: 104. 1931. Type: Java, *Thunberg s.n.* (UPS). Microfiche IDC card 515: 12275.

Rubus lasiocarpus Smith in Rees, Cyclop. 32: n. 6. 1819; Thw., Enum. Pl. Zeyl. 101. 1859; Hook. f., Fl. Br. Ind. 2: 339. 1878; Trimen, Handb. Fl. Ceylon 2: 138. 1894.
Rubus mysorensis Heyne in Roth, Nov. Pl. Sp. Ind. 235. 1821.
Rubus distans D. Don, Prod. Fl. Nepal 256. 1825.
Rubus micranthus D. Don, Prod. Fl. Nepal 235. 1825.
Rubus lasiocarpus Sm. var. *micranthus* (Don) Hook. f., Fl. Br. Ind. 2: 339. 1878.
Rubus lasiocarpus Sm. var. *membranaceus* Hook. f., Fl. Br. Ind. 2: 339. 1878, syn. nov.

Erect or semiscandent shrub, much branched at base. Stem glabrous, glaucous. Young branches villous. Prickles numerous, scattered, straight or slightly recurved. Leaves compound, imparipinnate; leaflets 5 to 9, membranous; lateral leaflets ovate oblong or elliptic, 2.5 to 5.5 cm long, 1 to 2.2 cm broad, cordate or truncate at base, acute, mucronate at tip, distantly serrate, serratures mucronate, slightly pilose above, greyish-white tomentose below, sessile; terminal leaflet broadly ovate, often lobed on one or both sides, long petiolulate. Rachis + petiole 5 to 15 cm long, slender, not perfectly furrowed above. Stipules lanceolate, up to 7 mm long, adherent to base of petiole, persistent. Inflorescence in many-flowered cymose panicles, chiefly terminal, sometimes reduced to 1 or 2-flowered cymes in highest leaf axils, densely villous. Flowers up to 1 cm across. Pedicel 1 cm long. Calyx lobes distant, triangular, acute-aristate, densely villous on both sides, not prickly. Petals broadly obovate or suborbicular, shorter than the calyx, hairy at the base inside, pink. Carpels about 50, borne on a rather low, villous receptacle. Fruit subglobose, less than 1 cm across, greenish-red, dark purple when ripe; drupelets flattened, dry or slightly fleshy, covered with a white tomentum, tipped with persistent style.

Distr. In temperate Himalaya, South India (Nilgiris, Mysore), Pakistan, Burma, Thailand, Laos, Vietnam, China (Yunan), Philippines and Java. Introduced in Kenya, Uganda, Tanzania and Sudan. In Ceylon, it is common in the lower montane zone, around Kandy and Matale Districts, occasional at higher elevation in the Nuwara Eliya District (Ramboda).

Ecol. Found in open places; border of patanas or by roadsides. Flower pink. Flowering from October to February, throughout the year according to Alston. The fruit is rather dry and has no flavour.

Vern. Rodu-Koetambila (S), according to Moon; woolly-berried bramble (E).

Illustr. Thunberg, Dissert. Rub.: f. 3. 1813; Focke, Sp. Rub. Biblioth. Bot. 17 (72): f. 75. 1911 (sub. *R. micranthus*).

Specimens Examined. KANDY DISTRICT: Castlereagh Estate Garden, *Alston 1087* (K); bet. Hunnasgiriya and Rangala, *Tirvengadum, Cramer & Balasubramaniam 67* (K, L, MAU, P, PDA, US); bet. Madulkele and Panwila, 16/12 Culvert, *Tirvengadum & Balasubramaniam 347* (G, K, L, MAU, P, PDA, US). MATALE DISTRICT: Rattota, Dambulla rock, *Tirvengadum, Dassanayake & Jayasuriya 6* (G, K, L, MAU, P, PDA, US); *8* (MAU, PDA, US). NUWARA ELIYA DISTRICT: N'Ellia 6000 ft alt., *Gardner 266* (K); Ramboda, *Tirvengadum & Jayasuriya 118* (G, K, L, MAU, P, PDA, US); Hakgala, cult., Botanic Garden, *Simpson 9119* (BM); Govinidahela, roadbank, 27.2.1906, *Silva s.n.* (PDA); Intérieur des forêts de Ceylan, *Leschenault s.n.* (P); Rikiligaskade, *Amaratunga 892* (PDA); *M.N.B. 613* (CAL). LOCALITY UNKNOWN: 5800 ft alt., *Pearson 240* (CGE); *s. coll. 1533* (BM, CAL, CGE, G, P, PDA); March 1829, *Macrae 621* (CGE).

5. **Rubus leucocarpus** Arn., Nova Acta Phys. Med. Acad. Caes. Leop. Carol. Nat. Cur. 18: 16 (334). 1834; Walp., Rep. 2: 19. 1843; Alston in Trimen, Handb. Fl. Ceylon 6: 103. 1931. Type: Ceylon, Col. *Walker* (according to Arnott l.c.).

Rubus lasiocarpus Smith var. *β. subglaber* Thw., Enum. Pl. Zeyl. 101. 1859; Hook. f., Fl. Br. Ind. 2: 339. 1878; Trimen, Handb. Fl. Ceylon 2: 138. 1894; Gamble, Fl. Pres. Madras 1 (3): 44. 1919. Type: Ceylon, *Thwaites C.P. 1534* (Holotype, K; Isotypes, BM, CAL, CGE, G, P, PDA).

Scrambling shrub. Stem glabrous, glaucous. Young branches villous. Prickles numerous, strong, large-based. Leaves compound, imparipinnate. Leaflets 5 to 7, membranous or coriaceous; lateral leaflets ovate-acuminate, 2.5 to 6.5 cm long, 1.5 to 3.5 cm broad, roundish or obtuse at base, acuminate at tip, finely serrate, serratures unequal, slightly villous above, glabrous or white lanate tomentose beneath, subsessile; terminal leaflet broadly ovate or cordate, sometimes lobed on one side or 3-lobed, petiolule 1 to 2 cm long. Rachis + petiole 8 to 25 cm long, stout, distinctly furrowed above. Stipules linear-subulate, up to 9 cm long, adherent to petiole base, persistent. Inflorescence in terminal and axillary 3-flowered corymbose cymes, rarely 2 or 1 in axils of lower leaves, densely velutinous. Flower slightly more than 1 cm across. Pedicel 1 or 2 cm long. Calyx lobes ovate-triangular, subulate at tip, velutinous inside and outside, armed with a few straight prickles. Petals orbicular, more or less equalling the calyx, deep pink. Carpels 35 to 40, borne on a slightly conical villous receptacle. Fruit globose, 1 to 1.6 cm across, greenish-white, purplish-blue when ripe; drupelets subglobular, fleshy, covered with a thick cottony white tomentum.

var. **leucocarpus**

Leaflets glabrous and green on both surfaces, 7, rarely 5, rather membranous.

Distr. Probably endemic.

Ecol. Found in open thickets or jungle paths or by road banks. Flower deep pink. Flowering from December to March, throughout the year according to Trimen.

Note. This species bears a close resemblance to the European *R. fructicosus* and to *R. pinnatus*, but the fruit is inedible and is covered with a woolly coat.

Hooker f. stated that Wight's Pulney Hills specimens of *R. racemosus* "have the leaflets more membranous, almost glabrous, except along the nerves, like *R. lasiocarpus* var. *subglaber*," but *R. racemosus* is a very distinctive taxon, characterised by a number of stipate glands present in various parts, and the leaf undersurface is not glabrous but covered with a yellowish tomentum.

Specimens Examined. NUWARA ELIYA DISTRICT: N'Ellia, 6000 ft alt., *Gardner 267* (BM, K); Pidurutalagala, *Tirvengadum & Cramer 25* (G, K, L, MAU, P, PDA, US); *26* (MAU, PDA, US); *30* (MAU, PDA, US); Kandapola Forest Reserve, *Tirvengadum & Cramer 145* (K, MAU, P, PDA, US); Hakgala, *Alston 134* (PAD); Hakgala Botanic Garden, Curator's compound, *Simpson 5033* (BM); *Amaratunga 411* (PDA); jungle at back of Hakgala, 1.3.1906, *s. coll. s.n.* (PDA); Ambawela, Blackpool, *Tirvengadum & Cramer 88* (MAU, PDA, US). BADULLA DISTRICT: Bandarawela, 3800 ft alt., *Pearson 747* (CGE). LOCALITY UNKNOWN: 4000 to 7000 ft alt., *s. coll. C.P. 1534* (BM, CAL, CGE, G, K, P, PDA); *C.P. 1533* (P); Regio temp. 5000 to 6000 ft alt., *Thomson s.n.* (K); 1845, 2000 to 6000 ft alt., *Walker 26* (K); *Walker s.n.* (K); *Gardner 267* (BM, K).

var. **tomentosa** Alston in Trimen, Handb. Fl. Ceylon 6: 104. 1931. Lectotype: Ceylon, *Alston 1243* (PDA).

Leaflets glabrous and green above, lanate tomentose underneath, silvery white at first, turning buff-coloured with age or on exposure, 5 to 7, coriaceous.

Distr. In Ceylon, in the upper montane zone, above 2000 m. Very common. Also reported in South India by Alston.

Ecol. Found in undergrowth of dense mossy forests, in openings in jungles and bushy thickets, borders of montane forests, edges of ravines and by road embankments.

Note. The stout stem and branches are covered with numerous strong, broad based, recurved prickles, looking like rosaceous thorns.

Specimens (*Tirvengadum & Cramer 281*) cited below, collected at Greater World's End, along borders of montane forests, bear smaller leaves with leaflets up to 2 cm long, up to 1.2 cm broad.

Specimens Examined. NUWARA ELIYA DISTRICT: Nuwara Eliya, 6000 ft. alt., *Gardner 268* (K); *s. coll.* 1883 *s.n.* (MH); Pidurutalagala, *Balasubramaniam 204* (PDA); *Tirvengadum & Cramer 24* (G, K, L, MAU, P, PDA, US); *29* (MAU, P, PDA, US); *31* (K, L, MAU, P, PDA, US); Kandapola Forest Reserve, *Tirvengadum & Jayasuriya 146* (G, K, L, MAU, P, PDA, US); bet. Nuwara Eliya and Hakgala, 51/7 Culvert, *Tirvengadum & Cramer 62* (MAU, P, PDA, US); *63* (MAU, PDA, US); Sita Eliya, 5800 ft alt., *Pearson 240* (CGE); Hakgala Peak, *Tirvengadum & Cramer 48*, *53* (MAU, PDA, US); Ambawela, in jungle, *Alston 1243* (PDA); *Tirvengadum & Cramer 87* (MAU, P, PDA, US); Horton Plains, North Entrance, *Fosberg & Mueller Dombois 50012* (K, PDA, US); Greater World's End, *Tirvengadum & Cramer 102* (K, MAU, P, PDA, US); *281* (G, K, L, MAU, P, PDA, US); BADULLA DISTRICT: Top of Namunukula, *Silva s.n.* (PDA). LOCALITY UNKNOWN: up to 6000 ft, alt., *s. coll. C.P. 1534* (BM, CAL, K, MH); Regio temp. 6000 ft, alt., *Thomson s.n.* (K); 2–3 July (6000) *Walker 27* (K); *Walker s.n.* (G); *s. coll. s.n.* (CGE).

6. Rubus ellipticus Smith in Rees, Cyclop. 30: n. 16. 1819; Hook. f., Fl. Br. Ind. 2: 336. 1878; Trimen, Handb. Fl. Ceylon 2: 137. 1894; Focke, Sp. Rub., Biblioth. Bot. 17 (72): 198. 1911. Type: Nepal, April 1802, *Buchanan s.n.* (LINN).

Rubus rotundifolius Wall., Cat. n. 730. 1828, nomen nudum. Type: Kumaon, *Wallich* n. *730* (Holotype, K; Isotypes, BM, CGE).

Rubus gowreephul Roxb., Hort. Beng. 39. 1814, nomen nudum; Roxb., Fl. Ind. 2: 517. 1832. Type: India, *Roxburgh* s.n. (K).

Rubus flavus Hamilton ex D. Don, Prod. Fl. Nepal 234. 1832; Thw., Enum. Pl. Zeyl. 101. 1859. Type: India, Coorg, *Heyne s.n.* (Holotype, K; Isotype, BM).

Rubus wallichianus Wight & Arn., Prod. Ind. Or. 298. 1834, syn. nov. Type: *Wight 1001* (K).

Suberect or scrambling shrub. Stem stout, flexuose, hispid. Young branches ferruginous hispid. Stem and branches covered with long, stiff, reddish bristles. Prickles scattered, strong, hooked. Leaves compound pinnate, trifoliate, coriaceous; leaflets elliptic or orbicular, 3 to 6 cm long, 2 to 5 cm broad, rounded or broadly obtuse at base, obtuse at tip, serrate, glabrous above, greyish lanate tomentose beneath, more or less sessile; terminal leaflet bigger, petiolulate. Petiole 2 to 4 cm long, slightly grooved above. Stipules subulate, up to 8 mm long, inserted on the petiole, often far from the base, occasionally with glandular hairs, persistent. Inflorescence in many-flowered axillary and terminal more or less condensed panicles, or in much branched corymbose panicles, hirtellous, with glandular or eglandular hairs. Flowers up to 1.2 cm across. Pedicel 0.5 to 1 cm long. Calyx lobes ovate, acute-acuminate, or obtuse-mucronate, erect in fruit. Petals obovate, slightly longer than the calyx lobes, white. Carpels less than 100, inserted on a rather low receptacle. Fruit globose, up to 1 cm in diameter, green at first, orange-yellow when mature; drupelets numerous. Stones rugose.

Distr. In India, Eastern and Western Himalaya, Kumaon, Assam, Nilgiris, Nepal, Pakistan, Burma, Thailand, Vietnam and the Philippines (Luzon). Naturalized in Queensland, Australia, West Java (Mt. Gede), Polynesia, East Africa, Nyasaland. In Ceylon, in montane regions above 1000 m.

Ecol. Met with at places of varied ecological conditions. Often found on red rocky lateritic soils, in jungle clearings or paths, along edges of jungles or of mossy forests, or by roadsides in villages. Flower pure white. Flowering all the year. The fruit breaks up easily on drying.

Vern. Nara-bute (S) according to Moon l.c. Oval-leaved bramble or false blackberry (E).

Illustr. Wight, Ic. Pl. Ind. Or. 1: Pl. 230/919. 1839. (sub. *R. gowreephul*).

Note. Although it is closely allied to the "Black-berries" of Europe, the fruit is non-edible. It is scarcely juicy and is not palatable at all. Gamble, however, reported that "the orange raspberry-like fruit is edible and good, it can be made into preserves".

Thuan (1970—Fl. Thailand 2 (1): 52) distinguishes two forms: f. *ellipticus* described above and f. *obcordatus* Franch., found in Indochina, China and India; f. *ellipticus* differs from the latter by its obovate leaves which are also emarginate at the tip.

Specimens Examined. KANDY DISTRICT: Kandy, *Moon 393* (BM); Adam's Peak, *Tirvengadum & Cramer 300* (MAU, P, PDA, US); NUWARA ELIYA DISTRICT: Nuwara Eliya, 4000 to 7000 ft alt., *s. coll. C.P. 2596* (BM, CAL, CGE, G, K, MH, P, PDA); 6000 ft alt., *Gardner 265* (K); 11.1883, *s. coll. s.n.* (MH); Surrounding hills near edge of village, *Van Beusekom 1426* (PDA); *Amaratunga 419 & 1549* (PDA); Nov. 1895, *Smith s.n.* (CGE); Lake Gregory, *Tirvengadum & Jayasuriya 133* (K, MAU, P, PDA, US); Pidurutalagala, along jungle borders, 2300 m alt., *Tirvengadum & Cramer 35* (MAU, P, PDA, US); Kandapola Forest Reserve, *Tirvengadum & Jayasuriya 147* (MAU, P, PDA, US); Sita Eliya, *Pearson 287* (CGE); Moon Plains, edge of jungle, *Alston s.n.* (PDA); Hakgala, open jungle, *Simpson 9063 & 9206* (BM); *Pearson 690* (CGE); Hakgala, edge of jungle towards N'Eliya, 26/2/06, *s. coll. s.n.* (PDA); 51/7, Culvert, Hakgala Road, *Tirvengadum & Cramer 65* (G, K, L, MAU, P, PDA, US): Blackpool, road to Ambawela, *Tirvengadum & Cramer 84* (G, K, L, MAU, P, PDA, US); Horton Plains, Pattipola Road, 1/4 mile downslope, 2100 m alt., *Comanor 977* (PDA); near Farr Inn, *Fosberg & Mueller Dombois*, 2100 m alt., *50079* (PDA). BADULLA DISTRICT: road Nanu Oya to Badulla, Mar. 1890, *Deschamps s.n.* (G); Haputale, 4000 ft alt., edge patana, *Pearson 395* (CGE). RATNAPURA DISTRICT: Kurupanawala, 2/5 mile marker, *Tirvengadum, Cramer & Balasubramaniam 185* (G, K, L, MAU, P, PDA, US). LOCALITY UNKNOWN: Regio Temp., 6000 ft alt., *Thomson s.n.* (PDA); 2000 to 6000 ft alt., *Walker 35* (K); *Walker s.n.*, part of *C.P. 2596* (K); March 1836, *Wight 911* (P); *Wight & Arnott 1001* (K).

7. **Rubus indicus** Thunb., Dissert. Rub. 5 & 7. 1813; Thunb., Fl. Ceil. 6. 1825; Alston in Trimen, Handb. Fl. Ceylon 6: 103. 1931; Kandy Fl. 25. f. 125. 1938. Type: Ceylon, *Thunberg s.n.* (UPS). Microfiche IDC card 515: 12269.

Rubus wightii Gamble, Fl. Pres. Madras 1 (3): 440. 1915. Type: India, Tinnevelly, Shewagherry Hills, *Wight 909* (K).
Rubus rugosus Smith var. *a* Thw., Enum. Pl. Zeyl. 101. 1859. Type: Ceylon, *Thwaites C.P. 2765* (Holotype, BM; Isotypes, CAL, CGE, G, MH, MPU, P, PDA).
Rubus rugosus auct. non Smith: Wight, Ic. Pl. Ind. Or. 1: Pl. 225. 1839.
Rubus moluccanus auct. non L.: Hook. f., Fl. Br. Ind. 2: 330. 1878, pro parte; Trimen, Handb. Fl. Ceylon 2: 136. 1894, pro parte.

Straggling shrub, woody at base. Stem stout, deeply ferruginous hirsute. Prickles numerous, scattered, reflexed with a large hairy base. Leaves simple, oval oblong, 6 to 18.5 cm long, 5.5 to 17 cm broad, subacutely 5-lobed, terminal one largest, lobulate, cordate at base, acute and acuminate at tip, finely crenate mucronate or coarsely dentate, membranous or subcoriaceous, rugose, almost flat, through sparsely rugose to coarsely and closely rugose and hirsute above, pale ferruginous or buff coloured, thinly lanate hirtellous beneath. Petiole stout, 3 to 6 cm long. Stipules joined to the stem, very different from bracts, up to 1.5 cm long, 1 cm broad, falcate, palmate laciniate or oblong lanceolate, incised, hirsute outside, glabrous inside, pinkish, soon deciduous. Inflorescence in lax terminal racemes or branched panicles, up to 40 cm long, and in axillary racemes, up to 15 cm long, rarely solitary or in clusters of 2 to 3 in the leaf axils, ferruginous hirsute. Bracts boat-shaped, up to 1 cm long, 0.9 cm broad, subentire, slightly or deeply toothed at apex, pinkish, deciduous. Pedicel 0.5 to 2 cm long. Calyx lobes triangular or ovate-lanceolate, with a distinct hyaline margin, up to 1 cm long, densely ferruginous hirsute outside, tomentose inside; outer sepals slightly larger and toothed at apex. Petals broadly obovate, up to 0.6 cm long, white. Carpels about 20, small, inserted on a slightly convex, hirsute receptacle. Fruit subglobose, up to 1.5 cm across, shiny scarlet red, translucent, with a mild watery acid taste; drupelets about 20, broadly ovoid, up to 1 mm long, fleshy. Stones very rugose.

Distr. In India (Tinnevelly, Shewagherry Hills), Thailand. In Ceylon, widespread at altitudes ranging from 300 to 2000 m in the montane zone.

Ecol. Found in montane forests, borders of secondary forests or by roadside jungles, on rocky road banks, on hilly slopes, sometimes in shady areas close to streams. Flower white. Flowering all the year.

Vern. Vel-buté (S).

Illustr. Wight, Ic. Pl. Ind. Or. 1: Pl. 225. 1839 (Sub. *R. rugosus* (Sm)); Alston, Kandy Fl. 25, f. 125. 1938.

Note. It is easily identifiable in nature by its pretty vinelike leaves. The fruit is glossy red, very attractive and is edible.

This species was probably collected by Thunberg from the low country (presumably Galle) according to Alston. Wight l.c. Pl. 225 is undoubtedly *R. indicus* Thunberg and not *R. rugosus* Smith or Linné. Both the Ceylon and Indian specimens examined confirm this. It is obviously a mistake committed by Wight and later authors, the two plants being so distinct from each other.

Specimens Examined. KANDY DISTRICT: Peradeniya, 1600 ft alt., *Gardner 261* (BM, CGE, K); Rangala, edge of jungle, *Alston 313* (K, PDA); Hunnasgiriya, *Tirvengadum, Cramer & Balasubramaniam 66, 68* (G, K, L, MAU, P, PDA, US); Adam's Peak, *Tirvengadum & Cramer 302* (MAU, P, PDA, US); Maskeliya, Moray Estate, *Tirvengadum & Cramer 308* (MAU, PDA, US); *309* (MAU, P, PDA, US); *310* (G, K, L, MAU, P, PDA, US); Madulkele, *Tirvengadum & Balasubramaniam 346* (MAU, P, PDA, US). MATALE DISTRICT: *Tirvengadum, Dassanayake & Jayasuriya 7* (K, MAU, P, PDA, US); *9* (G, K, L, MAU, P, PDA, US); *10* (MAU, P, PDA, US). N'ELIYA DISTRICT: N'Eliya, *Tirvengadum & Jayasuriya 115* (K, MAU, P, PDA, US); *117* (MAU, P, PDA, US); Ramboda, *Tirvengadum & Jayasuriya 120* (MAU, P, PDA, US); Sitha Eliya, *Tirvengadum & Jayasuriya 136* (G, K, L, MAU, P, PDA, US). RATNAPURA DISTRICT: Gilimale, *Tirvengadum & Cramer 179, 183* (G, K, L, MAU, P, PDA, US); Sinharaja, *Tirvengadum & Balasubramaniam 321, 322, 323* (MAU, PDA, US); Ginigathena road, roadside jungle, Rondura group, *Simpson 8551* (BM); Rasagala, Balangoda, *Cramer 2617* (PDA). BADULLA DISTRICT: near Bandarawela, Sept. 1890, *s. coll. s.n.* (PDA); Moretota, *Tirvengadum & Jayasuriya 152* (MAU, P, PDA, US). GALLE DISTRICT: Nelluwa, *Tirvengadum & Cramer 260* (MAU, P, PDA, US); Rikilagaskade, *Amaratunga 893* (PDA); Patana near Condegale, *Alston 870* (PDA). LOCALITY UNKNOWN: *s. coll. C.P. 2765* (BM, CAL, CGE, G, MH, MPU, P, PDA); *C.P. 1535* (G); *Ma-*

crae 1210 (CGE); *Walker 29* (G); *Walker s.n.* (G); *Walker 4*, part of *C.P. 134* (K); *s. coll. s.n.* (MH); *s. coll. 279* (CAL); Intérieur de Ceylan, bon terrain, *Leschenault s.n.* (P).

8. Rubus moluccanus L., Sp. Pl. ed. 1: 1197. 1753; ed. 3: 707. 1763; Thunb., Diss. Rub. 4. 1813; Thunb., Fl. Ceil. 6. 1825; Focke, Sp. Rub., Biblioth. Bot. 17 (72): 88. 1911. Type: *Rumphius*, Herb. Amboin. 5 (7): 88, f. 2. 1747.

Scrambling shrub up to 3 m long, ultimate branches horizontal. Stem slender, brownish, hirsute, older stem ultimately greyish-white. Prickles few, scattered, hooked, small. Leaves simple, oval oblong, distinctly 5-lobed, or broadly cordate, up to 10 cm long, 8 to 13 cm broad, deeply cordate at base, basal lobes more or less rounded, lateral and terminal lobes acute, acuminate, serrulate, membranous, loosely rugose and hirsute on the nerves above, thinly greenish lanate tomentose beneath (turning buff-coloured with age). Petiole slender, 3 to 6 cm long. Stipules joined to the stem, large and highly laciniate, densely hirsute outside, deciduous. Inflorescence in terminal and axillary branched lax panicles, densely hirsute. Bracts pectinately laciniate or pinnatifid towards the apex, 1.5 cm long, semipersistent. Pedicel 1 to 1.5 cm long. Calyx lobes acuminate, toothed at tip, tomentose outside, slightly so inside. Petals broadly obovate, white. Carpels about 30, inserted on a slightly convex hirsute receptacle. Fruit globose, red; drupelets fleshy. Stones rugose.

Distr. In India, Burma, Indo-china, Indonesia and the Malay Peninsula (native of Amboina). In Ceylon, collected in the montane zone at about 1800 m alt.

Ecol. It has a shrubby growth and is found growing on rocky road banks or close to stream sides. Flower white. Flowering October to January. Fruiting June.

Note. This very variable taxon has probably been recently introduced in the island. It bears close affinity with *R. hasskarlii* Miq. The specimens from Ceylon are probably one of the numerous forms of the complex *R. moluccanus*.

Specimens Examined. KANDY DISTRICT: Pussellawa, 20th mile marker, *Tirvengadum & Jayasuriya 114* (K, L, MAU, P, PDA, US); mile 18 from Kandy, *Hepper 4407* (K, P, PDA, US).

9. Rubus micropetalus Gardn. in Calcutta J. Nat. Hist. 8: 6. 1847; Focke, Sp. Rub., Biblioth. Bot. 17 (72): 84. 1911; Gamble, Fl. Pres. Madras 1 (3): 441. 1919; Alston in Trimen, Handb. Fl. Ceylon 6: 102. 1931. Type: Ceylon, *Gardner 263* (Holotype, K; Isotypes, BM, CGE).

Rubus rugosus Smith. var. *β*. Thw., Enum. Pl. Zeyl. 101. 1859, nomen

nudum; Trimen, Cat. 30. 1885, nomen nudum. Syntypes: Ceylon, *s. coll. C.P. 1536 & C.P. 1537* (BM, CAL, G, K, MH, P, PDA).
Rubus moluccanus auct. non L.: Hook. f., Fl. Br. Ind. 2: 330. 1878, pro parte.
Rubus glomeratus auct. non Blume: Hook. f., Fl. Br. Ind. 2: 329. 1878, pro parte; Trimen, Handb. Fl. Ceylon 2: 136. 1894, pro parte.

Shrub more than 6 m long. Stem, branches and young twigs slender, covered with cinnamomeous tomentum, older part becoming ultimately white or glabrous. Prickles few, scattered, hooked, small. Leaves simple, distant, hastate, 9 to 19 cm long, 8.5 to 15 cm broad, deeply or shallowly cordate at base, 3 (–5)-lobed, more constantly 3-lobed, the two basal lobes when present almost incipient, lateral lobes acute, terminal lobe very much larger, acute and very often acuminate, irregularly dentate, more or less coriaceous, distantly rugose, hirsute on the main nerves above, densely cinnamomeous lanate tomentose beneath. Petiole 3 to 5 cm long. Stipules joined to the stem, deeply laciniate, up to 1.5 cm long, with filiform, incised segments, glabrous inside, lanate tomentose outside, cinnamomeous at first, white later, deciduous. Inflorescence in short clusters on stout peduncles in axils of upper leaves and deeply pectinately-laciniate, persistent bracts, the whole forming an elongated cinnamomeous tomentose panicle. Pedicel 0.5 to 1 cm long. Calyx lobes ovate-acuminate, up to 7 mm long, entire or slightly lobed in the upper half, toothed at apex. Petals less than half the length of the calyx lobes, white. Carpels about 20 to 35, arranged on a short conical hairy receptacle. Fruit more or less spherical, up to 1 cm across, purplish to dark red; drupelets fleshy, with persistent style. Stones rugose.

Distr. In Ceylon, in the montane zone, above 1600 m (rather common according to both Trimen and Alston; now rather rare). Also in South India, Tinnevelly District, at Kativayapil (Gamble) and Malaya.

Ecol. Found in undergrowth of forests, under slight shade. Flower white. Flowering from June to October.

Illustr. Focke, Sp. Rub., Biblioth. Bot. 17 (72): f. 32. 1911.

Note. The stem tends to root at lower nodes. At the young stage it is erect and cinnamomeous tomentose, later sarmentose, whitish or glabrous.

Specimens Examined. NUWARA ELIYA DISTRICT: Nuwara Eliya Plains, 6000 ft alt., *Gardner 263* (BM, CGE, K); Hakgala, *s. coll. C.P. 1536* (BM, CAL, G, MH, P, PDA); Hakgala Peak, *Tirvengadum & Cramer 47* (K, L, MAU, P, PDA, US); *49* (MAU, PDA, US); *52* (K, MAU, P, PDA, US); May 1888, *s. coll. s.n.* (PDA); Ramboda, *s. coll. C.P. 1537* (K, PDA). LOCALITY UNKNOWN: Regio temp., 6000 ft alt., 1845, *Thomson s.n.* (K, P); Feb. 1829, *Macrae 619* (CGE); *s. coll. s.n.* (P).

10. Rubus glomeratus Blume, Bijdr. 1111. 1826; Miq., Fl. Ind. Bat. 1: 381

1855; Trimen, Handb. Fl. Ceylon 2: 136. 1894, pro parte. Type: Java, Gede, *Blume s.n.* (L).

Rubus fulvus Focke, Biblioth. Bot. 17 (72): 81. 1911, nom. illeg. (non Sudre, 1902); Gamble, Fl. Pres. Madras 1: 441. 1915, syn. nov.
Rubus moluccanus auct. non L.: Hook. f., Fl. Br. Ind. 2: 330. 1878, pro parte; Trimen, Handb. Fl. Ceylon 2: 136. 1894, pro parte.

Strong growing, subscandent shrub. Stem rather stout, terete, covered with fulvous tomentum. Prickles few, distant, recurved, minute. Leaves simple, ovate-cordate or broadly ovate, up to 10 cm long, up to 10 cm broad, 3 to 5-lobed, not lobulate, terminal lobe biggest; lateral and basal lobes obtuse to rounded, deeply cordate at base, acute acuminate at tip, finely crenate, subcoriaceous, rugose with small areoles, sparsely pubescent with hirsute nerves above, covered with fulvous tomentum beneath. Petiole rather stout, 3 to 6.5 cm long. Stipules joined to the stem, up to 1 cm long, laciniate or pinnatifid, deciduous. Inflorescence in terminal, spreading panicles or glomerules at the tip of long peduncles in the axils of upper leaves. Bracts narrow, more than 1 cm long, pinnatifid in the upper part, persistent. Pedicel about 0.5 cm long. Calyx lobes oval or triangular, slightly toothed at apex. Petals nearly as long as the calyx lobes. Carpels about 40. Fruit small, black purple; drupelets extremely small. Stones rugose.

Distr. In South India, 1000 to 2000 m alt. (West Nilgiris, Mysore, Kerala and Coimbatore) and the Malay Peninsula. In Ceylon, introduced probably from S. India, in the montane zone (Pusselawa ?). It is now extremely rare.

It has not been found by any collector since last reported by Trimen.

Specimens Examined. LOCALITY UNKNOWN: *s. coll. C.P. 1536* (K, P); *C.P. 2765* pro parte (MH).

11. Rubus fairholmianus Gardn., Calcutta J. Nat. Hist. 8: 6. 1847; Focke, Sp. Rub., Biblioth. Bot. 17 (72): 90. 1911; Gamble, Fl. Pres. Madras 1 (3): 440. 1919; Alston in Trimen, Handb. Fl. Ceylon 6: 102. 1931. Type: Ceylon, Bopatalawe Plains, bet. Adam's Peak and Nuwara Ellia, according to *Gardner* l.c. (not seen).

Rubus rugosus Smith var. γ. Thw., Enum. Pl. Zeyl. 101. 1858. Type: Ceylon, *s. coll. C.P. 134* (Holotype, K; Isotypes, BM, CAL, G, P, PDA).
Rubus moluccanus L. var. β. *fairholmianus* (Gardn.) Trimen, Handb. Fl. Ceylon 2: 137. 1894. Type: *s. coll. C.P. 135* (not seen).

Scrambling shrub up to 3 m long. Stem whitish pilose tomentose, old stem woody, glabrous. Prickles sparse, distant, weak, slightly recurved or falcate, conical with an aristate tip. Leaves simple, usually oval oblong, 5 to

13 cm long, 5 to 10 cm broad, shallowly 5 to 7-lobed, with obtuse lobes, lobulate, subcordate to truncate at base, subacute at tip, coarsely and closely rugose with hirsute nerves above, whitish pilose tomentose (tending to become greyish or dirty white with age) beneath. Petiole 2 to 5 cm long. Stipules joined to the stem, up to 1.5 cm long, deeply laciniate, or pinnately laciniate, pilose outside, glabrous inside, tend to split due to age or heavy showers, deciduous. Inflorescence in condensed axillary and terminal many-flowered corymbose cymes or panicles, pilose. Bracts similar to stipules, persistent. Pedicel up to 0.4 cm long. Calyx lobes triangular or oval-oblong, up to 1 cm long, dentate, pilose outside, tomentose inside. Petals obovate, 6 mm long, white. Carpels about 40, inserted on a shortly convex, hirsute receptacle. Fruit subglobose, up to 1.3 cm in diameter, orange turning glossy red when ripe; drupelets few, rather small. Stones pitted.

Distr. In Ceylon, in the upper montane zone, at an elevation of 2000 m. Also in South India (Pulney and Nilgiri Hills).

Ecol. Found in openings, on hilly slopes, borders of thickets, edges of ravines, clearings of jungles, sometimes by streams or ponds. Flower white. Flowering from November to March.

Illustr. Focke, Sp. Rub., Biblioth. Bot. 17 (72): 89. f. 36. 1911.

Note. The fruit which is often bright red, glossy and attractive is with little taste and rather insipid.

Specimens Examined. NUWARA ELIYA DISTRICT: N' Eliya, Lake Gregory, *Tirvengadum & Jayasuriya 134* (K, L, MAU, P, PDA, US); Hatton Road, *Tirvengadum & Cramer 83* (G, K, L, MAU, P, PDA, US); Hakgala, 5600 ft alt., *Pearson 639* (CGE); *Alston 454, 1242* (K, PDA); ascent to Hakgala Peak, *Tirvengadum & Cramer 42, 54* (MAU, P, PDA, US); road embankment, close to 51/7 culvert, *Tirvengadum & Cramer 64* (MAU, PDA, US). LOCALITY UNKNOWN: 5000 to 7000 ft alt., *s. coll. C.P. 134* (BM, CAL, G, K, P, PDA), *C.P. 1538* (K), *C.P. 134*, part of *C.P. 1536* (K); *Walker 4*, part of *C.P. 2765* (K).

12. Rubus gardnerianus Kuntze, Method. der Species und Rubus Monogr. 1: 42. 1879; Focke, Sp. Rub., Biblioth. Bot. 17 (72): 92. 1911; Gamble, Fl. Pres. Madras 1 (3): 441. 1919; Alston in Trimen, Handb. Fl. Ceylon 6: 102. 1931. Type: Based on *R. macrocarpus* Gardn.

Rubus macrocarpus Gardn., Calcutta J. Nat. Hist. 8: 7. 1847, nomen illeg. (non Bentham, 1844). Type: Ceylon, *Gardner 262* (K).

Rubus rugosus Sm. var. δ. Thw., Enum. Pl. Zeyl. 101. 1859, nomen nudum. Type: Ceylon, *s. coll. C.P. 1535* (Holotype, K; Isotypes, B, BM, CAL, CGE, G, P, PDA, W).

Rubus moluccanus L. var. *macrocarpus* (Gardn.) Trimen, Handb. Fl. Ceylon 2: 137. 1894, pro parte.

Rubus moluccanus L. var. *gardnerianus* Kuntze, Rev. Gen. 1: 222. 1891.
Rubus moluccanus auct. non L.: Hook. f., Fl. Br. Ind. 2: 331. 1878, pro parte.
Rubus moluccanus auct. non L.: Van Royen, Gen. Rubus, Phanerog. Monogr. 2: 98. 1969, pro parte.

Semiscandent shrub. Stem woody at base, glabrescent, dark brown. Branches ferruginous hirsute, densely so in younger parts. Prickles recurved, many, scattered, large, sometimes very minute. Leaves simple, suborbicular or broadly ovate, 9 to 20 cm long, 9 to 19 cm broad, broadly 5-lobed or obscurely 3-lobed, not lobulate, often shallowly cordate to truncate, rarely distinctly cordate at base, acute or acuminate at tip, crenate-dentate, membranous, slightly and distantly rugose with large areoles, hirsute on the main nerves above, thinly covered with a dark tawny tomentum (greenish white at first) beneath. Petiole 4 to 8 cm long. Stipules joined to the stem, deeply incised, sometimes bifid, broadly ovate, up to 2 cm long, 1.5 cm broad, palmatifid, hirsute outside, glabrous inside, deciduous. Inflorescence in few-flowered, axillary and terminal racemes, often lax, hirsute. Pedicel 1 to 2 cm long. Bracts similar to stipules, sometimes shallowly toothed, persistent, sometimes deciduous. Calyx lobes ovate-lanceolate or triangular, mucronate, up to 1 cm long, the exterior ones toothed at the apex, hirsute outside, lanate tomentose inside and on the margins outside. Petals obovate, up to 9 mm long, white. Carpels about 40, borne on a conical hirsute receptacle. Fruit hemispherical, up to 1.5 cm across, red, black when ripe; drupelets few, very large, up to 5 mm in diameter. Stones rugose.

Distr. In Ceylon, in the upper montane zone, at an elevation of about 2000 m. Also in South India (Nilgiri and Travancore Hills), Java and in the Philippines, according to Kuntze l.c.

Ecol. Found under slight shade, by margin of forests or close to stream. Flower white. Flowering from October to January.

Specimens Examined. NUWARA ELIYA DISTRICT: N'Eliya, 6000 ft alt., *Gardner 262* pro parte (K); surrounding hills, *Van Beusekom 1381* (PDA); Pedrotallagala, *Wawra 1103*, *1104* pro parte (W); Hakgala, *Tirvengadum & Cramer 41* (MAU, P, PDA, US); *43* (K, L, MAU, P, PDA, US); *56* (K, MAU, P, PDA, US); jungle above Hakgala, 4.4. 1906. *s. coll. s.n.* (PDA); Pattipola, *Tirvengadum & Cramer 90* (MAU, P, PDA, US). LOCALITY UNKNOWN: *s. coll. C.P. 1535* (B, BM, CAL, CGE, G, K, P, PDA, WU); part of *C.P. 2765* (G, MH, PDA); 2 to 3000 ft alt., *Thomson 29* (G, K, P); *s.n.* (G); *Walker 29* (K); *Maxwell s.n.* (K).

13. Rubus rugosus Smith in Rees Cyclop. 30: Rubus n. 34. 1819; Wall., Pl. As. Rar. 3: 19. 1832; Wight & Arn., Prod. Ind. Or. 1: 299. 1834; Focke, Sp. Rub. Biblioth. Bot. 17 (72): 92. 1911; Cardot, Fl. gen. Indoch. 2: 637. 1920. Type: Nepal, Sembu, *Buchanan s.n.* (LINN).

Rubus hamiltonianus Sér. in DC., Prod. 2: 556. 1825; Wall., Pl. As. Rar. 3: 19, t. 234. 1832 (*R. rugosus* in text). Type: *Wallich*, l.c. t. 234.
Rubus moluccanus auct. non L.: Hook. f., Fl. Br. Ind. 2: 330. 1878, pro parte.
Rubus rugosus var. *thwaitesii* auct. non Focke: Alston in Trimen, Handb. Fl. Ceylon 6: 102. 1931. Type: Ceylon, *s. coll. C.P. 1535* pro parte (K, P, PDA).

Strongly growing, scrambling shrub. Stem hispid, greyish-yellow or dark brown, old stem glabrous. Prickles few, scattered, almost straight, weak. Leaves simple, palmate, ovate or suborbicular, 7 to 14 cm long, 6 to 12 cm broad, or broadly ovate, up to 25 cm long, up to 25 cm broad, regularly 5-lobed, lobes rounded or subacute, scarcely lobulate, deeply cordate at base, acute-acuminate or rounded at tip, crenate or irregularly serrate, subcoriaceous to coriaceous, minutely rugose with very small areoles and hirsute on the main nerves above, covered with greyish or dark brown, hirsute tomentum beneath. Petiole 3 to 7 cm long. Stipules joined to the stem, palmatifid or flabellately laciniate, up to 1.5 cm long, 1.3 cm broad, often bifid, hirsute outside, glabrous inside, deciduous. Inflorescence in few flowered, compact axillary or extra-axillary racemes or in terminal panicles, 3 to 7 cm long, hispid tomentose. Bracts palmatifid or broadly ovate and entire, hispid tomentose, deciduous. Pedicel up to 2 mm long. Calyx lobes triangular, the three outer ones larger, toothed at apex, the smaller ones acuminate. Petals broadly obovate to orbicular, 7 mm long, white. Carpels about 40 inserted on a convex densely hirsute receptacle. Fruit globose, up to 2 cm in diameter, juicy, purplish-black when ripe; drupelets few, large, up to 3 mm in diameter. Stones rugose and pitted.

D i s t r. In Nepal (native of Sembu), South India (the Nilgiri Hills, common at Dodabetta at about 2500 m). Also in Thailand, Southeast Asia and the Malay Peninsula. In Ceylon, it is rather rare and confined to the upper montane zone, above 2000 m alt.

E c o l. Found under slight shade, in bushy thickets, along edges of jungles or in dense mossy forests. Flower white. Flowering from February to May; July; from November to December.

N o t e. The fruit is large and juicy and when ripe has a good flavour. This species is easily distinguished by its distinctive leaf outline and its peculiar tomentum.

It has several times been confused with *R. indicus* Thunb. Wight (1839— Ic. Pl. Ind. Or. t. 225), for example, gave a very illustrative figure of *R. indicus* Thunb., bearing the erroneous binomial *R. rugosus* Smith. (*R. rugosus* L., in the explanation of Plates.) This mistake practised by later authors caught the notice of Gamble (l.c.) who, however, gave it the epithet *R. wightii* (after Robert Wight, Superintendent of the Madras Botanic Garden, 1819–1853), evidently taking it for a new taxon, thereby committing another error. See *R. indicus* Thunb. for further explanation.

Specimens Examined. NUWARA ELIYA DISTRICT: Nuwara Eliya 6000 to 7000 ft alt. *s. coll. C.P. 1535* (K, P, PDA); *C.P. 262* pro parte; *C.P. 1538* (BM); Pidurutalagala, *Tirvengadum & Cramer 27, 271* (K, MAU, P, PDA, US); road to Hatton, 2 miles from Nuwara Eliya, *Tirvengadum & Cramer 82* (G, K, L, MAU, P, PDA, US); Kandapola Forest Reserve, *Tirvengadum & Jayasuriya 144* (K, L, MAU, P, PDA, US); Sita Eliya 5800 ft alt., *Pearson 354* (CGE); *Tirvengadum & Jayasuriya 135* (G, K, L, MAU, P, PDA, US); Moon Plains, *Alston 625* (K, PDA); *Balasubramaniam 212* (PDA); near Hakgala, May 1888, *s. coll.* (PDA); Horton Plains, North Entrance, 2100 m, *Fosberg & Mueller Dombois 50008* (PDA, US); World's End, 7000 ft, *Pearson 490* (CGE); Greater World's End, *Tirvengadum & Cramer 101* (K, L, MAU, P, PDA, US); *280* (G, K, L, MAU, P, PDA, US).

14. Rubus lineatus Reinw. in Blume, Bijdr. 17: 1103, 1108. 1826; Hook. f., Fl. Br. Ind. 2: 333. 1878; Focke, Sp. Rub., Biblioth. Bot. 17 (72): 47. f. 14. 1911. Type: Java, *Reinwardt s.n.* (according to Blume).

Rubus pulcherrimus Hook. f., Ic. Pl. 8: Pl. 729. 1848. Type: Hooker f., Pl. 729, based on a specimen of *Lobb s.n.*, from Java.
Rubus sp., Clarke, J. Linn. Soc. Bot. 15: 141. 1877.

Strong, suberect, nearly unarmed shrub. Stem sericeous-tomentose, old stem glabrous. Prickles 0, when present straight, few, scattered. Leaves compound palmate. Leaflets 3 or 5, elliptic-lanceolate, 14 cm long, 5.5 cm broad, cuneate at base, caudate-acuminate at tip, serrate-ciliate, often doubly serrate, coriaceous, glabrous above, densely silky-silvery tomentose or glabrous beneath, subsessile; nerves 20 or 30, straight, parallel, prominent, on each side of the mid rib, one to every tooth. Petiole about 6 cm long. Stipules joined to the stem, ovate oblong, or lanceolate, entire, membranous, slightly tomentose or glabrous, semipersistent. Inflorescence in short axillary clusters or heads and terminal elongate panicles, sericeous tomentose. Bracts large, concave, orbicular, about 1.5 cm broad, membranous, glabrous. Pedicel 1.5 cm long. Calyx lobes ovate, acuminate, sericeous tomentose. Petals obovate, shorter than the calyx, white. Carpels about 40, borne on a rather low hirsute receptacle; drupelets numerous, small, red, edible. Stones rugose.

f. lineatus

Leaf undersurface densely covered with a silky silvery white tomentum.

Distr. In India (Assam, Naga Hills, Sikkim, Darjeeling), Bangladesh, East Nepal, Burma, China, Vietnam, Malay Peninsula. Cultivated in Ceylon.

Illustr. Hook. f., Ic. Pl. 8: Pl. 729. 1848; Focke, Sp. Rub., Biblioth. Bot. 17 (72): 47. f. 14. 1911.

Specimens Examined. KANDY DISTRICT: Peradeniya, *s. coll. s.n.* (PDA).

f. **glabrior** Hook. f., Fl. Br. Ind. 2: 333. 1878. Type: Mishmi, *Griffith s.n.*, according to Hook. f., l.c.

Leaf undersurface glabrous, except for a few silky silvery hairs on the main nerves.

Distr. In India, Malay Peninsula and China?. Cultivated in Ceylon.

Specimens Examined. NUWARA ELIYA DISTRICT: Hakgala, cultivated, Botanic Garden, *Simpson 9118* (BM).

Note. Hooker f. distinguished a third form, forma *angustifolia*, in India. It is characterized by its smaller and narrower leaflets, tomentose beneath.

6. FRAGARIA

L., Sp. Pl. 494. 1753; L., Gen. Pl. ed. 5: 218. 1754. Type species: *Fragaria vesca* L.

Perennial herbs, rhizomatous, scapigerous, mostly stoloniferous. Leaves palmately 3-foliolate. Stipules adnate to the base of the petiole, often membranous and sheathing. Flowers in few-flowered cymes, borne on radical peduncles, mostly bisexual. Epicalyx and calyx 5-lobed; epicalyx smaller than the calyx. Petals 5, white. Stamens 15 to 20. Carpels many, free from each other, inserted on a convex receptacle; ovules solitary; style lateral, deciduous. Fruit of many achenes, dry and crustaceous, inserted on or slightly sunk in pits on the strongly enlarged, fleshy, deciduous receptacle. Seeds without endosperm.

A genus of about 20 species. In temperate and mountainous regions of the Northern Hemisphere, South America (Mexico, Chile, Brazil). About 10 species in Asia (temperate and subtropical Himalaya). One species is represented in Ceylon.

A few hybrids are cultivated for their delicious fruits (strawberries), e.g. var. "Cambridge vigour", introduced from Germany and grown successfully at Horton Plains, New Farm, at about 2600 m alt.

Fragaria vesca L. Sp. Pl. 494. 1753; Moon, Cat. 40. 1824; Hook. f., Fl. Br. Ind. 2: 344. 1878; Trimen, Handb. Fl. Ceylon 2: 138. 1894; Kalkman, Blumea 16 (2): 350. 1968. Type: Herbier Linné (LINN). Microfiche IDC card 342: 654.

Perennial herb with an erect, stout rootstock and slender creeping stolons. Leaves palmately 3-foliolate. Leaflets broadly ovate, 3 to 7 cm long, 2.5 to 6 cm broad, crenate serrate, soft herbaceous, bright green above, paler with very short inconspicuous glandular hairs beneath; terminal leaflet cuneate at base, petiolule up to 2 mm long; lateral ones very unequal and obtuse, sessile or with petiolule up to 1 mm long. Petiole 1 to 10 cm long. Stipules 1 to 3 cm long, tapering at end, entire, membranous, sheathing. Inflorescence usually a 3-flowered erect cyme, with foliaceous bracts, occasionally solitary, pilose. Bracts trifid at apex. Scapes equal to or longer than the leaves, pilose. Pedicel up to 1.5 cm long, becoming longer in fruit. Epicalyx lobes 3.5 cm long, acute; calyx lobes 5 to 7 mm long, acuminate; both pilose outside, more so in the fruit. Petals 3.5 to 5.5 mm long, obovate, spreading, white. Carpels numerous, placed on a high convex and fleshy receptacle, pale green when immature, pinkish-red and much enlarged and juicy when mature. Fruit obovoid, 2 to 2.5 cm long, 1 to 2 cm broad, bright red; fruiting epicalyx and calyx lobes reflexed, spreading in the ripe fruits; achenes numerous, smooth, seated in pits on the oblong or globose, red or pinkish-white receptacle.

Distr. Cultivated in most temperate regions of the world. In Ceylon, cultivated in Horton Plains and in Hakgala.

Vern. Strawberry (E).

Note. The Ceylon specimens of this species investigated and quoted below are all cultivated specimens. The badly dried specimens from Ceylon gathered at 2000 m, seen by Walker Arnott (1836—Pugil. Pl. Ind. Or. 16) and cited by Trimen, must have been lost or destroyed. The only recorded specimen I have seen at Peradeniya is a sheet dated March 1893 *s.n.* with the citation "introduced. Hakgala W. rock", presumably from Hakgala Botanical Garden. "The first wild strawberries came up in a *Cinchona* clearing made in 1864", according to Trimen. But, "the patches of so-called wild strawberries found occasionally about Nuwara Eliya", which he referred to as "escapes from cultivation" must have been completely destroyed and superseded by vegetable crops that are cultivated abundantly in those regions.

This species which is identical to the common form in Europe bears little affinity with *F. nilgerrensis* Schlecht. ex Gay, of Ootacamund, South India.

Specimens Examined. NUWARA ELIYA DISTRICT: Hakgala Botanic Garden, 2300 m alt. *Tirvengadum & Jayasuriya 128* (MAU, P, PDA, US), *129* (G, K, L, MAU, P, PDA, US); Rock Garden, *Tirvengadum & Cramer 291* (K, L, MAU, P); introduced, Hakgala, West rock, *s. coll. s.n.* (PDA); Horton Plains, New Farm, *Tirvengadum & Cramer 106* (var. Cambridge vigour) (G, K, L, MAU, P, PDA, US).

7. DUCHESNEA

Smith, Trans. Linn. Soc. London 10: 373. 1811. Type species: *Duchesnea indica* (Andr.) Focke.

Perennial herbs with creeping stolons. Leaves palmately 3 (–5) foliolate. Stipules adnate to the base of the petiole. Flowers axillary, solitary, bisexual. Epicalyx and calyx usually 5-lobed; epicalyx sometimes 4-lobed, larger than calyx. Petals 5, yellow. Stamens about 20. Carpels many, free, inserted on a convex receptacle; ovules solitary; style lateral, deciduous. Fruit of many achenes, minute, spread on the surface of the slightly enlarged spongy (pulpy when dry), deciduous receptacle. Seeds without endosperm.

A genus with 2 species in the Indo-Malayan region and the Far East (China and Japan). One species is represented in Ceylon and India.

The generic name *Duchesnea* was chosen by Smith (l.c.) to commemorate the merits of A.N. Duchesne (1747–1827), author of "Histoire Naturelle des Fraisiers" (1766). Smith separated the genus *Duchesnea* from *Fragaria* in 1811. This decision was not accepted by later authors, viz. De Candolle (1825), Bentham & Hooker (1865) and Baillon (1869). Focke (in Pflanzenfam. 3 (3): 33. 1888), however, reestablished it and made the correct combination for *Duchesnea indica* (=*D. fragiformis* Sm.; *Fragaria indica* Andr.). Transfer to *Potentilla* was proposed by Wolf (Biblioth. Bot. 16: 71. 1908) followed by Kalkman (Blumea 16 (2): 325. 1968), but most authors retained the species (1 or 2) in *Duchesnea*. It could therefore be argued that *Duchesnea* is a transitional form between *Fragaria* and *Potentilla*.

For a discussion of the genus and affinity with other species, see Vidal (Fl. Camb. Laos & Vietn. 6: 111–114. 1968) and Kalkman (Blumea 16 (2): 326. 1968).

Duchesnea indica (Andr.) Focke in Pflanzenfam. 3 (3): 33. 1888; Hara & Kurosawa, J. Jap. Bot. 34: 165. 1959; Vidal, Fl. Camb. Laos & Vietn. 6: 111. 1968; Vidal, Fl. Thail., 2 (1): 61. 1970. Type: Based on *Fragaria indica* Andr.

Fragaria indica Andr., Bot. Reposit. 7: Pl. 479. 1807; Hook. f., Fl. Br. Ind. 2: 343. 1878; Abeywick., Ceylon J. Sci., Biol. Sci. 2 (2): 168. 1959. Type: According to Vidal, the type specimen had not been found at (K) nor at (BM). Hence he proposed the colour plate 479 made from a plant originating from North East Bengal, cultivated in the garden of the Honourable C. Greville, and described by Andrews (l.c.) as type.

Duchesnea fragiformis Smith, Trans. Linn. Soc. London 10: 373. 1810; D. Don, Prod. Fl. Nepal 233. 1825, nom. illeg. Based on *Fragaria indica* Andr.

Potentilla indica (Andr.) Wolf in Asch. & Graebn., Syn. Mitt. Eur. Fl. 6 (1): 661. 1904; Kalkman, Blumea 16 (2): 344. 1968. Type: Based on *Fragaria indica* Andr.

For complete synonymy, see Vidal l.c.

Extensive runner with many slender, creeping stolons. Leaves palmately 3-foliolate, distant, many at each node. Leaflets rhomboid or obovate, scabrid above, glandular pubescent beneath; terminal leaflet 1.5 to 2.5 cm long, 1 to 2 cm broad, cuneate at base, serrate except in basal part; petiolule 0.5 to 1 mm long; lateral leaflets slightly smaller, cuneate, usually very unequal at base, crenate-serrate, sometimes with deeper incisions or deeply divided so that the leaf almost becomes 5-foliolate, petiolule short or more or less sessile. Stipules narrowly lanceolate, acute, membranous, hairy on the margins. Flowers 1.5 to 2.5 cm across. Bracts toothed or incised at tip. Pedicel 3.5 to 9 cm long, slender, pilose. Epicalyx 4 to 5, foliaceous, lobes 5 to 8 mm long, ovate-acuminate; calyx lobes smaller than epicalyx lobes; both slightly pilose on both surfaces. Petals 4 mm long, obovate, spreading, bright yellow. Carpels numerous, arranged on a convex receptacle, pink, slightly enlarging in fruit, not very juicy. Fruit subovoid, 1.5 cm long, 1.3 cm broad, insipid, scarlet red when ripe; epicalyx and calyx incurved over the fruit, later spreading or reflexed; achenes numerous, scattered on the outer surface of the receptacle. Dodecaploid species, 2n = 84.

Distr. In temperate and subtropical Himalaya (Darjeeling, Simla, Kumaon, Sikkim), South India, Nepal, Bhutan, Afghanistan, Laos, Vietnam, Thailand, China, Korea, Japan, Taiwan, Luzon, Malay Archipelago. Introduced in Europe, America, Africa, Azores (Sao Miguel) and in South Africa (Johannesburg). In Ceylon, in the montane zone at an elevation of about 2500 m. In Horton Plains, it is very common and abundant.

Ecol. Usually found in disturbed habitats like tea or way-side gardens. Rarely in damp places. Flowering and fruiting from December to February.

Vern. Indian strawberry (E).

Illustr. Andrews, Bot. Repos. 7: Pl. 479. 1807 (Sub. *F. indica*); Wight, Ic. Pl. Ind. Or. 3. Pl. 989. 1845 (*F. indica*).

Note. This plant is often cultivated along paths and alleys in gardens in Europe and America because of the lively yellow flowers and the brilliance of its scarlet red fruits. But the fruit is insipid, watery and entirely destitute of flavour. The achenes of *Duchesnea* are sometimes provided with distinct appendage as in certain *Potentilla* species.

D. chrysantha (Zoll. & Mor.) Miquel, Fl. Ind. Bat. 1: 372. 1855; Hara & Kurosawa (J. Jap. Bot. 34: 165. 1959), a species that bears close affinity with *D. indica* was described from Java by Zollinger & Moritzi as *F. chrysantha* Zoll. & Mor., Syst. Verz. 7. 1846. Blume, earlier, named the Javanese Plant "*F. indica* Andr.". This practice was followed in continental Asia, so that the Javanese and Asian plants, whether from India, Japan or China were usually considered conspecific and were mostly called *D. indica*. Hara & Kurosawa (l.c.), however, made a karyological investigation on the *D. indica* group and discovered in the Javanese species the existence of a diploid number (2n = 14), besides the dodecaploid one (2n = 84). The latter they called *D. indica* s.s., and the diploid one *D. chrysantha*.

Specimens Examined. NUWARA ELIYA DISTRICT: Hakgala Botanic Garden, behind office compound, *Tirvengadum & Jayasuriya 131* (K, MAU, P, PDA, US); Rock garden, *s. coll. s.n.* (PDA); by the side of pond,

Tirvengadum & Jayasuriya 126 (G, K, L, MAU, P, PDA, US); Horton Plains, New Farm, *Tirvengadum & Cramer 105* (G, K, L, MAU, P, PDA, US).

8. POTENTILLA

L., Sp. Pl. 495. 1753; L., Gen. Pl. ed. 5. 219. 1754. Type species: *Potentilla reptans* L.

Perennial or annual herbs. Leaves imparipinnate or palmately compound. Leaflets serrate. Stipules adnate to the base of the petiole, persistent. Flowers in leafy panicles or corymbose cymes, bisexual. Epicalyx and calyx 5-lobed, epicalyx equalling the calyx or smaller than the calyx. Petals 5, yellow. Stamens many, often about 20. Carpels many, free, inserted on a small, convex receptacle; ovules solitary; style terminal, deciduous. Fruit of several achenes, on a dry, raised but not enlarged, persistent receptacle. Seeds without endosperm.

A genus of about 300 species. In temperate and warm regions of the Northern Hemisphere (Asia, Europe), a few species distributed in the mountainous regions of South America; 175 species in Asia; 2 species are represented in Ceylon.

KEY TO THE SPECIES

1 Perennial herbs. Leaves imparipinnate. Flowers in panicles . . . . **1. P. polyphylla**
1 Annual herbs. Leaves palmately 3 or 5 foliolate. Flowers in cymes . . . . . . **2. P. sundaica**

1. Potentilla polyphylla Wall. [Cat. 1026. 1829, nomen nudum] ex Lehman, Nov. et Min. Cogn. Stirp. Pug. 3: 13. 1831; Wall., Monogr. Gen. Potent. Suppl. 1: 11, t. 5. 1835; Wolf, Biblioth. Bot. 16 (71): 143. 1908; Kalkman, Blumea 16 (2): 338. 1968. Type: From Gossain Than, *Wallich 1026* (Holotype, K; Isotype, CGE).

Potentilla mooniana Wight, Ic. Pl. Ind. Or. 1: Pl. 233, text p. 14. 1839; Lehm., Rev. Potent.: 54. 1856; Thw., Enum. Pl. Zeyl. 101. 1859; Hook. f., Fl. Br. Ind. 2: 349. 1878; Trimen, Handb. Fl. Ceylon 2: 139. 1894; Murata in Hara, Fl. East Himalaya: 124. 1966. Type: *Wight s.n.* from Nuwara Eliya, Ceylon, which is probably lost.

Potentilla barbata Wall., Cat. 1030. 1829, nomen nudum. Type: Bengal, *Wallich 1030* (Holotype, K; Isotypes, CGE, P).

Potentilla polyphylla Wall. ex Lehm. var. *barbata* Lehm., Rev. Potent. 54. 1856; Wolf, Biblioth. Bot. 16 (71): 145. 1908. Type: Based on *P. barbata* Wall.

See Kalkman (l.c.) for complete synonymy.

Erect perennial herb with a stout, woody, rootstock, giving off long slender stems, up to 1 m long, hispid, ascending, sometimes prostrate. Leaves imparipinnate, in a rosette at the apex of rootstock, distant on the branches. Rachis + petiole 10 to 50 cm long, hispid. Leaflets irregularly arranged, singly below and in pairs above, the paired ones up to 3 cm long, up to 1.5 cm broad, alternating with two intermediary smaller ones, broadly ovate or oblong, truncate or rounded at base, obtuse at apex, sharply serrate, hirtellous above, more so below, sessile. Stipules membranous, acuminate, basal ones entire, upper ones leafy, toothed or lobed. Flowering stems longer than rosette leaves, 15 to 60 cm long. Inflorescence an open, irregularly branched, lax panicle, hispid. Bracts leaf-like, toothed at apex. Pedicel slender, up to 3 cm long. Epicalyx lobes foliaceous usually 3-toothed at apex. Calyx lobes ovate, acute, more or less equalling the epicalyx lobes; both hispid outside. Petals up to 0.6 cm long, obovate-orbicular, yellow. Stamens about 20. Carpels about 50, arranged on a slightly elevated hairy receptacle. Fruit ovoid; achenes smooth, spread on a dry receptacle.

Distr. In India, temperate Himalaya (from Kashmir to Sikkim, Khasi Hills), Bhutan, China (Yunnan), Malaysia, Java (Mt. Papandajan). In Ceylon, confined to the upper montane zone at altitudes ranging between 2000 to 2500 m. Rather rare.

Ecol. In swampy habitats, boggy places, open marshy grounds among grass, wet patanas, often close to mountain streams. Flower yellow. Flowering July, August.

Illustr. Wight, Ic. Pl. Ind. Or. Pl. 233 l.c. (Sub. *P. mooniana*); Van Steenis, De Trop. Nat. 21: 101. f. 3–4. 1932 (Sub. *P. mooniana*).

Specimens Examined. NUWARA ELIYA DISTRICT: Pidurutalagala, 2500 m alt., *Tirvengadum & Cramer 272* (K, L, MAU, P, PDA, US); Sept. 1890, *s. coll. s.n.* (PDA); Horton Plains, at road from Pattipola just where it enters Horton Plains, 2175 m alt., *Mueller Dombois & Comanor 67070950* (PDA, US); at New Farm, *Cooray 68051716* (PDA, US); swampy patanas, *Silva s.n.* (PDA); Patana, 7200 ft alt., *Pearson 567* (CGE); Sept. 1890, *s. coll. s.n.* (PDA); *s. coll. s.n.* (CGE). LOCALITY UNKNOWN: *s. coll. C.P. 2767* (BM, CGE, MH, P, PDA).

2. **Potentilla sundaica** (Bl.) Kuntze, Rev. Gen. 1: 219. 1891; Backer & Bakh. f., Fl. Java 1: 518. 1963; Kalkman, Blumea 16: 343. 1968. Type: Based on *Fragaria sundaica* Bl.

Fragaria sundaica Blume, Bijdr. 17: 1106. 1826. Type: Java, *Blume s.n.* (L).

Potentilla wallichiana Delile [in Wall., Cat. 1022. 1829, nomen nudum] ex Lehman, Nov. et Min. Cogn. Stirp. Pug. 3: 30. 1831, nom. illeg. (non Sér. 1825); Rev. Potent. 34: 80. 1854; Wolf, Monogr. Gatt. Potent., Bibl. Bot.

16 (71): 411. 1908. Type: Nepal, *Wallich 1022* (Holotype, K; Isotypes, BM, CGE).

Potentilla kleiniana Wight & Arn. in Wight, Ill. Ind. Bot. 1: Pl. 85. 1831; Wight & Arn., Prod. 1: 300. 1834; Thw., Enum. Pl. Zeyl. 102. 1859; Hook. f., Fl. Br. Ind. 2: 359. 1878; Trimen, Handb. Fl. Ceylon 2: 139. 1894; Alston in Trimen, Handb. Fl. Ceylon 6: 104. 1931. Type: *Wight 914* (Holotype, K; Isotypes, L, NY) according to Kalkman. Vidal gives *Wight 1007* from Nilgiris, India as type.

Annual herb 45 to 60 cm tall. Stem decumbent at base, but not often rooting, pubescent. Leaves palmately 3 or 5-foliolate. Rosette leaves larger than cauline leaves. Petiole slender; radical ones 40 cm long; stem ones 5 cm long, pubescent. Leaflets unequal, terminal one larger, up to 2 cm long, up to 1 cm broad, obovate, acute at base, obtuse at apex, serrate dentate, glabrous above, pilose on nerves beneath, more or less sessile. Stipules membranous, lanceolate, rarely trifoliate. Inflorescence a compound corymbose cyme, spreading, with 2 (sub) opposite dichasial to monochasial branches under the terminal flower. Bracts leaf-like, sparsely pilose. Pedicel slender, 2.5 cm long. Epicalyx lobes oblong, mostly entire; calyx lobes triangular, slightly larger than the epicalyx. Lobes sparsely pilose outside. Petals obovate, 3 to 3.5 mm long, yellow. Stamens 15 to 20. Carpels 40 to 50, placed on a slightly hairy, low receptacle. Achenes arranged in a globose head on the dry receptacle.

Distr. In India (Punjab, Uttar Pradesh, Assam, the Nilgiris), Bhutan, Nepal, Burma, Malaya, China (Yunnan, Kiangsu), North Vietnam, Laos, Japan, Korea and Java. In Ceylon, in the upper montane zone at an elevation of 2000 to 2500 m. Reported to be rather common by Trimen, it has now become extremely rare or extinct due to extensive clearings and agricultural activities at Horton Plains, Pattipola and Hakgala.

Ecol. It seems to be a species from often damp to marshy places, banks of brooks and lakes. Flower yellow. Flowering April, from September to October, according to Trimen.

Kalkman observes that in the specimens he studied petals develop late and fall early.

Illustr. Wight, Ill. Ind. Bot. 1: Pl. 85/925. 1831. (Sub. *P. kleiniana*); Miquel, Fl. Ind. Bat. 1: 372, t. 6. 1855. (Sub. *Duchesnea sundaica* Bl.).

Specimens Examined. NUWARA ELIYA DISTRICT: Nuwara Eliya, 6000 ft alt., *Gardner 270* (BM, CGE, K); Hava Eliya, April 1920, *Alwis s.n.* (PDA); Hakgala, stream through patana, *Simpson 9497* (PDA); Horton Plains, patana, 7200 ft alt., *Pearson 566* (CGE); Pattipola, September 1890, *s. coll. s.n.* (PDA). LOCALITY UNKOWN: 6000 ft alt., *Thomson s.n.* (K); *Walker 147* (PDA); *94* (K), *s.n.* (K); *s. coll. C.P. 2768* (BM, CGE, K, P, PDA).

9. AGRIMONIA

L., Sp. Pl. 448, 1753; L., Gen. Pl. ed. 5. 206, n° 534. 1754. Type species: *Agrimonia eupatorium* L.

Perennial herbs. Leaves compound, interruptedly imparipinnate. Leaflets serrate. Stipules adnate to the base of the petiole, persistent. Flowers in terminal spike-like racemes, bisexual. Calyx tube grooved, spiny; calyx lobes 5, persistent. Epicalyx absent. Petals 5. Stamens 5 (–10) or more. Carpels 2, free, included in the calyx tube (concave receptacle); ovule 1; style terminal, exserted. Fruit of 1 or 2 achenes, enclosed by the persistent, dry, hardened calyx tube, clothed above with hooked spines. Seeds without endosperm.

A genus of about 25 species occurring in the temperate regions and the tropical mountains of the Northern Hemisphere and South America; one species in South Africa; nine in Asia; only one species is represented in Ceylon.

Agrimonia zeylanica Moon [Cat. 37. 1824, nomen nudum] ex Hook. f., Fl. Br. Ind. 2: 362. 1878; Trimen, Handb. Fl. Ceylon 2: 141. 1894. Type: Skalicky (in Herb. P) chose *C.P. 2769* as lectotype (BM, CAL, CGE, MH, P, PDA).

Agrimonia eupatorium auct. non L.: Arn., Nova acta Phys. Med. acad. Caes. Leop. Carol. Nat. Cur. 18: 334. 1836; Wight, Ic. Pl. Ind. Or. n 12, t. 224. 1839; Thw., Enum. Pl. Zeyl. 102. 1859.

Erect perennial herb up to 1 m tall. Stem with a strong rootstock, pilose, densely so in younger parts. Leaves interruptedly imparipinnate, radical ones larger than the stem ones. Leaflets 9 to 13, very unequal, larger ones obovate, 2.5 to 4 cm long, 2 to 3.5 cm broad, obtuse or rounded at base, coarsely serrate, slightly mucronulate, scabrid above, pilose beneath, almost sessile. Stipules semilunar, 1.5 to 2 cm long, 1 cm broad, deeply toothed, silky hairy on both sides. Inflorescence in terminal spike-like racemes 10 to 20 cm long, pilose, becoming longer in fruiting conditions. Bracts trifid, leafy, persistent. Pedicel 1 to 2 mm long, pilose. Calyx tube turbinate, provided below the limb outside with 2 whorls of erect and hooked spines; calyx lobes acuminate. Petals oblong, erect at first, then spreading, yellow. Stamens 5, glabrous. Carpels 2, enclosed within the calyx tube. Achenes included in the enclosed persistent calyx tube, armed above with the persistent ring of erect and hooked spines.

Distr. Endemic. It is quite common in Ambawela, but rather rare in Pattipola.

Trimen reported it to be rather rare, but locally plentiful, especially in Uva Province.

Ecol. In open patanas (grassy montane scrub); in bushy places or among grass in marshy places. Flower yellow. Flowering from February to April.

Illustr. Wight, Ic. Pl. Ind. Or. Pl. 224 l.c. (Sub. *A. eupatorium* L.).

Note. This species is remarkable in that it does not occur in the neighbouring South Indian Hills of Peninsular India. It is, however, close to *A. eupatorium* of Europe and approaches the tropical form of that plant, especially amongst the Sikkim and Khasi specimens.

Specimens Examined. NUWARA ELIYA DISTRICT: Nuwara Eliya, 6000 ft, *Gardner 272* (BM, CGE, K); *s. coll. C.P. 2769* (BM, CAL, CGE, MH, P, PDA); Sita Eliya, *Amaratunga 730* (PDA), *Pearson 246* (CGE), *Walker 97* (K); Ohiya Railway Line, *Simpson 9575* (BM); Hakgala, 5600 ft, *Pearson 691* (CGE), *Alwis s.n.* (PDA), *Silva s.n.* (PDA); Hakgala Garden, *Tirvengadum & Jayasuriya 127* (MAU, P, PDA, US); Ambawela, *Balasubramaniam 40* (PDA); by railway track, close to Ambawela Railway Station, *Tirvengadum & Cramer 91* (G, K, L, MAU, P, PDA, US); Pattipola, *Tirvengadum & Cramer 286* (G, K, L, MAU, P, PDA, US). LOCALITY UNKNOWN: 5000–7000 ft, *Beddome s.n.* (BM); *Gardner s.n.*; part of *Gardner 272* (K); *Walker s.n.* (K); *Mrs. Walker s.n.* (K).

10. SANGUISORBA

L., Sp. Pl. 116. 1753; L., Gen. Pl. ed. 5. 53. 1754. Type species: *Sanguisorba officinalis* L.

Perennial herbs, scapigerous. Leaves compound, imparipinnate. Leaflets serrate. Petiole sheathing at the base. Stipules sheathing, adnate to the petiole. Flowers in dense terminal capitula, bisexual or polygamous. Calyx tube deeply concave. Epicalyx absent. Calyx lobes 4, petaloid, deciduous. Petals absent. Stamens 4 or numerous. Carpels 2 (–1), included in the concave receptacle (calyx tube); ovules solitary; style terminal. Fruit consisting of 2 achenes enclosed by the 4-angled, winged, hardened, dry, persistent calyx tube. Seeds without endosperm.

A genus confined to the temperate parts of the Northern Hemisphere. Subgenus *Poterium* occurs in Europe, N. Africa and India. It is represented by only one species in Ceylon, which is endemic and rare.

Sanguisorba indicum (Gardn.) Tirvengadum, comb. nov. Type: Based on *Poterium indicum* Gardn.

Poterium indicum Gardn., Calcutta J. Nat. Hist. 8: 10. 1847; Thw., Enum. Pl. Zeyl. 102. 1859; Hook. f., Fl. Br. Ind. 2: 363. 1878; Trimen, Handb. Fl. Ceylon 2: 141. 1894. Type: *Alwis C.P. 2771* (PDA).

Erect perennial herb up to 1 m tall. Stems several arising from a rootstock, angular and striate, ferruginous villous especially in the lower part. Rachis with petiole up to 20 cm long, sheathing at base. Leaflets about 8 pairs, with an odd one, broadly ovate or ovate oblong, subcordate at base, coarsely inciso-serrate, shortly mucronate, glabrous above, sparsely villous underneath. Stipules foliaceous, inciso-dentate. Inflorescence in fairly conspicuous terminal globose head at the apex of elongated scapes, up to 45 cm long, erect. Flowers polygamous. Calyx lobes ovate-elliptical, glabrous. Stamens numerous. Carpels 2; style terminal; stigmatic papillae numerous, penicilliform, as long as the style. Achenes 2, enclosed in the 4-angled, hard and winged calyx tube.

Distr. Endemic.

Note. A very rare species collected twice on Adam's Peak by H. de Alwis in 1835 and J.C. Mackenzie in 1839, according to Trimen. Gardner (l.c.) based his description on Alwis' collections which are represented by one poor specimen in fruit (now *C.P. 2771*) in the Peradeniya Herbarium. Mackenzie's specimen which is also in fruit condition is in the Kew Herbarium. I have seen a third fertile specimen, without definite locality, bearing the following label (*J. Watson*, from Ceylon, *121*, received August, 1886) in the Calcutta Herbarium. Later botanists have searched for it in vain.

Nordborg (*Sanguisorba* L.: Delimitation and Subdivision of the genera, Op. Bot. 11 (2): 72. 1966) divides genus *Sanguisorba* into 4 subgenera: subg. *Sanguisorba*; subg. *Poteridium*; subg. *Poterium*; subg. *Dendriopoterium* and into 6 sections. *Sanguisorba indicum* probably belongs to subgenus *Poterium* (L.) A. Braun & Bouché (Ind. Sem. Hort. Bot. Berol. Append. 20. 1867). The three specimens examined tally with Gardner's description and bear greater affinity with subg. *Poterium* than with subg. *Sanguisorba*. The former is characterized by "the upper flowers of capitulum female, the middle and lower bisexual; stamens numerous; carpels 2; stigmatic papillae long".

Specimens Examined. KANDY DISTRICT: Adam's Peak, 1835, *Alwis C.P. 2771* (PDA). LOCALITY UNKNOWN: *Mackenzie s.n.* (K); Received Aug. 1886, *Watson 121* (CAL).

11. ALCHEMILLA

L., Sp. Pl. 123. 1753; L., Gen. Pl. ed. 5. 58. 1754. Type species: *Alchemilla vulgaris* L.

Annual or perennial herbs, decumbent to erect. Branches often covered with overlapping leaves. Leaves simple, numerous, palmate-lobed. Stipules adnate to the sheathing petiole. Flowers in axillary corymbs or cymes, sometimes paniculate, bisexual. Calyx tube urceolate, with constricted throat; calyx lobes 8 (–10), 2-seriate, inner (sepals) alternating with an outer epicalyx of small bracteoles. Corolla absent. Stamens 4 (–1). Carpels 1 (–4), free, enclosed in the concave receptacle (calyx tube); ovules solitary; style basal. Fruit 1 (–4) achenes enclosed by the persistent, membranous calyx tube.

A genus of about 250 species or more; natives chiefly of the American Andes, a few are European, North Indian and American. Only one species is represented in Ceylon.

Alchemilla indica Gardn., Calcutta J. Nat. Hist. 8: 8. 1847; Hook. f., Fl. Br. Ind. 2: 360. 1878; Trimen, Handb. Fl. Ceylon 2: 140. 1894. Type: Ceylon, *Gardner s.n.* (Holotype, K; Isotypes, BM, CGE).

Alchemilla vulgaris L. var. *sarmentosa* Thw., Enum. Pl. Zeyl. 102. 1859. Type: *Thwaites C.P. 2770* (Holotype, K; Isotypes, BM, CGE, P, PDA).

Alchemilla vulgaris auct. non L.: Arn., Nova Acta Phys. Med. Acad. Caes. Leop. Carol. Nat. Cur. 18: 334. 1836.

Herb with an erect, somewhat woody rootstock. Stems decumbent, fairly stout or extremely slender, pilose, clothed at the base with remains of old sheathing stipules. Leaves 6 to 9-lobed, spreading, orbicular, more or less cordate at base, dentate mucronate, pilose above, densely so below and at the margin, coriaceous; basal leaves 8.5 to 10 cm in diameter, sometimes extremely small, less than 2 cm in diameter; cauline leaves smaller than basal ones; young leaves plicate in bud condition. Basal petioles longer, 12 to 21 cm, pilose. Stipules membranous, connate at base into a tube surrounding the stem, with the free end oblong-lanceolate, or acute, slightly toothed at the apex. Inflorescence crowded into dense corymbs from axis of stem leaves or in paniculate cymes, villous. Bracts simple or trifid, acute. Peduncle 5 to 7 cm long, villous. Calyx lobes 8, in 2 whorls, inner lobes (sepals) ovate, acuminate, ciliate; outer ones smaller (epicalyx). Stamens 4. Carpel 1, enclosed within the calyx tube. Achene included in the membranous, persistent calyx tube.

Distr. A species confined to the upper montane zone, at an elevation of 2000 to 2300 m. Occurs also in South India in the Nilgiris and the Pulney Hills.

Ecol. Found at margins of forests, or along jungle paths; in open marshy patana grounds or plains, in peaty depressions and on moist banks. When very luxuriant, the leaves are larger with the lobes (9 to 11) greatly overlapping. Flower yellowish green. Flowering March to April.

Note. This species, often confounded with *A. vulgaris* L., is a totally distinct species. Gardner (l.c.) chiefly distinguished it by its great pilosity, its more or less orbicular leaves and semi-orbicular lobes, the overlapping of the basal leaves and the differently shaped stipules. The flowers are much larger, the venation of the calyx segments different.

var. **indica**

Stem fairly stout. Leaves up to 10 cm in diameter. Inflorescence in dense axillary corymbs.

D i s t r. That of the species.

I l l u s t r. Wight, Ic. Pl. Ind. Or. 1: Pl. 229. 1839 (Sub. *A. vulgaris*).

S p e c i m e n s E x a m i n e d. NUWARA ELIYA DISTRICT: Moon Plains, *Tirvengadum & Cramer 142, 143* (G, K, L, MAU, P, PDA, US); Ambawela, 26.3.06, *Silva s.n.* (PDA); Horton Plains "Old Farm" 2200 m alt., *Fosberg 49986* (K, PDA, US); near Black Bridge, 2100 m alt., *Hoogland 11557* (BM, BR, CAL, CGE, E, G, K, L, NY, PDA, US); along the roadside, near settlement, near Farr Inn, *C.F. & R.J. Van Beusekom 1479* (PDA, US); Ohiya Road, near the Old Farm, 7000 ft alt., *Koyama & Cooray 13, 534* (PDA, US); *Jayasuriya 177* (PDA, US); along a track parallel with the Belihul Oya between Slab Rock Fall and Baker's Fall, *Townsend 73/157* (K); *s. coll. C.P. 2770* (BM, CAL, CGE, MH, P, PDA); Jungle path on Horton Plains, *s. coll. s.n.* (PDA); World's End, *Tirvengadum & Cramer 95, 103* (G, K, L, MAU, P, PDA, US). LOCALITY UNKNOWN: *Gardner s.n.* (BM, K, CGE); Regio temp. 4000 ft alt., *Hooker & Thomson s.n.* (P); Common about 6000 ft (7200 ft), *Pearson 565* (CGE); *Walker 3* (K); *Walker s.n.* (PDA); *Watson s.n.* (PDA); *Mc. Rae s.n.* (K).

var. **sibthorpioides** Hook. f., Fl. Br. Ind. 2: 361. 1878; Trimen, Handb. Fl. Ceylon 2: 140. 1894. Type: Ceylon, *Gardner 273* (Holotype, K; Isotype, CGE).

Alchemilla ceylanica Moon, Cat. 37. 1824, nomen nudum.

Stem extremely slender. Leaves less than 2 cm in diameter. Inflorescence in paniculate cymes.

D i s t r. Endemic in Ceylon. Rare.

S p e c i m e n s E x a m i n e d. NUWARA ELIYA DISTRICT: Horton Plains, 25/5/1911, *Silva s.n.* (PDA); Patana, Ohiya Road, *Simpson 9546* (K). LOCALITY UNKNOWN: *Gardner 273* (K).

12. PRUNUS

L., Sp. Pl. 473. 1753; L., Gen. Pl. ed. 5. 213. 1754. Type species: *Prunus domestica* L.

Pygeum Gaertn., Fruct. 1788.

Evergreen or deciduous trees or shrubs. Leaves simple, entire or serrate with 2 (–4) glands at base or on petiole. Stipules deciduous, rarely persistent. Flowers solitary, in fascicles or racemes, usually bisexual. Calyx tube obconic or campanulate; calyx lobes 5 or 6, rarely 4 or more, deciduous. Petals 5 or 6, rarely 4 or more, sometimes hardly different from sepals, white or pink. Stamens 10 to 35. Carpels 1, rarely 2 or more, free, enclosed by the calyx tube; ovary superior; ovules 2; style terminal. Fruit a drupe; mesocarp dry or fleshy; endocarp bony or woody. Seeds without endosperm.

A genus of about 300 species, distributed in the temperate and subtropical regions of the Northern Hemisphere. About 200 species in Asia. About 34 species of Section *Mesopygeum* (subg. *Laurocerasus*) in tropical Asia, from India to New Guinea, the Solomon Islands and Queensland; only two species of that section are represented in Ceylon. They are spread in the montane zone or in more open vegetation from 100 to 2600 m. Two introduced species belonging to subgenera *Amygdalus* and *Cerasus* occur at high altitudes above 1500 m.

The genus *Pygeum* was first described by Gaertner in 1788 on some loose fruits sent to him from Ceylon, as *Pygeum zeylanicum*. It has always been considered to be closely related to, but different from *Prunus*. Thus Trimen (Handb. Fl. Ceylon 2: 134. 1894) described two species of *Pygeum* and Alston (Handb. Fl. Ceylon 6 (Suppl.): 100. 1931) recognized 5 species of which two he considered doubtful. Kalkman (Blumea 13 (1): 1–31. 50–55. 1965) reduces *Pygeum* to *Prunus*. Following Koehne (Bot. Jahrb. 51: 181. 1915) who divided *Prunus* into four subgenera: *Prunophora*, *Amygdalus*, *Cerasus*, *Padus*, and Rehder (Man. Trees & Shrubs, ed. 2: 538. 1940) who divided the latter into 2 subgenera: *Padus* and *Laurocerasus*; he further distinguishes three sections of subgenus *Laurocerasus* and places under section *Mesopygeum* the large majority of the species of the former genus *Pygeum*. He maintains in his taxonomic revision of the Ceylon species of "Pygeum" the two original species described by Trimen, viz. *P. zeylanicum* Gaertn. and *P. wightianum* Bl. The two species are now respectively called *Prunus walkeri* (Wight) Kalkm. and *Prunus ceylanica* (Wight) Miq. The former is endemic and is distinguished from the latter by the pubescence of its leaves and its hairy ovary and fruit. Many species are cultivated for their edible fruits: plums, peaches, apricots, cherries and a few for their edible seeds or kernels (e.g. almond). A large number are of horticultural importance.

KEY TO THE SPECIES AND VARIETIES

1 Margin of leaves serrate. Glands petiolar or on margin of leaf base. Flowers in fascicles or solitary. Leaves deciduous

2 Stipules linear, not laciniate. Flowers sessile or shortly pedicellate, solitary. Ovary and fruit tomentose .... **1. P. persica**

2 Stipules laciniate. Flowers pedunculate (umbelled or fascicled). Ovary and fruit glabrous .... **2. P. cerasoides**

1 Margin of leaves entire, not toothed. Glands not petiolar, but on base of leaf lamina. Flowers in solitary racemes or in clusters. Leaves persistent

3 Perianth not distinctly differentiated as sepals and petals. Stamens 12 to 17. Leaves, ovary and fruit woolly pubescent. Drupe more than 2 cm broad, transversely ellipsoid (500 to 1000 m alt.) .... **3. P. walkeri**

3 Perianth distinctly differentiated as sepals and petals. Stamens 10 to 20. Leaves, ovary and fruit glabrous. Drupe up to 1 cm broad, transversely ellipsoid to didymous (above 1000 m alt.) .... **4. P. ceylanica**

4 Leaves 15 cm long, 9 cm broad. Racemes more than 6 cm long. Pedicel up to 5 mm

long (up to 1900 m alt.; common and locally abundant)................var. **ceylanica**
4 Leaves 6.5 cm long, 3.5 cm broad. Racemes up to 3 cm long. Pedicel up to 1 mm long (up to 2500 m alt.; rare)..var. **parvifolium**

1. Prunus persica (L.) Batsch, Beitr. Entw. Pragm. Gesch. 1: 30. 1801; Redher, Man. Trees & Shrubs, ed. 2: 463. 1940; Bailey, Man. Cult. Pl. Rev. ed. 542. 1949. Type: Herb. Linn. (LINN). Microfiche IDC card 329: 639 (Sub. *Amygdalus persica* L.).

Amygdalus persica L., Sp. Pl. 472. 1753.
Persica vulgaris Mill., Gard. Dict. ed. 8. 465. 1768: Sér. in DC., Prod. 2: 531. 1825.

Deciduous tree up to 5 m tall. Young twigs glabrous. Leaves 8 to 13.5 cm long, 1.5 to 4.5 cm broad, oblong, lanceolate, cuneate at base with two sessile glands on the margin, long acuminate at tip, uniformly crenate-serrate, membranous, glabrous. Petiole 0.5 to 1.5 cm long. Stipules linear, subulate, glandular, deciduous. Flowers usually solitary, in axils of fallen leaves, 2.5 to 3.5 cm across. Bracts ovate, tomentose outside. Pedicel very short or sessile. Calyx tube campanulate; calyx lobes 5, oval elliptic, obtuse, pubescent outside. Petals 5, up to 2 cm long, much expanded, pink, showy. Stamens 20 to 40. Ovary tomentose; style villous at base, glabrous above. Drupe globose, 5 to 7 cm across, fleshy, tomentose. Stone deeply pitted and furrowed, very hard.

Distr. Native of China. Cultivated in temperate regions and in mountainous regions in tropical countries. In Ceylon, cultivated in botanic gardens and nurseries; naturalized at high altitude.

Vern. Peach (E).

Note. It was once believed to be from Persia, whence the name Persica. Many varieties and hybrids are cultivated for economic and ornamental purposes. The fruit is small and of a poor quality, the flower very showy.

Specimens Examined. NUWARA ELIYA DISTRICT: compound near Loreto Convent, *Tirvengadum & Cramer 81* (MAU, P, PDA, US); Hakgala Botanic Garden, *Tirvengadum & Jayasuriya 138* (MAU, P, PDA, US).

2. Prunus cerasoides D. Don, Prod. Fl. Nepal 239. 1825; Redher, Man. Trees & Shrubs ed. 2. 470. 1940. Type: Nepal, *Buchanan s.n.* (BM).

Cerasus puddum Roxb. ex Sér. in DC., Prod. 2: 537. 1825; Roxb. ex Wall., Pl. As. Rar. 2: 37, t. 143. 1831. Type: Nepal, *Wallich 4714* H (K).
Prunus puddum Roxb. ex Brandis, For. Fl. Br. Ind. 194. 1874; Hook. f., Fl. Br. Ind. 2: 314. 1878. Type: Based on *Cerasus puddum* Roxb.
See Vidal (1968—Fl. Camb. Laos & Vietn. 6: 169) for complete synonymy.

Deciduous tree 8 to 13 m tall. Young twigs glabrous. Leaves ovate or oblong lanceolate, up to 14 cm long, up to 5 cm broad, more or less rounded or broadly cuneate at base, acuminate at tip, sharply serrate, membranous, glabrous. Petiole 0.8 to 1.5 cm long, with a pair of circular glands at apex, grooved above and provided with scales at base. Stipules laciniate, glandular, deciduous. Inflorescence from lateral buds, fascicled or umbelled, often 3-flowered, glabrous. Bracts spatulate, 4 cm long, 3.5 cm broad, toothed at margin. Pedicel slender, 0.7 to 2 cm long. Flowers 2 to 3 cm across. Calyx tube campanulate; calyx lobes 5, triangular, acute or obtuse, expanded. Petals 5, obovate or suborbicular, 1 cm long, white or pink. Stamens 30 to 35 in 2 rows, outer ones longer. Ovary glabrous. Drupe ellipsoid, 1.5 cm long, 1 cm broad, glabrous, yellowish-red. Stone bony, rugose and furrowed.

Distr. Temperate Himalaya, South India, Nepal, mountainous regions of South China, Burma, Thailand, North Vietnam and North Laos. In Ceylon, common in forests and on the hills, at high altitude above 1500 m; often cultivated in parks and gardens.

Vern. Indian cherry (E).

Illustr. Wallich, Pl. As. Rar. 2: 37, t. 143. 1831 (Sub. *Cerasus puddum*); Brandis, Indian Trees f. 121. 1906 (Sub. *Prunus puddum*); Worthington, Ceylon Trees Pl. 226. 1959 (Sub. *Prunus puddum*); Vidal, Fl. Camb. Laos & Vietnam. 6 (1): 22, f. 11–13. 1968 (Sub. *Prunus cerasoides*).

Note. The fruit is non-edible, acid and somewhat astringent.

Specimens Examined. NUWARA ELIYA DISTRICT: Nuwara Eliya Park, *Wight s.n.* (PDA); *Tirvengadum & Jayasuriya 149* (MAU, P, PDA, US); circular lawn in front of Grand Hotel, Aug. 1922, *s. coll.* (PDA); October 9, 1921, *s. coll. s.n.* (PDA); *Tirvengadum & Cramer 292* (K, L, MAU, P, PDA, US); Deltota, cultivated, Hospital, *Appuhamy s.n.* (PDA); Halgran Oya, Glen Devon, *Worthington 6211* (PDA); Hakgala Botanic Garden, *Tirvengadum & Jayasuriya 137*, *139* (MAU, PDA, US).

3. Prunus walkeri (Wight) Kalkman, Blumea 13: 55. 1965. Type: Based on *Polydontia walkeri* Wight.

Polydontia walkeri Wight, Ill. 1: 203. 1840. Type: India, *Walker s.n.* I have not seen the holotype which is probably lost. Isotype, (E).

Pygeum walkeri (Wight) Blume ex C. Muell. in Walp., Ann. 4: 642. 1857; Koehne, Bot. Jahrb. 51: 181. 1913. Type: Based on *Polydontia walkeri* Wight.

Pygeum zeylanicum auct. non Gaertn.: Hook. f., Fl. Br. Ind. 2: 321. 1878 ("Ceylanicum"); Trimen, Handb. Fl. Ceylon 2: 135. 1894; Alston in Trimen, Handb. Fl. Ceylon 6: 101. 1931.

Evergreen tree 20 to 40 m tall. Young twigs ferruginous tomentose. Leaves ovate to ovate oblong, 12 to 19 cm long, 6 to 9 cm broad, obtuse to rounded at base, acute to slightly acuminate at tip, entire, membranous, usually glabrous above except on midrib, pubescent beneath, more so on the nerves; young leaves densely pubescent, covered with golden tomentum above and below; basal glands 2 to 5 (usually 2), spherical, flattened, light green. Petiole 1 cm long, lanate tomentose. Stipules 3 to 6 mm long, linear, densely tomentose, deciduous. Inflorescence in axillary racemes or solitary, 2.5 to 7 cm long, in fruit up to 10 cm long, pubescent. Bracts pubescent, deciduous. Pedicel 2 to 3 mm long. Calyx tube obconic; calyx lobes usually 6, subulate, pilose. Petals 6, hardly different from sepals, up to 1 mm long. Stamens 12 to 17. Ovary densely tomentose. Drupe transversely ellipsoid, 1.5 to 1.6 cm long, 2.2 to 2.4 cm broad, woolly pubescent. Young fruit globose, rufous tomentose, tipped by remnant of style; endocarp bony, glabrous.

Distr. Endemic in Ceylon. It is found mostly in the warm moist parts of the island up to an elevation of 1000 m. Common in Gilimale Forest Reserve and abundant in and about Sinharaja Forest. Reported in Ginigathena and Vattegoda.

Ecol. Found along fringes of secondary forests, in heavy jungles or heavily shaded forests, slopes of hill, borders of valley and road embankments. Flowers March to April.

Vern. Kunbumbal-Kétiya or Golu-Mora (S).

Illustr. Worthington, Ceylon trees Pl. 224. 1959.

Note. Wood hard, close grained, rather heavy, yellowish white, almond-scented. Leaf margin often tattered by wind. New leaf pale dirty green, old leaf dries copper brown; basal glands turn rusty with age. Flowers creamy white. Fruit smell of prussic acid. Timber used for rafters, planks, fuel.

Specimens Examined. KURUNEGALA DISTRICT: Badagama, 443 ft alt., *Worthington 4289* (BM). KANDY DISTRICT: Kandy, alt. 2000 ft, Rest Harrow, *Worthington s.n.* (PDA); Kandy Reservoir Catchment, 2250 ft alt., *Worthington 6886* (PDA); Roseneath Small Reservoir, *Worthington 6361* (PDA); Kadugannawa, Uduwela 1800 ft alt. in the heavy jungle, *Worthington 1438* (BM); Andiatanne, *Worthington 329* (BM); Madulkele, Arratenne, 1700 ft alt., *Worthington 2007* (BM, PDA); Haliganga (Road) *Worthington 1977* (PDA); between Panwila and Madulkele, *Tirvengadum & Balasubramaniam 344* (K, L, MAU, P, PDA, US); Hantane, 3000 ft, *Gardner 482* (BM); also part of *s. coll. C.P. 1532* (K); *s. coll. C.P. 1532* (BM, CAL, K, MH); Maskeliya, Moray Estate, *Kostermans 24267* (PDA, US); Near Ginigathena in jungle, *Alston 899* (K, PDA); mile 20, Gampola-N'Eliya Road, *Meijer, Dassanayake, Balasubramaniam 599* (PDA, US); Pusselawa, 20th mile post, *Tirvengadum & Cramer 116* (G, K, L, MAU, P, PDA, US). KALUTARA DISTRICT: Morapitiya Forest Reserve, Mile 15, Culvert 7, Maha Kirihembiliye, *Meijer 484* (PDA, US). KEGALLA DISTRICT: Nawalapitiya, *Worthington 589* (BM); RATNAPURA DISTRICT: Gilimale, *Balasub-*

ramaniam 599 (PDA, US); Gilimale Forest, alt. 350 ft, *Worthington 6454* (PDA); Forest Reserve, *Meijer 488* (PDA, US); between Ratnapura and Gilimale Road, 2/5 Culvert, *Tirvengadum, Cramer & Balasubramaniam 164, 165* (G, K, L, MAU, P, PDA, US). GALLE DISTRICT: Nakiadeniya, alt. 200–250 ft, *Worthington 2344, 5258* (PDA); Deniyaya-Ensalwatte Road, *Tirvengadum & Balasubramaniam 326, 327* (G, K, L, MAU, P, PDA, US); Sinharaja Forest, above Beverley Estate, *Worthington 2592* (PDA); *Tirvengadum & Balasubramaniam 329* (L, MAU, P, PDA, US); *331, 332* (G, K, L, MAU, P, PDA, US); Deniyaya-Rakwana Road, *Tirvengadum & Balasubramaniam 336* (G, K, L, MAU, P, PDA, US); Kuruwita Korale, Feb. 1892, *s. coll. s.n.* (PDA). LOCALITY UNKNOWN: *Wight s.n.*; *Macrae 229, 454* (BM).

4. **Prunus ceylanica** (Wight) Miq., Fl. Ind. Bat. 1 (1): 366. 1855; Kalkman, Blumea 13 (1): 52. 1965. Type: Based on *Polydontia ceylanica*, Wight.

Pygeum zeylanicum Gaertn., Fruct. 1: 218, t. 46. 1788. Type: *Koenig 6404* (fruits only). Holotype in Leyden, according to Kalkman.

Pygeum acuminatum Colebr., Trans. Linn. Soc. London 12: 360, t. 18. 1818; Hook. f., Fl. Br. Ind. 2: 318. 1878. Type: Bengal, at Silhet (not seen).

Polydontia ceylanica Wight, Ill. 203. 1840. Type: From above "Numbady" *Wight s.n.* (not seen). Kalkman believes that *Wight 293* (E) is probably an authentic specimen.

Pygeum wightianum Bl. ex Muell. in Walp. Ann. 4: 642. 1857; Hook. f., Fl. Br. Ind. 2: 319. 1878; Trimen, Handb. Fl. Ceylon 2: 134. 1894; Alston in Trimen, Handb. Fl. Ceylon 6: 101. 1931. Type: Based on *Polydontia ceylanica* Wight.

Pygeum gardneri Hook. f., Fl. Br. Ind. 2: 321. 1878; Gamble, Fl. Pres. Madras 3: 439. 1919. Lectotype: India, *Wight 908* (K), according to Kalkman.

Pygeum glaberrimum Hook. f., Fl. Br. Ind. 2: 319. 1878. Lectotype: India, from Chittagong, *Hooker & Thomson s.n.* (BM, CAL, K, L, P), according to Kalkman.

Pygeum plagiocarpum Koehne, Bot. Jahrb. 51: 180. 1913; Alston in Trimen, Handb. Fl. Ceylon 6: 101. 1931. Type: *Thwaites C.P. 638* (A, BP, BM, BR, CAL, K, MH, P, PDA).

Pygeum tenuinerve Koehne, Bot. Jahrb. 51: 180. 1913; Alston in Trimen, Handb. Fl. Ceylon 6: 100. 1931. Type: *Moon s.n.* (sterile). Holotype in L, according to Kalkman. A sterile specimen (in Herb. PDA) bearing the same nomenclature is probably an isotype.

Pygeum sisparense Gamble, Kew Bull. 238. 1918; Gamble, Fl. Pres. Madras 3: 438. 1919. Lectotype: *Gamble 14339* (K), according to Kalkman.

See Kalkman (l.c.) for complete list of synonymy and types.

Evergreen tree up to 25 m tall. Young twigs pubescent, older ones glabrescent. Leaves broadly oblong, ovate or ovate-lanceolate, 4 to 6.5 (– to 15) cm long, 1.5 to 3.5 (– to 9) cm broad, cuneate or rounded to truncate at base, usually decurrent, subobtuse or acute or sometimes acuminate at tip, entire, coriaceous or subcoriaceous, glabrous above, sometimes pubescent beneath when very young, but rapidly glabrescent; basal glands 2, flat, large, oval to orbicular. Petiole 0.5 to 1 (– to 1.5) cm long, pubescent when young, glabrescent later. Stipules 3 to 4 mm long, pubescent outside, more or less glabrous inside, margins sometimes glandular, deciduous. Inflorescence in solitary racemes, in axils of leaves or their scars, 2 to 3 (– to 7) cm long, pubescent. Bracts pubescent outside, more or less glabrous inside, deciduous. Pedicel 0.5 to 1 (up to 5) mm long. Calyx tube obconic, calyx lobes 5, rarely 4 or 6, triangular or subulate, tomentose. Petals 5, elliptic or oval, up to 1.4 mm long. Stamens 10 to 20. Ovary with a ring of hairs around insertion. Drupe transversely ellipsoid to didymous, very variable in size, 9 to 1.8 cm long, 1.4 to 2.5 cm broad, glabrous, tipped by remnant of style; endocarp bony, glabrous.

Distr. In India (Assam, Bengal) Bangladesh, Burma, Thailand, Cochin China, Laos, South Vietnam (Annam), Andaman Is. and in Java. In Ceylon, in the montane zone.

Ecol. Found in secondary montane forests, often in disturbed forests or jungles.

Note. Inner bark fibrous, darkening after exposure, distinctly almond scented, exuding an aromatic almond oil, ochre in colour. Wood white, soft. Flowers white, sweet scented. Fruiting from August to October. Fruit green, through purplish red to black. Kernel smelling and tasting like bitter almond. The ripe seeds when bruised have a strong odour of prussic acid. The leaves turn brown when dry.

var. **ceylanica**

Leaves broadly oblong, 10 to 15 cm long, 5 to 9 cm broad, coriaceous; petiole 1 to 1.5 cm long. Racemes 5 to 7 cm long; pedicel up to 5 mm long.

Distr. Reported from the montane zone and occasionally at lower elevation, by Trimen. I have not collected this species which has probably become very rare due to rapid deforestation in various localities.

Illustr. Colebr., Trans. Linn. Soc. 12: t. 18. 1818; Wight, Ic. Pl. Ind. Or. 1: Pl. 256. 1839. (Sub. *Polydontia ceylanica*); 3: Pl. 993. 1845 (Sub. *Pygeum acuminatum*).

Note. The dry leaves are darkish brown and leathery in the herbarium specimens examined.

Specimens Examined. NUWARA ELIYA DISTRICT: Ramboda, 5000 ft alt., *Gardner 275* (K); 6000 ft alt., *Worthington 1925, 1926* (PDA); N'Eliya, *Thomson* in 1845 *s.n.* (K); *s. coll. C.P. 672* (K); Near N'Eliya, Coll. *Beddome 2695* (BM); Sita Eliya Forest Reserve, 6250 ft alt., *Worthington*

2907 (BM); Hakgala, Coll. *Beddome 2694* (BM); Jungle below road, above Hakgala, 21.3.1906, *Silva s.n.* (PDA); Jungle above road, above Hakgala, 30.5.1906, *Silva s.n.* (PDA). LOCALITY UNKNOWN: Regio. temp., 5000 to 7000 ft alt., *Thomson s.n.* (K, P); *s. coll. C.P. 638* (BM, CAL, K, MH, P, PDA); *Moon*? *s.n.* (PDA).

var. **parvifolium** (Thw.) Tirvengadum, comb. nov. Type: Based on *Pygeum wightianum* var. *parvifolium* Thw.

Pygeum wightianum Blume *var. β. parvifolium* Thw., Enum. Pl. Zeyl. 102. 1859; Hook. f., Fl. Br. Ind. 2: 319. 1878; Trimen, Handb. Fl. Ceylon 2: 135. 1894. Type Nuwara Eliya, *Gardner 276* (Holotype K; Isotype BM).

Pygeum parvifolium (Thw.) Koehne, Bot. Jahrb. 51: 179. 1913; Alston in Trimen, Handb. Fl. Ceylon 6: 101. 1931.

Leaves oval, or oval lanceolate, 4 to 6.5 cm long, 1.5 to 3.5 cm broad, subcoriaceous; petiole 0.5 to 1 cm long. Racemes 2 to 3 cm long; pedicel 0.5 to 1 mm long.

Distr. Probably endemic in Ceylon. Common in the upper montane zone at an elevation of 2300 to 2500 m. Abundant in Pidurutalagala (Mt. Pedro) and at Totapolakande.

Ecol. Found along borders of secondary montane and submontane forests or in disturbed forests.

Note. The leaves turn brown when dry. The inner bark and the seeds when bruised have a strong smell of bitter almond.

Specimens Examined. NUWARA ELIYA DISTRICT: Nuwara Eliya, *Gardner 276* (K, type BM); Pidurutalagala 2400 m alt., *Tirvengadum & Cramer 269, 270* (G, K, L, MAU, P, PDA, US); Totapolakande, *Tirvengadum & Cramer 111* (G, K, L, MAU, P, PDA, US). LOCALITY UNKNOWN: *s. coll. C.P. 1596* (BM, CAL, K, MH, P, PDA); *Thomson*, part of *C.P. 672* (K); 4000 to 8000 ft alt., *s. coll. s.n.* (MH); *s. coll. s.n.* (PDA).

SABIACEAE

(by C.F. van Beusekom*)

Type genus: *Meliosma* Blume.

For description, see under the genus.

The Sabiaceae mainly occur in the tropics and subtropics of Southeast Asia and Central and South America. It is a small family consisting of about 40 species in three genera, of which only the largest one, *Meliosma* Bl., occurs in Ceylon, with two species, which both have a wide distribution in Southeast Asia. The highest species density of *Meliosma* is found in Malesia and in continental Southeast Asia; the two Ceylon species occur in the southwestern offshoot of the area. In Ceylon the family is restricted to the montane and the everwet regions.

The most outstanding character of *Meliosma* is found in its peculiarly irregular flowers, which are essentially 5-merous, but show reductions towards 3-merousness in sepals, petals, and stamens. The flowers open suddenly, by an exploding mechanism, after which petals and stamens soon drop off. Hence in a panicle mainly buds and passed flowers, but few open flowers occur. The profuse cream-coloured panicles are very showy and fragrant.

Meliosma is subdivided in two subgenera of which only subg. *Meliosma* sect. *Meliosma* occurs in Ceylon; the two Ceylon species each belongs to a different subsection of sect. *Meliosma*. For more information about the taxonomy and other aspects of the genus the reader is referred to the submonographic treatment of the Asian species of *Meliosma* in Blumea 19: 355–529. 1971.

MELIOSMA

Blume, Cat. 32. 1823; Blume, Rumphia 3: 196. 1849; Benth. & Hook. f., Gen. Pl. 1: 414. 1862; Warb. in Pflanzenfam. 3 (5): 371. 1895; Beus., Blumea 19: 429. 1971. Type species: *M. lanceolata* Bl.

Millingtonia Roxb. [Hort. Beng. 3. 1814, nomen] Pl. Corom. 3: 50, t. 254. 1820, non L. f., 1781, nec Dorn, 1807. Type species: *Millingtonia simplicifolia* Roxb.

***Rijksherbarium, Leiden.**

Wellingtonia Meisn., Pl. Vasc. Gen. (Comm.) 2: 207. 1840, in nota. Type species: *Wellingtonia arnottiana* (Wight) Meisn.

Trees. Leaves alternate, simple or imparipinnate with (sub) opposite leaflets without stipules. Inflorescence terminal, a pyramidal panicle. Bracts small to minute, bracteoles absent. Flowers numerous, small, bisexual, irregular. Sepals distinct, sometimes surrounded by several persistent small bracts indistinguishable from them, all together (4) 5–13. Petals 5, distinct, 3 outer ones larger, imbricate to subvalvate, convex, (sub) orbicular, 2 inner ones much smaller, bifid, opposite and more or less adherent to the base of the fertile stamens. Disk small, annular, more or less irregularly 5-dentate. Stamens 5; 2 fertile, inflexed in bud, with short, strap-shaped filaments dilated at the top into a wide cup-shaped connective bearing two globose-elliptic, transversely dehiscent anther-cells, 3 staminodial, opposite and more or less adherent to the base of the 3 outer petals, deformed, broad, irregularly shaped, with 1 or 2 holes near the top in which fit the anther-cells of the fertile stamens, often somewhat coherent and forming a cup over the ovary. Ovary globose to ovoid, 2-locular, with simple, short, conical style and minute stigma; ovules 2 (1) per locule, superimposed, attached to the partition, hemianatropous. Fruit a more or less oblique drupe, small, with one stone; stone 1-celled, 1-seeded, splitting in two valves; seed subglobose, without endosperm. Embryo with rather long, folded radicle and folded cotyledons.

Remarks. See under the family.

KEY TO THE SPECIES

1 Leaves simple .... **1. M. simplicifolia** (subsect. **Simplices**)
1 Leaves pinnate .... **2. M. pinnata** ssp. **barbulata** (subsect. **Pinnatae**)

1. Meliosma simplicifolia (Roxb.) Walp.

Millingtonia simplicifolia Roxb.
Millingtonia pungens Wall. ex W. & A.
Meliosma pungens (Wall. ex W. & A.) Walp.
Meliosma wightii Planch. ex Brandis.

For references and typification see under the subspecies.

Evergreen tree, small to medium-sized. Leaves elliptic to lanceolate or obovate-oblong to obovate-lanceolate, rather small to big, up to c. 50 by 18 cm, entire or dentate, chartaceous to coriaceous, glabrous to pubescent, especially beneath and on midrib and nerves; nerves ascending, often looped and joined near the margin; petiole short. Panicles spreading, (5–) 10–55 cm, nearly always with decrescent leaves subtending the lower primary axes, with (sub) sessile flowers. Mature buds 1.5–2.5 mm across. Sepals and bracts (round-) ovate, usually more or less unequal, entire, the outer ones often more or less pubescent, usually all with ciliolate margin. Inner petals more or less deeply bifid, 0.6–1 mm. Ovary 0.5–0.75 (–1) mm, glabrous. Style about as long as the ovary or somewhat shorter, glabrous. Fruit (sub) globose, 5–7 mm across; stone 3.5–5.5 mm across, with distinct median keel and reticulum on the surface.

Distr. Widespread and common in Southeast Asia; from Northwestern India to Japan, and southward to South India and Ceylon, and the West and Central Malesian islands.

Note. *M. simplicifolia* is very variable, and can be divided into eight well-marked subspecies, which for the most part cover separate, adjoining areas, but are often also restricted to different altitudinal zones, for instance in the Himalayas where three subspecies replace each other at different altitudes. In contact zones between subspecies transitional forms are usually found. In Ceylon the two following subspecies occur, which are restricted to different altitudinal zones.

KEY TO THE SUBSPECIES

1 Sepals (including one bract) (5) 6 .... 1. ssp. **simplicifolia**
1 Sepals (including several bracts) 8–13 .... 2. ssp. **pungens**

1. ssp. **simplicifolia** Beus., Blumea 19: 476. 1971.

Meliosma simplicifolia (Roxb.) Walp., Rep. 1: 103. 1842; Thw., Enum. Pl. Zeyl. 59. 1858; Beddome, Fl. Sylv. 3: 77. 1869; Brandis, For. Fl. 116. 1874; Hook. f., Fl. Br. Ind. 2: 5. 1876; Trimen, Handb. Fl. Ceylon 1: 315. 1893; Brandis, Ind. Trees 194. 1906; Worthington, Ceylon Trees 151. 1959.
Millingtonia simplicifolia Roxb. [Hort. Beng. 3. 1814. nomen] Pl. Corom. 3: 50, t. 254. 1820; Roxb., Fl. Ind. 1: 103. 1820; Wight & Arn., Prod. 1: 115. 1834; Anon., Ic. Roxb. 4: 40, t. 20. 1970. Lectotype: Roxburgh Ic. nr. 2104, in Kew Library (reproductions in Roxb., Pl. Corom., l.c., and Anon., Ic. Roxb., l.c.).

Leaves obovate-oblong to obovate-lanceolate, up to c. 50 by 18 cm, index (2–) 2.5–3.5 (–4), entire (but often dentate in leaves of water-shoots!), the base long cuneate to acute, the apex acute to short-cuspidate, above glabrous, beneath sometimes sparsely pubescent, often with tufts of hair in the nerve-axils; midrib and lesser so the nerves above nearly always more or less prominent, rarely flat or slightly impressed; nerves 8–23 pairs. Panicles rather lax, branched up to the 3rd or 4th order; axes slender, angular, sparsely to rather densely pubescent but not tomentose. Flowers more or less crowded to solitary; mature buds 1.5–2 mm across. Bracts 1 per flower. Sepals (4) 5.

Inner petals 0.6–0.75 mm, usually over halfway bifid; lobes more or less divergent, narrow, glabrous. Stone rather oblique, nearly triangular at central view; median keel usually very prominent.

Distr. Ceylon: Central and Southern Provinces. Outside Ceylon: South Peninsular and Northeast India, Bangladesh, Burma, Southwest China, North Vietnam, Thailand, Sumatra.

Ecol. In primary and scondary, moist evergreen tropical forests, found from c. 300 m up to c. 900 m altitude.

Vern. Elbedda, Elbetta or Albadda-gass (S).

Note. See under ssp. *pungens*.

Specimens Examined. KANDY DISTRICT: Peradeniya Bot. Gard. c. 500 m alt. *Van Beusekom 1656* (L, PDA, US); Peradeniya, *Macrae s.n.* ex Herb. *Gardner* (F); Gannoruwa forest, c. 500 m alt., *Van Beusekom 1656A* (L, PDA, US); Udawattekelle, 1700 ft alt., *Worthington 6565* (K); Hantane, 3000 ft alt., *Gardner 132* (BM, F); Tumpane valley, 1500 ft alt., *Worthington 1472* (K); Gonagampitiya in Ganga Palata, 1900 ft alt., *Worthington 6566* (K). RATNAPURA DISTRICT: Ambagama, 3000 ft alt., *s. coll. C.P. 703* (PDA). KEGALLE DISTRICT: Alagala, Dekkanda, 2500 ft, *Worthington 332* (K). DISTRICT UNKNOWN: s. loc. *Walker 1323* (E), *Wight 116* (E).

2. ssp. **pungens** (Wall. ex Wight & Arn.) Beus., Blumea 19: 466. 1971.

Millingtonia pungens Wall. ex Wight & Arn., Edinb. New Phil. J. 15: 178. 1833; Wight & Arn., Prod. 1: 115. 1834; Wight, Ic. Pl. Ind. Or. 3: t. 964/3. 1845. Type: *Wight 945* (Peninsular India) (E, P).

Millingtonia pungens (Wall. ex Wight & Arn.) Walp., Rep. 1: 423. 1842; Thw., Enum. Pl. Zeyl. 59. 1858; Bedd., Fl. Sylv. 1: t. 160. 1869; ibid. 3: 77. 1869.

Millingtonia wightii Planch. ex Brandis, For. Fl. 116. 1874; Hook. f., Fl. Br. Ind. 2: 4. 1876; Trimen, Handb. Fl. Ceylon 1: 314. 1893; Brandis, Ind. Trees 194. 1906; Gamble, Fl. Pres. Madras 1: 256. 1918. Type: '*M. wightii* Planch.' *s. coll.* et loco (Peninsular India or Ceylon) (K).

Leaves elliptic to oblong, sometimes lanceolate, up to c. 20 (–30) by 8 (–10) cm, index 2–3 (–4), without or with some distant teeth (leaves lower on the branches and on water-shoots are usually dentate), acute to rounded at the base, acute to acuminate at the apex, usually distinctly pubescent on midrib and sometimes on nerves above, sparsely to moderately pubescent beneath, usually with tufts of hair in the nerve-axils; midrib and nerves usually impressed above, sometimes flat or slightly prominent; nerves 7–18 pairs. Panicles lax to dense, branched up to the 2nd (3rd) order; axes rather coarse, terete to slightly angular, densely short-tomentose. Flowers crowded in dense glomerules; mature buds 2–2.5 mm across. Bracts (3) 4–6 (–8) per flower.

Sepals 5. Inner petals c. 1 mm, slightly bifid; lobes divergent, wide, glabrous. Stone not much oblique but sometimes rather irregular; median keel distinct, but not very prominent.

Distr. Ceylon: Central Province. Outside Ceylon: Southwest Peninsular India.

Ecol. Common in hill-forests at least from c. 1600 up to c. 2100 m altitude.

Note. This subspecies and ssp. *simplicifolia* are restricted to well-separate altitudinal zones, so that contact between them hardly or not occurs. Transitional forms between them have as yet not been found in Ceylon, but from South India two such cases are known. About the name of ssp. *pungens* there has been much confusion; the current epithet '*wightii*' should be replaced by the older name '*pungens*'. The flowering tree has the habit of a mango tree; in South India it is sometimes called 'hill mango'.

Specimens Examined. NUWARA ELIYA DISTRICT: N. Eliya, Maturata, *Thwaites C.P. 300* pro parte (F, K, L, P, PDA, W); Hakgala, c. 1650 m alt., c. 1900 m alt., *Van Beusekom 1441* (L, PDA, US), *Worthington 3006* (K), *A.M. Silva s.n.*, 22 IV. 1906 (PDA); Horton Plains, 7000 ft alt., *Worthington 1677* (K); Elephant Plains, 6000 ft alt., *Gardner 133* (BM); Haputale, *A.M. Silva s.n.*, II. 1890 (PDA). KANDY DISTRICT: Hunnasgiriya, *s. coll. C.P. 300* pro parte (K, P, PDA).

2. Meliosma pinnata (Roxb.) Maxim., Diagn. Pl. Nov. Jap. Mandsh. 4 & 5: 263. 1867, in obs.; Beus., Blumea 19: 494. 1971.

Millingtonia pinnata Roxb., Hort. Beng. Roxb., 3. 1814, nomen; Roxy., Fl. Ind. 1: 103. 1820. Type: Roxburgh, Ic. nr. 2105, in Kew Library (reproduction in Anon., Ic. Roxb. 4: 38, t. 19. 1970).

Meliosma arnottiana (Wight) Walp., Rep. 1: 423. 1842; Thw., Enum. Pl. Zeyl. 59. 1858; Beddome, Fl. Sylv. 1: t. 160. 1869; ibid. 3: 77. 1869; Hook. f., Fl. Br. Ind. 2: 6. 1876; Trimen, Handb. Fl. Ceylon 1: 315. 1893; Brandis, Ind. Trees 195. 1906; Gamble, Fl. Pres. Madras 1: 256. 1918; Worthington, Ceylon Trees 150. 1959.

Meliosma pinnata ssp. *arnottiana* (Wight) Beus., Blumea 19: 499. 1971, nom. illeg.

Millingtonia arnottiana Wight, Ill. 1: 144, t. 53. 1840. Type: *Wight 554* (Peninsular India) (E, K).

Meliosma rhoifolia Maxim, Diagn. Pl. Nov. Jap. Mandsh. 4 & 5: 262. 1867 Syntypes: *Oldham 85* (K) and *86/1* (A, BM, K, P, W) (both from Taiwan).

For description, see under the subspecies.

Distr. Widespread and common in SE Asia; all over Malesia up to

New Guinea, northward up to Japan, Central China and NE India, to the southwest in Peninsular India and Ceylon.

Note. *M. pinnata* is very variable; it is divided into nine well-marked subspecies, most of which have similarly replacing areas as *M. simplicifolia* (see there).

In Ceylon only one subspecies is found.

Ssp. **barbulata** Cufod., Oest. Bot. Z. 88: 254. 1939 (as *M. rhoifolia* ssp. *barbulata*). Type: *Handel-Mazzetti 12235* (China) (A, W).

Meliosma arnottiana (Wight) Walp.
Meliosma pinnata ssp. *arnottiana* (Wight) Beus.
Millingtonia arnottiana Wight.
Meliosma rhoifolia Maxim.

For references and typification, see under the species.

Evergreen tree, small to medium-sized. Leaves (2) 3–7 (8)-jugate, with up to c. 40 cm long rachis; leaflets ovate to ovate-oblong, or elliptic to oblong or lanceolate, especially the upper ones sometimes more or less obovate to obovate-oblong, more or less increasing in size towards the leaf-top, small to big, up to c. 15 by 5 cm, index 1.5–4, acute to rounded or truncate at the base, acuminate to cuspidate at the apex, entire to dentate, chartaceous to coriaceous, moderately pubescent beneath and on midrib above, often with tufts of hair in the nerve-axils beneath; nerves ascending, looped and joined near the margin or not; petiolules short, pubescent. Panicles 10–50 cm, spreading, rather dense, profusely branched up to the 4th order, with (sub) sessile, crowded flowers, pubescent, nearly always with decrescent leaves subtending the lower primary axes. Buds c. 2 mm across. Sepals (4) 5, and 1 bractlet, (round-) ovate, more or less unequal, entire, glabrous or the outer ones with a few hairs, all with ciliolate margin. Inner petals deeply bifid, c. 0.6 mm. Ovary c. 0.6 mm, densely pubescent. Style about as long as, or somewhat longer than, the ovary, glabrous. Fruit (sub) globose, 4–6 mm across; stone 3–4.5 mm across, with distinct median keel and reticulum on the surface.

Distr. Ceylon: Central Province. Outside Ceylon: South Peninsular and Northeast India, Burma, Thailand, Vietnam, China, South Japan, Philippines, Malay Peninsula, Sumatra, Java.

Ecol. Common in hill-forests from c. 900 m up to c. 2000 m altitude. In the upper montane zone often occurring together with *M. simplicifolia* ssp. *pungens*.

Vern. Wal-bilin or Nika-dawula (S); Kusavi (T).

Specimens Examined. NUWARA ELIYA DISTRICT: N. Eliya, 4000–7000 ft alt., *s. coll. C.P. 293* pro parte (F, K, L, P, PDA, W), *Worth-*

ington 1934 (BM); N. Eliya, Bulu Ela, 6250 ft alt., *Worthington 2914* (BM), *2915*, *2924*; Hakgala, 1850 m alt., *Van Beusekom 1443* (L, PDA, US); below Hakgala, *A.M. Silva s.n.* 24 V. 1911 (PDA). KANDY DISTRICT: Maskeliya, near Gartmore Estate, *J.M. Silva s.n.*, 9 IV. 1926 (PDA); Hunnasgiriya, *Thwaites C.P. 293* pro parte (K, PDA, W); Udawattekelle, 2900 ft alt. *Worthington 1483* (K); Madulkelle, 4285 ft alt., *Worthington 1974* (BM); Dunally Gap, Galaha, 3650 ft alt., *Worthington 2423* (BM). MATALE DISTRICT: Kandanuwara, Matale, 3100 ft alt.. *Worthington 2388* (BM). KEGALLE DISTRICT: Kellie Estate, Dolosbage, 3200 ft alt., *Worthington 1886* (K). RATNAPURA DISTRICT: Balangoda Estate, towards Denegama, 3250-4200 ft alt., *Worthington 3253*, *3254*, *3262*, *3274* (BM). BADULLA DISTRICT: Tonacombe, Namunukula, Uva, *Worthington 5419* (K). DISTRICT UNKNOWN: S. loc. *Walker 1285* (E), *Wight 115* (E).

SCROPHULARIACEAE

(by L.H. Cramer*)

Type genus *Scrophularia* L.

Unarmed, annual or perennial herbs or shrubs, sometimes root parasites; stem prostrate, decumbent or erect. Leaves usually opposite, occasionally verticillate, rarely alternate, exstipulate, entire to pinnatipartite, mostly penninerved. Flowers axillary, solitary, or in axillary or terminal spikes, racemes or cymes, often bracteate, bisexual, usually zygomorphic, occasionally subactinomorphic. Calyx generally tubular or campanulate, rarely spathaceous; lobes 4–5, variously divided to nearly free at base. Corolla limb subactinomorphic to 2-lipped; lobes 4–5, imbricate or valvate in bud. Stamens inserted on corolla tube, 2, with the anterior ones sometimes reduced to staminodes, or generally 4, the fifth staminodial or completely suppressed, the four usually didynamous; anterior filaments occasionally appendiculate at base; anthers free, less often connivent in pairs, cells distinct, rarely confluent or 1-celled, longitudinally dehiscent, rarely poricidal. Ovary superior, sessile, 2-locular, with axile placentation; ovules numerous in each loculus; style terminal; stigma capitate or bilobed. Fruit a capsule, septicidally, loculicidally or loculicido-septicidally dehiscent, generally 2-valved, less often 3–4-valved; seeds very small, endospermous; embryo straight or curved.

A large family considered to be tropical in origin but truly cosmopolitan in distribution, of 220 genera and about 3000 species, with representatives in Ceylon among 20 genera and 57 species. The Ceylon species have a wide distribution in the dry and wet zones of the country, from sea level to 2335 m. Of these the majority count as common annuals growing along the borders of tanks, ponds and in paddy fields of the dry zone and of the wet lowlands. Restricted to the dry zone are *Peplidium maritimum* (L.f.) Aschers. and *Verbascum chinense* (L.) Santapau; and to the montane region of the wet zone are *Torenia aerinea* Alston, the species of *Veronica* L., *Sopubia trifida* Buch.-Ham. ex D. Don, *Verbascum thapsus* L. and *Pedicularis zeylanica* Benth.

The Scrophulariaceae are most closely allied to the Solanaceae from which they are distinguished by the absence of bicollateral vascular bundles in the stem, the median position of the carpels, in most cases by the typically zygo-

*Flora of Ceylon Project, Peradeniya.

morphic corolla, and the reduction of the posterior stamen. Of the other allied families the Orobanchaceae and Gesneriaceae differ from it in their unilocular ovary with parietal placentation, the Bignoniaceae and Pedaliaceae in the absence of endosperm and the Acanthaceae in the internal structure of the capsule and the presence of usually conspicuous bracts.

The accepted classification of the Scrophulariaceae today is the one formulated by von Wettstein (1891). According to this system the family is divided into three subfamilies—the Verbascoideae (Pseudosolanaceae), Scrophularioideae (Antirrhinoideae) and Rhinanthoideae. Each of these is further subdivided into two or more tribes with their respective genera. The artificial key provided below has deviated from this system in an attempt to facilitate identification in the field.

KEY TO THE GENERA

1 Leaves usually all alternate
 2 Corolla subrotate; calyx tubular; fruiting calyx longer than capsule...... **1. Verbascum**
 2 Corolla bilabiate; calyx spathaceous; fruiting calyx shorter than capsule.. **2. Pedicularis**
1 Leaves all opposite or verticillate or at least the lower ones opposite
 3 Leaves sessile or subsessile
  4 Stem quadrangular or angular; usually terrestrial plants (except 7)
   5 Corolla limb widely cupular, shortly spurred; lower lip concave and tuberculate at base.. **3. Angelonia**
   5 Corolla limb flattened or narrowly bilabiate, not spurred; lower lip flattened and plain at base
    6 Corolla subrotate or subregular; filaments of stamens pubescent
     7 Lower leaves trifid pinnatisect; bracteoles filiform; calyx campanulate, entire; anther-cells not spurred at base.................... **4. Sopubia**
     7 All leaves entire; bracteoles oblong; calyx spathaceous, split anteriorly; anther-cells spurred at base............................ **5. Centranthera**
    6 Corolla bilabiate or subbilabiate; filaments of stamens glabrous
     8 Calyx ribbed or angular and after anthesis longer than capsule; anthers 1-celled
      9 Corolla tube abruptly incurved near the middle; bracteoles present; root parasites .... **6. Striga**
      9 Corolla tube straight; bracteoles absent; autophytic plants...... **7. Microcarpaea**
     8 Calyx smooth and after anthesis more or less equal to or shorter than capsule; anthers 2-celled
      10 Glabrous or simply pubescent herbs without a distinct scent; bracteoles absent; anther-cells sessile
       11 Stem higher than 45 cm, corolla more than 4 mm across towards base, longer than 2 cm....................................... **8. Artanema**
       11 Stem shorter than 45 cm, corolla less than 4 mm across towards base, shorter than 2 cm.. **9. Lindernia**
      10 Viscid-pubescent, aromatic herbs; bracteoles usually present; anther-cells stipitate.. **10. Stemodia**
  4 Stem terete or subterete; usually limnophytes (except 15)
   12 Perfect stamens 4, or, if 2, with staminodes
    13 Leaves entire

14 Calyx lobes very unequal; perfect stamens 4; capsule septicidal........**11. Bacopa**
14 Calyx lobes more or less equal; perfect stamens 2; capsule loculicidal............ ....**12. Dopatrium**
13 Leaves serrate or pinnatifid to laciniate
15 Sepals very unequal: anthers free, sessile............................**11. Bacopa**
15 Sepals more or less equal; anthers connivent in pairs, stipitate.....**13. Limnophila**
12 Perfect stamens 2, without staminodes
16 Leaves fleshy; capsule globose, rounded at apex; stamens included; anthers 1-celled ....**14. Peplidium**
16 Leaves thin; capsule compressed, emarginate at apex; stamens exserted; anthers 2-celled..**15. Veronica**
3 Leaves petiolate
17 Corolla subregular
18 Corolla throat densely hairy; bracts absent; capsule globose, rounded at apex; perfect stamens 4..**16. Scoparia**
18 Corolla throat glabrous; bracts present; capsule compressed, emarginate at apex; perfect stamens 2..**15. Veronica**
17 Corolla irregular
19 Leaves pinnately lobed; corolla lower lip calceolate; perfect stamens 2.......... ....**17. Calceolaria**
19 Leaves entire, serrate or pinnatifid; corolla lower lip simple; perfect stamens 4
20 Filaments of anterior stamens appendiculate at base
21 Calyx smooth; corolla less than 1.5 cm long, tube not dilated above......... ....**9. Lindernia**
21 Calyx winged or angular; corolla more than 1.5 cm long, tube dilated above ....**18. Torenia**
20 Filaments of anterior stamens inappendiculate at base
22 Flowers distinctly pedicellate; corolla lower lip bullately 2-lobed at base; flowering calyx cleft to middle; herbs without a distinct scent......**19. Mazus**
22 Flowers sessile or subsessile; corolla lower lip smoothly 3-lobed at base; flowering calyx cleft to base; usually aromatic herbs
23 Bracteoles present; 2 or all stamens with one anther-cell imperfect........ ....**20. Adenosma**
23 Bracteoles absent; all stamens with both anther-cells perfect
24 Adaxial calyx lobe more or less equal to the others; filaments of posterior stamens glabrous; terrestrial plants......................**10. Stemodia**
24 Adaxial calyx lobe much longer than the others; filaments of posterior stamens pilose; limnophytes.................................**13. Limnophila**

1. VERBASCUM

L., Sp. Pl. 177. 1753; L., Gen. Pl. ed. 5, 83. 1754; Endl., Gen. Pl. 670. 1839; Benth. & Hook., Gen. Pl. 2: 928. 1876 (omnes emend. Kuntze, Rev. Gen. Pl. 2: 468. 1891). Type species: *Verbascum thapsus* L.

Celsia L., Sp. Pl. 621. 1753; L., Gen. Pl. ed. 5. 272. 1754; Benth. & Hook. ibid. 929. Type species: *Celsia orientalis* L.

Large annual, biennial or perennial pubescent herbs, rarely undershrubs. Stem erect and unbranched, rarely weak and branched. Leaves usually soft, entire, crenate, sinnuate-dentate, pinnatifid or pinnatipartite. Spikes or racemes terminal, simple or branched. Pedicels when present ebracteolate. Calyx deeply 5-lobed, lobes imbricate, nearly equal. Corolla yellow or brown, subrotate; limb spreading; lobes 5, slightly subequal, the 2 posterior outside in bud. Stamens 5, or when 4, didynamously inserted at base of corolla tube; posterior filaments bearded; anthers 1-celled or 2-celled, cells confluent. Stigma dilated, compressed, slightly bilobed. Capsule globose, ovoid or oblong, septicidally 2-valved, valves usually bilobed; seeds numerous, rugose or warted.

A genus of temperate regions and the old tropics with 2 species in Ceylon. *Verbascum* and *Celsia* were placed in distinct classes in the Linnaean system. O. Kuntze relying on von Nees' evidence that the fifth stamen in *Verbascum* is often incomplete and sterile fused *Celsia* to *Verbascum*. In the lack of unanimity among present-day taxonomists regarding the validity of *Celsia* as a distinct genus, the opinions of Kuntze (1891), Pennell (1943) and Santapau (1950) are followed in treating these two as one genus.

KEY TO THE SPECIES

1 Indumentum lanate with stellately dendriform hairs; leaves entire to crenate; flowers in spikes; stamens 5 .... **1. V. thapsus**
1 Indumentum hirsute or strigose; leaves coarsely dentate to pinnatifid; flowers in racemes; stamens 4 .... **2. V. chinense**

1. Verbascum thapsus L., Sp. Pl. 177. 1753; Trimen, Handb. Fl. Ceylon 3: 241. 1895; Alston in Trimen, l.c. 6: 210. 1931. Type: No. 242. 1 (LINN.).

Annual. Stem erect, up to 2.8 m high, simple or slightly branched, subterete, stout below. Leaves sessile, oblong-lanceolate to oval-oblong, 12.8–54 × 2.8–16.4 cm, narrowed and decurrent at base down stem there becoming winged, lower ones acute, upper acuminate at apex, sparsely stellately dendriform above, densely so beneath, becoming bracteate above. Spikes 0.6–1.22 m long, simple, stiff. Bracteoles linear-lanceolate, up to 10 × 1.75 mm. Calyx tube campanulate, 4–5 mm long; lobes lanceolate, 8–10 × 2 mm, acuminate, stellate-tomentose without. Corolla subrotate, chrome-yellow; tube 7–8 mm long, glabrous on both surfaces; limb 2–2.7 cm across; lobes spreading, suborbicular, glabrous within, stellate-tomentose without. Stamens exserted; two anterior filaments longer than three posterior ones, glabrous; posterior filaments bearded. Ovary stellate-tomentose. Capsule broadly ovoid, 7–10 × 7–7.5 mm, bluntly 4-lobed, stellate-tomentose; seeds ovoid-oblong, up to 1 mm long, rugosely grooved, dark brown.

Distr. Native of Europe and temperate Asia including the Himalayas, spread through North India and the Deccan Peninsula to Ceylon.

Ecol. An introduced weed naturalized in waste ground and road embankments around Nuwara Eliya. The exact date of its introduction is not known. Till about the year 1956 it was reported to be common towards the Hakgala-Ambawela side of the Nuwara Eliya district; but extensive clearings on this side have rendered the plant there extinct. It is now observed, however, to be spreading in the direction of the Kandapola side of the district where it still continues to be common enough. Flowering October–March.

Specimens Examined. NUWARA ELIYA DISTRICT: Sita Eliya, Oct. 1906, *Silva s.n.* (PDA); Nuwara Eliya, in compound of house opposite Catholic Church, *Cramer 3724* (PDA); Kandapola, roadside near culvert 11/7, Nuwara Eliya Rd., *Cramer 3754* (PDA, US).

2. Verbascum chinense (L.) Santapau, Fl. Purandhar 90. 1958.

Scrophularia chinensis L., Mant. 2: 250. 177. Type: No. 773. 20 (LINN.).
Celsia coromandeliana Vahl, Symb. Bot. 3: 79. 1794; Thw., Enum. Pl. Zeyl. 217. 1860: Hook. f., Fl. Br. Ind. 4: 251. 1885; Trimen, l.c. 240. Type: "India orientalis," *Koenig* (?).
Verbascum coromandelianum (Vahl) Kuntze, Rev. Gen. Pl. 1: 468. 1891; Abeywick., Ceyl. J. Sci., Biol. Sci. 2 (2): 221. 1959.

Annual. Stem 60–100 cm high, stout, pubescent, branched at summit. Radical leaves on long petioles, compound or pinnatisect, with several small leaflets or segments at base and a large oblong-oval acute terminal one; cauline ones alternate, sessile, oblong-ovate, becoming bracteolate above; all coarsely dentate, pubescent on both surfaces. Flowers moderate-sized, in long erect narrow racemes terminating stem and branches. Bracts leafy. Pedicels slender, glandular-pubescent. Calyx lobes oblong, acute, glandular-pubescent. Corolla yellow; limb 1.2 cm in diam.; lobes rounded. Filaments equal, pubescent. Capsule globose, 8 mm broad.

Distr. "Very rare,"—Trimen. Afghanistan; India from the eastern Himalayas through the Deccan Peninsula and Ceylon to Burma; China.

Ecol. In the dry zone.

Specimens Examined. JAFFNA DISTRICT: Delft, *Gardner, C.P. 2021 & 2041* (PDA, K).

Note. Except for some modification given above the description is Trimen's. He says, "I have only seen the C.P. specimens which were collected by Gardner in the island of Delft ('Neduntivu')." The author himself could find no trace of the plant in Delft. It is very probable that this species is already extinct in Delft owing to grazing by sheep and goats that roam about untended in the island.

2. PEDICULARIS

L., Sp. Pl. 607. 1753; L., Gen. Pl. ed. 5. 266. 1754; Benth. & Hook., Gen. Pl. 2: 978. 1876; Wettst. in Pflanzenfam. 4 (3b): 104. 1891. Type species: *P. sylvatica* L.

Annual or perennial, often semi-parasitic herbs. Root-stock often well developed; roots fibrous or tuberous. Stem erect. Leaves usually alternate, occasionally verticillate, simple, crenate-serrate, pinnately lobed. Flowers often second, in terminal spikes or racemes, ebracteolate. Calyx spathaceously tubular-campanulate, anteriorly, less often posteriorly, cleft nearly to base, 2–5-toothed above. Corolla red or purple; tube cylindric, slightly dilated above; limb strongly 2-lipped, the posterior lobes inside in bud; upper lip erect, galeate, lower spreading or deflexed, 3-lobed. Stamens 4, didynamous, under upper lip of corolla; anthers connivent in pairs, 2-celled; cells equal, parallel. Style slender; stigma capitate. Capsule compressed, falcate or oblique, often beaked, loculicidally 2-valved, the valves coherent in the middle; seeds numerous, ovoid or oblong.

Over 600 species, mainly from the temperate regions of the Northern Hemisphere, with only one indigenous species in Ceylon.

Pedicularis zeylanica Benth., Scroph. Ind. 54. 1835; Benth. in DC., Prod. 10: 580. 1846; Thw., Enum. Pl. Zeyl. 221. 1860; Hook. f., Fl. Br. Ind. 4: 317. 1885; Trimen, l.c. 260; Abeywick., l.c. 222. Type: Ceylon, *Macrae s.n.* (K).

Annual with tuberous roots. Stem erect, up to 52 cm high, generally simple, occasionally branched, terete, hoary hirsute. Leaves usually crowded near base of stem, fewer and distant above, uppermost ones bracteiform, oblong to ovate-oblong, 2.4–4.6 × 1.1–1.6 cm, truncate at base, obtuse at apex, irregularly crenately lobed, scabrid on both surfaces; petioles 2–5 mm long. Racemes lax, 5–16 cm long. Bracts 6–15 × 2–10 mm. Calyx tube 3–10 mm long, split down anterior side, 4–5-ribbed, hirsute on ribs, pinkish-purple; lobes broadly ovate, 2–3 mm long, irregularly sublobed at apex, minutely hirsute on upper margin. Corolla bright pink, 2–2.5 cm long; tube 8–11 mm long, dilated above, faintly white, pilose without, villous within; upper lip erect, 8–9 mm long; lower lip 10–16 mm across, deeply lobed, lobes reflexed with age, midlobe slightly smaller than lateral ones. Filaments pilose below, glabrous above; anther cells shortly spurred at base. Style glabrous; stigma globose, faintly emarginate, papillose. Capsule ellipsoid-lanceolate, 12–15 × 5–7 mm, about $1\frac{1}{2}$ times as long as calyx, subfalcate, subcompressed, glabrous; seeds ellipsoid, 1.5 × 1 mm, scrobiculately reticulate, blackish-brown.

D i s t r. From the western Ghats of South India to Ceylon.

E c o l. In patana slopes of the upper hill country above 2000 m. The plant usually grows in association with *Satyrium nepalense* D. Don and at first sight can be easily mistaken for this orchid in the field but for its crenately lobed and penninerved leaves and the scrophulariaceous corolla. About April, after fruiting, the plants all wither and are not to be seen till shortly before the next flowering season. The whole plant blackens on drying. Flowering August–February.

Specimens Examined. NUWARA ELIYA DISTRICT: Galagame, *Gardner s.n.* (PDA); Sita Eliya, Oct. 1906, *Willis s.n.* (PDA), Sep. & Oct. 1906, *de Silva s.n.* (PDA); Moon Plains, Aug. 1926, *de Silva s.n.* (PDA); Horton Plains, Ohiya rd., below Rest House, *Comanor 418* (PDA), *Cramer 3140* (NBV, PDA, US); Ambawela, near culvert 4/1, Ambawela-Nanu Oya Rd., *Cramer 3255* (PDA).

3. ANGELONIA

Humb. & Bonpl., Pl. Aequin. 2: 92, t. 108. 1809; Benth., Gen. Pl. 2: 930. 1876; Wettst. in Pflanzenfam. 4 (3b): 53. 1891. Type species: *A. salicariifolia* Humb. & Bonpl.

Annual or perennial herbs, rarely shrubs; stem procumbent, weakly climbing, or erect, glabrous, pilose or viscid-pubescent. Leaves opposite or verticillate, or the upper ones alternate. Flowers solitary and axillary, becoming racemose above, ebracteolate. Calyx deeply 5-partite. Corolla subrotate to 2-lipped; tube short, expanded; limb 5-lobed, lobes spreading, rounded; throat broadened into a sac-like cup, often 2-tuberculate at base. Stamens 4; filaments short, curved; anthers 2-celled, cells divaricate. Stigma capitate. Capsule globose or ellipsoid, loculicidally 2-valved, valves entire, rarely indehiscent.

About 30 species, mostly native of tropical America.

Angelonia salicariifolia Humb. & Bonpl., Pl. Aequin. 2: 92, t. 108. 1812.

Viscid-pubescent herb with a perennial rootstock. Stem up to 1.30 m high, branched below, densely glandular-hispid. Leaves opposite throughout, lanceolate to lanceolate-oblong, 3–10 × 0.5–1.2 cm, narrowed at base, acute at apex, shallowly serrate at margins, penninerved, minutely glandular-hirsute on both surfaces. Flowers opposite, becoming racemose above. Bracts foliaceous, up to 4 × 0.6 cm, acuminate. Pedicels 12–15 mm, slender, cernuous, densely glandular-hispid. Sepals ovate to ovate-lanceolate, 4–5 × 1.5–2 mm, acuminate at apex, glandular-hispid without, closed in after anthesis. Corolla purplish-blue; tube very short; limb widely cupular, up to 2.2 cm across, faux spotted dark violet within; upper lip 4-lobed, lobes suborbicular, with yellow glandular hairs at bases on upper surface; lower lip up to 1.2 cm long, concave at base, with a white emarginate tongue-shaped appendix at lower margin of cavity and two small tubercles at upper. Filaments thickened, incurved, pilose; anthers pale blue. Style glabrous. Capsule globose, 3–6 mm in diam., tipped with withered style, longer than calyx, sparsely puberulent; seeds oblong-cuneiform, 1 mm long, 4–5-angular, reticulately foveolate.

Distr. Native of Brazil spread through cultivation in North America, West Indies, India & Ceylon.

Ecol. First introduced into cultivation in Ceylon in 1847, the plant is now becoming naturalised along the borders of ditches, ponds and tanks of the lowlands. The flowers are very conspicuous on account of their shape and colour and have a pleasant mint-like scent. Flowering throughout the year.

Uses. An ornamental plant in gardens.

Specimens Examined. KANDY DISTRICT: Peradeniya, Royal Botanic Gardens, Apr. 1921, *s. coll. s.n.* (PDA). GALLE DISTRICT: Koggala, *Cramer 3351* (PDA, US). TRINCOMALEE DISTRICT: Andakulam, by Tank, *Cramer 3007* (PDA). BATTICALOA DISTRICT: Kalkudah, close to culvert 20/1, Batticaloa Rd., *Cramer 3624* (PDA, US).

4. SOPUBIA

Buch.-Ham. in D. Don, Prod. Fl. Nepal 88. 1825; Benth. & Hook., Gen. Pl. 2: 970. 1876; Wettst. in Pflanzenfam. 4 (3b): 94. 1891. Type species: *Sopubia trifida* Buch.-Ham. ex D. Don.

Raphidophyllum Hochst., Flora 24: 666. 1941. Type species: *R. simplex* Hochst. (= *Sopubia dregeana* Benth.).

Glabrous or pubescent annual herbs. Stem erect, much branched. Leaves opposite below, alternate above, or verticillate, narrow, often pinnatisect with linear or filiform segments. Flowers solitary and axillary, or in terminal racemes, shortly pedicelled; pedicels 2-bracteolate. Calyx campanulate or tubular, 5-toothed or shortly 5-lobed. Corolla usually subrotate, more rarely subbilabiate limb spreading; lobes 5, subequal, the two posterior ones inside in bud. Stamens 4, didynamous, not completely included; filaments short, curved; anthers, at least one pair, connivent, 2-celled, one cell fertile, poricidal, the other smaller, stipitate, empty. Stigma thickened, tongue-shaped, obtuse. Capsule ovoid or oblong, loculicidally 2-valved; valves entire or 2-fid; seeds numerous, oblong or obovoid.

About 40 species, all palaeotropic.

KEY TO THE SPECIES

1 Lower leaves pinnatisect; corolla bilabiate, rose or pinkish-violet; stamens included; capsule longer than calyx, rounded at top .... **1. S. delphinifolia**

1 Lower leaves trifid; corolla subrotate, yellow with a purple eye; stamens exserted; capsule as long as calyx, emarginate at top .... **2. S. trifida**

1. Sopubia delphinifolia (L.) G. Don, Gen. Syst. 4: 460. 1837; Thw., Enum. Pl. Zeyl. 220. 1860; Hook. f., Fl. Br. Ind. 4: 302. 1885; Trimen, l.c., 257; Alston in Trimen, Handb. Fl. Ceylon 6: 217. 1931; Abeywick., l.c., 222.

Gerardia delphinifolia L., Cent. Pl. 2: 21. 1756. Type: "India orientalis," No. *764. 1* (LINN.).

Stem erect, up to 1 m high, with erecto-patent branches, glabrous, somewhat woody and dark purple towards base. Lower leaves in verticils of 3–5, 2–5.7 cm long, segments unequal, filiform, spreading, flexuous; upper ones opposite, linear-filiform, 1.5–2.2 cm long, entire. Flowers numerous, axillary, solitary, becoming racemose above. Bracts 1.2–2.4 cm long. Pedicels 2–6 mm long, slender, glabrous. Bracteoles 2–2.5 mm long. Calyx tube 4–6 mm long, 10-ribbed, glabrous; lobes linear-lanceolate, 4–6 mm long. Corolla tube 1.4–1.6 cm long, puberulous without, glabrous within; lower lip 8–13 mm across, lobes suborbicular, mid-lobe larger than lateral ones. Filaments inserted in middle of corolla tube, glandular-scabrid, pink, those of anterior pair arched. Style minutely strigose. Capsule oblong-ellipsoid, 5.5–6.5 × 3 mm, rounded at top, apiculate by withered style, glabrous, black; seeds oblong-cuneate, 0.5–0.8 mm long, truncate at broader end, striate, black.

Distr. India, from the Gangetic Plain through the Western Ghats and Deccan Peninsula to Ceylon.

Ecol. In open grassy places in the dry zone of the southern province and in the wet hill country from 600–1300 m; fairly common. Owing to its profusion of flowers the plant in full bloom affords a pleasant sight. Flowering December–February, August–September.

Specimens Examined. JAFFNA DISTRICT: Jaffna, Feb. 1890, *s. coll.* (PDA). KANDY DISTRICT: Hunnasgiriya, June 1926, *Silva* (PDA); Madugoda, *Simpson 9489* (PDA); Urugala, near culvert 22/2, Mahiyangana Road, *Cramer 3612* (PDA); Muruthalawa, Nelli Hill, *Cramer 2798* (PDA, US). NUWARA ELIYA DISTRICT: Maturata, *Gardner C.P. 2022* (K, PDA); Between Ramboda & Pusellawa, above Highway A 5, *Mueller-Dombois & Cooray 68011206* (PDA, US). BATTICALOA DISTRICT: Bintenne, Apr. 1923, *Silva* (PDA). BADULLA DISTRICT: Bandarawela, *s. coll.* Sep. 1890, *s. coll. s.n.* (PDA). HAMBANTOTA DISTRICT: Yala, Ruhuna National Park, Block 1, *Mueller-Dombois & Cooray 68012827* (PDA, US); Patanagala, *Cramer 3309* (PDA, US); Jamburagala, *Cramer 3313* (NBV, PDA, US).

2. Sopubia trifida Buch.-Ham. ex D. Don, Prod. Fl. Nepal 88. 1825; Benth. in DC., Prod. 10: 522. 1846; Thw., Enum. Pl. Zeyl. 220. 1860; Hook. f., Fl. Br. Ind. 4: 302. 1885; Trimen, l.c. 257; Abeywick., l.c. 222. Type: North Nepal, *Hamilton* (?).

Stem erect, up to 60 cm high, stiff, thin, sulcate and angular, glabrous, pale green above, purplish below, often branched. Leaves opposite or in verticils of 3–4; lower ones 2.8–3.5 cm long, segments linear, the terminal one longer than the lateral ones; upper ones entire, linear, up to 16 mm long; all glabrous above, puberulous beneath. Flowers in terminal racemes up to 16 cm long. Bracts linear, 1–2.8 cm long. Pedicels 5–8 mm long, slender, slightly cernuous. Bracteoles subulate, 3.5 mm long. Calyx tube 3–4.5 mm long, 10-ribbed; lobes ovate-deltoid, 2 × 1.5 mm, acute, glabrous without, lanate within along margins. Corolla tube 3–4.5 mm long, glabrous; limb 1.4–1.6 cm across, lobes orbicular, 6–7 × 5.6 mm. Filaments declinate, glabrous, purplish; lobes of perfect anther puberulent on inner margins. Style minutely scabrid. Capsule ellipsoid, 5–6 × 3 mm, obliquely emarginate at top, apiculate by withered style, glabrous, black; seeds oblong-cuneate, 0.7–1 mm long, truncate at one end, striate, blackish-brown.

Distr. India, from the Western Himalayas southwards through the Western Ghats and the Deccan to Ceylon.

Ecol. In patana slopes of the upper hill country above 3100 m; common. A pretty herb on account of its dissected leaves and yellow flowers. Flowering throughout the year.

Specimens Examined. NUWARA ELIYA DISTRICT: Maturata, *s. coll. C.P. 525* (K, PDA); Hakgala, Patana, Oct. 1906, *s. coll. s.n.* (PDA); Roadside below Plot Pl, *Mueller-Dombois & Cooray 68011429* (PDA, US); Boragas, by Boragas-Ambawela Road, *Cramer 3150* (PDA, US); *& 3246* (NBV, PDA, US). BADULLA DISTRICT: Bandarawela, Sep. 1890, *s. coll. s.n.* (PDA).

5. CENTRANTHERA

R. Br., Prod. 438. 1810; Endl., Gen. Pl. 692. 1839; Benth. & Hook., Gen. Pl. 2: 969. 1876; Wettst. in Pflanzenfam. 4 (3b): 94. 1891. Type species: *Centranthera hispida* R. Br. (*nom. illegit.*).

Razumovia Spreng., Mant. Fl. Hal. 1: 45. 1807 (non *Razumovia* Spreng. 1805). Type species: *R. tranquebarica* Spreng. (= *Centranthera tranquebarica* (Spreng.) Merr.).

Annual scabrid or hispid herbs with orange-coloured roots. Stem prostrate or erect, quadrangular or terete. Leaves opposite throughout or opposite below, alternate above, oblong or linear, entire. Flowers axillary, solitary or forming lax spikes or racemes; pedicels bibracteolate. Bracteoles oblong. Calyx asymmetrical, shortly 3–5-lobed. Corolla rose-coloured or purple; tube curved, widely infundibuliform; limb indistinctly bilabiate; lobes 5, spreading, subequal, two posterior ones inside in bud. Stamens 4, included, didynamous; filaments pilose; anthers connivent in pairs; cells spurred or mucronate at base, one perfect, the other small, empty or aborted. Stigma tongue-shaped. Capsule loculicidally 2-valved; valves entire; seeds numerous, oblong-cuneate.

About 9 species of the oriental tropics, reported to be root-parasites on grasses.

The earlier name *Razumovia* Spreng. employed for the genus by Britten, Alston and Merrill is a later homonym of *Razumovia* Spreng. in Allg. Lit.-Zeitung Intelligenzbl. 136. 1805, denoting a genus of the Compositae. Hence the name *Razumovia* Spreng. being already preoccupied cannot be applied again to the scrophulariaceous genus.

KEY TO THE SPECIES

1 Leaves oblong or ovate-oblong; corolla more than 15 mm long; filaments pilose........ **1. C. indica**

1 Leaves linear; corolla less than 15 mm long; filaments glabrous....**2. C. tranquebarica**

1. Centranthera indica (L.) Gamble, Fl. Pres. Madras 971. 1924; Abeywick., l.c. 222.

Rhinanthus indica L., Sp. Pl. 603. 1753. Type: Ceylon, Hermann, Mus. No. 20 (BM).

Centranthera hispida R. Br., Prod. 438. 1810; Thw., Enum. Pl. Zeyl. 220. 1860; Hook. f., Fl. Br. Ind. 4: 301. 1885; Trimen, l.c. 259; Abeywick., l.c. 222. Type: not known.

Centranthera procumbens Benth. in DC., Prod. 10: 525. 1846; Hook. f., Fl. Br. Ind. 4: 301. 1885; Trimen, l.c. 258; Petch, Ann. R. Bot. Gard. Peradeniya 6: 96. 1915. Type: "India orientalis," *Wight 2484* (?).

Razumovia indica (L.) Alston in Trimen, Handb. Fl. Ceylon 6: 217. 1931; Alston, Kandy Fl. 60, f. 325. 1938.

Razumovia hispida (R. Br.) Britten, J. Bot. 39: 69. 1901; Alston in Trimen, l.c.

Stem prostrate and diffusely branched from base or erect, up to 90 cm high, branched above, hispid. Leaves oblong to ovate-oblong, 1.2–3 × 0.4–0.7 cm, spreading or erect, truncate at base, subacute at apex, ciliate at margins, hispid on both surfaces. Flowers axillary, solitary or in terminal, lax spike-like racemes up to 15 cm long. Bracts linear-lanceolate to ovate-lanceolate, 4–7 × 1.5–4 mm, ciliate at margins. Pedicels 1–2 mm long. Bracteoles oblong to ovate-oblong, 3–3.5 × 1 mm. Calyx tube 5–19 mm long, obtusely 5-angular, scabrid or hispid without; lobes lanceolate. Corolla 18–31 mm long; tube 14–23 mm long, dotted with yellow glands and puberulous without, commonly purplish-red, less often rose-coloured or white; limb commonly white, less often rose-coloured. Style glabrous. Capsule ovoid, 5–8 × 4–5 mm, shorter than calyx, glabrous, dark brown; seeds oblong-cuneate, 1 mm long, striate.

Distr. South India through Ceylon and Burma to Malaysia. China and Australia.

Ecol. In open damp or dry grassy places of the wet lowlands and the lower hill country up to 1000 m, close to a source of water; common; rather rare in the dry region (Batticaloa). It may be a root parasite of grasses, but field observations have been unable to establish or dismiss this. Flowering throughout the year.

Specimens Examined. KANDY DISTRICT: Hunnasgiriya, *Gardner C.P. 2025* (K, PDA); Galagedera, Oct. 1882, *s. coll. s.n.* (PDA); Ganoruwa, Nov. 1927, *Alston s.n.* (PDA). COLOMBO DISTRICT: Colombo, *Gardner C.P. 2026* (K, PDA); Delgoda, Mar. 1919, *Lewis s.n.* (PDA). KALUTARA DISTRICT: Kalutara, *Gardner C.P. 2025* (K, PDA); Nagoda, by tarmac of abandoned airport, *Cramer 2790, 2791 & 3225* (PDA, US). RATNAPURA DISTRICT: Opata, Mar. 1881, *s. coll. s.n.* (PDA); Dewalagawa, behind Dewalagawa Vidyalaya, *Cramer 3098* (PDA, US) *& 3100* (NBV, PDA, US). KURUNEGALA DISTRICT: Kurunegala, by Tittawela Tank, *Cramer 3237* (PDA, US). GALLE DISTRICT: Koggala, by tarmac of airport, *Cramer 2332* (PDA) *& 3363* (PDA, US); Hiniduma, foot of Godekande forest, *Cramer 3345* (NBV, PDA, US). HAMBANTOTA DISTRICT: Yalewala, Ruhuna National Park, Block 2, *Mueller-Dombois 67062326* (PDA, US). BATTICALOA DISTRICT: Batticaloa, *Gardner C.P. 2026* (K, PDA); Kalkudah, *Cramer 2738* (PDA, US); Idaikuruppu, *Cramer 2751* (PDA, US).

Note. Indian floras generally use flower colour as one of the characters for delimiting this species from the earlier *Centranthera hispida* R. Br. They associate white corollas with *C. indica* and rose-coloured or purple ones with *C. hispida*. Previously, Trimen's observations were about the same though he did not use flower colour as the basis of his key. Close observations in the field, however, have indicated that such colour characters are not constant in these two species, the different colours occurring in either of the two. Besides, Pennell (1943) associates 4 villous ciliate filaments with the white- or yellow-flowered species and 2 each of glabrous and villous filaments with the rose or purple flowered ones. The specimens of *C. indica* examined by the author all had both pairs of filaments villous-ciliate. In the absence, therefore, of reliable and constant characters by which these two species can be separated, the author has followed Santapau (Fl. Khandala 184. 1967) in merging *C. hispida* R. Br. under *C. indica* (L.) Gamble.

2. Centranthera tranquebarica (Spreng.) Merr., 150th Anniv. Vol. R. Bot. Gard. Calcutta 55. 1941.

Razumovia tranquebarica Spreng., Mant. Fl. Hal. 45. 1807 & Syst. 2: 812.

1825; Britten, J. Bot. 29: 69. 1901. Type: not known.

Torenia lepidota Roth, Nov. Pl. Sp. 281. 1821. Type: "India orientalis."

Centranthera humifusa Wall. ex Benth., Scroph. Ind. 50. 1835 & in DC., Prod. 10: 525. 1846; Thw., Enum. Pl. Zeyl. 221. 1860; Hook. f., Fl. Br. Ind. 4: 301. 1885; Trimen, l.c. 259. Type: *Wall. Num. List 3883* (K-Wall.).

Razumoiva lepidota (Roth) Alston in Trimen, Handb. Fl. Ceylon 6: 217. 1931.

Stem erect, up to 15.5 cm high, often diffusely branched from base, slender, hispid on angles. Leaves linear, opposite below, alternate above, 0.9–2.8 × 0.2 cm, narrowed at base, subobtuse at apex, pustular scabrid on both surfaces. Flowers in axils of middle and upper leaves. Bracts oblong, 1–1.5 mm long. Bracteoles linear-oblong, 1–1.5 mm long. Calyx tube inflated, 4.5–5 mm long, hispid without. Corolla 10–11 mm long; tube 6–9 mm long, dorsally dull purple, ventrally yellowish; lobes of limb yellow; lower lip 7 mm across. Style glabrous. Capsule globose, 4 × 4 mm, slightly shorter than calyx, tipped with withered style, glabrous, brown; seeds oblong-cuneate, 0.8 mm long, reticulate.

Distr. From India through Ceylon to Malaysia.

Ecol. In open, damp grassy places close to watercourses or ponds or paddy fields; rather rare. Flowering May–February.

Specimens Examined. KURUNEGALA DISTRICT: Polgahawela, by Polgahawela-Narammala Rd., *Cramer 3269* (NBV, PDA, US). COLOMBO DISTRICT: Colombo, *Gardner C.P. 2888* (K, PDA). KALUTARA DISTRICT: Nagoda, in meadow opposite Catholic Convent, *Cramer 2974* (PDA, US). KEGALLE DISTRICT: near Kegalle, *Alston 977* (PDA). GALLE DISTRICT: Galle, *Gardner C.P. 3888* (K, PDA); Udugama, Panwila, *Cramer 2363* (PDA). HAMBANTOTA DISTRICT: Yala, Ruhuna National Park, Uraniya, *Mueller-Dombois 68050303* (PDA, US).

6. STRIGA

Lour., Fl. Cochinch. 22. 1790; Benth. in Hook., Comp. Bot. Mag. 1: 36. 1826; Endl., Gen. Pl. 685. 1839; Benth. & Hook., Gen. Pl. 2: 968. 1876; Wettst. in Pflanzenfam. 4 (3b): 96. 1891. Type species: *S. lutea* Lour.

Campuleia Thouars, Gen. Nov. Madagasc. 7. 1807. Type species: *C. coccinea* Hook. (= *Striga lutea* Lour.).

Annual, often semiparasitic scabrid herbs. Stem erect. Leaves opposite below, alternate above, generally linear to subulate, sometimes reduced to scales. Flowers axillary, solitary, sessile, in terminal bracteate spikes above, usually minutely bibracteolate. Calyx tubular, strongly 5–15 ribbed; limb 5-toothed or 5-fid. Corolla tube narrow, incurved near middle, two posterior lobes of limb inside in bud; upper lip usually emarginate or 2-fid; lower lip spreading, 3-lobed. Stamens 4, included, didynamous; anthers 1-celled, vertical, the connective sometimes mucronate. Stigma thickened or clavate. Capsule oblong, loculicidally 2-valved; valves entire, septiferous; seeds numerous, obovoid or oblong.

About 40 species, of the palaeotropics.

KEY TO THE SPECIES

1 Stem leafless, the leaves reduced to scales; calyx-ribs as many as number of lobes...... .... **1. S. gesnerioides**
1 Stem with ordinary leaves; calyx-ribs more than the number of lobes
 2 One calyx-rib terminating at the tip of each lobe, the rest terminating at the sinuses; flowers chrome-yellow.. **2. S. lutea**
 2 Three calyx-ribs terminating at the tip of each lobe; flowers white....**3. S. angustifolia**

1. Striga gesnerioides (Willd.) Vatke, Oesterr. Bot. Z. 25: 11. 1875; Alston in Trimen, Handb. Fl. Ceylon 6: 216. 1931; Abeywick., l.c. 222.

Buchnera gesnerioides Willd., Sp. Pl. 3: 338. 1800. Type: "India orientalis prope Pondepillj," not known.

Buchnera orobanchioides R. Br., Salt. Abyss. App. 64. 1814, nom. nud. & ex Endl., Bot. Zeit. 2: 388, t. 2. 1832; Benth., Scroph. Ind. 40. 1835. Type: not known.

Striga orobanchioides (R. Br. ex Endl.) Benth. in Hook., Comp. Bot. Mag. 1: 36, t. 19. 1836; Trimen, J. Bot. 23: 172. 1841; Hook. f., Fl. Br. Ind. 4: 299. 1885; Trimen, l.c. 255.

Stem erect, 19–42 cm high, leafless, moderately stout, terete, branched, scabrid, pale green or purplish, globosely tuberous at base. Scales triangular, 3–4 mm long, crowded and overlapping at base of stem, scattered above, becoming bracteate at flowering end. Flowers whorled, in erect, tapering spikes 15–20 cm long. Bracts ovate, 7 × 2.5 mm, subacuminate, concave, scabrid, pale green. Bracteoles linear-lanceolate, 3.5 × 1 mm, acuminate, scabrid. Calyx tube 2.5–3 mm long, scabrid without, 5-ribbed, ribs terminating at apex of lobes; lobes unequal, linear, 3–5 mm long, acuminate, pale green. Corolla white or pale pink; tube up to 11 mm long, stiff, glabrous; upper lip irregularly obovate, 3 mm long; lower lip 6–7 mm across, lobes cuneiform, 5 mm long; throat villous within. Filaments short, those of posterior pair 1 mm long. Style glabrous; stigma spathulate, bent. Capsule ovoid-globose, 5–7 × 3 mm, glabrous, apiculate by long, withered style, glabrous, black; seeds linear-oblong, 0.6 mm long, striate, black.

D i s t r. From tropical and South Africa through Arabia and Sind to Western India and Ceylon.

E c o l. Under shade of *Euphorbia antiquorum* L. and parasitic on its surface-feeding roots. "The whole plant is either pale green when the flowers are white, or deep reddish-purple when the flowers are dull pale purplish-pink. Dries black."—Trimen. Flowering August–February.

Specimens Examined. ANURADHAPURA DISTRICT: near Anuradhapura, in 1882, *s. coll. s.n.* (PDA). HAMBANTOTA DISTRICT: Hambantota, Dec. 1882, *s. coll. s.n.* (PDA); close to culvert 2/1, Tissa Rd., *Cramer 2818* (NBV, PDA, US); Bundala, Dec. 1882, *s. coll. s.n.* (PDA); Yala, Ruhuna National Park, Block 1, *Cooray 69111607* (PDA, US).

2. Striga lutea Lour., Fl. Cochinch. 22. 1790; Hook. f., Fl. Br. Ind. 4: 299. 1885; Trimen, l.c. 256; Alston in Trimen, Handb. Fl. Ceylon 6: 216. 1931; Alston, Kandy Fl. 60, f. 324. 1938. Type: Canton, *Loureiro* (P).

Buchnera asiatica L., Sp. Pl. 630. 1753, pro parte.

Buchnera hirsuta Benth., Scroph. Ind. 41. 1835; Thw., Enum. Pl. Zeyl. 220. 1860.

Stem up to 22.5 cm high, rigid, occasionally branched above, setulose. Leaves linear, 5–11 × 1–2 mm, rounded at base, subacute at apex, ciliolate at margins, setulose on both surfaces. Flowers in terminal, lax spikes up to 6 cm long. Bracts linear-lanceolate, 3–6 × 0.5–1 mm, minutely hispid on both surfaces. Bracteoles linear-subulate, 2–4 mm long, ciliolate at margins. Calyx tube 3–4 mm long, 10-ribbed, scabrid on ribs; lobes linear-lanceolate, 1.5–3 mm long, acuminate, scabrid without. Corolla chrome-yellow; tube glandular puberulous without; throat villous; upper lip suborbicular, 2-lobed above; lower lip 6–7 mm across, lobes obovate, 2–3 mm long, minutely dotted brown above, mid-lobe larger than lateral ones. Style glabrous; stigma 2-lobed. Capsule ovoid, 3.5–5 × 1.5–2 mm, obtuse at apex, glabrous, dark brown; seeds irregularly oblong-cuneate, 0.5 mm long, glabrous, black.

Distr. Tropical and South Africa, India, Ceylon and Southeast Asia.

Ecol. In open, grassy places of the wet lowlands and of the hill country; fairly common. Owing to its small size the plant can easily be missed in the field, but the chrome-yellow flowers help to spot it out. Probably parasitic on the roots of grasses. Flowering almost throughout the year.

Note. Some authors such as Merrill (1935) and Santapau (1950) have referred *Striga lutea* Lour. to *S. asiatica* (L.). Kuntze. *Buchnera asiatica* L., however, the basionym of Kuntze's combination, has been used for different species of *Striga* in different senses and, according to Art. 69 of the Code, "so has become a long-persistent source of error." Both *Buchnera asiatica* L. and *S. asiatica* (L.) Kuntze, therefore, become nomina ambigua and must be rejected. Thus according to the Rule of Priority *Striga lutea* Lour. becomes the valid name of the species.

Specimens Examined. KANDY DISTRICT: Kandy, *Gardner 647* (PDA); Peradeniya, *Gardner C.P. 2028* (K, PDA); Hantane, May 1924,

Silva s.n. (PDA). NUWARA ELIYA DISTRICT: Maturata, *s. coll. C.P. 2886* (K, PDA); Pussellawa, Patana close to culvert 32/9, Ramboda Road, *Cramer 3863* (PDA, US); Boragas, close to culvert 5/4, Boragas-Ambawela Rd., *Cramer 2955* (PDA, US) *& 3152* (NBV, PDA, US). GALLE DISTRICT: Kalegana, Diocesan Seminary Estate, *Cramer 1481* (PDA).

3. **Striga angustifolia** (Don) Saldanha, Bull. Bot. Surv. Ind. 5 (1): 70. 1963.

Buchnera asiatica L., Sp. Pl. 630. 1753, pro parte (?) nom. rej.; Vahl, Sym. Bot. 3: 81. 1794 (nom. illegit.). Type: 'India orientalis,' *Vahl* (C).

Buchnera angustifolia D. Don, Prod. Fl. Nepal 91. 1825. Type: Nepal, *Wallich* (?).

Buchnera euphrasioides Benth., Scroph. Ind. 41. 1835 (non Vahl 1794); Thw., Enum. Pl. Zeyl. 220. 1860; Hook. f., Fl. Br. Ind. 4: 299. 1885; Trimen, l.c. 246.

Striga euphrasioides (Benth.) Benth. in Hook., Comp. Bot. Mag. 1: 364. 1835; DC., Prod. 10: 503. 1846; Hook. f., Fl. Br. Ind. 4: 299. 1885; Trimen, l.c. 256.

Striga asiatica (L.) Kuntze, Rev. Gen. Pl. 466. 1891; Alston in Trimen, Handb. Fl. Ceylon 6: 216. 1931; Abeywick., l.c. 222. (nom. illegit.).

Stem up to 40 cm high, rigid, obtusely quadrangular, scabrid, often branched. Leaves linear, 10–20 × 1.5–2.5 mm, setulose on both surfaces, ciliolate and purplish at margins, becoming bracteate. Flowers in lax, erect spikes up to 12.5 cm long. Pedicels 1 mm long. Bracteoles subulate, 2 mm long, minutely strigose without. Calyx tube 5–6 mm long, 15-ribbed, scabrid on ribs; lobes linear-lanceolate, 3 × 1 mm, acuminate, faintly purplish at tips. Corolla white; tube linear, 10–11 mm long, minutely hirsute without; upper lip broadly orbicular, 3 × 5 mm, shallowly emarginate; lower lip 10–12 mm across, lateral lobes 3–4 mm long, mid-one broadly obovate, 6 × 3 mm. Style glabrous. Capsule ovoid, 5–6 × 3 mm, rounded at top, apiculate by withered style, glabrous; dehisced valves strongly reflexed; seeds cuneiform, 0.6 mm long, trigonous, glabrous, black.

Distr. India, Ceylon and Indonesia.

Ecol. Along roadsides and open country among short grass in the dry zone and in open grassy slopes of the wet lowlands and hill country; fairly common. The plant parasitizes the roots of grasses the commonest host being *Eleusine indica* (L.) Gaertn. Flowering August–October, December–May.

Note. For reasons given above under *Striga lutea* Lour., *Buchnera asiatica* L. is a nomen rejiciendum. The original *Striga euphrasioides* Benth., then, must be renamed according to the rules of priority. From the examination of type material Saldanha has shown that *S. euphrasioides* of Vahl (1794) and of

Bentham (1836) are really two different plants. Hence according to Art. 55 of the Code the name *S. euphrasioides* must be restricted to Vahl's plant, and Bentham's plant must be renamed. *Buchnera asiatica* Vahl, which is Bentham's plant, cannot be called *Striga asiatica* as then it becomes a later homonym of the Linnaean plant. Saldanha, accordingly, from the oldest legitimate epithet available in *Buchnera angustifolia* Don has made the combination *Striga angustifolia* (Don) Saldanha.

Specimens Examined. JAFFNA DISTRICT: Mullaitivu, *Alston 134* (PDA). KANDY DISTRICT: Peradeniya, 3/4 mile from Circuit Bungalow, *Mueller-Dombois 67103107* (PDA, US). BADULLA DISTRICT: Bintenne, *Gardner C.P. 3887* (K, PDA). AMPARAI DISTRICT: Lahugala, grassy ground, *Cramer 5045* (G, PDA, US). MANNAR DISTRICT: by culvert 138/1, Medawachchiya-Mannar Rd., *Cramer 3899* (NBV, PDA, US). COLOMBO DISTRICT: Negombo, Aug. 1930, *Silva s.n.* (PDA). HAMBANTOTA DISTRICT: Yala, Ruhuna National Park, Block 1, near Karangaswela, *Mueller-Dombois & Cooray 68012801* (PDA, US); Patanagala, at foot of Rock, *Cramer 3307* (PDA, US); Jamburagala, on top of Rock, *Cramer 3314* (PDA, US). BATTICALOA DISTRICT: Punanai, roadside close to culvert 64/7, Mannampitiya Rd., *Cramer 3633* (PDA).

7. MICROCARPAEA

R. Br., Prod. 435. 1810; Endl., Gen. Pl. 683, 1839; Benth. & Hook., Gen. Pl. 2: 957. 1876; Wettst. in Pflanzenfam. 4 (3b): 77. 1891. Type species: *Microcarpaea muscosa* R. Br. (=*M. minima* (Koenig) ex Retz.) Merr.

Very small herbs. Stem slender, diffusely branched. Leaves narrowly oblong. Flowers very small, axillary, solitary, ebracteolate. Calyx tubular, 5-angled or 5-ribbed, 5-fid. Corolla tube short, straight; limb subbilabiate; upper lip of 2 subconnate lobes; lower lip spreading, 3-lobed. Stamens 2, included; filaments filiform; anthers connivent, 1-celled. Style filiform, curved and slightly thickened at apex. Capsule ovoid to ellipsoid, loculicidally 2-valved; valves entire; seeds not very numerous.

A monotypic genus of the palaeotropics.

Microcarpaea minima (Koenig ex Retz.) Merr., Philipp. J. Sci. 7: 100. 1912; Alston in Trimen, Handb. Fl. Ceylon 6: 214. 1931; Alston, Kandy Fl. 60, f. 322. 1938; Abeywick., l.c. 222.

Paederota minima Koenig ex Retz., Obs. Bot. 5: 10. 1789. Type: *Koenig* (LD).
Microcarpaea muscosa R. Br., Prod. 426. 1810; Thw., Enum. Pl. Zeyl. 220. 1860; Hook. f., Fl. Br. Ind. 4: 287. 1885; Trimen, l.c. 254. Type: same as for *Paederota minima.*

Stem prostrate, slender, filiform, rooting at nodes, much branched, terete, with lacunae within, striate, glabrous. Leaves linear-oblong to linear-spathulate, 2.7–5.5 × 0.75 mm, slightly narrowed at base, subobtuse at apex, glabrous on both surfaces, punctate above. Flowers solitary in leaf-axils. Calyx tube campanulate, 1.5 mm long, lobes ovate, 1.5 × 0.75 mm, ciliate at margins, recurved with age. Corolla 2–3 mm long; pinkish-red; tube 1–1.5 mm long; upper lip slightly 2-lobed at apex; lobes of lower lip oblong, midlobe longer than lateral ones; all ciliate at margins. Stamens inserted towards top of corolla tube; filaments glabrous. Style glabrous; stigma shortly bifid. Capsule ellipsoid, 1.5–2 mm long, about half as long as calyx, glabrous, pale brown; seeds ellipsoid, 0.4 mm long.

Distr. Palaeotropical.

Ecol. In harvested paddy fields of the wet lowlands and lower wet hill country; not common. Owing to its creeping habit the plant covers large surfaces forming soft, pale green mats. At first sight this might be mistaken for a minute form of *Dentella repens* but is readily distinguished from the latter by the absence of interpetiolar stipules and its scrophulariaceous corolla. Flowering March–May; September.

Specimens Examined. KANDY DISTRICT: Ganoruwa, *Alston 1244* (PDA). KALUTARA DISTRICT: *s. coll. 2728* (PDA). GALLE DISTRICT: Bentota, Mar. 1887, *s. coll. s.n.* (PDA). RATNAPURA DISTRICT: Kitulgala, by Ganepalle-Kitulgala Rd., *Cramer 3338* (NBV, PDA, US).

8. ARTANEMA

D. Don in Sweet, Br. Fl. Gard. ser. 2, t. 234. 1829 nom. cons.; Benth. & Hook., Gen. Pl. 2: 953. 1876; Wettst. in Pflanzenfam. 4 (3b): 79. 1891. Type species: *Torenia fimbriata* Hook. ex Graham (= *Artanema fimbriata* (Hook. ex Grahm) D. Don).

Bahel Adans., Fam. 2: 210. 1768. Type species: *Bahel tsjulli* Rheede, Hort. Mal. 9: t. 87. 1698.

Achimenes Vahl, Symb. Bot. 2: 71. 1791 (non P. Br. 1756). Type species: as for *Bahel*.

Diceros Pers., Syn. 2: 164. 1807 (non Lour. 1970). Type species: *D. longifolius* (L.) Pers. (= *Artanema longifolium* (L.) Vatke).

Erect robust herbs or undershrubs. Stem stout, leaves entire or serrate. Flowers large, in terminal racemes, each in axil of a foliaceous bract; pedicels ebracteolate. Calyx 5-partite; lobes very imbricate. Corolla tube moderately wide; two posterior lobes of limb outside in bud; upper lip erect, broad, emarginate; lower lip broad. Stamens 4, didynamous; posterior filaments short, included, anterior ones with long arched filaments appendiculate at base, arching to become connivent under upper lip; anthers connivent. Stigma bilamellate. Capsule globose, septicidally 2-valved; valves entire; seeds numerous, foveolate rugose.

Four species, Indo-Malaysian and Australian.

Artanema longifolium (L.) Vatke, Linnaea 43: 207. 1880–82; Wettst. in Pflanzenfam. 4 (3b): 79. 1891 (*A. longiflorum*); Alston, Kandy Fl. 59, f. 311. 1938 (*A. longifolia*).

Columnea longifolia L., Mant. 1: 90. 1767. Type: Rheede Ic. in Hort. Mal. 9: t. 87. 1698.

Achimenes sesamoides Vahl, Symb. Bot. 2: 71. 1791. Type: as for *Columnea longifolia* L.

Diceros longifolius (L.) Pers., Syn. 2: 164. 1807. Type: as for above.

Artanema sesamoides (Vahl) Benth., Scroph. Ind. 39. 1835; Benth. in DC., Prod. 10: 408. 1846; Wight, Ic. Pl. Ind. Or. 1410. 1849; Thw., Enum. Pl. Zeyl. 219. 1860; Hook. f., Fl. Br. Ind. 4: 274. 1885; Trimen, l.c. 248.

Herb; stem up to 1.5 m high, stout, somewhat woody below, branched, acutely quadrangular, often purplish especially at nodes. Leaves oval-lanceolate, 4.5–15.5 × 0.8–3.2 cm, tapering to base and decurrent on petiole, acuminate at apex, fairly distantly serrate with warty asperities above at margins, scabrid on both surfaces, with prominent nerves beneath. Racemes lax, 15–42 cm long. Bracts linear-lanceolate, up to 5.5 × 2 mm. Pedicels 6–12 mm long. Calyx lobes ovate-lanceolate; two outermost up to 11 × 6 mm, inner ones shorter, narrower; all ciliolate. Corolla pale violet-purple, up to 2.5 cm long; tube subventricose, up to 1.6 cm long, glandular-hirtellous on both surfaces; limb 1.3–2.3 cm across; lobes orbicular. Filaments all glabrous; appendage of anterior ones tuberculate, 2-lobed, purplish. Style glabrous. Disk undulate. Capsule invested by persistent calyx lobes, 7–10 mm in diameter, tipped by withered style; seeds suborbicular, 5–7 mm in diam., flattened on one side, warty, dark brown.

Distr. From India and Ceylon eastwards to Malaysia.

Ecol. Under shade in moist localities towards forest borders, in waste places, and along borders of paddy fields of the wet lowlands and the hill country up to 1030 m; fairly common. Flowering May–November.

Specimens Examined. KANDY DISTRICT: Hantane, *s. coll. C.P. 2024* (K, PDA). GALLE DISTRICT: Hiriyamalkumbura, *Cramer 1457* (PDA).

9. LINDERNIA

All., Misc. Taur. 3: 178, t. 5, f. l. 1766; Wettst. in Pflanzenfam. 4 (3b): 79. 1891; Pennell, Acad. Nat. Sci. Philadelphia Monogr. 1: 137. 1935. Type species:

Anagalloides procumbens Krocker (=*Lindernia procumbens* (Krocker) Philcox).

Vandellia L., Mant. 89. 1767; Endl., Gen. Pl. 683. 1839; Benth. & Hook., Gen. Pl. 2: 955. 1876. Type species: *Vandellia diffusa* L. (=*Lindernia diffusa* (L.) Cramer, comb. nov.).

Ilysanthes Rafin., Ann. Nat. 13. 1820; Benth. & Hook., l.c.; Wettst., ibid. 80. Type species: *I. riparia* Rafin. (=*Lindernia pyxidaria* Pursh. (non L. 1771)).

Bonnaya Link & Otto, Ic. Pl. Selec. Berol. 25, t. 11. 1821; Endl., l.c.; Benth. & Hook., ibid. 956. Type species: *B. brachiata* Link & Otto (=*Lindernia ciliata* (Colsm.) Pennell).

Annual glabrous or pubescent herbs. Stem slender, erect or creeping, often quadrangular, sometimes terete. Leaves opposite, entire, crenate or serrate, penninerved or parallel-nerved. Flowers solitary in the axils of leaves, becoming terminal racemes, subsessile or pedicellate; pedicels often deflexed in fruit. Calyx 5-fid or 5-partite; lobes linear to subulate or narrowly lanceolate. Corolla shorter than 2 cm; tube cylindric or slightly widened above; upper lip (posterior lobe) outside in bud, erect, broad, entire, emarginate or 2-fid; lower broader, with 3 unequal lobes. Stamens 4 or 2, the posterior pair inserted near apex of corolla tube, the filaments short; anterior pair inserted at throat, the filaments long and arched with a basal, tooth-like, linear or clavate appendage, or reduced to staminodes; anthers 2-celled, connivent or subconnivent, cells divaricate. Stigma bilamellate. Capsule globose, ovoid, oblong to linear, septicidally 2-valved; seeds numerous, variously sculptured.

A pantropic genus of about 80 species predominantly palaeotropic.

Lindernia All., *Vandellia* L., *Ilysanthes* Rafin., and *Bonnaya* Link & Otto were considered distinct genera by the older authors. Haines (1922) reduced the four genera to two uniting *Bonnaya* with *Vandellia* and *Ilysanthes* with *Lindernia*. Later Alston (1931) and Pennell (1943) fused them all under the genus *Lindernia* All. This is now the generally accepted view of Mukerjee (1945), Santapau (1950) and Li (1961). The *Lindernia-Ilysanthes* complex, however, does not still seem to be clear and needs further investigation.

KEY TO THE SPECIES

1 Leaves penninerved
  2 Bracts absent; perfect stamens 4
    3 Flowers in racemes .... **1. L. viscosa**
    3 Flowers solitary in axils of leaves
      4 Capsule globose or ovoid, shorter or slightly longer than calyx; lower anther lobes of anterior stamens not spurred
        5 Stem scabrescent; flowering calyx cleft to middle, ribbed .... **2. L. crustacea**
        5 Stem hispidulous, flowering calyx cleft to base, smooth .... **3. L. pusilla**
      4 Capsule linear, twice as long as calyx or longer; lower anther lobes of anterior stamens spurred

6 Stem erect; leaves linear, sessile.................................**4. L. angustifolia**
6 Stem prostrate or decumbent; leaves ovate, petiolate..................**5. L. anagallis**
2 Bracts present; perfect stamens 2
7 Leaves linear, distantly serrulate; flowers solitary in axils of leaves..**6. L. tenuifolia**
7 Leaves oblong or obovate-oblong, distinctly serrate; flowers in racemes
8 Leaves sharply aristate-serrate; flowering calyx cleft to middle........**7. L. ciliata**
8 Leaves usually serrate or crenate-serrate; flowering calyx cleft to base..**8. L. antipoda**
1 Leaves parallel-nerved
9 Stem erect; bracts present
10 Leaves lanceolate; capsule longer than calyx; upper lip of corolla entire at margins...**9. L. hyssopioides**
10 Leaves ovate-elliptic or broadly ovate; capsule shorter than calyx; upper lip of corolla serrulate at margins.......................................**10. L. srilankana**
9 Stem creeping or decumbent; bracts absent
11 Flowering calyx cleft to base; perfect stamens 2; staminodes 2..**11. L. rotundifolia**
11 Flowering calyx cleft to middle; perfect stamens 4; staminodes 0..**12. L. nummularifolia**

1. Linderina viscosa (Hornem.) Boldingh., Zakfl. Landb. Java 165. 1916.

Gratiola viscosa Hornem., Enum. Pl. Hort. Hafn. 19. 1807. Type: *Hornemann* (probably C).

Vandellia hirsuta Buch.-Ham. ex Benth., Scroph. Ind. 36. 1835; Benth. in DC., Prod. 10: 44. 1846; Hook. f., Fl. Br. Ind. 4: 280. 1885; Trimen, l.c. 250. Type: Bangladesh, Sylhet, J.D. Hooker, Wall., Num. List 3942 C (K-WALL.).

Vandellia multiflora G. Don., Gen. Syst. 4: 549. 1837; Thw., Enum. Pl. Zeyl. 219. 1860. Type: not known.

Lindernia hirsuta (Buch.-Ham. ex Benth.) Wettst. in Pflanzenfam. 4 (3b) 79. 1891; Mukerjee, J. Ind. Bot. Soc. 24 (3): 131. 1945; Abeywick., l.c. 222.

Lindernia viscosa (Reichb.) Alston in Trimen, Handb. Fl. Ceylon 6: 213. 1931 (nom. illegit.).

Stem short, erect, with many long branches from the base, densely covered with spreading hair; leaves 2.5–5 cm, oval, tapering into wide petiole (the upper ones nearly sessile), subacute, dentate-serrate, slightly hairy on both sides; flowers on long slender pedicels, in lax viscous-pubescent axillary racemes; sepals distinct, linear-lanceolate, acuminate, hairy; capsule ovoid, shorter than calyx.

Distr. "In the low country; very rare."—Trimen. From India through Ceylon and Burma eastwards to the Philippines.

Ecol. Flowering July–September.

Specimens Examined. RATNAPURA DISTRICT: Getahette, Sep. 1865, *s. coll. s.n.* (PDA). DISTRICT UNKNOWN: Between Negombo & Kurunegala, *s. coll. C.P. 3309* (K, PDA).

N o t e. The description given above is Trimen's. In spite of a long and careful search in the field the author has been unable to find this plant. The two specimens in the Peradeniya Herbarium are poor and need to be supplemented.

2. Lindernia crustacea (L.) F. Muell., Cens. Austral. Pl. 97. 1882; Wettst. in Pflanzenfam. 4 (3b): 80. 1891; Mukerjee, l.c. 130.

Capraria crustacea L., Mant. 87. 1767. Type: Rumph. l.c. in Herb. Amb. 5: t. 170. 1750.
Torenia crustacea (L.) Cham. & Schlecht., Linnaea 2: 570. 1827; Alston in Trimen, Handb. Fl. Ceylon 6: 213. 1931.
Vandellia crustacea (L.) Benth., Scroph. Ind. 35. 1835; Benth. in DC., Prod. 10: 413. 1846; Hook. f., Fl. Br. Ind. 4: 279. 1885; Trimen, l.c. 250.
Gratiola lucida Willd., Sp. Pl. 1: 103. 1797; Moon, Cat. 4. 1824 (nom. nud.). Type: as for *Capraria crustacea* L.

Stem usually prostrate and rooting at lower nodes, with branches ascending up to 20 cm, sometimes caespitose, scabrescent. Leaves ovate to ovate-cordate, 0.5–1.5 × 0.4–1.2 cm, truncate to subcordate at base, sub-acute at apex, coarsely crenate-serrulate, glabrous above, minutely pubescent on veins beneath, often dull purple at margins; petioles 1–6 mm long. Flowers solitary; pedicels 0.5–1.6 cm long, sparsely puberulous. Calyx tube cleft to middle, 1.2–2.5 mm long, ribbed, lobes ovate-lanceolate, 1.5–2.5 mm long. Corolla purplish-blue; tube 4–7 mm long, puberulous without, often pale blue, pale yellow at base; upper lip shallowly 2-lobed; lower lip 6–7 mm across, mid-lobe slightly larger than lateral ones. Posterior filaments 2 mm long; anterior ones longer, with a linear bluish-purplish appendage at base; all glabrous. Hypogynous disk pale yellow. Style glabrous; lamellae of stigma broadly ovate. Capsule oblong-ovoid, 3.5–5 × 2–3 mm, slightly shorter than calyx, apiculate by remnant of style, glabrous; seeds oval, 0.5 mm long, tuberculate, tawny.

D i s t r. Palaeotropical; introduced into tropical America.

E c o l. In damp grassy places, paddy fields after harvest, and along banks of tanks in the wet and dry lowlands; very common. In nitrogenous soils as in meadows grazed by cattle the plant tends to be dwarfed. The small leaves and colourful flowers make it quite an elegant plant. Flowering throughout the year.

S p e c i m e n s E x a m i n e d. JAFFNA DISTRICT: Jaffna, *Gardner C.P. 2048* (K, PDA). PUTTALAM DISTRICT: Mattakotuwa, *Cramer 2610* (PDA, US). KURUNEGALA DISTRICT: Kurunegala, *Gardner C.P. 2048* (K, PDA); Kurunegala, by Tittawela Tank, *Cramer 3182* (PDA, US). KANDY DISTRICT: Peradeniya, Nov. 1927, *Alston s.n.* (PDA). KALUTARA DISTRICT: Nagoda, in meadow opposite Catholic Convent, *Cramer*

2776 (US); Pelawatte, on Ranawake Rock, *Cramer 2809* (PDA, US); Bentota, *Cramer 2911* (PDA, US). GALLE DISTRICT: Galle, in Green, *Cramer 2624* (US); Waidyahengande, along slopes of tea estate, *Cramer 2635* (US); Wakwelle, along banks of pond, *Cramer 2658* (PDA, US); Gonapinnuwela, Supemuyana Estate, *Cramer 2705* (PDA, US); Talgaswela, Talgaswela Estate, *Cramer 2850* (PDA). HAMBANTOTA DISTRICT: Yala, Ruhuna National Park, Block 1, *Mueller-Dombois 68050302* (PDA, US); Jamburagala, *Cramer 3316* (PDA). BADULLA DISTRICT: Ella, June 1924, *Silva s.n.* (PDA). BATTICALOA DISTRICT: Kalkudah, by railway track, *Cramer 2737* (PDA).

3. **Lindernia pusilla** (Willd.) Boldingh., Zakfl. Landbouwstr. Java 165. 1916; Alston in Trimen, Handb. Fl. Ceylon 6: 213. 1931; Alston, Kandy Fl. 59, f. 312. 1938.

Gratiola pusilla Willd., Sp. Pl. 1: 105. 1797. Type: India, specimen not known.

Vandellia scabra Benth., Scroph. Ind. 36. 1835 & in DC., Prod. 10: 414. 1846; Thw., Enum. Pl. Zeyl. 219. 1860; Hook. f., Fl. Br. Ind. 4: 281. 1885; Trimen, l.c. 251. Type: Luconia, *Chamisso & Eschscholtz*. (?).

Torenia hirta Cham. & Schlecht., Linnaea 2: 571. 1827. Type: as for *Vandellia scabra*.

Lindernia hirta (Cham. & Schlecht.) Pennell, J. Arnold Arbor. 24: 250. 1943; Mukerjee, l.c. 131; Abeywick., l.c. 222.

Stem prostrate, branches often widely spreading from the root-stock, or prostrate at base, then ascending up to about 25 cm, with long internodes, hispidulous. Leaves ovate to orbicular-ovate, 0.6–1.7 × 0.5–1.4 cm, truncate at base, obtuse at apex, crenate-serrate, with prominent nerves beneath, scabrid on both surfaces; petioles 1–3 mm long. Flowers 1–4 in axils of leaves. Pedicels 3–20 mm long, slender, glandular-pubescent, longer and deflexed in fruit. Calyx tube 1 mm long, cleft to base, smooth, lobes lanceolate, 2–2.5 × 1 mm, glandular-pilose without. Corolla white or pale mauve or pale blue; tube 4 mm long; upper lip erect, 3 mm long, 2-lobed at apex, dirty pale brown or mauvish-brown with purplish lines continued down tube; lower lip 4–6 mm across, mid-lobe broadly ovate, slightly larger than lateral ones, with a yellow blotch at base. Posterior filaments 2 mm long; anterior ones longer, geniculate at base; all glabrous. Hypogynous disk white. Style glabrous; lamellae of stigma ciliolate at margins. Capsule globose, 3–4 mm in diam., glabrous; seeds oblong-cuneiform, 0.5 mm long, truncate at both ends, longitudinally alveolate, tawny.

D i s t r. Palaeotropical.

E c o l. In grassy bunds of paddy fields and along grassy borders of ponds or tanks in the wet and dry lowlands and hill country up to 1000 m; very common. The general pilose nature of the indumentum and the dirty pale brown upper lip are good distinguishing characters of the plant in the field. Flowering throughout the year.

Specimens Examined. VAVUNIYA DISTRICT: Mullaitivu, *Alston 1281* (PDA). KURUNEGALA DISTRICT: Tittawela, close to Tank, *Cramer 3187* (US); Dewadduwa, *Cramer 3191* (US). POLONNARUWA DISTRICT: Habarana, border of pond beside Bahirawagala Rock, *Cramer 5125* (G, PDA, US). KALUTARA DISTRICT: Kalutara, *s. coll. C.P. 3885* (K, PDA), *Cramer 2708* (PDA, US); Matugama, by Kalutara-Matugama Rd., *Cramer 2963* (PDA, US). GALLE DISTRICT: Wakwelle, beside Ukwatte Rd., *Cramer 2661* (PDA, US), Kottapitiya, near marker 30, Matugama Rd., *Cramer 2915* (PDA, US); HAMBANTOTA DISTRICT: Yala, Ruhunu National Park, Mahamagulvihara, *Cramer 3325* (PDA, US). RATNAPURA DISTRICT: Elapata, *Cramer 3104* (NBV, PDA, US). BADULLA DISTRICT: Ella, Jun. 1924, *Alston s.n.* (PDA); Wakkumbara, *Cramer 3155* (PDA, US). BATTICALOA DISTRICT: Kalkudah, John Kennedy Estate, *Cramer 2746* (PDA, US).

4. Lindernia angustifolia (Benth.) Wettst. in Pflanzenfam. 4 (3b): 79. 1891; Alston in Trimen, Handb. Fl. Ceylon 6: 213. 1931; Mukerjee, l.c. 132; Abeywick., l.c. 222.

Vandellia angustifolia Benth., Scroph. Ind. 37. 1835; Benth. in DC., Prod. 10: 417. 1846; Thw., Enum. Pl. Zeyl. 219. 1860; Hook. f., Fl. Br. Ind. 4: 282. 1884; Trimen, l.c. 251. Type: Wall., Num. List 3951 (K-Wall.).

Stem erect, up to about 25 cm high, often much branched from base, with a large central spongy pith surrounded by a whorl of radiating lacunae, glabrous. Leaves linear to linear-lanceolate, 1.2–4 × 0.3–0.6 cm, slightly narrowed at base, subacute at apex, entire or distantly serrulate, with obscure lateral nerves, minutely punctate on both surfaces, glabrous. Pedicels up to 1.2 cm long, longer and deflexed in fruit, stiff, glabrous. Sepals linear-lanceolate, unequal, 2–4.5 × 1 mm, faintly spinous-serrate at margins, purplish at tips. Corolla 9–10 mm long, pale pinkish-violet; tube 5–6 mm long, glandular-puberulous without; upper lip broadly pyramidal, 4.5 mm long, entire at apex; lower lip 5–6 mm across, with a rounded yellow blotch at base of midlobe. Posterior filaments 2.5 mm long; anterior filaments longer, with a linear appendage from near base; all glabrous. Spur of anterior anther lobes 1.5 mm long. Hypogynous disk incomplete, pale green. Style glabrous. Capsule linear-cylindric, 10–13 × 1 mm, two or two and a half times longer than calyx, apiculate by withered style; seeds obovoid-oblong, 0.7 mm long, longitudinally grooved and foveolate.

D i s t r. From Ceylon and India eastwards to Southeast Asia, the Malayan Archipelago and Japan.

E c o l. In harvested paddy fields of the wet lowlands; rare. Flowering October–February.

S p e c i m e n s E x a m i n e d. COLOMBO DISTRICT: Colombo, *Ferguson C.P. 2884* (K, PDA). KALUTARA DISTRICT: Kalutara, *Moon C.P. 2051* (K, PDA). KURUNEGALA DISTRICT: Matiyagama, by culvert 16/5, Narammala-Dambadeniya Rd., *Cramer 3236* (NBV, PDA, US); Polgahawela, by culvert 3/6, Polgahawela-Narammala Rd., *Cramer 3272* (PDA, US).

N o t e. The plant grows in association with *Lindernia anagallis* (Burm. f.) Pennell which it closely resembles in floral structure.

5. Lindernia anagallis (Burm. f.) Pennell, J. Arnold Arbor. 24: 252. 1943 & Acad. Nat. Sci. Philadelphia Monogr. 5: 31. 1943; Mukerjee, l.c. 133 (excl. syn. *L. antipoda* Alston); Philcox, Kew Bull. 17 (3): 484. 1964.

Ruellia anagallis Burm. f., Fl. Ind. 135. 1768. Type: Java, *Kleinhof* (L.).

Gratiola cordifolia Colsmann, Prod. Desc. Grat. 15. 1793; Moon, Cat. 4. 1824, nom. nud. Type: Ceylon, *Koenig*?

Vandellia pedunculata Benth., Scroph. Ind. 37. 1835; Benth. in DC., Prod. 10: 416. 1846; Thw., Enum. Pl. Zeyl. 219. 1860; Hook. f., Fl. Br. Ind. 4: 282. 1885; Trimen, l.c. 251. Type: India, Wall., Num. List. 3939 C p.p., in Hb. Madras (K-Wall.).

Lindernia cordifolia (Colsmann) Merr., Enum. Philipp. Pl. 3: 437. 1923; Alston in Trimen, Handb. Fl. Ceylon 6: 213. 1931 & Kandy Fl. 59, f. 314. 1938.

Stem prostrate and rooting at lower nodes, then ascending up to 30 cm, with a central spongy pith surrounded by a whorl of radiating lacunae, glabrous, sometimes faintly purplish at nodes. Leaves ovate to ovate-oblong, 0.9–2.5 × 0.6–1.8 cm, truncate or subcordate at base, subacute or obtuse at apex, shallowly crenate-serrate, with distinct lateral nerves beneath; petiole 1–3 mm long. Pedicels 0.6–3.4 cm long, stiff, glabrous. Sepals linear-lanceolate, subequal, 4–6 × 1 mm, acute, glabrous on both surfaces. Corolla faintly mauvish-violet; tube 6–8 mm long, glandular-puberulous without, yellow towards base; upper lip oblong-pyramidal, 4–5 mm long, concave; lower lip 7–11 mm across, lobes orbicular, midlobe larger than lateral ones with a round yellow blotch at base. Anterior filaments each with a linear appendage from near base; spur of anterior anther cells 1 mm long. Hypogynous disk incomplete, pale yellow. Style glabrous; capsule linear-cylindric, 7.5–12 × 2 mm, narrowed and obtusely pointed at apex, glabrous; seeds ellipsoid, 0.5 mm long, longitudinally grooved and foveolate, tawny.

D i s t r. Tropical Asia, Malaysia and tropical Australia.

Ecol. In harvested paddy fields and along banks of ponds in the wet lowlands and the wet hill country up to 550 m; common. It usually grows in association with *Lindernia rotundifolia* (L.) Alston. Flowering throughout the year.

Note. The correct name of this plant is an alteration, for reasons of typification, of the hitherto long-known name *Lindernia cordifolia* (Colsm.) Merr. Philcox, on tracing the Kleinof material, the type of *Ruellia anagallis* Burm. f. as cited by Burman himself in his Flora Indica (1768), found that this material markedly differs from *Ruellia antipoda* L. with which it was identified all this time. It differs from the Linnaean plant "in having shortly petiolate, crenate-serrate leaves distinctly cordate at the base and more closely matches the description of *Gratiola cordifolia* Colsm."—Philcox, l.c. 483. Philcox rightly separated *Ruellia anagallis* Burm. f. from *R. antipoda* L., referring the former to *Gratiola cordifolia* Colsm. On transference to the genus *Lindernia* Burman's name, therefore, by priority becomes the legitimate basionym of the new name for the species till recently known as *Lindernia cordifolia* (Colsm.) Merr.

Specimens Examined. KANDY DISTRICT: Peradeniya, *Alston 1681 & 1690* (PDA); Kandy, *s. coll. C.P. 352* (K, PDA). PUTTALAM DISTRICT: Mattakotuwa, *Cramer 2607* (PDA, US). KALUTARA DISTRICT: Kalutara, *Gardner C.P. 652* (K, PDA); Haburugala, by Anduwa Rd., *Cramer 2912* (PDA, US). GALLE DISTRICT: Udalamatta, *Cramer 2415* (US); Wakwella, by Wakwella-Ukwatte Rd., *Cramer 2626* (PDA, US); by Kalegana-Wakwella Rd., *Cramer 2666* (PDA, US); Talgaswela, Talgaswela Estate, *Cramer 2852* (PDA, US). RATNAPURA DISTRICT: Kuruwita, *Cramer 3095* (NBV, PDA, US). BADULLA DISTRICT: Wakkumbura, *Cramer 3154* (PDA, US).

6. Lindernia tenuifolia (Colsmann) Alston in Trimen, Handb. Fl. Ceylon 6: 214. 1931; Mukerjee, l.c. 134; Abeywick., l.c. 222.

Gratiola tenuifolia Colsmann, Prod. Desc. Grat. 8. 1793; Vahl, Enum. Pl. 1: 96. 1805. Type: Ceylon, *Koenig*?

Bonnaya tenuifolia Spreng., Syst. Veg. 1: 42. 1825; Thw., Enum. Pl. Zeyl. 220. 1860; Hook. f., Fl. Br. Ind. 4: 286. 1885; Trimen, l.c. 254. Type: not known.

Stem tufted, erect, up to 40 cm high, slender, much branched from base, glabrous. Leaves linear to linear-oblong, 1.3–5.2 × 0.2–0.9 cm, narrowed at base, subacute at apex, entire or distantly serrulate, punctate above, with indistinct lateral nerves, glabrous on both surfaces. Bracts subulate, 2.5 mm long. Pedicels 0.6–1.7 cm long, longer and deflexed in fruit. Calyx tube 1 mm long; lobes linear, more or less equal, 3 × 0.8 mm, subacute and purplish at tips, hyaline at margins, glabrous on both surfaces. Corolla pale blue to bluish-violet; tube 4 mm long, uniformly linear, glabrous on both surfaces; upper lip oblong-pyramidal, entire at apex; lower lip 6–7 mm across, lobes oblong-orbicular, mid-lobe almost equal to lateral ones. Filaments glabrous. Staminodes exserted, linear, 1.5 mm long, straight. Capsule linear-cylindric, 5–12 × 1 mm, twice to thrice as long as calyx, apiculate by withered style, glabrous, purplish towards apex; seeds ellipsoid, 0.5–0.7 mm long, reticulately foveolate, tawny.

Distr. India, Ceylon, Burma and China.

Ecol. In harvested paddy fields of the wet lowlands moderately close to the sea-shore; occasionally in the wet hill country up to 500 m; rare. Flowering September–April.

Specimens Examined. KALUTARA DISTRICT: Bandaragama, *Cramer 2792* (US); Haburugala, near marker 3/5, Anduwa Rd., *Cramer 2913 & 3214* (PDA, US). KANDY DISTRICT: Peradeniya, *s. coll. C.P. 2883* (K, PDA).

7. Lindernia ciliata (Colsmann) Pennell, Brittonia 2: 182. 1936; Pennell, J. Arn. Arb. 24: 253. 1943; Pennell, Acad. Nat. Sci. Philadelphia Monogr. 5: 32. 1943; Abeywick., l.c. 222.

Gratiola ciliata Colsmann, Prod. Desc. Grat. 14. 1793; Vahl, Enum. Pl. 1: 97. 1805. Type: Java, specimen not known.

Bonnaya brachiata Link & Otto, Ic. Pl. Select. 25, t. 11. 1820; Thw., Enum. Pl. Zeyl. 426. 1864; Hook. f., Fl. Br. Ind. 4: 418. 1885; Trimen, l.c. 253. Type: Link & Otto, l.c.

Gratiola serrata Roxb., Fl. Ind. 1: 140. 1820. Type: India, Bengal, specimen not known.

Lindernia serrata (Roxb.) F. Muell., Census 97. 1882; Alston in Trimen, Handb. Fl. Ceylon 6: 214. 1931; Alston, Kandy Fl. 59, f. 316. 1938.

Stem erect, up to 13 cm high, often patently branched from base, or decumbent and rooting at base, strigose on angles. Leaves oblong-lanceolate or oblong-oval, 0.6–2.4 × 0.2–0.7 cm, rounded at base, subacute at apex, with inconspicuous lateral nerves, aristate-serrate, punctate above, glabrous on both surfaces. Racemes terminal, lax, 2.5–3 cm long. Bracts linear-lanceolate, 3.5–4 × 1 mm, setaceous at apex, ciliolate and purplish at margins. Pedicels 1–2.5 mm long, stiff. Calyx cleft to middle, tube 2 mm long; lobes linear-lanceolate, 3–4 × 0.7 mm, setaceous at apex, hyaline and ciliolate at margins, sparsely and minutely scabrid abaxially along mid-nerve. Corolla tube linear, 6–8 mm long, glandular-puberulent without, pale mauve or white; upper lip erect, oblong, 3–4 mm long; lower lip 4–5 mm across, lobes suborbicular, mid-one slightly smaller than lateral ones, all violet-purple at base. Filaments 1.5 mm long, glabrous. Staminodes exserted, linear, curved outwards at tip, purplish-mauve at base. Style glabrous; lamellae of stigma oblong-orbicular. Capsule linear-cylindric, 10–17 × 1 mm, twice as long as calyx, glabrous, faintly purple; seeds oblong to pyriform, 0.8 mm long, muricate, dark brown.

Distr. Widespread in Ceylon, India, Southeast Asia, and Malaysia.

Ecol. In wet places among short grass, particularly on damp rocks, in the dry lowlands and the wet hill country up to 650 m; common; occasionally in the wet lowlands. The leaves and flowers make it an attractive little plant. Flowering almost throughout the year.

Specimens Examined. KURUNEGALA DISTRICT: Polgahawela, along Polgahawela-Narammala Rd., *Cramer 3268* (NBV, PDT, US) & *3273* (US). MATALE DISTRICT: Dambulla Rock, *Cramer 3280* (CAL, PDA, US). KANDY DISTRICT: Haragama, *s. coll. C.P. 3084* (K, PDA); Muruthalawa, Nelli Hill, *Cramer 2799* (PDA, US). RATNAPURA DISTRICT: Rasagala, *Cramer 2619* (US). HAMBANTOTA DISTRICT: Yala, Ruhuna National Park, Rakinawala, *Fosberg 50272* (PDA, US), *Cramer 3295* (PDA, US); Ratanagala, at foot of Rock, *Cramer 3300* (US); Mahamagulvihara, close to pond, *Cramer 3324* (PDA, US).

8. Lindernia antipoda (L.) Alston in Trimen, Handb. Fl. Ceylon 6: 214. 1931; Alston, Kandy Fl. 59, f. 316. 1938; Philcox, Kew Bull. 17 (3): 484. 1964.

Ruellia antipoda L., Sp. Pl. 635. 1753. Type: Ceylon, *Hermann* no. *17* (BM).
Gratiola veronicifolia Retz., Obs. Bot. 4: 8. 1786. Type: not known.
Bonnaya veronicaefolia (Retz.) Spreng., Syst. Veg. 1: 14. 1825; Thw., Enum. Pl. Zeyl. 219. 1860 (excl. syn. *B. verbenaefolia* Spreng.); Hook. f., Fl. Br. Ind. 4: 285. 1885; Trimen, l.c. 253.
Ruellia anagallis auctt. (non Burm. f. 1768).

Stem decumbent or more usually prostrate and rooting at nodes with branches ascending up to 28 cm, glabrous. Leaves variable, oval-oblong, obovate-oblong or oblanceolate, 0.6–4.5 × 0.2–1.4 cm, cuneate or subacute at base, subacuminate to obtuse at apex, closely serrate with the serratures aristate, or distantly and shallowly serrate to crenate-serrate, with the lateral nerves obscure, punctate and glabrous on both surfaces. Racemes terminal or axillary, lax, 0.8–12.5 cm long. Bracts linear or linear-lanceolate to oblong-lanceolate, 3–7 × 1–2 mm, serrulate at margins and mid-dorsal surface towards apex, glabrous. Pedicels 2–17 mm long, longer and deflexed in fruit. Calyx cleft to base, lobes linear-lanceolate, subequal, 3–5 × 1–1.5 mm, serrulate at margins towards apex. Corolla 8–13 mm long, pale violet or violet-blue to pale blue or white, occasionally pale purple; tube 3–7 mm long, glabrous on both surfaces, whitish or pale yellow towards base; upper lip linear-oblong, 3–7 mm long, concave, entire at apex; lower lip 5–12 mm across, lobes orbicular-oblong or suborbicular, midlobe slightly larger than lateral ones with usually dark violet or purplish spots at base, all reflexed with age. Filaments 1.5–2 mm long, glabrous. Staminodes slightly porrect, linear, slightly curved outwards and pale blue or yellow at tips. Capsule linear-cylindric or subulate-cylindric, 5–13.5 × 1–1.5 mm, two to two and a half times as long as calyx, glabrous, often dull purple; seeds oblong to ellipsoid, 0.5 mm long, reticulate, tawny.

Distr. India, from the Gangetic Plain eastwards and southwards to Ceylon, and eastwards to Malaysia and the Philippines.

Ecol. Among short grass along banks of tanks and ponds, in exposed, humus-filled rocky hollows and in paddy fields of the wet and dry lowlands and the wet hill country up to 550 m; very common. A plant with intermediate forms on account of its variable habit, leaves, length of inflorescence and colour of flowers. In wet habitats the stem is generally fully prostrate, the racemes short and the colour of the flowers a pale violet or pink. In dry habitats as in paddy fields after harvesting the stem tends to be more erect, the racemes longer and the colour of the flowers generally a pale blue. Flowering throughout the year.

Specimens Examined. JAFFNA DISTRICT: Jaffna, *Gardner C.P. 2050* (K, PDA). PUTTALAM DISTRICT: Chilaw, Feb. 1881, *Mrs. Neville s.n.* (PDA). KURUNEGALA DISTRICT: Kurunegala, by Wenduru Wewa, *Cramer 3060* (NBV, PDA, US); by Tittawela Tank, *Cramer 3181* (CAL, PDA, US); Polgahawela, by Polgahawela-Narammala Rd., *Cramer 3274* (PDA, US). COLOMBO DISTRICT: Giriulla, near 2nd mile post, Negombo-Giriulla Rd., *Cramer 3201* (PDA, US). KALUTARA DISTRICT: Maggona, in estate of St. Vincent's Home, *Cramer 2975* (E, PDA, US). KANDY DISTRICT: Gannoruwa, *Alston 1682* (PDA). RATNAPURA DISTRICT: Dewalagawa, behind Vidyalaya, *Cramer 3099* (CAL, PDA, US); Goduwela, by Inner Circular Rd., *Cramer 3107* (PDA, US); Dehiowita, by culvert 44/16, Mahagamana Rd., *Cramer 3341* (PDA, US). GALLE DISTRICT: Galle town, in Green, *Cramer 2625* (PDA, US); Wakwelle, by Ukwatte Rd., *Cramer 2659* (PDA, US). HAMBANTOTA DISTRICT: Yala, Ruhuna National Park, Talgasmankade, near Circuit Bungalow, *Cramer 3328* (PDA, US). BADULLA DISTRICT: Ella, June 1924, *de Silva s.n.* (PDA). TRINCOMALEE DISTRICT: Periyakulam, by Tank, *Cramer 2990 & 2991* (PDA, US).

Note. The synonymous position of *Ruellia anagallis* Burm. f. under this species has long been the subject of much controversy. The controversy has been ended by the tracing of Burman's type material which has established the separation of his species from *Ruellia antipoda* L. For a fuller discussion of this subject see notes under *Lindernia anagallis* (Burm. f.) Pennell.

9. Lindernia hyssopioides (L.) Haines, Bot. Bihar-Orissa 635. 1922; Alston in

Trimen, Handb. Fl. Ceylon 6: 214. 1931; Alston, Kandy Fl. 59. f. 315. 1938; Mukerjee, l.c. 132.

Gratiola hyssopioides L., Mant. 2: 174. 1771. Type: India, Tranquebar, *Koenig* (?).

Ilysanthes hyssopioides Benth. in DC., Prod. 10: 419. 1846; Thw., Enum. Pl. Zeyl. 219. 1860; Hook f., Fl. Br. Ind. 4: 283. 1885; Trimen, l.c. 252. Type: as for *Gratiola hyssopioides*.

Stem erect, up to 30 cm high or more, simple or diffusely branched, sharply quadrangular, glabrous. Leaves 0.6–1.6 × 0.3–0.6 cm, narrowed or truncate at base, subacute at apex, distantly and faintly serrate, 3-nerved, glabrous on both surfaces, punctate above, pale green. Flowers solitary, axillary. Bracts lanceolate, 4–8 × 1.5–2.5 mm, with a single serration on either margin. Pedicels 1–3 cm long, filiform, deflexed in fruit. Calyx lobes divided to base, more or less equal, linear-lanceolate, 2–3 × 0.5–0.75 mm, subacute at apex, glabrous on both surfaces, occasionally purplish at tips. Corolla pale blue; tube somewhat flattened, 6–9 mm long, glabrous on either surface with a few bristly hairs on either side of throat, whitish-blue; upper lip obcuneate, 4–5 mm long, 2-lobed at apex, with dark blue lines continued down tube; lower lip 7–10 mm across, with dark blue blotches at bases of lobes, midlobe broader than lateral ones. Stamens inserted about middle of corolla tube; filaments 2.5 mm long, glabrous; anther-cells divergent, exserted, linear, curving out and clavate at tips, dark blue, decurrent on corolla tube as glandular-bristly, orange-yellow ridges. Style glabrous. Capsule obliquely ovoid, 5–7 × 2 mm, almost twice as long as calyx, glabrous; seeds oblong, 0.4 mm long, truncate at both ends, striate, tawny.

Distr. Southeast Asia, Malaysia and China.

Ecol. In paddy fields of the lowlands and of the wet hill country up to 550 m, particularly towards harvest time and afterwards; also along borders of ponds; very common. The stems are rather brittle and require careful handling when being pressed. The corolla is very caducous. Flowering throughout the year.

Specimens Examined. KEGALLE DISTRICT: near Kegalle, Oct. 1927, Alston *s.n.* (PDA). KANDY DISTRICT: Peradeniya, *s. coll. C.P. 2047* (K, PDA). KALUTARA DISTRICT: Matugama, beside culvert 1/8, Kalutara-Matugama Rd., *Cramer 2961* (US). KURUNEGALA DISTRICT: Dewadduwa, *Cramer 3190* (NBV, PDA, US); Maharagama, near 22nd mile post, Giriulla-Dambadeniya Rd., *Cramer 3207* (CAL, PDA, US). ANURADHAPURA DISTRICT: Kithulhitiyawa, near 49th mile post Dambulla-Kekirawa Rd., *Cramer 3030* (PDA, US). GALLE DISTRICT: Galle, Kaluwella, beside Galle-Matara railway track, *Cramer 33401* (US). HAMBANTOTA DISTRICT: Ranna, Mar. 1927, *de Silva s.n.* (PDA); beside culvert

152/4, Tissamaharama Rd., *Cramer 2820* (PDA, US); Yala, Ruhuna National Park, Talgasmankade, near Circuit Bungalow, *Cramer 3332* (PDA, US). BADULLA DISTRICT: Ella, June 1924, *de Silva s.n.* (PDA).

10. Lindernia srilankana Cramer & Philcox; Cramer, Ceylon J. Sci., Biol. Sci. 11 (1): 34–35. 1974. Type: Sri Lanka; Batticaloa District: Panichchankeni, damp ground among short grass along Valaichchanai-Trincomalee Rd., *Cramer 3629* (K, PDA, US).

Stem up to 13.5 cm high, branched from near base, quadrangular, sparsely glandular-hirtellous on angles. Leaves subsessile, 0.5–1.7 × 0.4–1.1 cm, truncate to rounded and 5-nerved at base, subacute at apex, crenate-serrate, ciliolate at margins, punctate on both surfaces, glabrous above, sparsely hirtellous on nerves beneath. Flowers solitary in axils of lower leaves, in whorls of 4 in upper part of stem. Pedicels 6–10 mm long, slender, hirtellous, deflexed in fruit. Bracts linear, 1.75–2.5 mm long, acute at apex, ciliolate at margins. Calyx lobes cleft almost to base, linear-lanceolate, 2.5 × 0.5 mm, subacuminate at apex, sparsely glandular-hirtellous without, spreading in fruit. Corolla 8–10 mm long; tube 4–5 mm long, subannulate at base, glabrous on both surfaces, dull violet-purple; upper lip erect, obspathulate, 4 mm long, subconcave, 2-fid at apex, serrulate, brownish-violet, striped darker violet; lower lip up to 10 mm across, prussian blue to pale blue, suborbicularly lobed, with two white blotches at base of the larger, more deeply violet central lobe. Perfect stamens 2, included within upper lip of corolla; filaments short, glabrous; anthers connivent, divaricately lobed. Staminodes 2, corniform, exserted, bent outwards, pale blue. Style glabrous; stigma 2-lamellate. Capsule septicidal, globose, 2.5–4 mm long, tipped with withered style; seeds ovoid-oblong, 0.7 mm long, truncate at both ends, 6-grooved, minutely alveolate, pale brown.

Distr. Endemic to Ceylon.

Ecol. In open grassy localities in the eastern and north-central parts of the dry lowlands. An elegant little annual that at first sight might be mistaken for *Lindernia pusilla* (Willd.) Boldingh., but on closer examination is distinguished from the latter by the parallel nervation of its leaves, its two perfect stamens and the two staminodes. Flowering February.

Specimens Examined. POLONNARUWA DISTRICT: Gunner's Quoin, open scrub among short grass, *Cramer 5078* (G, PDA, US). TRINCOMALEE DISTRICT: Somapura, open scrub among short grass, *Cramer 5114* (PDA, US).

11. Lindernia rotundifolia (L.) Alston in Trimen, Handb. Fl. Ceylon 6: 214. 1931; Alston, in Kandy Fl. 59. f. 313. 1938; Mukerjee, l.c., 132; Abeywick., l.c. 222.

Gratiola rotundifolia L., Mant. 274. 1767; Roxb., Pl. Corom. 3: 3, t. 204. 1811. Type: No. 30.4 (LINN.).

Ilysanthes rotundifolia Benth. in DC., Prod. 10: 420. 1846; Hook. f., Fl. Br. Ind. 4: 284. 1885; Trimen, l.c. 252. Type: "India orientalis," Roxb.

Stem decumbent and rooting at lower nodes, with branches ascending up to 20 cm high, grooved on one side, with cortex of 13–14 radiating lacunae, glabrous. Leaves broadly ovate to orbicular, 5–13 × 4–12 mm, truncate at base, obtuse at apex, distantly serrulate towards apex, 4–5-nerved, glabrous on both surfaces, minutely punctate above. Flowers solitary in leaf-axils. Pedicels 5–15 mm long, slender, glabrous. Calyx divided to base; lobes more or less equal, lanceolate, 2.5–2.75 × 1 mm, acute at apex, minutely glandular-hispid on both surfaces. Corolla 10–13 mm long, bluish-white; tube 9–10 mm long, glabrous on both surfaces, whitish at base; upper lip broadly pyramidal, 5 mm long, 2-lobed at apex; lower lip 7–8 mm across, lobes orbicular with purplish-blue blotches at bases, mid-lobe slightly broader than lateral ones; throat villous. Stamens 2, inserted at middle of corolla tube; filaments 1.5 mm long, glabrous. Staminodes linear, clavate and curved outwards at apex, dark-blue with age, decurrent on corolla tube as glandular-scabrid, yellow ridges. Hypogynous disk orange-yellow. Style glabrous; stigma bent, with lamellae minutely scabrid at margins. Capsule ovoid-globose, 2.5–3.5 mm long, equal to or slightly shorter than calyx, glabrous; seeds oblong, 0.5 mm long, with transverse bars along sides, pale brown.

Distr. Mauritius, Madagascar, Western and Southern India and Ceylon.

Ecol. In paddy fields of the lowlands and wet hill country up to 550 m, particularly after harvesting; here the plant usually grows in association with *Lindernia anagallis* (Burm. f.) Pennell; also along ditches in the hill country up to 1700 m and along borders of ponds and tanks in the lowlands; very common. Flowering throughout the year.

Specimens Examined. PUTTALAM DISTRICT: Wilpattu, along banks of Kalivillu, *Cramer 3161* (PDA, US); Mattakotuwa, *Cramer 2608* (PDA, US). KURUNEGALA DISTRICT: Kottukachchiya, along banks of Tank, *Cramer 2654* (PDA, US); Kurunegala, by Tittawela Tank, *Cramer 3185* (PDA, US). KALUTARA DISTRICT: Nagoda, by pond opposite Catholic Convent, *Cramer* 2985 (PDA); Maggona, St. Vincent's Estate, *Cramer 2976* (PDA, US). NUWARA ELIYA DISTRICT: Ambawela, Mar. 1922, *de Alwis s.n.* (PDA). GALLE DISTRICT: Hapugala, by Hapugala-Uluwitike Rd., *Cramer 2673* (NBV, PDA, US).

12. Lindernia nummularifolia (Don) Wettst. in Pflanzenfam. 4 (3b): 79. 1891; Mukerjee, l.c., 132.

Vandellia nummularifolia D. Don, Prod. Fl. Nepal 86. 1825; Benth. in DC.,

Prod. 10: 416. 1846; Hook. f., Fl. Br. Ind. 4: 282. 1885. Type: Nepal, *Wallich* (?).

Vandellia sessiliflora Benth., Scroph. Ind. 37. 1835 & in DC., Prod. 10: 416. 1846. Type: Burma, Toong Dang, Wall., Num. List. 3959 (K-WALL.).

Lindernia sessiliflora (Benth.) Wettst. in Pflanzenfam. 4 (3b): 79. 1891; Mukerjee, l.c. 132.

Stem usually erect, up to 17 cm high, occasionally decumbent and rooting at lower nodes, then ascending, diffusely branched from below, minutely hispid on angles. Leaves 1–2.5 × 1.1–2.1 cm, truncate at base, obtuse at apex, serrate and ciliolate at margins, 2–3-nerved from base, glabrous and minutely punctate above, scaberulous on nerves beneath; petioles 1–3 mm long. Flowers solitary in axils of leaves. Pedicels 1 mm long. Calyx tube 2–3 mm long; lobes unequal, linear-lanceolate, 2–4.5 mm long, acuminate, ribbed at middle with ribs continued down tube. Corolla 7–10 mm long; upper lip oblong-spathulate, 3 mm long, pale violet-brown; lower lip 6–6.5 mm across, lobes orbicular, midlobe slightly narrower than lateral ones with a pale yellow blotch at base. Anterior filaments arched, each with a geniculate base produced into a shortly blunt appendage. Style glabrous. Capsule narrowly ellipsoid, 10–11 × 1 mm, about twice as long as calyx, broadened towards base, glabrous; seeds ovoid-oblong, 0.5 mm long, obscurely longitudinally ribbed, transversely rugose in between ribs, dull orange-brown.

Distr. Tropical Africa, and from the subtropical Himalayas through South India to Ceylon, Burma, and Western China.

Ecol. Among short grass along bunds of paddy fields and in wet places of the wet lowlands and the wet hill country up to 550 m; not common. The capsule is typical as before dehiscence its wall is semitransparent and shows not only the outline of the seeds but even their sculpturing. Flowering almost throughout the year.

Note. The identity of this species has been determined by Mr. D. Philcox of the Kew Herbarium to whom I render sincere thanks. The species has not been reported before for Ceylon and is, therefore, a new plant record for the country.

Specimens Examined. KANDY DISTRICT: Peradeniya, Nursery Garden, Royal Botanic Gardens, *Cramer 3400* (US). KALUTARA DISTRICT: Bandaragama, beside Kalutara-Bandaragama Rd., *Cramer 2777* (PDA, US). GALLE DISTRICT: Talgaswela, Talgaswela Estate, *Cramer 2848* (PDA, US); Kottawa, by culvert 11/4, Galle-Udugama Rd., *Cramer 3017* (K, PDA, US); Wakwelle beside Wakwelle-Ukwatte Rd., *Cramer 2772* (US); Gonapinuwela, *Cramer 2701* (US).

10. STEMODIA

L., Syst. Nat., ed. 10. 1118. 1759, nom. cons.; Benth. & Hook., Gen. Pl. 2:

950. 1876 (pro parte); Wettst. in Pflanzenfam. 4 (3b): 74. 1891 (pro parte). Type species: *Stemodia maritima* P. Br.

Stemodiacra P. Br., Hist. Jamaica 261. 1756; Kuntze, Rev. Gen. Pl. 2: 467. 1891 (pro parte). Type species: as for *Stemodia.*

Annual, glandular-pubescent, often aromatic herbs. Stem often woody below, profusely branching. Leaves opposite or verticillate. Flowers in axils of leaves or bracts or in terminal leafy spikes or racemes. Pedicels 1–2-bracteolate or not. Calyx 5-partite; lobes narrow, imbricate, equal or subequal. Corolla bluish or purplish; tube cylindric; limb bilabiate; upper lip outside in bud, suberect, broad, entire or emarginate; lower lip spreading, 3-lobed. Stamens 4, didynamous, included; anther-cells distinct, stipitate, fertile. Style dilated at apex; stigma bilamellate. Capsule loculicidally, less often septicidally, 2–4-valved; seeds numerous, small.

A small genus of 30 species, pantropical.

The first author to validly describe the species *Stemodia maritima* was P. Browne in Hist. Jamaic. 1756. Linnaeus himself to whom its authorship had been generally credited cites this reference to P. Browne in his Systema Naturae l.c. Hence according to priority the authorship of *Stemodia maritima* should rightly be credited to P. Browne.

KEY TO THE SPECIES

1 Leaves sessile, amplexicaul or auriculate at base; stem quadrangular; flowers distinctly pedicelled . . . . **1. S. viscosa**

1 Leaves petiolate, not amplexicaul or auriculate at base; stem terete; flowers subsessile . . . . **2. S. verticillata**

1. Stemodia viscosa Roxb., Pl. Corom. 2: 33, t. 163. 1802; Roxb., Fl. Ind. 3: 94. 1832; Benth., Scroph. Ind. 24. 1835; Benth. in DC., Prod. 10: 381. 1846; Hook. f., Fl. Br. Ind. 4: 265. 1885; Santapau, J. Bombay Nat. Hist. Soc. 49 (1): 32. 1950. Types: India, *Roxburgh* & Roxburgh Ic. t. 163, l.c.

Stemodiacra viscosa (Roxb.) Kuntze, Rev. Gen. Pl. 2: 466. 1891.

Perennial herb. Stem erect, up to 55 cm high, with ascending branches from base, viscid-hirsute. Leaves oblong to oval-oblong, 3–6.2 × 1.3–2.5 cm, sub-acuminate at apex, serrate, penninerved, scabrid above, strigose along prominent nerves beneath. Flowers axillary, solitary or crowded in racemes 2–3.5 cm long. Pedicels 6–9 mm long, filiform, glandular-strigose. Bracteoles linear, 3 mm long. Calyx lobes subequal, linear-lanceolate, 3.6–5 × 1 mm, acuminate at apex, minutely glandular-hirsute without. Corolla up to 10 mm long, dull violet-blue; tube 8 mm long, subcompressed, strigose without; throat pilose; upper lip broadly orbicular, emarginate; lower lip up to 7 mm across, with orbicular-cuneiform lobes, midlobe broader than lateral ones. Filaments glabrous. Style glabrous. Capsule ovoid, 4–6 × 2 mm, equalling calyx, tipped by withered style, glabrous; seeds oblong-cuneiform, 0.3 mm long, truncate at both ends.

Distr. Afghanistan, throughout India and Ceylon.

Ecol. In cereal fields and in clayey soil among short grass close to ponds in the dry lowlands; rare. The plant has a distinctly pleasant, aromatic scent. Flowering February–March.

Note. This species has not been reported before for Ceylon, and must be considered a new plant record for the country.

Specimens Examined. MANNAR DISTRICT: Mannar, along Madawachchiya-Mannar Rd., *Cramer 2895* (L, PDA, US).

2. Stemodia verticillata (Mill.) Sprague, Kew Bull. 1921: 211. 1921; Alston in Trimen, Handb. Fl. Ceylon 6: 211. 1931; Alston, Kandy Fl. 59. f. 317. 1938; Abeywick., l.c. 221.

Erinus verticillatus Mill., Gard. Dict. (ed. 8) no. 5. 1768. Type: not known.
Stemodia parviflora Ait., Hort. Kew. ed. 2. 52. 1810. Type: not known.

Annual herb. Stem tufted, erect, up to 14 cm high, much branched, densely hirsute. Lower leaves opposite, upper ones in verticils of 3, broadly ovate, 0.7–1.5 × 0.4–1.4 cm, truncate at base, subacute at apex, coarsely crenate-serrate, penninerved, punctate above, scabrid along nerves beneath; petioles 3–10 mm long, densely and minutely glandular-hispid. Flowers axillary, solitary. Pedicels 1 mm long. Calyx lobes subequal, linear-lanceolate, 2.5–3.5 × 0.6 mm, acuminate at apex, minutely glandular-hispid without. Corolla 5.5 mm long; tube cylindric, 3–4 mm long, minutely glandular-pilose without, pale green, with longitudinal dull purple lines; upper lip 2-lobed at apex, with clavate hairs at base; lower lip 2 mm across, with obovate-orbicular lobes, midlobe broader than lateral ones. Filaments glabrous, anterior pair 1.5–2 mm long, posterior pair shorter; posterior anther-cells unequally paired. Style short; stigma bent. Capsule orbicular, 2.5–3 × 3 mm, biconvex, shorter than calyx, glabrous, 4-valved; seeds cuneiform, 0.5 mm long, truncate at both ends, longitudinally ribbed, reticulate in between ribs, dark brown.

Distr. Native of tropical America spread in Ceylon and Java.

Ecol. In cultivated ground in the hill country. Flowering July–December.

Specimens Examined. NUWARA ELIYA DISTRICT: Dolosbage, Sep. 1885, *Mohandrian s.n.* (PDA). KANDY DISTRICT: Peradeniya, Royal Botanic Gardens, Dec. 1925, *de Silva s.n.* (PDA), *Cramer 2706* (US), *2801* (US), *3402* (L, US).

11. BACOPA

Aubl., Pl. Guiana 1: 128, t. 49. 1775, nom. cons.; Wettst. in Pflanzenfam. 4 (3b): 76. 1891. Type species: *B. aquatica* Aubl.

Monniera B. Juss. ex P. Br., Hist. Jamaic. 269. 1756 (*Moniera*). Type species: *M. africana* Pers. (= *Bacopa monnieri* (L.) Pennell).
Bramia Lam., Enc. 1: 459. 1785. Type species: Brami in Rheede, Hort. Mal. 10: 27, t. 14. 1690 (non Rhumph. ex Adans. 1763).
Herpestis auct. (non Gaertn. f. 1807).

Annual, glabrous, marshy, sometimes aquatic herbs. Stem erect or repent. Leaves opposite, entire or crenate-serrate. Flowers axillary, solitary, pedicellate; pedicels bibracteolate or not. Calyx 5-partite; lobes much imbricate, posterior one largest, lateral ones usually narrow. Corolla subbilabiate; upper lip outside in bud, emarginate or 2-lobed; lobes of lower lip subequal. Stamens included, didynamous; anthers connivent in pairs or subdistinct. Style dilated at apex; stigma capitate or slightly 2-lobed. Capsule enclosed in calyx, septicidally 2–4-valved; seeds numerous, minute.

About 100 species, in warm countries.

KEY TO THE SPECIES

1 Stem prostrate; leaves spathulate or obovate-oblong .... **1. B. monnieri**
1 Stem erect; leaves linear-lanceolate .... **2. B. floribunda**

1. Bacopa monnieri (L.) Pennell, Proc. Acad. Nat. Sci. Philadelphia 98: 94. 1946; Santapau, J. Bombay Nat. Hist. Soc. 49: 30. 1950.

Lysimachia monnieri L., Cent. Pl. 2: 9. 1756. Type: not known.
Gratiola monnieria L., Amoen. Acad. 4: 306. 1759. Type: P. Br., Ic. in Hist. Jam. t. 28, f. 3. 1756.
Herpestis monnieria (L.) HBK., Nov. Gen. 2: 366. 1817 pro parte quoad tab. P. Br.; Benth. in DC., Prod. 10: 400. 1846; Trimen, l.c. 246.
Herpestis monniera Benth., Scroph. Ind. 30. 1835; Hook. f., Fl. Br. Ind. 4: 272. 1884.
Bacopa monnieria (L.) Wettst. in Pflanzenfam. 4 (3b): 77. 1891; Alston in Trimen, Handb. Fl. Ceylon 6: 211. 1931; Abeywick., l.c. 221 (*B. monniera*).

Stem creeping, rooting at nodes, diffusely branched, with numerous lacunae in cortex, glabrous. Leaves 0.9–2.0 × 0.3–6.5 cm, narrowed at base, obtusely rounded at apex, punctate above, thick, glabrous on both surfaces. Flowers solitary in leaf-axils. Pedicels 1–1.3 cm long, glabrous. Bracteoles subulate, up to 3 mm long. Sepals divided to base, outer ones broadly ovate, 6–7 × 4 mm, subacute at apex, inner linear-lanceolate, 5–6 × 1.5 mm, acuminate at apex. Corolla white or pale violet-blue; tube broadly campanulate, 5–6 mm long, glabrous; upper lip 2-lobed, lobes oblong, 5 × 3–4 mm; lower lip 13–15 mm across, lobes oblong-spathulate, subequal, 5–8 × 5 mm, midlobe longer than lateral ones; all with longitudinal lines impressed above. Stamens inserted at top of corolla tube, exserted; filaments glabrous. Style glabrous; stigma capitate, with a central, shallow depression. Capsule oblong-ovoid, 3–4 × 2 mm long, enclosed in calyx, subacuminate at apex, tipped with withered style, glabrous; seeds ovoid-oblong, 0.6 mm long, reticulate.

Distr. Palaeotropical.

Ecol. In open marshy places and along banks of ponds and creeks of the lowlands close to the coast; very common; occasionally along beaches and in the upper, wet hill country. A gregarious plant which because of its creeping habit covers large tracts of ground, particularly in marshy localities. Flowering throughout the year.

Uses. The extract of the plant boiled alone in water or together with the root tubers of *Maranta arundinacea* is taken as a purgative. A pure extract of the plant boiled in water is also employed as a lotion against skin eruptions.

Vern. Lunuwila (S).

Specimens Examined. MANNAR DISTRICT: Talaimannar, July 1916, *de Silva s.n.* (PDA). PUTTALAM DISTRICT: Puttalam, along beach, *Cramer 2648* (US); Wilpattu, along banks of Kali Villu, *Cramer 3163* (PDA, US). NUWARA ELIYA DISTRICT: Hakgala, *de Silva s.n.* (PDA). KALUTARA DISTRICT: Haburugala, by culvert 3/5, Anduwa-Elpitiya Rd., *Cramer 3213* (PDA, US). GALLE DISTRICT: Koggala, near tarmac of airport, *Cramer 1427* (US) *& 3352* (PDA, US); Gonapinnuwela, in Supemuyana Estate, *Cramer 2703* (PDA, US).

2. Bacopa floribunda (R. Br.) Wettst. in Pflanzenfam. 4 (3b): 77. 1891; Alston in Trimen, Handb. Fl. Ceylon 6: 211. 1931; Santapau, J. Bombay Nat. Hist. Soc. 49: 31. 1950.

Herpestis floribunda R. Br., Prod. 442. 1810; Benth. in DC., Prod. 10: 400. 1846; Thw., Enum. Pl. Zeyl. 218. 1860; Hook. f., Fl. Br. Ind. 4: 273. 1876. Trimen, l.c. 246. Type: Java, not known.

Monniera floribunda Kuntze, Rev. Gen. Pl. 463. 1891. Type: not known.

Stem erect, 10–30 cm, subquadrangular, glabrous, slightly branched at base; leaves lanceolate-linear, 1.8–3.1 cm, acute, entire, narrowly revolute at margins, finely gland-dotted; flowers 1 or 2 in an axil, on short slender pedicels; bracteoles filiform, sepals membranous, veiny, puberulous, outer ones broadly oval, obtuse.

Distr. Palaeotropical.

E c o l. In wet places in the lowlands; rare.

N o t e. The description given above is Trimen's. In spite of a long and diligent search in the field, the author has been unable to find this plant. This corroborates Trimen's own finding: "I have not met with this plant."

S p e c i m e n s E x a m i n e d. COLOMBO DISTRICT: *Ferguson C.P. 2881* (K, PDA). TRINCOMALEE DISTRICT: *Glenie C.P. 2881* (K, PDA).

12. DOPATRIUM

Buch.-Ham. ex Benth., Scroph. Ind. 31. 1835; Benth. & Hook. Gen. Pl. 2: 953. 1876; Wettst. in Pflanzenfam. 4 (3b): 75. 1891. Type species: *Gratiola juncea* Roxb. (= *Dopatrium junceum* (Roxb.) Buch.-Ham. ex Benth.)

Slender, glabrous, erect herbs. Leaves opposite, mostly radical and close together, cauline distant and smaller. Flowers solitary in leaf-axils or in terminal, lax racemes; pedicels ebracteolate. Calyx 5-fid; lobes imbricate, equal; anterior stamens reduced to 2 minute staminodes. Style short; stigma bilamellate. Capsule globose, loculicidally 2-valved; seeds small, finely tubercled or rugose.

A palaeotropical genus of about 20 species.

KEY TO THE SPECIES

1 Capsule linear-cylindric; fruiting pedicels deflexed .... **1. D. nudicaule**
1 Capsule globose or broadly ellipsoid; fruiting pedicels horizontal or erect
2 Flowers solitary in leaf-axils; fruiting pedicels erect, radical leaves less than 1.8 cm long; stem much branched towards base .... **2. D. junceum**
2 Flowers in elongate racemes or panicles, fruiting pedicels horizontal; radical leaves much more than 1.8 cm long; stems usually single, or at most 2–3 from base .... **3. D. lobelioides**

1. Dopatrium nudicaule (Willd.) Benth., Scroph. Ind. 31. 1835 & in DC., Prod. 10: 407. 1846; Thw., Enum. Pl. Zeyl. 218. 1860; Hook. f., Fl. Br. Ind. 4: 274. 1884; Trim., l.c. 247; Abeywick., l.c. 221.

Gratiola nudicaulis Willd., Ges. Naturf. Fr. Neue Schr. 4: 193. 1803. Type: not known.

Gratiola cuneifolia Roxb., Fl. Ind. 1: 142. 1832. Type: India, Coromandel Coast, (?).

Stem up to 25 cm high, with long, erect branches from near base, slender, glabrous. Leaves chiefly crowded at base of stem with 2–3 distant pairs above, broadly oblong to oval-oblong, 2–7 × 1–3.5 mm, subamplexicaul at base, obtuse to subacute at apex, indistinctly nerved. Flowers axillary or binate. Bracts linear-lanceolate, 1 × 0.5 mm, glabrous on both surfaces. Pedicels 7–10 mm long, slender, longer in fruit. Calyx tube 1 mm long; lobes linear-lanceolate to oval-lanceolate, up to 0.75 mm long. Corolla 6–7 mm long; tube 3–4 mm long, dull blue, glabrous on both surfaces; throat minutely hirsute with glassy hairs; upper lip obovate, 3 mm long, slightly 2-lobed at apex, brilliant dark purple-blue; lower lip 7–8 mm across, pale bluish-purple, midlobe broadly obovate, slightly 2-lobed, with a large, white blotch at base. Stamens included; filaments very short, pale green, with a cluster of glandular-hairs at base of anthers. Hypogynous disk pale yellow. Style glabrous. Capsule 3 mm long, bluntly pointed at apex, glabrous, with a transparent wall; seeds irregularly cuneiform, 0.3 mm long, rugose, black.

Distr. South India, Ceylon and the Philippine Islands.

Ecol. Among short grass in wet places of the lowlands and the wet hill country up to 500 m, especially on rocky surfaces, occasionally in paddy fields; common. Often grows in association with *Eriocaulon quinquangulare*. The brilliant purplish-blue flowers make the plant conspicuous when it grows in masses. Flowering November–July.

Specimens Examined. KANDY DISTRICT: Haragama, *Alston 1312* (PDA). MATALE DISTRICT: Dambulla, at foot of Dambulla Rock, *Cramer 3281* (L, PDA, US). KURUNEGALA DISTRICT: Polgahawela, near culvert 2/6, Polgahawela-Narammala Rd., *Cramer 3267* (CAL, PDA, US). HAMBANTOTA DISTRICT: Yala, Ruhuna National Park, Jamburagala, on top of Rock, *Cramer 3315* (PDA, US); Mahamagulvihara, near pond, *Cramer 3323* (PDA, US). BADULLA DISTRICT: Bintenne, *s. coll. C.P. 2040* (K, PDA). BATTICALOA DISTRICT: Inginiyagala, Mar. 1954, *Silva s.n.* (PDA). AMPARAI DISTRICT: Lahugala, in wet flush among short grass, *Cramer 5052* (BM, G, PDA, US).

2. Dopatrium junceum (Roxb.) Buch.-Ham. ex Benth., Scroph. Ind. 30. 1835 & in DC., Prod. 10: 407. 1846; Thw., Enum. Pl. Zeyl. 219. 1860; Hook. f., Fl. Br. Ind. 4: 274. 1884; Trimen, l.c. 247; Alston, Kandy Fl. 59. f. 310. 1931; Abeywick., l.c. 221.

Gratiola juncea Roxb., Pl. Corom. 2: 16, t. 127. 1798. Type: Roxb. Ic., l.c. t. 129.

Stem up to 30 cm high, rather thick and spongy towards base, gradually becoming thinner upwards, with erecto-patent branches. Leaves oblong to oval-oblong or obovate-oblong, 5–14 × 2–4 mm, upper distinct, smaller, becoming bracteate towards stem apex, parallel-nerved. Flowers in axils of middle and upper leaves, pedicellate or nearly sessile. Pedicels 6–12 mm long, very slender. Calyx tube campanulate, 1–1.5 mm long; lobes oval-lanceolate, 1 mm long. Corolla purplish-blue, 6–7 mm long; tube 3.5–4 mm long; lower lip up to 8 mm across, lobes rounded, midlobe with a white blotch at base. Style very short, glabrous. Capsule broadly ellipsoid, 3–4 mm long, glabrous;

seeds cuneately ovoid, 0.5 mm long, longitudinally ribbed, transversely rugose between ribs, blackish-brown.

Distr. Throughout the eastern tropics.

Ecol. In paddy fields of the wet lowlands and the hill country up to 550 m and along banks of ponds and tanks in the dry lowlands; common. Flowering June–October.

Vern. Bin-savan (S).

Note. In the field it is often difficult to distinguish this species from *D. lobelioides* (Retz.) Benth. But the erect fruiting pedicels and the much-branched nature of the basal part of the stem are distinctive. Its stem, in addition, bears the more general appearance of tapering towards the apex than that of *D. lobelioides.*

Specimens Examined. KANDY DISTRICT: Peradeniya, *Gardner C.P. 2044* (K, PDA). PUTTALAM DISTRICT: Wilpattu, National Park, along banks of Datarampu Villu, *Cramer 3167* (CAL, PDA, US). KURUNEGALA DISTRICT: Dambadeniya, near culvert 20/2, Giriulla-Dambadeniya Rd., *Cramer 3208* (PDA, US). POLONNARUWA DISTRICT: Habarana, *Gardner C.P. 2044* (K, PDA); ibid, in shallow depressions of Bahirawagala Rock, *Cramer 5093* (BM, G, K, PDA, US). HAMBANTOTA DISTRICT: Yala, Ruhuna National Park, Uraniya, near Buttawa, *Mueller-Dombois & Cooray 68012812* (PDA, US). AMPARAI DISTRICT: Lahugala, near Circuit Bungalow, *Cramer 3654* (PDA, US).

3. Dopatrium lobelioides (Retz.) Benth., Scroph. Ind. 31. 1835 & in DC., Prod. 10: 407. 1846; Thw., Enum. Pl. Zeyl. 219. 1860; Hook. f., Fl. Br. Ind. 4: 274. 1884; Trimen, l.c. 247; Abeywick., l.c. 221.

Gratiola lobelioides Retz., Obs. 4: 7. 1786. Type: not known.

Stem up to 80 cm high, spongy, thickened below. Radical leaves mostly crowded at base of stem, oblong-lanceolate to oblong, 3.2–5.5 × 1–1.7 cm, subacute at apex, cauline ones distant, 0.7–2 × 0.5–0.9 cm; all parallel-nerved. Panicles up to 22.5 cm long. Bracts ovate-oblong, up to 2 × 1 mm, becoming red with age. Pedicels 7–17 mm, slender, sparsely glandular-pubescent. Calyx tube 1.5 mm long, minutely and sparsely glandular-hirsute without; lobes oblong, 1.5 × 1 mm. Corolla 10–13 mm long, pale bluish-purple; tube 5 mm long, minutely glandular-hispid without; throat sparsely scabrid; lower lip up to 14 mm across, midlobe broadly obovate, faintly 2-lobed, often with a large white blotch at base. Stamens inserted at top of corolla tube; filaments 1 mm long. Hypogynous disk pale yellow. Style short; lamellae of stigma broadly triangular-rhomboid. Capsule globose, 1.5–4 × 1.5–3 mm, tipped by withered style, glabrous; seeds oblong-cuneiform, 0.6 mm long, longitudinally ribbed, transversely rugose between ribs, blackish-brown.

Distr. South India and Ceylon.

Ecol. In shallow water of tanks of the dry lowlands and in flooded paddy fields of the wet lowlands; common; often growing in populations. Flowering almost throughout the year.

Specimens Examined. JAFFNA DISTRICT: Tunakari, Feb. 1890, *s. coll. s. n.* Kutchchaveli, *Gardner C.P. 2042* (K, PDA). ANURADHAPURA DISTRICT: Wilpattu National Park, Kudapatessa, by villu, *Cramer 3039* (PDA, US) *& 3158* (PDA, US); Kali Villu, *Cramer 3163* (CAL, PDA, US). PUTTALAM DISTRICT: Vannativillu, by tank, *Simpson 9854* (PDA). KURUNEGALA DISTRICT: Kurunegala, by Wenduru Wewa, *Cramer 3059* (PDA, US); Kaligamuwa, by culvert 1/9, Kurunegala Rd., *Cramer 3199* (PDA, US); Giriullagangoda, by Totadeniya-Giriulla Rd., *Cramer 3204* (PDA); Maharagama, by Giriulla-Dambadeniya Rd., *Cramer 3206* (E, PDA, US). COLOMBO DISTRICT: Between Negombo & Kurunegala, *s. coll. C.P. 2042* (K, PDA). KEGALLE DISTRICT: Dedigama, *Lord 844* (PDA).

13. LIMNOPHILA

R. Br., Prod. 442. 1810, nom. cons.; Roth, Nov. Pl. 294. 1821; Benth. & Hook., Gen. Pl. 2: 950. 1876. Type species: *L. gratioloides* R. Br. (=*L. indica* (L.) Druce).

Ambulia Lam., Enc. 1: 128. 1783; Wettst. in Pflanzenfam. 4 (3b): 73. 1891. Type species: *A. aromatica* Lam. (=*Limnophila aromatica* (Lam.) Merr.)

Diceros Lour., Fl. Cochinch. 381. 1790. Type species: *D. cochinchinensis* (=*Limnophila chinensis* (Osb.) Merr.).

Annual or perennial, aquatic or marsh herbs, frequently aromatic when bruised, glabrous, pubescent or glandular. Stems erect, prostrate or creeping and rooting at nodes, simple or more usually branched. Submerged leaves in aquatic species verticillate, pinnatifid to capillary-multifid, glabrous; aerial leaves opposite to verticillate, sessile or petiolate, entire to serrate, laciniate or pinnately divided, where undivided pinnately or parallel-nerved, punctate. Flowers sessile or pedicellate, solitary-axillary or in lax or compact, terminal or axillary spikes or racemes, bracteolate or not. Calyx tubular; lobes narrow, subequal or the posterior one larger. Corolla bilabiate; tube cylindric; upper lip outside in bud, entire or 2-lobed; lower lip 3-fid. Stamens 4, included, didynamous, the posterior pair shorter; anthers free, cells stipitate. Stigma bilamellate. Capsule ellipsoid to globose, septicidally 4-valved; valves bifid; seeds small, numerous, angular or ribbed.

A genus of the palaeotropics, originating from tropical Asia. Of the 35 recognized species, 10 are indigenous to Ceylon.

KEY TO THE SPECIES

1 Leaves undivided throughout stem
 2 Leaves opposite throughout
  3 Leaves sessile or subsessile; posterior calyx lobe not enlarged
   4 Stem glabrous or subglabrous; flowers sessile........................**1. L. repens**
   4 Stem pilose or hirsute; flowers pedicellate
    5 Stem with ± straight, patent not flexuous hairs........................**2. L. laxa**
    5 Stem with soft, flexuous hairs, not distinctly patent..................**3. L. chinensis**
  3 Leaves distinctly petiolate; posterior calyx lobe enlarged...............**4. L. rugosa**
 2 Leaves verticillate throughout or opposite and verticillate
  6 Stem glabrous or glandular; capsule ellipsoid...........................**5. L. aromatica**
  6 Stem hirsute; capsule globose or subglobose.............................**3. L. chinensis**
1 Leaves on lower or submerged part of stem pectinate to capillary-multifid
 7 Aerial stem patent-hirsute; flowers pedicellate..........................**6. L. aquatica**
 7 Aerial stem not patent-hirsute; flowers usually sessile
  8 Bracts present; flowers in spikes.......................................**7. L. polystachya**
  8 Bracts absent; flowers solitary
   9 Bracteoles present...**8. L. indica**
   9 Bracteoles absent
    10 Upper leaves on aerial stem divided..................................**9. L. sessiliflora**
    10 Upper leaves on aerial stem undivided................................**10. L. heterophylla**

1. Limnophila repens (Benth.) Benth. in DC., Prod. 10: 387. 1846; Philcox, Kew Bull. 24 (1): 154. 1970.

Stemodia repens Benth. in Lindl., Bot. Reg. 17, sub t. 1470. sp. 11. 1832; Benth., Scroph. Ind. 23. 1835. Type: India, Wall., Num. List. 3835 C. (lectotype, K-WALL.).

Limnophila conferta Benth. in DC., Prod. 10: 387. 1846; Hook. f., Fl. Br. Ind. 4: 266. 1884; Trimen, l.c. 243; Alston, Kandy Fl. 60, f. 318. 1938; Abeywick., l.c. 221, nom. illegit.

Limnophila serrata Thw., Enum. Pl. Zeyl. 218. 1846. Type: Ceylon, *C.P. 2049* (K, PDA).

Stem decumbent, rooting at nodes, or erect, up to 30 cm, stout, spongy. Leaves rhomboid-oblong to ovate-elliptic, 0.6–2.4 × 0.4–0.9 cm, narrowed to cuneate at base, subacuminate to subacute at apex, crenate-serrate to serrate at margins, penninerved, with prominent nerves beneath, rather thick, glabrous, punctate on both surfaces. Flowers solitary in axils of leaves; occasionally in short, leafy spikes. Bracteoles linear-spathulate, 3–3.5 × 0.5–0.6 mm, sparsely ciliolate. Calyx tube 1.25 mm long; lobes subequal, linear-lanceolate, 4–5 × 1–1.5 mm, acuminate at apex, scabrid without. Corolla dull violet-pink to pinkish-purple; tube 6–8 mm long, finely pubescent without and at throat within, longitudinally striped deep purple down plane of midlobe, broader than lateral ones. Posterior filaments 2 mm long; anther cells horizontal. Style glabrous; stigma spathulate. Capsule ellipsoid, 3–4 mm long, invested by slightly longer calyx lobes, obtusely 4-lobed, glabrous, blackish; seeds 0.75 mm long, bluntly angular, shining, faintly arillate.

Distr. From Ceylon and Southeast Asia through Malesia eastwards to China and the Philippines.

Ecol. Widespread in paddy fields and marshy places of the wet lowlands and lower wet hill country; occasionally in the dry lowlands. The plant has a variable habit being prostrate under dry and erect under wet conditions. The leaves have a characteristic lemonaceous odour and taste. Flowering almost throughout the year.

Uses. In certain parts of the lowlands the local people chew the stem and leaves of the plant as an astringent with betel.

Specimens Examined. JAFFNA DISTRICT: *Gardner C.P. 2049* (K, PDA). KALUTARA DISTRICT: *s. coll. C.P. 2049* (K, PDA); Bandaragama, beside Bandaragama-Kalutara Rd., *Cramer 2784* (PDA, US); Matugama, near culvert 1/8, Kalutara-Matugama Rd., *Cramer 2960* (NBV, PDA, US). GALLE DISTRICT: Wakwelle, by Wakwelle-Ukwatte Rd., *Cramer 2667* (PDA); Gonapinuwela, *Cramer 2699* (PDA, US); Talgaswela, in Talgaswela Estate, *Cramer 2845* (PDA, US); Miriswatte, by culvert 6/6, Bentota-Elpitiya Rd., *Cramer 3215* (PDA, US). KEGALLE DISTRICT: Navata, near culvert 52/7, Yatiyantota Rd., *Cramer 3339* (CAL, PDA).

2. Limnophila laxa Benth. in DC., Prod. 10: 388. 1846 excl. *Stemodia diffusa* Benth.; Hook. f., Fl. Br. Ind. 4: 267. 1884; Philcox, l.c. 161. Lectotype: Malaya; Malacca, *Cuming 2294* (BM, K, L, M).

Limnophila pulcherrima Hook. f., Fl. Br. Ind. 4: 267. 1884, pro parte (quoad descr.).

Distr. From Ceylon through Burma and Sumatra eastwards to Thailand.

Specimens Examined. LOCALITY UNKNOWN: *Mrs. Walker s.n.* (K, not seen).

Note. This species is included here on the authority of Philcox, l.c. 161. The plant has not been met with in the field and there is no specimen of it in the Peradeniya Herbarium.

3. Limnophila chinensis (Osbeck) Merr., Amer. J. Bot. 3: 581. 1916; Merr., Interpr. Herb. Rumph. Amb. 47. 1917; Merr., Enum. Philipp. Pl. 4: 433. 1923; Merr., Trans. Amer. Philos. Soc. n.s. 24: 350. 1935; Alston in Trimen, Handb. Fl. Ceylon 6: 211. 1931; Abeywick., l.c. 221; Philcox, l.c. 151.

Columnea chinensis Osbeck, Dagb. Ostend. Resa 230. 1757. Type: not seen.

Stemodia hirsuta Heyne ex Benth. in Lindl., Bot. Reg. 17, sub t. 1970, Sp. 15. 1832. Type: India, Heyne in Wall., Num. List. 3930 A (K, lectotype).

Limnophila hirsuta (Heyne ex Benth.) Benth. in DC., Prod. 10: 388. 1846; Thw., Enum. Pl. Zeyl. 218. 1860; Hook. f., Fl. Br. Ind. 4: 268. 1884; Trimen, l.c. 244.

Stem stout, rather spongy, rooting at nodes below, with soft, flexuous hairs. Leaves opposite or often in whorls of 3, lanceolate, about 2.5 cm, acute or subacute, finely serrate, slightly hairy. Flowers on slender hairy pedicels, rather large, solitary or forming short axillary racemes; calyx lobes very acute, hairy. Corolla violet-pink, 1.2 cm, glabrous.

Distr. India and Ceylon.

Ecol. Wet places and paddy fields in the wet lowlands; not common. Flowering April–September.

Specimens Examined. KALUTARA DISTRICT: *Gardner C.P. 2040* (K, PDA). BADULLA DISTRICT: Uda Pusellawa, Uma Oya, Apr. 1854, *s. coll. s.n.* (PDA).

Note. The author has not met with this plant. The description given above, therefore, is Trimen's.

4. Limnophila rugosa (Roth) Merr., Interpr. Rumph. Herb. Amb. 466. 1917; Merr., Enum. Philipp. Pl. 3: 434. 1923; Money, Bot. Bihar & Orissa. Suppl. 97. 1950; Philcox, l.c. 135.

Herpestis rugosa Roth, Nov. Pl. Sp. 290. 1821. Type: India, *Heyne* (L.).
Stemodia menthastrum Benth. in Lindl., Bot. Reg. 17, sub t. 1470, Sp. 5. 1832; Benth., Scroph. Ind. 23. 1835. Type: Bangladesh, Sylhet, Wall., Num. List. 3927 (K).

Stem simple or branched, rooting at lower nodes, then ascending up to 16 cm high, adpressed hispid. Leaves ovate-lanceolate to ovate-elliptic, 1.6–4 × 0.7–1.7 cm, tapering to base, subacute at apex, entire to crenate-serrulate, penninerved, with nerves impressed above, glabrous to scabrid above, hirsute on nerves beneath; petiole 7–10 mm long, densely hirsute. Flowers solitary in leaf-axils, sessile. Calyx lobes divided to base, unequal, lanceolate, up to 7 × 1 mm, acuminate, minutely hispid without. Corolla up to 10 mm long, hirsute without. Capsule broadly ovoid, 5 mm long, invested by longer calyx lobes.

Distr. From India and Ceylon through Southeast Asia and Malesia northwards to the Philippines and eastwards to New Guinea.

Ecol. In marshy places in the wet lowlands; very rare. Flowering September.

Specimens Examined. GALLE DISTRICT: Hiniduma Pattu, Sep. 1924, *Livera s.n.* (PDA). KANDY DISTRICT: Peradeniya, tank near Orchid House, *Alston 1653* (PDA).

Note. There are two specimens of this species in the Herbarium, Peradeniya. Alston determined Livera's specimen as *L. rugosa* but failed to determine fully his own specimen. There is no doubt that both specimens belong to the same species. It is strange, however, that Alston failed to documentate this species in his Supplement to Trimen's Handbook. Livera's specimen must now be considered a new record for the species in Ceylon.

5. Limnophila aromatica (Lam.) Merr., Interpr. Rumph. Herb. Amb. 466. 1917; Alston in Trimen, Handb. Fl. Ceylon 6: 211, 1931; Alston, Kandy Fl. 60, f. 319. 1938; Abeywick., l.c. 221; Philcox, l.c. 145.

Ambulia aromatica Lam., Enc. 1: 128. 1783. Type: Rheede, Ic. in Hort. Mal. 10: t. 6. 1690.

Limnophila punctata Blume, Bijdr. 750. 1826; Thw., Enum. Pl. Zeyl. 218. 1864. Type: Java, *Blume 1495* (L.).

Limnophila gratissima Blume, Bijdr. 749. 1826; Benth., Scroph. Ind. 26. 1835; Hook. f. in Fl. Br. Ind. 4: 266. 1884; Trimen, l.c. 243. Type: Java, Batavia, *Blume 1176* (L.).

Stem simple or branched, copiously rooting at lower nodes, then erect up to 90 cm, inflated, striate, pale green, occasionally dull purple. Leaves linear-lanceolate to oblong-lanceolate, 1.3–7.3 × 0.3–2 cm, semiamplexicaul, acute to subobtuse at apex, crenate-serrate to serrate, penninerved, glabrous and densely punctate on both surfaces. Flowers solitary in leaf-axils or in terminal racemes 3–14 cm long. Pedicels 3–16 mm long, glandular-puberulous. Bracteoles subulate, 1–3 × 0.5 mm, glabrous on both surfaces. Calyx tube 1–2 mm long; lobes subequal, narrowly lanceolate, 4–6.5 × 1–1.75 mm, acuminate at apex, glandular-pubescent without. Corolla mauve-purple to pale violet; tube 9–12 (–19) mm long, finely glandular-pubescent without, densely villous at throat; lobes of limb oblong-orbicular; lower lip 7–12 mm across, midlobe slightly smaller than lateral ones. Posterior filaments 2–2.5 mm long, anterior ones longer; all glabrous. Style glabrous. Capsule broadly ellipsoid, 4–5 × 2–3 mm, enclosed by longer calyx lobes, obtusely 4-lobed, glabrous; seeds oblong-cylindric, 0.5 mm long, truncate at both ends, muricate, tawny.

Distr. From Southeast Asia through Malesia eastwards to China and Japan, northwards to the Philippines and southwards to Micronesia and Australia.

Ecol. In ditches, flooded paddy fields and marshy places of the wet lowlands and lower wet hill country; common. The plant has a turpentine smell. Flowering almost throughout the year.

Specimens Examined. KURUNEGALA DISTRICT: Kurunegala, along borders of lake, *Cramer 3062* (PDA, US). KALUTARA DISTRICT: Dodangoda, by Kalutara-Matugama Rd., *Cramer 2709* (NBV, PDA, US); Waskaduwa, along Bandaragama Rd., *Cramer 2716* (CAL, PDA, US).

GALLE DISTRICT: Kottawa, by culvert 11/10, Galle-Udugama Rd., *Cramer 3020* (PDA, US). RATNAPURA DISTRICT: Dewalagawa, near marker 42, Ratnapura Rd., *Cramer 3097* (E, PDA, US). KANDY DISTRICT: Kandy, *Appuhamy s.n.* (PDA); Gannoruwa, *Alston, s.n.* (PDA).

6. **Limnophila aquatica** (Roxb.) Alston, Ann. R. Bot. Gard. Peradeniya 11: 205. 1929; Alston, in Trimen, Handb. Fl. Ceylon 6: 211. 1931; Abeywick., l.c. 221.

Cyrilla aquatica Roxb., Pl. Corom. 2: 47, t. 189. 1798; Roxb., Fl. Ind. 3: 115. 1832. Type: Roxburgh, Icon. t. 189, l.c.
Diceros aquaticus (Roxb.) Moon, Cat. 45. 1824.
Limnophila racemosa Benth., Scroph. Ind. 26. 1835; Benth. in DC., Prod. 10: 390. 1846; Thw., Enum. Pl. Zeyl. 218. 1866; Hook. f., Fl. Br. Ind. 4: 271. 1884; Trimen, Handb. Fl. Ceylon 3: 245. 1895. Types: Wall., Num. list 3907 A & B (syntypes, K-WALL.).

Stem rooting at lower nodes, then erect and up to 40 cm high, with basal part usually submerged, thick, tumid at nodes. Upper leaves opposite or in verticils of 2–3, ovate-lanceolate to oblong-lanceolate, 2.8–6.6 × 0.8–1.9 cm, rounded and semi-amplexicaul at base, acuminate at apex, finely spinulose-serrate, strongly 3–5-nerved at base, glabrous on both surfaces, punctate above; lower ones in verticils of more or less 10–12, pinnately dissected, up to 5.8 cm long, often deflexed and root-like. Flowers numerous, in terminal racemes 5–18 cm long; peduncles finely glandular-hispid. Bracts ovate to broadly lanceolate, 5–7 × 3–4 mm, subacuminate at apex, finely glandular-pubescent on both surfaces. Pedicels 2–6 mm long, finely glandular-pubescent. Calyx tube 2–3 mm long; lobes more or less equal, ovate-lanceolate, 2–3 × 0.75 mm, long acuminate, scarious at margins. Corolla tube pale greenish-white to white, 7–10 mm long, puberulous without, villous within at throat; limb white or pale blue or pale mauve; lobes of upper lip broadly orbicular with a pale purple blotch at centre of each; lower lip 9–17 mm across, midlobe broader than lateral ones. Posterior filaments 2–3 mm long, anterior ones longer; anther cells horizontally placed. Style glabrous. Capsule globose, 3–4 × 2–3 mm, enclosed by longer calyx lobes, tipped by withered style, glabrous; seeds oblong-cuneate to globular, 0.5–0.7 mm long, quadrangular, muricate.

Distr. India and Ceylon.

Ecol. In marshy ponds, tanks and flooded paddy fields of the wet and dry lowlands; common. A gregarious plant; when in masses it affords a beautiful sight at flowering time. The flowers are faintly scented. Under clear water the fan-wise submerged leaves form a pleasing sight. The young stem and leaves form stout, green, mushroom-like buds. Flowering March–December.

Specimens Examined. JAFFNA DISTRICT: *Gardner C.P. 2036*

(K, PDA). ANURADHAPURA DISTRICT: Rambewa, *Simpson 9393* (PDA), Madawachchiya, near culvert 108/2, Madawachchiya-Murugan Rd., *Cramer 2883* (CAL, PDA, US). PUTTALAM DISTRICT: Nattandiya, by Marawila-Nattandiya Rd., *Cramer 2614* (PDA, US); Kottukachchiya, along borders of Tank, *Cramer 2646* (PDA, US); Wilpattu, along borders of Katarampu Villu, *Cramer 3169* (PDA, US). KURUNEGALA DISTRICT: Kurunegala, by Tittawela Tank, *Cramer 3183* (PDA, US); Dambadeniya, by culvert 19/3, Giriulla-Dambadeniya Rd., *Cramer 3212* (PDA). COLOMBO DISTRICT: Kotadeniya, by Colombo-Kotadeniya Rd., *Cramer 3203* (PDA). KALUTARA DISTRICT: Waskaduwa, by Bandaragama Rd., *Cramer 2713* (PDA, US); Bandaragama, *Cramer 2789* (PDA, US); Nagoda, by tarmac of abandoned airport, *Cramer 3228* (PDA, US). TRINCOMALEE DISTRICT: Periyakulam, along borders of Tank, *Cramer 2994* (E, PDA); Andakulam, along borders of Tank, *Cramer 3004* (PDA, US). BATTICALOA DISTRICT: Kalkudah, in pond close to culvert 20/3, Batticaloa Rd., *Cramer 3623* (PDA, US). GALLE DISTRICT: Wakwelle, by Wakwelle-Ukwatte Rd., *Cramer 2168* (PDA, US); Gonapinuwela, *Cramer 2700* (PDA, US).

7. Limnophila polystachya Benth., Scroph. Ind. 26. 1835; Benth. in DC., Prod. 10: 390. 1846; Hook. f., Fl. Br. Ind. 4: 269. 1884; Philcox, l.c. 131. Type: India, Madras, Wall., Num. List. No. 3909 A (K, holotype).

Stemodia aquatica Willd., Sp. Pl. 3: 346. 1801. Type: India, 'prope Tranquebariam' (not seen).

Distr. South India and Ceylon.

Specimens Examined. LOCALITY UNKNOWN: *Rottler 291* (M).

Note. This species is included here on the authority of Philcox. The author has been unable to find this plant in the field and there is no specimen of it in the Peradeniya Herbarium.

The earliest valid name for this species is the one proposed here. The combination *Limnophylla aquatica* would be illegitimate since this had been made earlier by Alston in 1929 for another species based on Roxburgh's *Cyrilla aquatica*.

8. Limnophila indica (L.) Druce in Rep. Bot. Soc. Exch. Club Br. Isles 1913, 3: 420. 1914; Alston in Trimen, Handb. Fl. Ceylon 6: 211. 1931; Abeywick., l.c. 221; Philcox, l.c. 115.

Hottonia indica L., Sp. Pl. ed. 2. 208. 1762 & Syst. Nat. ed. 10. 919. 1759. Type: Sp. No. 204.2 (LINN).

Gratiola trifida Willd., Sp. Pl. 1: 104. 1797, nom. illegit.; Moon, Cat. 4. 1824.

Limnophila gratioloides R. Br., Prod. 442. 1810; Benth., Scroph. Ind. 25. 1835;

Benth. in DC., Prod., 10: 389. 1846; Thw., Enum. Pl. Zeyl. 218. 1860; Hook. f., Fl. Br. Ind. 4: 271. 1884; Trimen, l.c. 245. Type: Australia, Queensland, Broadsound, *R. Brown 2652* (BM, holotype).

Stem rooting at lower nodes, then erect up to 20 cm high, occasionally tufted, slender, finely glandular-pubescent. Lower leaves pinnately dissected, root-like, upper ones 3-fid, uppermost often undivided, linear-spathulate to linear-lanceolate, 7–23 × 1.5–5 mm, tapering to base, subacute at apex, serrate, 1-nerved, glabrous on both surfaces. Flowers in leaf-axils. Pedicels 4–10 mm long, longer in fruit, divaricate, finely glandular-pubescent. Bracteoles divaricate, linear, 2–3 × 1 mm, glabrous and punctate above. Calyx tube 1.5–2 mm long, subglobose in older flowers; lobes linear-lanceolate, 2–3 mm long, glandular-scabrid without, acuminate. Corolla up to 8 mm long, pale yellow, occasionally pale purple; tube 4–6 mm long, glabrous without, villous within at throat, faintly striped purple down plane of upper lip; upper lip faintly 2-lobed; lower lip 6–7 mm across, lobes orbicular, midlobe broader than lateral ones. Stamens inserted at middle of corolla; filaments glabrous; anther-cells horizontal. Style glabrous. Capsule subovoid, 4–5 × 2.5 mm, tipped by withered style; seeds oblong-cuneiform, 0.5 mm long, 4–5 angular, muricate, shining, black.

D i s t r. From Southeast Africa through Iraq, Ceylon and Southeast Asia eastwards to Malesia and China, and Southwards to Australia.

E c o l. Widespread along borders of tanks and ponds in the dry lowlands, less so in the wet lowlands. The young stem and leaves form small, green, mushroom-like buds. The flowers begin to open by more or less 10 a.m., are fully open till about 11.30 a.m. and begin to close by more or less 3 p.m. Flowering January–October.

S p e c i m e n s E x a m i n e d. VAVUNIYA DISTRICT: *Simpson 9345* (PDA). ANURADHAPURA DISTRICT: Alan Kulam, near Anuradhapura, Mar. 1883, *s. coll. s.n.* (PDA); Madawachchiya, along borders of Mudaliyakulam, *Cramer 2882* (NBV, PDA, US); Wilpattu National Park, along borders of Kali Villu, *Cramer 3162* (CAL, PDA, US). PUTTALAM DISTRICT: *Gardner, C.P. 2038* (K, PDA); Dharmapala Wewa, *Cramer 3230* (PDA). KURUNEGALA DISTRICT: Kalugamuwa, Ginipenda Wewa, *Cramer 3200* (PDA, US). HAMBANTOTA DISTRICT: by culvert 152/4, Tissamaharama Rd., *Cramer 2992* (PDA). BADULLA DISTRICT: Uma Oya, *s. coll. C.P. 2038* (K, PDA).

9. Limnophila sessiliflora (Vahl) Blume, Bijdr. 849. 1826; Benth., Scorph. Ind. 25. 1835; Benth. in DC., Prod. 10: 389. 1846; Thw., Enum. Pl. Zeyl. 218. 1860; Hook. f., Fl. Br. Ind. 4: 270. 1884; Trimen, l.c. 244; Abeywick., l.c. 221; Philcox, l.c. 112.

Hottonia sessiliflora Vahl, Symb. Bot. 2: 36. 1791. Type: India, Burmann (not seen).

Stem submerged and rooting at lower nodes, then ascending up to more or less 20 cm, branched below, slightly pubescent above. Upper leaves on aerial flowering stem in verticils of 3–8, often pinnately divided, glabrous on both surfaces; submerged leaves in verticils of 10–11, up to 20 mm long, pinnatisect. Flowers sessile in axils of upper leaves. Calyx tube 3–4 mm long, minutely sessile-glandular without; lobes oval-lanceolate, 1.5 mm long. Corolla pale mauvish-pink; tube 5–6 mm long, with lilac longitudinal lines; lower lip 6–7 mm across. Style glabrous. Capsule globose, 3.5 × 2.5 mm, slightly longer than calyx, tipped by withered style, glabrous, pale brown; seeds oblong, 0.5 mm long, truncate at both ends, 5-angular, muricate, blackish-brown.

Distr. From Ceylon and Southeast Asia eastwards through Malesia to China, Korea and Japan.

Ecol. In wet places and flooded paddy fields of the wet and dry lowlands; common. Flowering March–September.

Specimens Examined. POLONNARUWA DISTRICT: Minneriya, *Gardner C.P. 2037* (K, PDA). ANURADHAPURA DISTRICT: *Alston 1310* (PDA). KURUNEGALA DISTRICT: Dambadeniya, near culvert 20/2, Giriulla Rd., *Cramer 3210* (NBV, PDA, US); near culvert 19/3, Giriulla Rd., *Cramer 3211* (PDA, US).

Note. Observations in the field have confirmed the correctness of Hooker's note: "This (species) strongly resembles a sessile-flowered form of *L. indica* (*L. gratioloides*), or a short pubescent form of *L. heterophylla* wanting the multifid submerged leaves." This observation forms a good means of distinguishing this species in the field from the allied *L. heterophylla* which is less common.

10. Limnophila heterophylla (Roxb.) Benth., Scroph. Ind. 25. 1835; Benth. in DC., Prod. 10: 390. 1846; Hook. f., Fl. Br. Ind. 4: 270. 1884; Trimen, l.c. 244; Abeywick., l.c. 221; Philcox, l.c. 124.

Columnea heterophylla Roxb., Fl. Ind. 3: 97. 1832. Type: Bengal, *Roxburgh* (BM).

Limnophila sessiliflora Thw., Enum. Pl. Zeyl. 218. 1860 (pro parte). Type: Ceylon, *C.P. 2389* (pro parte—K).

Stem submerged and rooting at lower nodes, then ascending up to about 60 cm high, simple or branched, spongy, aerial part slender, glabrous. Upper leaves in verticils of 3, linear-lanceolate to oval-oblong, 6–10 × 2–7 mm, subamplexicaul and 3-nerved at base, subacute at apex, serrulate, glabrous on both surfaces; intermediate leaves pinnatipartite to pinnatisect; lowest ones in verticils of 8–12, capillary-multifid, up to 42 mm long. Flowers in axils of upper leaves. Calyx tube 2–3 mm long, glabrous: lobes ovate-lanceolate, 1.5–2 × 0.75 mm, glabrous on both sides. Corolla pale pinkish-violet or pink; tube 6–7 mm long, sparsely pilose within, glabrous without; throat sparsely pilose; upper lip broadly orbicular, 3.5 mm long, 2-lobed at apex; lower lip 6–7 mm across, lobes more or less equal, ovate-oblong. Filaments glabrous, posterior ones 2 mm long, anterior ones longer. Anther-cells horizontal. Style glabrous. Capsule ovoid, 3–4 × 2 mm, glabrous, pale brown, with a translucent wall; seeds cuneiform, 0.5 mm long, 4–5-angular, muricate, tawny-brown.

Distr. From Ceylon and Southeast Asia eastwards through Malesia (except Indonesia) to China.

Ecol. In shallow water of ponds and in paddy fields of the wet and dry lowlands; not common. Flowering February-July.

Specimens Examined. MANNAR DISTRICT: Madawachchiya, near marker 139, Mannar Rd., *Cramer 2898* (CAL, PDA, US). KURUNEGALA DISTRICT: Polgahawela, close to culvert 5/2, Narammala Rd., *Cramer 3278* (PDA, US).

14. PEPLIDIUM

Delile, Fl. Egypt. 148. 1812; Endl., Gen. Pl. 682. 1839; Benth. & Hook., Gen. Pl. 2: 957. 1876; Wettst. in Pflanzenfam. 4 (3b): 77. 1891. Type species: *P. humifusum* Delile (= *P. maritimum* (L. f.) Aschers).

Small annual herbs. Stem prostrate or creeping, much branched, glabrous. Leaves opposite, ovate to spathulate, somewhat fleshy. Flowers very small, 1–3 in an axil, subsessile or shortly pedicellate. Calyx tubular, 5-ribbed, 5-fid, enlarged in fruit. Corolla tube short; lobes 5, more or less broad, subequal, 2 upper ones outside in bud. Stamens 2, included; filaments arched, subdilated at base; anthers 1-celled. Stigma spathulate, recurved. Capsule globose, or ovoid, septicidally 2-valved; valves entire; seeds fairly numerous, small.

A small genus of 2 palaeotropical species.

Peplidium maritimum (L. f.) Aschers. in Schweinfurth, Beitr. Fl. Aethiop. 1306. 1867; Wettst. in Pflanzenfam. 4 (3b): 78. 1891; Alston in Trimen, Handb. Fl. Ceylon 6: 214. 1931; Abeywick., l.c. 222.

Hedyotis maritima L. f., Suppl. 119. 1781. Type: India orientalis, Koenig (?).
Peplidium humifusum Delile, Fl. Egypt. 148. 1813; Thw., Enum. Pl. Zeyl. 426. 1864; Hook. f., Fl. Br. Ind. 4: 287. 1884; Trimen, l.c. 255.

Stem prostrate, filiform, rooting at nodes, often submerged or floating, much branched, glabrous. Leaves obovate or spathulate, 5–10 × 4–6 mm, tapering to base, rounded at apex, minutely punctate above, glabrous on both surfaces; petioles up to 3 mm long. Flowers 1–3 in axils of leaves. Pedicels 1 mm long. Calyx tube 1.75 mm long, glabrous, 5-ribbed, ribs continued into lobes; lobes shorter than tube, obtuse. Corolla pinkish-white. Capsule globose, 2.5–3 mm in diam., enclosed in enlarged membranous calyx, glabrous; seeds oblong to cuneate, 0.5 mm long, longitudinally striate, dark brown.

Distr. Egypt, South India, Ceylon and tropical Australia.

Ecol. Along edges of tanks and in water of shallow ponds in the dry lowlands; rare. Flowering January–June.

Specimens Examined. MANNAR DISTRICT: Mantai, near pit, Feb. 1840, *s. coll. s.n.* (PDA). TRINCOMALEE DISTRICT: *Glenie, C.P. 3789* (K, PDA). HAMBANTOTA DISTRICT: Yala, Kohombagaswela, *Meijer 236* (PDA, US).

15. VERONICA

L., Sp. Pl. 9. 1753; L., Gen. Pl. ed. 5. 10. 1754; Endl., Gen. Pl. 688. 1839; Benth. & Hook., Gen. Pl. 2: 964. 1876; Wettst. in Pflanzenfam. 4 (3b): 85. 1891. Type species: *V. officinalis* L.

Annual or perennial herbs or shrubs. Stem erect or often prostrate. Leaves opposite or alternate. Flowers axillary and solitary or in axillary or terminal racemes, bracteate but ebracteolate. Calyx 4–5-lobed; the posterior lobe often wanting. Corolla tube short; limb spreading, 4–5-lobed; adaxial lobe often of 2 united lobes, the lateral lobes usually narrower than the others. Stamens 2, inserted at top of corolla tube, exserted; anthers 2-celled, cells parallel or divergent, with confluent tips. Ovary pilose at apex; stigma subcapitate. Capsule compressed or turgid, loculicidally 2-valved, or loculicidal-septicidally 4-valved; seeds few or numerous.

A genus of about 300 species, mostly temperate.

KEY TO THE SPECIES

1 Stem prostrate; leaves distinctly petiolate
- 2 Flowers in terminal racemes; leaves shallowly crenate .... **1. V. serpyllifolia**
- 2 Flowers solitary; leaves sharply crenate-serrate .... **2. V. persica**

1 Stem erect; leaves sessile or subsessile
- 3 Leaves all opposite; racemes not leafy .... **3. V. javanica**
- 3 Upper or floral leaves alternate; racemes leafy .... **4. V. arvensis**

1. Veronica serpyllifolia L., Sp. Pl. 12. 1753; Benth. in DC., Prod. 10: 482. 1846; Hook. f., Fl. Br. Ind. 4: 296. 1884; Alston in Trimen, Handb. Fl. Ceylon 6: 215. 1931; Abeywick., l.c. 222. Type: Europe, No. *26.30* (LINN).

Stem rooting at nodes, simple or diffusely branched. Leaves ovate-orbicular or ovate-elliptic, 6–15 × 4–13 mm, rounded at base, obtuse at apex, shallowly crenate, 3-nerved at base, glabrous on both surfaces; petioles 3–5 mm long. Flowers in leafy terminal racemes up to 7.2 cm long. Upper bracts linear-oblong. Pedicels 2–3 mm long, puberulous. Calyx lobes obovate-oblong, 3–3.5 × 2 mm, glabrous. Corolla bluish-violet or white; limb 4–5 mm across; lobes obovate-cuneate, adaxial lobe with longitudinal, dark violet lines. Filaments glabrous. Style exserted, glabrous. Capsule broadly obcordate, 4 × 5.5 mm, shorter than calyx, glabrous, sparsely glandular-ciliate along upper margins; seeds oblong, 0.6 mm long, compressed, minutely puberulous.

Distr. Europe, America, Northern Africa, Northern and Southern Asia including Ceylon.

Ecol. A naturalised weed in lawns, cultivated fields, grassy roadsides and rocky hill tops in the upper hill country above 1800 m; common. Flowering throughout the year.

Specimens Examined. NUWARA ELIYA DISTRICT: Moon Plains, Feb. 1927, *Alston s.n.* (PDA); Pidurutalagala Peak, among rocky outcrops, *Cramer 3497* (PDA); Nuwara Eliya, Apr. 1915, *Petch s.n.* (PDA); Nuwara Eliya, Mahagastota, *Cramer 2855 & 2879* (PDA, US); Golf Links, *Cramer 2942* (PDA, US); Hakgala, close to marker 153, Hakgala Rd., *Cramer 2958* (PDA, US); along roadside close to culvert 53/13 Nuwara Eliya-Hakgala Rd., *Cramer 3241* (CAL, PDA, US).

2. Veronica persica Poir. in Lam., Enc. 8: 542. 1808; Hook. f., Fl. Br. Ind. 4: 294. 1884.

Stem pilose, often purplish; branches often ascending. Leaves broadly ovate, 8–28 × 6–15 mm, rounded or truncate at base, obtuse at apex, 5-nerved at base, scabrid on both surfaces; petioles 2–4 mm long. Flowers in axils of leaves. Pedicels 10–22 mm long, slender, pilose, decurved in fruit. Sepals oblong or lanceolate, 4–5 × 1.7–2.5 mm, pilose without at base, ciliate. Corolla tube 0.5 mm long; limb up to 9 mm across; adaxial and lateral lobes ovate-orbicular, 3.5 × 4 mm, purplish-blue with longitudinal dark blue lines, abaxial ones ovate-oblong, 3.5 × 2 mm, white. Filaments dilated at middle, glabrous. Capsule broadly orbicular, 6 × 7.5 mm, equal to calyx, minutely hirsute along upper margins, pale brown; seeds naviculate, 0.5 mm long, pitted.

Distr. From Northern Europe through North Africa to Western and Eastern Asia including Ceylon.

Ecol. A weed of cultivated fields above 2000 m; not common. Flowering March–August.

Note. The identity of this species has been determined by Mr. F. Adema of the Rijsherbarium, Netherlands, to whom the author acknowledges his sincere gratitude. The species has not been reported before for Ceylon and is, therefore, a new record for the country.

Specimens Examined. NUWARA ELIYA DISTRICT: Horton Plains, close to P.W.D. Depot, *Cramer 2948* (NBV, PDA, US); Nuwara Eliya, close to marker 153, Hakgala Rd., *Cramer 2957* (US).

3. Veronica javanica Blume, Bijdr. 742. 1826; Hook. f., Fl. Br. Ind. 4: 296. 1884; Alston in Trimen, Handb. Fl. Ceylon 6: 215. 1931. Type: Java, *Blume* (probably L).

Veronica deltigera Wall. ex Benth., Scroph. Ind. 45. 1835. Type: Nepal, Wall., Num. List. 402 (K-WALL.).

Roots fibrous. Stem about 15 cm, sparingly branched (in the Ceylon specimen), pubescent; leaves ovate, about 1.2 cm long, coarsely crenate-serrate, subsessile, pubescent. Flowers shortly stalked in axillary racemes. Calyx lobes linear-oblong, exceeding the capsule. Capsule obcordate, flattened.

Distr. From East Africa eastwards through India and Ceylon to Java, Tonkin, Siam and the Luchu Islands.

Ecol. In the jungle in the upper hill country above 2000 m; rather rare. Flowering December.

Note. The author's long search for this plant has been without avail. The description given above, therefore, is Alston's.

Specimens Examined. NUWARA ELIYA DISTRICT: Nuwara Eliya, Single Tree Hill, *Alston 1313* (PDA).

4. Veronica arvensis L., Sp. Pl. 13. 1753; Hook. f., Fl. Br. Ind. 4: 296. 1884. Type: Europe, No. *26.58* (LINN.).

Stem erect, up to 35 cm high, with ascending branches, glandular-hirsute. Lower leaves opposite, broadly ovate, 8–13 × 6–12 mm, truncate at base, obtuse at apex, coarsely crenate, 5-nerved at base, scabrid on both surfaces; upper or floral ones oblong-lanceolate, 7–8 × 2–2.5 mm. Racemes lax, terminal and axillary, up to 19.5 cm long. Pedicels 1 mm long. Calyx lobes linear-lanceolate, 2–4 × 1–1.25 mm, glandular hirsute without, ciliolate. Corolla bright blue; limb 2 mm across, lobes suborbicular. Style 0.6 mm long, glabrous. Capsule broadly obcordate, 3.5 × 3.5 mm, shorter than calyx, glandular-ciliolate along upper margins; seeds oval-oblong, 0.5 mm long, faintly striate.

Distr. From Europe through North Africa and North Asia to South India and Ceylon.

Ecol. A weed in cultivated fields of the upper hill country above 1900 m; not common. Flowering April–November.

Note. This species is an introduction having been probably transported from abroad along with imported seeds. The specimens cited have been checked with foreign material preserved in the Peradeniya Herbarium and their identity confirmed. This species has not been reported before for Ceylon, and is, therefore, a new record for the country.

Veronica didyma Tenore though mentioned by Alston has not been found in recent years. On the other hand there is no representative specimen of the plant in the Peradeniya Herbarium nor is there an appropriate description of it for Ceylon. In the absence of such evidence it has been considered better to await more tangible evidence of the occurrence of this species in Ceylon.

Specimens Examined. NUWARA ELIYA DISTRICT: Mahagastota, close to marker 153, Hakgala Rd., *Cramer 2956* (PDA, US); Horton Plains, beside Ohiya Rd., about 1/8 km from Farr Inn, *Balakrishnan 471* (PDA, US); Ambawela, pasture slope 1.5 km from Cattle Farm, *Cramer 3506* (PDA).

16. SCOPARIA

L., Sp. Pl. 116. 1753; L., Gen. Pl. ed. 5. 52. 1754; Benth. & Hook., Gen. Pl. 2: 959. 1876; Wettst. in Pflanzenfam. 4 (3b): 84. 1891. Type species: *Scoparia dulcis* L.

Erect herbs or much-branched undershrubs. Leaves opposite or verticillate, crenate-serrate, gland-dotted beneath. Flowers axillary, solitary or geminate, pedicellate; pedicels ebracteolate. Calyx 4–5-partite; lobes imbricate. Corolla rotate, 4-fid, densely pilose at throat; lobes subequal, obtuse, the posterior ones inside in bud. Stamens 4, subequal; anthers 2-celled, cells parallel or divaricate. Stigma notched or truncate. Capsule globose or ovoid, septicidally 2-valved or loculicido-septicidally 4-valved; valves entire; seeds numerous, scrobiculate.

A genus of about 20 species, mostly from the neotropics.

Scoparia dulcis L., Sp. Pl. 116. 1753; Hook. f., Fl. Br. Ind. 4: 289. 1884; Trimen, l.c. 255; Alston in Trimen, Handb. Fl. Ceylon 6: 215. 1931; Alston, Kandy Fl. 60, f. 323. 1938; Abeywick., l.c. 222. Type: Holland, living specimen in Cliff. Gard.

Gratiola micrantha Nutt., Amer. J. Sci. 5: 267. 1822. Type: N. America, Florida, *Ware s.n.* (PH).

Scoparia grandiflora Nash, Bull. Torrey Bot. Club 23: 105. 1896. Type: N. America, *Nash 2417* (NY).

Erect, perennial herb. Stem up to 60 cm high, divaricately branched, 5–6-angular, glabrous. Leaves ternately whorled, obovate-oblong to oblanceolate, 1.4–3.5 × 0.8–1.5 cm, tapering to base, subacute at apex, coarsely crenate-serrate from above base, penninerved, glabrous on both surfaces, punctate beneath; petioles up to 9 mm long. Pedicels 5–7 mm long, glabrous. Calyx lobes divided to base, oval-oblong, 2.5–3 × 1 mm, 3-nerved, glabrous within and without, ciliate at margins. Corolla white; limb 7–8 mm across; lobes spathulate, 3–3.5 × 2 mm, reflexed with age; hairs of throat erect, white. Filaments inserted at top of corolla tube, glabrous; anthers pale greenish-yellow. Capsule subglobose, longer than calyx, tipped by withered style, glabrous, pale brown; seeds oblong-cuneiform, 0.5 mm long, reticulate, dull brown.

Distr. A native of tropical America, now pantropic.

Ecol. A weed of cultivated fields and roadsides in the dry and wet lowlands; common. Flowering throughout the year.

Specimens Examined. KALUTARA DISTRICT: Maggona, St. Vincent's Estate, *Cramer 2978* (PDA, US). RATNAPURA DISTRICT: Nawata close to culvert 52/7, Yatiyantota Rd., *Cramer 3340* (NBV, PDA, US). GALLE DISTRICT: Kottawa, *Cramer 1291* (PDA). MATARA DISTRICT: near Matara, Feb. 1881, *s. coll. s.n.* (PDA). AMPARAI DISTRICT: Inginiyagala, Senanayake Samudra, *Mueller-Dombois & Comanor 67072609* (PDA, US).

17. CALCEOLARIA

L., Mant. 2: 143. 1771; Endl., Gen. Pl. 671. 1839; Benth. & Hook., Gen. Pl. 2: 929. 1876, pro parte; Wettst. in Pflanzenfam. 4 (3b): 55. 1891, pro parte; Kranzlin in Pflanzenr. 4. 257 c (Heft 28): 21. 1907. Type species: *Calceolaria pinnata* L.

Annual herbs or shrubs. Stem erect. Leaves opposite or verticillate. Flowers in axillary or terminal compound cymes. Calyx 4-partite. Corolla yellow, white or red; tube very short; limb bilabiate; upper lip outside in bud, hooded, entire or obscurely bilobed, much smaller than saccate lower lip. Stamens 2, inserted near base of corolla tube; anthers 2-celled, united by an elongated connective, posterior cell fertile, anterior one sterile or fertile. Style not thickened at apex; stigma small. Capsule septicidally 2–4-valved; seeds numerous; testa covered with rows of tubercles.

A large genus of about 300–400 species from Central and South America.

Calceolaria chelidonioides HBK., Nov. Gen. & Sp. 2: 378. 1818; Kranzlin in Pflanzenr. 28: 24, f. 4 A–C. 1907; Alston in Trimen, Handb. Fl. Ceylon 6: 210. 1931; Abeywick., l.c. 221.

Stem profusely rooting at lower nodes, then ascending up to 45 cm high, occasionally decumbent at base, quadrangular, glandular-pilose, reddish-purple. Leaves in unequal pairs, ovate in outline, 2.5–7 × 2–5.5 cm, pinnatipartite to pinnatisect; lower leaves 7-lobed, intermediate ones 5-lobed, uppermost ones 3-lobed; lobes serrate, scabrid and dark green above, glabrous or sparsely pubescent beneath along mid-nerve and pale green, acute at apex; petioles 0.8–4.2 mm long, grooved, subamplexicaul at base. Flowers axillary, solitary or in dichasial cymes above. Pedicels 1.0–2.2 cm long, slender, glandular-hispid. Calyx lobes broadly ovate, up to 5 × 3 mm, acute at apex, glandular-pilose without. Corolla chrome-yellow; tube 1.5 mm long, glandular-hispid without; upper lip oblong-spathulate, 3–5 mm long, finely pubescent without; lower one orbicular, 10–14 × 10–13 mm, with a small orifice. Sterile anther-cells each produced into a narrow appendage 2–2.5 mm long. Style short. Capsule broadly ellipsoid, 5–6 mm long, equal to or slightly longer than calyx, tipped with withered style, glandular-hirsute, 4-valved; seeds ellipsoid, 0.8 mm long, obtusely 4–5-angular, striate, orange-yellow.

Distr. Native of Ecuador naturalised in Ceylon.

Ecol. A gregarious plant in the upper hill country above 1650 m in exposed, damp ground, roadsides, grassy embankments close to a watercourse or ditch, or occasionally along drains by gravel estate roads; common. The shape and colour of the corolla make the flowers very conspicuous. Flowering December–August.

Specimens Examined. NUWARA ELIYA DISTRICT: May 1911, *de Silva s.n.* (PDA); Nuwara Eliya, close to culvert 110/114, Nanu Oya Rd., *Cramer 2861* (NBV, PDA, US); tea estate, *Cramer 2935* (CAL, PDA, US); close to culvert 40/5, Ramboda Rd., *Cramer 3117* (E, PDA, US). KANDY DISTRICT: Maskeliya, Adam's Peak Range, Mar. 1899, *s. coll. s.n.* (PDA), *Cramer 3790* (PDA, US).

18. TORENIA

L., Sp. Pl. 619. 1763; L., Gen. Pl. ed. 5. 270. 1754; Benth. & Hook., Gen. Pl. 2: 954. 1876; Wettst. in Pflanzenfam. 4 (3b): 79. 1891. Type species: *Torenia asiatica* L.

Nortenia Thouars, Gen. Nov. Madag. 9. 1806. Type species: *Nortenia thouarsii* Cham. & Schlecht. (= *Torenia thouarsii* (Cham. & Schlect.) Kuntze).

Pentsteira Griff., Not. 4: 118. 1854 pro parte.

Erect or trailing herbs. Stem quadrangular, glabrous to hirsute. Leaves opposite, crenate-serrate to dentate. Flowers solitary in leaf axils or in cymes or racemes. Calyx tube angular or winged; limb bilabiate or 3–5-toothed. Corolla tube often dilated towards apex; limb bilabiate; upper lip in bud outside, erect, broad, entire, notched or bifid; lower one broader, spreading, 3-lobed. Stamens didynamous, 2 posterior ones short; anterior ones longer with arched filaments meeting under corolla upper lip, apendaged at base; anthers connivent in pairs. Stigma bilamellate. Capsule included in calyx, septicidally 2-valved; valves entire; seeds numerous.

A genus of about 50 species, all palaeotropic.

KEY TO THE SPECIES

1 Calyx narrowly winged; midlobe of lower lip of corolla without a yellow spot at base; anterior filaments appendiculate above base
 2 Corolla more than 2 cm long; appendage of anterior filaments linear
  3 Lower half of corolla tube orange-yellow .... **1. T. travancorica**
  3 Lower half of corolla tube deep blue or violet .... **2. T. cyanea**
 2 Corolla less than 2 cm long; appendage of anterior filaments clavate ...... **3. T. aerinea**
1 Calyx broadly winged; midlobe of lower lip of corolla with a yellow spot at base; anterior filaments inappendiculate .... **4. T. fournieri**

1. Torenia travancorica Gamble, Fl. Pres. Madras 957. 1923; Hill, Bot. Mag. t. 9615. 1942; Stearn in Chittenden, Dict. Gard. 2127. 1951; Abeywick., l.c. 222. Type: India, Western Ghats (not known).

Torenia asiatica Hook. f., Fl. Br. Ind. 4: 277. 1884, pro parte; Trimen, l.c. 249 (non L. 1753).

Torenia leucosiphon Alston in Trimen, Handb. Fl. Ceylon 6: 212. 1931. Type: Ceylon, de Alwis Ic. ined. (PDA).

Stem decumbent, rooting at lower nodes, diffusely branched, quadrangular, glabrous, often tinged purple at upper nodes; internodes up to 6.5 cm long. Leaves ovate to ovate-deltoid, 1.5–5 × 1–3.2 cm, subcordate at base, acute or sometimes acuminate at apex, coarsely crenate-serrate, penninerved, scabrous on both surfaces; petioles 0.8–2.2 cm long. Flowers axillary, solitary. Bracts linear, 3–3.5 mm long. Pedicels 2.2–3.2 cm long, longer and deflexed in fruit, sparsely scabrous on angles. Calyx tube 9–11 mm long; lobes lanceolate, 6–8 mm long. Corolla 3–3.7 cm long; tube 2–2.4 cm long, orange-yellow in lower half, pale purple in upper half, glabrous within and without; upper lip oblong-orbicular, 10–14 mm long, dark purple; lower lip 1.6–1.8 cm across, pale bluish-purple. Posterior filaments 4 mm long; anterior ones longer; appendages 5–5.5 mm long; anther-cells divaricate. Hypogynous disk pale yellow. Style glabrous. Capsule linear-cylindric, 8–15 × 2 mm, tapering to apex, glabrous, pale brown; seeds ellipsoid, 0.6 mm long, pitted.

Distr. India and Ceylon.

Ecol. Under shade by foot-paths along forest borders or under shade of roadside hedges in forest areas of the wet lowlands or lower hill country; very common. Contrary to Alston's observations all specimens examined showed the corolla tube to be orange-yellow for a little more than half its length. This and the largeness of the corolla make the plant readily recognizable in the field. Flowering throughout the year.

N o t e. Since Linnaeus' original reference to the 'Indies' broadly connotes the Asian continent from India to China, the older taxonomists were led to consider *Torenia asiatica* L. an Indian plant. Modern taxonomists, however, now consider this species neither an Indian nor a Ceylonese plant. The author has followed Stearn in reducing *Torenia leucosiphon* Alston to synonymy as all specimens examined in the field, contrary to Alston's evidence, have shown the greater part of the corolla tube to be distinctly orange-yellow and not white.

S p e c i m e n s E x a m i n e d. KANDY DISTRICT: Hantane, *Gardner C.P. 2032* (K, PDA). KALUTARA DISTRICT: Matugama, *Cramer 2710* (PDA, US); Pelawatte, Ranawake Rock, *Cramer 2813* (PDA, US) *& 3177* (NBV, PDA, US). GALLE DISTRICT: Hiniduma, *Cramer 2683* (PDA, US); Talgaswela, Talgaswela Estate, *Cramer 2843* (E, PDA, US). RATNAPURA DISTRICT: Pasdun Korale, June 1883, *s. coll. s.n.* (PDA); Gallangama, near culvert 2/7, Ratnapura Rd., *Cramer 3115* (CAL, PDA, US).

2. Torenia cyanea Alston in Trimen, Handb. Fl. Ceylon 6: 212. 1931; Alston, Kandy Fl. 60, f. 320. 1938; Abeywick., l.c. 222. Type: Ceylon, Rangala, *Alston 901* (K).

Torenia asiatica Thw., Enum. Pl. Zeyl. 219. 1860 (non L. 1753), excl. *T. hirsuta* Willd. Type: Ceylon, *C.P. 2032* (K).

Torenia rubens Thw., Enum. Pl. Zeyl. 219. 1860 (non Benth. 1846). Type: Ceylon, *C.P. 2033* (K).

Stem creeping, diffusely branched, slender, sharply quadrangular, sparsely strigose on angles; internodes up to 9.6 cm long. Leaves ovate to ovate-deltoid, 1.2–4.2 × 1.1–2.8 cm, truncate or subcordate at base, acute at apex, crenate-serrate, penninerved, glabrous above, sparsely strigose on nerves beneath; petioles 0.8–2 cm long. Flowers axillary, solitary. Bracts linear, 3–5 mm long. Pedicels 1.5–4.8 cm long, slender, sparsely strigose on angles, deflexed in fruit. Calyx tube 8–14 mm long; wings strigose on angles; lobes lanceolate, 2–6 mm long. Corolla 2.2–3.2 cm long; tube 1.7–2.7 cm long, prussian blue; upper lip broadly orbicular, 7 × 9 mm, deep prussian blue; lower lip 0.7–1.7 cm across, lobes suborbicular, pale blue to white, midlobe larger than lateral ones with a purple blotch at base. Stamens inserted at top of corolla tube; posterior filaments 2–4 mm long, anterior ones longer; anther-cells divaricate, pale purplish-blue. Hypogynous disk pale greenish-yellow. Style glabrous, faintly purple. Capsule linear-oblong, 12–15 × 2.5 mm, glabrous; seeds irregularly orbicular, 0.5 mm long, foveolate.

D i s t r. Endemic to Ceylon.

E c o l. Under shade in wet and boggy places among short grass, often close to a watercourse or stream, in the wet lowlands above 160 m and chiefly in the wet hill country up to about 1800 m; common. The conspicuous colour of the corolla makes the plant very attractive and worthy of cultivation. Flowering throughout the year.

Specimens Examined. RATNAPURA DISTRICT: Rasagala, Diyawela, close to culvert 2/11, Balangoda Rd., *Cramer 2618* (PDA, US) & *3265* (CAL, M, PDA, US); Dehiowita, Mahagamana, close to culvert 44/16, *Cramer 3343* (US). KANDY DISTRICT: Muruthalawa, Nelli Hill, *Cramer 2800* (NBV, PDA, US); Rangala, *Alston 1736* (PDA). MATALE DISTRICT: Knuckles, June 1926, *de Silva s.n.* (PDA). NUWARA ELIYA DISTRICT: Sangilipalama, along border of paddy field, *Cramer 3855* (PDA); Labukelle, *Alston 1731* (PDA); close to culvert 41/10, Nuwara Eliya Rd., *Cramer 2859* (E, K, PDA, US); Hakgala, Sep. 1920, *de Alwis s.n.* (PDA); Blackpool, *Cramer 3250* (PDA, US). BADULLA DISTRICT: Haputale, Aug. 1927, *de Alwis s.n.* (PDA).

3. Torenia aerinea Alston in Trimen, Handb. Fl. Ceylon 6: 213. 1931; Abeywick., l.c. 222. Type: Ceylon, Nuwara Eliya, *Gardner* (PDA).

Torenia hirtella Hook. f., Fl. Br. Ind. 4: 277. 1884, pro parte; Trimen, l.c. 249 (excl. *T. rubens* Thw., non Benth. 1846).

Stem prostrate with ascending branches, slender, strigose at nodes; internodes up to 5 cm long. Leaves ovate to ovate-lanceolate, 1.6–3.3 × 1.0–1.8 cm, truncate at base, acute at apex, crenate-serrate with serrations ciliate, minutely scabrid above, punctate beneath with the nerves sparsely scabrid; petioles 4–5 mm long. Flowers axillary and solitary. Bracts linear, 3–4 × 0.5 mm, sparsely ciliate. Pedicels 1.8–3 cm long, slender, strigose on angles. Calyx tube 6–11 mm long; wings strigose on angles; lobes lanceolate, 3–4 mm long. Corolla 1.7–1.9 cm long; tube 10–14 mm long, glabrous within and without, pale violet-purple without, dark purple within; upper lip erect, broadly orbicular, 4–6 × 11 mm, emarginate at apex, pale violet-blue or whitish; lower lip 11–13 mm across, lobes rounded, pale blue, with purple spots at base. Posterior filaments 2 mm long, anterior ones longer; all glabrous; appendage 3 mm long, dark purple. Hypogynous disk greenish-yellow. Style glabrous. Capsule linear-cylindric, 10–12 mm long, glabrous; seeds ovoid-globose, 0.6 mm long, pitted.

Distr. Endemic to Ceylon.

Ecol. Under shade in damp grassy ground in the upper hill country above 1800 m, usually in outskirts of Nuwara Eliya town. Flowering throughout the year.

Specimens Examined. NUWARA ELIYA DISTRICT: Pidurutalagala, Aug. 1884, *s. coll. s.n.* (PDA); base of track to Peak, *Cramer 3121* (CAL, E, NBV, PDA, US); Pattipola, roadside, *Cramer 3474* (PDA, US);

Single tree hill, borders of jungle, *Cramer 3813* (PDA); Condegalla, *Alston 1922* (PDA); Ambawela, Mar. 1906, *de Silva s.n.* (PDA).

4. Torenia fournieri Lind. ex Fourn. in Ill. Hortic. 23: 129, t. 249. 1876.

Stem usually erect, up to 30 cm high, occasionally rooting at lower nodes, strigose on angles. Leaves ovate, 2.0–4.6 × 1.3–3 cm, truncate or rounded or subcordate at base, acute at apex, serrate, glabrous above, scabrid along midnerve beneath; petioles 0.6–1.5 cm long. Flowers axillary and solitary below, often in terminal or axillary racemes above. Bracts linear-lanceolate, up to 19 × 4 mm. Pedicels 1–1.6 cm long. Calyx tube 7–11 mm long; wings strigose, stiffened in fruit. Corolla up to 3 cm long; tube up to 2.1 cm long, expanded and pale blue above, narrowed and orange-yellow at base; upper lip broadly orbicular, up to 1.3 × 2 cm, distinctly carinate, pale blue; lower lip up to 2.5 cm across; lobes deep purple, filaments glabrous. Hypogynous disk pale yellow. Style glabrous. Capsule narrowly ovoid, 7–9 × 2 mm, glabrous, tipped by withered style; seeds orbicular-oblong, 1 mm long, foveolate.

Distr. Native of Indo-China cultivated in the warm tropics.

Ecol. Occasionally a weed in gardens in the wet low country; often cultivated on account of its showy flowers. Flowering March–June.

Specimens Examined. GALLE DISTRICT: Kaluwela, *Cramer 2910* (US) *& 1983* (PDA, US).

19. MAZUS

Lour., Fl. Cochinch. 385. 1790; Benth. & Hook., Gen. Pl. 2: 947. 1876; Wettst. in Pflanzenfam. 4 (3b): 72. 1891. Type species: *M. rugosus* Lour.

Hornemannia Willd., Enum. Hort. Berol. 653. 1809 (non Vahl 1791). Type species: *H. bicolor* Willd. (= *Mazus rugosus* Lour.).

Low herbs. Stem simple or stoloniferous, glabrous, tomentose. Leaves opposite below, generally alternate above, crenate-serrate to crenate-dentate. Flowers subsecund, in terminal racemes. Bracts very shortly stalked. Bracteoles present or absent. Calyx widely campanulate, 5-fid. Corolla tube very short; limb bilabiate; upper lip outside in bud, erect, entire or shortly 2-fid; lower lip larger, spreading, 3-lobed. Stamens 4, didynamous, inserted in corolla tube; anthers 2-celled, with the cells divergent. Capsule globose or compressed, loculicidally 2-valved; valves entire; seeds very numerous, minute.

A genus of about 20 species from East and Southeast Asia, Indo-Malaysia and Australia.

Mazus pumillus (Burm. f.) van Steenis, Nova Guinea n. s. 9: 31. 1958.

Lobelia pumila Burm. f., Fl. Ind. 186, t. 60, f. 3. 1768. Type: Burmann, Ic. t. 60, l.c.
Lindernia japonica Thunb., Fl. Jap. 253. 1784. Type: Japan (not known).
Mazus rugosus Lour., Fl. Cochinch. 358. 1790; Benth., Scroph. Ind. 27. 1835. Type: Indo-China (not known).

Small herb. Stem tufted, sometimes erect. Leaves rosetted below, sometimes opposite or alternate above, oblong-oval to obovate-spathulate, 1.0–2.5 × 0.7–1.7 cm, narrowed or acute at base, obtuse at apex, irregularly crenate-serrate, penninerved, glabrous on both surfaces or scaberulous only above; petioles 0.4–1.5 cm long, chanelled dorsally, minutely pilose. Racemes lax, up to 6 cm long; peduncles glandular-pubescent. Bracts lanceolate, 2 mm long, acuminate at apex. Pedicels 4–5 mm long, longer in fruit, glandular-puberulous. Calyx tube campanulate, 2.5–3 × 1 mm, lobes spreading. Corolla 6–8 mm long; tube 2.5–3 mm long, glabrous within and without, pale blue on adaxial side, white on opposite side; upper lip short, shallowly bifid, pale blue; lower lip 5 mm across, white, with the midlobe narrower than the lateral ones; lobes of palate yellow, with clavate hairs. Filaments glabrous. Style glabrous; lamellae of stigma broadly orbicular. Capsule obovoid, 2.5 × 2 mm, rounded at top, subcompressed, glabrous; seeds cuneate, 0.5 mm long, truncate at both ends, longitudinally faveolate.

Distr. From India and Ceylon eastwards through Burma to Malaysia, China and Japan.

Ecol. Under shade in cultivated land in the wet hill country; an uncommon weed. Flowering throughout the year.

Specimens Examined. KANDY DISTRICT: Peradeniya, in Nursery Garden, Royal Botanic Gardens, Aug. 1928, *Alston s.n.* (PDA), *Cramer 3399* (PDA, US).

20. ADENOSMA

R. Br., Prod. 442. 1810 (non Nees 1832); Benth. & Hook., Gen. Pl. 2: 949. 1876. Type species: *A. coeruleum* R. Br.

Pterostigma Benth., Scroph. Ind. 20. 1835. Type species: *P. grandiflorum* Benth. (= *Adenosma glutinosum* (L.) Druce).

Annual, glandular-pubescent or villous, often aromatic herbs. Stem erect or suberect. Leaves opposite, occasionally verticillate, rugose, crenate-serrate. Flowers axillary, solitary, or in leafy racemes or heads above. Calyx 5-partite; lobes imbricate, the posterior one largest. Corolla tube cylindric; limb bilabiate; upper lip outside in bud, erect, entire or emarginate; lower lip spreading, 3-lobed. Stamens 4, didynamous, included; anther-cells distinct, stipitate; one cell of the 2 anterior anthers or of anterior and posterior anthers reduced. Style dilated at apex; stigma entire or bilamellate. Capsule longitudinally 2-grooved, loculicido-septicidally 4-valved; seeds numerous, small.

A genus of 15 species predominantly Indo-Malaysian and Australian.

KEY TO THE SPECIES

1 Flowers axillary, solitary . . . . **1. A. subrepens**
1 Flowers in terminal racemes or dense heads
 2 Flowers in racemes, yellow; leaves more than 4 cm long . . . . **2. A. camphoratum**
 2 Flowers in dense heads, blue-purple; leaves less than 4 cm long . . . . . . **3. A. indianum**

1. Adenosma subrepens (Thw.) Benth. ex Hook. f., Fl. Br. Ind. 4: 263. 1884; Trimen, l.c. 241; Abeywick., l.c. 221.

Pterostigma subrepens Thw., Enum. Pl. Zeyl. 426. 1864. Type: Ceylon, Hewesse, *C.P. 3821* (K, PDA).

Stem decumbent at base and rooting at nodes, ascending, rather stout, quadrangular, villous-hairy; leaves 1.8–3.7 cm long, broadly ovate, obtuse, very coarsely crenate, slightly villous-hairy, becoming glabrous above; petiole 1.2 cm long, densely villous. Flowers very shortly stalked. Bractlets filiform, shorter than sepals. 3 outer sepals ovate, acute, reticulate-nerved, slightly hairy, 2 inner linear-lanceolate, glabrous. Capsule broadly ovate-ovoid, pointed.

Distr. Endemic to Ceylon.

Ecol. On borders of paddy fields in the wet lowlands; very rare. Flowering August.

Specimens Examined. RATNAPURA DISTRICT: Hewesse, *s. coll. C.P. 3821* (K, PDA).

Note. The author's attempts to find this plant have been without avail. The description given above, therefore, is Trimen's.

2. Adenosma camphoratum (Vahl) Hook. f., Fl. Br. Ind. 4: 264. 1884; Trimen, l.c. 241; Abeywick., l.c. 221.

Stemodia camphoratum Vahl, Symb. Bot. 2: 69. 1791. Type: Ceylon, *Hermann* in Mus. Zeyl. (BM).
Stemodia lutea Moon, Cat. 45. 1824 (nom. nud.).

Large herb. Stem up to ± 1 m high, obtusely quadrangular, faintly grooved on two opposite sides, sparsely glandular-hirsute on older parts, densely so on younger, with long branches. Leaves oval to oval-lanceolate, 5–8 × 2.2–4 cm, tapering to base, acute at apex, coarsely crenate-serrate, sparsely glandular-pubescent above, more so beneath; petiole short, obscure.

Racemes lax, up to 12 cm long. Bracts foliaceous, linear, up to 6 × 1.5 mm, glandular-hirsute without. Pedicels 2.5 mm long, curved, glandular-hirsute. Calyx lobes oval-lanceolate; posterior lobe up to 10 × 4.5 cm, others up to 8 × 2 mm. Corolla tube 8–9 mm long; upper lip entire; lower lip up to 14 mm across, lobes oblong-orbicular. Filaments glabrous. Lamellae of stigma spathulate. Capsule globose-conical, 6–7 × 4–5 mm, bluntly pointed, glabrous.

Distr. Endemic to Ceylon.

Ecol. In damp places in the wet lowlands below 700 m close to ditches or a watercourse; fairly common. Flowering March–December.

Vern. Kaha-gona-kola (S).

Note. *Pterostigma villosum* Benth., originally considered a synonym of this species, is reported to have a wider distribution than the Ceylon endemic species, being found also in Burma and Singapore. This cannot be, therefore, our Ceylon endemic species and must be considered a misidentification as follows: *Pterostigma villosum* auctt. (non Benth. 1835).

Specimens Examined. RATNAPURA DISTRICT: Adam's Peak, *Gardner C.P. 2041* (K, PDA); Kuruwita, Korale, Jan. 1892, *s. coll. s.n.* (PDA). KALUTARA DISTRICT: Pasdun Korale, near Moragoda, Mar. 1887, *s. coll. s.n.* (PDA). GALLE DISTRICT: Yatalamatta, *Cramer 2643* (US).

3. Adenosma indianum (Lour.) Merr., Trans. Amer. Philos. Soc. n. s. 24 (2): 351. 1935.

Manulea indiana Lour., Fl. Cochinch. 386. 1790. Type: Indo-China (not known).

Stemodia capitata Benth., Bot. Reg. 17: sub t. 1470. 1832. Type: Chile, *Cuming & Bridges*.

Adenosma capitatum (Benth.) Hance, J. Linn. Soc. Bot. 13: 114. 1873; Hook. f., Fl. Br. Ind. 4: 264. 1884; Trimen, l.c. 241; Abeywick., l.c. 221.

Stem up to 45 cm high, terete, densely villous, dotted with pale yellow sessile glands, with whorled ascending branches. Leaves subsessile, in verticils of 3–4, lower ones occasionally opposite, oblong-lanceolate to oval-oblong, 1.2–3.6 × 0.7–1.7 cm, rounded at base, obtuse at apex, crenate-serrate, penninerved, scabrid on both surfaces, copiously punctate beneath. Heads up to 2.5 cm long; peduncles opposite or whorled, 3–8.6 cm long. Bracteoles linear, up to 10 × 0.75 mm, long ciliate. Calyx tube 1.5 mm long; posterior lobe oblong-lanceolate, 4 × 1.5 mm, lateral ones linear-lanceolate, 3 × 0.25 mm, anterior one lanceolate, 3.5 × 1 mm; all hirsute without, dotted with pale yellow sessile glands. Corolla 9–10 mm long, striped with violet-purple lines; tube 5 mm long, glabrous within and without; upper lip orbicular, 4.5 × 4.5 mm, faintly emarginate, flattened over lower lip; lower lip 6 mm across, lateral lobes broader than midlobe. Filaments inserted at base of corolla tube, glabrous. Style glabrous; stigma bent. Capsule ovoid, 3.25–4 × 2–2.5 mm, enclosed by longer calyx, with translucent walls; seeds ovoid-oblong, 0.5 mm long, distinctly reticulate, tawny.

Distr. Widespread through the Indo-Malaysian region.

Ecol. In open sandy places in the wet lowlands below 350 m; frequently a weed in cinnamon estates. The flower has the general appearance of one of the Labiatae, but on closer examination the very divided nature of the calyx lobes and the capsular fruit reveal its scrophulariaceous identity. Flowering July–November.

Specimens Examined. COLOMBO DISTRICT: *Gardner C.P. 2880* (K, PDA); Katunayake, near Negombo, *Alston 2386* (PDA); Ja-Ela, Ekkala, near "The Hermitage," *Cramer 3420* (CAL, E, K, NBV, PDA, US). WITHOUT LOCALITY: *Walker s.n.* (PDA).

SONNERATIACEAE

(by William Macnae and F. Raymond Fosberg*)

Trees; leaves opposite, entire, thick, simple; flowers dioecious, in groups of two or three together at the summit of twigs, rather large, regular; calyx thickly coriaceous, slightly adnate to the base of the ovary, of 4–6 segments, often coloured inside; petals when present linear, deciduous; stamens very numerous, inserted on the calyx, filaments very long, red or white, pollen sticky; ovary superior, many-celled, with many ovules in each cell, style long, stigma capitate; fruit large, fleshy, indehiscent, resting on the calyx tube; seeds small, angular, liberated by the rotting of the fallen fruit.

The family Sonneratiaceae is represented in Ceylon by mangrove species only, belonging to the genus *Sonneratia*. Three occur. *Sonneratia alba* is characteristically found only in the seaward fringes of mangrove forests under conditions of high salinity. It was found only in a few localities in the Chilaw and Puttalam lagoons. *Sonneratia apetala* and *S. caseolaris* occur upstream in estuaries under strong fresh water influence. *S. caseolaris* is widespread and possibly encouraged because of its bitter-sweet fruits. *S. apetala* is rare, occurring only near Muttur where some six tall trees occur on the bank of a distributary of the Mahaweli Ganga.

SONNERATIA

L. f., Suppl. Pl. 38. 1781, nom. cons.; Backer & van Steenis in Fl. Mal. I. 4: 280. 1951 (full bibliography).

Trees with the characters of the family, with prominent upwardly growing pneumatophores emerging from the system of cable roots.

These are trees of mangrove areas and other coastal regions, and one species may penetrate some distance up estuaries.

*The manuscript for the Sonneratiaceae was submitted in draft form, very incomplete and in a somewhat different format than is usual in botanical literature. Professor Macnae died before the treatment could be brought into final form. Fosberg largely rewrote it, inserting much additional material, and brought into final form for use in the Flora of Ceylon. The descriptive material and taxonomic concepts are essentially those included by Macnae. A key has been added. The literature citations and nomenclature have been somewhat edited.

KEY TO THE SPECIES

1 Calyx normally 4-parted, 1.5–2 cm long; petals none; fruit about 2 cm wide; leaves oblong-lanceolate, obtuse, 6–12 cm long .... **2. S. apetala**
1 Calyx normally 6– or more parted, over 2 cm long; petals present; fruit 4–7 cm broad
2 Inside of calyx lobes red; leaves almost or quite as broad as long; fruit to 4 cm wide, lower part somewhat enclosed in turbinate calyx; petals linear, white, at least distally .... **1. S. alba**
2 Inside of calyx lobes greenish; leaves distinctly narrower than long; fruit 5–7 cm wide, calyx flattish, scarcely enclosing fruit; petals linear-lanceolate, dark red .... **3. S. caseolaris**

1. Sonneratia alba J. Sm. in Rees Cycl. 33. no. 2. 1819; Trimen, J. Bot. 23: 171. 1885; Trimen, Handb. Fl. Ceylon 2: 230. 1894; Backer & van Steenis in Fl. Mal. I. 4: 285. 1951 (full bibliography).

A bush or small tree, with smooth grey bark, twigs cylindrical, marked with prominent leaf sears; leaves 3–10 cm, oval-rotundate, slightly tapering at the base, rounded at the apex, very thick and fleshy, petiole very short; flowers large, on very thick pedicels, buds 6-angled, pointed; calyx-tube with 6 prominent ridges extending from between the segments to the base, segments 6, narrowly lanceolate, acute, longer than the tube; petals 5, linear; style 3–5 cm, stigma large; fruit supported on the much enlarged calyx, which forms a cup round the base with spreading segments, turbinate, about 4 cm wide, much depressed, flat on top, with the apiculate short style-base in the centre.

Distr. Mangrove swamps, common from Chilaw to Puttalam lagoon, Kalpitiya, elsewhere rare; flowers November, probably all the year; fruit at all stages of development in October. This species occurs from Inhambane (Lat. 23 1/2°S) in Central Mozambique to Innisfall in north Queensland, and eastward in the Pacific to the Marshall Islands. In eastern Asia it extends north to the southernmost Ryukyu islands.

Specimens Examined. PUTTALAM DISTRICT: Chilaw, *Nevill* in 1881 (PDA); Puttalam Lagoon near Kalpitiya, *Macnae* in 1968 (US).

2. Sonneratia apetala Buch.-Ham. in Symes, Embasy Ava 3: 477. 1800; Trimen, J. Bot. 27: 163. 1885; Trimen, Handb. Fl. Ceylon 2: 229. 1894; Backer & van Steenis in Fl. Mal. I. 4: 286. 1951 (full bibliography).

An erect tree, attaining 15 m, with slender drooping branches; leaves 6–12 cm, narrowly oblong-lanceolate, much tapering at the base into a short petiole, obtuse at the apex, glabrous, rather thick; flower solitary, terminal, large, articulated on a rather long pedicel; calyx tube cup-shaped, of 4 segments longer than the tube, oblong-ovate, acute, deflexed; petals 0; fruit about 2 cm wide, depressed-globular, tipped with style, smooth, white.

Distr. Mangrove swamps, very rare; only found near Muttur in the estuary of the Koddiyar river where some 5 or 6 tall trees are present. One suspects these to be the same trees mentioned by Trimen, but there seems to have been no further colonisation. Worthington (personal communication) has not seen it elsewhere in Ceylon. This species is distributed round peninsular India from Bombay to the Sunderbans but does not extend into Malaya.

Specimens Examined. TRINCOMALEE DISTRICT: Koddiyar, [*s. coll.*] in 1885; Muthur (Koddiyar), *Worthington 1312* (K); 67 mile Post, *Worthington 1392* (K).

3. Sonneratia caseolaris (L.) Engl. in Pflanzenfam. Nachtr. 261. 1897; Arulchelvam, Ceylon Forester 8: 75. 1968; Backer & van Steenis in Fl. Mal. I. 4: 283. 1951 (full bibliography).

Rhizophora caseolaris L. in Stickman, Herb. Amb. 13. 1754.

Sonneratia acida L. f., Suppl. Pl. 252. 1781; Thw., Enum. Pl. Zeyl. 123. 1859; Trimen, Handb. Fl. Ceylon 2: 230. 1894; Alston in Trimen, Handb. Fl. Ceylon 6: 129. 1931.

A shrub or small tree reaching 12 m, twigs quadrangular; leaves nearly sessile, 6–7 cm, oblong-oval or roundish, tapering to base, obtuse, sometimes apiculate at the apex, slightly fleshy; flowers large, terminal, solitary, on very short thick pedicel, buds not ribbed, blunt but apiculate at the apex; calyx tube quite without ribs, of 6 segments, lanceolate, acute, longer than the tube; petals also 6, linear, about 2.5 cm, rather longer than the calyx segments, persistent, red; stamens about 5 cm, erect, red; style considerably exceeding stamens; fruit broadly ovoid or subglobose, supported on enlarged calyx which forms a shallow cup, about 6 cm wide, rounded and slightly depressed on top, with the persistent tapering style-base forming a point 2.5 cm long in the centre.

Distr. Tidal estuaries, extending upstream into fresh-water-dominated stretches; rather common. Chilaw, Negombo, Panadure, Kalutara, south to Matara, Yala reserve, Batticaloa; flowers September, probably all the year. Also in Bengal, Burma, Thailand, Java eastwards to Solomon Is. and New Hebrides and North Australia.

Vern. Kirilla (S); Kinnai (T).

Note. Wood white, soft, light, but fine-grained. The curious erect root-branches are figured by Karsten in his 'Mangrove Vegetation', (Veget. Bild. 2: t. 10, f. 139. 1904); they were first described by Rumphius (Herb. Amb. iii, 112. 1743). As they attain 18 in. to 3 ft in height, and 3 in. in diameter, and have a soft, firm, even texture, they form a fine substitute for cork, and are cut into slices and used for entomologists' boxes and other purposes. Their wood is very much lighter than that of the stem. (See Templeton, Trans. Ent. Soc. London 3: 302. 1844).

S p e c i m e n s E x a m i n e d. COLOMBO DISTRICT: Negombo *s. coll. C.P. 1595* (PDA); Negombo, Blue Lagoon near mcuth of Dandugam Oya River, *Meijer* 585 (US), *588* (PDA); Thalahena, lagoon edge, *Wass 725* (US); Crow Island, Colombo, *Worthington 1340* (K). KALUTARA DISTRICT: Kalutara, *s. coll. C.P. 1595* (PDA); Panadura, Horana, *Worthington 2517* (K). GALLE DISTRICT: Dedduwa Lake, Undagatota, Bentota, *Worthington 2469* (K). Koggala, marshy place, *D.B. & D. Sumithraarachchi DBS 1032* (US). MATARA DISTRICT: Matara, 1 mile inland, *Worthington 2469* (K). TRINCOMALEE DISTRICT: Muttur, above Ganga Ferry, *Worthington 1662* (K). BATTICALOA DISTRICT: Batticaloa, *Macnae* in 1968 (US). AMPARAI DISTRICT: Kumana-Yala Sanctuary, *Balakrishnan 383* (PDA, US), *384* (US). HAMBANTOTA DISTRICT: Ruhuna National Park; Yala, *Worthington 5689* (K); edge of Katupila Ara, *Comanor 357* (US).

SYMPLOCACEAE

(by H.P. Nooteboom*)

Desf., Mem. Mus. Hist. Nat. Paris 6: 9. 1820, 'Symploceae'; D. Don, Prod. Fl. Nepal. 144. 1825 p.p., 'Symplocineae'; G. Don, Gen. Syst. 4: 1. 1837; Gurke in Pflanzenfam. 4, 1: 158. 1891; Brand in Pflanzenr. Heft 6: 13. 1901; Hallier f., Beih. Bot. Centralbl. 39-B: 78. 1921; Nooteboom, Revision of Symplocaceae of the Old World, in Leiden Bot. Series 1: 33. 1975.

Only one genus.

SYMPLOCOS

Jacq., Enum Pl. Carib. 5: 24. 1760; Jacq., Select. Stirp. Amer. Hist. 166. 1763; L., Gen. Pl. ed. 6. 272. 1764; Thw., Enum. Pl. Zeyl. 184. 1860; Clarke in Hook. f., Fl. Br. Ind. 3: 572. 1882; Trimen, Handb. Fl. Ceylon 3: 103. 1895; Alston in Trimen, Handb. Fl. Ceylon 6: 186. 1931.

Bobu Adans., Fam. Pl. 2: 204. 1763; DC, Prod. 3: 23. 1828; 'Bobua' Miers, J. Linn. Soc. Bot. 17: 302. 1879.

Eugeniodes '*Eugenioides*' L., Fl. Zeyl. 192. 1747; Kuntze, Rev. Gen. Pl. 2: 409, 975. 1891, nom. illeg.

Hopea L., Mant. 105. 1767; L., Syst. Nat. ed. 2. XII. 1767, nom. rejic., non Roxb., 1811.

Symplocos subg. *Hopea* Clarke in Hook. f., Fl. Br. Ind. 3: 572. 1882.

Small evergreen shrubs to—in Ceylon—small trees (c. 15 m). An indumentum of simple hairs can be present on all parts of the plant, at least in young stages. Leaves spirally arranged or pseudo-subverticillate, estipulate, simple, penninerved; petiole always present, sometimes very short. Inflorescence a spike, raceme or panicle, sometimes condensed to clusters, from the upper leaf axils or from the axils of fallen leaves, flowers rarely solitary. Flowers supported by a bract (exceptionally bractless in the axil of a leaf) and 2 (or more) bracteoles. Flowers actinomorphic, bisexual, fragrant. Ovary inferior, seemingly forming the calyx tube (in descriptions always called ovary), 2–5-celled with complete septation; style 1, with punctiform or peltate stigma. Ovules 2–4 in each cell, pendulous, anatropous-epitropous or amphitropous,

*Rijksherbarium, Leiden.

unitegmic, tenuinucellate. Calyx on top of the ovary, persistent on top of the fruit, (3–) 5 (–6) lobed, imbricate or valvate. Corolla gamopetalous, but divided nearly to the base in subg. *Hopea*, the lobes (3–) 5 (–6), quincuncially imbricate, whitish, bluish or pink, sometimes yellow. Stamens many, connate in a long monadelphous tube in subg. *Symplocos*, only connate at the very base in subg. *Hopea*, staminal tube always inserted on the (base of) corolla. Anthers globose, bilocular, introrse, dehiscing longitudinally. Intrastaminal disk glandular, sometimes inconspicuous. Fruit a monopyrenous drupe, crowned by the persistent calyx lobes, usually with thin mesocarp, when ripe frequently bluish or purple. Seed straight or curved, 1 in each developed cell, with copious endosperm. Embryo straight or curved, cotyledons very short, linear.

About 250 species. Eastern parts of the Old World, reaching from Bombay in the west to Fiji in the east and from Manchuria in the north to New South Wales and Lord Howe I. in the south. In the New World from the state of Washington to South Brazil.

The treatment of the Ceylonese species is rather different from that in my revision (1975), which, at least for India and Ceylon, was based on inadequate material. For the Ceylon species described by Thwaites it appeared necessary to study the original type material in PDA, because the CP numbers often consist of several collections, and not rarely of several species. The localities often are written with pencil on the PDA specimens only. Besides, I could study the species in the field and collect enough additional material to get a better insight in the delimitation of species. The species in *Symplocos* often are closely related to each other, and probably spontaneous hybridising occurs in the field, making discrimination difficult.

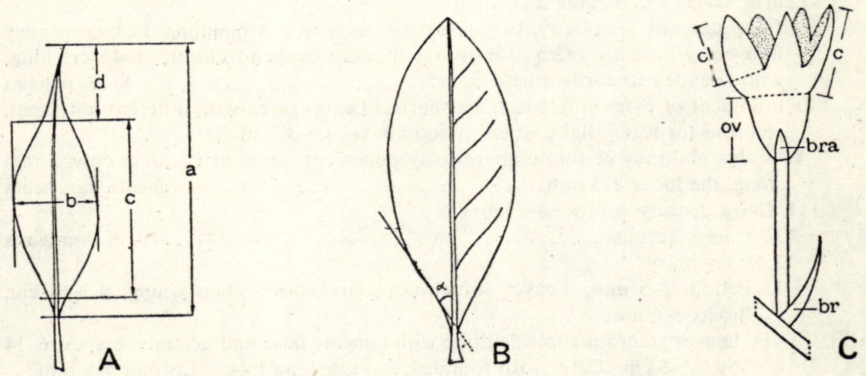

Fig. 1. Schemes elucidating descriptive terminology used in the text. A. *a* length of leaf, *b* width of leaf, *c* divided by *b* is leaf index, *d* length of acumen. B. way of expressing base angle α. C. deflorated flower; *br* bract, *bra* bracteole, *c* length of calyx, *cl* length of calyx lobes, *ov* height of ovary.

KEY TO THE SPECIES

1 Petals connate at least halfway up, but the margins free except the lowermost 1–3 mm. Staminal tube adnate to corolla except the upper 3–5 mm; anthers on 1–3.5 mm long filaments, inserted at various heights within the staminal tube. Fruit spindle-shaped, the enlarged calyx lobes surrounding the hairy conical persistent style base....**1. S. pendula**

1 Petals connate at the base only. Staminal tube very short. Fruit different

2 Twigs and leaves glabrous. Inflorescence a branched spike to 14 cm. Bracts and bracteoles appressedly pubescent or puberulous, persistent, enveloping the base of the ovary only, at most 2 mm long but mostly shorter. Ovary glabrous, 1–1.5 mm high. Calyx glabrous, consisting of 0.5–1 mm long lobes. Disk and style base glabrous. Fruit ampulliform with globose body, 6–8 by 4–6 mm...........**2. S. cochinchinensis**

2 Twigs and leaves glabrous or hairy. Inflorescence usually not branched. Bracts and bracteoles different. Ovary glabrous or hairy. Calyx glabrous or hairy, 1–5 mm. Disk and/or style base glabrous or more often hairy. Fruit different

3 Leaves subverticillate towards the end of the flushes

4 Inflorescence axis, bracts, ovary and calyx glabrous........**S. obtusa** var. **cucullata**

4 Inflorescence axis and bracts hairy. Ovary glabrous or hairy. Calyx hairy

5 Ovary glabrous (it is necessary to remove the bracteoles). Apex of leaves rounded or acute, sometimes shortly acuminate, acumen less than 5 mm....**3. S. cordifolia**

5 Ovary hairy. Apex of leaves acuminate, acumen mostly longer than 5 mm........ ....**4. S. coronata**

3 Leaves not subverticillate

6 Twigs and leaves glabrous

7 Bracts hairy. Ovary glabrous. Calyx hairy. Nerves in 10–18 pairs..**3. S. cordifolia**

7 Bracts glabrous. Ovary and calyx glabrous. Nerves in 5–11 pairs

8 Calyx 0.75–1.5 mm long. Leaves often obovate......................**5. S. obtusa**

8 Calyx 2–4 mm long. Leaves elliptic............................**6. S. bractealis**

6 Twigs hairy. Leaves hairy but sometimes glabrescent

9 Ovary glabrous. Calyx glabrous or nearly so, rarely some of the lobes more or less hairy. Leaves 2–9 cm. Petiole 2–9 mm.........................**7. S. elegans**

9 Ovary glabrous or hairy. Calyx lobes all hairy, at least when ovary glabrous. Leaves 4.5–33 cm. Petiole 2–25 mm.

10 Twigs patently long-hairy, in young twigs hairs to c. 3 mm long. Leaves sparsely long-hairy beneath, often also above, broadly ovate to elliptic, 6–15 cm long, with rounded to cordate base..................................**8. S. pulchra**

10 Indument of twigs different, hairs shorter. Leaves often with different indument, glabrous (or rarely hairy when young) above, 4.5–33 cm

11 Calyx glabrous or thinly appressedly pubescent, often at the base only, 3 mm long, the lobes 2–3 mm..............................**S. cuneata** var. **acuta**

11 Calyx densely appressedly hairy

12 Ovary glabrous..**10. S. kurgensis**

12 Ovary hairy

13 Petiole 2–6 mm. Leaves sometimes hairy above when young, 4.5–14 cm. Bracts 3–6 mm

14 Leaves glabrous above, elliptic with cuneate base and acuminate apex, 6–14 by 2.5–5 cm. Calyx with rounded, 2–4 mm long lobes. Corolla 4–5 mm.... ....**9. S. cuneata**

14 Leaves sometimes hairy above when young, (narrowly) ovate with cuneate to slightly cordate base and acuminate apex, 5–12 by 1–4 cm. Calyx with (narrowly) ovate to triangular 1.5–2 mm long lobes. Corolla c. 3 mm...... ....**11. S. diversifolia**

13 Petiole 5–25 mm. Leaves glabrous above, 6–33 cm. Bracts 3–10 mm
15 Twigs tomentose or tomentellous. Leaves usually densely tomentose on midrib and nerves, 7–20 by 2–11 cm with cuneate to slightly cordate base. Nerves 6–15 pairs... .... **12. S. macrophylla**
15 Twigs appressedly, or rarely patently, hairy. Leaves appressedly hairy especially on midrib and nerves, 6–14 by 2.5–5 cm with cuneate base. Nerves in 6–10 pairs....... .... **9. S. cuneata**

1. Subgenus SYMPLOCOS

1. Symplocos pendula Wight, Ic. Pl. Ind. Or. 4: 10, t. 1237. 1848; Thw., Enum. Pl. Zeyl. 184. 1860; Clarke in Hook. f., Fl. Br. Ind. 3: 587. 1882; Noot., Revision Sympl. 40. 1975.

Symplocos pauciflora Wight ex Benth. in Benth. & Hook., Gen., Pl. 2: 669. 1876, nomen ex Clarke in Hook. f., Fl. Br. Ind. 3: 587. 1882; Trimen, Handb. Fl. Ceylon 3: 111. 1895.
Symplocos sessilis Clarke in Hook. f., Fl. Br. Ind. 3: 587. 1882.
Symplocos henschelii (non Benth.) Clarke in Hook. f., Fl. Br. Ind. 3: 588. 1882, incl. var. *hirtistylis*, and Forbes & Hemsley, J. Linn. Soc. Bot. 26: 72. 1889, pro descr. et stirp, excl. syn Moritzi.
Symplocos confusa Brand in Pflanzenr. Heft 6: 88. 1901.
Cordyloblaste pendula Alston in Trimen, Handb. Fl. Ceylon 6: 188. 1931.

D i s t r. From Ceylon through India and Burma to China and Japan and throughout Malesia, except Java and the Sunda Is., with two varieties, of which only one in Ceylon:

var. **pendula**

Symplocos pendula Wight l.c.
Symplocos pauciflora Wight ex Clarke l.c.
Symplocos sessilis Clarke l.c.

Shrub (? or tree). Twigs glabrous or puberulous. Leaves glabrous, elliptic to obovate, 3–10 × 1.5–5 cm with cuneate base and rounded apex. Petiole 3–10 mm. Flowers solitary or few together in a short raceme; pedicel slender, 2–10 mm, with 2– several minute bracteoles. Ovary glabrous, conical, 3–5 mm high, semiinferior, the superior part densely greyish hairy. Calyx with very short, rounded ciliate lobes. Corolla tubular, trumpet shaped, 6–10 mm long. Petals 5, fleshy, connate halfway up but the margins free except the lowermost 1–3 mm, spathulate with rounded apex, white or pink in vivo. Staminal tube adnate to corolla except the upper 3–5 mm; free parts of filaments fleshy, 1–3.5 × 1 mm; anthers 30–50, inserted at various heights within the staminal tube. Fruit spindle shaped, 10–15 × 3–6 mm, the enlarged calyx lobes surrounding the hairy conical persistent style base.

Distr. Ceylon, India, China and Malay Peninsula.

Ecol. Rare in primary mountain forest and secondary shrubbery, alt. 1000–2300 m.

Specimens Examined. KANDY DISTRICT: Rangala hill, *s. coll. s.n.* (PDA); Mirisketiya, *Nooteboom 3402* (L, PDA), *Waas 1062* (PDA); Maskeliya, *J.M. Silva 31* (PDA); Adam's Peak, *s. coll. s.n.* (PDA). NUWARA ELIYA DISTRICT: Nuwara Eliya Town, *Cramer 4442* (PDA); Horton Plains, *Balakrishnan 1195* (PDA), *s. coll. C.P. 504* p.p. (PDA), *Theobald & Krahulik 2828* (PDA); Maturata, *s. coll. C.P. 504* p.p. (PDA). BADULLA DISTRICT: Top of Namunukula Kande, *Cramer 4357* (PDA), *Waas 862* (PDA).

2. Subgenus HOPEA

2. Symplocos cochinchinensis (Lour.) S. Moore, J. Bot. 52: 148. 1914.

Dicalyx cochinchinensis Lour., Fl. Cochinch. 1: 663. 1790 excl. syn. *Arbor rediviva* Rumph.
Myrtus laurinus Retz., Obs. Bot. 4: 26. 1786.
Laurus serrata floribus spicata Burman, Thes. Zeyl. 139, t. 62. 1737.
Eugenia Willd., Sp. Pl. 2: 967. 1799, p.p.
Bobua laurina DC., Prod. 3: 24. 1828.
Symplocos spicata var. *zeylanica* DC., Prod. 8: 254. 1844.
Eugenioides laurinum Kuntze, Rev. Gen. Pl. 2: 409. 1891, p.p.
Symplocos spicata Roxb., Hort. Beng. 40. 1814, nomen; Wall., Cat. 4417. 1831; Roxb., Fl. Ind. ed. Carey 2: 542. 1832; Clarke in Hook. f., Fl. Br. Ind. 3: 573. 1882, p.p., incl. var. *laurina*; Trimen, Handb. Fl. Ceylon 3: 104. 1895.
Symplocos laurina Wall. [Cat. 4416. 1831, nomen] ex G. Don, Gen. Syst. 4: 3. 1837.

Distr. Asia from Ceylon and India to China and Japan, southwards to Queensland and NS Wales, eastwards to the Solomon Is., New Hebrides and Fiji Is.

Four subspecies and many varieties. In Ceylon only one variety:

ssp. **laurina** (Retz.) Noot. var. **laurina**

All synonyms after *Arbor rediviva* Rumph.

Shrub, or tree to 12 m. Twigs and leaves glabrous except sometimes the youngest parts which can be appressedly hairy in innovations. Leaves broadly to narrowly elliptic, 6–17 × 2.5–8 cm; base cuneate, acuminate; apex blunt, acuminate; margin dentate, undulate or crenulate. Nerves in 6–10 pairs, anastomosing at some distance from the margin and often meeting in an intramarginal vein. Petiole 5–25 mm. Inflorescence a mostly compound spike to

14 cm; axis glabrous to more or less appressedly puberulous or pubescent. Bracts and bracteoles appressedly pubescent or puberulous, persistent, at most 2 mm long and broad, but mostly c. 1 mm, only enveloping the base of the ovary. Ovary glabrous, 1–1.5 mm high. Calyx glabrous, consisting of rounded ciliolate 0.5–1 mm long lobes. Corolla 3–5 mm. Stamens 30–100. Disk glabrous, more or less pulvinate. Style glabrous. Fruit ampulliform, the body globose, vertically ribbed when dry, the neck short cylindrical, 6–8 × 4–6 mm.

Distr. Asia, from Ceylon to China and Japan, southwards to Java, Borneo and Celebes.

Ecol. In Ceylon predominantly in the wet zone, from 0–2100 m, often in secondary vegetation.

Vern. Bobu, Bombu (S).

Specimens Examined. KURUNEGALA DISTRICT: Doluwa Kande, *Sumithraarachchi 630* (PDA, US). MATALE DISTRICT: Matale, *Worthington 1995* (PDA). KANDY DISTRICT: Kandy, *Meijer 693* (PDA, US), *Kostermans 24038* (PDA); Hantana, *Nooteboom 3042* (L, PDA, US); Knuckles, *Kostermans 25030, 25081, 23286* (L, PDA, US); Kadugannawa, Poilakane, *Cramer 4206* (PDA, US); Madugoda, *Alston 1962* (PDA); N. of Ulapane, *Maxwell 928* (PDA, US); Hunnasgiriya, *J.M. de Silva 54* (PDA); road Laxapana-Maskeliya, *Kostermans 24054, 24100* (L, PDA, US); upper part of Moray estate, *Kostermans 24527 A* (L, PDA); Ginigathena—Adam's Peak, *Balakrishnan 1189* (PDA, US). NUWARA ELIYA DISTRICT: Hakgala, *Balakrishnan 623* (PDA, US), *Cramer 3904* (PDA, US), *Cramer 4110* (PDA, US), *Nooteboom 3239* (L, PDA, US), *Sohmer 8484* (PDA, US), *Stone 11227* (PDA, US), *Tirvengadum & Cramer 39* (PDA), *Waas 836* (PDA, US); Horton plains, *van Beusekom 1518, 1399* (L, PDA, US), *Davidse 7631* (PDA, US), *Fosberg 50003* (PDA, US), *Hladik 725* (PDA, US), *Kostermans 23016* (PDA, US), *Mueller-Dombois 68051703* (PDA, US); Pidurutalagala, *Foster 5013* (PDA), *Meijer 54* (PDA, US); Kandapola f.r., *Sohmer 8362* (PDA, US). BADULLA DISTRICT: Lorana Watte, between Bandarawela and Welimada, *Wirawan 739* (PDA); Haputale, Adisham Monastery, *Kostermans 23417* (PDA). KEGALLE DISTRICT: Kitulgala, *Kostermans 849* (PDA, L), *Worthington 406* (PDA). KALUTARA DISTRICT: Moropitiya logging area, *Meijer 1087* (PDA, US); Kuruvita, *Sohmer 10245* (PDA); Hewesse, *Sohmer 10297* (PDA). GALLE DISTRICT: Nelowe, *Balakrishnan 238* (PDA, US). RATNAPURA DISTRICT: Belihul Oya, *Hepper 4572* (PDA, US), *Kostermans 23457* (PDA, L), *Worthington 447* (PDA); above Rassagala, *Kostermans 24643* (L, PDA, US); Ratnapura Hill, *Wheeler & Jayasuriya 12747* (PDA, US); Kukulawa Vihara Kande, *Waas 262* (PDA).

3. Symplocos cordifolia Thw., Enum. Pl. Zeyl. 187. 1860; Beddome, For. Man. Bot. 151. 1872; Clarke in Hook. f., Fl. Br. Ind. 3: 586. 1882; Trimen,

Handb. Fl. Ceylon 3: 110. 1895; Brand in Pflanzenr. Heft 6: 51, t. 5. 1901.

Lodhra cordifolia Miers, J. Linn. Soc. Bot. 17: 301. 1879.
Symplocos macrophylla Wall. ex DC., var. *cordifolia* (Thw.) Noot., Rev. Sympl. 233. 1975.

Shrub, or tree to 6 m. Twigs glabrous (or rarely patently long-hirsute), thick, often with ridges from the leaf insertions downwards. Leaves often, but not always, inserted at the end of the flushes, glabrous, or rarely patently long hirsute on midrib and nerves beneath, elliptic to obovate, 7–18.5 × 3–9 cm; base cordate to auriculate; apex rounded or acute, sometimes shortly acuminate, acumen less than 5 mm; margin sharply dentate or denticulate, often strongly revolute. Nerves 10–18 pairs, the lower ones very fine, pointing downwards; secondary veins many, transverse to nerves and midrib. Petiole 1–5 mm. Inflorescence: often many spikes crowded towards the end of the twigs; axis densely villous to long-pubescent, to 8 cm long. Bracts and bracteoles ovate to cordate, boat-shaped, usually fallen under the ripe fruit, appressedly sericeous (to hirsute), 6–12 and 3–8 mm. Ovary glabrous, c. 2 mm high. Calyx densely appressedly sericeous (to hirsute), consisting of rounded 3–5 mm long lobes. Corolla c. 8 mm. Stamens c. 100. Disk 5-glandular, surrounding the conical or semiglobose sericeous style base. Fruits ovoid to ellipsoid or cylindrical, 12–18 × 6–7 mm, the 3–5 mm long persistent calyx not included. Stone smooth or very shallowly lengthwise grooved.

Distr. Ceylon, in the central mountains.

Ecol. Mountain forest, often secondary shrubbery, also in moss forest, from 1200–2400 m, rather common.

Specimens Examined. RATNAPURA DISTRICT: Carney side of Adam's peak, *Bernardi 15795* (PDA). KANDY DISTRICT: Adam's Peak jungle above Moray Estate, *Jayasuriya 1112* (PDA, US), *Kostermans 24131, 24174, 24935* (L, PDA, US), *Nooteboom & Huber 3129, 3146* (L, PDA, US), *Sohmer 9937* (PDA); Adam's Peak range, Maratenne, *Nooteboom 3390* (PDA); Knuckles, Corbet's gap and surroundings, *Balakrishnan 201* (PDA, US), *Meijer 662* (PDA, US), *Nooteboom 3058, 3072* (L, PDA, US), *Waas 190, 1065* (PDA); Hunnasgiriya hill, *Sohmer 10589* (PDA). NUWARA ELIYA DISTRICT: Maturate, *s. coll. C.P. 271* p.p. (PDA, BM, FI, K, LE, MEL, P, US); Ramboda Pass, *Alston 626* (PDA); Wattekelle hill, *Becket 1124* (PDA); Mahacoodalugala Estate, *Sohmer 10146* (PDA); Pidurutalagala, *Nooteboom 3257* (PDA), *Waas 841* (PDA).

4. Symplocos coronata Thw., Enum. Pl. Zeyl. 187. 1860; Beddome, For. Man. Bot. 152. 1872; Clarke in Hook. f., Fl. Br. Ind. 3: 587. 1882 incl. var. *obovata*; Trimen, Handb. Fl. Ceylon 3: 11. 1895; Brand in Pflanzenr. Heft 6: 51, t. 6. 1901.

Symplocos apicalis Thw., Enum. Pl. Zeyl. 187. 1860, incl. var. *strigosa* and

var. *glabrifolia*; Beddome, For. Man. Bot. 151. 1872; Clarke in Hook., f., Fl. Br. Ind. 3: 586. 1882; Trimen, Handb. Fl. Ceylon 3: 110. 1895; Brand in Pflanzenr. Heft 6: 53. 1901; Alston in Trimen, Handb. Fl. Ceylon 6: 187. 1931.
Lodhra apicalis Miers, J. Linn. Soc. Bot. 17: 302. 1879.
Lodhra coronata Miers l.c.
Symplocos glabrifolia Brand in Pflanzenr. Heft 6: 53. 1901.
Symplocos macrophylla var. *apicalis* (Thw.) Noot. and var. *glabrifolia* (Thw.) Noot., Rev. Sympl. 233, 234. 1975.

D i s t r. Central and SW Ceylon.

1. var. coronata

Symplocos coronata Thw.
Symplocos apicalis Thw. incl. var. *strigosa*.

Treelet to 10 m. Twigs (sparsely) appressedly pilose to (long) pubescent, tomentellous or tomentose, inserted only towards the apex of the flushes. Leaves sparsely appressedly pilose (or only midrib and nerves so) to patently pubescent underneath, elliptic to obovate, tapering towards the base in var. *glabrifolia*, 13–36 (–45) × 5–16 cm; apex shortly acuminate, acumen 5–15 mm, rarely shorter; base (cuneate) to rounded or shortly cordate; margin nearly entire to denticulate. Nerves 11–16 pairs; secondary veins transverse to nerves and midrib, tertiary veins transverse to secondary ones. Petiole 5–15 mm. Inflorescence a spike or raceme from wood beneath the upper whorl of leaves, 5–13 cm; axis tomentose or pubescent; bract and bracteoles appressedly sericeous, broadly ovate, 4–7 and 3–4 mm respectively, often long persistent under the fruit. Pedicel to 1 mm. Ovary sericeous, c. 1 mm high. Calyx appressedly pubescent, 2–3 mm long, the lobes rounded, 1–2.5 mm. Corolla 5–6 mm. Stamens c. 100. Disk 5-glandular, with the style base sericeous. Style hairy only towards the base. Fruit ovoid to cylindrical, 8–14 × 6–7 mm; stone shallowly lengthwise grooved with a depression or transverse groove at one side near the base.

D i s t r. Ratnapura and Kandy distr.

E c o l. Tropical rain forest at low altitude, ascending in the lower mountain forest to 1200 m. In Moropitiya logging area becoming abundant after cutting of a new road.

S p e c i m e n s E x a m i n e d. KANDY DISTRICT: Knuckles, *Ferguson s.n.* (PDA); Hantane, *C.P. 127* (PDA, BM, CGE, FI, K, LE, P); ? Kalugamana, roadside of way to Galewatte, *J.M. de Silva s.n.* (PDA). KALUTARA DISTRICT: Moropitiya logging area, *Waas 895* (PDA), *Nooteboom 3275* (L, PDA, US). RATNAPURA DISTRICT: Kukulawa Vihara Kande, *Waas 278* (PDA); Rassagala, *Kostermans 23558* (L, PDA, US), *Nooteboom 3395* (L,

Fig. 2. *Symplocos coronata* Thwaites var. *coronata*.

PDA, US); Forest between Opanake and Wewelwatte, *Nooteboom 3294, 3309* (L, PDA, US); Sinharaja forest near Weddagala, *Nooteboom s.n.* (L, sterile). PRECISE LOCALITY UNKNOWN: Reigam Korala, *s. coll. C.P. 438* (PDA, BM, BO, FI, P).

N o t e. I accept the name *S. coronata* Thw., and not *S. apicalis* Thw. (as I did in my revision: *S. macrophylla* var. *apicalis* (Thw.) Noot.) because study in the field made it clear to me that the type of *S. apicalis* Thw., like a (sterile) collection I made myself in the Sinharaja forest (*Nooteboom s.n.*, L) might be of hybrid origin. At least these plants show characters (less hairy twigs and—in the C.P. collection—nearly glabrous leaves) of var. *glabrifolia*. In the sterile plant from the Sinharaja forest, innovations are present with nearly the same indument on the twigs as in the common var. *coronata*. The upper leaves possess the same shape as those of *C.P. 438*, the lower leaves have the shape that commonly occurs in var. *glabrifolia* (gradually tapering towards the base); they also become as large, 45 × 15 cm.

2. var. **glabrifolia** (Thw.) Noot.

Symplocos apicalis Thw. var. *glabrifolia*.
Symplocos glabrifolia Brand.

Twigs glabrous or the youngest ones hairy (hirsute). Leaves glabrous, or rarely patently long-hirsute on midrib and nerves, narrowly obovate, tapering towards the base, 11–45 × 6–15 cm; base rounded to cordate or auriculate; apex acuminate, acumen 7–15 mm; margin nearly entire. Nerves 13–17 pairs, secondary veins much prominent, transverse to the nerves, reticulation fine. Petiole 0–5 (–13) mm. Inflorescence a spike or raceme from the axis of the leaves or from wood; axis (patently) hirsute. Bracts and bracteoles hirsute, boat-shaped, elliptic to ovate, c. 5–15 and c. 3 mm long respectively, usually long-persistent under the fruit. Pedicel to 1 mm. Ovary appressedly sericeous, 1–2 mm high. Calyx (sparsely) appressedly pubescent, 2–4 mm long, the rounded lobes 1–4 mm. Corolla c. 6 mm. Stamens c. 100. Disk 5-glandular, shortly pilose to sericeous. Fruit as in var. *coronata*.

D i s t r. Ceylon, Sinharaja forest and adjacent area.

E c o l. Tropical rain forest at low altitude.

S p e c i m e n s E x a m i n e d. GALLE DISTRICT: Kottawa Kambala nat. arboretum, *Nooteboom 3198* (L, PDA, US); Kanneliya f.r., *Ashton 2050* (L, PDA, US). KALUTARA DISTRICT: Moropitiya f.r., *Nooteboom 3284, 3285* (L, PDA, US), *s.n.* (L), *Waas 1447* (PDA). RATNAPURA DISTRICT: Sinharaja Forest, *s. coll. C.P. 2933* (PDA, BM, FI, K, P), *Huber 557* (L, PDA, US), *Nooteboom 3269* (L, PDA, US).

N o t e. *Nooteboom 3198* is the only collection with hairy leaves (patently long-hirsute on midrib and nerves). Because it resembles the other material of var. *glabrifolia* in all other characters, I place it there.

Fig. 3. *Symplocos coronata* Thw. var. *glabrifolia* (Thw.) Noot. Already flowering at a juvenile stage.

5. Symplocos obtusa Wall., Cat. 4424. 1828, nomen ex G. Don, Gen. Hist. 4: 3. 1837; DC., Prod. 8: 255. 1844; Thw., Enum. Pl. Zeyl. 185. 1860, incl. vars *cucullata*, *obovata* and *major*; Clarke in Hook. f., Fl. Br. Ind. 3: 583. 1882; Trimen, Handb. Fl. Ceylon 3: 105. 1895; Alston in Trimen, Handb. Fl. Ceylon 6: 186. 1931; Noot., Rev. Sympl. 243. 1975 incl. var. *pedicellata*.

Symplocos laeta Thw., Enum. Pl. Zeyl. 184. 1860; Clarke in Hook. f., Fl. Br. Ind. 3: 584. 1882, incl. var. *pedicellata*.
Symplocos rufescens Thw., l.c.
Lodhra obtusa Miers, J. Linn. Soc. Bot. 17: 300. 1879.
Lodhra laeta Miers, op. cit. 301.
Lodhra cucullata Miers l.c.
Symplocos versicolor Clarke in Hook. f., Fl. Br. Ind. 3: 584. 1882; Trimen, Handb. Fl. Ceylon 3: 106. 1895; Livera, Ann. R. Bot. Gard. (Peradeniya) 10: 315. 1927.
Symplocos cucullata Brand in Pflanzenr. Heft 6: 56. 1901.
Symplocos major Brand, op. cit. 57.
Symplocos obovata Livera, Ann. R. Bot. Gard. (Peradeniya) 10: 317. 1927.
Symplocos suborbicularis Abeywick., Ceylon J. Sci., Biol. Sci. 2: 209. 1959.

D i s t r. South India and Ceylon.

KEY TO THE VARIETIES

1 Leaves with rounded base and very thick short petiole, subverticillate at the end of the flushes on very thick branches .... **2. var. cucullata**
1 Leaves not subverticillate, other characters different
2 Leaves obovate with cuneate base and rounded, rarely shortly acuminate apex........ .... **1. var. obtusa**
2 Leaves elliptic with cuneate base and acuminate apex .... **3. var. pedicellata**

1. var. obtusa

Symplocos obtusa Wall. ex Don, incl. var. *obovata* Thw. and var. *major* Thw.
Symplocos obovata Livera.
Symplocos major Brand.
Symplocos suborbicularis Abeywick.

Shrub, or tree to 15 m high and 30 cm diam., but mostly smaller. Twigs glabrous. Leaves glabrous, (broadly) obovate, with cuneate base and rounded, rarely shortly acuminate, apex (acumen shorter than 5 mm), evenly distributed along the twig, 2–18 × 1.2–10 cm; nerves in 5–10 pairs, meeting in a looped intramarginal vein. Petiole 2–20 mm. Inflorescence a 2–15 cm long spike or raceme, sometimes in small-leaved forms reduced to only 1 flower; axis glabrous; the (narrowly) elliptic or ovate bracts and the bracteoles glabrous but ciliate, caducous under the flower, 1.5–6 mm, sometimes the bracts leaf-like, and then inflorescence branched, thus forming a transition to the 1-flowered inflorescence; pedicels 0–10 mm, but usually only the lower flowers pedicelled. Flowers often fragrant. Ovary glabrous, 1.5–3 mm high. Calyx glabrous, divided into the semicircular 1–1.5 mm long ciliate lobes. Corolla white or yellow, 4–9 mm. Stamens 25–70. Disk glabrous, with some minute hairs to soft hairy. Style glabrous except the conical base which has the same indument as the disk. Fruit ellipsoid, 7–16 mm, the stone lengthwise ribbed.

Distr. India (Nilgiris), Ceylon.

Ecol. In mountain forest. Alt. 1000–2300 m.

Specimens Examined. KANDY DISTRICT: Kellie Estate, Dolosbage, *Worthington 1897* (PDA); Knuckles, *s. coll. s.n.* (PDA); path from Moray estate to Adam's Peak, *Kostermans 24245* (L, PDA, US), *Nooteboom 3380* (L, PDA, US), *Sohmer 8670* (PDA, US). NUWARA ELIYA DISTRICT: Pattipola road, *Comanor 458* (PDA, US); Horton plains, *Hladik 712* (PDA, US), *Hoogland 11528* (PDA), *Kostermans 23045*, *23065* (L, PDA, US), *Koyama 14073* (PDA), *Meijer 617*, *623* (PDA, US), *Nooteboom 3350*, *3356* (L, PDA, US), *Nowicke 225* (PDA, US), *Simpson 9526* (PDA), *Sohmer 8533* (PDA), *Waas 141* (PDA, US), *Worthington 1708* (PDA); 1 mile north of Horton Plains, *Gould 13575* (PDA, US); Ohiya, *Jayasuriya 1475* (PDA, US), *Cramer 4423* (PDA); Moon Plains, *Alston 623* (PDA); Sita Eliya, *Nooteboom 3250*, *3252* (L, PDA, US); Hakgala, above bot. gard., *Nooteboom 3244* (PDA), *Meijer 612* (L, PDA, US), *Jayasuriya 167* (PDA), *Cooray 70021108* (PDA), *Bernardi 15852* (PDA, US); Elk Plains, near Ambawela, *Kostermans s.n.* (L), *s. coll. C.P. 1820* p.p. (PDA); Dimbula, *s. coll. C.P.* 1820 p.p. (PDA); Totapola Kande, *Bernardi 15876* (PDA), *Worthington 1736*, *1737* (PDA); Pidurutalagala, *Nooteboom 3261* (L, PDA, US), *s. coll. C.P. 1819* (K, PDA); Lindula, *Kostermans 24614* (L, US); Bulu Ela, *Worthington 2910* (PDA); Ramboda Pass, *Worthington 5781* (PDA). RATNAPURA DISTRICT: Gongala, *Waas 1392* (PDA). MATARA DISTRICT: southern area of Sinharaja forest, *Waas 1478* (PDA); north of Deniyaya, *Huber 615* (L). BADULLA DISTRICT: Namunukula Kande, *J.M. de Silva s.n.* (PDA). LOCALITY UNKNOWN: *Wight 638* (K); *s. coll. C.P. 538* (PDA); *s. coll. C.P. 626* (L, PDA, US, etc.).

Note. *Worthington 1955*, from Madulkelle, in fruit, possibly belongs here, but the leaves are elliptic and acuminate. It might be a hybrid with var. *pedicellata*.

2. var. **cucullata** Thw.

Symplocos cucullata Brand.

Leaves elliptic to obovate, subverticillate on the very thick branches, with rounded base and very thick short petiole.

Distr. Ceylon.

Ecol. Mountain Forest.

Note. This is possibly a hybrid between var. *obtusa* and *S. cordifolia* Thw.

Specimens Examined. NUWARA ELIYA DISTRICT: Maskeliya, path to Adam's Peak, *Bernardi 15940* (PDA). KANDY DISTRICT: Ambagamuwa, *s. coll. C.P. 2835* (K, BM, LE, P, PDA).

3. var. **pedicellata** (Clarke) Noot., Rev. Sympl. 244. 1975.

Symplocos laeta var. *pedicellata* Clarke.
Symplocos rufescens Thw.
Symplocos versicolor Clarke.

Leaves elliptic, 3.5–9.5 × 1.5–3.5 cm with cuneate, attenuate base and acuminate apex (acumen 3–10 mm). Petiole thin, 3–(–20) mm. Ovary c. 2 mm. Calyx divided into 0.75–1 mm long lobes.

Distr. Ceylon.

Ecol. Mountain forest.

Note. *Kostermans 23662* from Deniyaya is aberrant because the flowers are smaller; the ovary is only 1 mm high, the calyx 0.5 mm long and the corolla c. 2 mm.

Specimens Examined. KANDY DISTRICT: Ambagamuwa, *s. coll. C.P. 3639* (PDA, K, LE, P); Adam's Peak, *s. coll. C.P. 165* p.p. (PDA); Knuckles, *W. Ferguson s.n.* (PDA); Rangala, *s. coll. s.n.* (PDA). NUWARA ELIYA DISTRICT: Horton Plains, *Theobald & Krahulik 2829* (US); Hakgala forest, *Balakrishnan 1206* (US); Nuwara Eliya, *s. coll. C.P. 165* p.p. (PDA), *W. Ferguson 15* (PDA). MATARA DISTRICT: Deniyaya, Enselwatte Estate, *Kostermans 23662* (L, PDA, US).

6. Symplocos bractealis Thw., Enum. Pl. Zeyl. 185. 1860; Beddome, For. Man. Bot. 151. 1872; Clarke in Hook. f., Fl. Br. Ind. 3: 584. 1882 incl. var. *revoluta*; Trimen, Handb. Fl. Ceylon 3: 106. 1895; Brand in Pflanzenr. Heft 6: 59. 1901.

Lodhra bractealis Miers, J. Linn. Soc. Bot. 17: 301. 1879.
Symplocos macrophylla Wall. ex DC. var. *revoluta* Noot., Rev. Sympl. 228. 1975.

Shrub or treelet to 12 m but mostly shorter; twigs glabrous. Leaves glabrous, mostly elliptic, 1.8 to 10 × 1.2 to 5 cm with cuneate to slightly cordate base and rounded to acute, rarely slightly acuminate apex, acumen if present shorter than 5 mm; margin dentate or denticulate, often strongly revolute; nerves in 7–11 pairs, meeting in an intramarginal vein which is often hardly distinct from the finely prominent reticulation beneath; petiole 5–7 mm. Inflorescence a spike to 3.5 (–5) cm, axis glabrous or sparsely hairy, rarely densely hairy; bracts glabrous but ciliate, circular to broadly elliptic, 4–10 mm long, with the 2–3 mm long bracteoles semipersistent, but usually fallen under the fruit. Ovary c. 1.5 mm high, glabrous. Calyx glabrous, 2–4 mm long, the lobes rounded, entirely glabrous or sparsely ciliate, 1.5–3 mm long. Corolla 5–7 mm. Stamens c. 100. Disk 5-glandular, (sparsely) soft hairy; style glabrous except the often soft hairy base. Fruit ovoid-ellipsoid, c. 12.5 × 6.5 mm; stone smooth or with shallow ridges.

Distr. Ceylon.

Ecol. Mountain forest, alt. 1170–2400 m.

Specimens Examined. KANDY DISTRICT: Adam's Peak region, *Bernardi 15909* (PDA), *s. coll. C.P. 174* (PDA, K, BM, P), *Cramer 4037* (PDA), *Tirvengadum & Cramer 294* (PDA, US). RATNAPURA DISTRICT: Path to Adam's Peak, *Bernardi 15785, 16083, 16092* (PDA, US). NUWARA ELIYA DISTRICT: Nuwara Eliya, path to top of Mount Pidurutalagala, *Tirvengadum & Cramer 266* (P, PDA, US), *Waas 182* (PDA), *Waas 839* (PDA, US), *Worthington 1927* (PDA), *Nooteboom 3255* (PDA, US, L); Sita Eliya, *Nooteboom 3248* (PDA, US, L); Galaha Kande near Hakgala, *Bernardi 15862, 15865* (PDA); Hakgala nat. res., *Cooray 70021113* (PDA, US), *Wirawan 749, 751* (PDA, US), *Nooteboom 3237, 3238* (PDA, US, L); Horton Plains, *Comanor 457, 960* (PDA), *Cramer 4024* (PDA), *Kostermans 23051, 23123* (PDA, L), *Meijer 638* (PDA), *Nowicke 260* (PDA, US), *Sohmer 9976, 10044, 10050* (PDA), *Tirvengadum & Cramer 274* (PDA), *Waas 852* (PDA, US), *Nooteboom 3345, 3338, 3340, 3349* (PDA, US, L).

7. Symplocos elegans Thw., Enum. Pl. Zeyl. 185. 1860, incl. var. *angustata*; Beddome, For. Man. Bot. 151. 1872; Clarke in Hook. f., Fl. Br. Ind. 3: 586. 1882; Trimen, Handb. Fl. Ceylon 3: 108. 1895; Brand in Pflanzenr. Heft 6: 62. 1901; Livera, Ann. R. Bot. Gard. (Peradeniya) 10: 318. 1927.

Symplocos hirs ta Wight & Gardner ex Thw., Enum. Pl. Zeyl. 185. 1860, non DC, incl. var. *minor* Thw.

Lodhra elegans Miers, J. Linn. Soc. Bot. 17: 301. 1879.

Symplocos nervosa DC var. *hirsuta* Beddome, For. Man. Bot. 149. 1872.

Symplocos latiflora Clarke in Hook. f., Fl. Br. Ind. 3: 585. 1882; Trimen, Handb. Fl. Ceylon 3: 108. 1895.

Symplocos minor Clarke in Hook. f., Fl. Br. Ind. 3: 586. 1882 incl. var. *eugenioides* and var. *glabrescens*; Trimen, Handb. Fl. Ceylon 3: 109. 1895; Brand in Pflanzenr. Heft 6: 64. 1901.

Symplocos angustata Clarke in Hook. f., Fl. Br. Ind. 3: 585. 1882; Brand in Pflanzenr. Heft 6: 64. 1901; Noot., Rev. Sympl. 125. 1975.

Symplocos glabrescens Brand in Pflanzenr. Heft 6: 64. 1901.

Symplocos amabilis Brand in Pflanzenr. Heft 6: 63. 1901.

Symplocos eugenioides Livera, Ann. R. Bot. Gard. (Peradeniya) 10: 317. 1927.

Symplocos macrophylla ssp. *hirsuta* (Wight & Gardner ex Thw.) Noot., Rev. Sympl. 235. 1975, incl. var. *elegans* and var. *minor*.

Distr. Ceylon, in the mountains of the wet zone.

Note. In this species 4 varieties are distinguished, although more have been described in the past. Some varieties even have been given the rank of species by some authors. I maintain only the varietal rank for taxa where the differences, although slight, are rather constant. Of var. *angustata* I am not certain whether this taxon still occurs, as no recent material has been collected at all, in the last 100 years.

KEY TO THE VARIETIES

1 Calyx 2.5–4 mm, the lobes glabrous or with some hairs towards the apex, sometimes some of the lobes hairy. Young twigs patently (stiff) long hairy. Leaves 4.5–9 cm long. Bracts 4–9 mm . . . . 3. var. **hirsuta**

1 Calyx 1–2 mm long, the lobes glabrous (or nearly so). Young twigs appressedly pilose or pubescent, patently long and soft villous, or ferruginously tomentose. Leaves 1–7 cm long. Bracts 0.5–3 mm

2 Young twigs ferruginously tomentose. Leaves more or less patently pilose beneath, with acute, shortly acuminate apex . . . . 1. var. **elegans**

2 Young twigs not tomentose. Leaves patently or appressedly pilose beneath, with rounded to acuminate or long acuminate apex

3 Inflorescence a very slender few flowered raceme to 5 cm long. Leaves (narrowly) ovate with long acuminate apex . . . . 2. var. **angustata**

3 Inflorescence a—not slender—spike or raceme to 3 cm. Leaves elliptic or ovate with rounded or acuminate apex . . . . 4. var. **minor**

1. var. elegans

Symplocos elegans Thw.
Symplocos amabilis Brand.
Symplocos macrophylla var. *elegans* (Thw.) Noot.

Shrub or treelet to 4 m. Young twigs ferruginously tomentose. Leaves elliptic or ovate, more or less patently pilose beneath, 2–6 × 1.2–3.5 cm; base acute to slightly cordate; apex acute, shortly acuminate, acumen 2–8 mm; margin recurved to revolute (leaves often cucullate), entire or denticulate. Nerves in 6–10 pairs, meeting in an intramarginal vein; reticulation fine. Petiole 2–5 mm. Inflorescence a spike or raceme to 3.5 cm; axis appressedly pilose to tomentose; bracts and bracteoles appressedly pilose, long persistent but fallen under the fruit, ovate, 1–3 mm long and ovate to nearly linear, 1–2 mm long respectively; pedicel very short. Ovary glabrous, c. 1 mm high; calyx glabrous, c. 2 mm long, the lobes rounded, 1.5–2 mm. Corolla 4–5 mm. Stamens c. 50. Disk 5-glandular, minutely hairy to pilose; style glabrous. Fruit ellipsoid, c. 12 × 6 mm.

Distr. Very local, probably only in the central mountains from Adam's Peak to Maratenna. On one *C.P. 23* sheet also Dimbulla is written.

Ecol. Mountain forests between 1000 and 1700 m.

Note. Brand has not seen the type of *S. elegans* Thw. in PDA. He based his *S. amabilis* on the fragments in European herbaria (K, P) of *C.P. 23*, that form the type of *S. elegans* Thw., according to his description and that of Thwaites. I therefore choose as lectotype for *S. elegans* Thw. and for *S. amabilis* Brand respectively a PDA sheet of *C.P. 23* and a fragment in K. of the same number that constitute the same taxon. Thus Brand's name is illegitimate.

Specimens Examined. KANDY DISTRICT: From Maskeliya to Adam's Peak and above Moray Estate, *s. coll. C.P. 23* p.p. (PDA, BM, FI, K, P), *Trimen s.n.* (PDA), *Nooteboom 3377* and *3378* (L, PDA, US). RATNAPURA DISTRICT: Between Carney and Adam's Peak, *Bernardi 16088* (PDA); mountains N. of Maratenna, *Huber 533* (PDA). ? NUWARA ELIYA DISTRICT: Dimbula, *s. coll. C.P. 23* p.p. (PDA).

2. var. **angustata** Thw.

Symplocos angustata Clarke.

Twigs appressedly pilose or pubescent, often a double indument of short fine hairs and long stiff hairs. Leaves (narrowly) ovate, appressedly pilose underneath, especially on midrib and nerves, 2–7 × 0.8–2.5 cm; base cuneate to rounded, apex long-acuminate, acumen 5–10 mm; margin recurved, denticulate. Nerves in 7–9 pairs, meeting in an intramarginal vein. Petiole 2–5 mm. Inflorescence a very slender raceme to 5 cm; axis sparsely appressedly pilose. Bracts and bracteoles appressedly pubescent, ovate to narrowly elliptic, persistent, 1–2 and 0.5–1.5 mm respectively. Pedicel slender, up to 2 mm. Ovary glabrous, 0.5–1.5 mm high. Calyx glabrous, 1–2 mm, the lobes rounded, 1–1.5 mm. Corolla 2–5 mm. Stamens c. 40– more than 100. Disk 5-glandular, with few hairs to shortly pilose. Style glabrous. Fruit ovoid-cylindric, c. 10 × 3 mm.

Distr. Ceylon, very rare.

Note. Mixed with this species on the sheets of *C.P. 78* is a hitherto undescribed taxon with hairy ovary and calyx. This taxon, *S. diversifolia* Brand var. *appressa* Nooteboom has not been collected until recently by S. Waas and myself, in Kurulugala.

Specimens Examined. KANDY DISTRICT: Hunnasgiriya, *s. coll. C.P. 78* p.p. (PDA); Galagama, *s. coll. C.P. 78* p.p. (PDA). NUWARA ELIYA DISTRICT: *Macrae s.n.* (K). LOCALITY UNKNOWN: *Moon 705* (K).

3. var. **hirsuta** (Wight & Gardner ex Thw.) Noot.

Symplocos hirsuta Wight & Gardn. ex Thw.
Symplocos nervosa DC var. *hirsuta* Bedd.
Symplocos latiflora Clarke.
Symplocos macrophylla ssp. *hirsuta* (Wight & Gardner ex Thw.) Noot. var. *hirsuta*.

Treelet 2.5–8 m. Young twigs patently (stiff) long hairy, hairs c. 2 mm long. Leaves long-pilose beneath, especially on midrib and nerves, glabrescent, (narrowly) elliptic, 4.5–9 × 1.5–4 cm; the base cuneate to slightly cordate, the apex rounded to acuminate, acumen 2–7 mm; margin dentate or denticulate. Nerves in 6–10 pairs, meeting in intramarginal vein, reticulation fine. Petiole c. 5 mm. Inflorescence a spike (or raceme) to 5 (! –14) cm; axis long-pilose; bracts and bracteoles persistent, but falling when the fruit ripens, long-pilose, (broadly) elliptic, 4–9 mm and (narrowly) elliptic, 2–5 mm respectively. Pedicel at most 1 mm. Ovary glabrous, 1.5–2 mm high. Calyx 2.5–4 mm, glabrous, or the rounded, 1.25–3 mm long lobes with some hairs towards the apex or some of the lobes hairy all over. Corolla 5–9 mm. Stamens c. 100. Disk 5-glandular, obscurely short-pilose. Style glabrous or with long pilose base. Fruit ovoid to ellipsoid, c. 15 × 7 mm.

Distr. Central Ceylon.

Ecol. In the mountains from 1200–1300 m (according to Thwaites to 2400 m).

Specimens Examined. MATALE DISTRICT: Campbells Lane, *Waas 964* (PDA). KANDY DISTRICT: Adam's Peak, *s. coll. C.P. 368* p.p. (PDA); Madulkelle, *s. coll. s.n.* (PDA); Dotulugala Kande, *Waas 1039* (PDA); Rangala, *s. coll. s.n.* (PDA), *Nooteboom 3414* (L, PDA, US); Between Rangala and Looloowatte, *Nooteboom 3059* (L, PDA, US); Hunnasgiriya, *s. coll. C.P. 368* p.p. (PDA). NUWARA ELIYA DISTRICT: Nuwara Eliya, *s. coll. C.P. 368* p.p. (PDA); Dimbula, *s. coll. C.P. 368* p.p. (PDA); Mool Oya, *Waas 1135* (PDA); Ludugala Kande, *s. coll. s.n.* (PDA). BADULLA DISTRICT: Haputale, *s. coll. C.P. 368* p.p. (PDA), *s. coll. s.n.* (PDA).

4. var. **minor** (Thw.) Noot.

Symplocos hirsuta Wight & Gardner ex Thw. var. *minor* Thw.
Symplocos minor Clarke, incl. var. *eugenioides* and var. *glabrescens*.
Symplocos glabrescens Brand.
Symplocos eugenioides Livera.
Symplocos macrophylla var. *minor* (Thw.) Noot.

Shrub or treelet to 3 m. Twigs densely patently soft and long-villous or more or less appressedly (long) pilose (or only so in innovations). Leaves c. elliptic or ovate, patently or appressedly pilose beneath, but often glabrous, 1–7 × 0.6–4 cm; apex rounded to acuminate, acumen 0–8 mm, base cuneate to slightly cordate; margin flat to strongly revolute, denticulate. Nerves in 7–11 pairs, meeting in an intramarginal vein. Petiole 2–9 mm. Inflorescence a spike or raceme to 3 cm, but often shorter; axis sparsely appressedly pilose or pubescent to densely long-villous; bracts and bracteoles appressedly hairy, persistent under the fruit or sometimes falling when the fruit ripens, (broadly) elliptic or ovate, 1–3 mm and (narrowly) ovate, 0.5–2 mm long respectively. Ovary glabrous, 1–1.5 mm high. Calyx glabrous, 1.75–2 mm long, the lobes rounded, 1.5–1.75 mm long. Corolla 4–6 mm long. Stamens c. 60 to more than 100. Disk 5-glandular, (minutely) hairy to densely pilose. Style glabrous, or pilose towards its base. Fruit ovoid or ellipsoid, 10–12 × 5–6 mm.

Distr. Ceylon, in the central mountains of the wet zone, in Nuwara Eliya and Badulla districts. Once recorded (*Ferguson s.n.*) from the Knuckles and Rangala in Kandy district. I doubt whether this is right.

Ecol. In the mountain forest, often as an ericoid shrub, usually above 1800 m, but I found it with much larger and thinner leaves in Haputale at 1300 m.

Note. This taxon is rather variable, often a local population shows distinct characters. On Mount Pidurutalagala the young twigs are patently very long villous, and the leaves patently pilose beneath; in Hakgala and Horton plains (and also Knuckles and Rangala, if it is really true that this plant is found in those places) the twigs are more or less appressedly villous, and sometimes very soon glabrescent. In the latter case the indument is only present in innovations and towards the very end of each twig. The leaves are appressedly pilose or glabrous. All plants from the Badulla district have large leaves which are often ovate. I suppose that they all grow at a lower altitude. The shape of the leaves, especially the apex being rounded or acuminate, as well as the leaf texture, in my opinion is not the result of genetic differences, but possibly of edaphic and climatological factors. At lower altitude the leaves are always acuminate (except on the Knuckles, see above) and rather thin. In the Horton plains and Hakgala all transitions are found, from coriaceous, often cucullate leaves with rounded apex to somewhat thinner leaves with flat margin and acuminate apex. Flower and fruit do not offer any differential characters at all. The small-leaved forms with acuminate apex formerly were often identified as *S. elegans* Thw., instead of as *S. minor* Clarke, or as *S. minor* Clarke var. *eugenioides* Livera. The same form with larger leaves sometimes is identified as *S. latiflora* Clarke, regardless its smaller flowers.

Specimens Examined.? KANDY DISTRICT: Knuckles and Rangala, *Ferguson s.n.* (PDA). NUWARA ELIYA DISTRICT: Pidurutalagala, *s. coll. C.P. 273* (PDA, BM, BO, FI, LE, P), *Davidse 8040* (US), *Nooteboom 3259* (L, PDA, US), *A.M. de Silva s.n.* (PDA), *Trimen s.n.* (PDA), *Waas 175* (PDA); Horton Plains, *s. coll. C.P. 2204* (PDA, BM, FI, K, P), *Kostermans 23046* (L), *Nooteboom 3351, 3352* (L, PDA, US), *Waas 137* (PDA); Hakgala

Mount, above Hakgala bot. gard., *Nooteboom 3226*, *3231*, *3232*, *3236* (L, PDA, US), *A.M. de Silva s.n.* (PDA), *Wirawan 749* (US); Mahakudalagala, *Waas 1143* (PDA). BADULLA DISTRICT: Jungle above Thotulagala Estate, *Sumithraarachhi 908* (PDA, US); Haputale, *Nooteboom 3396* (PDA); Namunukula Kande, *J.M. de Silva s.n.* (PDA), *Waas 742* (PDA).

8. Symplocos pulchra Wight, Ic. Pl. Ind. Or. 4: 10, t. 1230. 1848; Wight, Neilgherry Pl. 2: 39, t. 143. 1851; Clarke in Hook. f., Fl. Br. Ind. 3: 583. 1882; Brand in Pflanzenr. Heft 6: 63. 1901; Brandis, Indian Trees 440. 1906; Noot., Rev. Sympl. 266. 1975.

Lodhra pulchra Miers, J. Linn. Soc. Bot. 17: 300. 1879.
Symplocos hispidula Thw., Enum. Pl. Zeyl. 186. 1860.
Symplocos villosa Brand in Pflanzenr. Heft 6: 63. 1901.
Symplocos walkeri Brand in Pflanzenr. Heft 6: 63. 1901.

Distr. India (Deccan Peninsula) and Ceylon.

In Ceylon only one subspecies:

ssp. **hispidula** (Thw.) Noot.

Symplocos hispidula Thw.
Symplocos walkeri Brand.

Twigs patently long hairy when young, hairs to c 3 mm. Leaves sparsely (long) hairy on both surfaces, above soon, beneath later glabrescent, rarely glabrous above, broadly ovate to elliptic, 6–15 × 3.2–9.5 cm; base rounded to cordate; apex shortly acuminate; margin entire or denticulate. Nerves 6–10 pairs, meeting in an intramarginal vein. Petiole 5–8 mm. Inflorescence a raceme to 10 cm; axis patently long hairy, hairs c. 2 mm. Bracts villous, (narrowly) ovate, 5–8 mm, bracteoles hirsute, linear, c. 5 mm, both caducous under old flowers or under fruit. Ovary densely appressedly long hairy, c. 1.5 mm high. Calyx (sparsely) long hairy, 2–3 mm, the lobes rounded, 1.5–2.5 mm. Corolla 5–7 mm. Stamens 80–100. Disk 5-glandular, with the conical style base densely hairy. Fruit (obliquely) ellipsoid to flask-shaped, and then narrowed for the upper 4 mm, hairy, glabrescent, c. 15 × 5 mm.

Distr. Ceylon.

Ecol. Hill and lower montane zone forests.

Note. *C.P. 407* in PDA is mixed with a plant that possibly is a hybrid between *S. cuneata* Thw. var. *acuta* and *S. pulchra* Wight var. *hispidula*.

Specimens Examined. KANDY DISTRICT: Ambagamuwa, *s. coll. C.P. 407* p.p. (PDA). KALUTARA DISTRICT: Moropitiya logging area, *Kostermans 24686* (L), *Nooteboom 3277*, *3280* (L, PDA, US). RATNAPURA DISTRICT: Adam's Peak Jungle above Balangoda, *Kostermans 24467* (L);

Forest between Opanake and Wewelwatte, *Nooteboom 3392* (L, PDA, US); Rassagala, *Nooteboom 3394* (L, PDA, US); Bulutota Pass, *Nooteboom 3320 A* (L, PDA, US); Sinharaja Forest near Weddagala, *Nooteboom 3267 A* (L, PDA, US). LOCALITY UNKNOWN: *Wight 1693* (L).

9. Symplocos cuneata Thw., Enum. Pl. Zeyl. 186. 1860; Clarke in Hook f., Fl. Br. Ind. 3: 584. 1882; Trimen, Handb. Fl. Ceylon 3: 107. 1895; Brand in Pflanzenr. Heft 6: 61. 1901.

Symplocos acuta Thw., Enum. Pl. Zeyl. 186. 1860; Clarke in Hook. f., Fl. Br. Ind. 3: 584. 1882; Trimen, Handb. Fl. Ceylon 3: 106. 1895; Brand in Pflanzenr. Heft 6: 62. 1901.
Lodhra cuneata Miers, J. Linn. Soc. Bot. 17: 301. 1879.
Lodhra acuta Miers, l.c.
Symplocos macrophylla var. *cuneata* (Thw.) Noot., Rev. Sympl. 228. 1975.
Symplocos macrophylla var. *acuta* (Thw.) Noot., Rev. Sympl. 227. 1975.

Small tree to 15 m high and 20 cm diam. Twigs (densely) appressedly or rarely patently hairy. Leaves appressedly hairy, especially on midrib and nerves, glabrescent, elliptic, with cuneate base and acuminate, acute apex, 6–14 × 2.5–5 cm; nerves in 6–10 pairs, much prominent beneath, curved upwards and meeting in an intramarginal vein; secondary veins transverse to nerves much prominent; reticulation fine. Petiole 3–14 mm. Inflorescence a short raceme to 2.5 cm long; axis densely appressedly pubescent; pedicels 2 mm or shorter. Bracts and bracteoles caducous, densely appressedly pubescent, boat-shaped, 3–6 and 2–4 mm long respectively. Ovary densely appressedly pubescent to silky, 1 mm high. Calyx densely appressedly pubescent to glabrous, 3–4 mm, the lobes rounded, 2–4 mm long. Corolla 4–5 mm. Disk with few hairs to shortly pilose, 5-glandular. Stamens c. 100. Style glabrous except the base. Fruit nearly cylindrical, narrowed towards the apex, more or less hairy, 12–17 × c. 4 mm. Stone with shallow lengthwise grooves.

Distr. Ceylon.

Ecol. Prim. rain forest alt. 0–850 m.

Note. I reduced *S. acuta* Thw., because the only difference between the two taxa are the shorter, glabrous or thinly appressedly hairy calyx lobes of *S. acuta*. I consider this to be a variety, occupying the NE part of the species area, while var. *cuneata* occupies the SW part of it.

1. var. cuneata

Calyx 3–4 mm, the lobes 2–4 mm, densely appressedly pubescent.

Specimens Examined. RATNAPURA DISTRICT: Weddagala, *Waas 1466* (US) Hewesse, *s. coll. C.P. 2920* p.p. (PDA); Hillepee, *s. coll. C.P. 2920* p.p. (PDA). GALLE DISTRICT: Weerakande, Elpitiya, *Kostermans*

25605 (L, PDA); Neluwe Kande, *s. coll. C.P. 2920* p.p. (PDA), *Beddome 4951* (PDA); Kanneliya f.r. *Jayasuriya 2374* (P).

2. var. **acuta** (Thw.) Noot.

Calyx 3 mm, the rounded lobes 2–3 mm, glabrous, or thinly appressedly pubescent, often at the base only.

Note. In vivo the disk is orange coloured with white glands.

Specimens Examined. RATNAPURA DISTRICT: Heramitiyagala, near Rassagala, *Huber 541* (L); between Opanake and Wewelwatte (or between Pelmadulla and Rassagala) *Huber 504* (L), *Nooteboom 3314* (L, PDA, US); Palabaddala, *Waas 433* (US), *s. coll. C.P. 2681* p.p. (PDA); "Saffragam", *s. coll. C.P. 2681* p.p. (PDA).

10. Symplocos kurgensis Clarke in Hook. f., Fl. Br. Ind. 3: 576. 1882; Brand in Pflanzenr. Heft 6: 62. 1901; Brandis, Indian Trees 441. 1906.

Symplocos hohenackeri Clarke in Hook. f., Fl. Br. Ind. 3: 582. 1882; Brand in Pflanzenr. Heft 6: 90. 1901; Brandis, Indian Trees 439. 1906.
Symplocos hebantha Thw. ex Clarke in Hook. f., Fl. Br. Ind. 3: 586. 1882; Trimen, Handb. Fl. Ceylon 3: 109. 1895; Brand in Pflanzenr. Heft 6: 62. 1901.
Symplocos macrophylla Wall. ex DC var. *kurgensis* (Clarke) Noot., Rev. Sympl. 229. 1975.

Treelet to 10 m high. Twigs tomentose or appressedly pubescent. Leaves except the finely appressedly pilose midrib glabrous beneath, or patently pubescent or tomentose especially on midrib and nerves, c. elliptic, 5–14.5 × 2.5–6.5 cm; base cuneate to rounded; apex acuminate; margin entire or glandular dentate or denticulate. Nerves in 7–12 pairs, meeting in a looped intramarginal vein, secondary veins transverse to nerves. Petiole 6–20 mm. Inflorescence a raceme to 6 cm (much shorter in recent collections); axis patently soft hairy. Bracts and bracteoles soon caducous, appressedly (long) pubescent, circular or broadly ovate to elliptic, 4–7 mm and ovate to elliptic 3–5 mm long respectively. Pedicel c. 1 mm. Ovary glabrous, 1.25–2 mm high. Calyx densely appressedly silky, c. 3 mm long, divided into rounded lobes. Corolla c. 5 mm. Stamens c. 80. Disk 5-glandular, (shortly) pilose. Fruits ovoid-cylindrical, 13–20 × 6–8 mm, stone with shallow lengthwise grooves, depressed at one side towards the base.

Distr. India (Mysore, Mercara, Coorg) and Ceylon (Ratnapura district).

Ecol. Rain forest at low altitude, very rare.

Specimens Examined. RATNAPURA DISTRICT: Congala,

s. coll. C.P. 3918 p.p. (PDA); Kolonne Korale, *s. coll. C.P. 3918* p.p. (PDA, P); near Weddagala, *N. & S. Gunatilleke B 756*, *B 1628* (Bot. Departm, Univ. of Ceylon, L); Bambarabotuwa Forest Reserve between Pelmadulla and Rassagala, *Huber 505* (PDA).

Note. The recent collections (only two!) have much smaller leaves and a different indument on the twigs than the only other known collections from Ceylon, *C.P. 3918*, collected in 1866 and 1868.

11. Symplocos diversifolia Brand in Pflanzenr. Heft 6: 62. 1901.

Symplocos jucunda var. *diversifolia* Alston in Trimen, Handb. Fl. Ceylon 6: 187. 1931.

Symplocos macrophylla Wall. ex DC var. *macrophylla* p.p. Noot., Rev. Sympl. 227. 1975.

1. var. **diversifolia**

Shrub or treelet to ? 2 m. Twigs densely patently ferruginously pilose-tomentose. Leaves (narrowly) ovate, patently pilose underneath, 5–12 × 1–4 cm; apex acute, acuminate, base cuneate to slightly cordate; margin entire or denticulate; nerves in 5–10 pairs, meeting in an intramarginal vein. Petiole 2–6 mm. Inflorescence a fascicle, condensed spike or raceme to 5 cm; axis densely patently villous; bracts and bracteoles caducous, long hairy, narrowly ovate, c. 5 mm and linear, c. 2–3 mm respectively. Pedicel short. Ovary densely appressedly long hairy, 1.5 mm high. Calyx with same indument, c. 2 mm, the lobes (narrowly) ovate to triangular, 1.5–2 mm, sometimes apiculate by the excurrent midrib. Corolla c. 3 mm. Stamens 25–60. Disk 5-glandular, with the style base densely sericeous; style glabrous except the base. Fruit ovoid-ellipsoid, c. 10 × 5 mm, hairy, glabrescent.

Distr. Sinharaja Forest.

Ecol. Rain forest, alt. 550 m (once recorded). Very rare.

Specimens Examined. RATNAPURA DISTRICT: Deltotte and Saffragam, *s. coll. C.P. 2435* (PDA, K, BISH, BM, BO, FI, LE, MEL, P); S. of Weddagala, *Huber 552* (L). GALLE DISTRICT: Nellowe Kande, *s. coll. s.n.* (PDA).

2. var. **appressa** Noot. var. nov. Ramuli appresse pubescentes, folia infra appresse pilosa. Typus: *Nooteboom 3322*, Kurulugala, Bulutota Pass (L, isotype in PDA, US).

Shrub or tree to 5 m. Twigs appressedly ferrugineously pubescent. Leaves ovate to elliptic, finely appressedly pilose underneath, 4.5–8 × 1.5–2.5 cm; nerves in 6–8 pairs. Petiole c. 5 mm. Inflorescence with appressedly pubescent axis. Bracts and bracteoles caducous under the ripe fruit, appressedly pubescent, broadly and narrowly ovate respectively, both c. 3 mm long. Ovary appressedly pubescent, 1–2 mm high. Calyx appressedly pubescent, 2–3 mm, the lobes ovate, acute, 2–2.5 mm. Fruit c. elliptic, 10 × 5 mm, finely inconspicuously appressedly hairy, probably glabrescent.

Distr. Probably only from Kurulugala, Bulutota pass.

Ecol. Remnants of disturbed forest at about 1200 m. This locality probably will be entirely deforested in the near future.

Specimens Examined. RATNAPURA DISTRICT: Kurulugala, Bulutota pass, *Waas 1741* (PDA), *Nooteboom 3320*, *3322* (L, PDA, US). LOCALITY UNKNOWN: *s. coll. C.P. 78* p.p. (PDA).

Note. Some fragments of this taxon are part of several sheets of *C.P. 78* in PDA, and *C.P. 23* in P; the rest of these numbers belong to *S. elegans* Thw. See also the note under 12. *S. macrophylla.*

12. Symplocos macrophylla Wall., [Cat. 4431. 1831, nomen] ex DC. Prod. 8: 257. 1844; Clarke in Hook. f., Fl. Br. Ind. 3: 578. 1882; Brand in Pflanzenr. Heft 6: 61. 1901; Brandis, Indian Trees 439. 1906; Noot., Rev. Sympl. 222. 1975.

Symplocos gardneriana Wight, Ic. Pl. Ind. Or. 4: 10, t. 1231. 1848; Wight, Neilgherry Pl. 2: 39, t. 144. 1851; Clarke in Hook. f., Fl. Br. Ind. 3: 582. 1882; Brand in Pflanzenr. Heft 6: 1901; Brandis, Indian Trees 439. 1906.

Symplocos jucunda Thw., Enum. Pl. Zeyl. 186. 1860; Clarke in Hook. f., Fl. Br. Ind. 3: 585. 1882; Trimen, Handb. Fl. Ceylon 3: 107. 1895; Brand in Pflanzenr. Heft. 6: 61. 1901; Alston in Trimen, Handb. Fl. Ceylon 6: 187. 1931.

Symplocos marginalis Thw., Enum. Pl. Zeyl. 187. 1860; Clarke in Hook. f., Fl. Br. Ind. 3: 587. 1882; Trimen, Handb. Fl. Ceylon 3: 110. 1895; Brand in Pflanzenr. Heft 6: 50. 1901.

Lodhra macrophylla Miers, J. Linn. Soc. Bot. 17: 299. 1879.

Lodhra gardneriana Miers, op. cit. 300.

Lodhra jucunda Miers, op. cit. 301.

Lodhra marginalis Miers, l.c.

Symplocos ifsiana C.E.C. Fisher, Kew Bull. 37. 1940.

Treelet to 15 m high and 20 cm diam. Twigs densely tomentose or tomentellous. Leaves (narrowly) ovate or elliptic, usually densely tomentose on midrib and nerves, less densely hairy on the blade beneath, 7–20 × 2–11 cm, with cuneate to slightly cordate base and more or less acuminate apex; margin often revolute, entire to dentate or denticulate. Nerves 6–15 pairs, much prominent beneath, meeting in a looped intra-marginal vein; secondary veins transverse to midrib and nerves. Petiole 5–25 mm. Inflorescence a raceme or spike to 7 cm; axis densely pubescent or tomentose; bracts and bracteoles soon caducous, densely sericeous or pubescent, the lowermost up to 10 and 7 mm long respectively; pedicel stout, up to 4 mm in the lowermost flowers. Ovary appressedly sericeous or tomentose, c. 1.5 (–2) mm high. Calyx appressedly long pubescent or sericeous, 2–4 mm long, often entirely divided into the rounded 1.5–4 mm long lobes. Corolla 4–8 mm. Stamens 60—more than 100. Disk densely hairy. Style glabrous, except the very base, or sericeous halfway up. Fruit ovoid or ellipsoid to cylindrical, 13–20 × 6–8 mm, glabrescent.

Distr. India (Assam and S. India) and Ceylon.

Ecol. In forest, 0–2000 m.

Specimens Examined. KANDY DISTRICT: Ambagamuwa, *s. coll. C.P. 2688* (PDA); road Laxapana to Maskeliya, above double cutting, *Kostermans 24093* (L, PDA, US); Adam's Peak wilderness above Moray Estate, *Jayasuriya 1577* (PDA, US), *Nooteboom 3388* (L, PDA, US), *Waas 1695* (PDA, US); Deltota, *s. coll. C.P. 2435* (PDA, P); Maratenne, *Nooteboom 3391* (L, PDA, US). KALUTARA DISTRICT: Wallankanda forest, Pasdun Korale, *s. coll. s.n.* (PDA). RATNAPURA DISTRICT: Kurugala, Bulutota Pass, *Nooteboom 3319*, *3324* (L, PDA, US); S. side of forest between Opanake and Wewelwatte, *Nooteboom 3306* (L, PDA, US). MATARA DISTRICT: Kotapolla, near Deniyaya, *s. coll. s.n.* (PDA). LOCALITY UNKNOWN: *s. coll. C.P. 368* p.p. (P).

Note. *Symplocos diversifolia* Brand was based on *C.P. 3454*. The youngest leaves are hairy above. The leaves and the petioles are shorter than in the other collection cited by Thwaites under *S. jucunda* Thw., *C.P. 2435*. The last plant is chosen by Brand as the type of *S. jucunda*. *S. jucunda* var. *diversifolia* Alston was based on Brand's species. I reduce *S. jucunda* to *S. macrophylla*, because it belongs to the variety range of that species. In PDA there are three sheets that bear the number *C.P. 2435* only, which all belong to *S. jucunda* Thw. (*S. macrophylla* Wall. ex Don.).

UMBELLIFERAE

(by Joseph L. Krahulik and William L. Theobald*)

Juss., Gen. 218. 1789.

Plants herbaceous, rarely woody at base, acaulescent or caulescent, annual or perennial, slender or stout, erect, ascending, or prostrate; stems usually fistular and striate. Leaves alternate or basal, sometimes heteromorphic, compound, decompound, or sometimes simple, often much incised or divided; petioles usually dilated and sheathing at base; stipules absent (except in *Hydrocotyle*). Inflorescence of umbels, variously arranged; umbels sometimes simple or often compound with the primary rays terminated by an umbellet with pedicels; the umbel may be subtended by an involucre of bracts and the umbellet by an involucel of bractlets; flowers small, usually hermaphroditic. Calyx 5-merous, minute or absent. Corolla 5-merous, petals distinct, usually with inflexed apices, imbricate (sometimes valvate in *Hydrocotyle*). Stamens 5-merous, alternate with the petals; filaments filiform; anthers versatile, bilocular, dehiscing longitudinally. Ovary inferior; styles two, often swollen and spreading at the base, forming a stylopodium; disk around style sometimes present. Fruit a schizocarp composed of two separable, dry mericarps adjoined by their faces (commissure): flattened laterally (at right angles to the commissure), flattened dorsally (parallel with the commissure), to terete; each mericarp with five primary ribs: one median (dorsal rib), two on the margin nearest the commissure (lateral ribs), and two between the dorsal and lateral ribs (intermediate ribs), rarely with secondary ribs; the ribs filiform, thin, corky, or winged, sometimes spinescent; oil-tubes (vittae) usually present, filled with coloured, aromatic oils, commonly in the intervals between the ribs and on the commissural face, sometimes in the ribs; mericarps usually separating at maturity and sometimes attached at summit of a wiry, slender, central axis, the carpophore; mericarps unilocular, with solitary ovule; placentation axile. Seed pendulous, with small embryo; endosperm cartilaginous.

The family is a relatively large one of about 350 genera and 3000 species. They are predominantly North Temperate in distribution with the greatest concentrations being found in the Mediterranean region, Central Asia, and western North America. However, they are distributed throughout the

*Pacific Tropical Botanical Garden, Hawaii.

world, although in the tropics they are usually found at higher elevations. Ceylon representatives include 12 genera and 15 species. Of these, two are endemic and restricted to the higher elevations on the island. Several of these including *Eryngium foetidum* L., *Daucus carota* L., *Coriandrum sativum* L., *Apium leptophyllum* (Pers.) F. Muell., and *Trachyspermum roxburghianum* (Wall.) Craib appear to be introductions by man.

The most recent major treatment of the family is that by Drude 1897–1898, in Pflanzenfam. 3 (8): 63–128, 129–250. In more recent times a comprehensive review of the family has been prepared, which includes an extensive survey of the literature on various aspects of research underway within the group: Heywood, V. (ed.), Biology and Chemistry of the Umbelliferae, Academic Press, London.

KEY TO THE GENERA

1 Herbs prostrate, rooting at nodes; flowers in simple umbels; fruit strongly flattened laterally

2 Leaves crenate or lobed; petioles nonsheathing; stipules present; involucral bracts minute or absent; petals valvate; stylopodium present; fruit without evident secondary ribs or reticulations; carpophore absent........................... **1. Hydrocotyle**

2 Leaves entire to repand-dentate; petioles sheathing at base; stipules absent; involucral bracts conspicuous; petals imbricate; stylopodium absent; fruit with evident secondary ribs and reticulations; carpophore present............................. **2. Centella**

1 Herbs slender to stout, erect or ascending; flowers in heads, or simple or compound umbels; fruit terete, flattened dorsally, or slightly compressed laterally, never strongly flattened laterally

3 Leaves and involucral bracts very prominently spinescent; inflorescence capitate, the heads in dichasia.. **4. Eryngium**

3 Leaves and involucral bracts not spinescent; basic unit of inflorescence a compound umbel (appearing simple), or regularly compound umbel

4 Ovary and fruit armed with uncinate prickles or glochidiate spines

5 Leaves simple, palmately divided, lobes broad; flowers in compound umbels (appearing simple), the umbels in dichasia; carpophore absent............ **3. Sanicula**

5 Leaves pinnately decompound, ultimate divisions linear, filiform; flowers in regularly compound umbels; carpophore present................................ **5. Daucus**

4 Ovary and fruit unarmed, glabrous or variously pubescent

6 Leaves simple and entire, linear, grass-like; petals yellow............ **7. Bupleurum**

6 Leaves simple or variously compound, lobed, toothed, incised, or divided; petals white or variously tinted white

7 Fruit terete or slightly compressed laterally; lateral ribs not winged

8 Leaves heteromorphic; sepals large, conspicuous; vittae obscure.. **6. Coriandrum**

8 Leaves similar throughout; sepals minute; vittae conspicuous

9 Involucral bracts conspicuous............................ **9. Trachyspermum**

9 Involucral bracts minute or absent

10 Leaves pinnately decompound... **8. Apium**

10 Leaves simple or trifoliolate................................ **10. Pimpinella**

7 Fruit strongly flattened dorsally; lateral ribs conspicuously winged

11 Leaves triternate, ultimate divisions linear-lanceolate, glabrous; pedicels much longer than the fruit; ovary glabrous.. **11. Peucedanum**

11 Leaves ternately or pinnately compound, leaflets broad, variously cleft or lobed, pubescent; pedicels shorter than or equal to the fruit; ovary pubescent...... **12. Heracleum**

1. HYDROCOTYLE

L., Sp. Pl. 234. 1753.

Plants herbaceous, caulescent, perennial, slender, prostrate, glabrous or pubescent; stems creeping, rooting at the nodes. Leaves alternate, simple, ovate to orbicular-reniform, crenate, or lobed, venation palmate; petioles long, slender, nonsheathing; stipules present. Inflorescence usually a simple umbel; peduncles axillary, solitary or clustered, shorter or longer than the leaves; involucral bracts minute or absent; flowers sessile or pedicellate. Sepals minute or absent. Petals white or greenish-white, ovate, entire, acute, valvate. Stylopodium conspicuously conic to depressed. Fruit orbicular to ellipsoid, usually broader than long, strongly flattened laterally, the dorsal surface rounded or acute, commissure narrow; dorsal and intermediate ribs acute or absent, the lateral ones slender, acute, conspicuous or rarely absent; oil-bearing cells conspicuous to absent; carpophore absent; woody endocarp evident. Seeds ovate or ovate-oblong in transection, the seed face plane to concave.

A large and widespread genus of approximately 100 species found throughout temperate and tropical areas. Represented in Ceylon by two species.

KEY TO THE SPECIES

1 Mature leaves usually more than 3 cm broad, palmately lobed, lobes triangular; peduncles 2–6 at a node; umbels 10–30 flowered.. **1. H. javanica**

1 Mature leaves less than 2 cm broad, palmately cleft half-way down into 5–7 broad obtuse lobes; peduncles solitary at a node; umbels 4–8 flowered.......... **2. H. sibthorpioides**

1. Hydrocotyle javanica Thunb., Diss. Hydroc. no. 17, pl. 2. 1798; ed. Pers., 2: 415, pl. 3. 1800; Trimen, Handb. Fl. Ceylon 2: 275. 1894.

Hydrocotyle hirta R. Br. ex A. Rich., Ann. Gén. Sc. Phys. 4: 64. 1820.
Hydrocotyle nepalensis Hook., Exot. Fl. 1: pl. 30. 1823.
Hydrocotyle capitata Moon, Cat. 22. 1824. nom. nud.
Hydrocotyle hispida D. Don, Prod. Fl. Nepal. 183. 1825.
Hydrocotyle globata Blume, Bijdr. 15: 883. 1826.
Hydrocotyle hirsuta Blume, Bijdr. 15: 884. 1826.
Hydrocotyle heyneana Wall., Cat. 563. 1829.
Hydrocotyle zeylanica DC., Prod. 4: 67. 1830.

Hydrocotyle strigosa Ham. in Wall., Cat. 7219. 1832.
Hydrocotyle polycephala Wight & Arn., Prod. 1: 366. 1834.

Plants low; stems 10-55 cm long, succulent. Leaves orbicular-reniform in outline, rather broader than long, 0.5–3 cm long, 0.75–6 cm broad, palmately lobed, lobes triangular, margins very coarsely crenate, base very deeply cordate, shiny on both surfaces, especially above, strigose; petioles slender, ascending, 1–25 cm long, densely hirsutulous; stipules large, membranaceous. Umbels dense, globular, many-flowered; peduncles 2–6, clustered at nodes, 0.1–2 cm long, much shorter than the leaves, hirsutulous; involucral bracts absent or inconspicuous; pedicels up to 2.5 mm long; flowers 10–30, very small. Sepals absent or inconspicuous. Petals white to greenish-white, valvate in bud. Ovary orbicular, glabrous; styles filiform, divaricate. Fruit suborbicular, c. 1 mm long, c. 2 mm broad, glabrous; ribs visible; oil-bearing cells absent.

Distr. Between elevations of 650–2300 m in the central mountains of the Kandy, Badulla, and Nuwara Eliya Districts. Japan, south into southern India and Ceylon and Indonesia.

Ecol. Moist open places under and among trees and in disturbed areas. Flowering May to October, and apparently throughout the year.

Vern. Maha-gotukola (S).

Note. Trimen notes that the large stipules completely enclose the end of the growing branch and young leaf. He also noted that at the higher elevations the petioles and peduncles are more woolly, and the flowers are on longish pedicels.

Trimen mentions its use medicinally and refers to Mr. E. Green, of Pundaluoya, who stated that the Tamil labourers employed it as a fish poison.

Specimens Examined. BADULLA DISTRICT: top of Namunukula, *J.M. Silva s.n.*, 29-IV-1907 (PDA). KANDY DISTRICT: Dehigama and Murutalawa, *Appuhamy s.n.*, 16-VI-1955 (PDA); below summit of Hantane Mt. No. 1, Kandy, *Theobald & Grupe 2333* (A, BISH, K, L, PDA, UC, US). NUWARA ELIYA DISTRICT: Hakgala Botanic Garden, jungle stream at south boundary, *A.W. Silva s.n.*, 28-II-1906 (PDA); north entrance, Horton Plains, Dieter's plot No. P24, *Fosberg & Mueller-Dombois 50029* (BISH, PDA, US); along primate survey transect in cloud forest just north of Horton Plains Rest House, *Theobald & Krahulik 2742* (BISH, K, MO, PDA, UC, US); along trail from Galagama Falls to Horton Plains, *Theobald & Krahulik 2837* (BISH, L, P, PDA, UC, US); Maturata, *s. coll. s.n.*, in 1858, part of *C.P. 2812* (PDA). DISTRICT UNKNOWN: *Col. Walker s.n.* (PDA); *s. coll. s.n.*, in 1858, part of *C.P. 2812* (PDA); *s. coll. C.P. 1634* (PDA).

2. Hydrocotyle sibthorpioides Lam., Enc. 3: 153. 1789; Alston in Trimen, Handb. Fl. Ceylon 6: 137. 1931.

Hydrocotyle rotundifolia Roxb., Hort. Beng. 21. 1814. nomen; Roxb., Fl. Ind. ed. 2 (2): 88. 1832; Trimen, Handb. Fl. Ceylon 2: 275. 1894.
Hydrocotyle nitidula A. Rich., Ann. Gén. Sc. Phys. 4: 60. 1820.
Hydrocotyle tenella D. Don, Prod. Fl. Nepal. 183. 1825.
Hydrocotyle ranunculoides var. *incisa* Blume, Bijdr. 15: 884. 1826.
Hydrocotyle zollingeri Molkenb. in Miq., Pl. Jungh. 91. 1851.
Hydrocotyle perexiqua Hance in Walp., Ann. 2: 691. 1852.
Hydrocotyle puncticulata Miq., Fl. Ind. Bat. 1 (1): 732. 1856.

Plants low; stems filiform, branching diffuse, glabrous. Leaves small, nearly orbicular in outline, 0.25–1.25 cm long, 0.5–1.5 cm broad, palmately cleft half-way down into 5–7 broad obtuse lobes, margins crenate, base cordate, the basal lobes almost in contact, but not overlapping, glabrous and shining above, with scattered coarse hairs beneath; petioles ascending, filiform, 0.75–5 cm long, glabrous; stipules membranaceous. Umbels very small, few-flowered, compact, globular; peduncles solitary at a node, 0–1 cm long, usually shorter than the leaves, slender, glabrous; involucral bracts minute, oblong, as long as the flowers; pedicels very short to absent; flowers minute, 4–8. Sepals absent or inconspicuous. Petals greenish-white, valvate in bud. Ovary orbicular, glabrous; styles filiform, divaricate. Fruit suborbicular, c. 0.75–1 mm long, c. 1.25–1.5 mm broad, glabrous; ribs blunt; oil-bearing cells absent.

Distr. Between elevations of 1400–2300 m in the mountains of the Badulla and Nuwara Eliya Districts. Japan south and widespread throughout tropical Asia. Now introduced into the New World.

Ecol. Sunny or shaded, damp, fertile localities. Flowering throughout the year.

Specimens Examined. BADULLA DISTRICT: Bandarawela, patana stream, Craig Estate, *A.M. Silva s.n.*, 25-III-1906 (PDA). NUWARA ELIYA DISTRICT: North entrance, Horton Plains, Dieter's plot No. P24, *Fosberg & Mueller-Dombois 50023* (BISH, PDA, US); Nuwara Eliya, *s. coll. s.n.*, IV-1856, *part of C.P. 1635* (PDA).

2. CENTELLA

L., Gen. Pl. ed. 6. 485. 1764.

Plants herbaceous, caulescent, perennial, slender, prostrate, glabrous or pubescent; stems creeping, rooting at the nodes. Leaves alternate, simple, orbicular to reniform (in ours), margins entire to repand-dentate, venation palmate; petioles sheathing; stipules absent. Inflorescence a loose to subcapitate simple umbel; peduncles axillary, shorter or longer than the leaves; involucral bracts 2–3, ovate, conspicuous; pedicels very short, slender to absent. Sepals absent. Petals white or rose-tinged, orbicular with acute, inflexed apices, imbricate. Styles short; stylopodium absent. Fruit orbicular to ellipsoid, strongly flattened laterally, commissure narrow, glabrous; primary and secondary ribs filiform, conspicuous; vittae absent or occasional, small; carpophore entire; woody endocarp evident. Seeds elliptic to ovate-oblong in transection; the seed face plane.

A small genus of approximately 20 species, especially abundant in South Africa and also found in Australia, New Zealand, and extending into the northern hemisphere. One representative in Ceylon.

Centella asiatica (L.) Urban in Mart., Fl. Bras. 11 (1): 287. 1879; Alston in Trimen, Handb. Fl. Ceylon 6: 137. 1931.

Hydrocotyle asiatica L., Sp. Pl. 1: 234. 1753; Trimen, Handb. Fl. Ceylon 2: 276. 1894.
Hydrocotyle wightiana Wall., Cat. 7220. 1832.
Hydrocotyle lurida Hance in Walp., Ann. 2: 690. 1852.

Plants glabrous at maturity; stems one to several dm long. Leaves 1–6, clustered on a short shoot at each node, orbicular to reniform in outline, 0.5–2 cm long, 1–3 cm broad, apex rounded, margins entire, crenate, or usually repand-dentate, the sinuses usually broad, base cordate, glabrous and shiny above, paler beneath; petioles 1–10 cm long. Umbels opposite the leaves, mostly 3-flowered, dichasial; peduncles 0–2 cm long, much shorter than the leaves; involucral bracts 2, ovate, 2–3 mm long, small, conspicuous; pedicels very short, 0.5–1 mm long. Petals white or rose-tinged, minute. Ovary flattened laterally, glabrous; styles filiform, divaricate. Fruit orbicular to ellipsoid, 2–3 mm long, 3–4 mm broad; vittae absent.

D i s t r. Very abundant from sea-level to the highest elevations all over the island. Also widespread throughout the tropics and subtropics of the Old and New Worlds.

E c o l. A very common weed with great capacity for holding earth against erosion. Flowering May to October and probably throughout the year.

V e r n. Gotukola, Hin-gotukola (S).

U s e s. The plant is widely used as a salad and Trimen notes that the powdered dry leaves are an official drug in the Indian Pharmacopoeia.

S p e c i m e n s E x a m i n e d. ANURADHAPURA DISTRICT: Horawupotana, *s. coll. s.n.*, VIII-1856, part of *C.P. 2810* (PDA). KANDY DISTRICT: crest of ridge between Hantane Mts. No. 1 & 2 south of Kandy, *Theobald & Krahulik 2771* (A, BISH, K, L, LE, MO, P, PDA, UC, US). KEGALLE DISTRICT: Alagalla, *s. coll. s.n.*, VIII-1853, part of *C.P. 2810* (PDA). NUWARA ELIYA DISTRICT: patana, Hakgala, *A.M. Silva s.n.*, 25-II-1906 (PDA); Horton Plains and Nuwara Eliya, *s. coll. s.n.*, in 1857, part of *C.P. 2810* (PDA); Horton Plains, along path in cloud forest between Small World's End and Big World's End, *Theobald & Krahulik 2729*

(BISH, BM, CAL, NY, PDA, SING, UC, US); along the road from Pundaluoya to Hwy. A-5 and Nuwara Eliya, c. 2 miles west of intersection with A-5, *Theobald & Krahulik 2822* (BISH, PDA, US).

3. SANICULA

L., Sp. Pl. 235. 1753.

Plants herbaceous, biennial or perennial, low, slender, erect or decumbent, scapose or branching, glabrous or pubescent; stems with branches usually pseudodichotomous or dichasial, with a short rootstock or fusiform or tuberous roots. Leaves alternate, membranaceous to subcoriaceous, simple and palmately divided (in ours), the divisions variously toothed, lobed, pinnately dissected, or entire; petioles sheathing. Basic unit of inflorescence a compound umbel, appearing simple, the umbels in dichasia (in ours); involucral bracts entire to lobed, longer or shorter than the umbels; umbels polygamous, or unisexual with only staminate flowers; flowers hermaphroditic or staminate, the staminate flowers often prominently pedicellate, persistent or deciduous, the perfect flowers short-pedicellate or subsessile. Calyx synsepalous, the lobes setiform, linear-lanceolate, or obtusely ovate, nearly distinct to strongly connate, concealed by the fruit armature when mature. Petals white, greenish-white (in ours), spatulate to ovate, emarginate, with acute, inflexed apices. Styles shorter than to longer than the calyx lobes, spreading, recurved, or coiled; stylopodium absent or flattened and disk-like. Fruit oblong-ovoid to globose, solitary to many in each umbel, pedicellate or subsessile, somewhat compressed laterally, commissure broad, densely covered with prickles, spicules, squamae, or tubercles, the prickles often dilated or bulbous at base, straight or uncinate; ribs absent, rarely costate; vittae large or small, regularly or irregularly arranged, three to many under the dorsal and lateral surfaces, usually two on the commissure; carpophore absent. Seeds subterete to compressed laterally or dorsally in transection; the seed face plane, concave, or sulcate.

A small genus of approximately 37 species found throughout the world except New Guinea and Australasia. One representative in Ceylon.

For a monograph of the genus see: Shan, R.H. and Constance, L. 1951. The genus *Sanicula* (Umbelliferae) in the Old World and the New. Univ. Calif. Publ. Bot. 25: 1–78.

Sanicula elata Ham. ex D. Don, Prod. Fl. Nepal. 183. 1825.

Sanicula hermaphrodita Ham. ex D. Don, Prod. Fl. Nepal. 183. 1825.
Sanicula javanica Blume, Bijdr. 15: 882. 1826.
Sanicula montana Reinw. ex Blume, Bijdr. 15: 882. 1826.
Sanicula europaea var. *elata* Boiss., Bull. Soc. Bot. France 53: 421. 1906.

Sanicula europea sensu Trimen, Handb. Fl. Ceylon 2: 276. 1894.

Plants caulescent, perennial, erect; stems 20–70 cm tall, furrowed, glabrous, dichasially branched above, with a thick rootstock bearing fibrous roots. Leaves mostly in a basal rosette, cauline ones reduced upwards, simple, very broadly ovate to orbicular-cordate in outline, 3–6 cm long, 4–8 cm broad, deeply palmately divided into 3–5 elliptic lobes, segments obovate to broadly elliptic in outline, margins crenate-serrate, glabrous on both surfaces, dark green above, paler beneath; petioles erect, 5–20 cm long, glabrous. Peduncles 2–3, 2–10 mm long, unequal; involucral bracts linear, acuminate, minute; flowers 4–7 in each umbel; staminate flowers 1–4, pedicellate, pedicels 1–2 mm long; perfect flowers 3, short-pedicellate or subsessile. Sepals linear-lanceolate, acute, persistent. Petals greenish-white, obovate. Ovary ovoid, bristly; styles about twice as long as sepals, divaricate, with capitate stigmas; stylopodium flattened and disk-like. Fruit ovoid, c. 3 mm long, c. 2.5 mm broad, covered with stiff, slender, uncinate prickles; ribs obscure; vittae polymorphic: small, solitary one in each primary rib, some very slender ones scattered in the pericarp, and two larger ones on the commissure. Seed face slightly concave.

Distr. Very rare and only known from a few localities between 1000–2300 m in the Nuwara Eliya District. However, otherwise widespread and found from central China south to Indonesia and west to Southwest China, India and Eastern Africa to South Africa and Madagascar.

Ecol. In open woods. Flowering April to June.

Specimens Examined. NUWARA ELIYA DISTRICT: Horton Plains, *s. coll. s.n.*, IV-1854, part of *C.P. 2813*, (PDA); between Maturata and Nuwara Eliya, *s. coll. s.n.*, VIII-1853, part of *C.P. 2813* (PDA); Maturata, *s. coll. s.n.*, 17-IV-1859, part of *C.P. 2813* (CAL). DISTRICT UNKNOWN: *Col. Walker 91* (PDA).

4. ERYNGIUM

L., Sp. Pl. 232. 1753.

Plants herbaceous, biennial or perennial, erect, spinescent, usually glabrous; stems branching, with rootstocks bearing fibrous roots. Leaves mostly in a basal rosette or the upper ones sometimes alternate, coriaceous or membranaceous, simple, commonly lanceolate to oblanceolate, entire to pinnately or palmately lobed to divided, often ciliate to spinose, venation parallel or reticulate; petioles sheathing, sometimes septate. Inflorescence capitate, the heads solitary, or in dichasia (in ours); involucral bracts linear to ovate-lanceolate, softly to pungently spinescent, subtending the head; rays absent; involucel bractlets subtending the flowers entire or lobed; flowers sessile. Sepals ovate to lanceolate, acute to obtuse, entire or rarely spinescent, conspicuous. Petals greenish-white, or white to purple, ovate to oblong with variously inflexed and lobed to fimbriate apices. Styles shorter than to exceeding the sepals; stylopodium absent. Fruit globose to obovoid, nearly terete, scarcely compressed laterally, commissure broad, variously covered with scales, tubercles or papillae; ribs absent; vittae mostly five, inconspicuous; carpophore absent. Seeds subterete in transection; the seed face plane or slightly concave.

A large genus of approximately 230 species found throughout temperate and tropical areas except Tropical and South Africa. One introduced representative in Ceylon.

Eryngium foetidum L., Sp. Pl. 1: 232. 1753; Alston in Trimen, Handb. Fl. Ceylon 6: 138. 1931.

Plants caulescent, biennial, slender, erect, aromatic; stems 15–40 cm tall, furrowed, mostly solitary, dichasially branched above. Leaves oblanceolate-oblong in outline, 4–12 cm long, 1–2.5 cm broad, apex obtuse, margins crenate to finely spinosely-serrate, base cuneate, sessile, glabrous, venation pinnately reticulate. Inflorescence capitate, heads in dichasia; involucral bracts 5–7, linear-lanceolate, 1–3 cm long, 0.3–0.7 cm broad, greatly exceeding the heads, foliaceous, spreading and reflexed, with a few spiny teeth; heads cylindrical, 4–10 mm long, 3–5 mm broad, terminal on the branches or on short peduncles in the forks; involucel bractlets linear or lanceolate, 2–3 mm long, exceeding the fruit; flowers numerous. Sepals erect, lanceolate, longer than the petals, persistent, apex acute. Petals greenish white, oblanceolate to obovate, erect, clawed. Ovary ovoid; styles short, slightly exceeding the sepals, filiform, divaricate, with capitate stigmas. Fruit ovoid, 1–1.5 mm long, 0.5–0.75 mm broad, densely papillose.

D i s t r. Introduced weed found from 550–800 m about the Royal Botanic Garden, Peradeniya and Gammaduva. Only rarely reported. A native of the tropics and subtropics of the New World, now also widespread throughout tropical Africa and Asia.

E c o l. Probably will be found in shaded or sunny, fertile localities, disturbed areas, and along forest edges and stream banks. Flowering April, June, December and probably throughout the year.

V e r n. Andu (S).

N o t e. The plants have a very strong (unpleasant) smelling aromatic oil containing the aldehyde dodecen-2-al-1 (Kool-Haas, Rec. trav. chim. Pays-Bas 51: 460. 1932) and are used as a condiment.

S p e c i m e n s E x a m i n e d. KANDY DISTRICT: Royal Botanic Garden, Peradeniya, *Trimen s.n.*, IV-1882 (PDA), *Alston s.n.*, 8-XII-1925 (PDA). MATALE DISTRICT: Gammaduva, *Alston 667* (PDA).

5. DAUCUS

L., Sp. Pl. 242. 1753.

Plants herbaceous, caulescent, annual or biennial, erect, variously hispid; stems branching, with taproots. Leaves alternate, 2–3 times pinnately decompound, the ultimate divisions small, narrow, linear; petioles sheathing. Inflorescence a regularly compound umbel, subcompact by the incurving of the rays after anthesis; peduncles terminal and axillary; involucral bracts foliaceous, equalling or exceeding the rays, pinnately decompound into short, linear or lanceolate segments, or absent; rays few to numerous, spreading or the outer often connivent over the shorter inner ones; involucel bractlets linear, about equalling the pedicels, or absent; pedicels spreading, unequal. Sepals minute or absent. Petals white, or the central flowers of each umbel purple, obcordate, the outer often radiant, with inflexed apices. Styles short; stylopodium conic. Fruit oblong to ovoid, more or less flattened dorsally; primary ribs slender, bristly; secondary ribs winged, the wings with a single row of prominent barbed prickles; vittae solitary in the intervals, two on the commissure; carpophore free, entire or bifid at the apex. Seeds flattened dorsally in transection; the seed face shallowly concave.

A genus of approximately 60 species found throughout Europe, Africa, Asia, and America. One introduced representative in Ceylon.

Daucus carota L., Sp. Pl. 1: 242. 1753.

Plants biennial; stems 1.5–12 dm tall, solitary, glabrous to retrorsely papillate-hispid. Leaves oblong in general outline, 5–15 cm long, 2–7 cm broad, 2–3 pinnately decompound, the ultimate divisions linear to lanceolate, 2–12 mm long, 0.5–2 mm broad, apex acute, mucronate, margins entire or few-cleft, glabrous to hispid especially on the veins and margins; petioles 3–10 cm long. Umbels with flat or round surface; peduncles 2.5–6 dm long, retrorsely hispid; involucral bracts usually reflexed, filiform, pinnately divided or rarely entire, 3–30 mm long, scarious-margined; rays 15–30, 3–7.5 cm long, unequal, compact in fruit; involucel bractlets 5–7, equalling or exceeding the flowers, acuminate, entire or rarely pinnate, more or less scarious. ciliate; pedicels 20–30, 3–10 mm long. Sepals triangular, teeth acute. Petals white, unequal, obovate. Ovary ovoid, styles divaricate. Fruit ovoid, 3–4 mm long, c. 2 mm broad, broadest at middle; primary and secondary ribs evident, secondary ones with prominent long spines, glochidiate.

D i s t r. Abeywickrama notes it in his checklist, although we are unaware of any recent collections of wild material. However, it is a common weed in both the Old and New Worlds.

E c o l. Cultivated as a garden flower and vegetable and probably a garden escape at higher elevations on the island.

V e r n. Carrot (E).

6. CORIANDRUM

L., Sp. Pl. 256. 1753.

Plants herbaceous, caulescent, annual, slender, erect, glabrous; stems branching, with slender taproots. Leaves alternate, trifoliolate to ternately-pinnately decompound, heteromorphic: leaflets of basal leaves ovate and incised; ultimate divisions of cauline leaves linear; petioles sheathing. Inflorescence a loose, regularly compound umbel; peduncles terminal and lateral, sympodial; involucral bracts usually absent; rays commonly few, spreading-ascending; involucel bractlets few, filiform, lanceolate, inconspicuous; pedicels spreading. Sepals acute, often unequal, conspicuous. Petals white or rose, oblong, conspicuously unequal, the outer usually radiant, with inflexed apices. Styles slender, spreading; stylopodium conic. Fruit orbicular, terete, not constricted at the commissure, glabrous; ribs filiform to obscure; vittae obscure; carpophore 2-cleft to the base. Seeds flattened dorsally in transection; the seed face concave.

A very small genus of approximately two species found throughout the Mediterranean area. One introduced representative in Ceylon.

Coriandrum sativum L., Sp. Pl. 1: 256. 1753; Alston in Trimen, Handb. Fl. Ceylon 6: 138. 1931.

Cuminum cyminum Wall., Cat. 594. 1829.

Plants entirely glabrous; stems 2–7 dm tall. Basal leaves trifoliolate, leaflets ovate in outline, flabelliform, 1–2.5 cm long, 1–2.5 cm broad, apex rounded, margins variously incised, base truncate to cuneate; cauline leaves biternately decompound, ovate in general outline, the ultimate divisions linear-filiform, 3–10 mm long, 0.5–1.5 mm broad, apex obtuse, margins entire; petioles 1–6 cm long. Rays 3–5, 1–2.5 cm long; involucel bractlets 3–5, linear, 2–4 mm long; pedicels 4–8, 2–4 mm long. Sepals ovate-lanceolate. Petals white, pinkish-white or rose. Ovary ovoid, glabrous. Fruit orbicular, terete, 2–5 mm long, 1.5–3 mm broad.

D i s t r. Cultivated on the island and weedy at elevations between 550–2000 m. Reported wild about the Royal Botanic Garden, Peradeniya and Ambewela. A native of the Mediterranean region now widespread as both a cultivated and weedy species in both the Old and New Worlds.

V e r n. Kottha-Malli (S and T); Coriander (E).

U s e s. The dried ripe fruits of coriander are widely used as a spice and a seasoning.

Specimens Examined. KANDY DISTRICT: Royal Botanic Gardens, Peradeniya, *s. coll.*, II-1898, *s.n.* (PDA), *Livera s.n.*, 30-II-1926 (PDA), *F.W. de Silva 268* (PDA). NUWARA ELIYA DISTRICT: Ambewela, by roadside, *Alston s.n.*, 21-II-1927 (PDA). DISTRICT UNKNOWN: *Trimen s.n.*, XII-1893 (PDA).

7. BUPLEURUM

L., Sp. Pl. 236. 1753.

Plants herbaceous, or somewhat woody at base, caulescent or rarely acaulescent, annual to perennial, low or slender, erect or spreading, glabrous and often glaucous; stems branching, with woody or fibrous taproots. Leaves alternate, simple, linear, grass-like, venation nearly parallel, basal leaves with sheathing petioles; cauline leaves usually sessile and clasping, auriculate, or perfoliate. Inflorescence a loose, regularly compound umbel; peduncles terminal and lateral, sympodial; involucral bracts foliaceous, conspicuous or absent; rays few, spreading-ascending to divaricate; involucel bractlets foliaceous, often connate, rarely coloured, sometimes longer than the flowers and fruit; pedicels spreading. Sepals absent. Petals yellow (or rarely purple), oblong to orbicular with acute, inflexed apices. Styles short; stylopodium depressed-conic. Fruit ovoid to orbicular or ellipsoid, slightly compressed laterally, constricted at the commissure; glabrous or roughened; ribs filiform, narrowly winged to obscure; vittae numerous and continuous about the seed cavity, or several in the intervals and on the commissure, or obscure; carpophore 2-cleft to the base. Seeds subterete in transection; the seed face usually plane.

A large genus of approximately 150 species found throughout Europe, Asia, Africa, and North America. One representative in Ceylon.

Bupleurum mucronatum Wight & Arn., Prod. 1: 370. 1834.

Bupleurum nervosum Moon, Cat. 22. 1824, nom. nud.
Bupleurum ramosissmum Wight & Arn., Prod. 1: 370. 1834.
Bupleurum virgatum Wight & Arn. Prod. 1: 370. 1834, non. Wall. 1829; Trimen, Handb. Fl. Ceylon 2: 277. 1894.
Bupleurum falcatum sensu Dalz. & Gibs., Bombay Fl. 108. 1861, non. L. 1753.
Bupleurum mucronatum var. *virgatum* Clarke in Hook. f., Fl. Br. Ind. 2: 676. 1879.
Bupleurum mucronatum f. *virgatum* H. Wolff in Pflanzenr., Heft 43: 145. 1910.
Bupleurum wightii Koso-Polj., J. Russe Bot. 6: 7. 1913; Alston in Trimen, Handb. Fl. Ceylon 6: 138. 1931.

Plants herbaceous, caulescent, perennial, slender; stems 25–50 cm tall. Leaves oblong to linear, 3–12 cm long, 0.5–1 cm broad, apex acuminate, mucronate, margins entire, base amplexicaul, glabrous, glaucous. Involucral bracts 3–6, lanceolate-linear, 3–7 mm long, abruptly acuminate, glabrous; rays 3–6, 1–2 cm long; involucel bractlets 4–5, 2–4 mm long, about equal to the flowers; pedicels 6–8, 1–1.5 mm long. Petals yellow. Ovary oblong, terete, glabrous; styles filiform, divaricate. Fruit ellipsoid, 2–4 mm long, 1–2.5 mm broad, shiny, glabrous; ribs prominent; vittae broad, conspicuous, solitary in the intervals, two on the commissure.

Distr. Found between 650–2300 m from Belihul-oya to Nuwara Eliya and Horton Plains. Also known from the southern region of the Indian Peninsula.

Ecol. Quite common in rocky open places and patanas. Flowering December to August and probably throughout the year.

Vern. Wal enduru (S).

Specimens Examined. NUWARA ELIYA DISTRICT: Ambewela, in patanas, *Alston 1256* (PDA); patana beyond Lab. Hakgala, *J.M. Silva s.n.*, 24-II-1906 (PDA); Horton Plains, patana on Bogawantalawa Road, *J.M. Silva s.n.*, 26-I-1906 (PDA); Horton Plains, near Ohiya Creek below Rest House, growing in patana near forest edge, *Mueller-Dombois 68051701* (PDA); Horton Plains, patana at Small World's End, *Theobald & Krahulik 2727* (A, BISH, E, K, L, LE, PDA, UC, US); Horton Plains, patana at Big World's End, *Theobald & Krahulik 2730* (A, BISH, BM, BO, CAL, K, L, MO, NY, P, PDA, SING, UC, US); Horton Plains, rocky edge at Small World's End, *Theobald & Krahulik 2733* (BISH, PDA, UC, US); along trail from Galagama Falls to Horton Plains, *Theobald & Krahulik 2836* (A, BISH, BM, BO, CAL, E, K, L, LE, MO, NY, P, PDA, SING, UC, US); Nuwara Eliya, *Gardner 320* (PDA), *Gardner s.n.*, VII-1845, part of *C.P. 225* (PDA); between Nuwara Eliya and Maturata, *s. coll.* VIII-1853, part of *C.P. 225* (PDA); Nuwara Eliya, *s. coll.* IV-1854, part of *C.P. 225* (PDA). RATNAPURA DISTRICT: Galagama (=Belihul-oya), *Gardner s.n.*, VII-1845, part of *C.P. 225* (PDA).

8. APIUM

L., Sp. Pl. 264. 1753.

Helosciadium C. Koch, Nova Acta Acad. Leop.-Carol. 12: 125. 1825.
Ciclospermum Lag., Amen. Nat. 101. 1821.

Plants usually herbaceous, caulescent, annual to perennial, glabrous; stems usually branching, with taproots or rootstocks rooting at the nodes. Leaves alternate, simply pinnate to ternately-pinnately decompound, the ultimate divisions suborbicular to filiform (in ours); petioles sheathing. Inflorescence a loose to subcompact, regularly compound (or rarely simple by reduction) umbel; peduncles terminal and lateral, sympodial, or some or all the umbels sessile; involucral bracts minute or absent; rays usually rather few, spreading-ascending; involucel bractlets minute or absent; pedicels spreading or ascending. Sepals minute or absent. Petals white or light greenish-white, ovate to suborbicular with acute, inflexed apices. Styles short, spreading or divaricate; stylopodium short-conic to depressed. Fruit ovoid to orbicular or ellipsoid, slightly compressed laterally, somewhat constricted at the commissure, glabrous; ribs filiform, subequal, obtuse, conspicuous; vittae solitary in the intervals, two on the commissure; carpophore entire, bifid at the apex, or more or less deeply 2-cleft. Seeds subterete in transection; the seed face plane.

A genus of approximately 30 species found worldwide. One introduced representative in Ceylon.

Apium leptophyllum (Pers.) F. Muell.; Benth. & Muell., Fl. Austral. 3: 372. 1866; Alston in Trimen, Handb. Fl. Ceylon 6: 138. 1931.

Sison ammi L. sensu Jacq., Hort. Vindob. 2: 95, pl. 200. 1773, non L. 1753.
Pimpinella leptophyllum Pers., Syn. Pl. 1: 324. 1805.
Helosciadium leptophyllum DC., Mém. Soc. Phys. Nat. Hist. Genève 4: 493. 1829.
Apium ammi Urban in Mart., Fl. Bras. 11 (1): 341. 1879.
Apium tenuifolium Thell. in Hegi, Ill. Fl. Mittel-Eur. 5 (2): 1140. 1926, non *Cnidium tenuifolium* Moench 1794.

Plants herbaceous, annual, slender, erect to ascending; stems 40–50 cm tall, alternately branching above. Leaves oblong-ovate in general outline, 2–3 times ternately-pinnately decompound, the ultimate divisions linear-filiform, 2–10 mm long, 0.5–1 mm broad, apex narrowly obtuse, margins entire, glabrous; petioles up to c. 1 cm long, the sheath scarious-margined. Umbels regularly compound, sessile or rarely pedunculate; involucral bracts absent; rays 3–5, 0.5–1 cm long; involucel bractlets absent; pedicels 3–10, 2–4 mm long. Sepals absent. Petals white. Ovary ovoid, glabrous. Fruit ovoid, c. 1.5 mm long, c. 1 mm broad; carpophore shortly 2-cleft.

Distr. Only known from 1700 m about Haputale. Otherwise, widespread throughout the New and Old Worlds.

Ecol. Alston notes that it appeared in waste ground in 1917.

Specimens Examined. BADULLA DISTRICT: Haputale, *Petch s.n.*, XI-1917 (PDA).

9. TRACHYSPERMUM

Link, Enum. Hort. Berol. 1: 267. 1821.

Plants herbaceous, annual, erect; stems glabrous. Leaves simply ternate to ternately-pinnately decompound or 1–2 pinnate, the ultimate segments narrow to oblanceolate, glabrous; petioles present. Inflorescence a regularly compound umbel; peduncles terminal and lateral; involucral bracts 2–6; rays present; involucel bractlets 3–8; pedicels present; flowers few to numerous. Sepals inconspicuous or absent. Petals white, obcordate, with broadly inflexed obtuse apices. Styles divaricate; stylopodium low conic. Fruit narrowly ovoid, slightly compressed laterally, constricted at the commissure, glabrous to more or less hirtellous; ribs filiform; vittae three in the intervals, 4 on the commissure; carpophore 2-cleft to the base. Seeds terete in transection; the seed face sulcate to nearly plane.

A genus of approximately 20 species found throughout Tropical and Northeast Africa, east through Central Asia to India and West China. Represented in Ceylon by one native species and one introduction.

KEY TO THE SPECIES

1 Plants greater than 15 cm tall; involucral bracts 3–10 mm long; rays 1–2.5 cm long; fruit hirtellous..**1. T. roxburghianum**
1 Plants low, less than 12 cm tall; involucral bracts 1–4 mm long; rays 0.5–1 cm long; fruit minutely scabrous..**2. T. stictocarpum**

1. Trachyspermum roxburghianum (Wall.) Craib, Fl. Siam. En. 1: 788. 1931.

Apium involucratum Roxb. ex Flem., Ind. Med. Pl. in As. Res. 11: 157. 1810.
Athamanta roxburghiana Wall., Cat. 571.1829.
Ptychotis roxburghiana DC., Prod. 4: 109. 1830.
Phellandrium sp. Wall., Cat. 7217. 1832.
Pimpinella involucrata Wight & Arn., Prod. 1: 369. 1834, non Miq. ex Clarke 1879.
Carum roxburghianum Benth. ex Clarke in Hook., Fl. Br. Ind. 2: 682. 1879; Trimen, Handb. Fl. Ceylon 2: 278. 1894.
Trachyspermum involucratum H. Wolff in Pflanzenr. Heft 90: 89. 1927.
Trachyspermum roxburghianum H. Wolff in Pflanzenr. Heft 90: 129. 1927, nom. altern.

Plants caulescent; stems 15–70 cm tall, sympodially branched. Leaves alternate, ternately-pinnate or 1–2 pinnate, ultimate divisions linear-oblong to lanceolate, entire to lobed, 3–10 mm long, 1–2 mm broad, apex narrowly acute to acuminate; petioles sheathing, 0.5–1.5 cm long. Involucral bracts 2–5, linear-lanceolate, 3–10 mm long, acute; rays 3–9, 1–2.5 cm long; involucel bractlets 5–8, 2–3 mm long; pedicels 5–15, 2–6 mm long. Sepals absent. Petals white. Ovary ovoid. Fruit ovoid, 1–2 mm long, 1–2 mm broad, hirtellous; ribs prominent; vittae large.

Distr. Reported only from area around Royal Botanic Garden, Peradeniya at elevation of 570 m. It is also unknown in the wild anywhere else and Hooker (Fl. Br. Ind. 2: 682. 1879) notes that this species is "probably a cultivated form of *C. stictocarpum* which it exactly resembles except as to the fruit." However, on the basis of the Ceylon populations of these two species, they are both distinctive.

Ecol. An occasional weed and cultivated plant. Flowering probably throughout the year.

Specimens Examined. KANDY DISTRICT: Royal Botanic Garden, Peradeniya, *Alston 2015* (PDA), *F.W. de Silva 297* (PDA).

2. Trachyspermum stictocarpum (Clarke) H. Wolff in Pflanzenr. Heft 90: 89. 1927.

Carum stictocarpum Clarke in Hook. f., Fl. Br. Ind. 2: 681. 1879; Trimen, Handb. Fl. Ceylon 2: 278. 1894.

Plants short, caulescent; stems 4–12 cm tall. Leaves alternate, ternate, leaflets ovate, deeply cleft into oblanceolate-oblong, apiculate lobes, the lobes c. 5 mm long, c. 2 mm broad; petioles sheathing, 0.5–2 cm long. Involucral bracts 3–6, linear to lanceolate, 1–4 mm long, acute; rays 3–5, 0.5–1 cm long; involucel bractlets 3–6, linear-lanceolate, 1–2 mm long; pedicels 6–10, 1–2 mm long, filiform. Sepals minute to absent. Petals white. Ovary ovoid. Fruit ovoid, 1–2 mm long, 0.5–1 mm broad, minutely scabrous; ribs slender; vittae small.

Distr. Rare, only known near sea-level about Keerimalai, Jaffna District. Trimen notes "that is possibly an accidental introduction." Widespread throughout India.

Ecol. Dry zone, very rare. In sheltered crevices of the rocky boulders of coral-stone forming the headland (Trimen). Flowering February.

Note. These populations appear to be depauperate forms of those found in India.

Specimens Examined. JAFFNA DISTRICT: Keerimalai, in sheltered crevices of rocks on the coast, *Trimen s.n.*, 6-II-1890 (PDA). DISTRICT UNKNOWN: *Ferguson s.n.* (PDA).

10. PIMPINELLA

L., Sp. Pl. 263. 1753.

Plants herbaceous, or somewhat woody at base, caulescent, annual, biennial or perennial, low or slender, erect, ascending or procumbent, glabrous to hirsute; stems branching, with slender taproots. Leaves alternate, simple, or ternately or pinnately compound or dissected, the divisions various; petioles sheathing. Inflorescence a loose, regularly compound umbel; peduncles terminal and lateral; involucral bracts usually minute or absent; rays few to numerous, spreading-ascending; involucel bractlets small, or absent; pedicels spreading. Sepals usually minute or absent. Petals white, broad with acute, inflexed apices, glabrous to hirsute, all subequal, or nearly so. Styles slender; stylopodium conic or low-conic. Fruit oblong to orbicular, ellipsoid, or ovoid, narrower at apex, rounded or cordate at base, somewhat compressed laterally, constricted at commissure, glabrous to pubescent; ribs equal, filiform or indistinct; vittae numerous, continuous around the seed, or 1-several in the intervals; carpophore entire, or 2-cleft to the middle or to the base. Seeds pentagonal or terete in transection; the seed face concave to nearly plane.

An Old World genus of approximately 150 species found throughout Eurasia and Africa. Represented in Ceylon by two species.

KEY TO THE SPECIES

1 Leaves pinnately trifoliolate; rays glabrous; fruit glabrous.............**1. P. heyneana**
1 Leaves simple; rays pubescent; fruit pubescent.....................**2. P. leschenaultii**

1. Pimpinella heyneana Wall., Cat. 566. 1829; Trimen, Handb. Fl. Ceylon 2: 279. 1894.

Sium triternatum Moon, Cat. 22. 1824. nom. nud.
Helosciadium heyneanum DC., Prod. 4: 106. 1830.
Anethum trifoliatum Roxb., Fl. Ind. ed. 2 (2): 96. 1832.
Seseli zeylanicum Herb. Rottl. and Klein ex Clarke in Hook. f., Fl. Br. Ind. 2: 685. 1879.

Plants herbaceous, annual, glabrous; stems 15–70 cm tall, with ascending branches. Leaves pinnately trifoliolate, very broadly ovate in outline, 2–5 cm long, 2.5–8 cm broad, leaflets ovate to lanceolate in outline, 1.5–4 cm long, 0.75–1.75 cm broad, apex acuminate, margins serrate to deeply parted near base, base acute to rounded, glabrous; petioles 2–7 cm long. Involucral bracts absent; rays 6–12, 2–4 cm long, slender, glabrous; involucel bractlets 1–2, linear, 0.5–1.2 cm long, unequal; pedicels 4–10, 2–10 mm long. Sepals absent. Petals white, ovate-lanceolate. Ovary ovoid, glabrous; styles short, divaricate. Fruit ovoid, slightly compressed laterally, c. as broad as long, c. 1–1.5 mm long, c. 1.2–2 mm broad, glabrous; ribs prominent; vittae solitary in the intervals, two on the commissure.

Distr. Found at elevations between 70–170 m in the dry zone districts of Anuradhapura, Polonnaruwa, Vavuniya, and also reported from the Matara District. Thwaites states that it is very abundant in Uva up to 1300 m. Also known from India and Burma.

Ecol. Moist areas in grass under trees, waste ground, and roadsides.

V e r n. Wal asamodagam (S).

S p e c i m e n s E x a m i n e d. ANURADHAPURA DISTRICT: Nuwara Weva, on bank of tank, *Balakrishnan 340* (PDA, US). MATARA DISTRICT: Morawaka Korale, *s. coll. s.n.*, 28-II-1881 (PDA). POLONNARUWA DISTRICT: Polonnaruwa, edge of road, *Alston 573* (PDA). VAVUNIYA DISTRICT: Mankulam, open place in jungle, *Alston 1432* (PDA). DISTRICT UNKNOWN: *Col. Walker s.n.* (PDA); *s. coll. s.n.*, IV-1854, part of *C.P. 182* (PDA); *Alston 36* (PDA); Punnava Weva, jungle path near bank, *Simpson 9236* (PDA).

2. Pimpinella leschenaultii DC., Prod. 4: 122. 1830; Trimen, Handb. Fl. Ceylon 2: 279. 1894.

Plants herbaceous, perennial, with basal rosette when young, obscurely pubescent; stems 15–60 cm tall, weak, procumbent. Leaves simple, cordate-ovate in outline, 1.5–6 cm long, 2–8 cm broad, apex acute to obtuse, margins coarsely repand-dentate or crenate, base very deeply cordate with a narrow sinus and often with the rounded lobes slightly overlapping, glabrous above, more or less pubescent below and ciliate on margins, veins conspicuous beenath; petioles 5–20 cm long, sparingly hairy. Involucral bracts absent; rays 10–20, 1.5–3 cm long, pubescent; involucel bractlets 1–3, linear-lanceolate, 3–5 mm long; pedicels 6–10, 3–5 mm long. Sepals absent. Petals ovate-lanceolate. Ovary ovoid; styles long, erect, divaricate. Fruit ovoid, slightly compressed laterally, tapering upwards, 2–4 mm long, 2–3 mm broad, obscurely pubescent; ribs obscure; vittae small, three in the intervals, six on the commissure.

D i s t r. Rare. Found only at elevations between 650–2300 m between Belihul-oya and Horton Plains. Also known from the mountains of Southern India.

E c o l. At edge of forest along patanas. Flowering October to May and probably throughout the year.

S p e c i m e n s E x a m i n e d. NUWARA ELIYA DISTRICT: Horton Plains, *Gardner s.n.*, part of *C.P. 515* (PDA); Horton Plains, by the stream, *Trimen s.n.*, 21-II-1882 (PDA); Horton Plains, above Nonpareil, *Trimen s.n.*, 24-II-1882 (PDA); Horton Plains, edge of precipice at World's End, *A.M. Silva s.n.*, 3-V-1906 (PDA); Horton Plains, edge of forest at Small World's End, *Theobald & Krahulik 2728* (A, BISH, CAL, K, L, MO, PDA, UC, US); along trail from Galagama Falls to Horton Plains, *Theobald & Krahulik 2835* (BISH, BM, L, LE, P, PDA, UC, US). RATNAPURA DISTRICT: Galagama (=Belihul-oya), *Gardner* part of *C.P. 515* (PDA). DISTRICT UNKNOWN: *s. coll.* II-1857, part of *C.P. 515* (PDA).

11. PEUCEDANUM

L., Sp. Pl. 245. 1753.

Plants herbaceous, caulescent, perennial, slender to stout, usually erect, glabrous (in ours) to tomentose; stems with stout taproots. Leaves alternate, ternate, or ternately-pinnately or pinnately compound, the leaflets broad or narrow, distinct or more or less confluent, serrate or dentate to incised or lobed; petioles sheathing, cauline sheaths little to somewhat inflated. Inflorescence a loose, regularly compound umbel; peduncles terminal, or terminal and lateral; involucral bracts absent or one to several; rays few to numerous, spreading-ascending; involucel bractlets few to numerous, narrow, entire, or absent; pedicels slender, spreading-ascending, frequently webbed; flowers polygamous. Sepals prominent to absent. Petals white or reddish-white, oval to obovate with acute, inflexed apices, glabrous to pubescent dorsally. Styles short to long, spreading or recurved; stylopodium conic to low-conic. Fruit oblong-oval to orbicular, strongly flattened dorsally, commissure broad, glabrous to pubescent; dorsal and intermediate ribs filiform, unwinged, the lateral ones winged, narrower than the body; vittae solitary to several in the intervals, two to several on the commissure; carpophore 2-cleft to the base. Seeds flattened dorsally in transection; the seed face slightly concave to plane.

A large and very variable genus of approximately 120 species found throughout temperate Eurasia, Tropical Africa and South Africa. Probably not wholly natural. One endemic representative in Ceylon.

Peucedanum ceylanicum Gardn., Calcutta J. Nat. Hist. 8: 14. 1847; Trimen, Handb. Fl. Ceylon 2: 280. 1894 (‘*zeylanicum*’).

Palimbia ramosissima sensu Thw., Enum. Pl. Zeyl. 131. 1859, non DC. 1830.

Plants glabrous; stems 1–1.5 m tall. Leaves mostly basal, triternate, ultimate divisions numerous, linear, 5–40 mm long, 1–2 mm broad, terminal division much the longer, apex narrowly acuminate, margins entire, glabrous, glaucous; petioles 7–15 cm long. Involucral bracts absent; rays 10–12, 2–6.5 cm long; involucel bractlets c. 5, filiform, c. 3 mm long; pedicels 10–12, filiform, 1–2 cm long, much longer than the fruit. Sepals minute or absent. Petals white, obovate. Ovary ovoid, glabrous; styles long, divaricate; stylopodium thick-conic. Fruit oblong-elliptic, 5–8 mm long, 3–5 mm broad, glabrous; wings narrower than the body; vittae solitary in the intervals, 2–6 on the commissure.

D i s t r. Endemic. Very rare. Found above Belihul-oya on way to Horton Plains. Only collected once, in 1846 by Gardner.

E c o l. Open grassy rocky places. Flowering February, March.

Specimens Examined. RATNAPURA DISTRICT: Galagama (=Belihul-oya), *Gardner*, III-1846, part of *C.P. 135* (PDA).

12. HERACLEUM

L., Sp. Pl. 249. 1753.

Plants herbaceous, caulescent, biennial or perennial, erect, pubescent; stems with taproots or fibrous roots. Leaves mostly basal, ternately or pinnately compound, the leaflets broad, ovate to orbicular, serrate to variously cleft and lobed; petioles sheathing and usually conspicuously dilated. Inflorescence a regularly compound umbel; peduncles terminal and axillary; involucral bracts absent or few, narrow; rays numerous, unequal, spreading-ascending; involucel bractlets narrow, entire, or rarely absent; pedicels spreading-ascending; flowers polygamous. Sepals minute or absent. Petals white, oval to obcordate, the outer ones radiant and often 2-cleft, with inflexed apices. Styles short, erect or recurved; stylopodium conic. Fruit orbicular to obovate or elliptic, strongly flattened dorsally, commissure broad, glabrous or pubescent; dorsal and intermediate ribs filiform, the lateral ones broadly thin-winged and nerved near the outer margin of the wings; vittae large, visible on the dorsal surface, solitary in the intervals, 2–4 on the commissure, usually extending only part way from the stylopodium to the base of the carpel (all the way in ours); carpophore 2-cleft to the base; woody endocarp evident. Seeds flattened dorsally in transection; the seed face plane.

A genus of approximately 70 species found throughout the north temperate zone of Europe and Asia and tropical mountains with one species extending into North America. One endemic representative in Ceylon.

Heracleum ceylanicum Gardner ex Clarke in Hook. f., Fl. Br. Ind. 2: 716. 1879; Trimen, Handb. Fl. Ceylon 2: 280. 1894 ('*zeylanicum*').

Plants perennial, finely pubescent; stems 30–70 cm tall, slightly branched. Leaves mostly basal, pinnate, leaflets 3–7, broadly ovate in outline, 2–7 cm long, 2–7 cm broad, very deeply, pinnately cut into broadly oblong, coarsely spinous-serrate lobes, pubescent on both sides, apices obtuse-acute; petioles 2–30 cm long. Involucral bracts 2–3, linear, c. 4 mm long; rays 7–15, 1–6 cm long; involucel bractlets 6–7, filiform, c. 2 mm long; pedicels 10–20, shorter than or equal to the fruit, 2–10 mm long. Sepals minute, linear-lanceolate, acute. Petals white, obovate, attenuate. Anthers purple; pollen white. Ovary ovoid, pubescent. Fruit orbicular to obovate-elliptic, 6–12 mm long, 5–10 mm broad, glabrous; wings c. as broad as body.

Distr. Endemic. Found at elevations between 2000–2700 m about Moon Plains, Horton Plains, Adam's Peak, and Pidurutalagala.

E c o l. In patanas of upper montane zone, rather common. Flowering February to July.

N o t e. The only polyploid member (n = 22) of the genus so far known (L. Constance, personal communication).

Trimen notes that the form of the lower leaves is very variable, and a luxuriant form occurs on Horton Plains with very much cut leaves, which are nearly glabrous.

The entire complex of South Indian species, including *H. rigens* Wall. and *H. sprengelianum* Wight. & Arn., is in need of further investigation. *Heracleum ceylanicum* is definitely closely allied to these taxa and may represent a subspecific variant within a more widespread species.

S p e c i m e n s E x a m i n e d. KANDY DISTRICT: Adam's Peak, *Gardner*, part of *C.P. 145* (PDA). NUWARA ELIYA DISTRICT: Horton Plains, *Trimen s.n.*, 21-II-1882 (PDA), *J.M. Silva s.n.*, 26-I-1906 (PDA); Horton Plains, patana at World's End, *A.M. Silva s.n.*, 3-V-1906 (PDA); Horton Plains, 1/4 mile below Rest House at Ohiya Road, *Comanor & Mueller-Dombois 67070904* (PDA); Horton Plains, patana at Small World's End, *Theobald & Krahulik 2726* (BISH, PDA, US); bog area 0.2 miles from Pattipola Railroad Station, along road to Horton Plains, *Theobald & Grupe 2305* (A, BISH, BM, K, L, PDA, UC, US); Moon Plains, *J.M. Silva 144* (PDA); Nuwara Eliya, *Gardner s.n.*, part of *C.P. 145* (PDA); Pidurutalagala, *s. coll.*, IV-1852, part of *C.P. 145* (PDA); patana, Sita Eliya, *A.M. Silva s.n.*, 3-X-1906 (PDA). DISTRICT UNKNOWN: *Col. Walker s.n.* (PDA); *s. coll.* in 1857, part of *C.P. 145* (PDA).

